# Management of Hazardous Energy

Deactivation, De-Energization,
Isolation, and Lockout

# Management of Hazardous Energy

## Deactivation, De-Energization, Isolation, and Lockout

# Thomas Neil McManus

## CRC Press
Taylor & Francis Group
Boca Raton London New York

CRC Press is an imprint of the
Taylor & Francis Group, an **Informa** business

CRC Press
Taylor & Francis Group
6000 Broken Sound Parkway NW, Suite 300
Boca Raton, FL 33487-2742

First issued in paperback 2017

© 2013 by Taylor & Francis Group, LLC
CRC Press is an imprint of Taylor & Francis Group, an Informa business

No claim to original U.S. Government works

ISBN-13: 978-1-4398-7836-1 (hbk)
ISBN-13: 978-1-138-07211-4 (pbk)

**Visit the Taylor & Francis Web site at**
**http://www.taylorandfrancis.com**

**and the CRC Press Web site at**
**http://www.crcpress.com**

# Dedication

This book is dedicated to the memory of those people who either lost their lives or were seriously injured in accidents involving hazardous energy. May we learn from these tragedies.

# Contents

# Preface

The harnessing of energy from energy sources and materials containing energy is intrinsic to activity that occurs in all societies, regardless of their state of development. The misuse and inappropriate use of energy sources and materials containing energy have caused countless injuries and deaths. Much of the modus vivendi of the safety profession over the years has focused on minimizing the risk of harm arising from use of energy sources and materials containing energy.

Hazardous energy is the type and level of energy that is likely to cause rapid harm. Rapid harm can include serious traumatic injury and even death. Statistics suggest that on average, hazardous energy is involved in about 68% of fatal occupational accidents in the U.S. Faulty, flawed, or absent control of hazardous energy through safeguarding, lockout, and tagout is a major cause of citations issued by Occupational Safety and Health Administration (OSHA).

There is a serious dearth of information about hazardous forms of energy outside specialized industrial sectors and settings. As well, there is no recognizable path for individuals, especially safety practitioners, to be able to gain this knowledge in an efficient manner to a level sufficient to be able to contribute effectively to discussion in these areas.

One of the most important capabilities of the safety practitioner is the ability to ask the right questions. The safety practitioner must be able to anticipate what could go wrong, recognize when something is going wrong, evaluate the significance and severity of what could occur, and be able to recommend measures to eliminate or at least minimize the risk posed by these situations.

The right questions include what does the equipment, machine, or system do; how does it do it; what can happen; when can events happen; where can they happen; and most importantly, why and how can they happen? The safety practitioner must possess a functional level of knowledge about how things work and about how they fail in order to be able to ask the right questions.

This level of knowledge exceeds that possessed by the beginner in the field and sometimes by the operator of the equipment, machine, or system but is less than that possessed by a technical specialist. This knowledge must focus on hazardous conditions and risks known to be associated with a particular endeavor. This knowledge also should include consequences of actions taken in the past in response to purported innovations. In learning about these, the safety practitioner becomes knowledgeable about what could go wrong in the general context of the activity.

Resources needed to gain this level of knowledge are difficult to find. Perhaps of necessity, they are sector specific. On the one hand, current books and periodicals aimed at safety practitioners focus on regulatory requirements. In doing so, they fail to provide the background information needed to explain the technical rationale behind the requirement or they cover these areas at a level below what is needed. On the other hand, publications aimed at technical specialists in these sectors provide information at a level and in a manner that presumes that the reader possesses considerable background knowledge. As well, trade periodicals aimed at these sectors emphasize only the benefits perceived to accrue from adoption of new technologies and provide little, if any, discussion about potential consequences.

This book attempts to bridge this gap by providing sector specific information gathered over a period of many years. This information will assist safety practitioners interacting with technical specialists in these sectors in asking the "right" questions, thereby enabling effective contribution to discussion about hazard assessment and management.

A critical aspect of any discussion about hazardous energy is the role of computers and software in control of equipment, machines, and systems. Computerization of control has dramatically changed the manner in which operation of equipment, machines, and systems occurs and the manner in which operators and maintenance personnel interact with them. While on the one hand, this advancement produces tremendous benefit, cost accompanies every change, even for late adopters. Understanding about control of equipment, machines, and systems through which expression of hazardous energy occurs, is essential in order to enable consistent and knowledgeable approaches to deactivation, de-energization, isolation, lockout, and verification that address these realities.

This book also provides practical tools for assessing the input, transformation, storage, and output of hazardous energy in equipment, machines, and systems for creating concise, defensible procedures for deactivation, de-energization, isolation, lockout, and verification. These procedures provide the basis for ensuring safety for workers affected by hazardous energy and for complying with regulatory requirements.

# Author

**Thomas Neil McManus,** CIH, ROH, CSP, is a practicing industrial hygienist with more than 30 years of broad-spectrum experience "in the trenches." He holds certifications from the American Board of Industrial Hygiene, the Canadian Registration Board of Occupational Hygienists, and the Board of Certified Safety Professionals. He is an active volunteer in the health and safety profession and the local community, and has presented numerous times at conferences in Canada and the U.S., and in Brazil and Taiwan.

Mr. McManus is the chair of the ANSI Z9.9 Committee on Portable Ventilation Systems, and past chair of the Computer Applications and Confined Space Committees of the American Industrial Hygiene Association. He is a fellow of AIHA, the first non-U.S. citizen elected to this honor.

Mr. McManus has written numerous articles and short publications (available for download at www.nwohs.com). He is the author of *Safety and Health in Confined Spaces* and *Portable Ventilation Systems Handbook*; he is coauthor of *The Hazcom Training Program* and *The WHMIS Training Program*. Together with *Management of Hazardous Energy: Deactivation, De-Energization, Isolation, and Lockout,* these books form a unified system for the assessment and management of occupational health and safety risks in maintenance and other non-production activities.

Mr. McManus has an M.Sc. in radiation biology, an M.Eng. in occupational health and safety engineering, a B.Sc. in chemistry, and a B.Ed. specializing in chemistry and biology, all from the University of Toronto.

# 1 Hazardous Energy and Fatal Accidents

## INTRODUCTION

The harnessing of energy from energy sources and materials containing energy is intrinsic to the activity that occurs in all societies, regardless of the state of development. The misuse and inappropriate use of energy sources and materials containing energy have caused countless injuries and deaths. Much of the *modus vivendi* of the safety profession focuses on minimizing the hazardous use of energy sources and materials containing energy.

Much of industrial equipment and industrial endeavor is concerned with the production and extraction, storage, transfer and transport, and use or conversion of energy from energy sources and materials containing energy. (Energy cannot be created or destroyed, but can be converted from one form to another.) The actual outcome from these endeavors may be production of a tangible item or substance that may seemingly have nothing to do with energy. Yet, production of industrial products and chemical substances involves energy, and energy has everything to do with the hazardous conditions that can occur in certain workspaces. To illustrate, the sole purpose for many industrial structures is to confine processes and activities that involve energy. Entry into these structures before the energy is discharged and maintained at nonhazardous levels represents a pre-accident condition.

Energy sources and materials containing energy are present in the industrial environment as point sources (Figure 1.1), line sources (Figure 1.2) and distributed sources (Figure 1.3). Further, they occur in situations in which the energy source is confined (Figure 1.4), semi-confined (Figure 1.5), semi-unconfined (Figure 1.6), and unconfined (Figure 1.7, in open work areas and confined spaces (Figure 1.8).

Systems in which the energy source is confined employ barriers to contain the energy source or the material containing the energy. Examples include nuclear and chemical reactors and piping systems, food irradiators, and enclosed machinery, among many others. The boundary surface prevents contact between people and the energy source or the material containing the energy. Systems in which the energy source is semi-confined include welding operations in normal shop operations and sand piles and unshored excavations and trenches, and unguarded machines. Systems in which the energy source is unconfined include nuclear detonations and powerful weather systems, such as hurricanes.

Confined spaces prevent or impede the outward radiation of energy and hinder movement of individuals during emergency situations. Confined spaces occur across the spectrum of industry. Potentially any structure in which people work could be or could become a confined space.

The term confined space normally is used to indicate danger in a particular structure or workspace. What the term confined space actually does is to encode hazardous conditions that can occur in a workspace, rather than the workspace itself. The enigma of confined spaces is that under some conditions a particular workspace may pose no extraordinary hazard. Yet, following seemingly minor change, the conditions become life-threatening. On the other hand, the interior of some structures poses serious hazards under almost all conditions. The dangers are easily recognizable and these spaces receive due recognition and attention. In other spaces, hazardous conditions develop only under a few circumstances. The hazards that develop can be equally serious, yet receive no recognition. This, then, explains the difficulty in managing the hazards posed by confined spaces. The hazardous conditions may be transient and subtle, and therefore difficult to recognize and to address. As well, the attributes of the space can exacerbate the hazards.

This book uses the terms energy sources and materials containing energy. Energy sources refer to finite objects that may or may not be under the control of the user. A welding torch, for example, is a source of considerable energy that is directly under the control of the operator, as is the beam from a radioisotope source used in nondestructive testing of metal structures. An uninsulated overhead powerline is an energy source that is not under the direct control of utility linemen or other workers affected by it. Flowable solid materials, such as a pile of sand or coal, or the wall of an unshored trench, are ma-

Figure 1.1. Point source. The flame of an oxy-fuel torch creates an intense point source of hazardous energy in a very small volume in space. This source of energy is unconfined by boundary surfaces. Reliance on personal protective equipment is necessary to protect against exposure.

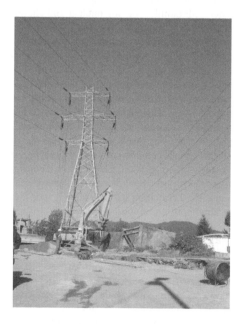

Figure 1.2. Line source. A high voltage transmission line is one of the few examples of line sources of hazardous energy. This source of energy is unconfined by boundary surfaces. Reliance on distance is necessary to protect against exposure.

Figure 1.3. Distributed source. The boiler of this locomotive, the Duchess, is an example of a distributed source of energy  This energy source is partly confined by a boundary surface. Shielding using an insulating layer both retains heat and improves performance.

Figure 1.4.  Confined energy source. The boundary surfaces of this equipment confine the energy sources (rotating equipment and flowable solid materials). This photo details the scene of a fatal accident involving an engulfment.

Figure 1.5. Semi-confined energy source. An embankment undergoing excavation at the exposed face is an example of a semi-confined energy source. The action of the excavating machinery or natural forces, such as an earthquake, undermines stability of the face of the the the embankment. (Source: MSHA.)

Figure 1.6. Semi-unconfined energy source. A pile of bulk, free-flowing granular material is confined only by the base on which it rests.

Figure 1.7.  Unconfined energy source. A hurricane is an energy source unconfined by boundary surfaces. This energy source is able to move in a manner unaffected by human intervention. (Source: National Oceanic and Atmospheric Administration.)

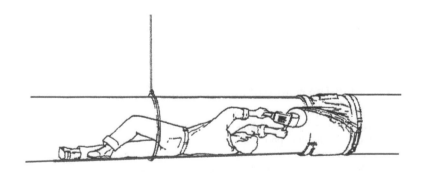

Adapted from OSHA

Figure 1.8.  Confined space. The boundary surfaces of the workspace limit the position of the energy source. This limitation on position can considerably increase the severity of injury, as would occur following shattering of a grinding wheel. (Source: Modified from OSHA.)

terials that contain energy. This energy can convert rapidly from one form to another (potential to kinetic) during slumpage (collapse) of the pile.

Another element in this discussion is the manner in which work occurs. Industry operates under two recognizable modes: production and non-production. Production mode often implies relatively invariate or steady-state conditions, day after day. Production mode occurs for hours, days, months, or even years. Predictability and repeatability characterize tasks that occur during production mode. Non-production modes include construction, inspection, maintenance, modification and renovation, and rehabilitation. Entire segments of industry, and organizational units within individual organizations, operate in non-production mode. Activities occurring during non-production mode usually are neither repetitive, nor are the outcomes fully predictable. Variability characterizes these activities and the conditions under which they are carried out.

Work occurring during non-production modes can be short in duration, and sometimes is highly hazardous. This work often occurs irregularly, and not uncommonly, during off-shifts. Complicating matters further is the pressure to complete the work in orderly fashion, often in the shortest possible time. Time pressure extends to all aspects of work during non-production modes, including hazard identification and assessment. The first line of defense against workplace hazards in these situations usually is personal protection, supplemented by portable ventilating equipment in the case of work occurring in confined spaces. Complicating things further, the work process and individual workers are forced to accommodate to realities necessitated by conditions and availability of equipment on site. Additional equipment may not be available until after completion of the work.

The nature of work also influences the likelihood of encounters with energy sources and materials containing energy. In some segments of industry encounters are infrequent, while in others, they are routine and daily. Similarly, occupation influences the likelihood of encounter with energy sources and materials containing energy. Some workers encounter energy sources and materials containing energy rarely, while others do so all the time.

In many organizations, the classic safety structure "owned" the problem of hazardous energy sources and materials containing energy. This "ownership" seemed logical, since the hazards associated with work involving energy sources and materials containing energy were perceived to be safety-related. The safety profession thus assumed leadership in this area. This led to the development of technical standards (ANSI 1975, ANSI 1982, ANSI/ASSE 2004, ANSI/ASSE 2008). Yet, hazardous energy poses multi-disciplinary problems and issues, since it can exist in many different forms. Whole professions, such as health physics and process safety management, have developed to address specialized parts of the bigger picture. Similarly, specialized groups within classic safety have developed to address hazards involved with electrical generation, distribution and usage, mining, manufacturing, and chemical handling. While the challenges posed by specific types of hazardous energy may seem different, methods for assessment and strategies to prevent or control contact (exposure) share many similarities.

## Hazardous Energy: The Nature of the Problem

In beginning an examination of this subject, one needs to comprehend the full meaning and implications of the term hazardous energy. Hazardous energy includes energies generated by physical, chemical, nuclear, mechanical, and other processes.

To the uninitiated, hazardous energy might connote high-voltage electrical energy. While correct as far as this goes, this notion is overly restrictive, because it focuses on high levels of energy. These far exceed the energy needed to injure or kill. Electrical energy at 50 to 80 Volts containing even low current flow, applied appropriately is capable of fatally injuring humans.

The presence of the boundary surfaces in a confined space exacerbates the hazard from energy sources contained within it. This results from the ability of many types of hazardous energy to reflect off walls of the structure, rather than being absorbed by them.

A point needing considerable emphasis here is that the purpose for many structures or equipment, the interior of which can be confined spaces, is to contain a source of hazardous energy or materials containing energy. Entry, especially entry during operation, was neither presumed nor intended to occur by designers of these structures or manufacturers of this equipment. Entry, therefore, can defeat the purpose for the containment by enabling contact with the energy source or materials containing energy. An implication inherent in the function of the structure as a barrier is that conditions inside can be considerably different from those encountered in normal workspaces. These conditions can develop prior to opening the structure during preparation for entry or as a result of work activity.

The logical starting point for beginning an exploration of this subject is the picture provided by summary accident statistics. The Bureau of Labor Statistics in the U.S. maintains detailed statistics of industrial experience. Table 1.1 provides the average for the years 1992 to 2002 inclusive (11 years) for each type of event or exposure. These data originate from a search performed on the BLS website (www.bls.gov). The data in the BLS reports shift slightly from year to year. This shift can occur for many reasons including the state of economic activity and the emphasis placed by regulatory and enforcement agencies on a particular causative element listed in the table.

If hazardous energy is definable as the level of energy capable of interacting with biological tissue needed to produce immediate and serious injury, the information provided by Table 1.1 indicates that hazardous levels of energy are responsible for the vast majority (about 92%) of fatal injuries sustained by workers on the job.

Obtaining more detailed information about accidents caused by hazardous levels of energy is a difficult task. This necessitates assembly of accidents that have occurred by task or in a particular sector of industry or for a particular reason. Over the years, the effort needed to do this has resulted in a number of government-sponsored reports and some reports created independently. Despite the age of some of the documents, the data and underlying lessons that they contain are as pertinent and valuable today as the day they were issued.

The Division of Safety Research of the National Institute for Occupational Safety and Health commissioned one of the first quantitative analyses of fatal accidents involving hazardous energy during maintenance and servicing (NIOSH 1983a). This document resulted from analysis of 300 accident scenarios found in 14 different literature sources. Of these, the authors selected 59 accidents to examine in more detail. The document provides very brief summaries of these accidents.

The authors had planned to perform statistical analysis on the data in order to identify factors contributing to the type of accident under consideration. This proved to be impossible due to the absence of the specifics in the accident reports needed to identify factors contributing to or causing the accidents. Absence of specifics is a common theme in this area of endeavor. The investigator is forced to "read between the lines" when performing an investigation. This results partly from the absence of detail, relevance and consistency in accident reports. This information is used primarily for processing of claims for workers' compensation. These focus on the nature of the injury, and not the accident. Attempts to gain information through trade groups, companies, unions, and insurance carriers similarly proved fruitless due to its proprietary nature or liability issues.

The authors categorized the causes of accidents in the following manner:

- Failure to de-energize the equipment or system or to control hazards associated with energy sources prior to initiating maintenance activities
- Inadequate blockage or isolation of energy sources
- Failure to dissipate residual (potential) energy
- Unintended or inappropriate activation of energy sources

The authors used these findings to illustrate the need for adequate methods of energy control.

The Division of Safety Research published a second report immediately on safety in grain elevators and feed mills (NIOSH 1983b). This document provided a focused, albeit now hopefully outdated, look at accident statistics for this industry. A review of the literature provided in the document indicated that of all dust explosions occurring in the U.S. from 1900 to 1956, about 48% occurred in industries handling grain, feed and flour (Theimer 1973). Chiotti and Verkade (1976) studied dust explosions that occurred in grain elevators and feed and cereal mills during the period from 1958 to 1975. Those occurring in grain elevators were the most frequent and caused the greatest number of injuries and amount of property damage. Dust explosions occurred in 137 grain elevators and 50 feed and

## Table 1.1

## Summary Statistics for Fatal Occupational Accidents 1992 to 2002

| Event or Exposure | Percent within Category | Total |
|---|---|---|
| Transportation | | 42 |
| Highway | 22 | |
| Other | 20 | |
| Contact with Objects and Equipment | | 16 |
| Strike by | 9 | |
| Caught in | 7 | |
| Assaults and Violent Acts | | 18 |
| Homicide | 14 | |
| Other | 4 | |
| Falls | | 11 |
| Falls to lower level | 10 | |
| Other falls | 1 | |
| Harmful Substances or Environments | | 9 |
| Electrocution | 5 | |
| Other | 4 | |
| Fires and Explosions | | 4 |

Source: www.bls.gov website report.

cereal mills in the U.S., and resulted in 336 injuries and 51 deaths. The U.S. Department of Agriculture compiled a later list of 434 explosions covering the period from 1958 to 1982 (USDA 1982). These resulted in 776 injuries and 209 deaths. Injuries and deaths varied considerably by year.

An earlier report listed probable ignition sources for 250 explosions (USDA 1980). The most frequent was welding and cutting, This was followed by electrical failure, presence of tramp metal, and fire other than welding and cutting. By far, the most probable location for the primary explosion was the bucket elevator. This was followed by grinding equipment and storage bins.

Currently there are considerable interest and attention focused on explosions involving combustible dust. The latter consists of dust of industrial and agricultural origin, although not flour.

The Occupational Safety and Health Administration (OSHA) of the U.S. Department of Labor published a series of analytical reports on fatal accidents in the 1980s (OSHA 1982a, OSHA 1982b, OSHA 1983, OSHA 1985, OSHA 1988, OSHA 1990). Each report focused on a specific type of hazardous condition (fires and explosions, lockout/tagout, atmospheric hazards in confined spaces), type of work (welding and cutting) or work environment (grain handling, ship repair), and provided a summary for each accident used in the investigation.

The first OSHA report examined the link between fatal accidents and fires and explosions in confined workspaces (OSHA 1982a). Associated with the 50 accidents selected for the study during the period 1974 to 1979 were 76 fatal injuries. Heat and the energy generated by the fire and/or explosion were the major causative agents. Multiple fatalities occurred in many of these accidents. Some of the accidents involved process hazards.

The second OSHA report examined fatal accidents involving energized equipment (OSHA 1982b). Associated with the 83 cases selected for the study during the period 1974 to 1980 were 83 fatalities. Although the involvement of confined spaces was not considered a factor during this study, the geometry of many of the workspaces would satisfy criteria generally accepted at the present time.

The OSHA study of grain handling reported on 105 fatal accidents that occurred during the period 1977 to 1981 (OSHA 1983). Associated with these accidents were 126 fatal injuries. Although the involvement of confined spaces was not considered a factor during this study, the geometry of many of the workspaces would satisfy criteria generally accepted at the present time.

The fourth OSHA report examined the relationship between fatal accidents in confined workspaces and toxic and asphyxiating (oxygen-deficient) atmospheres and other causes (OSHA 1985). While most of the accidents involved hazardous atmospheric conditions, some were caused by other hazards. Associated with the 122 accidents selected for the study during the period 1974 to 1982 were 173 fatalities.

The fifth report published by OSHA examined the relationship between fatal accidents and welding and cutting (OSHA 1988). Associated with the 217 accidents selected for the study during the period 1974 to 1985 were 262 fatal injuries. In 22% of the accidents, the welder was working in a workspace that would satisfy criteria generally accepted today for inclusion as confined spaces.

The OSHA study of shipbuilding and repairing reported on 151 fatal accidents that occurred during the period 1974 to 1984 (OSHA 1990). Associated with these accidents were 176 fatal injuries. In 36% of the accidents, the victim was working in a workspace that would satisfy criteria generally accepted today for inclusion as confined spaces.

The OSHA reports repeatedly commented about the lack of completeness of the data used in these studies. This occurred because the system for coding causal factors in accidents contained no reference to the parameters examined in these studies. The forms were injury based for compensation purposes, rather than accident based. As well, data were not available from all of the U.S. states. Despite these shortcomings, the summaries from individual accidents contained in these reports are an invaluable resource for further enquiry.

During the same period, NIOSH initiated the Fatality Assessment and Control Evaluation (FACE) program (formerly Fatal Accident Circumstances and Epidemiology) (NIOSH 1994, NIOSH 2000). FACE investigated 423 accidents involving the deaths of 480 workers from 1983 to 1993. The purpose of FACE was to identify factors that increase the risk of work-related fatal injury. While previous studies utilized only written reports or forms as the source of information, FACE included on-site visits. These visits provided the opportunity for first-hand observation and investigation of the worksite by a team of specialists and interviews with witnesses. The intent of the team approach was to provide continuity and consistency from one investigation to another and to reconstruct the accidents in a repeatable manner. FACE currently focuses on accidents involving falls from heights, contact with sources of electrical energy, entry into confined spaces and contact with machinery. With the publication of *Worker Deaths in Confined Spaces* (NIOSH 1994) and *Worker Deaths by Electrocution* (NIOSH 2000), NIOSH has made available a major tool for researching the etiology and causative factors in these accidents.

*Worker Deaths in Confined Spaces* documented worker deaths in confined spaces from 1983 to 1993 (NIOSH 1994). This document provided summaries for 70 investigations that resulted in 109 fatalities. This NIOSH report also included summary statistics from the NTOF (National Traumatic Occupational Fatalities) database for the period 1980 to 1989. The NTOF database provides a comprehensive view of fatal injuries by industry, reason for entry, and other epidemiologically significant categories.

The NTOF data quoted in the NIOSH report provide information about engulfment in trenches and other excavations, as well as confined spaces. Depending on the jurisdiction, trenches and excavations may or may not be considered confined spaces. OSHA reportedly set a depth of 5 feet (1.5 m) as the dividing line between a trench and a confined space (NIOSH 1994). This arbitrary division and the separate treatment of trenches and excavations by NIOSH and other researchers means that engulfment by soils following cave-in by walls of excavated holes in the ground may or may not have received thorough inquiry.

Engulfment in loose materials in confined spaces during the period 1980 to 1989 resulted in 227 of the 670 deaths reported in the NTOF database. Of these, 124 deaths occurred in grain and 26 in other agricultural products, 25 in sand, 22 in gravel, cement, or clay, 11 in sawdust, 11 in other materials, and 8 in unknown material. Although not stated in the report, the maximum number of accidents was 227 of the 585 total, assuming one fatality per accident.

The NTOF database also indicates that 576 accidents and 606 fatalities occurred in trenches and excavations during the period 1980 to 1989. Of the fatalities, 468 occurred in construction, and no more than 28 occurred in any other industrial sector. NIOSH cautions that this estimate should be considered a minimum due to problems in interpreting information on death certificates, the main source of information for the database.

More recently, NIOSH published a document outlining the details of worker deaths by electrocution (NIOSH 2000). Electrocutions are the fifth leading cause of death in U.S. workplaces, accounting for 7% (411 deaths per year) of all fatalities according to the NTOF database. A total of 5,348 workers were electrocuted in 5,180 accidents during the period from 1980 to 1992. This estimate is believed to represent about 80% of the true picture due to limitations on the information-gathering process. Of the accidents, 153 (3%) resulted in multiple fatalities. That is, 140 accidents caused 2 fatalities, 11 caused 3, and 2 accidents caused 4 fatalities. The report also provides detail about annual occurrence of accidents, age of the victims, and industrial grouping. The FACE program investigated 224 of the electrocution accidents that resulted in 244 fatalities from 1982 to 1994.

The Mine Safety and Health Administration (MSHA) compiled summaries of fatal accidents in mining (MSHA 1988). This report primarily detailed accidents resulting from the fluid behavior of bulk solids under conditions of slumpage. MSHA documented 44 fatalities from 38 accidents that occurred during the period 1980 to 1986. MSHA subsequently reported that 57 fatalities had occurred by the end of 1993 (MSHA 1994). MSHA noted that the majority of the fatalities resulted from collapse of bridged and caked material in bins, silos, and hoppers. Some of these accidents occurred in workspaces that fit the description of confined spaces.

The fatal accident summaries in the OSHA, MSHA, and NIOSH reports provide the starting point for obtaining additional information about these accidents (OSHA 1982a, OSHA 1982b, OSHA 1983, OSHA 1985, OSHA 1988, OSHA 1990; MSHA 1988; NIOSH 1994, NIOSH 2000). Most of the reports provide cryptic comments about the accidents from forms compiled by investigators at the scene.

Table 1.2 summarizes the data presented in the OSHA, MSHA, and NIOSH reports. A note of caution is needed before entering into overly enthusiastic use of this data. These reports span partially overlapping time-frames of varying length. They also are potentially dated because of their age compared to the current era. Some of the accidents were utilized in more than one report, for example, welding and cutting, and fire and explosion accidents in shipbuilding and repair. This is a natural outcome from the overlap between work activity and work environment. As well, not all of the accidents believed to have occurred were available to the researchers who created these documents. That said, this is the extent of the information base that exists from which any analysis can occur.

Accidents in categories, such as fire and explosion, are focused compared to those in shipbuilding and repair that report on one segment of industry. While accidents due to welding and cutting, fire and explosion, and toxic and asphyxiating atmospheres occurred in shipbuilding and repairing, so also did other types. Similarly, while accidents due to welding and cutting, fire and explosion, and toxic and asphyxiating atmospheres occurred during grain handling, so also did accidents due to lockout and engulfment.

## Table 1.2
## Summary of Data Contained in NIOSH, OSHA, and MSHA Reports

| Action/Condition | Period | Accidents | Fatalities | F/A |
|---|---|---|---|---|
| Fire/explosion (OSHA 1982a) | 1974-1979 | 50 | 76 | 1.5 |
| Lockout/tagout (OSHA 1982b) | 1974-1980 | 83 | 83 | 1.0 |
| Grain handling (OSHA 1983) | 1977-1981 | 105 | 126 | 1.2 |
| Toxic/asphyxiating atmospheres (OSHA 1985) | 1974-1982 | 122 | 173 | 1.4 |
| Welding/cutting (OSHA 1988) | 1974-1985 | 217 | 262 | 1.2 |
| Shipbuilding/repair (OSHA 1990) | 1974-1984 | 151 | 176 | 1.2 |
| Mining (MSHA 1988) | 1980-1986 | 38 | 44 | 1.2 |
| Confined spaces (NIOSH 1994) | 1983-1993 | 70 | 109 | 1.6 |
| Electrocution (NIOSH 2000) | 1982-1994 | 224 | 244 | 1.1 |

F/A is fatalities per accident.

Sources: OSHA 1982a, OSHA 1982b, OSHA 1983, OSHA 1985, OSHA 1988, OSHA 1990;
MSHA 1988; NIOSH 1994, NIOSH 2000.

The ratio of fatalities to accidents provides a measure of the severity of accidents occurring in each category. The ratio of fatalities to accidents exceeded unity in all categories except lockout and tagout. This means that more than one fatal injury occurred during many of these accidents.

Table 1.3 illustrates the role of hazardous energy in the OSHA, MSHA, and NIOSH reports (OSHA 1982a, OSHA 1982b, OSHA 1983, OSHA 1985, OSHA 1988, OSHA 1990; MSHA 1988; NIOSH 1994, NIOSH 2000). This classification reflects the interpretation of information provided in the anecdotal summary for each accident. To the extent possible in this and subsequent analysis, duplication of accident summaries that appeared in more than one report was eliminated. The most prominent use of data from other sources occurred in the report on shipbuilding and repair.

Accidents selected for use in this and further analysis reflected situations where energy stored in materials or in sources manipulated by the victim or other workers caused the accident. This selection excluded sources, such as welding equipment, that are necessary to performance of the work. Acci-

## Table 1.3
## Role of Hazardous Energy in Fatal Accidents

| Action/Condition | Accidents | | | Fatalities | | | F/A |
|---|---|---|---|---|---|---|---|
| | $A_{HE}$ | $A_T$ | % | $F_{HE}$ | $F_T$ | % | |
| Fire/explosion (OSHA 1982a) | 8 | 50 | 16 | 16 | 76 | 21 | 2.0 |
| Lockout/tagout (OSHA 1982b) | 72 | 83 | 87 | 72 | 83 | 87 | 1.0 |
| Grain handling (OSHA 1983) | 51 | 105 | 49 | 56 | 126 | 44 | 1.1 |
| Toxic/asphyxiating atmospheres (OSHA 1985) | | | | | | | |
| • Toxic | 17 | 55 | 31 | 21 | 80 | 34 | 1.6 |
| • Asphyxiating | 13 | 46 | 28 | 21 | 71 | 30 | 1.6 |
| • Other | 7 | 21 | 33 | 7 | 22 | 32 | 1.0 |
| • Total | 37 | 122 | 30 | 55 | 173 | 32 | 1.5 |
| Welding/cutting[*] (OSHA 1988) | | | | | | | |
| • Interior | 7 | 164 | 4 | 7 | 190 | 4 | 1.0 |
| • Exterior | 57 | 164 | 35 | 67 | 190 | 35 | 1.2 |
| Shipbuilding/repair[@] (OSHA 1990) | 12 | 151 | 8 | 14 | 176 | 8 | 1.2 |
| Mining (MSHA 1988) | 31 | 38 | 82 | 37 | 44 | 84 | 1.2 |
| Confined spaces (NIOSH 1994) | 8 | 70 | 11 | 8 | 109 | 7 | 1.0 |
| Electrocution (NIOSH 2000) | 220 | 224 | 98 | 240 | 244 | 98 | 1.1 |

$A_{HE}$ means fatal accidents associated with hazardous energy.

$A_T$ means total accidents.

$F_{HE}$ means fatalities associated with hazardous energy/hazardous energy sources.

## Table 1.3 (Continued)
## Role of Hazardous Energy in Fatal Accidents

$F_T$ means total fatalities.

@ means accident summaries utilized in other reports are not used in these statistics.

* The data on welding and cutting accidents used in this table reflect analysis of individual accident summaries contained in the OSHA report (OSHA 1988). The report claims that this abridged group (164 cases and 190 associated fatalities) reflects the total (217 cases and 262 associated fatalities).

Sources: OSHA 1982a, OSHA 1982b, OSHA 1983, OSHA 1985, OSHA 1988, OSHA 1990; MSHA 1988; NIOSH 1994, NIOSH 2000.

dents, where energy sources, such as welding equipment, activated energy stored in chemical substances and caused fires and explosions, were included for consideration. The secondary impact, that is, the fire or explosion, occurred because of the primary source, the welding arc or cutting flame. Accidents considered for use in this analysis occurred in open workspaces, as well as confined spaces.

The term hazardous energy as applied to this analysis refers to energy sources associated with the surroundings, rather than energy sources under direct control of the victim, for instance a welding torch. The use of the term in this context is intended to refer to energy sources needing assessment and control in order to perform the job safely. The victims presumably had some level of control over energy sources, such as welding equipment, that they used to perform the job.

According to this analysis, hazardous energy associated with the surroundings played the biggest role in electrical accidents, as measured by percentage of the total number, followed by lockout/tagout (primarily entanglement in mechanical equipment), mining (mainly engulfment in bulk material), and grain handling (mainly engulfment in bulk material) accidents. The smallest association with energy sources associated with the surroundings occurred in welding and cutting accidents occurring in confined spaces and in shipbuilding and repair.

The ratio of fatalities to accidents was highest for accidents occurring in confined spaces (fires and explosions, and toxic and asphyxiating atmospheres) according to the OSHA data. This is not supported by statistics derivable from the NIOSH data for the succeeding decade. This former would tend to confirm the increase in severity for accidents occurring in confined spaces compared to the overall group as detailed in other texts (McManus 1999).

## Control of Hazardous Conditions

A major issue in the causation, and ultimately prevention of accidents of this type, is control over hazardous conditions. That is, who has control over the work to be performed, and does that person recognize hazardous conditions posed by the environment and the work activity? Given that knowledge, is that person in a position to exercise control, and what happens when that occurs? In the circumstances to be examined here, all of this is past tense, since the sources of information are fatal accidents in which at least one person perished. In order to discuss this properly and to understand the implications of the analysis to follow, some insight is needed in how organizations function.

In many organizations, the equipment on which work is to occur is "owned" by production. When assistance is needed, production calls maintenance. Maintenance may be an internal function or one that is outsourced. Outsourcing of maintenance has become increasingly common with the downsizing or right-sizing of many operations.

Internal service providers in a small organization may know intimately the equipment to be serviced. On the other hand, with many pieces of equipment and inadequate staffing due to downsizing and retirement, those servicing the equipment may be completely unfamiliar with it. This also could occur due to infrequency of repairs to be made on a particular piece of equipment. There also may be continuity issues, in that only one individual works on a particular piece of equipment, or for that matter, all equipment in the organization. In a large organization, many people potentially may work on a particular piece of equipment. As well, this work can span more than one shift. This transference to more than one individual or even the possibility of transference through unanticipated or unauthorized action of another individual could change the status quo of a situation as expected by the initiating individual.

An option used by many organizations is outsourcing of maintenance to external service providers. This could apply to single pieces of equipment or to the entire operation. In either event, an individual who is unfamiliar with the intimate workings of the operation is required to work on a particular piece of equipment, that, in turn, is under the influence of other equipment both in the vicinity and remote to it.

This examination of the issue of control (Table 1.4) reflects the situation perceivable from summaries of fatal accidents from the perspective of the person performing the action that led to the accident.

In many types of accidents, persons who were external to the process or who had little reason to come into contact with the hazardous energy were unable to exercise control over exposure to it. The data suggest that this is the case even when the victim worked in the affected workspace. This situation is especially true for accidents involving electrocution (death due to exposure to electricity). Inadvertent contact between conductive structures, such as aluminum ladders and mobile augers, and uninsulated power lines, for example, occurs all too frequently.

## Temporal Development of Hazardous Conditions

The temporal relationship between development of the hazardous condition and onset of the accident is an important one. Hazardous conditions could exist prior to the start of work or they could develop during the performance of work activity.

The OSHA, MSHA and NIOSH reports provide a means to investigate this question. (OSHA 1982a, OSHA 1982b, OSHA 1983, OSHA 1985, OSHA 1988, OSHA 1990; MSHA 1988; NIOSH 1994). Table 1.5 compiles data obtained from a review of the accident summaries contained in these reports. The data are compiled according to type of hazardous condition. This information was compiled by topic from all of the reports, since small numbers of accidents were involved in some classifications. Some of the reports contained more than one type of hazardous condition. This is especially true for reports based on industries, such as grain handling, and shipbuilding and repair. To the extent possible, individual reports are used only once. The data reflect subjective estimation of the situations described in the accident summaries, and of course, suffer from this limitation. Second-guessing the actions of accident victims where eyewitnesses were not present, of course, is not possible.

The data in this compilation suggest that certain types of hazardous condition are linked to fundamental aspects in the way that work is performed. For example, hazardous conditions linked to engulfment and fires and explosions are most likely to develop as a result of work activity. This is analogous to sending the worker into a cage to stir up a sleeping tiger. Electrocution and entanglement hazards are most likely to develop prior to the start of work. In the case of electrical accidents, this is strongly so. Entanglement hazards involving mostly mechanical equipment depend on the nature of work. In this case, machinery often was deactivated to a shutdown or standby status while some action occurred. The victim or a third party reactivated the energy source while the victim was still in contact with it. The actions of the victim in these circumstances were inadvertent. In some cases, the actions of the third party were highly deliberate. This, by the way, is not to be construed as meaning that the third party intended to harm the victim. Rather, this means that the third party did not know that the victim was in contact with the energy source or believed that the victim was no longer in contact with the energy source.

## Table 1.4

## Role of Control of Hazardous Conditions in Fatal Accidents Involving Hazardous Energy

| Action/Condition | Number of Accidents | | | |
| --- | --- | --- | --- | --- |
| | External | Internal | Control | No Control |
| Fire/explosion | 2 | | 0 | 2 |
| (OSHA 1982a) | | 6 | 1 | 5 |
| Lockout/tagout | 4 | | 1 | 3 |
| (OSHA 1982b) | | 68 | 36 | 32 |
| Grain handling | 7 | | 0 | 7 |
| (OSHA 1983) | | 44 | 14 | 30 |
| Toxic/asphyxiating atmospheres | | | | |
| (OSHA 1985) | | | | |
| • Toxic | 8 | | 2 | 6 |
| | | 9 | 4 | 5 |
| • Asphyxiating | 4 | | 2 | 2 |
| | | 9 | 7 | 2 |
| • Other | 1 | | | 1 |
| | | 6 | 3 | 3 |
| • Total | 13 | | 4 | 9 |
| | | 24 | 14 | 10 |
| Welding/cutting[*] | | | | |
| (OSHA 1988) | | | | |
| • Interior | 14 | | 8 | 6 |
| • Exterior | | 50 | 22 | 28 |
| Shipbuilding/repair[@] | 5 | | 3 | 2 |
| (OSHA 1990) | | 7 | 3 | 4 |
| Mining | 2 | | 0 | 2 |
| (MSHA 1988) | | 29 | 6 | 23 |
| Confined spaces | 4 | | 1 | 3 |
| (NIOSH 1994) | | 4 | 2 | 2 |
| Electrocution | 90 | | 7 | 83 |
| (NIOSH 2000) | | 130 | 64 | 66 |

**Table 1.4 (Continued)**

**Role of Control of Hazardous Conditions in Fatal Accidents Involving Hazardous Energy**

External means that the normal work routine of the victim did not involve contact with the hazardous energy involved in the accident.

Internal means that the work routine of the victim normally did involve contact with the hazardous energy, and that effects of exposure and protection against it would be discussed as part of normal procedural and safety training.

Control means that the victim could have exercised control over exposure to the hazardous energy, and failed to do so, either by deliberate choice or subconscious one.

No control means that the victim had no option for exercising control over exposure to the hazardous energy. The victim may not have recognized the existence of the hazardous condition.

@ means accident summaries utilized in other reports are not used in these statistics.

* The data on welding and cutting accidents used in this table reflect analysis of individual accident summaries contained in the OSHA report (OSHA 1988). The report claims that this abridged group (164 cases and 190 associated fatalities) reflects the total (217 cases and 262 associated fatalities).

Sources: OSHA 1982a, OSHA 1982b, OSHA 1983, OSHA 1985, OSHA 1988, OSHA 1990; MSHA 1988; NIOSH 1994, NIOSH 2000.

Additional elaboration of this information is possible (Table 1.6). Several subdivisions exist for the temporal relationship between the victim and the energy source at the onset of the accident.

**Preexisting Hazardous Conditions**

• Knowledgeable Defiance: The victim attempted to perform the task without adherence to procedures or use or proper use of protective measures, despite the possibility of contact, with the full knowledge that the energy source was active. Full knowledge can arise from visual or other cues from the senses or from knowledge gained through training. This classification also covers actions taken in defiance of procedures or direct orders, or requests from supervisors and others. An example that illustrates this classification is intentional interaction with a machine during its operating cycle or knowingly working without protection on electrical circuits or uninsulated, energized electrical conductors.

• Knowledgeable Helplessness: The victim recognized the presence of the energy source, but performed the task regardless. The victim had no training in control of the energy source, or perceived the ability to perform the task without undue risk or perceived the lack of availability of protective measures. Protective measures include shutdown of the energy source, or installation of shielding between the source and the person, or use of personal protective equipment to prevent contact with it.

• Lack of Knowledge (Unknowing): The hazard of the energy source was not perceivable through the senses or recognizable by other common means, or the person lacked the knowledge needed to recognize its presence. An example that illustrates this classification is electrocution that occurred during performance of unrelated work around live electrical wires or wiring.

## Table 1.5
## Hazardous Conditions and the Work Cycle

**Number of Accidents**

| Condition | Electrocution | Engulfment | Entanglement | Fire/explosion | Process |
|---|---|---|---|---|---|
| Pre-existing | 234 | 11 | 39 | 4 | 35 |
| Work activity | 15 | 56 | 2 | 53 | 5 |
| Resurrected | 3 | – | 40 | 4 | 3 |

Pre-existing means that the hazardous condition existed prior to the action performed by the victim.

Work activity means that the activity performed created the hazardous condition.

Resurrected means that the hazardous condition arose from a circumstance independent of the work activity that did not exist prior to or result from the action performed by the victim. Resurrected conditions result from unexpected reactivation of equipment and energy sources.

Process means situations involving the movement of bulk materials usually through piping at or above normal atmospheric pressure into structures in which the victim was present. These could include gases or vapors, liquids, slurries and flowable solid materials.

Sources: OSHA 1982a, OSHA 1982b, OSHA 1983, OSHA 1985, OSHA 1988, OSHA 1990; MSHA 1988; NIOSH 1994, NIOSH 2000.

---

**Work Activity**

Actions of the victim precipitated the accident through use of an energy source associated with the task, or through deliberate destabilizing of energy contained in materials, such that transformation could occur. Entry into a silo for the express purpose of initiating or improving flow of bulk material is an example of such a situation. The material in the silo and the energy it contains are stable until disturbed.

**Resurrected Hazardous Conditions**

- Self-activated: The energy source was inactive at the time of onset of the task. The victim through some action inadvertently re-energized it.

- Other-activated: A third party activated the energy source while the victim performed the task that led to contact with it. In some circumstances, this occurred with the full knowledge of the victim.

This compilation of data from the accident summaries suggests that with some hazardous conditions, namely electrical, entanglement, and process hazards, failure to follow procedures by individuals trained in hazard recognition and personal protection was a significant factor in the causation of these accidents. In the case of electrical hazards, inability to do anything feasible about the hazard, labeled as helplessness, was the dominant factor. Lack of recognition was a factor in electrical and process-related accidents. Activation of the energy source after being "stabilized" into the standby or even deactivated state was mainly a factor in entanglement accidents.

There is a danger in reporting summary statistics that the sum introduces a bias toward a particular industry or type of work. Table 1.7 provides the summaries by individual source report.

## Table 1.6
## Temporal Development of Hazardous Conditions

Number of Accidents

| | Electrocution | Engulfment | Entanglement | Fire/Explosion | Process |
|---|---|---|---|---|---|
| **Pre-Existing Hazardous Conditions** | | | | | |
| Defiance | 60 | 1 | 35 | 3 | 17 |
| Helplessness | 105 | 1 | 4 | – | 2 |
| Unknowing | 69 | 9 | – | 1 | 6 |
| **Work Activity** | 15 | 56 | 2 | 53 | 5 |
| **Resurrected Hazardous Conditions** | | | | | |
| • Self-activated | 2 | – | 23 | 1 | – |
| • Other-activated | 1 | – | 17 | 3 | 3 |

Sources: OSHA 1982a, OSHA 1982b, OSHA 1983, OSHA 1985, OSHA 1988, OSHA 1990; MSHA 1988; NIOSH 1994.

As would be expected, certain industries and activities dominate particular hazardous conditions. Small numbers of accidents were reported in unrelated reports.

The preceding reports reflect, almost exclusively, conditions that existed in the 1970s and 1980s. With the implementation of regulations in the US to control hazardous energy (OSHA 1989) and work in confined spaces (OSHA 1993, OSHA 1994), downward trend in occurrence would be expected.

The nature of accidents that occurred under these headings reflects the culture of the workplace and the nature of the work itself. This is the subject of more detailed examination in the next sections. For example, in industries where engulfment occurs in storage and material-handling structures, entering to dislodge hang-ups and to break bridges and rat-holes to improve flow was accepted practice. This practice may continue to occur in the industrialized world, despite the imposition of regulations. Entry into a storage structure or material-handling structure where material exists in an unstable state of energy equilibrium is akin to entering a cage at the zoo containing a tiger. The behavior of the tiger is semi-predictable until the animal is provoked (maybe). Entering the same structure during energy conversion, as occurs during attempts to increase flow, is akin to entering a tank full of sharks during a feeding frenzy.

## Nature of Hazardous Conditions

This section provides further elaboration about the nature of fatal accidents contained in NIOSH, MSHA, and OSHA reports. This examination considers the situation from the perspective of the victim; that is, from the perspective of knowledge, training, and experience. What should be second nature to individuals in some job classifications could be completely foreign to individuals in others.

## Table 1.7
## Fatal Accidents by Source Report

| Condition | F/E<br>n = 8 | L/T<br>n = 72 | GH<br>n = 51 | CS<br>n = 37 | W/C*<br>n = 64 | S/R<br>n = 12 | Min<br>n = 31 | NIOSH<br>n = 8 | EI<br>n = 221 |
|---|---|---|---|---|---|---|---|---|---|
| **Number of Accidents** | | | | | | | | | |
| **Electrocution Accidents** | | | | | | | | | |
| Pre-existing hazardous conditions | | | | | | | | | |
| • Knowing | – | 3 | – | 1 | 1 | 1 | – | 1 | 53 |
| • Helpless | – | 1 | 9 | – | 1 | 1 | – | – | 93 |
| • Unknowing | – | 3 | – | – | 3 | 2 | – | 2 | 59 |
| Work activity | – | – | – | – | 1 | – | – | – | 14 |
| Post-existing hazardous conditions | | | | | | | | | |
| • Self-activated | – | – | – | – | – | – | – | – | 2 |
| • Other-activated | – | – | – | – | – | 1 | – | – | – |
| **Engulfment Accidents** | | | | | | | | | |
| Pre-existing hazardous conditions | | | | | | | | | |
| • Knowing | – | – | – | – | – | – | 1 | – | – |
| • Helpless | – | – | – | – | – | – | 1 | – | – |
| • Unknowing | – | – | – | – | – | – | 9 | – | – |
| Work activity | – | – | 28 | 3 | – | – | 20 | 5 | – |
| Post-existing hazardous conditions | | | | | | | | | |
| • Self-activated | – | – | – | – | – | – | – | – | – |
| • Other-activated | – | – | – | – | – | – | – | – | – |
| **Entanglement Accidents** | | | | | | | | | |
| Pre-existing hazardous conditions | | | | | | | | | |
| • Knowing | – | 26 | 7 | 1 | 1 | – | – | – | – |
| • Helpless | – | 4 | – | – | – | – | – | – | – |
| • Unknowing | – | – | – | – | – | – | – | – | – |

## Table 1.7 (Continued)
## Fatal Accidents by Source Report

| Condition | F/E $n = 8$ | L/T $n = 72$ | GH $n = 51$ | CS $n = 37$ | W/C* $n = 64$ | S/R $n = 12$ | Min $n = 31$ | NIOSH $n = 8$ | EI $n = 221$ |
|---|---|---|---|---|---|---|---|---|---|
| **Number of Accidents** | | | | | | | | | |
| Work activity | – | – | – | – | – | – | – | – | – |
| Post-existing hazardous conditions | | | | | | | | | |
| • Self-activated | – | 22 | – | – | 1 | – | – | – | – |
| • Other-activated | – | 13 | 1 | – | 2 | 1 | – | – | – |
| **Fires and Explosions (fuels)** | | | | | | | | | |
| Pre-existing hazardous conditions | | | | | | | | | |
| • Knowing | – | – | – | – | 3 | – | – | – | – |
| • Helpless | – | – | – | – | – | – | – | – | – |
| • Unknowing | – | – | – | – | 1 | – | – | – | – |
| Work activity | – | – | 6 | – | 47 | – | – | – | – |
| Post-existing hazardous conditions | | | | | | | | | |
| • Self-activated | – | – | – | – | – | 1 | – | – | – |
| • Other-activated | – | – | – | – | – | 1 | – | – | – |
| **Process Accidents** | | | | | | | | | |
| Pre-existing hazardous conditions | | | | | | | | | |
| • Knowing | – | – | – | 16 | – | 1 | – | – | – |
| • Helpless | – | – | – | 2 | – | – | – | – | – |
| • Unknowing | – | – | – | 13 | 1 | 2 | – | – | – |
| Work activity | 5 | – | – | – | – | – | – | – | – |
| Post-existing hazardous conditions | | | | | | | | | |
| • Self-activated | – | – | – | – | – | – | – | – | – |
| • Other-activated | 1 | – | – | 1 | – | 1 | – | – | – |

F/E is fires and explosions.
L/T is lockout/tagout.

GH is grain handling.

CS is hazardous non-atmospheric conditions in confined spaces.

**Table 1.7 (Continued)**

**Fatal Accidents by Source Report**

W/C is welding and cutting.

S/R is shipbuilding and repair.

Min is mining.

NIOSH is data from the FACE study.

EI is electrical accidents.

* The data on welding and cutting accidents used in this table reflect analysis of individual accident summaries contained in the OSHA report (OSHA 1988). The report claims that this abridged group (164 cases and 190 associated fatalities) reflects the total (217 cases and 262 associated fatalities).

Similarly, the nature of the work performed by individuals in different job classifications and the manner by which they come into contact with the hazardous condition likely will be completely different. Predicting what motivated an individual to perform an activity in a certain way is beyond the scope of this endeavor and potentially highly disrespectful to the memory of the deceased, about whom this type of discussion occurs in the abstract.

**Electrocution**
Electrocution is a fatal injury caused by electricity (NIOSH 2000). Electrocution is one of four main types of electrical injury, the others being electric shock, burns and falls. Electrocution in real-world accidents results from contact with energized conductors. Electrocution can occur at voltages less than 110 V (50 to 60 V), although 110 V and upward is what appears in accident records. The most productive study of accident records is based on voltage. Table 1.8 summarizes the nature of contact involved in job-related electrocution accidents.

High-Voltage Accidents Involving Electrical Workers
High-voltage electrical circuits usually mean electrical circuits operating at greater than 600 V AC (Volts Alternating Current) between conductors or between a conductor and ground (NFPA 1999). Note that this distinction is not universal in North America, as some jurisdictions use 750 V AC as the boundary between low and high voltages (WCB 1998). However, this distinction has no relevance to this discussion. These voltages are usually encountered in electrical generation, and in transmission and distribution systems in lines mounted at the top of wooden utility poles and high-voltage tubular and metal frame towers.

Table 1.9 summarizes information on fatal accidents involving electrocution of electrical workers during work on high-voltage transmission and distribution lines (NIOSH 2000). The term electrical worker, as used here, means persons whose employment normally brings them into potential contact with high-voltage lines and equipment. This can occur through employment by an electrical utility or through employment by a contractor performing work for an electrical utility. The distinction is that workers in these organizations likely have received training about hazardous conditions associated with this work and precautionary measures that are required. Organizations included in this discussion

### Table 1.8
### Nature of Contact Involved in Fatal Occupational Electrocutions

| Type of Contact | Number | Percent |
|---|---|---|
| Overhead power line | 139 | 40 |
| Wiring, transformers, other electrical components | 94 | 27 |
| Machine, tool, appliance, light fixture | 55 | 16 |
| Lightning | 15 | 5 |
| Underground/buried power line | 5 | 1 |
| Other/unspecified | 37 | 11 |

Source: BLS 1997.

include electrical utilities and their construction and electrical maintenance partners, as well as companies that perform ancillary functions including painting, demolition, and other tasks that take them with permission onto property owned by the utility.

Table 1.9 indicates that most of the victims were journeymen linemen or electricians performing similar tasks. Victims also included other members of crews performing this type of work. These included apprentices, support workers, and first and second level supervisors. Victims also included individuals employed by contractors performing work on utility property unrelated to line installation and maintenance. These included painters and construction laborers. Most victims were performing what appears to be normal work. Electrical work of this nature would be expected to include training about the hazards of high-voltage electricity. This logically would be expected as a condition for anyone to be permitted to work on utility property.

As would be expected, most of the fatal accidents occurred during line installation and maintenance. A substantial proportion of the accidents occurred at ground level. Accidents at heights in elevated work platforms (buckets on booms of bucket trucks) involved mostly linemen. Accidents occurring at ground level involved a much broader range of occupations.

Major factors in causation of the accidents included failure to use protective equipment, failure to use isolation equipment, and failure to ground equipment. Minor factors included failure to test or inappropriate test, loss of control over intended action, miscommunication, and complexity of the work task.

Work at heights typically occurs on an elevated work platform (bucket truck) or on the pole (reached by climbing using belt and spurs). The bucket truck utilizes a hydraulically positioned boom. Positioning of the boom and bucket requires finesse. Lack of skill in positioning led to some of the electrical contacts, possibly because of jerky movements. While the bucket(s) are nonconductive, the boom may be or become conductive. This is more likely the case with older equipment. In this case, the boom and the truck can become electrically live through contact with an energized conductor or by induction. Contact with electrically live mobile equipment led to numerous ground-level fatalities.

The modus operandi of work around high-voltage lines and equipment is to consider the circuit live, until proven otherwise. Sometimes even that cannot be assumed, as illustrated by fatal accidents involving downed distribution lines energized by backfeed from domestic portable generator sets, the latter being used to provide temporary power to a home or business while repair work occurred. In these cases, the main breaker in the building remained closed, thus creating a continuous path from the dwelling to the distribution line.

## Table 1.9
## Fatal Accidents: Electrical Workers and High-Voltage Conductors

| Element | Number of Accidents (n = 61) |
|---|---|
| **Occupation of Victim** | |
| Laborer | 7 |
| Lineman/electrician | 43 |
| Other trade | 2 |
| Supervisor | 8 |
| Manager | – |
| Other/unknown | 1 |
| **Work Activity** | |
| Usual | 54 |
| Unusual | 7 |
| **Nature of Work Activity** | |
| Installing line/fittings | 9 |
| Installing pole/fittings | 4 |
| Repairing/replacing equipment | 39 |
| Demolition | 2 |
| Other | 7 |
| **Work Height** | |
| Elevated | 41 |
| Ground level | 20 |
| **Other Factors** | |
| No test/inappropriate test | 6 |
| Equipment malfunction/failure | 7 |
| Loss of control over intended action | 7 |
| Miscommunication | 5 |
| Perspiration | 1 |

## Table 1.9 (Continued)
## Fatal Accidents: Electrical Workers and High-Voltage Conductors

| Element | Number of Accidents (n = 61) |
|---|---|
| Task complexity | 4 |
| Ground not installed | 16 |
| Isolation not installed | 28 |
| Isolation inappropriate/inadequate | 1 |
| Protective equipment inappropriate | 1 |
| Protective equipment not used | 26 |

Source: NIOSH 2000.

Tasks to be performed require strict adherence to procedures and control of conditions. Failure to adhere or inability to adhere to the procedural aspects of this work led to numerous electrocutions during elevated and ground-level work. This aspect of these accidents can be considered under loss of control over action or complexity of the procedure. This begins with positioning of the bucket or establishing position on the pole. Lack of finesse with operation of the controls can lead to jerky rather than smooth movement of the bucket. Considerable skill involving spatial perception and planning is needed to position the bucket in situations where several sets of wires intersect on a pole, or where the pole has several sets of wires at different levels on cross-arms. Jerky motion of the bucket can lead to unintended contact with energized conductors.

Handling and placement of rubber insulators is critical to the success of the procedure. Difficulty in working with this equipment has already received discussion. Additional difficulty has arisen when the end of a piece of tie wire has snagged the glove system and exposed the skin, causing contact. This caused at least one electrocution.

Further aspects involving lack of procedural control include handling of jumpers, unexpected motion of conductors and guy wires, failure of positioning hardware and cross-arms, and failure of wire connections. Unexpected and unpredictable motion represented loss of control over the spatial relationship between various components. These failures led to numerous ground-level electrocutions, resulting from contact between energized conductors and guy wires with vehicles, wire trailers, tensioning devices, and other equipment. Although not mentioned in these accidents, induction is an additional potential source of electrification. None of this equipment was grounded at the time. Victims were inside cabs, or in contact with the equipment through leaning or sitting on it. None of these individuals was protected or protectable through the use of any system of personal protection.

### High-Voltage Accidents Involving Non-Electrical Workers

The second major class of high-voltage electrical accident involved individuals not normally considered to be electrical workers. The work that these individuals perform in the course of their occupations, while unrelated to electricity, has the potential to bring them into contact with energized high-voltage conductors. Electrical workers who work on high-voltage lines and equipment receive extensive training about the hazardous nature of this energy source and measures for protection against it. Training of non-electrical workers, whose occupation can lead to contact with high-voltage lines and equipment, in these matters is much less certain.

Table 1.10 details fatal accidents occurring in occupations considered to be nonelectrical due to contact with energized, high-voltage conductors (NIOSH 2000; OSHA 1983, OSHA 1990).

As expected, most of the accidents involved hands-on workers (laborers and trades). Non-electrical workers identified in this section primarily include farmers and farm laborers, tree trimmers, painters and roofers, sign technicians, cable television installers, and structural construction workers. Truck drivers making deliveries of materials and equipment with trucks containing built-in cranes are an important subgroup. A high proportion of these contacts occurred during performance of normal duties. Most of the work at the time of the accident occurred at ground level. This included primarily the delivery and on-site handling of materials and equipment through use of cranes and related mobile equipment. Ground-level work also involved handling of protrusions, including ladders, poles, and mobile augers and related machinery used to extend reach above the surface of the ground. Some of this work occurred at elevation on elevated work platforms, ladders, and rooftops.

The data strongly suggest that victims had no control over the conditions under which the activity was to be performed. Coupled with this was the complexity of the task being performed.

Some accidents involved underground power lines. With increasing use of trenchless technology, the likelihood of contact with underground high-voltage cable increases considerably. This situation and the need for protection of individuals working overhead puts an increasing burden on utilities to assist with either shielding or shutting down this equipment.

## Low-Voltage Electrical Accidents

The third category of accidents examined here involved low-voltage electrical sources (less than 600 Volts). This category includes voltages encountered in a wide variety of industrial, commercial, and institutional settings, as well as the home (Table 1.11). (Homes are workplaces for some tradespeople.) Sources of information for this table were the NIOSH and OSHA reports (NIOSH 1994, NIOSH 2000; OSHA 1982b, OSHA 1985, OSHA 1988, OSHA 1990).

As expected, most of the victims were hands-on workers in the trades (laborers, electricians, and others). The predominant trade was electrical. A large proportion of the victims were performing work that would be normal to them. This included primarily the installation, testing, and maintenance and repair of electrical equipment. Victims performing unrelated work often were directly affected by the quality of work performed by electrical workers. These individuals, laborers, machine operators, cleaners, supervisors, and managers, whose daily activity would ordinarily not lead to contact with electrical equipment, were affected by wiring errors and sloppy work practices, including failure to close electrical junction and switch boxes. In some instances, machine operators, and in others, senior individuals, including supervisors and managers, attempted to troubleshoot faulty equipment by gaining access to control cabinets. The latter contain uninsulated conductors that operate at voltages well above 110 V. Most of the work occurred at ground level. Elevated work occurred primarily on ladders and sometimes on rooftops.

Testing to determine the presence of energized circuits was not performed in these accidents. Testing would have alerted many of the victims to the presence of live circuits or faulted circuits containing leakage paths. This presumably would have warned them not to venture forward.

Wiring errors occurring during initial installation and during renovations were responsible for numerous electrocutions. A subsidiary contributor was the removal of ground prongs on plugs. This led to loss of ground protection. Perhaps even more horrifying was substitution of improper plugs in order to fit certain receptacles. This led to mismatching of conductors and prongs and subsequent energization of outer cases of equipment. Use of plugs containing damaged fittings and equipment containing exposed conductors also contributed to the electrocutions. Owners of equipment failed to conduct periodic inspection to determine correctness and integrity of wiring.

Many of the victims involved in testing, installation, and renovation accidents worked "hot" as a matter of routine. Some individuals resorted to the use of nonconductive fiberglass ladders in the belief that these would isolate them and prevent establishing a path to ground. This strategy failed to account

## Table 1.10
## Fatal Accidents: Non-Electrical Workers and High-Voltage Conductors

| Element | Number of Accidents (n = 96) |
|---|---|
| **Occupation of Victim** | |
| Laborer | 34 |
| Truck driver | 17 |
| Other trade | 35 |
| Supervisor | 3 |
| Manager | 2 |
| Other/unknown | 5 |
| **Work Activity** | |
| Usual | 82 |
| Unusual | 14 |
| **Nature of Work Activity** | |
| Material/equipment delivery | 21 |
| Material handling | 15 |
| Equipment handling | 42 |
| Other | 18 |
| **Work Height** | |
| Elevated | 13 |
| Ground-level | 83 |
| **Other Factors** | |
| Equipment malfunction/failure | 3 |
| Loss of control over intended action | 81 |
| Miscommunication | 4 |
| Task complexity | 79 |
| Ground not installed | 3 |
| Other | 4 |

## Table 1.10 (Continued)
## Fatal Accidents: Non-Electrical Workers and High-Voltage Conductors

Sources: NIOSH 2000; OSHA 1983, OSHA 1990.

for the presence of conductive paths in metal structures (structural beams and the frames of suspended ceilings) with which they would come into contact.

Several individuals performing unfamiliar work at elevation contacted bare conductors used to supply power to overhead cranes. These individuals were unfamiliar with the means of powering the crane and either stepped on or grasped the conductors as a means of support while performing unrelated tasks.

Perspiration or water was a factor in accidents involving submersed equipment, and especially in welding accidents. Perspiration or water often is associated with accidents involving equipment containing improper or failed wiring where conductors are inappropriately exposed. Situations resulting from this violate the requirement inherent in welding processes for the welder to remain electrically isolated (McManus 1999). Water and perspiration increase the conductivity of insulating materials and the skin. Maintaining an isolated condition is especially critical when the welder is working inside a conductive structure. Conductive metals, including steel, stainless steel and aluminum, are the materials of construction of many industrial structures.

Electrical energy was a causative agent in fatal accidents other than electrocutions, namely fires and explosions. This subject is explored in more detail in other resources (McManus 1999). Almost all of the ignition sources in the nonwelding accidents were either light bulbs or arcs involving electrical equipment or conductors. The light bulbs either shattered at the time of ignition or acted as hot point sources.

### Engulfment

Engulfment is the process of burial in a flowable solid material. Engulfment can occur rapidly, as when the base material collapses rapidly, or more slowly through an action similar to that of quicksand where the base material disappears downward more quickly than a person can step up to remain on top of it. Engulfment also can occur when material adhering to the sides of a structure slumps downward. The latter is what occurs during collapse of the walls of a trench or excavation.

Engulfment that occurred during fatal accidents was reported by MSHA, NIOSH, and OSHA (MSHA 1988; NIOSH 1994; OSHA 1983, OSHA 1985). Table 1.12 summarizes information on fatal accidents involving engulfment.

Most of the victims were laborers. Laborers commonly entered bins, silos, and other storage structures to promote flow or to break up hang-ups on walls, and to aid in funneling residual materials toward collectors during clean-up. Equipment operators also were prominent among the victims. This reflects the occurrence of these accidents in mining operations. Mining operations involve large quantities of flowable solid material that are stored in large piles. Bulldozers and front-end loaders are used to move these materials toward the inlets to hoppers. On occasion, these inlets block, and operators have used this equipment in attempts to break open the blockage. This is, in many respects, a large-scale version of the situation involving hoppers and silos used to store sand, grain, and other materials. On occasion, slumpage of storage piles near which they had parked their vehicles has engulfed truck drivers and loader operators.

Despite the nature of the situation involving deliberate contact between individuals and an energy source (the energy stored in bulk materials), most of the victims appeared to be performing normal work.

Engulfment accidents cause death through suffocation and acute traumatic injury. During flow-induced engulfment, the material surrounds the chest during exhalation and prevents normal

**Table 1.11**

**Fatal Accidents: Contact with Low-Voltage (750 V or less) Circuits**

| Element | Number of Accidents (n = 95) |
|---|---|
| **Occupation of Victim** | |
| Laborer | 25 |
| Electrician | 31 |
| Welder | 9 |
| Other trade | 21 |
| Supervisor | 2 |
| Manager | 5 |
| Other/unknown | 2 |
| **Work Activity** | |
| Usual | 75 |
| Unusual | 20 |
| **Nature of Work Activity** | |
| Testing electrical equipment | 7 |
| Installing electrical equipment | 24 |
| Repairing/replacing electrical equipment | 19 |
| Welding | 7 |
| Unrelated activity | 38 |
| **Work Height** | |
| Elevated | 22 |
| Ground-level | 73 |
| **Other Factors** | |
| No test/inappropriate test | 50 |
| Equipment malfunction/failure | 14 |
| Loss of control over intended action | 10 |
| Miscommunication | 7 |

## Table 1.11 (Continued)
## Fatal Accidents: Contact with Low-Voltage (750 V or less) Circuits

| Element | Number of Accidents (n = 95) |
|---|---|
| Water/perspiration | 14 |
| Task complexity | 7 |
| No ground or ground removed | 12 |
| Contact with energized bare conductor | 18 |
| Incorrect/faulty wiring | 42 |
| Other | 1 |
| Protective equipment not used | 26 |

Sources: NIOSH 1994, NIOSH 2000; OSHA 1982b, OSHA 1985, OSHA 1988, OSHA 1990.

expansion during inhalation. In some circumstances, the victim suffocated despite not being buried above chest level. Acute traumatic injury results from contact with rapidly falling coalesced material during accidents involving hang-ups and the break-up of bridged or caked structures.

Contents of bins and storage structures usually are drawn from the bottom. This occurs either by gravity or through use of draw-off equipment, such as augers or conveyors. The bottom geometry of the space is either flat with one or more draw-off openings in the floor or tapered.

Most of the accidents occurred during attempts to promote flow. This type of work requires the victim to stand on the flowable solid material. In some cases, the process of engulfment, that is drawing the victim into flowing material, began as soon as the victim stepped onto the contents. In other cases, this process began with the collapse of a bridge or caked material.

Another important cause of engulfment was traumatic injury and burial during break-up of hang-ups on walls of silos and other storage structures, or the slumpage of flowable solid material. Hang-ups occur through coalescence of the material. The material can fall intact or in fragments, or can slump as an amorphous mass. Falling intact can lead to fatal traumatic injury due to the mass of material and distance through which the fall occurs. Slumpage can lead to almost instantaneous burial.

Slumpage also was an important factor in engulfment accidents that occurred during clean-out of residual material. In these cases, almost all of the contents of the space were removed previously by gravity. The remaining material forms a structure that slopes toward the collector, often an opening to an auger located in the floor. Clean-out entails creating a path for flow of this material into drainage channels. Slumpage and rapid burial can occur in circumstances where walls of material are high and unstable, and are undermined by attempts to promote flow.

During storage, coalescence of material sometimes results from the presence of moisture or biological action. Coalescence can lead to deposition on vertical surfaces of the structure and bridging across the horizontal plane of the material. Dislodging coalesced material on vertical surfaces can produce an avalanche of large, hardened, sharp fragments. Material that has not coalesced can flow from the underside of the bridge, thus creating a hollow. As well, "rat holes" (vertical channels in the coalesced material) can develop. "Rat holes" permit some flow to occur. This flow may hide the existence of a bridge. Entry with the express purpose of breaking a bridge was uncommon. However,

## Table 1.12
## Elements of Fatal Accidents Involving Engulfment

| Element | Number of Accidents (n = 67) |
|---|---|
| **Occupation of Victim** | |
| Laborer | 25 |
| Trade | 1 |
| Equipment operator | 18 |
| Supervisor | 3 |
| Manager | 5 |
| Other/unknown | 15 |
| **Work Activity** | |
| Usual | 51 |
| Unusual | 16 |
| **Reason for Action** | |
| Break bridge | 5 |
| Promote flow | 28 |
| Remove hang-up from wall | 6 |
| Clean out loose residual material | 13 |
| Inspect | 5 |
| Other | 10 |
| **Factors in Pre-Existing Hazardous Conditions** | |
| Flow occurring at entry | 27 |
| Unstable base (quicksand effect) | 28 |
| Wall collapse/slumpage | 19 |
| Bridge collapse | 16 |
| Other | 2 |

Sources: MSHA 1988; NIOSH 1994; OSHA 1983, OSHA 1985.

the breaking of bridges was a major cause of engulfment accidents during attempts to open a channel for flow.

### Entanglement in Moving Machinery (Mechanical Hazard)

Entanglement in moving parts of machinery and equipment causes traumatic injury through the tearing of skin and tissue, breaking of bones, and amputation and dismemberment.

Table 1.13 summarizes information on fatal accidents involving entanglement (OSHA 1982b, OSHA 1983, OSHA 1985, OSHA 1988, OSHA 1990).

Most of the victims were hands-on people (operators, tradespeople, and laborers). Mostly they were performing normal work associated with operation and maintenance of the equipment. Work tasks included cleaning and unblocking, repairing and replacing, and adjusting, lubricating, and inspecting the equipment.

In most of these situations, the victim deliberately entered the space while the hazardous condition was present. In some cases, the equipment was operating when the victim entered. The latter situations put the victim fully into the zone for contact with moving parts. More frequently, the victim shut off the equipment or put it into standby mode.

In the latter modes, start-up occurred through inadvertent actions of the victim, or inadvertent or inappropriate actions of others. In some cases, the victim made contact with the start switch or tripped position sensors, thus activating the equipment. Activation also occurred externally to the victim through the actions of others. This occurred for a number of reasons starting with inadvertent contact with start switches, and activating incorrect start switches. The instruction to activate a particular start switch led for reasons of illiteracy to activation of the incorrect switch.

Lack of redundancy of control was a factor in accidental activation. Most of the accidents occurred because of activation through a single level of control (McManus 1999). Multiple levels of control, however, do not necessarily provide greater security, especially when other factors, such as miscommunication, are operating.

Miscommunication was an important factor in these accidents. This is troubling when the ability to maintain safe conditions of work depends on the spoken and written word and its comprehension. Miscommunication led to misperception about the location of the victim and status of occupancy of spaces within the physical structure and operating envelope of equipment. This brings up issues of illiteracy (already mentioned) and continuity of working relationships. Shift change and absence are two important factors in this discussion. Without written indication about status, critical information is lost following shift change or absence of a key individual. In most situations, formal control procedures, such as lockout/tagout, were not utilized. However, one accident occurred despite the correct use of a lockout procedure. The operator removed the locks and activated several switches in the mistaken belief that everyone had vacated the equipment, and that start-up was safe to initiate.

### Fires and Explosions

Fires and explosions reflect chemical potential energy stored in the molecules of fuel. A fire is a gradual release of this energy, whereas an explosion is an abrupt release. Explosion also can occur due to energy stored by compression of gases in piping and structures. These are considered in the section addressing processes.

Table 1.14 summarizes information on fatal accidents involving fires and explosions reported by OSHA (OSHA 1982a, OSHA 1983, OSHA 1988, OSHA 1990). The focus of this investigation was the release of energy caused by activating a latent energy source (igniting fuel stored in a container), rather than development of the energy source through work activity. An explosion caused by development of an ignitable mixture of solvent mist and vapor in a structure during spraypainting is an example of the latter. In this case, the work activity introduces the energy source (the fuel), where prior to the start of work none existed. The latter subject is the focus of investigation concerning work in confined spaces (McManus 1999). Most of the fatal accidents involved welders and individuals having different titles, but performing the same or similar work (welding, cutting, and heating).

The major proportion of the work was normal, and involved repair, replacement and installation, and demolition. By far, the greatest proportion of the work involved welding and cutting.

## Table 1.13
## Fatal Accidents: Entanglement in Moving Machinery

| Element | Number of Accidents (n = 81) |
| --- | --- |
| **Occupation of Victim** | |
| Laborer | 11 |
| Trade | 21 |
| Equipment operator | 33 |
| Supervisor | 4 |
| Manager | 2 |
| Other/unknown | 10 |
| **Work Activity** | |
| Usual | 45 |
| Unusual | 36 |
| **Reason for Action** | |
| Clean, unblock | 27 |
| Repair, replace | 15 |
| Adjust, lubricate | 11 |
| Inspect | 8 |
| Other | 17 |
| Unknown | 4 |
| **Factors in Pre-Existing Hazardous Conditions** | |
| Equipment malfunction | 4 |
| **Factors in Self-Activation of Equipment** | |
| Standby condition | 10 |
| Placement of START switch | 7 |
| Equipment failure | – |
| Loss of control over action | 2 |

## Table 1.13 (Continued)
## Fatal Accidents: Entanglement in Moving Machinery

| Element | Number of Accidents (n = 81) |
|---|:---:|
| **Factors in Activation of Equipment by Another Individual** | |
| Standby condition | 5 |
| Miscommunication | 10 |
| Control remote from equipment | 20 |
| Automated control | 5 |
| Manual control | 24 |
| Failure of intended action | 1 |

Sources: OSHA 1982b, OSHA 1983, OSHA 1985, OSHA 1988, OSHA 1990.

Structures on which this work occurred included mostly tanks containing ignitable materials in quantities ranging from bulk liquids to traces present in vapor form. Most of the contents of these containers had high flash points (examples being asphalt, transformer oil, and diesel fuel). These require considerable heating to generate sufficient vapor for ignition, especially where bulk quantities are present. Knowledge about these properties may have been a consideration in the decision to proceed with the work. Welding and cutting processes develop intense localized sources of heat. These could have created localized sources of ignitable and explosible vapor. The intensity and concentration of energy associated with the arc and flame override considerations about flash point as a measure of hazard.

Many accidents involved so-called empty containers, usually 55 gallon (200 L) drums and other small containers. Small quantities of residue remaining on metal surfaces can easily vaporize to form an ignitable atmosphere. This is especially the case with 200 L drums which have a relatively high surface area-to-volume ratio.

Flushing the container to remove residues is no guarantee of success. In one case, the container, an automotive gasoline tank, reportedly was flushed several times with water. Despite this, an ignitable atmosphere still developed. A tanker truck used to transport spent caustic solution for scrubbing gasoline contained sufficient residual hydrocarbon for development of an explosible mixture. These examples highlight the difficulty of removing hydrocarbons from steel containers.

### Process Hazards

Process hazards refer to hazardous conditions created by chemical and biochemical processes. These processes occur in closed systems, such as process reactors. The outcome from process hazards includes runaway chemical reactions. As well, this category includes the movement and storage of bulk liquids and gases in piping and storage vessels. Bulk movement and storage of liquids and gases in piping occur under pressure. This category also includes hazardous conditions that develop in open systems, such as sewers. These can contain hazardous liquids and gases and vapors due to deliberate and accidental dumping, as well as the hazards posed by flow of bulk liquid.

## Table 1.14
## Elements of Fatal Fires and Explosions (fuel-related)

| Element | Number of Accidents (n = 61) |
|---|---|
| **Occupation of Victim** | |
| Laborer | 9 |
| Welder | 30 |
| Other trade | 8 |
| Operator | 1 |
| Supervisor | 5 |
| Manager | – |
| Other/unknown | 8 |
| **Work Activity** | |
| Usual | 48 |
| Unusual | 13 |
| **Reason for Action** | |
| Clean, unblock | 3 |
| Repair, replace, install | 39 |
| Demolish | 18 |
| Other | 1 |
| **Factors in Pre-Existing Hazardous Conditions** | |
| Equipment malfunction | 3 |
| Contents of container | 56 |
| **Ignition Source** | |
| Welding | 29 |
| Cutting | 25 |
| Heating | 2 |
| Other | 3 |

Sources: OSHA 1982a, OSHA 1983, OSHA 1988, OSHA 1990.

Table 1.15 summarizes information about the elements of accidents involving process hazards in the OSHA reports (OSHA 1982a, OSHA 1985, OSHA 1988, OSHA 1990). These primarily cover accidents involving open systems and closed systems not directly involved in chemical processing.

Most of the victims were hands-on workers (laborers, trades, and operators) who were performing normal work. These accidents occurred because of lack of isolation of the workspace from the causative agent or because of faulty or inappropriate isolation. An important minor causal factor in these accidents was failure of equipment. Equipment involved included piping, and valves and other components in pressurized steam systems. Steam systems seem to be especially prone to sudden failure in service. This results from embrittlement of materials of construction subjected to long periods of service at high temperature and pressure. Water hammer and unexpected impacts with ladders and other equipment has led to failure of piping and components, and subsequent rapid release of steam.

Other accidents occurred because of escape of liquids and gases in piping systems following failure of plugs and other components used for isolation. These accidents typify the concern for hazards in process systems in confined spaces that remain active during entry and work. Namely, the system of isolation could fail during occupancy and discharge the contents into the space. The victim then is unable to escape from the space before being affected by the process emission.

In providing detailed summaries, the MSHA, NIOSH, and OSHA reports, of necessity, highlighted the role of victims and on-site activity in fatal accidents. Process accidents differ from the others discussed in this section because by their very nature, chemical and biochemical processes are likely to occur in closed and increasingly automated systems. Chemical plants contain few operations people. Operators spend most of their time in the control room, except when inspecting the status of the unit firsthand. Video surveillance eliminates much of the need to leave the control room for an inspection. The energy source or materials containing energy are totally enclosed within metal structures and equipment. The only normal source of exposure is leakage at points that are incompletely sealed.

As a result, accidents involving chemical operations differ from those involving other sources of hazardous energy. For the moment, this discussion will stress primarily the role of people in these events to be consistent with what appears previously. Authors such as Duguid (2001) and the Mary Kay O'Connor Process Safety Center (Mannan et al. 2001) have published reports on this subject. These are part of larger considerations under the purview of chemical process safety. The focus of these reports is accidents that have occurred in operations involved in chemical manufacturing, distribution, storage, and use.

Duguid (2001) examined 562 incidents that occurred from 1950 to 1995. The average age of the incidents was 20 years. More than 60% of the incidents reported in this study occurred in the oil industry, 20% in chemical industries, and 10% in petrochemicals. This breakdown reflects the source of the incidents rather than actual industry experience. The causes of the accidents and responsibility for them followed similar patterns in the three industries. Duguid also commented on the likelihood of under-reporting of potential versus actual incidents.

He makes some interesting comments about the applicability of findings in the database. Only 1% of incidents could have happened in the 1950s and 1960s, as these did not involve failure of computer control. Only 12% of the incidents were caused by such an unusual combination of events that they were unlikely to be repeated. Even here, the individual causes and responsibilities were the same as identified in other accidents. Only 17% of the incidents were specific to the type of process unit. That is, 83% could have happened on a variety of process units. As a result, change in processes and equipment will not invalidate the outcome of findings arising from this work into the foreseeable future.

Major change in the status quo, detailed in Table 1.16, accounted for more than half of the incidents. During start-up, for instance, equipment is cold. Metals have not fully expanded to fill in gaps. The reverse occurs during shutdown. As gaps open or prior to closing, potential for leakage increases dramatically over steady-state conditions. Only 1% of incidents involved recently introduced equipment, and only 3% involved obsolete equipment or processes.

Causes of process accidents in chemical industries are considerably more equipment oriented than those of accidents occurring in other sectors (Table 1.16), as reported in previous sections. Or per-

**Table 1.15**

**Elements of Fatal Accidents Related to Industrial Processes**

| Element | Number of Accidents (n = 43) |
|---|:---:|
| **Occupation of Victim** | |
| Laborer | 12 |
| Other trade | 15 |
| Operator | 7 |
| Supervisor | 4 |
| Manager | 4 |
| Other/unknown | 1 |
| **Work Activity** | |
| Usual | 33 |
| Unusual | 10 |
| **Reason for Contact** | |
| Equipment failure | 10 |
| Overpressure | 5 |
| Water hammer | 2 |
| Valve leak | 3 |
| Lack of isolation | 16 |
| Faulty/improper isolation | 10 |
| **Causative Agent** | |
| Steam | 5 |
| Wastewater and gases | 12 |
| Purge gases | 6 |
| Carbon monoxide (process) | 2 |
| Hydrogen sulfide | 2 |
| Nitrogen | 8 |
| Cutting | 25 |

## Table 1.15 (Continued)
## Elements of Fatal Accidents Related to Industrial Processes

| Element | Number of Accidents (n = 43) |
|---|---|
| Heating | 2 |
| Other | 3 |

Sources: OSHA 1982a, OSHA 1985, OSHA 1988, OSHA 1990.

haps this should be expressed in terms of the fact that the people involved act remotely from the point of contact with the source of hazardous energy at the design, construction, and commissioning stages, rather than at hands-on operations and maintenance. There is little hands-on activity in these industries, since most of the hands-on work, such as controlling operation of pumps, adjusting the position of valves, and taking readings of process variables (temperature, pressure, pH, and so on), is performed by electrical and mechanical operators controlled by computerized process control. For the most part, the human operator is a spectator to this process.The Mary Kay O'Connor Process Safety Center reviewed five databases for the years following 1998 (Mannan et al. 2001). Nonfatal and fatal injuries resulting from chemical incidents show a decreasing trend over this period. The review indicated that injuries produced by small releases are proportional to those produced by large releases. Injuries are widely distributed across the chemical industry, regardless of the size of the facility. That is, the numerous small incidents balance the large ones. Mechanical failure is more likely to be the initiating event than human error. This finding is consistent with that reported by Duguid (Duguid 2001). Again paralleling the findings of Duguid, about half of the releases occur in industries that store, distribute, and use chemical products, rather than at the producer level (chemical manufacturers and petroleum refiners).

### Ionizing Radiation Accidents
The energy emitted by the nucleus of radioactive atoms as particles and electromagnetic waves has been utilized commercially, industrially, and medically for many years in both open and closed systems. The administration of radiopharmaceuticals and diagnostics orally and by injection are examples of an open system. A nuclear reactor is an example of a closed system. Naturally occurring radioactive materials contain insufficient radioactivity to cause the acute injury which is the focus of this book. However, artificial sources are more than capable of causing short-term, and sometimes almost immediate, fatal injury.

González (1999) reported on 134 major accidents that occurred world-wide since 1945. These included 21 accidents involving reactors and criticality, 89 involving radioactive sources, 23 involving radiation-producing machines, and 3 in which the source was not identified. This list is known not to be inclusive (Lubenau and Strom 2002).

"Orphan source" is the term used to describe radioactive sources that have escaped institutional control. These have been involved in 60 severe radiological accidents and caused 39 fatalities (Yusko 2001). Interestingly, one feature that distinguishes accidents involving "orphan sources" is overexposure of "members of the public" rather than "radiation workers" (Lubenau and Strom 2002).

Due to the public profile of ionizing radiation, applications in commercial, industrial, and medical environments generally are subject to considerably more stringent control than other energy sources, such as those mentioned earlier in this chapter. Radioactive sources are distributed for use under the terms of licenses in many jurisdictions. Among other things, the license stipulates conditions

## Table 1.16
## Summary of Findings by Duguid

| Cause | Percent |
| --- | --- |
| **Major Change in the Status Quo** | |
| Shutdown | 13 |
| Start-up | 14 |
| Maintenance | 10 |
| Abnormal operation | 14 |
| **Background Causes** | |
| Inadequate design for safety | 34 |
| Inadequate temporary or permanent procedures | 33 |
| Operator error | 22 |
| Inadequate inspection during construction | 15 |
| Maintenance failure | 10 |
| Faulty mechanical design | 10 |
| Inadequate supervision | 10 |
| **Total (average incident had more than one cause)** | 156 |

Source: Duguid 2001.

for use and requirements for control of exposure. These include the stipulation for training of operators of equipment and users of non-tethered radioactive sources. Adherence to the terms of the license is subject to auditing by regulatory authorities. The system of licenses and the requirements for adherence are not universal in scope. Despite this approach to control, fatal accidents have happened and continue to occur.

Accidents involving ionizing radiation have multiple causes (LANL 2000; Ortiz et al. 2000).

### Industrial Radiography
One of the first practical applications of ionizing radiation (starting in the early 1900s) was industrial radiography (IAEA 1998). Ionizing radiation is sufficiently energetic to penetrate metals and other materials, and in so doing, provides a means for recording irregularities (defects) onto film. The concept utilizes X-ray equipment and radioisotope sources. The radioactive source initially utilized was radium. With the appearance of $^{60}$Co and $^{192}$Ir the 1940s (a result of the creation of fission products during nuclear reactions), application of industrial radiography grew rapidly.

The radioactive source is held in a shielded case (lead, depleted uranium, or other dense material). Some devices have a controllable opening, such as a flap or guillotine blade. This allows an intense beam of gamma energy to pass in a defined direction, similar to a beam of light produced by a flashlight. There is one important difference in this analogy, since gamma radiation from a source radiates in all directions. This means that low levels pass through the shielding. In other units, a mechanism moves the source from the protection of the shielded storage to an exposure position.

The history of industrial radiography contains accidents that have resulted in nonfatal and fatal injuries (IAEA 1998). Nonfatal injuries include burns, blistering, and sometimes loss of a digit or limb. Most of the victims were members of the public and workers not associated with use of the source (Ortiz et al. 2000). Common factors (IAEA 1998) in incidents involving sources used in industrial radiography include failure to use radiation monitoring equipment, failure to respond to safety alarms, disconnecting or defeating safety interlocks and alarm systems, entering the path of the beam, damage to or mechanical failure of the source positioning assembly, inadequate measures in positioning equipment, and miscommunication. In some situations, the source was lost or removed from the containment of the shielded container. Many of these reflect deliberate or accidental failure to follow operating procedures. Inadequate or inappropriate training was an issue in some situations. Operators have attempted procedures, such as dismantling equipment to examine or reattach or recover unshielded sources, that are considerably beyond the mandate provided by normal training. Leakage of soluble forms of the source material has occurred on occasion due to failure of the sealing weld under conditions of corrosion.

In a typical accident, the source fails to move to the shielded position on command (Ortiz et al. 2000). This occurs because of jamming in the guide tube or detachment from the drive cable. The operator then must decide what to do. Actions begin with attempts to move the source back and forth along its path of travel. When this fails, at some point the operator ups the ante. Actions have included removal of dosimeters and shutting down of survey instruments, and dismantlement of the equipment. The latter invariably leads to detachment of the source from the equipment. The operator or another individual then handles the source, and attempts to reposition it into the equipment, and even on occasion, puts it into a pocket.

A common feature of these accidents is separation of the source from its shielded container. The unshielded source often is not labeled, and for this reason, not perceived to pose any hazard. On a number of occasions sources became lost, and were "found" and taken home by bystanders. Industrial radiography is repetitive and highly competitive. Most of the work involves positioning equipment to maintain safety, especially that of bystanders. Safety-related work includes positioning collimators, barriers, and shielding materials, installing barrier tapes, patrolling the area, using the survey meter to ensure that leakage of the beam is not occurring, and wearing alarming and other types of dosimeters. Shielding materials are heavy due to their high density. These actions add substantially to the cost of performing the work, which is often done during off-shifts. Industrial radiographers often work alone, under poor lighting, in inconvenient positions, and without supervision. Under these conditions and tight production deadlines, the occurrence of incidents is not surprising. The reported ones likely represent only the tip of the iceberg.

## Industrial Irradiation Facilities

One industrial application of ionizing radiation is the irradiation facility. This uses high energy gamma photons and electron beams (IAEA 1996). Currently, there are more than 160 gamma irradiation and 600 electron beam facilities in operation in countries that are members of the International Atomic Energy Agency (IAEA). The most common uses of these facilities are to sterilize medical and pharmaceutical products, to preserve foods, to synthesize polymers, and to eradicate insects.

Fatal accidents have occurred in both developing and developed countries. This is an ever-present consequence of improper operation and the high dose rate produced by these energy sources. The first of these units were put into service in the late 1950s for food preservation. Nonfatal accidents occurred as early as 1965 in the U.S. The first fatal accident occurred in 1975 in Italy, and the second in Norway in 1982. Between 1975 and 1994 five fatal accidents were reported, and since 1989, one reported fatal accident has occurred per year.

These accidents occur for several reasons. The first is mistrust of warning signals (Ortiz et al. 2000). Warning signals provided by instruments are an essential component in the strategy for protection against exposure to energy sources that provide no warning to the senses. Conflict between information reported by radiation sensors and sensors that indicate the position of the radiation source has led to misadventure. That is, the radiation sensor indicates a high level, while the source position sensor indicates that the source is in its shielded container. Despite the value that practitioners in health and safety put on instrumental response, operator experience can be completely contrary, especially after they experience false instrument readings.

Operator overconfidence is another important factor. This has led to deliberate bypassing of interlocks and ignoring of procedures and warning signals. A factor related to this is jamming of materials handling equipment located in the unit. Materials to be irradiated enter the sensitive volume of the unit on a conveyor or other similar materials' handling device. Materials to be irradiated are packaged in bags or in boxes. On occasion, the materials handling equipment that moves these items into the path of the beam jams. This can result from mispositioning of the item, so that it comes into contact with structural elements of the facility. The jam requires entry into the material path and corrective action by the operator. In the case of jamming that occurs frequently, untrained operators and operators willing to take the risk have devised means to defeat interlocks intended to prevent entry and to maintain safe levels of exposure.

## Radiotherapy

Medical uses of ionizing radiation intentionally expose patients sometimes to high levels of radiation. This is especially the case in radiotherapy where doses are extremely high and a small deviation from the prescribed dose could have severe or fatal consequences. Even a dose significantly below that prescribed can have severe consequences to the patient (Ortiz et al. 2000). Radiotherapy treatment is a complex process involving many professional participants. Treatment involves many variables. The practitioner may see a large number of people per day, each receiving treatment that varies only slightly, but possibly significantly. Treatment preparation and delivery require intense concentration. This can suffer from the distractions in a busy area having a high workload. As a result, there is considerable potential for error.

Radiotherapy includes external beam therapy, brachytherapy, and nuclear medicine involving diagnostics and radiopharmaceuticals. (Brachytherapy is radiotherapy administered by implanting wires, ribbons, or tubes containing grains of highly radioactive material into or close to a tumor.) The aim of radiation therapy is to deliver a dose and dose distribution that is sufficient for destruction or control of tumors, while minimizing complications in normal tissue (IAEA 2000). Hence, demands for precision and accuracy are high.

## External Beam Therapy

Factors associated with accidents involving beam units include equipment faults and mistakes by operators (IAEA 2000). These have an immediate outcome on the efficacy of the treatment. Equipment faults usually involve mechanical problems, including failure of components whose function is to control dose delivery and provide protection against overexposure. However, one situation was caused by faulty software used to control operation of the equipment. Mistakes by operators range from errors in estimating delivered dose, the result of faulty calibration and incorrect calculations, to fabrication of records. Mistakes have occurred in confirming the identity of patients, so that the wrong patient received the treatment. Operational errors in use of the equipment also have occurred.

## Brachytherapy

Brachytherapy puts tissue of the human patient directly into contact with a source capable of producing an intense localized dose (Ortiz et al. 2000). Brachytherapy requires handling of the source by the practitioner and implanting into the body of the patient. Sources include ribbons and needles containing pellets of radioactive material. Brachytherapy involved many of the same types of errors mentioned in the previous section on external beam therapy. Despite the seeming difference of approach to administering the therapy, many of the same types of incidents occurred, namely errors in calculating dose delivery, delivery to the wrong site, and even treatment of the incorrect patient, and failure of equipment. Incidents characteristic of brachytherapy and source handling included use of incorrect source and isotope, source leakage, incorrect application and handling technique, and failure to implant all of the intended sources.

Concern also exists with exposure to persons helping to comfort the patient. These have resulted following failure to follow rules, and even release of the therapeutic device from the body of the patient. Sometimes the patient leaves the treatment facility with the therapeutic device still inplace. This has led to continued exposure and eventual death. The therapeutic device can cause exposure to persons unrelated to the patient and to members of the public in these circumstances.

Accidental occupational exposures to medical staff have resulted from unintended expulsion of a brachytherapy source from a patient.

## Radiopharmaceuticals and Diagnostics

Radiopharmaceuticals and diagnostics are unsealed sources. These are administered orally or by injection. Administration involves handling of the source by the practitioner. Despite the seeming difference of approach to administering the therapy, many of the same types of incidents occur, namely errors in calculating dose delivery, delivery to the wrong site, and even to the wrong patient. Spillage is a concern during administration of radiopharmaceuticals and diagnostics.

## Decommissioning, Source Disposal and Dispersal

Exposure of the public following decommissioning of equipment containing radioactive sources is an emerging concern. This is especially the case with radiotherapy units and industrial irradiation equipment (IAEA 1998, IAEA 2000). This compounds the situation that has existed for a considerable period of time over the loss or theft of sources used in industrial radiography.

Normal practice is to license the radioactive source to the owner of the equipment. However, this was not always the case. Conditions of the license and compliance with regulatory statutes require periodic inspection, often by the manufacturer and more certainly by the regulator. At the end of service of the equipment, the source is to be returned to the manufacturer for disposal. Common practice in industrial radiography, for example, is to allow the source to decay in the source holder prior to return to the manufacturer for disposal.

Return of the source to the manufacturer during decommissioning does not always happen. In some situations, the equipment was abandoned either in situ or elsewhere, or sold for disposal as scrap (Ortiz et al. 2000). The source is contained within a shield of a very dense material, such as lead or depleted uranium. The actual radioactive material may be present as a metal slug, or as pellets, or powder, all readily dispersible. Scavenging of scrap material, melting down, and remanufacturing has led to widespread dispersal of source material. In addition to injury caused by exposure, these events have led to extensive contamination of the environment. Some of the contamination problem has arisen from disposal of radioactive level gauges containing $^{60}$Co, $^{137}$Cs, and $^{241}$Am.

Most of the fatal accidents involving loss of control of sources led to fatal injury to members of the public (Ortiz et al. 2000). Typically, the source became detached from its enclosure inside the equipment, and was recovered by someone not connected with its use, put in a pocket, and taken home, as described previously.

## Criticality Accidents

Criticality is a condition in the atomic nucleus when at least one of the several neutrons emitted during a fission process causes a second nucleus to split (Cember 1996). Not all systems that are fissionable can go critical. The chain reaction dies out when more neutrons are lost by escape from the system or are absorbed by nonfissionable materials (impurities or poisons) than are produced by fission. A criticality accident occurs when fissionable material is present in conditions that permit a nuclear chain reaction to happen. A criticality event releases a very large amount of energy in a very short time. Criticality depends on the supply of neutrons having energy appropriate to initiate fission, and on the availability of fissile atoms. This depends, in turn, on the size and geometry of the fissile assembly. Much of what is known about criticality control is the result of trial and error and happenstance: the circumstances described in the earliest criticality accidents.

Criticality is a by-product of activity that started in the 1940s. Naturally occurring radioactive materials cannot become critical except under specific circumstances involving use of heavy water (deuterium oxide), such as employed in CANDU and similar types of reactors. (The CANDU reactor is a Canadian-designed product that uses natural uranium as fuel and heavy water as the "moderator".) Criticality is a serious concern following concentration of naturally occurring radioactive materials, such as $^{235}U$ by enrichment processes, and extraction and processing of synthetic radioactive elements, such as the plutonium series, produced through irradiation of naturally occurring ones.

Early criticality accidents occurred during handling of bomb components. Exactly when the first criticality accident happened is not recorded (LANL 2000). Many were non-routine events that formed part of the learning curve during the race against time to create the first and subsequent atomic weapons. Criticality accidents differ from chemical accidents. In criticality accidents the event often occurs in an instant. That is, in an instant, the victim has often received a lethal dose of gamma and neutron radiation. Chemical accidents involving reaction processes require considerably more time to release energy.

Associated with each criticality accident typically was about $10^{17}$ fissions. To put meaning into this number, each fission produces about 190 MeV (Million electron Volts) of energy. Since there are $1.6 \times 10^6$ erg/MeV and 1 J (Joule)/$10^7$ erg, the energy released during a typical criticality accident is $3.0 \times 10^6$ J or 3.0 MJ (million Joules). The Watt, the more familiar unit of power is 1 J/s. Hence, this burst of energy, released over one second of time, if harnessed, would produce 3.0 MW of power.

In systems containing water, some of this energy is trapped in the water and materials of construction of the vessel. This deposition of energy causes rapid heating and sometimes boiling, and in some accidents, has caused a steam explosion. In systems containing only metals, the energy heats the metal rapidly, causes warping, and on occasion, melting. Most of the energy produced in these accidents radiates outward in all directions and interacts with matter in its path only at considerable distance. Some of the energy and particles passes through the tissues of victims present nearby and causes death through radiation sickness.

*Nuclear Process Accidents*
Of the accidents reviewed in the Los Alamos report, 22 involved process operations (LANL 2000). Of these, 21 occurred with the fissile material present in solutions or slurries. One occurred with metal ingots. None occurred with powders. Eighteen of the accidents occurred in manned, unshielded facilities. There were nine fatalities. Three of the survivors lost limbs through amputation. None of the accidents occurred during transportation or storage. No equipment was damaged. Measurable contamination by fission products (slightly above natural levels) beyond the plant boundary occurred in only one accident. Only one accident resulted in measurable exposures (well below allowable worker annual exposures) to members of the public.

A training course in August 1997 yielded unexpected disclosure of accidents in Russia and the former Soviet Union (Vargo 1999). These accidents resulted in seven fatalities. The earliest accidents occurred through use of equipment having unsafe geometry. Violations of operating practices and configuration controls on equipment were major factors. Other factors included inadequate maintenance and surveillance, and inaccurate or incomplete results.

Criticality accidents in process facilities are considered to be unexpected. Small quantities are handled in bench-scale or pilot-scale, batch processes and equipment. Processes involving fissile materials are applications of elementary chemical reactions. Misapplication of chemical or physical processes was partly responsible for accumulation of critical quantities of the fissile material in the appropriate geometry, and hence the conditions needed for occurrence of some of these accidents. Additional complications included subtleties associated with the physical characteristics of turbulence during start-up of mixing of contents containing a layer of organic liquid floating on a water-based layer. Manifestation of the accident typically included rapid production of heat, followed by boiling and evolution of vapor and steam. This energetic release often ruptured fittings and expelled liquid and solid material from the equipment.

Process facilities carrying out operations with fissile material attempt to avoid criticality accidents through physical and administrative controls. These controls are intended through design and procedures to prevent occurrence of critical or near-critical configurations in the facility, since under normal working conditions, operators are required to work in close proximity to potentially critical configurations. Typical requirements include the need to operate control switches, remote robotic handling equipment, and to adjust valves and other fittings.

Factors associated with these accidents include those unique to the activity: inadequacy of design/layout of equipment, unfavorable geometry of vessels, inadequacy of process design, poor operating condition of equipment, and change in routine, such as shutdown for material audit, start-up of new processes, one-of-a-kind operations, and equipment malfunction due to inadequate maintenance.

Also associated with these accidents were qualities typically associated with human failings:

• Calculational errors

• Miscommunication

• Loss of continuity in communication between shifts

• Failure to follow procedures/orders

• Lack of training

• Lack of familiarity with the procedure

• Panic/instinctive reaction during the excursion

• Fatigue and distraction

As with other types of radiation accidents, none of the process-related accidents was caused by a single failure (LANL 2000). That is, each involved multiple causes. Equipment failure or malfunction was either a minor or a non-contributing factor in all of the accidents. No accident occurred with fissile material in storage. These operations are relatively simple. Controlling criticality is also simple. No accident occurred with fissile material during transport. National and international transport regulations specify defense in depth for criticality safety that goes far beyond what would be practical and cost-effective for plant operations. No accident resulted in significant radiation consequences beyond the facility, either to people or the environment. Criticality accidents are similar to small, benchtop-scale, chemical explosions in their personnel and environmental consequences. That is, they pose issues of worker safety.

The most obvious and significant characteristic of these accidents is that all but one of them involved solutions or slurries (LANL 2000). This is attributable to several factors:

• The relatively small quantity of fissile material needed for criticality when well moderated

• High mobility of solutions and the ease with which they adapt to changes in vessel shape

• Potential for changes in concentration

• Ability of fissile materials to exchange between aqueous and organic phases

The accidents that occurred in Russia and the former Soviet Union provided similar findings (Vargo 1999).

The time profile for these accidents indicates that lessons have been learned and applied (LANL 2000). For about the first decade of operations with significant quantities of fissile materials, no accidents were reported. This likely was due to the relatively small scale of individual operations and amount of fissile material (almost exclusively plutonium and enriched uranium) available for use. From the late 1950s to the mid-1960s, there was about one accident per year in both the U.S.S.R. and the U.S. During this time, a very large increase in the production of fissile material and the scale of operations occurred at process sites. Since the mid-1960s, the frequency of accidents dropped by a factor of 10, to about one per 10 years.

This drop is attributable to several factors. The first was application of significant lessons learned from the earlier accidents, a major one being the need to avoid use of vessels having unfavorable geometry. The second was a significant increase in management attention to criticality safety, particularly the employment of staff devoted specifically to controlling this hazard. Recognition about the importance of the lessons provided by these accidents prompted study about critical mass and operational best practice. The first compilations of data began appearing in the late 1950s, and the first national standards in the mid-1960s.

### Criticality Experiments/Operations

Thirty-eight of the accidents mentioned in the LANL report involved criticality experiments or operations involving reactors (LANL 2000). Reactor and critical experiment research facilities purposely plan and achieve near-critical and critical configurations. The intent for these experiments is to determine the physical characteristics of fissibility: geometry, mass, physical form, and concentration. This information aids in development of applications for nuclear energy. This, of course, includes the handling of fissile materials in various physical and chemical forms. Personnel operating this equipment usually are experts in criticality physics. Although they carry out hands-on operations with fissile material under restrictions similar to those found in process facilities, the planned near-critical and critical configurations are performed under shielded or remote conditions. Because criticality is expected, lessons learned from this section do not contribute directly to process criticality safety and operation of commercial nuclear power reactors.

Of the 38 accidents studied, five occurred in what are categorized as working reactors, and 33 in critical facilities where the properties of the assemblies themselves were being investigated. These accidents usually resulted in no damage to equipment, but sometimes caused warped or melted fuel assemblies, and even steam explosions that led to extensive damage or even destruction of the equipment.

Factors associated with these accidents included those unique to the activity:

• Lack of information about outcomes from deliberate challenges to determine limits of tolerance of criticality criteria

• Lack of anticipation of conditions

• Loss of control over conditions

Accompanying these factors were qualities typically associated with human failings: errors in calculation, failure to follow procedures, overconfidence, operator error, equipment failure, and faulty design.

Perhaps surprisingly to outsiders to the nuclear industry, more fatalities occurred during criticality experiments and small reactor accidents than during processing of fissile material (12 versus 9). This difference in perception likely is due to the greater visibility associated with processing activities.

## Factors in Accidents Involving Hazardous Energy

This chapter has provided an overview of accidents involving sources of hazardous energy and materials containing energy that have occurred across a broad spectrum of industry. While the industrial sectors differ widely in terms of products and services, there are commonalities regarding the manner in which nonfatal and fatal accidents have occurred. Accidents involving hazardous levels of energy occur in two ways: direct contact with the energy source or energy released by it, and exposure to energy released during conversion from one form to another. Energy sources in the first category include nuclear and electrical sources, and moving components of equipment. Energy release during conversion from one form to another occurs in materials and objects, including components of machines and equipment that store energy by virtue of position once operational motion has ceased.

In many industrial sectors, hazardous conditions exist within equipment and structures. The sole purpose for many of these structures is to provide containment. Containment provides a physical barrier or a boundary surface, to use the language of confined spaces (McManus 1999). This prevents interaction between the worker and hazardous conditions within. The critical point comes in the decision about how to perform required work. The option that provides the closest assurance to safety is to deactivate, de-energize, isolate, lock out, and then to verify by testing that no energy is present. The latter determines and confirms the effectiveness of the process prior to opening the equipment or structure to perform work. This approach is a component of the concept known as defense in depth that incorporates layers of actions to ensure safe conditions and therefore, safety. Despite the availability of this approach, which is codified in standards and regulations, it is not always followed. This failure results from lack of knowledge, insufficient or inappropriate leadership, and even defiance of existing procedures.

In some industrial sectors, workers interact in close proximity with materials containing energy. The purpose for this interaction is to initiate or promote conversion of energy from one form to another, the outward manifestation of this interaction being, for example, to initiate or increase the flow of materials. Fatal accidents happen when the rate of energy conversion exceeds the ability of the worker through expenditure of energy to counteract the flow to maintain position on the surface of the material.

In some industrial sectors, work occurs in close proximity to energy sources that are not shielded against contact. These situations place heavy reliance on avoidance of contact, personal and other forms of protective equipment, and disciplined adherence to procedures. However, many workers who potentially come into contact with these energy sources have no means of protection other than avoidance.

A critical question to answer in the approaches taken in different industrial sectors is whether they provide equivalent protection to workers in potential contact with the energy source or material containing energy. A society built on an inherent sense of fairness would insist that workers in different industrial sectors should face similar risk of performing work. This concept is difficult to quantify, since many factors are involved. These include the risk of performing procedures, number of workers affected, frequency of performing the procedure, and so on. Summary statistics, such as number of workers, hours of work and frequency of occurrence of injuries and fatalities, fail to convey the information needed to resolve this issue.

Despite precautions available to prevent contact with energy sources and materials containing energy, nonfatal and fatal accidents continue to happen. Some of the factors intrinsic to these accidents relate to equipment and surroundings, while others pertain to human factors extrinsic and intrinsic to them. Analysis of accidents in some sectors by industry experts indicates that these accidents typically have at least one cause. This observation probably is extendable across industry.

Factors related to equipment and surroundings include poor design and layout, breakdown and malfunction, equipment or processes inappropriate to the task, loss of control over conditions, inadequate lighting, high noise level, and uncomfortably hot or cold temperature.

Human factors manifested in these accidents through physiological and emotional aspects. Physiological factors include fatigue from long hours of work, inability to see due to low light level, forgetfulness (short-term memory) leading to failure to complete a critical task, substance abuse/misuse, overexertion, and inability to process required information due to overcomplexity of procedures, and other influences including commotion, distraction, and high noise level. Emotional factors include panic, fear of breaking equipment, fear of being caught for making a mistake, responsibility-taking during an emergency, acting beyond limits of knowledge, and frustration with malfunctions of equipment and processes.

## References

ANSI: Safety Requirements for Sand Preparation, Molding and Coremaking in the Sand Foundry Industry (ANSI Z241.1-1975). New York: American National Standards Institute, 1975.

ANSI: American National Standard for Personnel Protection – Lockout/Tagout of Energy Sources – Minimum Safety Requirements (ANSI Z144.1). New York: American National Standards Institute, 1982.

ANSIASSE: Control of Hazardous Energy: Lockout/Tagout and Alternative Methods (ANSI Z144.1-2003). Des Plaines, IL: American Society of Safety Engineers, 2004.

ANSIASSE: Control of Hazardous Energy: Lockout/Tagout and Alternative Methods (ANSI Z144.1-2003, R2008). Des Plaines, IL: American Society of Safety Engineers, 2004.

BLS: National Census of Fatal Occupational Injuries, 1995 by G. Toscano and J. Windau. In: Fatal Workplace Injuries in 1995: A Collection of Data and Analysis – April 1997 (Report 913). Washington, DC: Bureau of Labor Statistics, Census of Fatal Occupational Injuries: Reports (1997). 12

pp. (Also published as Toscano, G. and J. Windau: National Census of Fatal Occupational Injuries, 1995. *Comp. Work. Cond.* September 1996. pp. 34-45.)

Cember, H.: *Introduction to Health Physics*, 3$^{rd}$ Ed. New York: The McGraw-Hill Companies, Inc., 1996.

Chiotti, P. and M. Verkade: Literature Survey of Dust Explosions in Grain Handling Facilities: Causes and Prevention (Project 400-24-04). Ames, IA: Energy and Mineral Resources Research Institute, Iowa State University. March 25, 1976.

Duguid, I.: Take This Safety Database to Heart. *Chem. Eng. 108 (7)*: July 2001. pp. 80-84.

González, A.J.: Strengthening the Safety of Radiation Sources and the Security of Radioactive Materials: Timely Action. *IAEA Bull. 41*: 2-17 (1999).

IAEA: Lessons Learned from Accidents in Industrial Irradiation Facilities. Vienna: International Atomic Energy Agency, 1996.

IAEA: Lessons Learned from Accidents in Industrial Radiography (Safety Reports Series No. 7). Vienna: International Atomic Energy Agency, 1998.

IAEA: Lessons Learned from Accidental Exposures in Radiotherapy (Safety Reports Series No. 17). Vienna: International Atomic Energy Agency, 2000.

LANL: A Review of Criticality Accidents, 2000 Revision (LA-13638) by T.P. McLaughlin, S.P. Monahan, N.L. Pruvost, V.V. Frolov, B.G. Ryazanov, and V.I. Sviridov. Los Alamos, NM: U.S. Department of Energy, Los Alamos National Laboratory, 2000.

Lubenau, J.O. and D.J. Strom: Safety and Security of Radiation Sources. *Health Phys. 83*: 155-164 (2002).

McManus, N.: *Safety and Health in Confined Spaces*. Boca Raton, FL: Lewis Publishers/CRC Press, 1999.

Mannan, M.S., M. Gentile, and T.M. O'Connor: Chemical Incident Data Mining and Application to Chemical Safety Analysis. In *Proceedings of the CCPS 2001 International Conference and Workshop*, Toronto, Ontario, Canada, October 2-5, 2001, pp. 137-156.

MSHA: Think "Quicksand": Accidents Around Bins, Hoppers and Stockpiles, Slide and Accident Abstract Program. Arlington, VA: U.S. Department of Labor, Mine Safety and Health Administration, National Mine Health and Safety Academy, 1988.

MSHA: Hazard Information Alert: Confined Space Fatalities – 57 since 1980. Arlington, VA: U.S. Department of Labor, Mine Safety and Health Administration, 1994.

NFPA: National Electrical Code (ANSI/NFPA 70). Batterymarch Park, MA: National Fire Protection Association, 2011.

NIOSH: Guidelines for Controlling Hazardous Energy During Maintenance and Servicing (DHHS (NIOSH) Publication No. 83-125). Morgantown, WV: National Institute for Occupational Safety and Health, 1983a.

NIOSH: Occupational Safety in Grain Elevators and Feed Mills (DHHS/NIOSH Publication No. 83-126). Morgantown, WV: National Institute for Occupational Safety and Health, 1983b.

NIOSH: Worker Deaths in Confined Spaces (DHHS/PHS/CDC/NIOSH Pub. No. 94-103). Cincinnati, OH: National Institute for Occupational Safety and Health, 1994.

NIOSH: Worker Deaths by Electrocution (DHHS (NIOSH) Publication 2000-115. Cincinnati, OH: National Institute for Occupational Safety and Health, 2000. CD-ROM.

Ortiz, P., M. Oresegun, and J. Wheatley: Lessons from Major Radiation Accidents. *Proc. 10th Congr. Internat. Radiat. Prot. Assoc. Hiroshima, Japan, May 2000*, 2000.

OSHA: Selected Occupational Fatalities Related to Fire and/or Explosion in Confined Work Spaces as Found in OSHA Fatality/Catastrophe Investigations. Washington, DC: U.S. Department of Labor, Occupational Safety and Health Administration (U.S. DOL/OSHA), 1982a.

OSHA: Selected Occupational Fatalities Related to Lockout/Tagout Problems as Found in Reports of OSHA Fatality/Catastrophe Investigations. Washington, DC: U.S. Department of Labor, Occupational Safety and Health Administration (U.S. DOL/OSHA), 1982b.

OSHA: Selected Occupational Fatalities Related to Grain Handling as Found in Reports of OSHA Fatality/Catastrophe Investigations. Washington, DC: U.S. Department of Labor, Occupational Safety and Health Administration (U.S. DOL/OSHA), 1983.

OSHA: Selected Occupational Fatalities Related to Toxic and Asphyxiating Atmospheres in Confined Work Spaces as Found in Reports of OSHA Fatality/Catastrophe Investigations. Washington, DC: U.S. Department of Labor, Occupational Safety and Health Administration (U.S. DOL/OSHA), 1985.

OSHA: Selected Occupational Fatalities Related to Welding and Cutting as Found in Reports of OSHA Fatality/Catastrophe Investigations. Washington, DC: U.S. Department of Labor, Occupational Safety and Health Administration (U.S. DOL/OSHA), 1988.

OSHA: "Control of Hazardous Energy Sources (Lockout/Tagout); Final Rule," *Fed. Regist. 54*: 169 (1 September 1989). pp. 36644–36696.

OSHA: Selected Occupational Fatalities Related to Ship Building and Repairing as Found in Reports of OSHA Fatality/Catastrophe Investigations. Washington, DC: U.S. Department of Labor, Occupational Safety and Health Administration (U.S. DOL/OSHA), 1990.

OSHA: "Permit-Required Confined Spaces for General Industry; Final Rule," *Fed. Regist. 58*: 9 (14 January 1993). pp. 4462–4563.

OSHA: "Confined and Enclosed Spaces and Other Dangerous Atmospheres in Shipyard Employment; Final Rule," *Fed. Regist. 59*: 141 (25 July 1994). pp. 37816–37863.

Theimer, O.F.: Cause and Prevention of Dust Explosions in Grain Elevators and Flour Mills. *Powder Technol. 8 (3-4)*: 137-147 (1973).

USDA: Prevention of Dust Explosions in Grain Elevators – An Achievable Goal. Washington, DC: U.S. Department of Agriculture, 1980.

USDA: List of Explosions. Washington, DC: U.S. Department of Agriculture, 1982.

Vargo, G.J.: A Brief History of Nuclear Criticality Accidents in Russia—1953–1997. *Health Phys. 80:* 505-511 (1999).

WCB: Part 19: Electrical Safety. In Occupational Health & Safety Regulation (BC Regulation 296/97, as amended by BC Regulation 185/99). Vancouver, BC: Workers Compensation Board of British Columbia, 1999.

Yusko, J.G.: The IAEA Action Plan on the Safety of Radiation Sources and Security of Radioactive Material. In *Radiation Safety and ALARA Considerations for the 21st Century*. Madison, WI: Medical Physics Publishing, 2001. pp. 137–142.

# 2 Occurrence of Hazardous Energy

## INTRODUCTION

Hazardous energy is the cause of many nonfatal and fatal accidents according to reports published by the International Atomic Energy Agency (IAEA), Mine Safety and Health Administration (MSHA), the National Institute for Occupational Safety and Health (NIOSH), and the Occupational Safety and Health Administration (OSHA), among others (IAEA 1996, IAEA 1998, IAEA 2000; MSHA 1988; NIOSH 1979, NIOSH 1983a, NIOSH 1983b, NIOSH 1994, NIOSH 2000; OSHA 1982a, OSHA 1982b, OSHA 1983, OSHA 1985, OSHA 1988, OSHA 1990).

Transmission of energy and transformation of energy from one form to another are responsible for many kinds of accidents. Indeed, statistics compiled by the (U.S.) Bureau of Labor Statistics on their website (www.bls.gov) for the years 1992 to 2002 indicate that 'strike by' accidents which average about 9% of the total and electrocutions, about 6%, are the fourth and fifth ranking causes of fatal occupational injury. They are the first and second ranked causes due to materials containing energy or an energy source with which workers routinely come into contact during the course of work activity. (The other leading causes of fatal occupational injury are highway traffic incidents (about 21% of the total), workplace violence (about 16%), and falls from heights (about 10%).) Statistics compiled by NIOSH indicate that engulfment by loose materials was the leading cause of fatalities in accidents in workspaces that fit the technical description of confined spaces (NIOSH 1994). These workspaces include trenches, excavations and ditches in which engulfment occurred. Trenches, excavations and ditches normally are not considered to be confined spaces. However, engulfment is engulfment. The mechanism of engulfment in a trench, excavation or ditch is similar to what occurs in a structure, such as a silo or hopper. When considered without the imposition of regulatory constraints, engulfment easily would become the first ranking cause of fatal accidents in these workspaces. Please note that these rankings are subject to change created by the emphasis placed by regulatory authorities on individual classes.

The data suggest that most of the work performed by victims, when considered from their perspective, was normal. Many of the hazardous conditions existed prior to the start of work. Coincident with this are the following subsets of circumstance.

- First, the victim knew about the hazardous condition through training and education or safety talks provided by supervisors or others, but attempted to perform the task without following procedures, or use or proper use of protective measures.

- Second, the victim recognized the presence of the energy source, but had no training in control, or perceived the ability to perform the task without undue risk, or perceived the lack of availability of protective measures.

- Third, the hazard of the energy source was not perceivable through the senses or recognizable by other common means, or the person lacked the knowledge needed to recognize its presence. In some circumstances, workers inactivated the energy source prior to the onset of the task. Through some action, the victim or a third party inadvertently or sometimes intentionally re-energized the energy source. This sometimes occurred with the full knowledge of the victim.

- Lastly, in some situations, the victim precipitated the accident through use of an energy source associated with the task, or through deliberately destabilizing materials, such that transformation of energy trapped within them could occur. The relative importance of a particular scenario depends on the nature of the work and the industry in which it occurs.

Contact with energy sources produces a range of effects. At the one extreme, the effect is immediately fatal, or is fatal within a short span of time following the contact. These effects include visible impacts, such as severe traumatic injury, but they also can occur as injury that is completely invisible. At the other extreme, the effects are minor annoyances to the overall daily routine. There also can occur as delayed effects, such as cancers or other diseases that have a long period of latency. The book focuses mainly on acute and subacute effects. These are the outcomes that have the greatest impact on the safety or actions of victims, or individuals who could become victims. These are the outcomes that must be prevented in order to enable people not to be victims of accidents.

## Physical Concepts

In the abstract world of the physics textbook, energy is defined as the capability to do work (Weast 1968). Energy is conserved and cannot be either created or destroyed, but can be transformed from one form to another. Work is the change produced in a body by a force acting through a resistance. Force is that which changes or tends to change the state of rest or motion in matter or an object. Work is measurable as the product of the force and the distance through which the object moves. That is, work is done or energy expended when a force is exerted through some distance (Cember 1996). Power is the rate at which work is done per unit of time.

These definitions hardly describe the encounters that we experience with energy and its effects. Understanding the outcome from collision between the human body and a moving object is readily comprehensible at an intuitive level. The object has motion. A moving object hits a stationary one with a whack. The whack can destabilize or change the position of a stationary object, or cause evident damage or destruction. On the other hand, visualizing the outcome from the collision between a rapidly moving sub-nuclear particle and an organelle in a cell in the body is completely impossible. Such is one of the difficulties in conceptualizing about the many forms of hazardous energy.

Energy is a highly intangible "something" with which we all have familiarity. Energy is intrinsic to certain objects, and can be transferred to them and from them to others, and makes things happen. Energy contained in some materials and systems is :"stable" and resists release, while in others is unstable and is released with minimal disturbance.

## Motion

A discussion about the mathematical relationship between these words will help clarify their meaning and relationships. The startingpoint for this discussion is a statement of the equations of motion. The following equations provide the relationship between velocity, acceleration, and time.

$$v = u + at \tag{2-1}$$

$$d = ut + (0.5)at^2 \tag{2-2}$$

$$v^2 = u^2 + 2ad \tag{2-3}$$

where:
u = initial velocity (m/s).
v = final velocity (m/s).
d = displacement (m).
a = acceleration (m/s$^2$); acceleration due to gravity is 9.8 m/s$^2$.
 t = time (s).

Force is the product of mass and acceleration (Equation 2-4) and has the unit Newton (N).

$$F = ma \tag{2-4}$$

where:
F = force (N = [kg][m/s$^2$]).
m = mass (kg).

Work is the product of force and distance (Equation 2-5) and has the unit Joule (J).

$$W = Fd \tag{2-5}$$

where:
W = work (J = [kg][m/s$^2$][m] = [kg][m$^2$/s$^2$]).
d = distance (m).

Energy exists in two main forms: potential energy and kinetic energy. Potential energy is the energy of storage or position. Potential energy is the energy that is available to be expressed. Potential energy is the energy "stored" in the nucleus of the atom, in chemical bonds, in fluids under pressure, in objects that can fall to a lower level, and in the geometric configuration of flowable solid materials. Gravitational energy of position is given by:

$$E_{potential} = mgh \tag{2-6}$$

where:
$E_{potential}$ = potential energy due to gravity (J = [kg][m/s$^2$][m] = [kg][m$^2$/s$^2$]).
m = mass of the object (kg).
g = acceleration due to gravity (9.81 m/s$^2$).
h = height above the reference plane (m).

Note from Equation 2-6 that the unit for both work and energy is the Joule; that is, it is the same. The unit of power is the Watt (1 W = 1 J/s).

Kinetic energy is the energy associated with moving objects or particles. Kinetic energy is the energy expended in process of doing something. The kinetic energy of slow-moving (not affected by relativistic considerations) objects or particles is given by:

$$E_{kinetic} = (0.5)mv^2 \tag{2-7}$$

where:
$E_{kinetic}$ = kinetic energy (J = [kg][m$^2$/s$^2$]).
m = mass of the object or particle (kg).
v = velocity of the object or particle (m/s).

## Photoelectric Energy

A considerable part of the effort in physics in the late 1800s and early 1900s concerned study of matter at the atomic and nuclear level. One of the first phenomena demonstrated was the photoelectric effect. This concerned emission of electrons from metal surfaces irradiated by light. These experiments produced surprising results that related the energy of the emitted photons to the wavelength and frequency of the incident light through a constant (Planck's constant) and the speed of light. The speed of all electromagnetic waves in a vacuum is the same: $3.0 \times 10^8$ m/s, the speed of light. Planck's equation (Equation 2-8) forms the link between the wavelength and the energy of photons

$$E = h\nu = \frac{hc}{\lambda} \tag{2-8}$$

where:
E = energy (Joules or J).
h = Planck's constant = $6.627 \times 10^{-34}$ Joule•second (J•s).
$\nu$ = frequency of the emission, waves/second (1/s).
c = speed of light = $3.00 \times 10^8$ meters/second (m/s).
$\lambda$ = wavelength of the emission, meters/wave (m).

**Electrical Energy**
The Joule is the unit of energy commonly used in electrical circuits. The following equation (Equation 2-9) describes the energy required to power an electrical circuit over a period of time. The use of power in the previous sentence is colloquial, since power is the rate of "using" electrical energy, expressed in Watts as the "use" of energy at the rate of one Joule/second (1 W = 1 J/s).

$$E = VIt \qquad (2\text{-}9)$$

where:
E = the energy related to current flow, Joules (J).
V = the impressed voltage, Volts (Volt = Joule/coulomb = J/C).
I = the current, Amperes (A = coulomb/second = C/s).
t = time, seconds (s).

The second equation of importance (Equation 2-10) expresses electrical energy stored in a capacitor. A capacitor is a fundamental component in electrical circuits. Capacitors sometimes arise by happenstance in unintended ways and have caused many accidents due to unintended accumulation and discharge of electrical charge.

$$E = 0.5\ QV \qquad (2\text{-}10)$$

where:
E = energy stored in the capacitor, Joules (J).
Q = charge separated in the capacitor, coulombs (C).
V = voltage impressed to separate the charge, Volts (V = J/C).

The third important equation(Equation 2-11) relates energy converted to heat (also a form of energy) in an electrical circuit.

$$E = I^2Rt \qquad (2\text{-}11)$$

where:
E = energy expended against the resistance of the circuit, Joules (J).
I = electrical current flowing in the circuit, Amperes (A = coulombs/second).
R = resistance to flow of current, Ohm ($\Omega = J\bullet s/C^2$).
t = time, seconds (s).

This equation relates the amount of energy (or power) "consumed" in operating an electrical appliance, such as a toaster or electric stove, when cooking.

**Heat (Thermal Energy)**
Heat is the ultimate end-point for energy converted (transformed) from other forms. Heat (thermal energy) is the destination for energy converted from potential to kinetic forms.
The classic unit for measuring heat was the calorie. The calorie is the amount of heat needed to raise the temperature of water 1° C (Celsius). There is considerable confusion with the Calorie, the unit for expressing the energy content in food, which has and continues to have widespread use. Equation 2-12 provides the relationship between the calorie, the Calorie and the Joule.

$$1\ cal = 0.001\ Cal = 4.18\ J \qquad (2\text{-}12)$$

where:
cal = calorie (cal).
Cal = Calorie, the unit of energy provided by food (Cal).

Equation 2-13 relates energy radiated from a surface as the temperature increases.

$$E_{radiant} = \sigma e T^4 \tag{2-13}$$

where:
$E_{radiant}$ = energy radiated from the surface, Joules (J).
$\sigma$ = Stefan–Boltzmann constant, $5.67 \times 10^{-8}$ Joules/(metre)$^2$/second/($^\circ$Kelvin)$^4$, (J/m$^2$/s/$^\circ$K$^4$).
e = a factor that depends on the material and the temperature.
T = surface temperature, degrees Kelvin ($^\circ$K = $^\circ$C + 273).

## Conservation and Transformation (Conversion) of Energy

Conservation of energy in the universe is a fundamental principle. That is, transformation of energy from one form to another or transfer from one object to another can occur, but creation or destruction of energy cannot occur. Ultimately, energy is transformed to heat (also a form of energy), which is expressed through the motion of atoms and molecules.

Transformation is the change of energy from one form to another, for example from potential to kinetic. Transformation of energy from one form to another is a seamless process, as illustrated by movement of the pendulum of a clock. The energy stored in the pendulum is entirely potential at the high point of the swing. The energy is entirely kinetic at the low point of the swing. Another example of transformation of energy is the partition between static pressure and velocity pressure, as occurs inside the casing of a fan, inside the components of a ventilation system, and the interior of temporary enclosures created for asbestos and mold remediation (McManus 2000, McManus 2006).

Transmission of energy is transfer from one object to another. Transmission of energy occurs, for example, when one railcar strikes another and induces motion.

Transformation and transmission of energy in the previous examples affect the entire object, and not its minutest components, the individual atoms and molecules.

Transformation and transmission of energy in an unintended or uncontrolled manner are responsible for many fatal accidents. Gaining an understanding and an appreciation about the workplace in energy terms, therefore, is essential to forming and implementing strategies for control.

As the focus of this book is hazardous energy, there is a need to determine the intended meaning for this term. The intent to focus on effects that can significantly impair the ability of a victim to lead a productive life following the contact narrows the scope of the definition. For the purposes of this book, hazardous energy is the level of energy interacting with biological tissue capable of producing injury, and secondarily, disease.

Energy sources are the points of emission of things containing energy. Sources occur in various geometries, starting with the point source. Examples of point sources of intense emissions include the sun, welding arcs, and cutting flames, and searchlights, to name a few. Line sources include electrical transmission lines and radio broadcast antennas. Distributed sources include the heated surfaces (walls and roof) in a bake oven.

The previous examples illustrate sources that emit energy. Hazardous levels of energy also are stored in materials. Many materials are stored or configured in a manner that can allow rapid transformation from one form of energy to another. While each component of the material, a seed of grain or a crystal of sand or molecule of highly unstable, explosive material itself may have insignificant energy, the collective aggregate is what poses the harm during energy transformation. Some examples of materials containing energy include a sandpile, snow on the crest of a mountain, and sand in a silo. In each case, slumpage can rapidly change the geometry of the material, and in the process transform potential energy to kinetic energy.

Critical to understanding the physical and biological action of physical bodies is the concept of energy transfer. Energy transfer occurs at the nuclear, atomic, and molecular level, as well as at the gross level that is part of our normal experience. Accidents characterized as "struck by" or "struck against", or "fall from height" or "fall from the same level" for example, all involve collision between the human body or some part thereof and another object, and hence energy transfer.

One mechanism for energy transfer is collision (Cember 1996). Collision is either elastic or inelastic. During an elastic collision, the sum of the kinetic energy of the two bodies and the sum of the momentum of the two bodies remain the same. That is, kinetic energy and momentum are conserved (Beer and Johnston 1987). Elastic collision acts to redistribute both kinetic energy and momentum. The bodies need not make physical contact during an elastic collision. "Like" poles of two magnets or objects carrying high levels of "like" electrostatic charge need not come into physical contact to collide elastically.

Momentum of slow-moving (not affected by relativistic considerations) objects or particles is given by:

$$M = mv \qquad\qquad (2\text{-}14)$$

where:
M = momentum ([kg][m/s].
m  = mass of the object or particle(kg).
v  = velocity of the object or particle(m/s).

An inelastic collision occurs when some of the energy is expended in freeing a tethered object. To illustrate, some nuclear particles remove electrons from orbitals in atoms. Removing the electron from the status quo (electrostatic attraction with the nucleus of the atom) requires some of the energy carried by the colliding particle. Hence, the energy shared between the electron and colliding particle after the collision is less than the total energy carried by the particle and the electron originally.

Energy and matter are inter-convertible, as predicted by Einstein in 1905 (Equation 2-15):

$$E = mc^2 \qquad\qquad (2\text{-}15)$$

where:
E = the energy equivalent of matter (J).
m = mass converted to energy (kg).
c = the speed of light in a vacuum ($3 \times 10^8$ m/s).

That is, E is the total energy equivalent to a piece of matter of mass, m. According to the theory of relativity, all matter contains potential energy by virtue of its mass, since mass is directly convertible to energy.

Energies resulting from nuclear processes are measured in electron Volts (eV). The conversion between the SI (Système Internationale) unit of energy, the Joule (J) and the eV is:

$$1 \text{ eV} = 1.60 \times 10^{-19} \text{ J} \qquad\qquad (2\text{-}16)$$

## Nuclear Energy

The atomic nucleus measures about $5 \times 10^{-13}$ cm in diameter (Cember 1996). Study of the atomic nucleus and its energetic emissions dominated much of the scientific effort of the 20th century, starting with their discovery by Becquerel in 1896 (Martin and Harbison 1979; Shapiro 1981). Subsequent discoveries by physicists and some chemists showed that the nucleus contains two fundamental particles: protons and neutrons. The proton contains an electrical charge of $4.80294 \times 10^{-10}$ sC (statCoulombs), nominally designated a value of + 1 units, while the neutron has no charge (Table 2.1). The neutron is slightly heavier than the proton. The number of protons (atomic number) determines the chemical identity of the atom. Surrounding the nucleus are clouds of electrons in orbitals. These are equal in number in atoms having no net charge to the number of protons in the nucleus. The electron has an electrical charge equal to and opposite that of the proton (Table 2.1). The electron is considerably lighter than the proton and neutron.

Protons are positively charged, and therefore repel one another. They co-exist with one another and with the neutrons that have no electrical charge in the nucleus in an at times uneasy relationship.

## Table 2.1
## Major Atomic Particles

| Particle | Charge | Mass |
|---|---|---|
| Proton | + 1   (+ 4.80294 × 10⁻¹⁰ sC) | 1.6724 × 10⁻²⁴ g |
| Neutron | 0 | 1.6747 × 10⁻²⁴ g |
| Electron | − 1   (− 4.80294 × 10⁻¹⁰ sC) | 9.1086 × 10⁻²⁸ g |

Source: Cember (1996).

The fact that there are many stable chemical elements and isotopes within those elements indicates that this relationship does exist. We also know, based on the poking and prodding that the nucleus has received over the years during scientific experimentation, that this stability is substantive. Stable elements, starting with helium, which have more than one proton, have at least one neutron. That is, there is at least one neutron in the nucleus per proton. The ratio of neutrons to protons increases as the atomic number increases (Cember 1996). An increasing proportion of neutrons is needed to help maintain stability in the nucleus as the number of protons increases, since the protons, being very close together, repel one another with considerable electrical force.

The nucleus of the atom, being stable, therefore contains attractive forces having considerable strength. That is, the attractive forces must have sufficient strength to overcome the repulsive electrical forces. Nuclear forces act across the very short distances between the particles in the nucleus. Electrical repulsive forces between protons act across much longer distances. This is the reason for the increasing number of neutrons needed to maintain stability in the nucleus as atomic number increases.

Certain numbers of neutrons represent stable configurations with a fixed number of protons, hence the occurrence of stable isotopes (atoms having the same number of protons but different numbers of neutrons). Stability occurs within a narrow range of the neutron-to-proton ratio. Hence, addition of additional neutrons to a nucleus does not ensure or enhance stability. These configurations are so energetically unstable that they do not occur, and still other configurations have fleeting stability ranging from fractions of seconds to thousands of years. These are the unstable or radioactive isotopes. They become stable through the spontaneous process of radioactive decay.

Nuclear processes occur in the keV to MeV range. Energies of this magnitude, deposited in the appropriate manner in a biological system, are capable of causing injury, up to and including fatal injury. Even so, these quantities seem vanishingly small in a world that operates with energies measured in Joules. (A standard fluorescent tube consumes 35 to 40 J of electrical energy per second to produce light and some heat.)

### Radioactivity
Radioactivity is a spontaneous change in the composition of the nucleus. Radioactivity usually results in formation of a new element. The type of radioactive process that occurs in a specific situation depends on the neutron-to-proton ratio and the mass-energy relationship between the parent nucleus, nucleus of the progeny, and the emitted particle (Cember 1996).

### Alpha Particle Formation
An alpha particle has the same composition as the nucleus of a helium atom: two protons and two neutrons. The alpha particle is the heaviest of the particles emitted from the nucleus of the atom. It also carries an electrical charge of +2. Alpha particles possess considerable kinetic energy, the result of large mass and the high speed of ejection needed to overcome binding forces in the nucleus. Alpha emitters

almost exclusively have atomic numbers greater than 82 (Cember 1996). This results from the balance between cohesive and repulsive forces in the nucleus. The energy of the alpha particle results from a matter-to-energy conversion and occurs in the MeV range.

The energy of the matter-to-energy conversion apportions to the alpha particle and the nucleus of the resultant atom. The nucleus recoils in the direction opposite to that of the emitted alpha particle. Some of the energy may remain in the nucleus. This condition is termed an excited nucleus. The nucleus later sheds the surplus energy through emission of a gamma ray photon to return to the non-excited (ground) state. (A photon is a packet of energy. Photons have wave-like and particle-like characteristics, and are describable in terms of wavelength, frequency, or energy.) Energies of gamma photons range from keV to MeV. Generally alpha particle emissions are not accompanied by gamma emissions. The resulting nucleus may still be unstable due to the preponderance of neutrons (Martin and Harbison 1979).

### Beta Particle Formation

A beta particle is the same in every characteristic as an electron (Johns and Cunningham 1974). It is created in the nucleus during conversion of a neutron to a proton, the beta particle, and a particle of vanishingly small mass called a neutrino. Beta particle formation is a means of establishing the balance between cohesive and repulsive forces. Beta formation occurs in nuclei having a surplus of neutrons to protons. Mass surplus to the conversion of the particles converts to energy. This apportions to the nucleus of the new atom as recoil energy, to the beta particle and to the neutrino. As a result, beta particle energy generally ranges from a maximum measured in keV to MeV. In some cases, the nucleus of the progeny atom retains some of the energy. In these situations, the excited nucleus emits a gamma photon in order to return to the non-excited state.

### Positron Formation

Positron formation occurs when the neutron-to-proton ratio is too low for stability and emission of an alpha particle is not energetically possible (Cember 1996; Johns and Cunningham 1974). A positron is the same as an electron in every characteristic except charge. A positron has a charge of $+4.80294 \times 10^{-10}$ sC or $+1$ units of charge. A positron results during transformation of a proton to a neutron, the positron, and a neutrino. Because a neutron has greater mass than a proton, energy must convert to matter in the process. The number of protons decreases by one in the transformation. Hence, the nucleus has a surplus of cohesive force due to the reduction of electrostatic force. The energy unused during the energy-to-matter transformation apportions between the nucleus of the progeny, the positron, and the neutrino. The nucleus loses some energy during recoil from the positron, but may remain excited, in which case emission of a gamma photon occurs. Positron emissions occur in the keV to MeV range. The neutrino carries surplus energy.

### Electron Capture

An alternate process for addressing neutron deficiency is electron capture (Cember 1996; Johns and Cunningham 1974). Electron capture refers to the capture of an electron from one of the orbital clouds that surround the atom. Usually the electron captured is one that spends the largest proportion of its time near the nucleus, and hence is more likely than others to be available for capture. In this process the electron enters the nucleus and unites with a proton to form a neutron and a neutrino. The nucleus emits a gamma photon and a mono-energetic neutrino (a neutrino having only a single level of energy). The energies involved are in the keV to MeV range.

Accompanying the process of electron capture is emission of an X-ray photon. This is the energy released during migration of one of the outer electrons to the position formerly occupied by the electron converted to the neutrino. The energies of X-ray photons are in the keV range.

### Gamma Ray Photons

Gamma ray photons are packets of energy emitted from the nucleus of atoms undergoing radioactive decay processes (Shapiro 1981). Gamma emissions occur at specific energy levels. That is, an emission from a particular isotope has a characteristic energy. As mentioned previously, these energies are in the keV to MeV range.

Photons have both particle- and wave-like properties. That is, a particular photon has a trajectory in space. A population of photons emits randomly in all directions from source nuclei, one at a time. The wave-like characteristics of photons are similar to other forms of electromagnetic waves, starting with X-rays. (Electromagnetic waves have magnetic and electrical components oriented perpendicularly to each other.) The only difference between the electromagnetic wave properties of gamma photons and those of electromagnetic waves from other sources is the wavelength.

## Internal Conversion

Internal conversion is an alternate mechanism for dissipation of energy during the transition from the excited to the non-excited level (Cember 1996). This process involves transfer of the energy to an electron in the cloud surrounding the nucleus with subsequent ejection of the electron from the atom. The energy of the ejected electron is that left after overcoming the attractive forces of the nucleus for electrons. This energy is in the keV to MeV range.

Accompanying the process of electron ejection is emission of an X-ray photon. This is the energy released during migration of one of the outer electrons to the position formerly occupied by the ejected electron. The energies of X-ray photons are in the keV range.

## Fission Processes

Fission is the process of splitting the nucleus into two fragments (Cember 1996). The nuclei of some isotopes of thorium and uranium split spontaneously. These contain large numbers of protons and even larger numbers of neutrons. Despite the fact that fission of the nuclei of some isotopes can occur, spontaneous fission is a rare event. This is the case because energy is needed to excite the nucleus in order to cause the occurrence of the fission event.

One way to promote fission is to bombard the nucleus with slow-moving (thermal) neutrons. (Thermal neutrons have the same kinetic energy as gas molecules in the same environment. At room temperature, this is about 0.025 eV.) Capture of the neutron releases binding energy. Nuclei that capture neutrons but do not split release the excess binding energy through emission of a gamma photon. Other nuclei will split when bombarded with fast neutrons having energies exceeding 1.1 MeV.

Fission produces two nuclear fragments, several neutrons and considerable energy (about 190 MeV per event). This energy apportions mostly to the kinetic energy of the nuclear fragments, with much smaller amounts distributed to neutrons, gamma ray photons, and beta particles and neutrinos produced during subsequent radioactive decay processes. Hence, most of the energy is released as heat represented by the movement of particles.

A major concern in radiation protection involving fissionable material is criticality. As mentioned in the previous chapter, criticality is the condition where at least one of the neutrons emitted during a fission process causes a second nucleus to split. Uncontrolled fission leads to rapid and uncontrolled release of energy. To put meaning into this situation, each fission produces about:

$$190 \text{ MeV/fission} \times 1.6 \times 10^{-19} \text{ J/eV} \times 10^6 \text{ eV/MeV} = 3.0 \times 10^{-11} \text{ J/fission} \qquad (2\text{-}17)$$

Since a criticality accident typically produces about $1 \times 10^{17}$ fissions (LANL 2000), the amount of energy released is:

$$3.0 \times 10^{-11} \text{ J/fission} \times 1 \times 10^{17} \text{ fissions} = 3.0 \times 10^6 \text{ J or 3 MJ} \qquad (2\text{-}18)$$

The energy released during a typical criticality accident, $3.0 \times 10^6$ J, is 3.0 MJ (megaJoules or million Joules). The Watt, the more familiar unit of power, is 1 J/s. Hence, this burst of thermal energy, released over one second of time, if harnessed, would produce 3.0 MW (MegaWatts) of power. No wonder these events lead to rapid heating, boiling, and vaporization of process solutions, as well as stressing and buckling of metals, and destruction of structures, as reported in the accident summaries.

Particle Emission Processes

Fission was one of the first of the atomic processes induced by particle bombardment. Neutrons are produced by bombardment of the nucleus of target elements in particle accelerators, such as the cyclotron. The products of bombardment are unstable and achieve stability through emission of neutrons. Another technique for producing neutrons is to bombard nuclei by alpha particles in a powdered, thoroughly admixed mixture. One such example of a neutron source is a sealed capsule containing a mixture of finely powdered beryllium and radium, polonium, or plutonium. Neutrons emitted in these reactions have energies in the MeV range (Shapiro 1981).

## Atomic (Electron) Processes

Atomic processes involve electrons. Electrons normally are found in orbital "clouds" that surround the nucleus of the atom.The shapes of these "clouds" found in chemistry textbooks, derive from complex mathematical equations. In an atom having no net charge, the number of protons (atomic number) equals the number of surrounding electrons. Almost all atoms occur in nature in chemical combination, either as combined elements or chemical compounds involving one or more other elements. Atoms in the inert gases (helium, neon, argon, krypton, xenon, and radon) exist singly. Their outer electrons already exist in the configuration which other atoms strive to achieve. As a result, atoms of inert gases have no need to form chemical bonds with other atoms.

Because of their insular nature, inert gases provide the opportunity to study the electronic arrangement in the atom without perturbation from chemical bonds. Electrons exist in specific energy levels. Associated with these main levels are sub-levels (Barrow 1966). Both are related to the arrangement of electron orbital clouds around the nucleus. Providing energy to these atoms causes electrons to move from stable energy states to those at higher energy. The transition from lower to higher energy states physically means that an energized electron spends less time near the nucleus and more time at distances remote from the nucleus. On removal of the exciting energy, the electrons return to the lower energy levels. The amount of energy absorbed in a specific jump from a lower to a higher level or vice versa is specific to the jump, and repeatable from episode to episode. The energy released by electrons during transitions emits as photons. Energies associated with transitions by electrons between energy levels occur in the eV to keV range.

Photons emitted by electrons during transitions between energy levels have characteristics similar to those of gamma ray photons. That is, they have both particle and electromagnetic wave-like properties. They move at the speed of light and differ from each other only in wavelength. Hence, wavelength is directly related to the energy of the photon.

The energy of the photon endows it with certain characteristics. The first of these is the ionization potential (IP). The ionization potential is the energy needed to remove the least tightly held electron from an atom. These values range from a low of 3.9 eV (cesium) to a high of 17.4 eV (fluorine) (Weast 1968). Removal of electrons from inner orbital clouds in large atoms where the electron is found near the nucleus a high proportion of the time requires thousands of electron volts (keV). Removal of an electron from an inner cloud leads to replacement by an electron from an outer cloud. This electron sheds energy during the transition in the form of a photon. Replacement of the first electron can create a cascade of replacements. Each electron in the cascade sheds energy through emission of a photon.

Scientists have put different labels onto photons having specific energy. The photons having the greatest energy in the process are called X-ray photons. These occur in the keV range. X-rays have characteristics of electromagnetic waves. Hence, X-rays are indistinguishable from gamma rays having the same energy. The only difference is the process of production. Photons produced in less energetic transitions are given the label of UV (ultra-violet) photons. UV photons have energies ranging from 12.4 eV to 3.1 eV. UV photons behave as electromagnetic waves. The preceding energies correspond to the wavelengths, 100 to 400 nm (nanometers) (Johns and Cunningham 1974; Repacholi 1992). The choice of 100 nm is arbitrary and does not correspond to the ionization potential of a particular element, as noted above.

UV photons are capable of removing electrons from some elements, primarily metals. This is the basis for the photoelectric effect. The ultraviolet region ends at the point at which the retina can detect visible light. This varies in individuals between 380 nm to 400 nm (nanometers).

### Chemical Bonds and Ionizing Radiation

Most elements occur in nature as chemical compounds. At the one extreme are ionic compounds, whose atoms acquire one or more electrons (becoming negative ions), or donate one or more electrons (becoming positive ions) to each other. They do this in an attempt to satisfy their chemical need to have the electrons in specific arrangements. This drive to achieve chemical stability supersedes the need to balance the number of charges between the nucleus and the surrounding electrons. At the other extreme are covalent compounds whose atoms mutually share certain electrons with one another, so that these shared electrons satisfy the chemical needs of each atom for the complete electron configuration for part of the time. Such is the nature of chemical bonds.

Covalent chemical bonds are thought to involve orbital clouds similar in concept to those associated with atoms (Barrow 1966). Bond formation represents a way of lowering energy content. Excitation of electrons forming the chemical bond can occur through absorption of energy. Excitation energies range from 2.2 eV to 6.6 eV (Willard et al. 1981). Energies are quantized; that is, they are specific to the transition. Most molecules have excited electronic states, many of which are not stable. Elevation to these states can cause dissociation of the molecule and severing of the bond. This energy is re-emitted from excited molecules that do remain intact when the electrons return to their original positions. The levels of these emissions occur in the UV and visible regions of the electromagnetic spectrum. These emissions are the mechanism behind fluorescence by chemical compounds, especially following irradiation by ultraviolet light (March 1985; Roberts and Caserio 1964). Some energy is lost through collisions and by increasing vibration and rotation of the bond.

Chemical bonds between atoms are capable of being ionized through interaction with energy sources. The ionization potential of compounds forming covalent bonds (shared electrons) ranges from about 7.4 eV to 14.8 eV. The ionization potential for some diatomic gaseous elements ranges as high as 17.8 eV. More typically, ionization potentials range from 8.5 eV to 11.5 eV (RAE Systems 2000). Instruments containing photoelectric detectors exploit ionization induced by irradiation from a UV source.

Ionizing radiation also can break chemical bonds. Bond strengths range from 1.6 eV to 11.2 eV, but more typically from 2.2 eV to 4.8 eV (Morrison and Boyd 1966; Roberts and Caserio 1964). Bond dissociation energies are in the same range as bond ionization energies, as noted above.

## Interaction of Nuclear Emissions with Matter

Previous discussion has established that nuclear emissions involve charged particles (alpha, beta, electrons, and positrons), uncharged particles (neutrons and neutrinos), and photons. These emissions interact with both electron orbital clouds (Figure 2.1(a)) and atomic nuclei (Figure 2.1(b)). The nature and capabilities of the interaction depend on the energies involved and charges carried by the particles (Pizzarello and Witcofski 1967). Previous discussion has also established that atoms and molecules can undergo excitation or ionization, and that chemical bonds between atoms can be broken. Nuclei and other atomic particles undergo recoil when struck by a moving particle. The energy carried by emissions from radioactive atoms ultimately dissipates as heat through these mechanisms.

### Alpha Particles

Alpha particles lose energy by electron excitation and ionization (Shapiro 1981). Each ionization in air or soft tissue requires about 35 eV. The amount of energy expended is considerably greater than that needed strictly for ionization. Some of this energy converts to the kinetic energy of the emitted electrons and positive ions. Alpha particles create tens of thousands of ion pairs per centimeter of travel through the air.

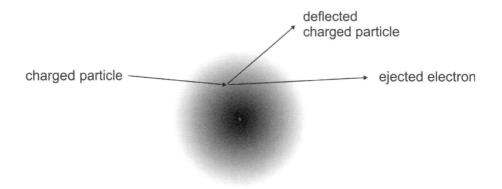

Figure 2.1(a). Particle interaction with orbital electrons. Charged alpha particles (helium nuclei), protons, and beta particles (electrons) moving at high velocity dissipate energy through removal of electrons from the electron cloud distributed around the nucleus of the atom (hydrogen shown here for simplicity).

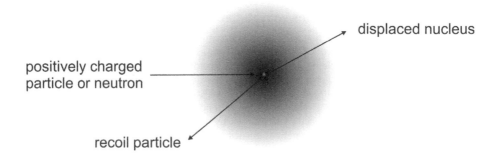

Figure 2.1(b). Particle interaction with the nucleus. Collision with the nucleus by atoms dissipates the energy of neutrons and charged particles. The nucleus presents a small target to particles compared to the electron cloud.

## Beta Particles

Beta particles lose energy by electron excitation and ionization through long-distance interaction between the electric field of the particle and that of electrons and nuclei in atoms. This is similar to the interaction between two magnets whose south poles, or north and south poles, face each other. Expenditure of energy in ejecting an electron from an atom is about two to three times the ionization potential. The average loss per ionization in air is about 35 eV. This quantity decreases for absorption occurring in hydrocarbon gases. Energy not used in ionization is expended in recoil energy provided to the particles. Some interactions transfer a large quantity of energy to a secondary displaced electron. This electron, in turn, can transfer energy to tertiary electrons, and so on.

Absorption of energy from beta particles depends on the density of electrons in the path (Cember 1996). This would suggest that the most efficient absorbers of beta particles are dense materials. In practical terms, materials having high atomic number are not used as beta absorbers. Materials having low atomic numbers, not greater than 13 (the atomic number of aluminum), are used instead to minimize production of Bremsstrahlung. Bremsstrahlung are X-ray emissions produced when high speed electrons undergo rapid acceleration when near a large nucleus. Hence nuclear charge is a major issue in minimizing the production of secondary radiation during stopping of high speed electrons.

## Positrons

Positrons are identical to electrons in every characteristic, except charge (Shapiro 1981). Positrons have the same charge as protons. Positrons are anti-matter. That is, they cannot co-exist with matter. Matter and anti-matter annihilate when in contact, in a matter-to-energy conversion. In this case, the matter with which the positron annihilates is an electron. This annihilation gives rise to two gamma ray photons that move in opposite directions, each having the energy equivalent to at least the mass of the electron (0.51 MeV at rest).

## Neutrons

Neutrons lose energy through collisions with atomic nuclei (Cember 1996). Neutrons in collision with a nucleus undergo scattering. In the process, the neutron may transfer some of its energy, creating an excited nucleus. The excited nucleus, in turn, emits a gamma photon to regain stability. The more likely transfer of energy creates recoil movement by the colliding nucleus. The energy of the neutron may be completely transferred to the colliding nucleus. This transfer is most effective with light nuclei, such as those in hydrogen atoms. The struck nucleus becomes an ionizing particle, and dissipates its kinetic energy through excitation and ionization. Once slowed down to thermal energy levels, neutrons may undergo capture by a nucleus. This is followed by emission of a photon or another particle. Boron and cadmium readily absorb neutrons, hence their use in neutron control rods in nuclear reactors. The boron reaction results in emission of an alpha particle. The cadmium reaction results in emission of a high energy gamma photon. The nuclei of other elements also absorb neutrons. This means that in an accident, such as a criticality accident, involving production of neutrons, peripheral materials and equipment could become radioactive.

## Neutrinos

Neutrinos are believed to have little interaction with matter, and for this reason, are not considered to have any significance to humans or other interactions with matter (Shapiro 1981). They are extremely difficult to study. To illustrate the difficulty of studying neutrinos, a laboratory was constructed deep in a hardrock mine near Sudbury, Ontario, Canada. Neutrinos reach the laboratory detection chamber, a plastic sphere containing thousands of liters of heavy water (deuterium oxide), after traveling through more than one thousand meters of solid rock. Another neutrino study facility located in Antarctica uses the entire planet as a shield. Detectors are mounted in thick ice.

## Photon Interactions That Ionize Matter

Photons that can ionize matter include gamma and X-rays. Their energy spectra overlap. As well, they both have particle and wave-like characteristics. Hence, at the same level of energy, they are indistinguishable from each other. Interposition of material into a beam reduces the intensity. Energy dissipation by photons occurs by several mechanisms (Figure 2.2). The one that predominates depends on energy level (Cember 1996; Johns and Cunningham 1974). More than one mechanism can participate in dissipation of the energy content of a particular photon.

Photoelectric Effect
The photoelectric effect involves interaction between the incident photon and an orbital electron. The photon annihilates, contributing its entire energy to the electron. The electron escapes from the atom and dissipates its energy through excitation and ionization. The atom may emit secondary photons as other electrons cascade inward to fill the vacancy. The photoelectric effect is highly dependent on atomic number (number of protons). Hence, heavy elements, such as lead are highly effective at shielding X-rays.

Compton Scattering
Compton scattering refers to a nondestructive collision between the photon and an orbital electron. This results in ejection of the electron and emission of a secondary photon of less energy. The target atom may emit secondary photons as other electrons cascade inward to fill the vacancy. The scattered electron dissipates its energy by excitation and ionization. High energy photons lose a greater fraction of their energy through this mechanism than low energy photons. The probability of Compton scattering increases with decreasing quantum energy and decreasing atomic number of the absorber. Compton scattering is the main mechanism of interaction in light elements (low atomic number). Scattering properties depend on density of electrons.

Pair Production
Pair production refers to an energy-to-matter conversion process in which the photon annihilates as it passes near the nucleus and a positron and an electron appear. Almost all of the surplus energy of the

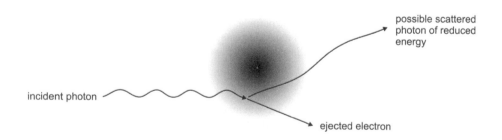

Figure 2.2. Interaction of high-energy photons with orbital electrons. Transfer of energy from photons to orbital electrons leads to ionization (photoelectric effect, Compton scattering) and pair production.

photon apportions as kinetic energy to the two particles. The probability of this mechanism is related to atomic number. Hence, pair production is more important in absorbers having high atomic number. Pair production requires at least 1.02 MeV of energy (0.51 MeV for production of each particle). The positron annihilates with an electron, as discussed previously.

Annihilation
Annihilation refers to capture of the photon by the nucleus of the absorbing atom. In most cases, this is followed by emission of a neutron. Annihilation is a high-energy reaction, generally requiring greater energy than is available from decay of radioisotopes. This reaction is a concern with electron accelerators, where the requisite high energies are achieved. An important exception to this is beryllium. Beryllium activation requires only 1.66 MeV.

## Nonionizing Radiation and the Electromagnetic Spectrum

In practice, the electromagnetic spectrum covers the range of photon energies from BeV ($10^9$ eV) to feV ($10^{-15}$ eV). These include energies already discussed that are created through nuclear processes, and that impact nuclei and electronic structure of target atoms and electronic structure of molecules, as well as regions not yet discussed. Commercially these regions are utilized in illumination, radiant heating, radar, television and radio broadcasting, and the transmission of electric power through conductors. Photons at some of these energies produce effects on the nucleus and electron clouds involved in chemical bonds (Barrow 1966; Willard et al. 1981).

### Visible Region
Visible light is the quality that we assign to photon energies to which the retina of the eye is sensitive. Depending on the individual, these range from 3.2 eV to 1.6 eV and correspond to wavelengths 385 to 780 nm (Repacholi 1992). Energy in this range corresponds to transitions involving outer orbital electron clouds (those that spend less of their time near the nucleus). Photon emissions from individual atoms or molecules consist of lines or bands of color, rather than the continuous spectrum extending from violet to red with which we are familiar. This reflects the single energies or narrow bands of energy involved in the transitions.

### Infra-Red Region
Molecular motions requiring the next lower input of energy after electron transitions are rotation and vibration of the atoms with respect to each other (Barrow 1966). Energies required to induce changes in rotation are usually about 1/100,000 of those needed for electron transitions. Energies needed to induce change in vibration are about 1/20 of what is needed for electron transitions. Such energies occur in the infra-red (IR) region of the electromagnetic spectrum. Photon energies range from 1.6 eV to 1.2 meV (millielectron Volts). These correspond to wavelengths ranging from 780 nm to 1 mm (Repacholi 1992). Vibrations of chemical bonds include symmetric and asymmetric bending and stretching modes. Bending modes refer to change in the angular relationship between atoms in the chemical bond. Stretching modes refer to change in the distance between atoms involved in the bond. Increasing the motion and rate of motion of components of a molecule is a manifestation of heat.

Infra-red emitters usually are heated surfaces, objects or point sources, such as light bulbs. Heating a material causes transitions to occur in the orbital electrons. This leads to emission of photons having energies in the infra-red region (McKinlay 1992). Emission of photons does not occur uniformly across the range of wavelengths, and is not temperature dependent. That is, the preponderance of wavelengths of emitted photons follows a pattern that is independent of the temperature of the emitter (Figure 2.3). This means that the increase in emissions is greater for some wavelengths than others for the same increase in temperature. The peak of the emission is temperature-dependent and shifts to lower wavelength (whiter light) as temperature increases. Infra-red radiators typically operate in the thousands of degrees Kelvin (K = C + 273). The filament of an incandescent lamp, for example, operates at 2850° K (Willard et al. 1981).

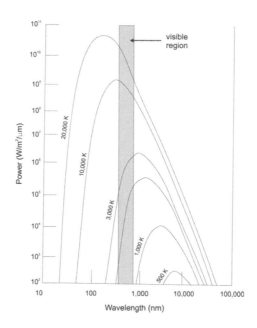

Figure 2.3. The thermal emission curve shows the relationship between temperature and wavelength of emission of photons from hot surfaces. Note the shift in the spectrum of emissions as the temperature increases.

### Microwave Region

Orbital electrons spin and generate a magnetic field as do the nuclei of most atoms (Willard et al. 1981). Electrons behave as magnets with the poles along the axis of rotation. The axis of motion is off-set from the vertical, as is the case with the planet Earth. Imposition of an external static magnetic field causes the spin to precess around the vertical, and establishes energy levels. This precession is similar to the motion of a top. Introduction of an electromagnetic field can cause the electron to jump between energy levels. Energy absorbed by orbital electrons occurs in the microwave region of the electromagnetic spectrum. This region covers photon energies from 1.2 meV to 1.2 $\mu$eV. These correspond to wavelengths of 1 mm to 1 m (Repacholi 1992).

Microwaves are generated in a klystron or magnetron tube similar to what is present in a microwave oven (Willard et al. 1981). A beam of electrons oscillates in pulses back and forth in the cavity between the cathode and a reflector filament. The electrons shed kinetic energy lost in deceleration as photons. Microwaves are produced only in specialized equipment.

### Radiofrequency Region

In about half of the isotopes, the nucleus is known to spin on an axis similar to the planet Earth (Willard et al. 1981). This spin generates a magnetic field. An external magnetic field applied to the atom causes the spin to precess in a manner similar to a spinning top. Increasing the strength of the field increases the rate of precession. Application of a weak radiofrequency field perpendicular to the uniform magnetic field at the precession frequency causes the nucleus to absorb energy and jump to a higher energy level. This is a resonance phenomenon. Energy absorbed by the nucleus occurs in the radiofrequency (RF) region of the electromagnetic spectrum. This region covers photon energies from 1.2 $\mu$eV to 1.2 neV. These correspond to wavelengths of 1 m to 1 km (Repacholi 1992).

## Chemical Bonding and Reaction Processes

As mentioned previously most elements occur in nature combined together or in chemical compounds containing atoms of two or more elements. Gold is one of the few metallic elements to occur in nature where the atoms are combined together. Hydrogen, nitrogen, and oxygen and some of the nonmetallic elements exist as combinations of atoms in molecules.

Atoms exhibit the need to be chemically complete; that is, to have the electron configuration of the nearest inert gas in the periodic table. Achieving this configuration occurs through two main bonding processes: ionic bonding and covalent bonding. An understanding of the rudiments chemical bonding is essential for understanding about energy stored in chemical bonds and interaction between different chemical entities.

### Ionic Bonding

Atoms become negative ions when they acquire one or more electrons, and positive ions when they donate one or more electrons to other atoms. They do this in an attempt to satisfy their chemical need to have the electrons in the specific arrangement of the nearest inert gas. This drive to achieve chemical stability supersedes the need to balance the number of charges between the nucleus and the surrounding electrons. Ions of opposite charge attract each other in arrangements that form balanced charge structures (no net charge) (Barrow 1966). The coming together of ions of opposite charge makes the overall process of ion formation energetically possible. Charge transfer can occur through metal structures (such as wires in a storage battery).

### Covalent Bonding

Covalent bonding involves the sharing of outer electrons between atoms. In this way, the shared electrons satisfy the chemical needs of each atom for completion of the electron configuration for part of the time (Barrow 1966). Covalent bonds are thought to involve the formation of molecular orbital clouds similar in concept to those associated with atoms. Molecular orbitals are mathematical combinations of atomic orbitals. Each combination leads to a bonding orbital of lower energy and an antibonding or excited orbital at a higher energy level. Each orbital can accommodate two electrons, each having spin opposite that of the other. Bond formation is a way of lowering energy content.

Most covalent bonding involves elements that are not identical. Nuclei of different elements have different affinity for shared electrons. As a result, the electron spends more of its time closer to one nucleus than the other. Electronegativity is a means of expressing the strength of attraction of an atom for electrons. The difference in the electronegativity of the two elements is a useful measure of the equality of sharing of electrons between nuclei of different elements. A difference of zero or a small number implies equality or near equality of sharing electrons. A large difference indicates that the bond has highly ionic character.

Complex bond structures provide stability to molecules. This is a reflection of energy released during formation.

### Metallic Bonding

The third type of primary bonding of atoms in materials is the metallic bond (Shackelford 1988). This involves sharing orbital electrons and is nondirectional, similar to ionic bonds. In this case, the electrons are delocalized. This means that they are highly mobile throughout the solid, and equally likely to be associated with any of a large number of atoms. The metallic bond is an unnatural situation, since few metallic elements exist in uncombined form in nature. Almost all exist as oxides or sulfides. Metals of commerce are reduced from the oxide or sulfide to the elemental state. The metallic bond provides a means of delocalizing electrons to satisfy the chemical requirement of the atoms to achieve the configuration of the nearest inert gas for at least part of the time.

The ready mobility of the valence band of electrons is the basis for the electrical and thermal conductivity of metals. Electrical transmission involves the movement of a cyclic energy disturbance along the electron cloud. This is similar in concept to transfer of pull in a wave along a line of railway cars when slack exists in the couplings and the train starts to move.

## Chemical Reaction Processes

Chemical reaction describes interactions between atoms, molecules, ions and chemical entities that lead to the formation of new substances. Reaction rate depends on the concentration of reacting species and temperature (Barrow 1966). Reaction mechanism refers to the manner in which the interaction occurs.

Activation energy is the energy needed to raise the reacting entity to an excited state (activated complex) in which the reaction actually occurs. The activated complex can decompose back to the reactants or proceed forward to the products. Reactants and products are both at stable minima of potential energy, although one minimum is usually higher than the other. The activated complex corresponds to the energy at the top of the potential energy barrier (Moore 1962). For reactions that are exothermic (release heat to the surroundings), the energy of the products is less than that of the reactants. For reactions that are endothermic (absorb heat from the surroundings), the energy of the products is greater than that of the reactants.

Rate of reaction is a concept of considerable interest in laboratory and production chemistry. One influence on rate is temperature. The reaction rate constant increases exponentially with temperature. Activation energies range typically from 0.2 eV to 2 eV (Moore 1962). For those in the range of 0.9 eV, the rate doubles or triples for each increase in temperature of $10°$ C (Barrow 1966). Increasing temperature provides the energy needed to create more activated complexes.

Another means of energizing the reaction is to provide energy in the form of photons from external sources. This is the subject of photochemistry. These photons include energies in the visible, ultra-violet and gamma/X-ray regions of the electromagnetic spectrum. Energies associated with photons in the visible range are sufficient in some circumstances to produce an activated complex. Photons in the ultra-violet region can break chemical bonds. Adverse photochemical reactions can occur on the skin and in the retina of the eye.

A catalyst lowers the activation barrier, and by so doing, increases the rate of reaction (Moore 1962) by providing an alternate route for the reaction to follow. The catalyst may react with reactants or provide a surface on which reaction can occur. The catalyst is not changed in the overall process.

## Chemical Processing Hazards

Chemical processing is a major activity in even primitive societies. In advanced ones, it consumes large proportions of investment and effort. Chemical processing starts at the test-tube level in the lab and ranges upward to pilot plants, and finally to major industrial operations. Serious accidents have occurred in laboratory-scale operations where the worker is physically close in proximity to the reacting material. For reasons of lack of visual impact, these accidents do not attract the attention of those occurring in large facilities where damage is not readily cleaned up and is obvious to all spectators. Yet, they can be just as consequential to the individuals involved for reason of proximity to the source of energy release. In the laboratory, the causes of injury include projectiles of shattered glass produced by over-pressurizing of containers, burns from hot surfaces and from burning liquids and solids expelled from containers, and overexposure to chemical contents.

The hazards in these situations reflect the release of energy in nuclear and chemical reactions. Energies at the level of the nucleus and chemical bond are very small. However, what is readily forgotten is that a few grams of highly unstable and reactive material can contain Avogadro's number (6.02 × $10^{23}$) of atoms and molecules, and several times that number of chemical bonds.

Physical and chemical processes rearrange atoms and molecules in relation to one an other. The action of so doing liberates energy though heating the surroundings or consumes energy through cooling them. Associated with the overall process are likely to be various sub-processes and energy associated with each (Table 2.2).

Design of chemical operations must consider all of the applicable factors to ensure that situations that lead to release of energy occur only in a predicted and controlled manner. Situations in which unpredicted and uncontrolled release of energy can happen include mixing of incompatible chemicals, such as acids and bases (caustics or alkalis), and runaway reactions, among others. Capable design and pre-construction reviews intend to assure that these situations (hopefully) cannot occur. Yet they do, and they have. This is not necessarily the result of oversight by designers. It can and has resulted from

## Table 2.2
## Energetic Physical and Chemical Processes

| Process Type | Energetic Process |
| --- | --- |
| Physical | Heat of solution |
|  | Heat of crystallization |
|  | Heat of fusion |
|  | Heat of vaporization |
|  | Heat of transition (from one crystal structure to another) |
|  | Heat of sublimation |
|  | Heat of adsorption |
|  |  |
| Chemical | Heat of reaction |
|  | Heat of decomposition |
|  | Heat of substitution |
|  | Heat of oxidation-reduction |
|  | Heat of hydrolysis |
|  | Heat of neutralization |
|  | Heat of combustion |
|  | Heat of afterburn |
|  | Heat of explosion |
|  | Heat of polymerization |
|  | Heat of formation |
|  | Heat of dissociation |
|  | Heat of dilution (acids) |

incompleteness of information available to designers on substances, such as explosives, which are deliberately designed to release stored chemical energy in an instant.

Reactivity Hazards
Reactive hazards involve release of energy at rates or in quantities too large to be absorbed by the immediate surroundings (Breatherick 1990). Most chemical reactions involving reactivity hazards are exothermic. However, reactivity hazards are not limited to exothermic reactions. Some endothermic reactions also pose reactivity hazards. Reactions of concern range from decomposition of a single reactant to rearrangements involving complex multi-component systems (Pohanish and Greene 1997).

Reactivity incidents occur under extraordinary conditions with reactions that under normal circumstances pose no extraordinary hazard. To illustrate, some incidents occurred during inadequate mixing of reactants contained in multi-phase systems. This can happen when mixing ceases and reformation of the phases occurs or where mixing speed is inadequate. Concentration of reactants also impacts the rate of reaction. Increasing concentration beyond the normal recipe in a process can profoundly increase reaction rate (Breatherick 1990). As well, addition of additional catalyst beyond what is specified or insufficient addition of reaction inhibitor also can have the same effect.

Rasmussen (1987) studied 190 unwanted reactions (Table 2.3) during chemical processing and indicated that they occurred for a limited number of reasons. Accidents involving batch reactors were much more frequent than those occurring in continuous flow processing units.

Instability in single compounds or high reactivity in mixtures is attributable to particular groupings of atoms or structures, such as those containing a high proportion of oxygen or nitrogen. Hypergolic mixtures are so reactive that they ignite spontaneously on contact or can be made to do so with additives (Breatherick 1990). These mixtures are utilized as rocket propellants. Compounds used in these reactions are exotic oxidizers, and fuels that are only slightly less so.

Redox (reduction-oxidation) reactions are an especial concern in any discussion of reactivity hazards. Oxidizing agents (electron acceptors) are the most commonly involved class of molecule in these accidents (Breatherick 1990). Reducing agents (electron donors) involved with oxidizing agents produce highly energetic reactions. Some substances react spontaneously with air, while others (alkali metals like sodium) do so in water.

Pyrophoric materials are spontaneously combustible or combustible following slight heating in air. These substances include elemental phosphorus, finely divided metal powders, metal oxides and sulfides, and an array of exotic chemical compounds. Research performed by Duguid (2001) indicates, however, that pyrophoric materials were involved in only 1.2% of the 562 incidents that occurred in process operations between 1950 and 1995.

Finely divided metal powders become pyrophoric when the surface area exceeds a critical value (Breatherick 1990). Some of these are shock sensitive. Factors influencing pyrophoricity of metal powders include particle size, heat of combustion, thermal conductivity, and presence of moisture, among others. Alloys of reactive metals are often more pyrophoric than the parent. Some catalysts are pyrophoric. These are particularly reactive following adsorption or absorption of hydrogen onto the surface of the catalyst, either before a hydrogenation reaction or after separation of the catalyst from the reaction mixture. Susceptible catalysts contain finely divided metal powders.

Iron-sulfur compounds are particularly susceptible to pyrophoric activity. Rusty scale in tanks and vessels involved in sulphide removal from petroleum fractions has been involved in many fires and explosions. Disturbing the iron sulfide/rusty scale can produce sparks and persistent incandescence. Residual hydrocarbons provide the fuel.

Another redox reaction of concern is the thermite reaction. Thermite reactions are most common between aluminum or magnesium and metallic oxides, such as rusty iron. The thermite reaction is used to generate molten iron for filling in the gap between rails in railway operations. In these reactions, an igniter creates sufficient heat to cause aluminum powder to remove oxygen from iron oxide. This reaction, the oxidation of aluminum and reduction of the iron oxide generates sufficient heat to melt the reduced metal.

Runaway Chemical Reactions

The factor of greatest concern in reactivity is temperature (Breatherick 1990). As mentioned previously, for many reactions, increasing the temperature by $10°$ C doubles or triples the rate. In the case of exothermic reactions, release of energy into the reaction mixture can cause further increase in rate in an escalating process that has overcome the capacity of the cooling system. This situation can lead to a runaway reaction. Rate of release of energy is the critical factor, as this determines extent of damage. High rates of energy production are associated with explosions.

The accelerating rate calorimeter is among the most sophisticated means of determining the hazard potential of mixtures and reactions. This device contains the heat of reaction inside the reaction chamber, and then measures increases in temperature and pressure. These provide a means of predicting what will occur under actual conditions where control of temperature is lost.

Polymerization reactions are subject to runaway conditions. Polymerization is the formation of long-chain molecules (polymers) from single molecules (monomers). Each monomer contains a reactive chemical group capable of participating in chain formation. These reactions are mediated through formation of free radicals (unpaired, unbonded electrons) on the growing end of the polymer. Free radicals are highly reactive. Runaway reactions and explosions have occurred with 9 of the 12 monomers

## Table 2.3
## Reactivity Hazards: Unwanted Chemical Reactions

| Element or Condition | Percent of Accidents |
| --- | --- |
| Incorrect storage or handling of materials | 24 |
| Contamination or catalytic impurities | 20 |
| Use of incorrect chemicals | 19 |
| Incorrect charging or processing conditions | 19 |
| Insufficient agitation | 13 |
| Other | 5 |

Source: Rasmussen 1987.

most significant in industrial processes: acrylic acid, acrylonitrile, 1,3-butadiene, ethylene oxide, methyl methacrylate, styrene, vinyl acetate, vinyl chloride, and vinylidene chloride.

Duguid (2001) surveyed 562 incidents in the process industries in the period 1950 to 1995. His research indicates that runaway reactions were responsible for 18% of the incidents from 1950 to 1969, 6% from 1970 to 1979, and 15% from 1980 to 1995.

## Combustion Processes

Combustion is an oxidative chemical process. It produces energy as heat, and often as light. Combustion is called fire when the oxidative process occurs fast enough to be self-sustaining. Combustion that produces sudden and violent release of energy is called an explosion (Meyer 1989). Combustion differs from slow oxidative processes, such as rusting. The difference is the rate of release of heat. Temperature near the surface of slow oxidation increases only slightly — never more than 1° C above that of the surroundings. Combustion occurs so rapidly that heat production occurs faster than heat dissipation. Temperature increases by hundreds and often thousands of degrees Celsius above the surroundings. Heating is so intense that light is emitted (Friedman 1991).

Combustion occurs in two modes: flaming (including explosions) and surface (including glow and deep seated glowing embers) (Haessler 1986). These modes occur singly or in combination, but are not mutually exclusive.

Flammable liquids and gases burn in flaming mode only, as do plastics that melt and vaporize on heating. Flaming combustion occurs only when the fuel is present in vapor state. Vapor formation can occur through vaporization from liquid, or through melting and vaporization or pyrolysis of solids. Pyrolysis is the process of breakdown of complex molecular structures into smaller pieces. Pyrolysis products include ignitable gases and vapors.

Surface combustion occurs directly on the surface of the material as with pure carbon and readily oxidizable metals and nonmetals. Flaming and surface combustion occur during the burning of coal, sugars and starches, wood, straw and vegetable matter, and certain plastics that do not melt. Early stages of combustion of these materials involve thermal decomposition and destructive distillation.

Considered from simplest perspectives, combustion is a sustaining chemical reaction involving three components: fuel (a substance capable of being oxidized under the conditions available), an oxidizer, and a source of energy. The energy source initiates the reaction (ignites the fuel), and sustains and maintains the reaction. Energy produced by oxidizing the fuel may enable the combustion to be self-sustaining, or an external source (such as a lighted match) may be necessary. Removing any of these elements extinguishes the fire.

In chemical terms, an oxidizer is a substance capable of accepting electrons during a chemical reaction. Oxidizers usually are gases, although they need not be. The oxidizer most commonly encountered in fire situations is oxygen. Gaseous oxidizers that occur in industrial applications include: halogens and halogen oxides, nitric oxide and nitrogen dioxide, oxygen and ozone. Some liquid and solid substances readily release oxygen during chemical or thermal decomposition. Liquid oxidizers include hydrogen peroxide, per- and peroxy- acids, per- and peroxy- compounds, and nitro- compounds, among others. Solid oxidizers include peroxides, hydroperoxides, and peroxidates, hypohalites, halites, halates and perhalates, metal oxides, metal peroxides, chromates, permanganates, and nitrates and nitrites, among others (Breatherick 1990).

Historically, the components considered essential for the sustenance of combustion — oxidizer, fuel, and heat — were represented as a triangle: the fire triangle (Figure 2.4). All three components must be present simultaneously in a mutually beneficial relationship. The fire triangle is a valid representation of flameless surface combustion (including glow and deep seated glowing embers). However, other situations indicate that the process is more complex. That this is true, is attributed by the ability of magnesium, aluminum, and calcium to "burn" in nitrogen, or zirconium in carbon dioxide. These processes occur only under highly specific conditions. This is an important caveat since nitrogen and carbon dioxide are commonly used as extinguishing agents or fire suppressants under normal conditions. Further, some materials such as hydrazine, nitromethane, hydrogen peroxide, and ozone, when heated, will decompose emitting light and heat (Haessler 1986).

Flaming combustion is mediated through the elements contained in the fire triangle plus chemically reactive species called free radicals (Haessler 1974). A free radical is a molecular fragment having one or more unpaired electrons but no electrical charge. Free radicals are formed in many kinds of chemical processes. Free radicals are highly reactive, and therefore short-lived. Reactivity means that

## combustion products

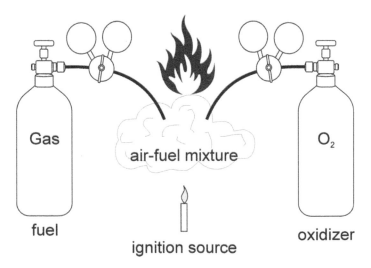

Figure 2.4. The fire triangle represents idealized conditions for combustion. Combustion requires the simultaneous presence of a fuel-air mixture in appropriate proportion, an oxidizer, and a source of heat. (Sources: McManus and Green 1999, McManus and Green 2001.)

free radicals actively seek to form chemical bonds having paired electrons. This occurs as part of the process of combustion.

The burning of methane illustrates the importance of free radicals in combustion. The combustion of methane occurs through a series of elementary reactions. The product of one reaction is a reactant in one or more others. The reaction mechanism thus is a series of elementary steps. The reaction mechanism for the combustion of more complex organic substances is considerably more complicated (Meyer 1989).

Free radicals are fundamental to the process of flaming combustion. Destroy the free radicals and flaming combustion ceases. That is, the fire is extinguished and the explosion prevented. This technique is effective for flaming combustion but not surface (glow) combustion. Extremely effective extinguishing agents, such as halogenated ethanes and methanes, exploit this principle. These substances readily form free radicals through loss of a halogen atom. The free radicals formed from this decomposition unite with intermediates of combustion, which are also free radicals, thereby interrupting and terminating the process. Free radicals add a fourth component to the three-sided fire triangle. The result is the so-called fire tetrahedron (Haessler 1974).

Combustion produces heat. Heat of combustion is the energy released during the reactions that constitute combustion. Heat released during combustion is essential to sustaining the combustion reaction. Heat activates molecules of both fuel and oxidizer by raising their internal energy. For combustion to be self-sustaining, the rate of production of activated molecules must exceed the rate of decay back to the unexcited state (Meyer 1989). Once ignition has occurred, combustion will continue until one of the following has occurred:

- The reaction consumes all available fuel or oxidant
- Cooling extinguishes the flame
- Some agent reduces the number of reactive species below sustaining levels (Drysdale 1991)

Combustion of substances containing carbon results in formation of two possible gaseous products: carbon monoxide and carbon dioxide. Carbon monoxide, the product of incomplete combustion, can undergo further oxidation to carbon dioxide. Carbon dioxide is the product of complete combustion. Unburned substances can form smoke (solid particles containing elemental carbon), sludge, or ash.

The different forms in which matter can exist – gas, vapor, mist, pure liquid and solution, bulk and powdered solids, and aerosols (airborne solid and liquid particulates) – complicate the study of combustion. Different measures of combustion apply to matter in its different forms. Liquids that can burn, and solutions containing them, themselves do not burn or explode. Combustion occurs only in the vapor space above the liquid. A vapor-air mixture having appropriate characteristics of concentration must form before combustion can occur. A similar process must occur in the vapor space above volatile solids.

Molecules of gas or vapor are constrained in their motion only by the boundaries of the containment and are free to intermingle. Given sufficient time, molecules of a gas or vapor introduced into a container of air as a discrete cloud will migrate to form a uniform mixture. Airborne dusts and liquid aerosols possess considerably enhanced surface area at which vapor formation can occur compared to bulk solids and liquids.

## Fire

Fire occurs in situations where the fuel and oxidizer (usually oxygen in air) have not undergone thorough mixing. As a result, combustion proceeds slowly and potentially is delayed by the need for mixing of fuel and oxidizer. Combustion rate is controlled by the rate at which mixing occurs, rather than the rates of chemical reactions occurring within the flame. Conditions conducive to explosion require thorough mixing of oxidizer and fuel prior to ignition. Burning rate is much greater during explosions than in fire situations (Drysdale 1991).

## Gases

Gas is the physical state of a substance that takes the shape and volume of the container in which it is confined. Gases exist exclusively in the gaseous state at ambient temperature and pressure (21° C and 101 kPa or 760 mm Hg). Liquids have a vapor pressure not exceeding 40 psia at 100° F (275 kPa or 2 070 mm Hg at 38° C) (NFPA 1991a). Vapor is the gaseous state of a substance that exists in the liquid or solid state under normal conditions of temperature and pressure. Conversely, by this definition, the vapor pressure of gases exceeds 40 psia at 100° F (275 kPa or 2 070 mm Hg at 38° C).

Flammable gases burn in the normal concentration of oxygen in air. Gases that do not burn in any concentration of air or oxygen are nonflammable. Some gases support combustion of other substances. These are oxidizers. Gases that suppress combustion are called inerting gases. By this definition, under normal conditions, inerting gases include carbon dioxide and nitrogen, as well as the chemically inert gases, helium, neon, argon, krypton, xenon, and radon. However, both carbon dioxide and nitrogen will support combustion of some metals under special circumstances (Lemoff 1991).

Flammable gases burn in air in the same manner as the ignitable vapors. Ignitable vapors arise from liquids and some solids. That is, flammable gases burn only within a range of concentration in air (the flammable range) and ignite only above a specific temperature (the ignition temperature). A fire involving a flammable gas is an aborted explosion. Fires usually occur in outdoor locations where containment by surroundings is minimal. Massive release and partial confinement by structures instead could lead to an explosion. The intent of continuously burning a pilot flame in gas-burning equipment, such as ovens, water heaters, boilers, and furnaces, is to cause a gas fire instead of permitting development of conditions that could lead to a more destructive explosion.

## Volatile Liquids and Solids

A key property of some liquid and solid substances is the ignitability of vapors emitted from the surface. (The terms ignitable, flammable and combustible have specific meaning related to the temperature at which sufficient vapor is present for ignition.)

Vaporization and condensation occur continuously at the surface of volatile liquids and solids. At any particular temperature, an equilibrium establishes between vaporization and condensation, given sufficient time. As temperature increases, vaporization increases relative to condensation, thus producing an ever increasing concentration of vapor. At a critical temperature, the concentration of vapor is sufficient to be ignitable (burst into flame) by an energetic ignition source. When heated slowly in the absence of a source of ignition, the vapor will self-ignite at a higher critical temperature, assuming that breakdown of the substance does not occur.

### Flash Point

The minimum temperature of liquid or solid at which there is sufficient vapor in air to be ignited by an energetic source is the flash point (Figure 2.5). Practical measurement of flash point is a difficult process. Replicating measurements requires meticulous attention to detail and procedure, and use of standardized apparatus (Wierzbicki and Palladino 1994). The measured value depends on the specific conditions created by each type of testing apparatus and the environment in which it is used. The configuration of the interface between the chamber in which the sample is held and the airspace in which the ignition source is situated reflects two fundamental design philosophies.

Closed-cup instruments contain an enclosed chamber containing a small opening in the top for escape of vapor. Methods using these instruments attempt to simulate ignitability in a semi-closed environment, such as the vent on a container. Open-cup instruments provide much larger surface area for emission of vapor. Methods using these instruments attempt to simulate ignitability in open environments, such as an open-topped container. The Cleveland tester is completely open, whereas the other testers have openings ranging between 5.4% to 6.4% of the surface area of liquid. Flash point measured using open cup testing equipment generally are higher than those measured using the closed cup units. For most liquids, the open cup flash point is 10 to 20% higher in degrees Fahrenheit than the closed cup flash point (NFPA 1991a). Flash point testing creates its own hazards. These include burn hazard from the flash and potential for overexposure to vapor and combustion products.

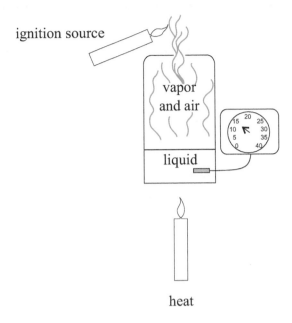

ignition source

vapor
and air

liquid

heat

Figure 2.5. Flash point is the temperature at which the surface of a liquid or solid emits sufficient vapor for ignition to occur. (Sources: McManus and Green 1999, McManus and Green 2001.)

The National Fire Protection Association created the most widely used system for classification (Table 2.4) (NFPA 1991b). This system has been adopted in various forms by many jurisdictions.

The division of classes is somewhat arbitrarily based on temperatures found in normal situations. Class I refers to temperatures that could occur at some time during the year. The cut-off of 100° F (38° C) also is body temperature. Stated another way, volatile liquids, such as gasoline, having flash points considerably below room temperature (–45° F or –43° C), pose a considerable ignitability hazard. Given an energetic source of ignition, gasoline vapor is easily ignitable at room temperature. The same also applies to the vapors of other industrial solvents in common use. Substances having flash points in Class I have the potential to be very hazardous in normal circumstances of use. Substances having flash points in Class II require moderate heating to generate sufficient vapor for ignition. Moderate heating of industrial solvents other than in process situations involving closed systems is not common. Fires involving substances having flash points in this region are less likely to occur in normal circumstances. Class III refers to substances requiring considerably more heating than provided by ambient sources to generate sufficient vapor for ignition.

Flash point provides an indication of several things: ability or inability of vapors to burn, relative level of concern appropriate to ignitability, and susceptibility to ignition. Flash point is a key indicator of the concern posed by liquids that can burn. Flash point is a predictor about the potential for vapor formation. The lower the flash point relative to the temperature at which the liquid or solid is used, the more likely that vapor will be present. Data tabulated in reference sources have practical limitations. These apply to pure substances. Commercial products are seldom as highly purified (NFPA 1991a).

This limitation highlights a broader concern about flash point. Actual experience from real-world use of chemical products bears little relationship to properties determined in the controlled environment of a laboratory testing instrument.

The most important application of flash point is comparison of ignitability (ability to burn) between substances. In order to optimize the utility of these comparisons, benchmarks are required, since

## Table 2.4

## NFPA Classification of Ignitability (Flammability and Combustibility)

| Classification | Flash Point | Boiling Point |
|---|---|---|

**Flammable Liquid:** Flash point below 100° F (38° C) and vapor pressure not exceeding 40 psia (275 kPa or 2 070 mm Hg) at 100° F (38° C).

| Classification | Flash Point | Boiling Point |
|---|---|---|
| Class IA | <73° F (23° C) | <100° F (38° C) |
| Class IB | <73° F (23° C) | ≥100° F (38° C) |
| Class IC | 73° F (23° C  FP < 100° F (38° C) | |

**Combustible Liquid:** Flash point ≥ 100° F (38° C).

| Classification | Flash Point | Boiling Point |
|---|---|---|
| Class II | 100° F (38° C) ≤ FP < 140° F (60° C) | |
| Class IIIA | 140° F (60° C) ≤ FP < 200° F (93° C) | |
| Class IIIB | ≥ 200° F (93° C) | |

**Note:** NFPA defines a liquid as a fluid having a vapor pressure not exceeding 40 psi absolute (275 kPa or 2 070 mm Hg) and fluidity greater than that of 300 penetration asphalt (a unit of viscosity).

Source: NFPA (1991b).

flash point temperatures vary from very low to very high. Benchmarks provide reference points for putting individual values into perspective.

Mixtures containing both ignitable and nonignitable liquids complicate the concept of ignitability. Addition of a volatile nonignitable substance to an ignitable one can suppress ignitability. The mixture as initially formulated may be nonignitable. This occurs because vapor from the ignitable and nonignitable components contribute to the total vapor pressure of the mixture. An example of this situation and its potential ramifications is a blend containing carbon tetrachloride and gasoline. Carbon tetrachloride is more volatile than some of the components in gasoline. On standing in an open container, the carbon tetrachloride will evaporate more rapidly than the gasoline. The residual fluid thus becomes enriched in hydrocarbons. Over time, the residual liquid will exhibit a high flash point that progressively decreases. The flash point of the final 10% of the original mixture approximates that of the heavier fractions of the gasoline (Sly 1991).

A mixture of flammable and combustible components represents a variation on this concept. Unequal evaporation would lead to enrichment of the vapor by the more volatile component(s). This would result in a decrease or increase in the flash point depending on the ignitability of the more volatile components. An example of this situation is a high-flash nonvolatile mixture (such as motor oil) that contains a volatile low-flash substance (such as toluene). Preferential evaporation by the toluene produces an unexpectedly low flash point for the mixture. This type of mixture easily could arise as a waste liquid in a garage operation. Toluene can vaporize readily from this mixture. The flash point of the mixture is the temperature at which sufficient toluene vapor is present for ignition. This temperature would be considerably less than the flash point for the oil. Waste oil products containing small amounts of solvent can constitute major fire hazards.

Appreciable evaporation and consequent change in ignitability can occur under conditions involving prolonged exposure of solvent blends to the atmosphere. Examples include solvents used in dip tanks and degreasers and waste solvents whose history is not known. Fractional evaporation tests conducted at room temperature in open vessels provide a means to evaluate the fire hazard of these mixtures. Flash point can be determined on fractions containing 10, 20, 40, 60 and 90% of the original sample. The results of such tests indicate the ignitability grouping into which to place the liquid. The open cup test method may give a more reliable indication of the ignitability hazard than closed cup methods in such situations (NFPA 1991b).

Flammable organic substances can occur in aqueous solutions or emulsions. Vapor from the organic component still can pose an ignitability hazard. Examples of such mixtures include spent caustic solution containing traces of gasoline, or commercial products containing aqueous solutions of alcohols. Raoult's law can be used to estimate the closed-cup flash points of these mixtures (Johnston 1974). This method assumes that Raoult's law applies and that the vapor space above the liquid contains water, solvent vapor, and air. The temperature of the mixture must be sufficient to generate the vapor concentration at the flash point in dry air. This temperature is greater than the flash point of the pure liquid. For substances having high flash points, the contribution of water vapor to the mixture in the vapor space becomes critical at elevated temperatures. At elevated temperatures, water vapor can suppress vapor formation by the organic component, so that a flammable mixture cannot develop.

Decreasing total atmospheric pressure decreases the flash point relative to sea level. For example, the flash point of toluene measured in Denver, CO (altitude = 1.6 km or 1 mi; standard atmospheric pressure = 627 mm Hg = 83.6 kPa) is 1° C less than the value at sea level (Bodurtha 1980). This results from increased vaporization relative to condensation as total pressure decreases. Increasing total atmospheric pressure increases the flash point. This property has practical significance. The fire hazard associated with flammable solvents increases slightly with increasing altitude or decreased atmospheric pressure and decreases slightly with depth, as in a deep mine. In the real world, these effects are negligible.

## Mist

Mist is a suspension of droplets in air. Mist can develop by condensation of vapor or by active processes, such as spraying and spraypainting, pouring, aerating, grinding and burning, sieving, splashing and nebulization, among others. Another source of liquid aerosols is foam. The bursting of bubbles in a froth propels liquid droplets into the air. These processes increase surface area by breaking the mass into smaller particles.

Droplets dramatically increase the area of the surface on which vaporization can occur. Vapor formation occurs on the surface of each droplet. This can provide conditions for ignitability at temperatures below the flash point. This property depends on particle size. The concept of flash point is predicated on passive generation of vapor at the surface of a volatile liquid or solid. Insufficient concentration of vapor for ignition exists at temperatures below the flash point.

Processes that aerosolize liquids dramatically increase ignitability by loading the air with an ignitable mixture at a temperature below the flash point of the liquid. This is an especial concern with combustible liquids. The lower ignitable limit for mists containing fine droplets plus accompanying vapor is approximately 48 g/m³ at 0° C and 101 kPa (760 mm Hg), regardless of the flash point of the liquid. This concentration of small droplets (smaller than 10 μm) corresponds to a very dense mist. (Burgoyne 1957, Zabetakis 1965). A 100 Watt bulb is visible only for a few centimeters (inches) at this concentration.

## Dust

Most finely divided combustible solids are hazardous (Schwab 1991). Deposits of combustible dust on beams, machinery, and other surfaces are subject to flash fires and explosions. Two types of fires occur in dust deposits: smouldering (surface) and flaming (Lees 2001).

Smouldering involves very slow combustion. Smouldering results from restricted air access and heat retention. Dust layers 2 mm and thicker can sustain smouldering for periods measured in years. Smouldering may give no readily detectable effects, such as smoke or odor. Upon reaching the surface of a layer of dust a smouldering fire can burst into flame. Smouldering rate commonly is determined from rate of travel along a dust "train" of stated dimensions. Typical smoldering rates for dusts are approximately 5 cm/hr for wood and 20 cm/hr for coal in a bed having a depth of 1 cm. Magnesium has an extremely high rate of smolder of 14 m/hr. Characteristics of dust that influence the type of fire include volatile content, melting point, and particle size. Detection of dust fires is difficult. The effects of smoldering are difficult to detect with the sensors normally used in fire protection systems.

In a flaming fire heat from the flame causes volatilization of volatile substances from the surface of dust particles. The flame may spread across the surface of the dust while smoldering occurs underneath. Heat produced in a smoldering fire liberates volatile compounds from the dust. If sufficient volatile material is present, the flame may travel rapidly across the surface. Otherwise, the smoldering rate determines the rate of propagation over the surface. Large particles reduce packing density, thereby permitting airflow and flame propagation through the dust. Combustion occurs rapidly. Large particles smolder poorly.

**Ignitability Limits**
An energetic source of ignition can ignite flammable gases and ignitable vapor from flammable and combustible liquids and volatile solids only within specific limits of concentration (Figure 2.6). Under equilibrium conditions for liquids and volatile solids, these concentrations correspond to a range of temperature in air or oxygen.

The lower limit of concentration at which ignition by an energetic source can occur is the lower flammable limit (LFL) for flammable gases and vapors and lower combustible limit (LCL) for combustible vapors. Under equilibrium conditions for volatile liquids and solids, these limits occur at the

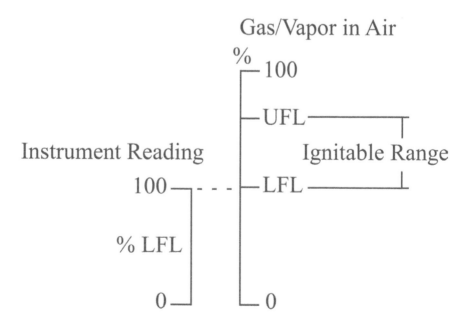

Figure 2.6. Ignitable Limits. The lower limit is the concentration at which there is enough vapor to be ignited by an energetic source of ignition. The lower limit occurs at the temperature of the Flash Point. (Sources: McManus and Green 1999, McManus and Green 2001.)

flash-point temperature. A flame does not propagate away from the source of ignition at concentrations below the lower flammable limit.

The corresponding upper limits of concentration are the upper flammable limit (UFL) and upper combustible limit (UCL). Generally, these limits are known only by the terms LFL and UFL, although strictly speaking the other terms apply. The difference between these concentrations is the ignitable range, consisting of the flammable range and combustible range, respectively. Values of the LFL and UFL for many chemical substances are available from standard sources (Kuchta, 1985; NFPA 1991a; NIOSH 2005; Zabetakis, 1965).

Adding further to the confusion is the fact that the lower flammable limit is often referred to as the lower explosive limit (LEL), the upper flammable limit as the upper explosive limit (UEL), and the flammable range as the explosive range. Strictly speaking, these terms are not identical. The LEL occurs at a higher concentration than the LFL and the explosive range is narrower than the ignitable range. Explosive mixtures will burn. Not all flammable mixtures will explode. The difference rests with the amount of energy needed to cause an explosion, rather than a fire, and the limits of concentration within which the mixture will explode. Stated in popular terms, a mixture whose concentration is less than the LFL is too "lean" to burn or explode. A mixture whose concentration is greater than the UFL is too "rich" to burn or explode.

While commonly expressed in volume percent, when expressed on a mass/concentration basis, LFLs for hydrocarbons, and therefore hydrocarbon mixtures, generally lie in the range, 40 to 50 $g/m^3$ of air at $0°$ C and 101.325 kPa. Consequently, ventilation rates needed to reduce concentrations of equal masses of different hydrocarbons to a specified percent of the lower limit are approximately the same (Bodurtha 1980). The lack of uniformity in LFL values across chemical classes is a major disincentive to the calibration of testing instruments by one substance for assessment of hazardous atmospheres containing other substances. This lack of uniformity becomes an especial concern in the assessment of unknowns and mixtures of different classes of substances.

The general effect of increasing temperature is to decrease the LFL and increase the UFL, thus broadening the ignitable range (Zabetakis 1965). This outcome is very specific to equipment and conditions. An increase in temperature also can cause a previously non-ignitable mixture to become ignitable by increasing vapor formation. A decrease in temperature can cause a previously ignitable mixture to become non-ignitable by decreasing vapor formation (Drysdale 1991).

Variation in pressure at normal atmospheric level has little impact on ignitable limits. The effect of larger pressure is specific to each mixture. Decreasing the pressure below atmospheric can increase the LFL and reduce the UFL until the ignitable limits coincide, thus rendering the mixture non-ignitable. Increasing the pressure above atmospheric levels can widen the ignitable range. The effect is more marked on the UFL than the LFL. However, in some cases, an increase in pressure narrows the ignitable range (Zabetakis 1965).

One aspect about flash point and ignitable limits that cannot be overemphasized is the problem posed by "empty" containers. "Empty" containers were involved in 16% of the fatal welding and cutting accidents reported by OSHA (OSHA 1988). Methods for the safe handling of "empty" containers are the subject of several safety standards (ANSI/AWS 1988; NFPA 1993). No empty container should be presumed to be clean or safe. Containers that have held hazardous substances can be made safe by following appropriate measures.

An "empty" container (Table 2.5) easily can confine an ignitable atmosphere under sub-ambient, ambient and super-ambient conditions of temperature. The governing factors are temperature of the surroundings and flash-point temperature of the liquid. All that is required for formation of an ignitable or explosible atmosphere is sufficient residual liquid on interior surfaces and time. Time is needed for the formation of the vapor-air mixture. Containers also may contain coated surfaces, a liner or sludge. When heated, the coating, liner, or sludge could release vapors or liquids trapped in or between the coating or liner and the container wall. Table 2.5 outlines conditions that could exist in "empty" containers prior to the application of heat.

Heating a container holding only vapor that is below or within the flammable range creates the potential for fire or explosion. Heat applied from a cutting torch creates a hot spot on the surface. This pressurizes the gaseous contents in a sealed container and would cause outflow from any opening in-

### Table 2.5

### Possible Conditions in "Empty" Containers Prior to Application of Heat

| Contents | Ignitability | Comments |
| --- | --- | --- |
| Gas or vapor | Non-ignitable | Concentration below ignitable range |
| | | Concentration above ignitable range |
| Gas or vapor | Ignitable | Concentration within ignitable range |
| Liquid + vapor | Non-ignitable | Vapor concentration below ignitable range |
| | | Vapor concentration above ignitable range |
| Liquid + vapor | Ignitable | Vapor concentration within ignitable range |

Source: McManus (1999).

cluding that created by the torch. Other less dramatic heat sources capable of creating hot spots that could ignite vapors include saws and grinders.

The ignitability of the atmosphere inside a container holding both liquid and vapor depends on the temperature of the containment and the contents. Ignitability could change with the season, since radiant heating from the intense summer sun could sufficiently heat a high-flash liquid to create an ignitable mixture. On the other hand, heat from the summer sun could put the vapor concentration from a low-flash liquid, such as gasoline, above the ignitable range. The reverse could apply under winter conditions. The key to assessing this concern is the interrelationship between vapor pressure and ambient or localized temperature and flash-point temperature.

The relationship between a liquid and its vapor in a container at any point in time reflects two possible conditions: equilibrium and non-equilibrium (disequilibrium). Equilibrium is a precise condition in which the rate of vaporization and rate of condensation are equal. Equilibrium mixtures can develop in the vapor space of all containers from large and small tanks to small-sized containers used for everyday use to "empty" containers that contain just sufficient liquid to form an ignitable mixture. Attainment of equilibrium is fully possible only in closed containers where vapor is prevented from escaping and temperature remains constant. Equilibrium is unlikely to be attained in containers in which the vapor is not fully confined. This is especially problematic for low-flash liquids in summer conditions, since they are volatile enough for the vapor to escape from containment.

Disturbance of equilibrium occurs following raising or lowering the temperature or changing the liquid content in the container. The relationship between temperature and vapor concentration depends on the kinetics of attainment of equilibrium. Where equilibration is rapid, the lag time in reaching the new condition is short. This situation is less problematic for liquids having high volatility.

The question then becomes whether the air-vapor mixture that develops in non-equilibrium situations can be ignited. The answer is related to the concentration of vapor in the region in which the energetic source of ignition is located. The mixture should be ignitable provided that the concentration exceeds the LFL and lies within the flammable range. Conditions under which a non-equilibrium mixture develops are extremely common. To summarize, this then begs the question whether equilibrium is needed to have an ignitable mixture. The answer is no.

**Sources of Ignition**
Fires and explosions occurring in industry usually involve a source of ignition, rather than slow heating and self-ignition (autoignition). Sources of ignition are many and varied. They may be obvious or unobtrusive. Recognizing that they are present is critical to addressing the hazard that they pose. Sources of ignition (Table 2.6) are related to an existing process, introduced as part of a maintenance activity, or associated with the actions of individuals (Lees 2001).

The preceding sources are obvious to the trained observer. Raising awareness about less obvious sources is critical to addressing this problem. All fire-producing and spark-producing sources should be considered as potential sources of ignition.

Ignition sources in the industrial environment are capable of heating the vapor to its ignition temperature in the presence of the surrounding air. Under many conditions, the ignition source also must provide sufficient energy to volatilize the fuel. Hot surfaces and flames can be a source of ignition if large enough and hot enough. The smaller the surface, the hotter it must be to cause ignition. As well, vapor from the liquid or solid must remain close to the source of ignition to enter the ignitable range prior to being carried off by convection currents (Sly 1991).

Chemical Energy
Rapidly released chemical energy can act as a source of ignition. Rapid release of chemical energy can occur during different classes of chemical reaction under appropriate conditions. These include reduction-oxidation (redox) reactions of which combustion reactions and reactions by chemical explosives are subgroups, acid-base neutralization, hydration of salts, and formation of solutions. Runaway chemical reactions have caused considerable devastation because of rapid release of energy.

Situations that provide the conditions for rapid generation of energy through redox reactions include metallic smearing of pyrophoric iron sulfide. Rusty steel smeared with metallic aluminum or aluminum paint or magnesium is an example of a metallic smear. Struck by a heavy metallic object, a strong spark results. Energy that creates the spark is partly mechanical and partly chemical. The mechanical component arises from the sudden impact. The chemical component arises from the release of chemical energy during the oxidation-reduction process.

Pyrophoric iron sulfide scale forms in the reaction between hydrogen sulfide and iron in exposed steel. This reaction is a concern in industrial situations where sulfur and sulfides, including hydrogen sulfide, are generated or otherwise potentially present. Under dry, warm conditions, the scale may glow red and act as a source of ignition. Pyrophoric iron sulfide is best removed by dampening the surface and scraping to remove the scale. No attempt to remove it by scraping should occur prior to dampening (Lees 2001).

Mechanical Energy
Mechanical energy is the source of ignition for a significant number of fires each year. Frictional heating and friction sparks, and heat of compression are the source of ignition in most of them (Drysdale 1991).

*Frictional Heating*
Frictional heating results from conversion of energy used to overcome resistance to motion when two solids are rubbed together. The significance of this hazard reflects availability of mechanical energy, and the rate of heat generation and dissipation.

Friction sparks result from the sudden impact of two hard surfaces. One surface usually is metal. In workplace situations, friction sparks often arise in a casual manner. Examples include dropping of steel tools onto a concrete floor or machinery (or piping), ricocheting of tramp metal inside a grinding mill, and impact between shoe nails and a concrete floor. Heat, generated by impact or friction, initially heats a metal particle. The freshly exposed surface may oxidize at the elevated temperature. The heat of oxidation may increase the temperature of the particle until it becomes incandescent. The extent of this process depends on the ease of oxidation and the heat of combustion of the particle. The incandescence temperature of most metals is well above the ignition temperature of ignitable materials. The ignition potential of a spark, however, depends on total heat content. Thus, the practical danger from

**Table 2.6**

**Sources of Ignition**

| Type | Examples |
|---|---|
| Related to existing processes | Open flames (burners, heaters, furnaces and flare stacks) |
| | Hot soot (ships and diesel locomotives) |
| | Unlagged surfaces (process equipment and piping) |
| | Distressed machinery (faulty or damaged bearings and seals) |
| | Electrostatic discharge (bulk materials handling) |
| | Compression (diesel effect) |
| | Abrasion (buffing, sanding, scraping) |
| Introduced as part of maintenance activities | Electric arc (welding, cutting, burning, electrical equipment) |
| | Hot metal and slag (welding, cutting, grinding and burning) |
| | Engines/exhaust systems (vehicles and mobile equipment) |
| | Electrostatic discharge (abrasive blasting, spray painting, steam cleaning, power washing) |
| Unrelated to the activity | Open flames (cigarettes, matches and lighters) |
| | Electrostatic discharge (nylon and static-generating fabrics in clothing, nails in soles and heels of footwear, tramp metal) |

Source: Lees (2001).

---

mechanical sparks is limited by their small size and low heat content. They cool quickly and start fires only under highly favorable conditions.

*Heat of Compression*
Heat of compression is the heat released when a gas is compressed. Heat of compression has found practical application in the operation of the diesel engine. In a diesel, air is compressed in the cylinder by motion of the piston, and then an oil spray is injected into it. The heat contained in the compressed air is sufficient to vaporize and ignite the oil spray.

Electrical Energy
Electrical energy is one of the most common forms of energy with which we come into contact. Electrical energy occurs in three forms: static electricity, lightning, and current electricity. Electrical energy converted to heat is an important source of ignition energy (Drysdale 1991). This conversion can occur through several mechanisms.

Electrical systems are inherently different from hydraulic and pneumatic systems. Unlike the latter, where leakage of liquid and gas is readily detected, leakage from electrical systems is not obvious. Leakage of electrical current readily causes fatal accidents. Detecting leakage of electrical energy is the function of insulation testing and ground fault circuit interrupters (GFCIs) at 110 V and equipment leakage circuit interrupters (ELCIs) at higher voltages.

*Static Electricity*

Static electricity is a surface phenomenon produced generally by the contact and separation of dissimilar materials (Lees 2001). During contact, electrons migrate from one material to the other. As a result, one surface acquires a positive electrical charge, and the other, a negative charge. If the materials are good conductors, the electrons move freely and neutralize the charge separation at last contact before physical separation occurs. If one or both of the materials are poor conductors, the electrons do not move freely, and the bodies retain charge after separation.

When two insulators come into contact, charge transfer occurs because of the different states of their surface energy (Menguy 1998). Electrostatic charge transferred to the surface of an insulator can migrate deeper into the material. Humidity and surface contamination affect the behavior of electrostatic charge. Surface humidity increases surface conductivity. This favors charge recombination, and facilitates electron mobility. The polarity (relative positivity and negativity) of two insulators in contact depends on the electron affinity of each material.

When an isolated piece of metal and an insulator come into contact, electrons transfer to the insulator. The amount of charge transferred is proportional to the electron affinity of the metal. During the separation the insulator is left with a surplus of electrons and the metal with a deficit. The increase in potential caused by contact and separation ranges from 1 V (Volt) to 1,000 V.

A charged body can induce charge and charge separation in a neutral body, even without physical contact. This occurs because of the electromagnetic field that surrounds the charged object. The electromagnetic field radiates in all directions (Palmquist 1988). Charge induced in a conductor by an oppositely charged insulator could be considerably more consequential than the inducing charge that accumulated on the insulator.

Static electricity is characterized by low current and high voltage. The increase in potential caused by contact and separation easily can exceed 10,000 V. Process materials, equipment, and the human body all can become electrostatically charged. A person wearing insulating shoes is a common example of an insulated conductor (Menguy 1998). The human body is an electrostatic conductor, with a typical potential of up to 30 kV. People experience electrostatic discharge, such as produced when a hand approaches a door handle or other metal object, as a noticeably uncomfortable sensation.

Accumulation of charge increases the strength of the associated electrical field. Accumulation beyond a critical value leads to electrostatic discharge. This discharge is similar in some respects to lightning. Electrostatic discharge through the air occurs by several mechanisms: complete electrical breakdown (spark discharge) or partial breakdown (corona discharge). A spark discharge occurs in a very small fraction of a second and gives a short, sharp, crackling sound. A corona discharge occurs over a longer time and may give a faint glow and a hissing sound. Both corona and spark discharges are capable of igniting an ignitable mixture. The corona discharge usually is less hazardous than a spark discharge

Many industrial processes (Table 2.7) involve contact, movement, and separation of poorly conducting materials (Lees 2001). These processes involve solid, liquid, or gaseous phases, singly or mixed. During discharge, each of the preceding processes can generate electrostatic charge sufficient to ignite an ignitable mixture of gas or vapor.

Electrostatic effects in liquids are explainable by the classical model of the electrical double layer at an interface (Lees 2001). According to this model a layer of positive ions and a layer of negative ions accumulate at the interface between two immiscible liquids, or the liquid and the wall of a container. Movement of the liquid produces unequal distribution of the ions and a resulting electrostatic charge. Situations in which charge separation occurs, include pipeline flow, especially of immiscible liquids such as oil and water; settling of water droplets through oil; splash filling of tanks involving free fall;

## Table 2.7
## Electrostatic Generating Actions

| Action | Examples |
| --- | --- |
| Fluid handling | Pipeline flow, tank filling, agitation in process vessels |
| Dust and powder handling | Grinding, sieving, pneumatic conveying, unloading bags of powdered material |
| Sprays and mists | Steam cleaning, steam leaks, power washing, spray application of coatings |
| Abrasive blasting | Dry ice (undergoes a phase change), solid media |
| Moving equipment | Conveyor belts, bucket elevators, web-handling of rolled materials |

agitation of liquids in tanks,. and splashing of oil droplets on the side of a tank. Filters, valves, and other constrictions in pipelines are points of generation of high levels of electrostatic charge.

The extent of charge separation depends on the resistivity of the liquid. Appreciable charge separation without immediate recombination can occur in liquids having high resistivity. Electrostatic accumulation and discharge are major concerns in the handling of these liquids. Liquids having high resistivity include gasoline, kerosene, naphtha, benzene and other 'white oils'. Liquids having low resistivity include water, ethanol and crude oil.

Electrostatic discharge is a potential source of ignition wherever ignitable substances are present. Electrostatic discharge is believed to be the cause of many apparently mysterious explosions that have occurred in process plants (Lees 2001). To illustrate this situation using a recent example, fires have ignited spontaneously during filling of portable fuel containers in the backs of pickup trucks containing plastic bed liners or cars containing carpeted surfaces. In this situation, the plastic of the liner or the carpet acted as the dielectric material in a capacitor formed by the metal of the gasoline can and the metal of the vehicle (NIOSH 2002). The electrostatic arc jumps between the metal of the pump nozzle and the gasoline container.

Electrostatic arcs normally do not produce sufficient heat to ignite ordinary combustible materials such as paper. Some, however, are capable of igniting flammable vapors and gases, and clouds of combustible dust.

The ability of an electrostatic arc to cause ignition depends on the delivery of energy to the flammable mixture (Scarbrough 1991). The minimum voltage needed to produce an arc across the shortest measurable gap of 0.01 mm under ideal conditions is about 350 V. However, once established, 80 V is all that is required to maintain it, hence, the voltage used in welding equipment.

Quenching by associated equipment and the surrounding atmosphere precludes ignition under this condition. However, incendive arcs induced across gap lengths in the range of 0.5 mm to 1.5 mm (which exceed the quenching distance) with voltages in the range of 1,500 to 5,000 V can produce ignition under ideal conditions. Most hydrocarbon mixtures require about 0.25 to 0.5 mJ (milliJoules) of energy for ignition under ideal conditions. Ignition energy is discussed in the next section. Ionization potential of air is about 30,000 V/cm.

Arcs from good conductors are more incendive than those from poor conductors or insulators. Charge flows along the contiguous surface of a conductor during discharge. On the other hand, charge flows only from a localized region of an insulator during discharge. For this reason, discharge from insulators or poor conductors usually is not implicated in fires and explosions.

Discharge occurs more readily when one, or both, of the oppositely charged objects are pointed (Lees 2001). Thus, a metal object protruding into a tank and electrostatically bonded to it can act as a discharge path. Objects in process operations that can act as discharge paths include grounded probes, dipsticks and ullage tapes, tank washing machines and metallic objects that float on the surface of a liquid.

Actions that introduce pointed conductors into containers holding ignitable liquids are especially hazardous. Activities during which this can occur include level gauging and drum sampling. Discharge between the charged liquid and the conductive object may occur in two ways. First, the object may act as a discharge path from liquid to ground directly or via a person. Alternately, discharge may occur from the liquid to the tank. Antistatic additives or time allowance for charge relaxation dramatically reduce the hazard of charged liquids. Nonconductive ullage tapes and permanently fixed and grounded sounding pipes that extend to within a few centimeters of the bottom of the tank reduce the hazard from level gauging.

Relative, as compared to absolute, humidity influences the accumulation and stability of electrostatic charge (Scarbrough 1991). Most fires attributed to electrostatic discharge occur indoors in winter when relative humidity is less than 30%. Electrostatic charge tends to dissipate when the relative humidity is maintained above 50%. Under the latter condition surface conductivity increases sufficiently to prevent static accumulation. To illustrate, the surface conductivity of plate glass increases 1,000-fold when relative humidity increases from 20% to 50%. Humidification is effective only when surfaces are held at room temperature. Humidification does not affect surfaces heated above room temperature, as in the case of dry textiles or other materials moved over heated surfaces.

Most industrial dusts and powders are poor electrical conductors. Production and transport of these substances tend to generate electrostatic charge. Operations in which this occurs include micronizing, grinding, mixing, sieving, gas filtration, pneumatic conveying, and mechanical transfer. Factors influencing the rate of generation include conductivity of the material, turbulence, and interfacial surface area between materials and presence of impurities. High generation rates are common in dispersing operations and when materials are mixed, thinned, combined, or agitated. Hazardous arcing can occur between poorly conductive materials and the agitator blade in a mixer or the conductive fill pipe in a ball or pebble mill.

In piping systems, the generation and accumulation of electrostatic charge are functions of the materials, flow rate, velocity, and pipe dimensions (Gregg 1996). In filling operations, high rates of flow, turbulence in splashing or free-falling liquids and powder fines contribute to electrostatic charge accumulation. Disconnection of hoses and valves can result in electrostatic discharge. Filters can generate levels of electrostatic charge 200 times higher than in pipe alone. The most serious mistake in powder and dust handling is failure to bond and ground equipment made from conductive material.

Process gases also can cause problems with static electricity (Scarbrough 1991). Steam, carbon dioxide, compressed air, and other process gases can generate electrostatic charge under certain conditions. The electrostatic hazard from steam arises from the formation of water droplets through condensation. Strong electrification can occur at the orifice during escape of steam containing droplets of condensed water. When escaping under high pressure from an orifice as a liquid, carbon dioxide immediately changes to gas and solid (snow). The opposite situation occurs during use of dry ice (solid carbon dioxide) in abrasive blasting.

The process of phase change can result in generation and accumulation of electrostatic charge on the discharge device and the receiving container or surface. Carbon dioxide should not be used for the rapid inerting of ignitable atmospheres by injection under high pressure for this reason. Compressed air or other process gases containing solid or liquid impurities also can produce strong electrification on escaping through an orifice. Bombardment of a conductive body by gas contaminated with dust, mist, scale, and metallic oxides can produce strong electrification of conductive fittings that are not bonded and grounded. Gas contaminated in this manner should not be used for purging or cleaning.

*Lightning*

Lightning is a form of electrical discharge. Usually lightning occurs between charged clouds, and between charged clouds and the Earth. Lightning is typically associated with storms; however, lightning also is associated with emissions from volcanoes (Newcott 1993). Lightning is a hybrid between static and current electricity. The energy that supplies the output remains in stasis until the moment that current flows (Davis 1991; Frydenlund 1966; Whitehead 1983). The destructive potential of lightning, of course, is well-known. Less well appreciated is the fact that current electricity is considered to be able to flow without the guidance provided by wires from source to destination.

On average, the surface charge of the Earth is negative while the upper atmosphere carries a positive charge (Grandolfo 1998). The resulting electrostatic field near the surface of the Earth has a strength of about 130 V/m. This field decreases with height to about 100 V/m at 100 m elevation, 45 V/m at 1 km, and less than 1 V/m at 20 km. Actual values vary widely, and depend on local temperature and humidity, and the presence of ionized contaminants. (To put these values into perspective, typical levels of man-made electrostatic fields range from 1 to 20 kV/m in offices and households; these fields are frequently generated around high-voltage equipment, such as TV sets and video displays, or by friction.)

Beneath thunderclouds, for example, and even as thunderclouds approach, large variations in the electrostatic field occur at ground level, because normally the lower part of a cloud is negatively charged, while the upper part is positively charged. Charge separation occurs due to collisions between rising ice crystals and falling hailstones within the cloud. In addition, there is a space charge between the cloud and ground. As the cloud approaches, the field at ground level may first increase and then reverse, with the ground becoming positively charged. During this process, fields of 100 V/m to 3 kV/m are observable even in the absence of local lightning. Field reversals occur rapidly, usually within 1 min. High field strengths can persist for the duration of the storm.

Lightning involves the flow of 2000 to 200,000 A (Amperes) of current (median 30,000 A) at a potential difference of 10 million to 100 million Volts for periods lasting several microseconds (Davis 1991; Frydenlund 1966; Whitehead 1983). Peak temperature can reach 33,000° C (60,000° F) for several microseconds. Cloud-to-ground flashes can extend to 14 km (9 mi), and cloud-to-cloud flashes, 140 km (90 mi) (Uman 1999).

A lightning strike is most likely to the object in a group that is the best conductor, nearest the approaching cloud or the most prominent. The lightning bolt generally follows a metallic path to ground. Fulfilling this mandate can entail jumps from one source of metal to another through direct and indirect paths. The lightning stroke, in effect, is an arc between two plates of a capacitor, each having the opposite charge (Uman 1999).

A cloud-to-ground flash comprises at least two strokes: a leader stroke and a return stroke (EB 1998; Newcott 1993; Uman 1999). The leader stroke carrying negative charge normally passes from cloud to ground. The leader stroke is not very bright, is often stepped, and has many branches extending out from the main channel. Each step of the leader is about 46 m in length and lasts about 50 $\mu$s (microseconds). The leader creates an ionized path, but involves movement of only a small current. The flash of cloud-to-ground lightning is initiated by the neutralization of the small net-positive charge in the lowest region of the cloud. As it nears the ground, it induces an opposite charge, concentrated at the point to be struck, and a return stroke carrying a positive charge from ground to cloud is generated through the ionization column. The two strokes generally meet about 50 m above the ground. At this moment, a highly luminous return stroke of high current passes through the column to the cloud. The leader stroke reaches the juncture point or the ground in about 20 ms (milliseconds), and the return stroke reaches the cloud in about 70 $\mu$s. The return stroke from earth to the charged cloud through the ionized path carries the major part of the discharge current.

Direct lightning strikes can damage overhead transmission lines, transformers, outdoor substations and other electrical installations. Electrical transmission equipment can also pick up induced voltage and current surges that can enter buildings. Fires, damage to equipment and serious interruption to operations may result. Surge arresters divert these voltage peaks to ground through effective grounding.

Thunder, the noise associated with lightning, is caused by the rapid heating of air to high temperatures along the whole length of the lightning channel. Air thus heated, expands at supersonic speeds, but within 1 to 2 m, the shock wave decays into a sound wave, which is then modified by the intervening air and topography. The result is a series of claps and rumbles.

*Current Electricity*
Current electricity is the flow of an electrical disturbance along the surface of a metal wire. Electrons in the wire move only slightly. What moves is the disturbance. The disturbance moves at the speed of light $3 \times 10^8$ m/s (in a vacuum). Current electricity occurs in two main forms: direct current (DC) and alternating current (AC) (Palmquist 1988). Voltages associated with direct current in a circuit are either positive or negative, but not a mixture of both. These voltages can be constant or embody a waveform. The presence of a waveform is indicative that current flow varies in magnitude, but not direction. DC circuits are found only in specialized equipment and industrial applications. Alternating current flows in both directions in a circuit, usually in a waveform. This results in voltage levels that fluctuate between positive and negative according to the waveform and direction of flow of the electrons. The time-varying nature of the waveform is the important factor with regard to properties of electrical energy. Almost all AC circuits operate at 50 or 60 Hz (Hertz). (One Hertz is the passage of one complete waveform per second.) Higher frequencies are employed in specialized equipment.

Associated with waveform disturbances involving movement of electrons in wire is the production of an electromagnetic field. This field has an electrical component and a magnetic component. These components are oriented at 90° to each other and perpendicular to the direction of the conductor. The electrical wave and magnetic wave are completely in phase with each other. The time variation of these waves is the property that enables the transportation and utilization of electrical energy, because what is being transported and utilized is the disturbance, not bulk motion of electrons.

Some industrial facilities contain DC systems that utilize energy from storage batteries. Batteries usually are considered part of back-up systems. In some industrial facilities, banks of batteries are integral to the process. Electrical generation systems, for example, utilize batteries to produce the magnetic field in the rotor. Other electrical installations, such as electric transit systems, also utilize DC current provided from batteries (as well, of course, as DC provided by rectifiers).

Resistive heating occurs during flow of current through a conductor. The power required to produce current flow is the resistance of the conductor times the square of the current. Less power is needed for passage of current through conductors, such as copper and silver, that have low resistivity. Heat generated by filaments in incandescent bulbs and infrared heaters is an example of resistive heating.

Dielectric heating results from distortion of atomic or molecular structure caused by externally applied time- and direction-varying electric potential. A rapidly varying external potential can cause considerable heating of a dielectric (a good insulator).

Induction heating occurs in a conductor subjected to a fluctuating or alternating magnetic field. This also occurs in a conductor moved across the lines of force of a magnetic field. Rapidly changing or alternating potentials produces heat due to mechanical and electrical distortion of the atomic or molecular structure. Heating increases with the frequency of alteration. Microwave heating is an application of this principle.

Leakage current also is a source of heat. Leakage current is current flow through insulators that are subjected to substantial voltages. The heat produced by leakage current becomes important only in unusual situations involving breakdown of insulation.

Arcing is a phenomenon of current electricity. Arcing occurs when current flow in an electric circuit is interrupted. Arcing is especially severe when motors or other inductive circuits are involved. The temperature of the arc is very high. In some circumstances the arc may melt the conductor and scatter the molten metal. Energy released by an intrinsically safe electrical circuit during a fault condition such as arcing must be less than that needed to ignite the hazardous atmosphere in which the circuit is located.

## Ignition Energy

Ignition is the process of initiating self-sustained combustion. Ignition requires energy. The energy needed for ignition varies from substance to substance and depends on conditions. In general, molecules of both fuel and oxidizer require activation before they can undergo chemical reaction. Activation requires energy. The energy needed for activation can be obtained in two ways: an external source or an increase in the internal energy of the mixture as reflected by its temperature. As the temperature of the mixture increases, the amount of supplemental energy needed from the external source decreases. At a sufficiently high temperature, the mixture ignites spontaneously. No supplemental source of ignition is required. This process is called self- or autoignition. Ignition occurring through supply of energy from an external source — a flame, spark, or glowing object (ember) — is called piloted ignition. The temperature of the fuel-air mixture at which piloted ignition can occur is considerably lower than the autoignition temperature (Drysdale 1991).

The energy required to ignite a flammable mixture varies according to composition. This passes through a minimum in the middle of the ignitable range (Zabetakis 1965). The minimum ignition energy is the minimum energy needed from an external source, usually a spark discharge to cause ignition.

Flammable mixtures of hydrogen and acetylene require only 0.02 mJ (milliJoules) to cause ignition. This compares to about 0.25 to 0.5 mJ required to ignite other hydrocarbons. These quantities are extremely small. Excellence in design of equipment is the reason that acetylene and hydrogen cause so few problems in everyday use. Equipment handling these gases contains features to control electrostatic and other electrical discharges.

The energy stored by the human body following accumulation of electrostatic charge ranges around 10 mJ. This assumes a potential of 10,000 V (Volts) and capacitance of 200 pF (picoFarads). Much higher voltages can develop in a dry environment. This build-up depends on clothing worn, humidity, materials of construction, and grounding (Lees 2001). A person charged to 15 kV could provide a discharge of 22.5 mJ. This level of energy is within the range produced by a spark plug (20 to 30 mJ) used to detonate the gasoline-air mixture in the automotive engine. Thus, commonplace sparks and arcs can ignite flammable vapor and gas mixtures with energy to spare. More energy is required to ignite dusts (Bodurtha 1980).

Tabulated values for minimum ignition energies from spark discharges reflect optimized conditions, for example, a fixed gap between electrodes and fixed electrodes versus breaking a contact. A spark gap of 2.5 mm requires less energy than a wider gap. The minimum ignition energy for arcs generated by broken electrical contacts is considerably higher than for fixed electrodes. Oxygen concentration and pressure also influence minimum ignition energy. Increasing the oxygen concentration and total pressure reduces the minimum ignition energy.

## Spontaneous (Self-) Heating

Exposed to the normal atmosphere, practically all organic substances undergo oxidation, even at ambient temperature. The rate of oxidation usually is so slow that heat of combustion is transferred to the surroundings as rapidly as it is produced. The usual result is no appreciable increase in temperature of the substance (Drysdale 1991).

Under certain conditions, materials retain heat. Spontaneous or self-heating results in an increase in temperature of the material without input from the surroundings. If enough energy remains in the material, spontaneous heating can produce spontaneous ignition. Several factors control the severity of the process: nature of the substance, rate of heat generation, air supply, and geometry and insulating properties of the immediate surroundings (Drysdale 1991; Lees 2001).

Bulk materials in process, storage, or transport may undergo spontaneous heating. Examples include bulk materials handled in process driers, stored in piles in warehouses, or transported in large containers, such as ships. Well known is the spontaneous combustion of coal stored in piles or bunkers (Lees 2001). Self-heating is a significant source of ignition of dust. Dust in a pile has high surface area and controlled air circulation, both of which favor self-heating. Storage of a large quantity of dust at a high initial temperature creates the greatest hazard. A typical example is the discharge of a hot dusty product from a drier into a hopper (Bodurtha 1980).

Spontaneous heating is not likely to occur in volatile substances having access to the atmosphere. Vaporization removes the material, as well as the accumulating heat. Spontaneous heating is restricted to materials, such as oils, that have low volatility and combustible dusts.

Sufficient air must be available for oxidation, yet heat must not be removed by convection. Tightly packed materials might provide the necessary insulation, yet impede air movement. Predicting the occurrence of spontaneous heating under a given set of conditions is very difficult because of the many possible interactions between air supply and insulation (Drysdale 1991).

Substances undergoing oxidation first form intermediate products of oxidation. These may catalyze further oxidation. To illustrate, oil that has become rancid oxidizes faster than fresh oil due to the presence of intermediate products of oxidation.

Supplemental heat provided by other sources can promote the process by increasing the rate of oxidation. An additional source of supplemental heat is bacterial metabolism. This is a common cause of heating in agricultural crops stored in silos or wood chips stored in piles. Continued heating beyond temperatures generated by bacterial action occurs due to chemical oxidation. Moisture promotes bacterial activity. However, evaporation of the moisture removes heat. Agricultural products having a high content of oxidizable oils, such as cornmeal feed, linseed, rice bran, and pecan meal are susceptible to spontaneous heating.

A concern of major importance in process operations arising from self-heating is a lagging fire (Lees 2001). Lagging is the insulation applied to piping and process equipment. Lagging on plant equipment frequently becomes impregnated with oils and other liquids. Heat transferred from the equipment to the lagging can cause preheating of the liquid and spontaneous combustion. The temperature which can be attained in the lagging depends on geometry and temperature of the underlying equipment. Typical leakage points at which a lagging fire can occur include pumps, flanged joints, and sample and drain points.

The most important factor in a lagging fire is the liquid that impregnates the lagging. Significant self-heating requires several conditions (Table 2.8).

Several precautions can prevent lagging fires. The most obvious is to prevent leakage into the lagging. This requires not only a high standard of operation and maintenance, but also the action of knowledgeable personnel. Additional measures include leaving bare known points of leakage and protecting lagging at critical points with metal collars. Another approach is to use insulating materials less prone to fire. This type of solution often introduces additional cost. A final alternative is not to insulate.

## Oxygen Enrichment

The principal oxidizer in common experience is oxygen. Oxygen occurs in the normal atmosphere at a concentration of 20.9% and partial pressure of 21 kPa (159 mm Hg) at sea level. Oxygen in process situations can occur at concentrations and pressures considerably higher than those encountered under ambient conditions. Oxygen sometimes is used as a process gas, for example in steelmaking and the bleaching of pulp. Processes could involve transport within pressurized lines. Small quantities of compressed oxygen, relative to process applications, are used, sometimes with disastrous consequences, in medical applications and in oxy-fuel torches.

The likelihood of ignition and the rate of flame propagation of a combustible are greatly influenced by the oxygen content of the atmosphere (Figure 2.7). Oxygen is a nonflammable gas, meaning it does not burn under normal conditions. In general, oxygen enrichment significantly increases the ignitability of materials, although this is not guaranteed (Frankel 1991). Oxygen enrichment widens ignitability limits by decreasing the LFL and increasing the UFL (Zabetakis 1965). The LFL in oxygen is not markedly affected, since the concentration of oxygen in the normal atmosphere already exceeds combustion requirements. The UFL increases markedly in oxygen-rich atmospheres, tending to be above 50%.

The fire hazard in an oxygen-enriched atmosphere is significantly greater than that in a normal atmosphere (Frankel 1991). This is due in part to the reduction in minimum energy needed for ignition

## Table 2.8
## Conditions Required for Self-Heating

| Condition | Comments |
|---|---|
| Physical state | Liquid having low volatility. |
| Chemical reactivity | Intrinsic reactivity, such as unsaturated chemical bonds. |
| Fire resistance | Low fire resistance — Fire-resistant liquids containing antioxidants or hydraulic fluids containing water can become combustible if preferential loss of the antioxidant or water occurs in the warm lagging. |
| Leakage rate | Sufficient to provide new source of fuel. |
| Thickness of lagging | The relationship between pipe temperature and lagging thickness is the reverse of that required for heat insulation. |
| Time | Time needed to attain critical temperature may be considerable, possibly months. Frequently, there is an induction period because of the presence of natural antioxidants in many materials. |
| Insulation | A good insulating material has low thermal conductivity and porous structure of low density. Surface area, porosity, and heat retention favor self-heating. |
| Protective covering | Impervious coatings, such as cement finishes, bituminous materials, or aluminum foil, greatly reduce diffusion of oxygen into the insulation. |

Source: Lees (2001).

and the greater rate of flame spread. That is, combustible materials ignite more easily and burn more rapidly in an oxygen-enriched atmosphere. Generally, ignition energy decreases with increasing oxygen concentration. The rate of flame spread increases with the increase in the oxygen concentration at constant pressure or with the increase in the total pressure at constant percentage of oxygen (increased oxygen partial pressure).

The autoignition temperature (AITs) of most hydrocarbons tends to be slightly less in oxygen than in air. In the case of lubricants and hydraulic fluids, this decrease is significant. These products pose significantly greater fire hazard in oxygen-enriched atmospheres than under normal conditions. Minimum ignition energy decreases significantly in an oxygen-enriched atmosphere. Ignition occurs more readily in an oxygen-enriched atmosphere.

In an enriched atmosphere even at normal pressure, oxygen adsorbs onto fabrics and the skin, thus increasing combustibility. In general, ignition temperature and flame resistance are lower in an oxygen-enriched atmosphere. OSHA documented fatal accidents in which oxygen enrichment occurred through inadvertent or deliberate release of pressurized oxygen (OSHA 1985). The resulting fires indicated that risk of combustibility is greatly enhanced even at normal atmospheric pressure. Almost all materials will burn in an atmosphere of pure oxygen. This would change the NFPA classification of

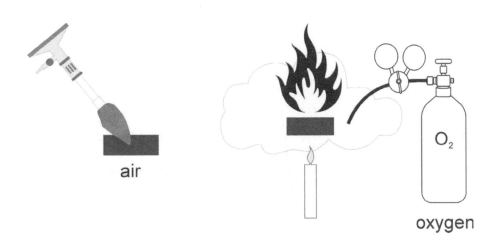

air

oxygen

Figure 2.7.  Oxygen enrichment. Many substances that will not burn in air will burn readily in an oxygen-enriched atmosphere. Oxygen at concentrations above that occurring in the normal atmosphere clings to the skin and clothing.

some materials from non-ignitable to ignitable. This situation can seriously challenge presumptions about safety in selection of materials for use in oxygen service.

Lubricants and hydraulic fluids are the most sensitive of the types of substances for which information is available. In the case of lubricants, this sensitivity changes from oxygen-deficiency through normal concentrations through oxygen enrichment. The lowest of the tested partial pressures corresponded to a concentration of 31% oxygen relative to the sea level dry atmosphere.

## Explosion

An explosion is an event characterized by the sudden release of energy (Cruice 1991). Explosions result from chemical and physical processes. Detonation of an explosive or rapid combustion of a cloud of flammable vapor or gas are examples of explosions resulting from chemical processes. Rupture of a container caused by overpressure or rapid vaporization and uncontrolled escape of a liquid or vapor are examples of explosions resulting from physical processes. The effects produced by an explosion occur due to at least one of the following processes: production of gases, rapid expansion of gases, and motion of projectiles.

One of the main characteristics of an explosion is the shock wave or blast wave (Figure 2.8). The shock wave is characterized by rapid increase in atmospheric pressure due to rapid expansion of gases. The shape of the pressure profile depends on the type of explosion. The shock wave produced during an explosion moves outward at subsonic or supersonic velocity, depending on the type of explosion.

During an explosion, atmospheric pressure rises almost instantaneously to a peak and gradually decreases to the starting level. As the shock wave travels outward, the height of the peak of the pressure wave decreases at the shock front with increasing distance. The shock wave in air usually is referred to as a blast wave because a strong wind may accompany it. The peak wind velocity behind the shock front depends on the peak overpressure. Overpressure is pressure above the normal atmospheric level. Overpressures generated by the blast wave can injure or kill people, and damage or destroy equipment

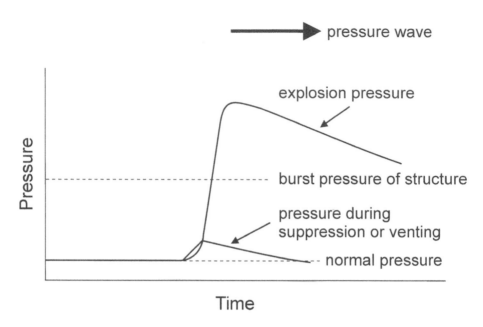

Figure 2.8. Explosion. The pressure wave created in an explosion moves outward from the central point. The height of the peak decreases with increasing distance. This is the same as the wave created when a pebble falls into still water. The wave contains a region of overpressure and a region of underpressure.

and buildings. A region of negative pressure, or underpressure, follows the overpressure in the shock wave. The underpressure usually is quite weak and usually does not exceed about 100 kPa (760 mm Hg) gauge (Lees 2001; Bodurtha 1980). The high pressure components of the shock wave move outward at higher velocities. Initially, the shock wave from a detonation travels at supersonic speed. As the intensity of the wave subsides, it becomes sonic (Lees 2001, Bodurtha 1980).

**Deflagration**
A deflagration is a type of explosion involving ignitable gas/vapor-air mixtures. A deflagration is an exothermic reaction that propagates from the burning gases to unreacted material by conduction, convection, and radiation. The combustion rate during a deflagration is less than the velocity of sound. A subsonic shock wave results from a deflagration. During a deflagration, the pressure equalizes at the speed of sound following the overpressure (Zabetakis 1965). The pressure drop across the flame front is relatively small.

The pressure rise associated with a deflagration in an unvented vessel is approximately 8 atmospheres (800 kPa) (Zabetakis 1965; Bodurtha 1980). The increase may be as much as 20 atmospheres (2,000 kPa) in fuel-oxygen systems.

**Detonation**
A detonation is an exothermic reaction characterized by the presence of a supersonic shock wave. A detonation can involve compressed gases, ignitable gas/vapor-air mixtures, and solids. In the case of detonations involving combustion, the shock wave establishes and maintains the reaction. A rapid exothermic chemical reaction occurs just behind the shock front (Cooper 1996). The reaction zone propagates at a rate greater than the velocity of sound. The principal mechanism of heating is shock compression. The temperature increase is directly associated with the intensity of the shock wave,

rather than thermal conduction. The wave speed in the forward direction is constant after initiation. Supersonic propagation of the flame from the source of ignition occurs during a detonation explosion. The rate of pressure equalization at the flame front is less than the rate of propagation. This results in considerable pressure drop across the flame front (Zabetakis 1965).

**Physical Explosions**
Physical explosions result from physical events, rather than rapid chemical reactions (Cruice 1991). Mechanical or other physical action is the source of all of the high-pressure gas that undergoes rapid expansion.

Vapor Explosions
A vapor explosion results from rapid heating (superheating) and rapid vaporization of a liquid in a confined geometry (Bodurtha 1980). Superheating results from contact between a hot liquid and a cold liquid. The energy source is the sensible heat plus the heat of fusion of the hot liquid. A liquid may become superheated under hydrostatic pressure or in a boiling regime that favors superheating.

This occurs on a stove when water is put onto a pot of burning oil during a cooking fire. The water sinks to the bottom of the pot and is surrounded by oil at a much higher temperature. The pressure exerted by the oil against the water droplets suppresses release of vapor from the surface of the globule. Eventually, the outward force overcomes the containing force and an explosion occurs, sometimes with disastrous consequences. Normally, however, the release of superheat energy occurs without great violence through generation of nucleated bubbles, as in normal heat transfer to a boiling liquid. In some cases the release of vapor is considerably more violent.

In industrial explosions of this type, the cold liquid generally is water. The hot liquid often is molten metal. Molten salts also can produce vapor explosions if they are quenched or come into contact with water. The usual mode of occurrence follows from a spill of molten metal onto a puddle of water on the floor. A vapor explosion also can occur when water is poured onto the melt. A layer of water at the bottom of a tank of hot oil produces vapor in the same manner. This causes boil-over, slop-over, and froth-over during some serious fire situations involving burning liquids (Sly 1991).

Water acts as the hot liquid after coming into contact with relatively cold liquids, such as chlorodifluoromethane or LNG (Liquified Natural Gas) that have low boiling points. In the latter case the superheat energy may release explosively, although without ignition.

Another type of vapor explosion occurs when a hot pressurized liquid is suddenly depressurized. The bulk liquid may undergo spontaneous nucleation with sudden vaporization and development of a shock wave (Bodurtha 1980).

Containment Rupture
Containment rupture occurs when the internal pressure exceeds the strength of the containment (Cruice 1991). Overpressure can result from heating of liquids and gas/vapor in the container. Vessel failure generally occurs at four times the allowed working pressure. Failure occurs at the weakest point. This can result in expulsion of projectiles or flinging open of walls. Gas release is extremely rapid under these circumstances. The pressure wave and projectiles from the vessel are highly directional. Damage potential is approximated by the product of the volume of gas and the pressure at time of failure in units of TNT equivalent (the amount of TNT needed to produce a shock wave of equivalent intensity).

Boiling Liquid, Expanding Vapor Explosions (BLEVE)
Another important type of physical explosion is the boiling liquid, expanding vapor explosion (BLEVE). A BLEVE occurs when a container is heated so that the containment ruptures and discharges a stream of boiling liquid. In the lower pressure environment outside the containment, the liquid vaporizes. This is especially critical where the boiling point of the liquid is low, as in the case of liquefied gases (superheated liquids), compared to ambient temperatures.

The aftereffects of a BLEVE depend on the ignitability of the liquid in the vessel. In all cases the initial explosion may generate a blast wave and missiles. If the liquid is ignitable, a fire always occurs.

The boiling liquid vaporizes and forms a vapor cloud, giving rise to a second explosion (Lees 2001). The definition of the term BLEVE indicates that mechanisms for the release of energy are strictly physical, including the effects of flying missiles and blast. Combustion following release of a flammable or combustible liquid is a secondary occurrence (Cruice 1991).

In most BLEVEs, failure of the containment originates in the metal of the vapor space (Lemoff 1991). The metal stretches and thins. A longitudinal tear develops and progressively lengthens to a critical length. At this point, the metal fatigues from the heat of the impinging fire and pressure exerted by contained superheated liquid. The failure propagates at sonic velocity in both longitudinal and circumferential directions. As a result, the container often ruptures in two or more places.

In a fire situation, a BLEVE will occur under one of the following conditions. First is the absence of a pressure relief device or presence of an undersized one. The second condition is impingement of the fire onto the metal above the liquid in the container. The magnitude of a BLEVE depends on the quantity of liquid that vaporizes when the container fails and the weight of the pieces. Most BLEVEs involving liquefied gases occur when the container is half to three-quarters full. Under these conditions, the blast can propel pieces up to 1 km.

Storage of liquefied gases occurs in containers at temperatures above their boiling points at normal temperatures and pressures. This pressure ranges from less than 7 kPa for some cryogenic gases to more than 1000 kPa for noncryogenic liquefied gases. Container failure reduces the internal pressure to atmospheric pressure. Rapid vaporization of a portion of the liquid occurs. The amount is related to the difference in temperature between the liquid at the instant of container failure and the normal boiling point. For many liquefied flammable gases, this can result in vaporization of a considerable fraction of the liquid. The liquid remaining is refrigerated by "self-extraction" of heat and cooled to near its normal boiling point.

Accompanying the vaporization is the liquid-to-vapor expansion. This expansion provides the energy for explosion of the container, atomization of the remaining liquid and rapid mixing of the vapor and air, resulting in the characteristic fireball upon ignition. Many of atomized droplets burn as they fly through the air.

### Chemical Explosions
Chemical explosions result from the evolution of gaseous products during a rapid chemical reaction (Cruice 1991). The products differ substantially from the reactants.

Ignitable Gas/Vapor Explosions
Fire occurs in situations where fuel and oxidizer are initially unmixed. Mixing of fuel and oxidizer is controlled by the combustion process itself. Burning rates are restricted primarily by the supply of fuel and oxidizer, rather than the rate of elementary chemical reactions occurring within the flames. These reactions generally occur so fast that they immediately consume all available fuel and oxidizer. The basic gas-phase process usually occurs along thin flame sheets, called "diffusion flames." These separate regions rich in fuel vapor from regions rich in oxidizer. Fuel vapor and oxidizer diffuse toward the flame sheet. Combustion products and heat, in turn, diffuse away from the flame sheet (Drysdale 1991).

Explosions involving ignitable mixtures generally occur only where fuel and oxidizer have mixed intimately prior to ignition (Cruice 1991). As a consequence, combustion occurs very rapidly. Two types of mixtures can result: homogeneous, uniform mixtures and heterogeneous, nonuniform mixtures. The components of a homogeneous mixture are intimately and uniformly mixed. A small sample is representative of the entire mixture. Examples of substances able to form homogenous mixtures include gases and vapors. Other types of flammable or combustible mixtures are heterogenous. These include mists, foams, and dusts.

Gas- and vapor-air mixtures have both ignitability limits (deflagration) and explosive limits. Explosive limits depend upon the initiating stimulus and the environment. They usually are slightly narrower than ignitability limits. The most intense explosions of vapor-air mixtures occur near the middle of the ignitable range. Explosions involving ignitable vapor-air mixtures most frequently occur in geometries that confine the atmosphere (Bodurtha 1980). The violence of the explosion depends upon the nature of the substance, the quantity of the mixture, and the enclosure confining it.

Explosions in Contained Systems
While not common, fires and explosions have occurred in process equipment, such as gas compression systems (Anonymous 1959; Burgoyne and Craven 1973; Perlee and Zabetakis 1963). The causes relate to the process of compression or to operation of a unit, such as a compressor. Compression of gas or vapor-air mixtures in the presence of air can lead to explosion. Air should be purged from systems handling flammable gases or vapors with inert gas to prevent this occurrence. Location of intakes is critical to avoid entraining air containing hydrocarbon vapors. Compression of endothermic gases, such as acetylene, requires special care and consideration. Another potentially dangerous condition occurs when a centrifugal pump, containing an ignitable liquid and an air pocket, operates deadheaded. Adiabatic compression of the gas pocket may develop temperatures above the autoignition temperature or cause decomposition of the liquid.

Another cause of compressor explosions is the compressor itself (Lees 2001). Generally, the type of compressor involved in these explosions is the oil-lubricated reciprocating compressor. Carbonaceous deposits in the vicinity of final-stage outlet valves and oil films appear to be the principal sources of ignitaible materials. Excessive carbon deposits result from an improper oil feed rate.

A suspension of finely divided drops of ignitable liquid can form in gas compression systems during condensation of saturated vapor or atomization of liquid by mechanical forces (Lees 2001). Explosions in compressed air systems are essentially oil mist explosions. An explosion occurs because of the presence of oil mist and condensed oil in the system. Often the compressor has a history of faulty operation and high outlet temperature prior to the explosion. The risk of explosion is not completely linked to abnormally high oil consumption.

The explosion that occurs in a compressed air line having a thin film of ignitable oil is unique. Typically, the air line ruptures at intervals along its length. The explosion is a detonation. Dispersing the oil film into a mist by some primary shock or explosion creates the conditions for a more powerful secondary explosion. High outlet temperature at the compressor can vaporize and ignite the oil. Sudden release at high pressure can cause simultaneous formation of mist and ignition. Carbonaceous residues can undergo self-heating and ignite. Oil film explosions have occurred in the compressed air starter systems of large diesel engines. In these cases, the initiator of the explosion generally appears to be the diesel engine rather than the air compressor.

Crankcase explosions in engines and compressors involve oil mist. The suspension of oil mist in a crankcase during normal operation originates from two sources: mechanical spray and condensed mist from the lubricated parts. Lubricated parts are hotter than the average temperature in the crankcase. Serious overheating accelerates mist formation, a condition often referred to as "smoke." Ventilation and injection of inert gas suppress ignitability.

Explosions in the Surroundings: Vapor Cloud Explosions
Previous discussion has centered on gases and vapors confined in experimental apparatus or process or industrial equipment. Gases and vapors that have escaped from containment also can pose considerable explosion hazard. The more familiar cause of explosion involves ignition of ignitable mixtures that have escaped from containment as a cloud. The majority of vapor cloud explosions documented in recent years resulted from ignition of denser-than-air vapors. These tend to stratify near the ground and resist dispersion. Release could occur from vent stacks above ground level, pumps, valves, and other sources at ground level or below grade in pits, pump rooms, sumps, chemical sewers, and other contained structures. Ignitable concentrations from releases of dense vapors at the ground seldom extend to appreciable heights. Less dense gases rise in the atmosphere as a result of buoyancy and generally do not accumulate at low levels.

Vapor cloud explosions usually result from leaks of flashing liquids — liquids stored under pressure at temperatures above their atmospheric boiling points. These liquids vaporize rapidly at the higher temperature and lower pressure outside containment (Kletz 1977). The source of ignition can be a considerable distance from the point of release.

Rate of release, rather than quantity, is the primary criterion for assessing potential hazard of continuous releases. Concentration remains constant once steady-state is achieved. High concentrations can occur at large distances from the release point of instantaneous emissions. These concentrations are short-lived at any fixed location because of the brevity of the emission.

An unconfined vapor-cloud explosion or turbulent wind could disperse more concentrated sections of the cloud to form puffs. A puff expands as it moves downwind. This could dilute unreacted vapor previously above the upper flammable limit into the flammable range. Total quantity of vapor in the puff is the critical factor (Bodurtha 1980). When ignited, the mixture burns rapidly and produces heat rapidly. The heat is absorbed by the fuel, combustion products, and other gases in the vicinity of the flame. Confinement by structures restricts the ability of the heated gases to expand, thereby increasing pressure and the severity of the explosion. Oxygen enrichment further increases explosion pressure.

Rate of emission, density of the gas/vapor, and ventilation rate all affect accumulation of gas or vapor in a structure. Classic laws of diffusion are inapplicable under actual conditions because the combination of extremely slow release rates and airtight structures is seldom encountered. The ignitable mixture occupied less than 25% of the structure in most explosions. In addition, most flammable mixtures are approximately 90% air. The density of the mixture is approximately the same as the density of air, regardless of the density of the gas or vapor. Hence, density of the gas or vapor is seldom a significant factor in gas explosions involving structures (Lemoff 1991).

## Dust Explosions

The risk of a dust explosion exists wherever ignitable dusts are handled. Historically, industries particularly affected by dust explosions include flour milling, grain processing and storage, and coal mining and processing (Lees 2001). Stringent regulations that control releases into the environment have introduced new opportunities for dust explosions. To illustrate, there is a growing trend toward installation of dry collectors, such as baghouses and electrostatic precipitators to handle these materials. This equipment can contain large quantities of combustible dust at high airborne concentration. Explosions and fires have occurred in these installations (Bodurtha 1980). Combustible dust currently has become the subject of considerable attention in the US following a number of major explosions that have led to multiple fatalities.

A dust explosion can occur only when the dust is dispersed in air. Transition from a dust fire to a dust explosion or vice versa can occur. Burning particles from an explosion can act as the source of ignition for a fire of other flammable materials (Lees 2001).

The ability of detonation explosions to occur in normal industrial situations has not been established. Combustion in a dust explosion is very rapid. The flame speed is comparable with that in gas deflagrations. Maximum explosion pressures often are close to theoretical values calculated assuming no heat loss during the explosion. Most of the evidence for detonation relates to coal mines where a strong ignition source initiates the explosion. Industrial plants generally have weak ignition sources. The general practice for protection against dust explosions in industrial plants is to assume the occurrence of deflagration, rather than detonation. This procedure has proved to be satisfactory in practice (Palmer 1973).

An industrial dust explosion often involves two stages: a primary and a secondary explosion. The shock from the primary explosion dislodges settled dust from horizontal surfaces or ruptures dust handling and storage equipment in adjacent areas. Burning material produced during the primary explosion acts as the ignition source for the newly formed, unexploded dust cloud. The quantity of dust in the secondary explosion often exceeds that in the primary one. Secondary explosions often are more destructive than the primary one. The possibility of a highly destructive secondary explosion makes dust explosions rather unpredictable (Lees 2001). Venting dust collecting equipment inside a building could create serious repercussions in the progress of a primary explosion. Where feasible, installation of dust handling and processing equipment should occur outdoors or at least venting should occur outside through as short a duct as possible (Bodurtha 1980).

As with flammable gases and vapors, dusts explode within a specific range of concentration (DI 1954, DI 1962, DI 1965). Generally, the lower explosible limit (LEL) of fine solid organic materials in

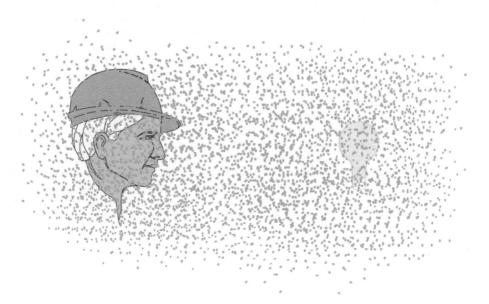

Figure 2.9. Explosible dust. A cloud of combustible material forms an explosible mixture in air when a person cannot see a light bulb at a distance of 1.5 m (5 ft).

air is about 20 g/m³. This loading in air is described as a very dense fog (Figure 2.9). The general "rule of thumb" is that an explosible mixture is present when one cannot see an operating light bulb at a distance of 1.5 m (5 ft).

Particle size is extremely important in this determination. The test specified by the U.S. Bureau of Mines used dust that passes through a 200-mesh screen (74 $\mu$m particle size or smaller). Dusts coarser than 200-mesh have high LEL. The effect is negligible for finer dusts (DI 1961). In general, dusts having particle size larger than 500 $\mu$m do not explode (DI 1968). Sample purity, oxygen concentration, strength of ignition source, turbulence of dust cloud, and uniformity of dispersion also affect the LEL (Schwab 1991). Determination of upper explosible limits (UEL) for dust clouds has not occurred due to experimental difficulties. However, this information would have limited applicability in real-world situations. UELs are estimated to be about 4 kg/m³ (Bodurtha 1980). Dust suspensions in industrial equipment often occur above the UEL. Using the UEL as a design parameter is not generally practical, since the LEL is the concentration of prime interest (Lees 2001). The most violent dust explosions occur at concentrations slightly above that required for reaction with all of the oxygen in the atmosphere. At lower concentrations, less heat is generated and smaller peak pressure develops. At higher concentrations absorption of heat by unreacted dust apparently reduces explosion pressure (Schwab 1991).

The minimum spark ignition energy (MIE) of combustible dusts is about 100 times that of flammable vapors or gases. MIE is the minimum spark energy from a condenser required to produce flame propagation 100 mm or longer in the test apparatus. MIEs are determined at the most easily ignitable concentration, generally 5 to 10 times the LEL for the dust. Consequently, ignition energies for marginally explosible concentrations are much higher than published MIEs. MIE values are affected by many factors, including particle size and the presence of ignitable vapors. Dusts wetted by ignitable solvent or suspended in a vapor-air mixture is more easily ignitable. Sources in industrial settings having sufficient energy to ignite dust with an MIE > 500 mJ are uncommon (DI 1960).

Ignition temperature of a dust cloud is determined in a Godbert-Greenwald furnace. The ignition temperature measured by this determination is the hot-surface temperature in the furnace, not the lower (and unknown) temperature of the dust-air mixture. Thus, ignition temperature, while not the autoignition temperature, is similar to the hot-surface ignition temperature for vapors and gases. Natural substances tend to have lower ignition temperatures than synthetic products (DI 1960, DI 1961). Ignition temperatures required for dust explosions are much lower than the temperatures and energies available in most common sources of ignition. As with flammable vapors and gases, common sources of high temperature can ignite dust clouds. These include open flames, light bulbs, smoking materials, electric arcs, hot filaments of infra-red heaters, friction sparks from grinding and other processes, high pressure steam lines and other hot surfaces, welding and cutting operations, and other common sources of heat. Not surprisingly, all common sources of ignition have caused dust explosions (Schwab 1991).

Trouble lamps on extension cords or bare bulbs are extremely hazardous in dusty locations. Ignition of dust by electric light bulbs has occurred often. Bulbs overheat when coated with dust. The hot glass can ignite the dust before the filament burns out. Moreover, a loose bulb presents a double hazard, as it may arc, as well as produce heat (LeVine 1972). Hot surfaces also result from distress in machinery such as pumps and motors. Another important source of ignition is mechanical sparks. Dust may block equipment and cause overloading. Sparks can arise from foreign materials, such as tramp metal, in rotating machinery (Lees 2001).

Moisture in dust particles raises the ignition temperature of the dust cloud. Heat is required to vaporize the moisture. Moisture in air surrounding a dust particle has no significant effect on the course of a deflagration once ignition occurs. There is a direct relationship between moisture content and minimum energy required for ignition, lower explosibility limit, maximum pressure, and rate of pressure increase. With spark ignition, the moisture content in dust needed to prevent ignition varies from 16% for coal to 35% for paper (DI 1964). In practice, moisture is not explosion preventive, since most ignition sources provide more than enough heat to vaporize the moisture. In order for moisture to prevent ignition, the dampness in the dust would preclude formation of a dust cloud (Schwab 1991).

Electrostatic discharge is a source of ignition for dusts, accounting for 10% of dust ignitions (Eichel 1967). Electrostatic charge is generated on dust particles when dust clouds are produced and conveyed through transfer equipment. The particles, often highly insulating, can retain charge for considerable time (Jones and King 1991). Pneumatic transport produces the highest level of charge. Thus, airborne dust can become the source of electrostatic charge which is a source of ignition. A high level of electrostatic charge can develop in dust clouds. This is a concern in industrial situations only in dust clouds at least the size of a small house. The density of electrostatic charge in the dust cloud at the bottom of a chute in a receiver, such as a silo, can be relatively high. An electrostatic discharge from the charged powder to a grounded pointed object in the silo could cause an explosion involving dust or associated flammable vapors. Avoidance of protrusions from the top into containers receiving dusty bulk solids should occur for this reason (Cooper 1953).

Electrostatic discharge between an isolated electrical conductor charged by the dust and nearby grounded equipment (capacitive discharge) has caused the vast majority (90%) of dust explosions caused by electrostatic discharge (Glor 1985). An electrostatic arc from an isolated conductor rapidly dissipates the entire stored energy in a single discharge. The voltage needed for discharge is approximately 350 V; 100 V is used as a standard to provide a margin of safety in the design of this equipment (Gibson 1979). Electrostatic arcs from a charged insulator are relatively weak. They contain only a portion of the energy stored on the insulator, since discharge occurs only locally. They also can be relatively long in duration.

Other types of electrostatic discharge can occur in powder handling systems (Glor 1985). While these thus far have had little impact on the occurrence of dust explosions, some types of discharge have sufficient energy to ignite ignitable vapors. Brush discharge can occur between a conductor having a radius of curvature in the range 5 to 50 mm (0.2 to 2 in) and either another conductor or a charged insulating surface. Brush discharges, while unlikely to ignite dust, can ignite ignitable vapors (Maurer 1979).

Corona discharge involves electrodes having a radius of curvature less than 5 mm (Glor 1985). These include sharp edges, protrusions, such as threaded bolts, welded seams, and brackets inside vessels. Corona discharge is too weak to ignite gases and vapors.

Propagating brush (Lichtenberg) discharge occurs when a thin insulating layer inside a vessel, such as a lining, that is backed by a conducting layer, becomes charged. These are the most energetic of all discharges and the most potentially destructive. The thin insulating layer acts as a broadly distributed capacitor. For this reason, nonconductive linings inside vessels, bins, and pipes must be avoided.

Maurer discharge is a form of brush discharge. It occurs along the conical surface of the powder during filling operations, in conjunction with sliding or avalanching of particles (Maurer 1979). Maurer discharges possess sufficient energy to cause a dust explosion.

Lightning-like discharges emanating from the dust cloud suspended above the pile to a grounded vessel wall are considered to be possible, but have yet to be reported.

Low humidity promotes accumulation of electrostatic charge because of the decrease in moisture content and conductivity of the particles of dust. This condition is associated with the higher frequency of fires and dust explosions in winter than in summer (Kissell et al. 1973).

Dust explosions are more destructive than those involving flammable gases and vapors (Cruice 1991). The destructiveness of a dust explosion depends on a number of factors, although primarily on the rate and duration of pressure increase. Other contributing factors include maximum pressure, confinement, concentration of oxygen, and quantity of dust compared to the quantity of heat generated during combustion.

Closely associated with rate of pressure increase and maximum pressure as indicators of destructiveness is the duration of excess pressure. The area under the time-pressure curve determines the total impulse exerted. Total impulse rather than instantaneous force determines the amount of destruction. This partly explains the reason that dust explosions, which generally have slower rates of pressure rise than gas and vapor explosions, are more destructive. The destructiveness in many dust explosions does not reach full potential because the dust is not uniformly dispersed throughout the cloud. A dust explosion in real-world situations rarely occurs under optimum conditions.

Particle size has a profound effect on several characteristics of dust explosions. The lower explosible limit, ignition temperature, and energy necessary for ignition decrease as particle size decreases. Fine dust ignites more easily than coarse dust. Rate of pressure increase and maximum pressure are greater for fine dusts. The exposed surface per unit weight increases as particle size decreases. Also, fine dust disperses more easily and remains in suspension longer. Particle size also affects the rate of pressure increase during the explosion. Explosions involving fine dusts have a higher rate of pressure increase.

### Explosives

An explosive is a substance or mixture of substances that, when subjected to the appropriate stimulus, undergoes an exceedingly rapid self-propagating reaction (Porter 1991). Most explosives are organic compounds containing carbon, hydrogen, nitrogen, and oxygen (Cooper 1996). There are also some inorganic explosives. The process undergone by an explosive following detonation is a redox (reduction-oxidation) reaction in which the explosive acts as both oxidizing and reducing agent. Gases produced during decomposition of explosives can include carbon dioxide, carbon monoxide, nitrogen, nitric oxide, oxygen, steam, and sulphur dioxide (Meyer 1989). The oxygen needed for combustion is generated during the decomposition of the solid or liquid material that constitutes the explosive, and does not originate from the atmosphere. Further, atmospheric oxygen is unnecessary for the decomposition. The gases produced following the detonation expand to 10,000 times the original volume of the material at the elevated temperature and pressure of the reaction.

The products of the reaction usually include stable gases, heat, and a blast wave. The explosive release of these gases occurs in microseconds. The blast wave results from the expansive action of heat on the gaseous products. Commercial explosives exploit the properties of instability inherent in some materials in a controlled (read safe) manner.

Burning or deflagration can occur in an explosive (Cooper 1996). Detonation occurs when the decomposition reaction occurs at shock velocities. Initiation of detonation versus deflagration depends on the rate at which the reaction occurs. Detonation requires projection of a pulse of energy through the material. This can occur through collision between the explosive and some other object. The shock created by the pulse of energy must travel a finite distance through the explosive before detonation will occur. This depends on the peak input shock pressure.

Greatest heat production occurs when there is sufficient oxidizer contained in the molecules to react with all of the reactants. Inexact proportioning leads to unreacted material that can absorb energy. These explosives are completely self-contained and do not need atmospheric oxygen for completion of chemical reactions. Final products from a completely self-contained reaction include $N_2$, $H_2O$, and $CO_2$. More typically, however, products of explosion include NO and $NO_2$, and CO. These products of explosion can pose serious exposure hazards in enclosed spaces. In situations where complete combustion does not occur (where insufficient oxygen is present in the molecules), the products of the explosion themselves act as fuels. These can ignite in air to form an energetic fireball (afterburn). Heat liberated in the afterburn can be more than twice the energy liberated during the detonation.

Explosive compounds include a wide range of structures and substituents. These are broadly classed into fuel contributors and oxidizer contributors. Fuel contributors are common structures from organic chemistry, plus amino ($-NH_2$), ammonium ($NH_4^+$), and imino ($-NH$) groups. Oxidizer contributors include nitrate ($-ONO_2$), nitro ($-NO_2$) and nitroso ($-NO$) groups, hydroxyl ($-OH$), carbonyl (CO) and carboxyl (COOH) groups, halogens, and halogen-substituted amino groups. Some structures are both fuel and oxidizer contributors. Other high energy bonds used in explosives include azides ($-N_3$) and diazo ($-N_2-$) compounds.

Inorganic explosives include mercury and silver fulminate (cyanate), lead and silver azide, and ammonium nitrate. Sodium azide is the explosive used to inflate automotive air bags. Silver cyanate is shock sensitive. Silver azide is light sensitive.

Initiating explosives, such as mercury fulminate and lead azide, are used to trigger explosion by more stable explosives. This is the concept behind the blasting cap.

Detonation produces a shock wave that moves at velocities ranging from 7 km/s to 9 km/s for pure explosive materials. For mixtures containing stabilizers and other ingredients, the velocity is reduced somewhat. The speed of sound in air at room temperature is about 0.33 km/s. Hence, the shock wave moves at considerably greater velocity than the speed of sound.

The other parameter of interest with explosives is detonation pressure. This is approximately:

$$P = \frac{\rho D^2}{4}$$                                                                           (2-19)

where:
P = detonation pressure (GPa = $10^{12}$ Pa).
$\rho$ = density of the unreacted explosive (g/cm$^3$); range is 1.65 g/cm$^3$ to 1.94 g/cm$^3$.
D = detonation velocity (km/s).

For a typical explosive having density of 1.8 g/cm$^3$ and detonation velocity of 8 km/s, detonation pressure is approximately 28.8 GPa. (Normal atmospheric pressure at sea level is 101.325 kPa.) Detonation pressure is very large compared to normal atmospheric pressure.

## Hazardous Energy: Physical Effects In Context

This chapter has presented a wide-ranging discussion about the quantitative aspects of some sources of energy capable of and known to have caused harm to workers and other people, thereby making them hazardous.

This discussion introduced several units of historical importance or importance to specialists in particular areas of endeavor. Historical units for reporting of energy suited the needs of the investigators of the time. Their need was a shorthand that minimized the effort of communication between them. Table 2.9 summarizes this information using the Joule as the common unit of energy.

## Table 2.9
## Summary of Hazardous Energy Sources

| Example | Wavelength | Energy | |
|---|---|---|---|
| | m | eV | J |
| **Nuclear Emission** | | | |
| Fission | NA | $190 \times 10^6$ | $3.0 \times 10^{-11}$ |
| Alpha particle | NA | $1.0 \times 10^6$ | $1.6 \times 10^{-13}$ |
| | | $9.0 \times 10^6$ | $1.4 \times 10^{-12}$ |
| Beta particle | NA | $0.018 \times 10^6$ | $2.9 \times 10^{-15}$ |
| | | $16 \times 10^6$ | $2.6 \times 10^{-12}$ |
| Positron | NA | $0.3 \times 10^6$ | $4.8 \times 10^{-14}$ |
| | | $18 \times 10^6$ | $2.9 \times 10^{-12}$ |
| Neutron | NA | $0.5 \times 10^6$ | $8.0 \times 10^{-14}$ |
| | | $11.5 \times 10^6$ | $1.8 \times 10^{-12}$ |
| Gamma ray photons | $0.12 \times 10^{-9}$ | $0.01 \times 10^6$ | $1.6 \times 10^{-15}$ |
| | $0.12 \times 10^{-12}$ | $10.2 \times 10^6$ | $1.6 \times 10^{-12}$ |
| X-ray photons | $100 \times 10^{-9}$ | 12.4 | $2.0 \times 10^{-18}$ |
| | $9.4 \times 10^{-12}$ | $0.134 \times 10^6$ | $2.1 \times 10^{-14}$ |
| **Electronic Emissions** | | | |
| Ultra-violet photons | $385 \times 10^{-9}$ | 3.1 | $5.0 \times 10^{-19}$ |
| | $100 \times 10^{-9}$ | 12.4 | $2.0 \times 10^{-18}$ |
| Visible light photons | $780 \times 10^{-9}$ | 1.6 | $2.6 \times 10^{-19}$ |
| | $385 \times 10^{-9}$ | 3.1 | $5.0 \times 10^{-19}$ |
| Infra-red photons | $1 \times 10^{-3}$ | $1.2 \times 10^{-3}$ | $1.9 \times 10^{-22}$ |
| | $780 \times 10^{-9}$ | 1.6 | $2.6 \times 10^{-19}$ |
| Microwave | 1 | $1.2 \times 10^{-6}$ | $1.9 \times 10^{-25}$ |
| | $1 \times 10^{-3}$ | $1.2 \times 10^{-3}$ | $1.9 \times 10^{-22}$ |
| Radiofrequency | 1000 | $1.2 \times 10^{-9}$ | $1.9 \times 10^{-28}$ |
| | 1 | $1.2 \times 10^{-6}$ | $1.9 \times 10^{-25}$ |
| **Applications** | | | |
| Television picture tube | | $20 \times 10^3$ | $3.2 \times 10^{-15}$ |

## Table 2.9 (Continued)
## Summary of Hazardous Energy Sources

| Example | Wavelength | Energy | |
|---|---|---|---|
| | m | eV | J |
| **Electronic Absorption** | | | |
| Air or tissue ionization, average deposition | NA | 35 | $5.6 \times 10^{-18}$ |
| Pair production, minimum deposition | NA | $1.02 \times 10^6$ | $1.6 \times 10^{-13}$ |
| Ionization energy (atoms) | NA | 3.9 | $6.2 \times 10^{-19}$ |
| | | 17.4 | $2.8 \times 10^{-18}$ |
| Excitation of electrons involved in bond formation | NA | 2.2 | $3.5 \times 10^{-19}$ |
| | | 6.6 | $1.1 \times 10^{-18}$ |
| Ionization of chemical bonds | NA | 7.4 | $1.2 \times 10^{-18}$ |
| | | 17.8 | $2.8 \times 10^{-18}$ |
| Strength of chemical bonds (organic molecules) | NA | 2.6 | $4.2 \times 10^{-19}$ |
| | | 11.2 | $1.8 \times 10^{-18}$ |
| Activation for chemical reaction | NA | 0.2 | $3.2 \times 10^{-20}$ |
| | | 2.0 | $3.2 \times 10^{-19}$ |
| **Other Examples** | | | |
| Ignition, most hydrocarbon vapors in air | NA | | $0.25 \times 10^{-3}$ |
| | | | $0.5 \times 10^{-3}$ |
| Falling body, 70 kg, 3 m starting from rest | NA | | 2060 |
| 1000 kg vehicle traveling at 100 km/h | NA | | $386 \times 10^3$ |

**Notes:**

Nuclear emissions refer to non-accelerated particles.

The average energy of beta particles and positrons is about 40% of the maximum for the emission.

Sources: Barrow 1966; Cember 1996; Johns and Cunningham 1974; March 1985; Moore 1962; Morrison and Boyd 1966 ; RAE Systems 2000; Repacholi 1992; Roberts and Caserio 1964; Shapiro 1981; Weast 1968; Willard et al. 1981).

. Where Table 2.9 reports two values, these are the lower and upper limits of the range of energies reported in the literature. This consideration applies to observed quantities (nuclear particles and atomic properties and chemical bonds), and to limits that define classes of energy, such as the range that constitutes visible light. Table 2.9 mainly examines energy resulting from phenomena that occur naturally in our environment, but also considers energy associated with machines, such as vehicles.

The feature that stands out in Table 2.9 is that the quantities of energy involved in causing harm range from the unimaginably small to the unimaginably large. Another factor is the overwhelming amount of energy produced during nuclear processes compared to that needed to break chemical bonds, the target in closest proximity to the source. This acknowledges the importance of proximity of the source to the target of the action and the mechanism by which the action of harm occurs.

## References

Anonymous: Maintenance Notes. *Maintenance*: December 1959. pp. 11-12.

ANSI/AWS: Recommended Safe Practices for the Preparation for Welding and Cutting of Containers That Have Held Hazardous Substances (ANSI/AWS F4.1-88). Miami, FL: American National Standards Institute/American Welding Society, 1988.

Barrow, G.M.: *Physical Chemistry*, 2nd Ed. New York: McGraw-Hill, Inc., 1966.

Beer, F.P. and E.R. Johnston, Jr.: *Mechanics for Engineers, Statics and Dynamics*, 4th ed. New York: McGraw-Hill Book Co., 1987.

Bodurtha, F.T.: *Industrial Explosion Prevention and Protection*. New York: McGraw-Hill Book Co., 1980.

Breatherick, L.: *Breatherick's Handbook of Reactive Chemical Hazards*, 4th Ed.London: Butterworth & Co. (Publishers) Ltd., 1990.

Burgoyne, J.H.: Mist and Spray Explosions. *Chem. Eng. Progr. 53*: 121M-124M (1957).

Burgoyne, J.H. and A.D. Craven: Fire and Explosion Hazards in Compressed Air Systems. In *Proceedings of the Chemical Engineering Progress 7th Loss Prevention Symposium*. New York: 1973. pp. 79-87.

Cember, H.: *Introduction to Health Physics*, 3rd Ed. New York: The McGraw-Hill Companies, Inc., 1996.

Cooper, P.W.: *Explosives Engineering*. New York: VCH Publishers, Inc., 1996.

Cooper, W.F.: The Practical Evaluation of Electrostatic Hazards. *Brit. J. Appl. Phys. 4*: Suppl. 2, S71-S77 (1953).

Cruice, W.: Explosions. In *Fire Protection Handbook*. 17th Ed. Cote, A.E. and J.L. Linville (Eds.). Quincy, MA: National Fire Protection Association, 1991. pp. 1-56 to 1-71.

Davis, N.H., III: Lightning Protection Systems. In *Fire Protection Handbook*. 17th Ed. Cote, A.E. and J.L. Linville (Eds.). Quincy, MA: National Fire Protection Association, 1991. pp. 2-293 to 2-304.

DI: Laboratory Explosibility Study of American Coals by I. Hartmann, M. Jacobson and R.P. Williams (Report of Investigation, 5052). Pittsburgh: U.S. Department of the Interior/Bureau of Mines, 1954.

DI: Laboratory Equipment and Test Procedures for Evaluating Explosibility of Dusts by H.G. Dorsett, M. Jacobson, J. Nagy.,and R.P. Williams (Report of Investigation, 5624). Pittsburgh: U.S. Department of the Interior/Bureau of Mines, 1960.

DI: Explosibility of Agricultural Dusts by M. Jacobson, J. Nagy, A.R. Cooper, and F.J. Ball (Report of Investigation, 5753). Pittsburgh: U.S. Department of the Interior/Bureau of Mines, 1961.

DI: Explosibility of Dusts Used in the Plastics Industry by M. Jacobson, J. Nagy, A.R. Cooper, and F.J. Ball (Report of Investigation, 5971). Pittsburgh: U.S. Department of the Interior/Bureau of Mines, 1962.

DI: Pressure Development in Laboratory Dust Explosions by J. Nagy, A.R. Cooper, and J.M. Stupar (Report of Investigation, 6561). Pittsburgh: U.S. Department of the Interior/ Bureau of Mines, 1964.

DI: Float Coal Hazard in Mines: A Progress Report by J. Nagy, D.W. Mitchell, and E.M. Kawenski (Report of Investigation, 6581). Pittsburgh: U.S. Department of the Interior/Bureau of Mines, 1965.

DI: Dust Explosibility of Chemicals, Drugs, Dyes and Pesticides by H.G. Dorsett, Jr. and J. Nagy (Report of Investigation, 7132). Pittsburgh: U.S. Department of the Interior/Bureau of Mines, 1968.

Drysdale, D.D.: Chemistry and Physics of Fire. In *Fire Protection Handbook*. 17th Ed. Cote, A.E. and J.L. Linville (Eds.). Quincy, MA: National Fire Protection Association, 1991. pp. 1-42 to 1-55.

Duguid, I.: Take This Safety Database to Heart. *Chem. Eng. 108(7)*: July 2001. pp. 80-84.

EB: Lightning. In *Encyclopaedia Britannica* CD 98. Chicago: Encyclopaedia Britannica Inc., 1997. [CD-ROM].

Eichel, F.G.: Electrostatics. *Chem. Eng.*: March 13, 1967. pp. 153-167.

Frankel, G.J.: Oxygen-Enriched Atmospheres. In *Fire Protection Handbook*. 17th Ed. Cote, A.E. and J.L. Linville (Eds.). Quincy, MA: National Fire Protection Association, 1991. pp. 3-160 to 3-169.

Friedman, R.: Theory of Fire Extinguishment. In *Fire Protection Handbook*. 17th Ed. Cote, A.E. and J.L. Linville (Eds.). Quincy, MA: National Fire Protection Association, 1991. pp. 1-72 to 1-82.

Frydenlund, M.M.: Modern Lightning Protection. *Fire J. 60*: 10-15 (1966).

Gibson, N.: Electrostatic Hazards in Filters. *Filtr. Separ. 16*: 382-386 (1979).

Glor, M.: Hazards due to Electrostatic Charging of Powders. *J. Electrostat. 16*: 175-181 (1985).

Graham, F.D.: *Power Plant Engineers Guide* (rev. by C. Buffington). Indianapolis, IN: The Bobbs-Merrill Co. Inc.: 1983.

Grandolfo, M.: Static Electric and Magnetic Fields. In *Encyclopaedia of Occupational Health and Safety*, 4th ed. Stellman, J.M. (Ed.). Geneva: International Labour Office, 1998. [CD-ROM].

Gregg, B.: Generation and Control of Static Electricity. *Plant Serv. 17(6)*: June 1996. pp. 83-87.

Haessler, W.M.: *Extinguishment of Fire*, Rev. Ed. Quincy, MA: National Fire Protection Association, 1974.

Haessler, W.: Theory of Fire and Explosion Control. In *Fire Protection Handbook*. 16th Ed. Cote, A.E. and J.L. Linville (Eds.). Quincy, MA: National Fire Protection Association, 1986. pp. 4-42 to 4-47.

IAEA: Lessons Learned from Accidents in Industrial Irradiation Facilities. Vienna: International Atomic Energy Agency, 1996.

IAEA: Lessons Learned from Accidents in Industrial Radiography (Safety Reports Series No. 7). Vienna: International Atomic Energy Agency, 1998.

IAEA: Lessons Learned from Accidental Exposures in Radiotherapy (Safety Reports Series No. 17). Vienna: International Atomic Energy Agency, 2000.

Johns, H.E. and J.R. Cunningham: *The Physics of Radiology*, 3rd Ed. Springfield, IL: Charles C Thomas Publisher, 1974.

Johnston, J.C.: Estimating Flash Points for Organic Aqueous Solutions. *Chem. Eng. 81*: no. 25, Nov. 25, 1974. p. 122.

Jones, T.B. and J.L. King: *Powder Handling and Electrostatics*. Chelsea, MI: Lewis Publishers, Inc., 1991.

Kissell, F.N., A.E. Nagel, and M.G. Zabetakis: Coal Mining Explosions: Seasonal Trends. *Science 179*: 891-892 (1973).

Kletz, T.A.: Unconfined Vapor Cloud Explosions. In *Proceedings of the Chemical Engineering Progess 11th Loss Prevention Symposium, Houston, Texas, 1977*. pp. 50-58.

Kuchta, J.M.: *Investigation of Fire and Explosion Accidents in the Chemical, Mining, and Fuel-Related Industries — A Manual* (Bull. 680). Pittsburgh, PA: U.S. Department of the Interior, Bureau of Mines, 1985.

LANL: A Review of Criticality Accidents, 2000 Revision (LA-13638) by T.P. McLaughlin, S.P. Monahan, N.L. Pruvost, V.V. Frolov, B.G. Ryazanov and V.I. Sviridov. Los Alamos, NM: U.S. Department of Energy, Los Alamos National Laboratory, 2000.

Lees, F.P.: *Loss Prevention in the Process Industries*, 2nd Ed., Vol. 2. London: Butterworths-Heinemann, 2001.

Lemoff, T.C.: Gases. In *Fire Protection Handbook*. 17th Ed. Cote, A.E. and J.L. Linville (Eds.). Quincy, MA: National Fire Protection Association, 1991. pp. 3-63 to 3-82.

LeVine, R.Y.: Electrical Safety in Process Plants ... Classes and Limits of Hazardous Areas. *Chem. Eng. 79*: 51-58 (1972).

March, J.: *Advanced Organic Chemistry,* 3rd Ed. New York: John Wiley & Sons, 1985.

Martin, A. and S.A. Harbison: *An Introduction to Radiation Protection*, 2nd Ed. London: Chapman & Hall Ltd., 1979.

Maurer, B.: Discharges due to Electrostatic Charging of Particles in Large Storage Silos. *Ger. Chem. Eng. 2*: 189-195 (1979).

McKinlay, A.: Introduction to Optical Radiation. In *Non-ionizing Radiation. Proceedings of the 2nd International Non-ionizing Radiation Workshop, Vancouver, British Columbia, 1992 May 10-14*. M. Wayne Greene (Ed.). London: International Radiation Protection Association. pp. 227-251.

McManus, N.: *Portable Ventilation Systems Handbook*. London: Taylor & Francis, 2000.

McManus, N.: Differential (Negative) Pressure. *Cleaning Restor. 43*: No. 10 (October) 2006. pp. 40-46.

McManus, N. and G. Green: *The HazCom Training Program*. Boca Raton, FL: Lewis Publishers, 1999.

McManus, N. and G. Green: *The WHMIS Training Program*. North Vancouver, BC: Training by Design, Inc. 2001.

Menguy, C.: Static Electricity. In *Encyclopaedia of Occupational Health and Safety*, 4th ed. Stellman, J.M. (Ed.). Geneva: International Labour Office, 1998. [CD-ROM].

Meyer, E.: *Chemistry of Hazardous Materials*, 2nd Ed. Englewood Cliffs, NJ: Prentice-Hall, Inc., 1989.

Moore, W.J.: *Physical Chemistry*, 3rd Ed. Englewood Cliffs, NJ: Prentice-Hall, Inc., 1962.

Morrison, R.T. and R.N. Boyd: *Organic Chemistry,* 2nd Ed. Boston, MA: Allyn & Bacon, 1966.

MSHA: Think "Quicksand": Accidents around Bins, Hoppers and Stockpiles, Slide and Accident Abstract Program. Arlington, VA: U.S. Department of Labor, Mine Safety and Health Administration, National Mine Health and Safety Academy, 1988.

Newcott, W.R.: Lightning. *National Geog. 184*: 83-103 (1993).

NFPA: *Fire Protection Guide to Hazardous Materials,* 10th Ed. Quincy, MA: National Fire Protection Association, 1991a.

NFPA: Basic Classification of Flammable and Combustible Liquids. (NFPA 321-1991). Quincy, MA: National Fire Protection Association, 1991b.

NFPA: Cleaning or Safeguarding Small Tanks and Containers. (NFPA 327-1993). Quincy, MA: National Fire Protection Association, 1993.

NIOSH: Criteria for a Recommended Standard, Working in Confined Spaces (DHEW (NIOSH) Pub. No. 80-106). Cincinnati, OH: DHEW/PHS/CDC/NIOSH, 1979.

NIOSH: Guidelines for Controlling Hazardous Energy During Maintenance and Servicing (DHHS (NIOSH) Publication No. 83-125). Morgantown, WV: National Institute for Occupational Safety and Health, 1983a.

NIOSH: Occupational Safety in Grain Elevators and Feed Mills (DHHS/NIOSH Publication No. 83-126). Morgantown, WV: National Institute for Occupational Safety and Health, 1983b.

NIOSH: Pocket Guide to Chemical Hazards (DHHS (NIOSH) Pub. No.2005-149). Cincinnati, OH: Department of Health and Human Services//Centers for Disease Control and Prevention/National Institute for Occupational Safety and Health, 2005.

NIOSH: Worker Deaths in Confined Spaces (DHHS/PHS/CDC/NIOSH Pub. No. 94-103). Cincinnati, OH: National Institute for Occupational Safety and Health, 1994.

NIOSH: Worker Deaths by Electrocution (DHHS (NIOSH) Publication 2000-115. Cincinnati, OH: National Institute for Occupational Safety and Health, 2000. CD-ROM.

NIOSH: Fire Hazard from Filling Portable Gas Cans in Pickup Trucks and Cars. *Appl. Occup. Environ. Hyg. J. 17*: 242-243 (2002).

OSHA: Selected Occupational Fatalities Related to Fire and/or Explosion in Confined Work Spaces as Found in OSHA Fatality/Catastrophe Investigations. Washington, DC: U.S. Department of Labor, Occupational Safety and Health Administration (U.S. DOL/OSHA), 1982a.

OSHA: Selected Occupational Fatalities Related to Lockout/Tagout Problems as Found in Reports of OSHA Fatality/Catastrophe Investigations. Washington, DC: U.S. Department of Labor, Occupational Safety and Health Administration (U.S. DOL/OSHA), 1982b.

OSHA: Selected Occupational Fatalities Related to Grain Handling as Found in Reports of OSHA Fatality/Catastrophe Investigations. Washington, DC: U.S. Department of Labor, Occupational Safety and Health Administration (U.S. DOL/OSHA), 1983.

OSHA: Selected Occupational Fatalities Related to Toxic and Asphyxiating Atmospheres in Confined Work Spaces as Found in Reports of OSHA Fatality/Catastrophe Investigations. Washington, DC: U.S. Department of Labor, Occupational Safety and Health Administration (U.S. DOL/OSHA), 1985.

OSHA: Selected Occupational Fatalities Related to Welding and Cutting as Found in Reports of OSHA Fatality/Catastrophe Investigations. Washington, DC: U.S. Department of Labor, Occupational Safety and Health Administration (U.S. DOL/OSHA), 1988.

OSHA: Selected Occupational Fatalities Related to Ship Building and Repairing as Found in Reports of OSHA Fatality/Catastrophe Investigations. Washington, DC: U.S. Department of Labor, Occupational Safety and Health Administration (U.S. DOL/OSHA), 1990.

OSHA: Concepts and Techniques of Machine Safeguarding (OSHA 3067, Revised). Washington DC: U.S. Department of Labor, Occupational Safety and Health Administration, 1992.

Palmer, K.N.: Dust Explosions and Fires. London: Chapman & Hall, Ltd., 1973.

Palmquist, R.E.: Electrical Course for Apprentices and Journeymen, 3rd ed., rev. by J.A. Tedesco. New York: Macmillan Inc., 1988.

Perlee, H.E. and M.G. Zabetakis: Compressor and Related Explosions. (U.S. Bureau of Mines Report of Investigation, 8187). Washington, D.C.: U.S. Department of the Interior, 1963.

Pizzarello, D.J. and R.L. Witcofski: Basic Radiation Biology. Philadelphia: Lea & Febiger, 1967.

Pohanish, R.P. and S.A. Greene: Rapid Guide to Chemical Incompatibilities. New York: Van Nostrand Reinhold, 1997.

Porter, S.J.: Explosives and Blasting Agents. In Fire Protection Handbook, 17th Ed., Cote, A.E. and J.L. Linville (Eds.) Quincy, MA: National Fire Protection Association, 1991. pp. 3-92 to 3-100.

RAE Systems: Applications and Technical Guide (2000/2001 Ed.). Sunnyvale, CA: RAE Systems Inc., 2000.

Rasmussen, B.: Unwanted Chemical Reactions in the Chemical Processing Industries (Report M-2631). Roskilde, Denmark: Riso National Laboratory, 1987.

Repacholi, M.: Non-ionizing Radiation. In Non-ionizing Radiation. Proceedings of the 2nd International Non-ionizing Radiation Workshop, Vancouver, British Columbia, 1992 May 10-14. M. Wayne Greene (Ed.). London: International Radiation Protection Association. pp. 3-13.

Roberts, J.D. and M.C. Caserio: Basic Principles of Organic Chemistry. New York: W.A. Benjamin, Inc., 1964.

Scarbrough, D.R.: Control of Electrostatic Ignition Sources. In Fire Protection Handbook. 17th Ed., Cote, A.E. and J.L. Linville (Eds.). Quincy, MA: National Fire Protection Association, 1991. pp. 2-284 to 2-292.

Schwab, R.F.: Dusts. In Fire Protection Handbook, 17th Ed., Cote, A.E. and J.L. Linville (Eds.). Quincy, MA: National Fire Protection Association, 1991. pp. 3-133 to 3-142.

Shackelford, J.F.: Introduction to Materials Science for Engineers, 2nd Ed. New York: Macmillan Publishing Company, 1988.

Shapiro, J.: Radiation Protection, A Guide for Scientists and Engineers, 2nd Ed. Cambridge, MA: Harvard University Press, 1981.

Sly, O.M., Jr.: Flammable and Combustible Liquids. In Fire Protection Handbook. 17th Ed., Cote, A.E. and J.L. Linville (Eds.). Quincy, MA: National Fire Protection Association, 1991. pp. 3-43 to 3-53.

Uman, M.A.: Lightning. In World Book Millenium 2000, Deluxe Ed. Chicago: World Book Inc., 1999. [CD-ROM].

Weast, R.C.(Ed.): Handbook of Chemistry and Physics, 49th Ed. Cleveland, OH: The Chemical Rubber Company, 1968.

Whitehead, E.R.: Lightning. In *Encyclopaedia of Occupational Health and Safety,* 3rd rev. Ed. Vol. 2
(L-Z). Parmeggiani, L. (Ed.). Geneva: International Labour Organisation, 1983. pp. 1231-1234.

Wierzbicki, V. and D. Palladino: Standard Flash-Point Testing Methods. *Environ. Testing Anal. 3:*
30-37 (1994).

Willard, H.H., L.L. Merritt, Jr., J.A. Dean, and F.A. Settle, Jr.: *Instrumental Methods of Analysis,* 6th
Ed. Belmont, CA: Wadsworth Publishing Company, 1981.

Zabetakis, M.G.: *Flammability Characteristics of Combustible Gases and Vapors* (Bull. 627). Washington, DC: U.S. Department of the Interior/Bureau of Mines, 1965.

# 3 Biological Effects of Hazardous Energy

## INTRODUCTION

Hazardous energy is the level of energy that is likely to cause bodily harm. Contact with hazardous levels of energy occurs during accident situations. Contact with nonhazardous levels of energy can occur as a routine part of work experience. To take this even further, nonhazardous levels of some types of energy, for example ionizing radiation, are part of the normal environment in which we live. In fact, there are some parts of the Earth where the normal background exceeds levels that would prompt investigations in workplace situations.

The previous chapters examined the types of fatal accidents that occur in real-world situations and mechanisms by which energy is produced and interacts with matter, starting at the nuclear level and progressing to the macro level with which we are most familiar. This discussion indicated that nuclear events, and events involving chemical bonds produce seemingly inconsequential quantities of energy. Yet, these same levels of energy are capable of causing serious traumatic injury or even death. The key to understanding why this is the case is to consider the mode of delivery and the critical target.

Few of the energy sources encountered in accident situations are capable of causing immediate death (Table 3.1) (BLS 1997a). Table 3.1 suggests that, on average, hazardous energy is involved in 68% of fatal occupational accidents in the U.S. There is no evidence to argue for the lack of applicability of the information in this statement over the intervening years. Of this total, kinetic energy involving moving objects associated with normal workplace activities, as opposed to those associated with workplace violence, is involved in 41%, or 60% of the group caused by hazardous energy. The victim was struck by a moving object, including vehicles, and falling and flying objects in 12% of the accidents, or 18% of the group caused by hazardous energy. Transportation accidents constitute the largest group at 29% of the total. This is followed by workplace violence (15%), electrocutions (8%), struck by accidents (6%), fires and explosions (3%), caught in equipment or machinery (2%), compressed by equipment or machinery (2%), caught in or crushed by collapsing materials (2%), and temperature extremes (1%).

Most of the energy provided by these sources during an accident situation is not sufficiently intense as to cause immediate death. In fact, much of the effort of trauma units in hospitals is to prevent death, and to prolong life following such accidents, so the true impact of hazardous levels of energy encountered during accident situations will never be known. More commonly, death will come following a period of stabilization, albeit possibly a short one.

Another important matter in this type of discussion is the role of hazardous levels of energy in injuries serious enough to cause time loss from work (Table 3.2) (BLS 1997b). Although Table 3.2 refers to private industry, this provides a basis for comparison across the spectrum of industrial activity. Again, there is no evidence to argue for the lack of applicability of the information in this statement over the intervening years.

The document on lost-time injuries provides considerably less detail about the cause of injury compared to the document on fatal injuries. As a result, comparison between the two is not as exact as it could be. The causes of lost-time injury involving energy sources differ considerably from those of fatal injury. That is, the type of injury that kills workers is different from what sends them to the hospital or other medical facility, and leads to loss of time only. Hazardous energy plays a smaller role in lost-time accidents (38.8% of the total) than in fatal ones (68% of the total). Kinetic energy is by far the most prominent form involved in lost-time accidents. Transportation incidents and workplace violence play a small role in lost-time accidents compared to fatal ones according to these statistics. Slightly more than half (52%) of the total lost-time accidents involving energy occur due to struck by or against objects. Accidents involving equipment (caught in or compressed by) comprise 17% of the energy-related, lost-time accidents.

## Table 3.1

## Occupational Fatal Accidents Related to Sources of Hazardous Energy

| Major Class | Type of Energy | Percent |
|---|---|---|
| **Transportation Incidents** | | |
| Collision involving vehicle containing victim | kinetic | 23 |
| Struck by vehicle | kinetic | 6 |
| **Workplace Violence** | | |
| Shooting | kinetic | 12 |
| Bombing | chemical, kinetic | 3 |
| **Contact with Objects and Equipment** | | |
| Struck by falling object | kinetic | 5 |
| Struck by flying object | kinetic | 1 |
| Caught in running equipment or machinery | kinetic | 2 |
| Compressed by equipment or objects | kinetic | 2 |
| Caught in or crushed by collapsing materials | kinetic | 2 |
| **Harmful Substances or Environments** | | |
| Electrocution | electrical, heat | 8 |
| Temperature extremes | heat | 1 |
| Fires or explosions | chemical, kinetic, heat | 3 |
| **Total** | | 68 |

Source: BLS 1997a.

The focus of this chapter is biological effects of hazardous levels of energy; that is, how energy interacts with atoms, molecules, cells, tissues, and organ structures. The preceding discussion indicates that kinetic energy involving large bodies is responsible for by far the largest proportion of fatal and nonfatal accidents. Yet, probably less information is available about the mechanism of the damage than for other forms.

### Ionizing Radiation

The previous chapter indicated that the major nuclear particles are the proton and neutron. Electrons in equal number to protons populate the region around the nucleus in atoms carrying no net charge. Atoms containing unstable nuclei can achieve stability by emitting alpha particles (equivalent to helium nuclei) and neutrons. Neutrons can change into protons and electrons, and protons can change into

## Table 3.2

**Occupational Lost-Time Injuries Related to Sources of Hazardous Energy**

| Major Class | Type of Energy | Percent |
|---|---|---|
| **Transportation Incidents** | Kinetic | 3.6 |
| **Workplace Violence** | Kinetic | 1.1 |
| **Contact with Objects and Equipment** | | |
| Struck by object | Kinetic | 13.2 |
| Struck against object | Kinetic | 7.0 |
| Caught in equipment or object | Kinetic | 4.6 |
| Compressed by equipment or objects | Kinetic | 2.0 |
| Caught in or crushed by collapsing materials | Kinetic | 2.0 |
| **Harmful Substances or Environments** | | 5.1 |
| Fires or explosions | Chemical, kinetic, heat | 0.2 |
| **Total** | | 38.8 |

Source: BLS 1997b.

neutrons and positrons (positive electrons.) These particles emit from the nucleus at high velocity carrying large quantities of kinetic energy.

Experimentation has shown that creating an ion pair requires expenditure of 30 to 35 eV (electronVolts) of energy, regardless of the material (Barrow 1966). Particles carrying kinetic energy in the MeV range (millions of electronVolts) can create dense paths of ionizations and recoil of nuclei resulting from collisions. These particles interact with other atoms primarily by collision with the nucleus, by capture into the nucleus and by ejecting electrons. These actions cause the nucleus of the target atom to move in recoil and to sever chemical bonds between atoms. Bond breakage is especially consequential when the area of deposition of energy is the DNA molecule in a chromosome or a key protein. Chromosomal damage is especially consequential to the future functioning of the affected cell.

The cell is largely composed of water. Hence, the preponderance of interactions involves water molecules, and for this reason, water plays a large role in radiation protection. The radiolytic decomposition of water is an important first step in biological response at the biochemical level. This occurs according to the following postulated reactions (Barrow 1966):

$$H_2O + h\upsilon \rightarrow H_2O^+ + e^- \qquad\qquad (3\text{-}1)$$

$$H_2O^+ \rightarrow H^+ + \cdot OH \qquad\qquad (3\text{-}2)$$

$$H_2O + e^- \rightarrow H_2O^- \rightarrow H\cdot + OH^- \qquad\qquad (3\text{-}3)$$

This mechanism suggests that the transfer of energy to water molecules leads to formation of the highly energetic species, $H_2O^+$, which decomposes to form the highly reactive hydroxyl radical, $\cdot OH$. (Radicals are molecular species. They are highly reactive by virtue of the unpaired electron.) Hydroxyl radicals are second only to fluorine in oxidation potential, and as such are highly reactive. Similarly, the electron expelled from the water molecule in Equation 3-1 can generate a hydrogen atom through reaction with a water molecule. The hydroxyl radicals and hydrogen atoms can chemically attack macromolecules, such as DNA, proteins, and membranes that are essential for proper cellular function. As well, the reaction produces hydrogen ($H^+$) and hydroxide ions ($OH^-$).

Another possible reaction involves combination of hydroxyl radicals to form hydrogen peroxide (Moore 1962):

$$\cdot OH + \cdot OH \rightarrow H_2O_2 \tag{3-4}$$

Production of hydrogen peroxide depends on the availability of hydroxyl radicals in close proximity. This is more likely to occur with alpha particles that deposit a large quantity of energy in a small volume (Cember 1996). Hydrogen peroxide, while a highly reactive chemical species, is much less so than the hydroxyl radical. As a result, molecules of hydrogen peroxide are able to migrate within the aqueous environment and attack susceptible molecules in an area beyond that where they were formed.

Cellular fluids also contain dissolved oxygen (Pizzarello and Witcofski 1967). Oxygen reacts with the hydrogen atom (radical) formed in reaction 3-3:

$$O_2 + H\cdot \rightarrow HO_2\cdot \tag{3-5}$$

The hydroperoxyl radical is not as reactive as the hydroxyl radical. The hydroperoxyl radical combines with the hydrogen atom, as formed in Equation 3-3, to form additional hydrogen peroxide.

$$HO_2\cdot + H\cdot \rightarrow H_2O_2 \tag{3-6}$$

This situation means that oxygen in cellular fluids, which is essential for life, enhances the damage of radiation energy deposited in the cell. Antioxidants, including vitamin E, sulfur-containing amino acids, cobalt, and some chelating agents are radioprotective. These substances administered before (if warning is available) and after an acute (short-term, high-level) dose of radiation can enhance survivability.

Of considerable concern to biologically important molecules is the formation of radicals. This can occur following a direct hit in the manner described previously for water in Equation 3-1 (Pizzarrello and Witcofski 1967).

$$RH + h\upsilon \rightarrow R\cdot + H\cdot \tag{3-7}$$

where:
RH is a molecule containing a carbon-hydrogen bond.
$R\cdot$ and $H\cdot$ are both free radicals.

The free radical $H\cdot$ can, in turn, react with a molecule of dissolved oxygen to form a peroxyl free radical, and by so doing, an entirely different species:

$$R\cdot + O_2 \rightarrow RO_2\cdot \tag{3-8}$$

The peroxyl free radical is also a reactive species. This can engage in chemical reactions with additional neutral molecules.

$$RO_2\cdot + RH \rightarrow RO_2R + H\cdot \tag{3-9}$$

$$RO_2\cdot + RH \rightarrow RO_2H + R\cdot \tag{3-10}$$

These reactions indicate that this process can be highly disruptive to the normal order at the bio-chemical level.

An interesting follow-up to these reactions is the interest currently expressed about free radicals and the aging process. Ionizing radiation is one of several means of creating free radicals in the cellular medium. Another is ozone, $O_3$. Ozone is a highly reactive molecular species that can enter the body through the lung. The mechanisms of free radical production are similar, as are the means for combating them. These currently include advocacy for consumption of antioxidants as part of the diet.

Actions at the molecular level are the starting point for consideration about the actions of ionizing radiation on the molecular components of the cell. These interactions express themselves ultimately at the organismal level through the ionizing radiation syndromes.

**Ionizing Radiation Syndromes**
Some effects of ionizing radiation follow dose response relationships. As a result, damage by dose is readily predictable. These effects result from deposition of large quantities of radiation within a short period of time, as would occur during an accident situation. This deposition can result from internal or external sources.

To put these effects into perspective, an introduction to units used in radiation protection is necessary. Units for radiation dose have their roots in the early part of the 20th century. The SI (Système Internationale) unit for absorbed dose is the Gray (Gy).

$$1 \text{ Gy} = 1 \text{ J/kg} = 100 \text{ rad} \tag{3-11}$$

That is, one Gray represents the deposition of one Joule of energy per kilogram of tissue. The impact of different types of ionizing radiation differs due to the efficiency of deposition along the path through the tissue. Alpha particles, for example, are densely ionizing, whereas neutrons are much less so (Wilkening 1991a). This situation is resolved through use of the Quality Factor which is based on the rate of energy deposition in keV/$\mu$m of path. The Quality Factor differs with the type of radiation.

$$(\text{Absorbed Dose}) \times (\text{Quality Factor}) = \text{Dose Equivalent} \tag{3-12}$$

This gives rise to the unit of dose equivalent, the Sievert (Sv).

$$1 \text{ Sv} = 1 \text{ J/kg} = 100 \text{ rem} \tag{3-13}$$

The Sievert (and its historical predecessor, the rem) is the unit of direct comparison between effects due to ionizing radiation.

The lethal, whole body dose to 50% of the human population ($LD_{50}$) is estimated to be 4 to 6 Gy, or 4 to 6 Sv, assuming a quality factor of 1.0 (Johns and Cunningham 1974). Acute overexposure to ionizing radiation affects all organs and systems of the body (Cember 1996). The acute effects are classified into three syndromes in order of increasing dose: hematopoietic or blood marrow (BM) syndrome, gastrointestinal (GI) syndrome, and central nervous system (CNS) syndrome (Table 3.3). Effects that are common to all of these syndromes include nausea and vomiting, malaise and fatigue, increased temperature and blood changes. The appearance of symptoms depends on the dose and the syndrome. None of the symptoms is immediate, as would be the case with chemical overexposure. Doses of ionizing radiation that cause death in 30 days are considered to be immediately lethal, and the cause, acute overexposure (Pizzarello and Witcofski 1967).

The CNS (Central Nervous System) syndrome appears at high levels of dose (50 to 100 Gy) (Pizzarello and Witcofski 1970). At these levels of exposure, the victim could die during delivery of the dose. This level of dose is completely lethal in any event. The victim becomes agitated and irritable, followed by lethargy. Loss of equilibrium, spasms, convulsive seizures, prostration, coma, and death follow. Mean (average) survival time is dose dependent. Death is attributable to neuronal damage due to inflammation of the walls of the small blood vessels, swelling, and increased intracranial pressure.

## Table 3.3
## Effects of Penetrating Ionizing Radiation

| Dose | Syndrome | Effect |
|---|---|---|
| 50 to 100 Gy, acute, whole body | CNS | Failure of the central nervous system; death in <3 days |
| 10 to 100 Gy, acute, whole body | GI | Death of stem cells that replenish the lining of the gut; loss of absorptive capacity; death about 6 days later |
| 2 or 3 to 10 Gy, acute, whole body | BM | Death of stem cells in bone marrow that replenish blood cell lines; death about 45 days later |
| 4 to 6 Gy, acute, whole body | | Estimated lethal dose to 50% of humans in 30 days |
| 3 Gy, acute, skin | | Reddening of the skin (erythema) |
| 3 Gy, acute, local to the ovaries | | Temporary sterility, women |
| 2 to 5 Gy, acute, local to the lens of the eye | | Cataract development |
| >0.25 Gy, acute, whole body | | Gene mutations proportional to dose |
| 0.25 to 0.5 Gy, acute, whole body | | Observable changes in blood |
| 0.3 Gy, acute, local to the testes | | Temporary sterility, men |
| 0.0008 to 0.79 Gy | | Background level due to geologic and cosmic sources, no predictable effect |
| 0.020 Sv | | Annual permissible dose for atomic radiation workers |

Sources: Cember 1996; ICRP 1990; Pizzarello and Witcofski 1967; UNSCEAR 2000.

Exposure to 10 to 100 Gy of whole body radiation causes the GI (Gastro-Intestinal) syndrome (Pizzarello and Witcofski 1970). Onset of the GI syndrome is independent of dose. The GI syndrome results from damage to and death of the cells lining the GI tract (Figure 3.1) and the bone marrow. The GI syndrome is characterized by loss of appetite, sluggishness and inertia, infection and dehydration, loss of appetite, prolonged retention of food in the stomach, and decreased absorption in the gut. This leads to weight loss and dehydration. Cells of the villi in the gut that normally are sloughed off are not replaced. This is due to death of stem cells in the crypts at the base of the villi. Cells produced in the crypts normally undergo specialization and maturation as they migrate up the crypt. At the same time, the villi become flattened. Initial non-permanent damage to cells in the crypts occurs at 1 Gy. At levels of whole-body irradiation needed to induce the GI syndrome, the bone marrow syndrome also occurs.

The bone marrow (BM) syndrome appears at the lowest level of whole-body dose (2 to 10 Gy). The initial sign of damage is disturbance of function of the GI tract. A subsequent period follows in

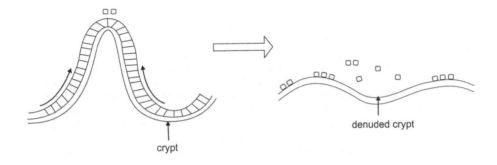

Figure 3.1. Gastrointestinal (GI) syndrome. Ionizing radiation kills stem cells located in the crypts at the base of the villi in the small intestine. Progeny of the stem cells normally migrate up the villi and slough off into the gut. Loss of these cells reduces absorptive capacity and enables entry of disease-causing bacteria into the body. (Modified from Pizarello and Witcofski 1967.)

which outward symptoms are absent. However, what has happened is that the radiation has killed stem cells in the bone marrow. These stem cells are precursors of the various cell lines found in the blood. Loss of mature blood cells and failure of the bone marrow to replenish them precede signs and symptoms leading to death. The latter include weight loss, diarrhea, fluid imbalance, loss of hair, and depression of sperm and egg formation.

Treatment for some forms of cancer involves administration of a whole body dose of ionizing radiation sufficient to destroy stem cells in the bone marrow, followed by repopulation using marrow removed from a donor. This treatment literally means administering a lethal level of whole body radiation and then gambling that the donated marrow containing the stem cells will able to survive in the recipient.

The effects reported in Table 3.3 are direct, semi-quantitative, and dose-related. These effects have appeared in victims of radiation accidents and survivors of atomic bomb blasts (LANL 2000). Many of these effects follow a dose-response curve. That is, a particular dose produces a particular effect. There is evidence to argue in favor of a threshold; that is, a dose that produces zero effect (Cember 1996).

Ionizing radiation also produces delayed effects (Cember 1996). Those of greatest concern include cancer, shortening of life span, and cataracts. Normally, delayed effects result from long-term, low-level chronic exposure, rather than a single, high-level acute one. Radiologists (physicians who use radiation in their practice), for example, show significantly higher incidence of cancer than other physicians. A notable exception to the chronic exposure situation is the higher than expected occurrence of leukemia in survivors of the atomic bomb who were within 1500 m of the epicenter. This is the result of a single acute dose. There is an ongoing debate about whether radiation-induced leukemia follows a threshold of occurrence.

Significant variation exists in the natural background to which humans are exposed at various locations on the Earth. This variation ranges close to 1000-fold. The highest measured absorbed doses are considerably higher than the recommended permissible annual dose of 0.020 Sv per year (ICRP 1990). Residents of areas having high background radiation show no difference in chromosomal abnormalities compared to the control group, and no increase in incidence of cancer or leukemia. There is evidence of an adaptation to these conditions (Ghiassi-nejad et al. 2002).

Sources of ionizing radiation other than background levels have two main origins: naturally occurring and man-made.

Naturally occurring sources of ionizing radiation result from substances containing compounds of uranium, radium, and thorium (Shapiro 1981). These substances occur in low concentration in the Earth's crust and at substantially higher concentration in certain minerals, such as the monazite sands, phosphate-containing rock, and pitchblende (Metzger et al. 1980, Hewson and Fardy 1993). Industrial processes, such as combustion of coal and oil, can concentrate compounds containing these elements to considerably higher concentrations (de Santis and Longo 1984, Cohen 1985). The net result is that deposits containing high concentrations of these substances can form on the inside (boundary) surfaces of processing equipment. Such deposits develop during extraction and processing of phosphates in minerals, the production of fertilizer, and in the processing of oil and natural gas (Wilson and Scott 1992). Airborne dust from these deposits can cause significant exposure to the lungs during work in these structures. Uranium, radium, and thorium decay through a series of unstable isotopes (Shapiro 1981). These isotopes emit alpha, beta, and gamma radiation in varying amounts. Radiation from these substances is unlikely to produce symptoms of acute exposure.

Radon, the heaviest of the inert gases, is an exception to the other naturally occurring radioactive substances which are solids (Shapiro 1981). Radon-222, the most prevalent form of radon, originates from the radiological decay of $^{226}$Rn radium. Radon can diffuse into sub-surface concrete structures from surrounding soil or from materials of construction containing $^{226}$Rn.

Radon-222 can accumulate in spaces in which air exchange is low. Radon-222 has a half-life of 3.8 days. This means that half of the original number of atoms disappear in 3.8 days. Radon acts chronically. High levels, as encountered in uranium mining, are a causative agent in lung cancer. There is no evidence to support any concern about injury from acute exposure, as might occur during entry into a poorly ventilated structure containing accumulated radon. The initial purge of subsurface structures containing accumulated $^{222}$Rn removes all traces (Wiegand and Dunne 1996).

Gamma-emitting sources have been used industrially for many years in level and thickness gauges, as well as for radiographic nondestructive testing. When used as a level gauge, the source is mounted externally to the vessel. The beam passes through the walls of the vessel and the substance under measurement to a detector. The extent of attenuation is an indication of level, or thickness, in the case of a thickness gauge. The source contains a shutter that is to be closed and locked out during entry and servicing of the vessel. Damage to the shutter mechanism has occurred on occasion in corrosive environments. In these circumstances, the shutter failed to close when the arm was moved to the closed position. In other cases involving poor housekeeping, burial of the source by powder has occurred. In these situations, the source was undetectable to outside contractors.

Industrial radiography utilizes gamma sources to produce an image on photographic film or other image receptor. The source is located on one side of the object to be examined, and the film on the other. Prevention of unnecessary exposure of bystanders and the radiographer is a function of the skill used in planning the exposure and bystander control. In these situations, in the absence of a confirmatory check using a gamma-detecting, radiation survey instrument, work could occur under conditions of partial to full irradiation.

In the previous applications, the source is external to the enclosure. In many other applications, the enclosure is designed to confine energy from the source. The concept of using shielding for confinement of energy of radiological significance has formed the basis for radiation protection for many years. Application of this concept ranges from the walls that shield the diagnostic X-ray machine in the dental office to the thick high-density concrete used to shield high-energy gamma sources used in irradiation of food and sterilization of equipment. In the former case, there are no constraints to prevent the operator from entering the field of the X-rays during operation of the equipment. In the latter case, interlocks on doors shut down or otherwise deactivate the source at time of unsanctioned entry.

## Nonionizing Radiation

Nonionizing radiation, gamma radiation, and X-rays all are forms of electromagnetic energy. Electromagnetic energy has both wave-like and particle-like characteristics. The wavelength, the length of one complete wave, is related to the energy of the particle, through Planck's constant. Hence, the wavelength differentiates between the nature of the interaction with matter and the biological effects produced, and the name given to the type of energy.

The distinction between ionizing and nonionizing electromagnetic energy is the ability to ionize (remove electrons from) atoms and molecules. Electromagnetic energy, classed as nonionizing radiation, carries insufficient energy to cause ionization. The basis for the distinction between ionizing and nonionizing radiation is the energy needed for ionization of atomic oxygen and hydrogen, 10 to 12 eV, hence, the slightly arbitrary choice of 10 eV.

Contrary to ionizing radiation, levels of nonionizing radiation encountered during accident situations are incapable of causing death. This results from the difference in action on biological tissue between photons that can ionize and those that cannot. This difference does not influence the ability to penetrate into matter and tissue, as some types of nonionizing radiation are highly penetrating.

### Ultra-Violet Region

The ultra-violet (UV) region of the electromagnetic spectrum contains the highest energies of nonionizing radiation. These energies cause excitation of electrons in atoms and molecules, but not outright ionization. UV energy has considerable application. As a result, UV sources have widespread use in equipment and processes.

The UV spectrum is subdivided into three regions based on biological effect: UVC, 100 to 280 nm (nanometers); UVB, 280 to 315 nm; and UVA, 315 to 380–400 nm (CIE 1987). Peak absorption of DNA occurs at 260 nm, and proteins at 280 nm.

The sun is the major source of UV exposure at the Earth's surface (Wilkening 1991b). UV emissions are most significant when the sun is high in the sky, as would occur from late morning to early afternoon (Sliney 1992a). This results from the lack of attenuation by the atmosphere at that position in the sky. At lower positions, the atmosphere absorbs the UV. This is the reason that one can drive facing into the sun in late afternoon without damaging the eyes.

Ozone in the upper atmosphere absorbs UV emissions at wavelengths less than 290 nm. These normally do not reach the surface of the Earth. One exception is the hole in the ozone layer that exists over Antarctica. Significant industrial sources of UV include mercury-containing discharge lamps, high-intensity discharge lamps, plasma torches, and welding arcs.

A major industrial application of UV energy is polymerization of monomers by activation of photosensitive initiators (Moss 1980). Applications of this process include the curing of inks, adhesives, and wood and metal coatings. The UV sources used in these processes produce significant output in the 300 to 400 nm region. This is within the region (180 to 400 nm) that produces biological effects (INIRC 1985a). Most energy in the UV region is not visible to the eye; hence, workers who enter areas containing unshielded UV sources may not be aware of the presence of this energy. The only indication may be emission of blue-colored light and fluorescence of certain types of clothing. In such circumstances, warning devices and interlocks are essential at entrances to the space. There is no guarantee, however, that these devices will function as intended, or that tampering to defeat them cannot or will not occur.

Another major source of UV energy is welding. Welding emissions cover the range from the ultra-violet to the infra-red (Hinrichs 1980). In general, UV emissions from electric welding processes decrease in the following order: Gas Metal Arc Welding > Shielded Metal Arc Welding > Gas Tungsten Arc Welding > Plasma Arc Welding. Welding arcs produce emissions at wavelengths below 300 nm. As a consequence, these are sources of ozone production.

The critical targets for UV energy are the eye and the skin. The cornea absorbs all optical radiation having wavelength less than 300 nm (McKinlay 1992; Zuclich 1980). Hence, UVB and UVC are more hazardous to the cornea than the retina. Electrical arcs produced by welding and electrical short circuits are the main source of exposure to UVC in the work environment. UV exposure of the cornea causes photokeratitis (inflammation of the cornea). This condition, known as welder's flash, feels like sand in the eye. This condition, fortunately, is self-correcting. The extent of damage is inversely proportional to the delay in onset. UV exposure also can cause conjunctivitis (inflammation of the membranes in the eyelids) and photophobia (avoidance of light).

The lens absorbs UVB with a peak centered around 365 nm. Cataract formation is involved with a narrow band of wavelengths below 325 nm (Pitts 1986). Intense UVA centered at 365 nm can cause thermal damage to the anterior surface of the lens (Zuclich 1980).

Outdoor situations also can cause photokeratitis. Normally, the geometry of the eye in the skull protects against exposure by the direct path (Sliney 1992a). Reflection from surfaces can cause photokeratitis. This depends on reflectivity of the surroundings. Highly reflective surroundings include fresh snow, sea surf, and beach sand (in descending order). Snow blindness is an example of photokeratitis.

A highly intense flash and exposure, as could occur during an accident situation involving open electrical arcs, could cause temporary visual difficulty or perhaps outright blindness. Onset of photokeratitis and photophobia could be rapid, thereby compounding the difficulty of escape or seeking help.

UVB is the principal region of the UV absorbed by the skin. UVB penetrates deeper into the epidermis than UVC. UVC is absorbed primarily in the surface layers of dead cells of the epidermis (Urbach 1980). Hence, UVB produces a more severe reaction. Maximum sensitivity of the skin occurs at 295 nm (Sliney 1992a). This produces reddening (erythema). Erythema appears up to 12 hours after exposure. This can lead to blistering and peeling of the skin. Absorption of UV radiation can cause skin cancer. Exposure to UVB and UVA is believed related to premature aging of the skin.

Another concern regarding skin exposure to UV is phototoxicity induced by chemicals on the skin (Urbach 1980). These produce erythema at wavelengths of UVA that normally are harmless. Phototoxic chemicals include furocoumarins found in the rind of lemons and limes, in celery, and in some perfumes. Fluorescent compounds in coal tar are highly phototoxic. A related reaction, photoallergy, involves the immune system. Compounds responsible for this condition were found to be mostly halogenated carbanilides and bacteriostatics of the type formerly added to soap. ACGIH lists additional compounds associated with photoallergy (ACGIH 2001a). These include antibiotics (tetracycline and sulphathiazole), antidepressants (imipramine and sinequan), and some diuretics, cosmetics, and antipsychotic drugs.

The most illustrative real-world example of exposure to UV from an uncontrolled source occurred in a school gymnasium containing a damaged mercury vapor lamp (Andersen 1980). Mercury vapor lamps contain an inner lamp enclosed in quartz and an outer glass envelope. The inner lamp continued to function following damage of the outer envelope. Without the intact glass envelope, the lamp emitted considerable UV energy. In general lighting situations, the glass envelope of mercury vapor and other types of lamps effectively absorbs UV energy (McKinlay 1992). High-intensity discharge lamps, such as used in photopolymerizers and sun lamps, can emit considerable UV energy. The gymnasium incident resulted in severe corneal photokeratitis, conjunctivitis, and erythema.

**Visible Region**
The energies contained in the narrow band of the electromagnetic spectrum known as visible light are detectable by the retina. Visible light covers the region from 380 to 400 nm to 760 to 780 nm (CIE 1987). Wavelengths of light in the visible region pass through the structure of the eye and are focused by the lens onto the retina. An added factor in admission of light into the eye is the action of the iris and pupil. Pupil size is maximum in the dark-adapted eye at 7 mm (ACGIH 2001a).

Visible light only recently has received recognition as being hazardous at levels incapable of causing retinal burn. "Blue light" in the region 400 to 550 nm (peak 440 nm) can cause photochemical injury to the retina (Sliney 1990). The discovery of photochemical damage from blue light provided an explanation for unanswered questions about effects, such as eclipse blindness (Sliney 1983). The principal effect from viewing a source containing "blue light" is photoretinitis. The onset of photoretinitis is dose dependent. That is, blue-light retinal injury can result from viewing an intense source for a short time or a weak one for a longer time. Duration of exposure must exceed 100 seconds. Injury appears several hours to 2 days following the exposure (Sliney 1992b). A complicating factor in risk of injury from blue light is removal of the lens from the eye (aphakia) as a result of cataract surgery. Removal of the lens increases the risk from the retinal blue light hazard.

Energy levels capable of causing blue-light injury can vary more than 1,000-fold. High intensity discharge lamps are potential sources of blue light hazard, as is the sun (McKinlay 1992). Solar irradiation to "blue light" is influenced, as in the case of ultra-violet exposure, by height in the sky. The atmosphere attenuates blue light, so that this hazard exists primarily when attenuation is least, from late morning to early afternoon. Welding arcs are another potential source of exposure to blue light. Reflections from surfaces of structures can cause indirect exposure to these sources.

The greatest potential for exposure to "blue light" (either indoors or outdoors) arises during contact with light reflected from fresh snow, especially that from late morning to early afternoon. This light also contains a significant contribution from the UV. Failure to wear eye protection under these conditions can result in erythropsia (red vision) and photokeratitis (snow blindness) (Sliney 1992b). Erythropsia results from excessive illumination of the cornea.

Flash blindness is a temporary condition resulting from exposure to high-intensity sources (Wilkening 1991b). This condition causes bleaching of the pigments and an "afterimage". An afterimage is a temporary blind spot in the field of vision. Persistence of the afterimage depends on intensity of the light source and duration of exposure. A high-intensity source can cause temporary blindness.

Thermal injury of the retina can occur at energies considerably greater than those discussed thus far from sources of intense visual radiation. Energy at visible wavelengths is used, for example, in lasers used to repair retinal conditions (Allen 1980). Other sources include flashbulbs, spotlights, welding arcs, and carbon arcs (Wilkening 1991b). Thermal damage to the retina is possible because of the focusing ability of the lens to concentrate light onto the retinal surface. The primary mechanism for retinal damage is production of a steam bubble or explosion, plus the resultant mechanical trauma. Thermal damage can result when the rate of radiation of heat generated in the interaction with retinal pigments and processes is less than the rate of production. At this point, accumulation of heat occurs in the local area. Damage results when the build-up exceeds tolerance levels of the cellular structure. Thermal damage occurs when the energy deposition ranges around $10 \text{ J/cm}^2$ and temperature increases to 45° C where coagulation of protein occurs (ICNIRP 1997; Tengroth et al. 1980).

### Infra-red Region
Infra-red sources span the range from 760 to 780 nm to 1 mm (1,000,000 nm) (CIE 1987). The IR spectrum is subdivided into three regions based on biological effect: IRA, 760–780 nm to 1,400 nm; IRB, 1,400 nm (1.4 $\mu$m) to 3.0 $\mu$m; IRC, 3.0 $\mu$m to 1 mm (1,000 $\mu$m).

IR energy from 770 to 1,400 nm (IRA) focuses onto the retina. Energy absorbed by the lens and iris is believed to be involved in cataract formation. Aversion reflexes offer protection against retinal injury caused by, for example, prolonged staring at the sun (McKinlay 1992). Normally, the geometry of the eye in the skull protects the cornea against exposure by the direct path. At sufficient intensity, a retinal burn, similar to that occurring with visible energy, can develop (Allen 1980; Sliney 1983). Injury to the retina at wavelengths above 550 nm in the visible and IR regions occurs principally from thermal injury (Sliney 1992b). However, this appears likely only from pulsed sources. Thermal damage occurs when energy deposition ranges around $10 \text{ J/s/cm}^2$ and temperature increases to 45° C at which coagulation of protein occurs (ICNIRP 1997; Tengroth et al. 1980). The cornea absorbs IR energy at wavelengths longer than 2 000 nm (IRB and IRC).

The skin is also affected by IR radiation. Solar radiation causes more adverse effects than other sources (McKinlay 1992).

Conventional sources of IR energy, such as molten glass and metal, apparently do not produce the irradiance necessary to cause either acute or chronic effects beyond "dry eye" (Sliney 1983). Incandescent (heated filament) lamps pose no hazard to the retina. However, infra-red heat sources that provide no strong visual stimulus could pose a retinal hazard during prolonged staring (McKinlay 1992).

Sources of IR energy are present in industrial workplaces. Warning devices and interlocks are essential in situations where the source provides no strong visual or other warning. There is no guarantee, however, that these devices will function as intended or that tampering to defeat them cannot occur. The impact of IR sources in some confined spaces is exacerbated by the reflective interiors and boundary surfaces. IR sources also are present outside confined spaces. Those operating at wavelengths containing no visible component could pose a serious optical and skin contact hazard.

**Microwave Region**

The microwave region spans wavelengths from 1 mm to 10 m (30 MHz to 300 GHz) (Wilkening 1991b). (MHz is megaHertz, and GHz is gigahertz; these are units of frequency, meaning the number of waves that pass by a fixed point in one second.) The microwave region and radiofrequency radiation are part of the same continuum of the electromagnetic spectrum. The continuum extending from the far IR (IRC) is the beginning of the region in which the individual components of the electromagnetic wave, the electrical and the magnetic, produce the effect on tissue. The photon energy at microwave frequencies is too low to ionize molecules, regardless of the number absorbed. As a result, ionization effects are excluded.

Microwave frequencies are used in radar, for cooking and food processing, industrial sealing, heating and gluing, medical diathermy, satellite communications, cellular telephones, police, fire, and ambulance communications, and television and FM radio broadcasting (Glaser 1980; Loral 1992). Natural sources include cold weather fronts and solar radiation (Wilkening 1991b).

Absorption of microwave energy depends on the size, configuration, shape, orientation, and dielectric properties of the object being irradiated (Glaser 1980). In humans, absorption of microwave and radiofrequency energy occurs in the 60 to 100 MHz frequency range, with a peak at 70 MHz (Gandhi 1975). Under grounded conditions, these can decrease to 30 MHz. This range coincides with frequencies used in FM radio and VHF television broadcasting, and police, fire, and ambulance communications. Local absorption peaks exist for the head, neck and legs, and arms. The range of frequencies at which these occur are well within the boundaries of the microwave region. The outer surface of the skin absorbs wavelengths less than 3 mm. Wavelengths ranging from 3 mm to 10 cm penetrate 1 to 10 mm. Wavelengths of 10 to 20 cm penetrate deeply enough that potential for damage to internal organs is a consideration. The body is transparent to wavelengths greater than 500 cm.

The main mechanisms of interaction between microwave and RF energies and cellular components are polarization of bound charges, orientation of permanent dipoles, and displacement of free charge carriers (Bernhardt 1992). Energy dissipation can occur from friction losses occurring during alignment of dipoles, such as water molecules, and induced oscillation and rotation occurring within molecules. Orientation of dipoles occurs in the range from 1 to 100 GHz. Maximum absorption by bound water, peptides, proteins, or side chains of large molecules occurs in the range 1 to 20 GHz. In the frequency range, 100 kHz to 100 MHz, the entire cell acts as a dipole.

Microwave ovens heat food through interactions between microwaves at a particular frequency (2450 MHz) and molecules of water. (This frequency corresponds to a wavelength of 12 cm). Wavelength is calculated by dividing the speed of light ($3 \times 10^8$ m/s) by the frequency $2450 \times 10^6$ waves/s. Excitation of the -OH bond is sensed as heat. Since water is the major constituent of cells and tissues, and -OH bonds are present in many other molecules, excitation produces a profound internal effect.

Of the many possible effects in this region of the electromagnetic spectrum investigated over the years (now well more than 5,000 articles), most relate to heating of the exposed subject (Cleary 1980). This continues to be the perspective about exposure to electromagnetic radiation in this region (COMAR 2002). Thermal damage occurs at a power density of 100 mW/cm$^2$ (Wilkening 1991b). Energy deposition at high levels, as could occur in an accident situation, would be felt as heating. Operators of equipment have remarked on these sensations.

Some level-sensing devices that are used in storage hoppers, bin, tanks, and so on operate in the microwave region. High levels of microwave energy are utilized in other industrial applications, such as sealers, welding, and heating processes. Entry into areas containing microwave equipment must be restricted, as there is no indication to the senses of overexposure, except at the very high level noted here. In such circumstances, warning devices and interlocks at entrances to these workspaces are essential. There is no guarantee, however, that these devices will function as intended or that tampering to defeat them cannot occur.

Nonthermal effects are also considered possible. These effects primarily result from interactions with the central nervous system (Cleary 1980). This interaction is expressed through behavioral end-points and signs and symptoms (KlimKova-Deutschova 1974). The latter include circulatory and digestive upsets, sleeplessness and headache, sleeplessness, fatigue, and changes in activity of brain waves.

Some individuals can "hear" microwave pulses (Frey 1962). The "sound" is sensed as a buzzing or ticking, depending on pulse rate. The current explanation for this effect is induction of thermoelastic waves in the head. To date, there is no evidence of injury from this effect. Microwave energy also can cause cataracts under special conditions of exposure (Wilkening 1991b).

### Radiofrequency Region
Radiofrequency radiation includes frequencies from 300 Hz to 30 MHz and wavelengths ranging from 10 m to 1,000 km (Stuchly 1992). Radiofrequency and microwave energies represent an artificial division of the same continuum. Radiofrequency applications include CB radio, heat sealers, vinyl welders, diathermy units, dielectric heaters, RF communications, plasma etching and sputtering, AM radio broadcasting, induction heaters, radio navigation, ship-to-shore communications, video displays, and aids to navigation (Loral 1992). Natural sources include lightning (NCRP 1981).

Little absorption of energy occurs in humans in the radiofrequency region (Glaser 1980). There are reports of ocular changes, changes in neuroendocrine function, behavioral, and CNS (central nervous system) changes, alteration of immunologic function, and decreased sperm production.

In the frequency range 100 kHz to 100 MHz, the entire cell acts as a dipole. For frequencies below 100 mHz, interaction with excitable tissue is the primary concern. Above 100 kHz, the threshold for interaction with excitable tissue is so high that thermal interactions become important (Bernhardt 1992).

### Extremely Low Frequency Region
Extremely low frequency (ELF) ranges from 30 to 300 Hz and wavelength from 1,000 to 10,000 km (Stuchly 1992). The most important application of energies at these frequencies is electrical power generation, transmission, distribution and utilization (50 or 60 Hz).

Associated with electromagnetic waves in space is a time-varying electric and magnetic field (Morgan 1989). Electrical energy transmitted in a conductor also utilizes a time-varying electromagnetic field. This field is present in and emanates from all conductors (insulated and uninsulated), from high-voltage transmission lines to home appliances operating on household voltage (110 V).

This field typically decreases rapidly with distance from line sources (such as electric wires) and does not radiate readily, as do electromagnetic fields of higher frequency (BPA 1986). The effects of the electric and magnetic fields generated at these frequencies have received considerable attention in recent years. The wavelengths are very long compared to the dimensions of the human body.

Electric and magnetic waves exhibit a characteristic of waves in general: addition and cancellation (BPA 1986). The net strength of a wave in a particular region in space is the vector sum of the contributions from each wave. Waves that have similar phase are additive, while those that are out of phase are subtractive. This affects the magnitude of the energy field surrounding conductors having multiple phases, such as are found in transmission and distribution systems. Addition and cancellation can occur. This also is the case with the hot conductor and the grounded neutral conductor. When side by side, the external energy fields almost cancel.

Electric and magnetic fields of extremely low frequency occur throughout nature and in all living things (Morgan 1989). They are necessary for operation of the nervous system. The electric field component involved in the electrical system can induce currents in the human body. However, the currents induced by power frequency fields are very small, on the order of $10^{-5}$ to $10^{-7}$ that of the external inducing field (Tenforde and Kaune 1987).

These induced currents appear to have the ability to interact with receptors on cell surfaces to change the rate of production of hormones, enzymes, and other proteins. Additional effects include changes in rate of production of DNA (the genetic material), and rate of error generation occurring in transcription of RNA from the DNA template. Effects are known to occur at the level of expression of these molecules regarding circadian rhythms (daily biological cycles) and neurotransmitters, and rate of growth and division of some cells.

The body and other materials perturb the electric field component. Hence, shielding is readily achieved. Measurement of the electrical field requires use of remote measuring probes in order to avoid perturbation by the presence of the human body.

The body perturbs the magnetic field component only to a minimal extent. Hence, measurement of the magnetic field component of the energy field requires no special precautions. Dosimeters are available for measuring magnetic fields. The body is essentially transparent to the magnetic field. Providing portable shielding to nullify the magnetic field from an isolated conductor would be extremely difficult, if not technically impossible. To date there is no evidence for any short-term effect from fields to which humans normally are exposed in industry or the home (Bernhardt and Matthes 1992).

The electric field radiating from a conductor induces a field in conductive objects. Discharge to ground can occur through a conductive path. The threshold in humans to sense an electric field is 2 to 10 kV/m (INIRC 1990). The threshold for sensing spark discharges is about 3 kV/m

The principal man-made sources of exposure to magnetic fields in the 50/60 Hz region are transmission lines and high-voltage electrical facilities. These operate at high voltage to minimize current flow. The principal source of industrial exposure is equipment utilizing high current. Examples include welding machines, electroslag refining, electric furnaces, induction heaters, stirrers, and electrolytic processes (Bernhardt and Matthes 1992). Exposure to magnetic fields generated by industrial sources can easily exceed that from transmission lines due to proximity to people.

Present guidelines for occupational and public exposure attempt to prevent acute effects (muscle or neural stimulation), such as reflex reactions, associated with induced or conducted currents produced by the electrical or magnetic field components of the electrical wave (Kavet et al. 2001). These occur at an induced current density of $10$ mA/m$^2$ (Sheppard et al. 2002). Alterations of cognitive function and stimulation of the visual system in humans also begin to occur at this level (Saunders and Jefferys 2002). Effects on the visual system include phosphenes (AIHA 2002). Phosphenes are flickering sensations in the eye. They appear to result from stimulation of the retina. Phosphenes also result from pressure, mechanical shock, chemical substances, and sudden fright.

There is no support at this time for a causal link between cancer and exposure to 50/60 Hz electromagnetic fields. This remains an open question.

Lasers
Laser is an acronym for Light Amplification by Stimulated Emission of Radiation (WHO 1982; ACGIH 1990). Lasers have widespread utilization in industrial applications (Court and Courant 1992). Applications include alignment, welding, cutting, drilling, heat treatment, distance measurement, entertainment, advertising, and surgery (ICNIRP 1996; INIRC 1985b). Some industrial applications can produce exposure to highly intense beams through direct viewing, as well as reflected paths.

Adverse health effects from laser radiation are particularly a concern from 400 to 1,400 nm (visible to IRA) (INIRC 1985b). Retinal damage can occur in this range. However, biological effects can occur across the optical spectrum of wavelengths from 180 nm to 1 mm. The biological effects induced by optical radiation at any wavelength are essentially the same for coherent (laser) and incoherent sources. These range from bleaching of pigments in the retina to destructive damage (Wilkening 1991b). The increased concern with lasers results from the intensities and irradiances that are produced. As well, lasers are highly monochromatic (single wavelength) (Figure 3.2). Some lasers produce their emissions in pulses. Pulsed emissions amplify the hazard. The eye and skin are the critical organs for laser exposure, as for other types of optical radiation (ICNIRP 1996).

Damaging levels of laser energy are utilized in many industrial applications, including welding and cutting, heating processes, optical processes, electronic processes, photochemical processes, and so on. Provision of a strong visual or other stimulus as a warning inside spaces where boundary surfaces serve as shielding provide limited benefit. In such circumstances, warning devices and interlocks at entrances to the space are essential. There is no guarantee, however, that these devices will function as intended or that tampering to defeat them cannot occur.

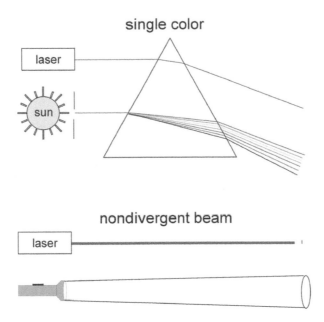

Figure 3.2. Lasers are monochromatic sources of energy which range from the ultra-violet to the visible to the infra-red regions of the spectrum. Laser emissions broaden only slightly with distance.

## Hot Surfaces and Heat-Related Injury

Many workplaces contain heated surfaces and sources of heat. These include surfaces of process equipment, as well as portable equipment and processes used during work activity. Direct contact and at-a-distance irradiation of unprotected skin and heated surfaces can cause thermal burns.

The skin is a complex heterogeneous organ. An understanding about the anatomy of the skin is needed to appreciate the ramifications of damage caused by hot and cold surfaces. The skin is predominantly two layers: the outer epidermis and the inner dermis (Figure 3.3).

The epidermis is contiguous with the dermis and acts as the outer cover of the cushion of connective and elastic tissue that guards the blood and lymphatic vessels, nerves, secretory glands, hair shafts, and muscles (Birmingham 1991). The epidermis consists of the stratum corneum or keratin layer, and the epidermal cell layer. The keratin layer contains multiple layers of dead cells. These form a shield of strong, tightly packed fibrous protein. This shield provides the protective function of the skin against liquids and physical abrasion. Cells of the epidermal cell layer replace the cells of the sacrificial keratin layer. The epidermal cells themselves are replaced by germinal cells resident in the lowermost layer of the epidermis.

The dermis is much thicker than the epidermis (Birmingham 1991). The dermis contains connective tissue, composed of collagen and elastic tissue, structures that house the sweat glands and ducts, hair follicles, sebaceous glands, blood and lymphatic vessels, some muscle tissue, and nerve endings. The dermis is an extremely strong fibrous envelope (Olishifski 1988). This is the part of the animal skin that becomes leather. The dermis acts as the main protective layer against trauma. When injured, the dermis repairs itself by forming new tissue, a scar.

Beneath the dermis is the subcutaneous layer of fat. This layer also contains the lower parts of some of the structures found in the dermis. The fat layer acts as an insulator and cushion (Tucker and Key 1982). Underlying the subcutaneous fat layer is the musculature.

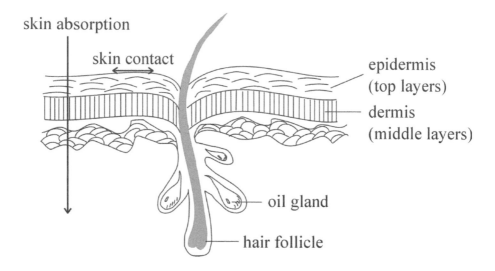

Figure 3.3. The skin. The layers of the skin provide the first line of protection, both physical and chemical, against external agents. As with other tissues, burn, synonymous with coagulation of protein, occurs at temperatures exceeding 45° C. Full-thickness burns require grafting. (Sources: McManus and Green 1999, McManus and Green 2001).

Thickness of the skin and its various layers varies with location on the body (Tucker and Key 1982). As experience would suggest, the thickest layer of the stratum corneum is found at the base of the plantar in the foot, and the thinnest in the scrotum (Feldman and Maibach 1967).

A shear force, such as a scrape, removes the epidermis and the uppermost layer of the dermis (Tucker and Key 1982). Much larger force is required to shear the middle and lower layers of the dermis. A gel between the bundles of collagen provides compressibility to the skin.

The thermal threshold of pain in the skin occurs at 45° C (Hardy 1958; Wertheimer and Ward 1952). This temperature is critical for producing skin burn and scalds. Pain threshold is related to skin temperature. The extent of a burn injury is affected by the temperature of the surface and duration of contact. Duration of contact is subject to the location on the body on which contact occurs. Contact with skin containing few sensory nerves could occur without warning longer than contact with skin in the face and neck that contains many sensory nerves. The nature of the contact also influences the outcome. Contact can occur through grasping the hot surface, through momentary brushing or grazing against it, or through prolonged contact following an uncontrolled action, such as a fall onto the surface.

Depth, size, and location determine severity of a burn injury (Wilkerson 1985). The terms partial and full thickness burn have replaced the older terms: first, second, and third degree burn. First and second degree burns now are considered partial thickness burns. Partial thickness burns are less destructive. Under the old terminology, a first degree burn produced only redness of the skin (erythema) and no death of tissue. Second degree burn damaged the upper layers of the skin and produced blisters. Recovery would occur without the need for grafting. Third degree burns now are classed as full-thickness burns. Full-thickness burns require skin grafts. Deeper burn injury can involve underlying muscle, as well as the bone, blood and lymphatic vessels, and other structures.

Until recently, few individuals survived full-thickness burns covering more than 50% of the skin surface area (Wilkerson, 1985). This observation applied even when treatment occurred in specialized burn centers. Under proper care, few burns covering less than 15 to 20% are fatal.

The more commonly occurring outcome involving thermal energy is heat stress. Heat stress is the impact of the total net heat load on the body (ACGIH 1991). Heat stress causes an increase in the core temperature of the body. An increase to 41° C (106° F), as occurs during heat stroke, is almost certainly fatal unless cooling and medical attention are administered promptly. This temperature is considerably less than the 45 to 50° C applied to the skin needed to produce scalds and burns. The Threshold Limit Value was set to prevent the core temperature from rising above 38° C (100° F). Considerable compensatory action occurs with even a seemingly slight increase in core temperature above 37° C (98.6° F). This temperature represents consensus among work physiologists for minimizing risk of heat-induced illness.

The contribution to the heat load from the external environment is a serious concern in workspaces, such as metal clad structures, under some conditions. The surface solar intensity (insolation) at noon in early July in the northern U.S. from radiation at wavelengths less than 3.5 $\mu$m is about 3.3 J/cm$^2$/sec (Moran and Morgan 1989). The pyranometer is a standard instrument used for measuring the intensity of solar radiation striking a horizontal surface. Insolation varies with time of day, month of the year, latitude and presence of clouds. Some materials of construction of the boundary surface, such as conductive metals, absorb and then retransmit solar energy into the interior of the structure. They also can retain heat produced by processes used during work activity. Heat transfer from boundary surfaces into the interior of the structure occurs by the standard mechanisms: conduction, convection, and radiation. Heat transfer to the air inside the space can increase the interior temperature to a level considerably above that outside.

## Cold Surfaces and Cold–Related Injury

Cold is a serious workplace hazard in many parts of the world. Temperatures below the range of comfort challenge the ability of the body to accommodate. Fatal accidents involving humans almost always have resulted from exposure to low environmental air temperatures or immersion in water (ACGIH 1991). The most important life-threatening aspect from exposure to cold is a decrease in the deep core temperature of the body. Cold stress is the condition resulting from loss of heat from the body or a portion, such as feet, hands, limbs or head. The Threshold Limit Value for cold stress was set to prevent the deep core temperature from decreasing below 36° C (96.8° F). Considerable compensatory action occurs with even a seemingly slight decrease in core temperature from 37° C (98.6° F). Prolonged exposure to cold air or immersion in cold water at temperatures above freezing can cause this condition.

Some workspaces contain very cold surfaces. These can result from refrigeration processes used to maintain temperature, but more probably from weather conditions. Cold weather conditions pose an especial concern in small, unheated structures located in the outside environment. Mobility is restricted in small spaces. Restricted mobility can lead to reduced ability to maintain body heat through physical activity. Boundary surfaces can provide a shield from the wind, but they cannot protect against the effects of low temperature. Ventilation used in the space to control airborne contaminants can exacerbate heat loss by convection.

Cold surfaces can damage or destroy unprotected skin. The most familiar example is the "freezer burn" on the skin of a frozen chicken or turkey. Superficial or deep tissue freezing of human skin occurs at temperatures less than −1° C (30° F) (ACGIH 1991). Unprotected contact with air should not occur at wind chill equivalent temperature of −32° C (−26° F). Contact frostbite with surfaces can occur at temperatures less than −7° C (19° F). Rapid freezing can occur. Severe freezing injuries also can occur from contact with liquids, such as gasoline, alcohol, or some solvents that remain liquid at temperatures far below 0° C (Wilkerson 1986a). Liquids at these temperatures can absorb considerable heat during evaporation.

Frostbite results from formation of ice crystals in the fluids and underlying soft tissues of the skin (McDonald 1986). Frostbite damage results mostly from disruption of cellular activity caused by extraction of water and obstruction of the blood supply. Obstruction of the blood supply results from leakage of serum into the tissues (Wilkerson 1986a). Frostbite is ranked according to extent of damage. First degree frostbite includes freezing without blistering or peeling. Second degree frostbite includes blistering and peeling. Third degree frostbite includes freezing and possibly death of skin and deeper tissues (McDonald 1986). The most common sites of frostbite are the extremities: toes, fingers, ear-lobes, and the nose (Wilkerson 1986a).

In addition to frostbite, damage to tissue can occur from immersion injury (Wilkerson 1986b). Trench foot (immersion foot) results from prolonged wetting of the feet in near freezing temperatures. Prolonged cooling damages the tissues. Vasoconstriction also occurs to preserve heat in the core of the body. Greatest damage occurs to the nerves. This causes pain, prickling or tingling sensations (paresthesia) or the total anesthesia that may result. Damage also occurs to the skin and other tissues.

## Energized Electrical Conductors

A previous section examined biological effects due to 50/60 Hz electromagnetic fields. These are the energy fields that result when an electrical disturbance moves through a wire. These energy fields are at-a-distance from the affected individual. Of considerably greater concern and impact are electrical contact accidents. These usually entail contact with a bare, energized conductor.

Conductors are energized through failure to shut down the circuit of which the conductor is a part. Conductors become energized because switches are closed and fuses inserted in the circuit in which the circuit is a part. Conductors also become energized through unintended contact between an unrelated hot conductor and the conductor under consideration. Conductors become energized because of backfeed from portable generators connected into the distribution network. The latter usually follows the downing of above-ground power lines and the attempt to restore service to a dwelling or business. Conductors also become energized because of electrostatic and electromagnetic induction involving energized conductors. Electrostatic and electromagnetic induction do not require physical contact between the energizing and energized conductor.

High-voltage transmission and distribution lines lack insulation. Low-voltage lines usually contain an insulating covering, which is sometimes described only as providing weatherproofing and not electrical protection. The bared part of an insulated wire in contact with electrical fittings can make live other metal parts of the fitting.

As detailed in Chapter 1, electrical contact accidents occur during electrical work involving high and low-voltage energized circuits, and unintended contact with energized conductors during non-electrical work. The statistics indicate that fatal outcomes routinely occur with electrical contact involving standard North American household voltages of 110 to 120 V. Water or perspiration (a conductive liquid) was a factor in some of these accidents. In the U.S., about 10% of fatal occupational injuries result from electrocution. This translates into about 700 lives lost per year (CDC 1984).

The less common waveform disturbance with which we are familiar is found in DC (Direct Current) circuits. DC disturbances contain voltage only greater than zero or less than zero relative to some point of reference. A battery provides constant voltage. Hence, the waveform in a battery-powered DC circuit is a horizontal line at the indicated voltage. Other waveforms are possible, depending on how the electronics massage the signal. Waveforms could be half of a sine wave, square wave, saw-tooth, or flat. Direct current is provided in battery systems used in reliable power circuits, for energizing electrical generators, in steelmaking furnaces, electrolysis and electroplating, and for powering municipal transit systems, and railway locomotives, among other applications. DC is used in some electrical transmission systems. These operate as high as $\pm$ 500 kV and contain two conductors (a positive and negative). (Two wires are characteristic of DC transmission systems. Some operating circuits, such as street car systems, use only a single wire and rely on grounding through the wheels and the rails to complete the circuit.)

The more common waveform is the sine wave used in AC (Alternating Current) circuits. The voltage relative to a zero point of reference is positive for half the period of the wave and negative for the other half, but changes continuously. The common frequencies used in electrical transmission and distribution are 50 or 60 Hz. At one time, 25 Hz was the frequency used in these systems. Higher frequency AC systems are found in some electronic circuits. These circuits can operate at high voltages. Transmission systems in 50/60 Hz circuits operate as high as 750 kV. These systems use three wires, carrying disturbances having a phase difference of 120° relative to each other. That is, the position of the disturbances in the sine waves differ from each other in time by 120°. This difference is important in powering large electric motors and other heavy industrial equipment that operate under considerable load.

Electrocution means death by electricity, and not electrical shock in general (Bernstein 1991). Electrocution results from electric shock and severe tissue damage. Electric shock depends on the waveform of the electrical disturbance, frequency, current, duration of the contact, and pathway through the body (Casini 1998; Wolff 1999). Electrical injuries result from electrical shock (direct contact), arc flash (nonionizing radiation), blast effects (explosive) and burns (electrical, flash and flame).

**Electric Shock**
The most prevalent electrical contact is 60 Hz. The effects of normal 60 Hz electrical energy reflect the root mean square of the current and the duration of contact. Voltage is important through its involvement in Ohm's law with impedance (resistance) of the body for calculating current flow.

Resistance to current flow varies with the part of the body. Impedance, rather than resistance, is the term used to determine current flow in AC circuits. Both terms use the ohm ($\Omega$) as the unit of resistance. The body is a complex system. The impedance of each component through which the current passes differs. The various impedances include skin contact impedance, and impedances for bone, muscle, nerves, the heart, the brain, and various body tissues and fluids (Geddes and Baker 1967). The resulting impedance is a series combination of the skin contact and internal body impedances in the path taken by the current (IEC 1984). Table 3.4 summarizes body impedances (Geddes and Baker 1967; IEC 1984). Note that moisture, in particular conductive solutions, considerably reduces the impedance of the body.

Electrical contact produces a small entry wound and a much larger exit wound, similar to a gunshot (DeVaul 2002). The entry wound may be difficult to detect. The entry and exit wounds are the only visible signs of damage and injury. The path taken by current (Table 3.5) is a critical determinant about whether the electrical shock will be fatal. However, the current path is unpredictable in individual cases. A current path involving passage through the heart or brain considerably increases the risk that the shock will cause a fatal outcome.

The side(s) of the body that form the path are another critical factor in determining the outcome of the shock (Rekus 1997; Wolff 1999). The heart is usually located in the left side of the chest. A current path involving the heart is more likely to be fatal than one that does not. For example, a current path from right hand to the right foot is less likely to lead to fibrillation than a path from left hand to left foot. Fibrillation is the uncoordinated beating of the heart. During fibrillation, the heart is unable to pump blood through the body.

The current paths mentioned in Table 3.5 and illustrated in Figure 3.4(a), Figure 3.4(b), and Figure 3.4(c) have important implications for humans and animals in contact with energized surfaces. Voltage decreases with distance from the source. Grazing animals are especially sensitive to potentials caused by contact with energized surfaces. These contacts produce responses due to what are known as step and touch potentials. A path through the mouth to the ground through the hooves involving a potential difference creates a touch-step potential. This can produce a tingling sensation in the mouth resulting in refusal to eat or drink. This situation can be fatal to grazing animals. Similarly, a step potential can develop because of the relationship of the four legs of the animal.

A person standing on energized ground created as a result of an accident can become affected by step potential when the feet are separated. Keeping the feet together and hopping away from the source, or shuffling so that the heels do not pass the toes will prevent the occurrence of a current path through the body due to contact with surfaces at differing potential. Electric shock can produce a number of types of injury (Table 3.6) (DeVaul 2002).

## Table 3.4
## Body Impedances

| Condition/Description | Impedance ohm |
|---|---|
| Lowest resistance across the chest | 50 to 100 |
| Foot immersed | 100 to 300 |
| Hand immersed | 200 to 500 |
| Human body, excluding the skin | 200 to 1000 |
| Broken, ruptured skin | 500 |
| Minimum impedance between limb extremities, such as hand to hand, or hand to foot | 500 |
| Wetted skin, clothing | 1000 |
| Dry, intact skin | >100,000 |

Sources: Geddes and Baker 1967; IEC 1984.

Low Voltage (< 600 V)
Current flow through the body depends on adequate voltage. Voltage of 50 V or less at 60 Hz is regarded as not particularly hazardous (NFPA 1990). Few authenticated electrocutions have occurred at these voltages (Kouwenhoven 1949). Similarly, welding units regulate open circuit voltage to 80 V or less (UL 1987). Electrocutions at this voltage have occurred when the operator made good skin contact with the electrodes or established a current path across the chest. Hence, the lower cutoff of voltage capable of causing electrocutions is about 80 V. At higher voltages, especially household voltages (110 to 120 V), electric shock is routinely fatal.

## Table 3.5
## Current Paths through the Body

| Current Path | Example |
|---|---|
| Touch to touch | Hand to hand, hand to head, elbow to shoulder |
| Touch to step | Hand to foot, elbow to foot, hand to knee, head to foot |
| Step to step | Foot to foot, knee to foot |

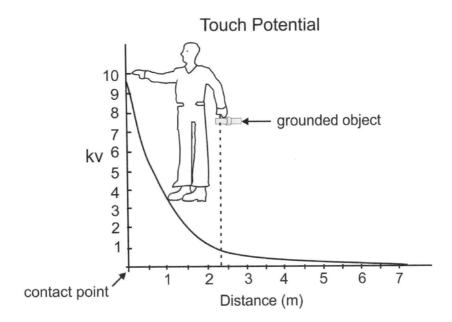

Figure 3.4(a). Touch potential results when the hands or head and hand contact surfaces at different potential. If the difference in potential is sufficiently large, current flows through the body. Current flow in the region of the heart is potentially lethal.

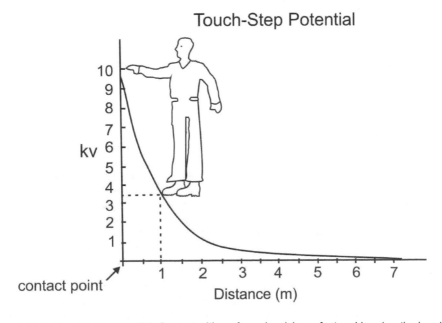

Figure 3.4(b). Touch-step potential. Contact with surfaces involving a foot and hand or the head can subject the body to different potentials. If the potential difference is sufficiently large, current flows through the body. Contact involving the left hand routes current through the chest near the heart.

## Step Potential

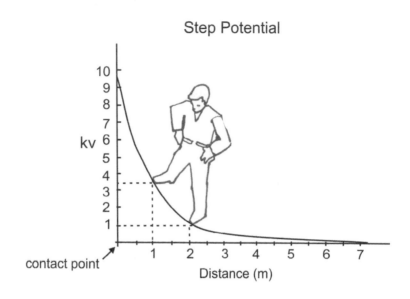

kv

contact point

Distance (m)

Figure 3.4(c).  Step potential. Contact of the feet with surfaces at different potential can lead to flow of current through the body.


Current that enters the body is capable of producing deleterious effects. These depend on factors already discussed, but also on voltage and contact time. Table 3.7 summarizes the effects of electrical current at 60 Hz for voltages less than 600 V.

The threshold of perception for a finger tapping contact at 60 Hz is approximately 0.2 mA. The perceptual level for half of a group of men is 0.36 mA, and 0.24 mA for half of a group of women (Kahn and Murray 1966). No uncontrolled startle reaction occurred when the shock current was below 0.5 mA, even though perception of a shock might occur. Women seem to have greater sensitivity to shock than men (Smoot and Stevenson 1968). As a result, the threshold level for a startle reaction was established to be 0.5 mA.

DC levels that produce the perception or startle reaction are about three times the current required at 60 Hz (Dalziel and Massoglia 1956). DC does not pose the hazard of AC. This is reflected in other measures of harm, such as let-go and fibrillation current (see following discussion).

Higher levels of current cause the hand to close involuntarily and to grasp the electrified contact that the palm or fingers were touching (Dalziel and Massoglia 1956). The individual cannot "let go" and remains trapped unless the power is turned off or is physically removed from contact with the circuit. Shocks at the let-go current level are quite painful, although usually not lethal. The let-go current threshold is a "go-no go" situation (Figure 3.5). Once immobilized or frozen to a circuit, an individual either will break the contact and live, or will not be able to break the contact. If the contact is not broken, the person's skin contact resistance may decrease, allowing the current to increase to the lethal level, ultimately causing death. To break contact and become free of the circuit, the individual must overcome involuntary muscular contraction while enduring the painful shock.

Let-go current for women is lower than that for men. Dalziel and Massoglia (1956) found that 0.5% of women could not let go at 6 mA, whereas 0.5% of men could not let go at 9 mA. As a result, 6 mA was set as the safe value for let-go current because shocks at this level will not freeze individuals to energized circuits. The Ground Fault Circuit Interrupter (GFCI) detects current imbalances greater

## Table 3.6
## Electric Shock Outcomes

Low-voltage contact wounds

High-voltage contact wounds from entry and exit of current

Burns

Respiratory difficulty due to swelling of the tongue

Respiratory damage due to inhalation of vaporized metal

Infectious complications

Injury to bone through falls, heat necrosis and muscle contraction (a cause of fracture, especially in the neck)

Injury to the heart (ventricular fibrillation, cardiac arrest

Injury to internal organs

Injury to the eyes (cataracts)

Source: DeVaul 2002.

than 6 mA between supply and return currents and disconnects the circuit in sufficient time to prevent electrocution (UL 2006).

Let-go increases exponentially with frequency, although there is little difference between 10 Hz and 100 Hz.

Let-go for direct current is the level at which test subjects refuse to let go because of the anticipated severe jolt (Dalziel and Massoglia 1956). Direct current causes severe pain when the circuit is completed or broken, but little while current is maintained. Let-go current is about the same in AC and DC systems.

Asphyxiation occurs when the passage of current through the chest cavity causes constant contraction of the chest muscles (Cabanes 1985). The individual cannot breathe or break the contact. Current levels at which this occurs are less than what is needed to trigger ventricular fibrillation.

Respiratory arrest can occur when current passes through the respiratory center (Lee 1966). The respiratory center is located in the medulla. (The medulla is situated slightly above a horizontal line extending from the back of the throat.) Thus, shocks passing from the head to a limb or between two arms could lead to respiratory arrest. Shocks at current levels above those that cause ventricular fibrillation can produce respiratory arrest, even though the primary path is not through the respiratory center (Hodgkin et al. 1973).

The usual cause of death due to electrocution is interference with control of heart muscle contraction (Bernstein 1991a). Ventricular fibrillation is the uncoordinated, asynchronous contraction of ventricular muscle fibers. Unconsciousness occurs in about 10 seconds because circulation ceases, and irreversible brain damage occurs in 3 to 6 minutes. An electrical shock passing through or across the chest can cause ventricular fibrillation.

For shocks shorter than a cardiac cycle, the electrical current causing fibrillation is large and must occur during the vulnerable period (Bernstein 1991a). Shocks lasting longer than a cardiac cycle lower the shock threshold current by causing premature ventricular contractions. Using these concepts, the following safe current limits were proposed: 500 mA for shocks less than 0.2 seconds and 50 mA for shocks longer than 2 seconds (Biegelmeier and Lee 1980). Cardiopulmonary resuscitation, promptly administered, can provide some circulation of oxygenated blood to the brain and heart until defibrillation occurs (NIOSH 1986a). The only way to restore heart rhythm is to use a defibrillator. A defibrillator applies a pulse shock to the chest.

## Table 3.7
## Effects of 60 Hz AC Electrical Current

| Reaction | Current mA |
|---|---|
| Perception of faint tingle. | 0.5 |
| Slight shock: disturbing, not painful, tingling in the affected area, slight discoloration of the skin, minor burns. Average individual can let go of energized conductor. Injury can occur from involuntary reaction. | 5 |
| Severe shock: painful, numbness or temporary paralysis of body parts, muscle spasms or loss of control, possible loss of consciousness, breathing difficulties; let-go range. | 6 to 25 (women) 9 to 30 (men) |
| Respiratory paralysis, stoppage of breathing. | 30 (men) |
| Deadly shock: extreme pain, respiratory arrest, severe muscle contractions. Individual cannot let go, second to third degree burns, possible destruction of nerves, tearing of muscle, shattering of teeth and breakage of bone. Death possible. | 50 to 150 |
| Fibrillation threshold 0.5%. | 75 (men) |
| Fibrillation threshold, 99.5%. | 250 (men) |
| Ventricular fibrillation, muscle contraction and nerve damage after short duration. Death likely. | 1000 to 4300 |
| Cardiac arrest, severe burns, probable death. | >10,000 |

Sources: Lee 1971; after OSHA 1986.

Immediate symptoms shown by victims of electric shock include confusion, amnesia, headache, cessation of breathing, cessation of heartbeat, and burns (Wolff 1999). Survivors of injuries displaying these symptoms in former times were declared cured. Recent study indicates that symptoms can persist into the longer term. Subacute to subchronic symptoms can include paralysis, muscular pain, abnormal vision, swelling, headache, and cardiac irregularities. Chronic (long-term) symptoms can include paralysis, impaired speech or writing, loss of taste, and numerous other disorders.

Pulse and impulse-type shocks also can cause damage to the heart. These shocks have a time duration much shorter than the heart cycle. Pulse and impulse-type shocks result from capacitor, electrostatic, and lightning discharges. The hazard due to an impulse-shock results from the electrical energy in the discharge. Any pulse shock having an energy content of 50 J probably is hazardous. Shocks below 0.25 J, while disagreeable, probably are not hazardous (Dalziel 1971). The annoying electrostatic discharge shock produced by walking across a carpet is about 10 mJ. Defibrillators have a maximum output in the range of 200 to 400 J (Bernstein 1991a).

The DC current needed to induce ventricular fibrillation during short contact (0.2 seconds) is the same as AC (both about 500 mA) (Dalziel and Massoglia 1956). For shocks lasting 0.5 seconds or

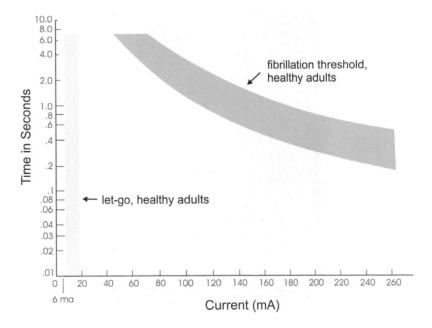

Figure 3.5. Time-Current Relationships. Time and current control the severity of an electric shock, and the manner of preventing this harm. This is the basis of the design of the ground fault circuit interrupter. (Modified from Dalziel and Massoglia 1956.)

longer, the DC current needed is 400 mA. The corresponding value of AC current is 75 mA. This again illustrates the greater hazard posed by AC current.

High Voltage
High voltage (above 600 V) produces different effects from those discussed thus far (Bernstein 1991). The more likely result from passage of current through the heart is asystole, rather than fibrillation. Asystole is cardiac standstill; that is, the heart does not beat. Current flow above 1 A through the chest cavity can cause asystole. Such high currents are associated with high voltage. Hence, asystole rather than ventricular fibrillation is the usual cause of death in accidents at voltages greater than 1,000 V. Unlike ventricular fibrillation, asystole may revert to a normal heart rhythm, once the triggering voltage is removed. As a result, individuals have survived asystole following exposure to high-voltage, high-current shock, while others have died from ventricular fibrillation following a low-current, 120-V shock. This is the basis for concern expressed in NIOSH Alert documents about 110-120 V electrical contacts (NIOSH 1986a, NIOSH 1986b).

**Arc Flash**
An electric arc is an intense source of ultra-violet, visible, and infra-red energy. An arc flash (Figure 3.6) can result without contact between the individual and the energized conductor. An arc flash lasts from 0.01 to more than 1 second. Cataracts are one result from exposure to ultraviolet energy during an arc flash (Bussmann 2000). The heat generated by an electric arc flash can produce temperatures up to 19,000° C (35,000° F) (Lee 1982). This temperature is almost four times that at the surface of the sun. This temperature can liquefy and vaporize metal, producing a rapidly expanding plasma of ions. These emissions can cause extensive damage and serious injury. Their full potential for injury and damage have yet to be determined.

Figure 3.6. Arc flash. An arc flash and the accompanying arc blast caused when a circuit is disconnected under load is one of the most fearsome and feared events during electrical work.( Image reproduced with the permission of Salisbury by Honeywell © 2011 Honeywell International Inc. All rights reserved.)

Blast Effects

A high-energy arc fault generates a blast wave (Lee 1987; Floyd et al. 2001). This blast wave has saved many individuals by hurling them from the blast zone. Pressures generated by the blast could be 3.5 MPa (500 lb/in$^2$) or more measured at 0.6 m (2 ft) from the arc. Blast pressure results from liquefying and vaporizing metal, and expanding atmospheric gases. Copper, for example, expands 67,000 times its volume during this process. The blast wave can emanate at supersonic velocity. This rapid expulsion of an individual from the blast zone can cause other injuries including concussion, hearing loss, and the traumatic injury resulting from being hurled.

Burn

As mentioned, temperatures generated in an arc are extremely high. At least 80% of the thermal radiation is available to cause thermal burns (Lee 1982; Wolff 1999). Temperature exceeds 96° C (205° F) in 0.1 second, in which time irreversible skin burns occur.

Surface burns on the body can result from the heat produced by the arc, heating at the point of contact by a high-resistance contact or large currents, spatter of molten particles from damaged conductors or equipment, and from burning of fabrics worn at the time of the accident or the skin itself (Bernstein 1991; Floyd et al. 2001; Wolff 1999). A shock from contact with energized equipment can cause external or internal burns. The magnitude of the current, and type and duration of contact determine the severity of the injury. Electrocution burns progress from the interior of the body to the exterior.

**Lightning**

Lightning is a form of direct current electrical energy capable of causing serious injury and death (Grim 2002). A lightning strike can generate 200,000,000 V and current flow of 20,000 to 40,000 A. Lightning can overcome communication in the nervous system causing the heart to stop and the brain to "short out". After the strike, the heart may begin normal function, although recovery of breathing requires more time. Numbness in the limbs and feathery patterns in the skin are common symptoms. The latter result from damage to capillaries beneath the skin. These mark the path of current through the surface of the body. Other short-term symptoms include memory loss, vertigo, and deafness.

Long-term problems can include cataract formation in the eyes. Long-term problems can affect any organ system, but the nervous system seems especially at risk both physiologically and psychiatrically. Neuropsychiatric complications involving the nervous system are the most devastating. These include sleep disturbance, anxiety, pain, nerve damage, fear of storms, and depression. Problems can appear years afterward.

## Sound

Sound is the form of energy resulting in a medium (solid, liquid, or gas) from the transfer of kinetic energy from a surface that is vibrating. That is, unlike the transfer of nuclear and electromagnetic energy, the transfer of the disturbance requires the presence of atoms or molecules. The denser the packing of atoms or molecules, the more effective is the transfer of energy. The disturbance moves through the medium by causing a coherent change in position of some of the atoms or molecules. The coherence of the change transfers to subsequent particles as the disturbance moves through the medium. Note that the disturbance moves through the medium, and not the particles through mass displacement. The particles remain in their average position.

### Infrasound

Infrasound refers to frequencies below those detectable by the human ear. This generally means frequencies less than 20 Hz. Infrasound has only recently become the subject of attention by standard-setters (ACGIH 2001b). The documentation indicates that infrasound down to 1 Hz can be heard or at least sensed. The sensation is usually described as a "chugging" or "motor-boat" sound.

Damage to hearing is considered to be a definite possibility. This occurs through damage to the eardrum, and possibly temporary and permanent threshold shifts (ACGIH 2001b). Pain or pressure reported in the middle ear is attributed to displacement of components beyond tolerance limits.

Resonance of the whole body occurs at 4 to 8 Hz. This is a result of compression and rarefaction of airspaces within the body. This is the main response of concern due to exposure to infrasound and forms the basis for the TLV for whole body vibration.

### Audible Frequencies

Audible frequencies range from 20 Hz to 20,000 Hz. The upper limit of this range depends on age, gender, and history of exposure to noise. Generally, youths and women have sensitivity to the highest frequencies in the audible range. (Berger et al 1986, Berger et al 2000).

The main effect of acute exposure to noise is the temporary threshold shift (Hétu and Parrot 1978). The temporary threshold shift (TTS) is a temporary shift in the sensitivity of the auditory system to sound signals at different frequencies (Figure 3.7). The shift appears following noise exposure and disappears following a prolonged period of relative quiet, optimistically provided during sleep and leisure hours.

The temporary threshold shift results from the manner in which the detection mechanism of the ear functions (Ward et al. 2000a). The detection mechanism functions through vibrations imparted to liquid fluid in which is immersed a frequency-sensitive membrane. Liquid fluids are not compressible. Hence, inward movement of the oval window caused by movement of the bones in the middle ear causes outward movement of the round window. At the same time, the basilar membrane that forms the separator in the middle of the channel, along which the fluid moves, deflects in response to the disturbance caused by movement of the fluid. The amount of this deflection depends on distance from the oval window. Microprojections in sensor cells ("hair cells") that are immersed in the fluid deflect against the tectorial membrane. This deflection causes a neural signal to be transmitted to the brain. Back and forth deflection beyond a particular magnitude causes reduced response from the hair cells (fatiguing). This fatiguing translates into reduced signal strength to the brain (threshold shift), and the

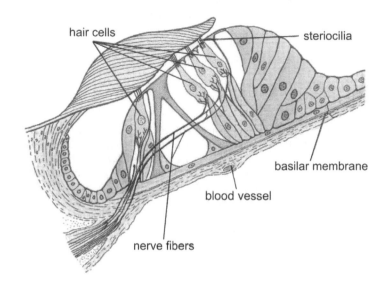

Figure 3.7. Hearing apparatus. Vibration of the end and round windows induces vibration in the fluid. Vibration of the fluid induces vibration in the membranes, which, in turn, induces vibration in the sensory cilia. Vibration of the cilia cause a neurological signal. (Modified from Rasmussen 1943.)

need for additional deflection to produce the same response. Restoration of ambient sensitivity requires a period of quiet.

Permanent threshold shift results from destruction of the hair cells. This can result from prolonged exposure to high levels of continuous noise or from brief exposure to exceedingly loud sources (impactive or impulsive noise).

Temporary threshold shift received considerable attention in previous years (Ward 1969). Growth in TTS (measured in dB) is almost linear with the logarithm of time. TTS begins at noise levels of 80 dB and increases linearly to at least 130 dB. This means that TTS produced between 80 and 85 dB will be about the same as that produced between 90 and 95 dB. Considerably higher temporary threshold shifts than those mentioned above can occur. Moderate TTS decreases exponentially with time during the period of recovery. At these levels, complete recovery is expected after 16 hours of quiet. When the TTS exceeds 40 dB, recovery becomes linear with time. As a result, full recovery may require days or weeks in quiet conditions. Hence, TTS of 40 to 50 dB is regarded as a threshold for prevention of permanent damage.

TTS depends on frequency. As indicated by observation, TTS is less for equal levels of noise occurring in low frequencies compared to high frequencies. That is, a rumble is less damaging than a screech. TTS from noise energy concentrated in a narrow range of frequencies occurs at half to a full octave higher.

Intermittency is another important factor in development of TTS. TTS is proportional to the fraction of the time that the noise is present. That is, short bursts are tolerated better than continuous noise.

In the long term, an exposure to noise producing a daily TTS of a particular magnitude is expected to produce a permanent threshold shift (loss) of the same magnitude (Ward et al. 2000b). Noise exposures in the range of 85 to 95 dBA were shown to produce TTS values of 10 dB or lower at frequencies of 1000 Hz or lower, 15 dB at 2000 Hz, or 20 dB at 3000 Hz in a person whose hearing is normal (Ward 1966). The actual TTS in a particular circumstance depends on the actual spectrum of the noise source.

The (U.S.) National Academy of Sciences/National Research Council Committee on Hearing Bioacoustics and Biomechanics (CHABA) examined the problem of temporary and permanent threshold shift (Kryter 1963). CHABA offered the view that the limits below which handicap in speech reception would not occur are:

- not greater than 10 dB at frequencies up to 1 kHz
- 15 dB at 2 kHz
- 20 dB at 3 kHz and above

The temporary shift in hearing sensitivity compounds the effect of permanent losses incurred following previous exposure to loud noise plus loss due to aging plus loss due to use of hearing protection. The net result for an individual can be considerable loss in hearing sensitivity. The impact of this loss can result in the inability to hear warning signals.

Vibrating objects and surfaces cause minute changes in atmospheric pressure. These changes impinging on the eardrum create mechanical movement, that, in turn, leads to vibrations in the fluid in the inner ear that induces motion in the hair cells, as mentioned previously. The energy radiated by the vibrating object or surface usually is expressed in Watts, the unit of power. The Watt is the "use" of 1 Joule/second ($W = J/s$). Acoustic power in human experience spans a wide range from nanoWatts or nW ($10^{-9}$ W) to megaWatts or MW ($10^6$ W).

Sound pressure level (Magrab 1975) at any point, r, from a point source is given by:

$$L_p = L_w + 10 \log \left[ \frac{Q}{4\pi r^2} + \frac{4}{R} \right] \quad \text{dB re 20 } \mu\text{Pa} \tag{3-14}$$

$$R = \frac{S\alpha}{1-\alpha} \tag{3-15}$$

where:
$L_p$ = the sound pressure level, dB (re 20 $\mu$Pa).
$L_w$ = the sound power level, dB (re 1 pW).
$Q$ = a geometry factor: $Q = 1, 2, 4,$ or 8, for sphere, hemisphere, quarter sphere, and eighth sphere.
$r$ = the distance from the source, m.
$R$ = the absorption constant of boundary surfaces.
$S$ = the surface area of the boundary surface, m$^2$.
$\alpha$ = the absorption coefficient for the boundary surfaces. In this case, the surfaces are assumed to be the same material; otherwise, the weighted average is used.

Note in Equation 3-14 that sound power level, $L_w$, and sound pressure level, $L_p$, are both expressed in decibels (dB). Yet, the reference units for comparison are pW (picoWatts or $10^{-12}$ W) for sound power level and $\mu$Pa (microPa or $10^{-6}$ Pa) for sound pressure level. The distance factor, r, relates to point sources suspended in space. There is a corresponding equation for decrease in $L_p$ with distance for line sources. In the latter case, $L_p$ decreases with the first power of r; that is, $r^1$ rather than $r^2$ as in the case of point sources. The R factor accounts for absorption of sound energy by materials of construction of the space.

The net result from Equation 3-14 is that, in a structure constructed from highly reflective material, the sound pressure level approaches the sound power level with losses due only to distance from the source. This is the case in many confined spaces where the sound pressure level receives contributions from the direct and reflected paths.

Table 3.7 provides some examples of sound power and sound power levels, and corresponding sound pressure levels. The threshold of pain occurs at sound pressure levels around 130 dB. This level is measured at the position of the receiver versus the power radiated by the source.

**Table 3.7**

**Acoustic Power**

| Source | Sound Power W | Sound Power Level dB re $10^{-12}$ W | Sound Pressure Level dB re 20 μPa |
|---|---|---|---|
| Saturn rocket | 25 to 40 x $10^6$ | 195 | |
| Ramjet engine | $10^5$ | 170 | |
| Turbojet engine (7000 lb of thrust) | $10^4$ | 160 | |
| Four-engine propeller aircraft | $10^3$ | 150 | |
| | $10^2$ | 140 | |
| 75-piece orchestra, peak output | 10 | 130 | |
| Large chipping hammer | 1 | 120 | |
| "Blaring" radio | 0.1 | 110 | |
| Auto on the highway | $10^{-2}$ | 100 | |
| Voice: shouting | $10^{-3}$ | 90 | 90, measured at 1 m |
| | $10^{-4}$ | 80 | |
| | $10^{-5}$ | 70 | |
| Voice: normal conversation | $10^{-6}$ | 60 | 60, measured at 1 m |
| | $10^{-7}$ | 50 | |
| | $10^{-8}$ | 40 | |
| Voice: very soft whisper | $10^{-9}$ | 30 | |

Source: Peterson 1980.

The reference pressure of 20 μPa is the threshold of human hearing in the frequency range of 1000 to 4000 Hz. This produces a value in the frequency range of 1000 to 4000 Hz of 0 dB in the calculation of sound pressure level from sound pressure.

Low-frequency audible sound ranges from 20 to 80 Hz (ACGIH 2001b). Sound in these frequencies can cause resonance effects involving body parts containing airspaces. Compression and rarefaction occur in air contained in these cavities. This can lead to displacement of tissue. Air-containing spaces in the body occur in the lung and external and middle ear. Tickling and choking in the throat leading to coughing and the sensation of nausea occur at frequencies just above 20 Hz. Frequencies in the 20 to 50 Hz range cause vibration of the chest wall, gag sensations, and changes in respiratory rhythm and fatigue. Sounds in the 50 to 60 Hz range cause resonance in the chest. This can cause annoyance and discomfort. Frequencies in the 50 to 100 Hz range cause headache, choking, coughing, visual blurring, and fatigue.

**Ultrasound**

Ultrasound includes frequencies above the normal audible range (greater than 20 kHz). Ultrasound has received little attention in the mainstream industrial hygiene literature (Wiernicki and Karoly 1985). Upper limits for ultrasound applications range to 10 Hz in gases and $10^9$ Hz in liquids and solids. Intensities range as high as $10^5$ W/cm$^2$. Ultrasonic energy is used in a wide variety of industrial applications (INIRC 1984). These include cleaning, drilling, mixing, and emulsification. Ultrasound emitted by intrusion alarms and some computer equipment is audible to some people.

Ultrasound is also used in medical applications (O'Brien 1992a; Wiernicki and Karoly 1985). Medical ultrasound equipment operates in the range 1 to 15 MHz. Therapeutic equipment operates around 1 MHz. Diagnostic equipment used for abdominal examinations operates in the range from 2.25 to 7.5 MHz where deep penetration is required. For situations where shallow penetration is required, frequencies of 7.5 to 15 MHz are used.

Most common industrial applications occur in the range 20 to 50 kHz, although some extend to 100 kHz. Sources utilize continuous and pulsed output. The main effects of high-energy exposure to ultrasound include heating of tissue, mechanical disruption of membranes, formation of gas bubbles, and damage to tissues and organ systems. Many of these high-energy effects occur in the brain and central nervous system.

Adverse health effects have been reported at frequencies up to 100 kHz. Nearly all occur below 50 kHz (INIRC 1984). These include headache, fatigue, nausea, tinnitus (ringing in the ears), pressure in the ears, tingling in the scalp and cheeks, or an unpleasant sensation of fullness. Some data indicate hearing loss at high levels. The ear is the main target organ for airborne ultrasound. Some is absorbed by the skin and the hair. The basis for the INIRC guidelines and the Threshold Limit Value is prevention of possible hearing loss from sub-harmonics of those frequencies (ACGIH 1991; INIRC 1984). Sub-harmonics of these frequencies can cause temporary threshold shifts (Parrack 1966). The INIRC guidelines apply to third octave frequency bands ranging from 20 to 100 kHz (INIRC 1984). The TLV covers the range in third octave bands from 10 to 50 kHz (ACGIH 1991).

Exposure to airborne ultrasound is considerably less hazardous than contact exposure, since air readily attenuates transmission. As well, the signal is directional and easily stopped by barriers constructed from materials normally encountered in workplaces in furniture and furnishings.

When propagated through tissue, an ultrasonic wave loses energy through absorption or scattering (O'Brien 1992b). These losses can produce thermal effects, cavitation, and mechanical disruption in tissue (Wiernicki and Karoly 1985). The thermally induced lesions are similar to those produced by a resistance wire implanted into tissue. Absorption involves conversion of energy to heat. Scattering involves change of direction. The American Institute for Ultrasound in Medicine review of bio-effects of mammalian exposure to ultrasound concluded that no significant bio-effects occur in mammalian tissues exposed *in vivo* at unfocused exposures below 100 W/cm$^2$ or 1 W/cm$^2$ for focused exposures (AIUM 1987). (Low frequency ultrasound is the most penetrating.)

Ultrasonic disturbances can cause cavitation in liquids (O'Brien 1992a). Cavitation results from vapor and consequent bubble formation due to reduced pressure. Cavitation is observable as a boiling action that produces considerable audible noise. Cavitation in tissue occurs at increasing frequencies starting at 10 kHz, which is in the audible region. Cavitation can produce heating, mechanical stress, and ionization (Dunn and Fry 1971). Cavitation can produce shear stresses leading to stretching, twisting and rupturing of membranes, and production of free radicals. This occurs at frequencies around 1 MHz.

Greatest human exposure to ultrasound occurs during medical procedures, particularly diagnostic ones (ter Haar and Hill 1989). Ultrasonic devices utilize continuous output or short pulses (1 to 10 $\mu$s). There is evidence to support a hypothetical risk of injury due to heating and cavitation at levels of energy used during diagnostic procedures, despite lack of examples during actual use (O'Brien 1992b). There is an especial concern regarding fetal exposure.

Therapeutically, ultrasound, like radiofrequency and microwave energy, is used to produce deliberate internal heating in tissues. Most units operate around 1 MHz. The range extends from 0.75 to 3 MHz (Bly and Harris 1992). Ultrasound produces heat at moderate depth in a localized volume. Ultrasound is also used surgically to treat urinary kidney stones and gallstones. Another application is the treatment of glaucoma through generation of focal lesions. Additional medically oriented applications include ultrasonic baths for cleaning patients. Non-occupational sources include ultrasonic humidifiers and ultrasonic occupancy sensors.

Some of these effects could occur in individuals working around sources of airborne ultrasound. More serious are the potential effects that could result from physical contact with these sources. Physical contact could occur accidentally or through deliberate grasping, or through a coupling medium, such as water.

## Vibration

Vibration is the periodic back and forth movement of an object. A vibratory system contains a means for storing potential energy (a spring), a means for storing kinetic energy (mass or inertia) and a means by which energy is gradually lost (a damper) (Magrab 1975). Vibration of a system involves alternate transfer of energy between kinetic and potential forms. A system containing damping dissipates some energy during each cycle. For vibration to continue, replacement of energy from an external source must occur. A single physical structure can perform all three of these functions. The simplest vibratory system contains a mass attached by a spring and a viscous damper to an immovable support.

Vibration in machinery occurs for two basic reasons: out-of-balance forces, and impact and shock (Bramer et al. 1977). Out-of-balance forces affect rotating and reciprocating machinery. Rotating machinery vibrates when out of balance. Reciprocating machinery is inherently out of balance. Some machines, such as tampers, vibrating screens, conveyors, and bowl feeders, produce and utilize vibration as a deliberate part of their operation.

Imbalance in rotating equipment can occur both statically and dynamically. A simple rotor is statically imbalanced when the center of mass of the rotating element does not turn on the same axis as the rotating assembly (Jones 1994a). This creates an out-of-balance force that rotates at rotor speed. In practical terms, this means that the rotating element will tend to stop in the same place when allowed to stop from free rotation.

Dynamic imbalance only occurs during rotation (Bramer et al. 1977; Jones 1994a). One condition caused by rotation is skew. Skew causes alternating forces on bearings. Skew can result from inadequate hold-down of bearings. Flexure of the shaft will cause bowing at a particular rotational frequency. This can lead to permanent bending of the shaft. Aerodynamic imbalance of fan blades can cause both static and dynamic imbalance. This can result from improper manufacture, but also can result from erosion of the blades and loading with particulate material. Dynamic imbalance of fan blades also can result from overly ambitious application of grease and incomplete cleaning. Impellers in pumps wear unevenly or suffer damage through passage of foreign materials. Sheaves (pulleys) used on motors can be out of round. A cracked or broken weld in a small component can cause dynamic imbalance.

Additional causes of dynamic imbalance include misalignment of shafts and couplings, belts and sheaves, and chains and sprockets. Misalignment is responsible for 70 to 75% of vibration problems (Jones 1994a). Misalignment causes abnormal wear on bearings. Misalignment can be parallel or angular. A small face-to-face misalignment of 0.025 cm (0.01 in) can result in considerable cost in terms of wear and operating expense. Looseness in parts is another major cause of vibration. Looseness can result from loosened nuts and bolts, broken bolts, worn and undersized shafts, and installation of incorrect or inappropriate components.

Reciprocating equipment adds inherently unbalanced reciprocating masses to static and dynamic imbalances (Bramer et al. 1977). This leads to complex forces and couples at various frequencies. The number of cylinders and crank arrangement in engines and reciprocating compressors influence forces and couples. Six- and eight-cylinder, in-line engines and compressors possess inherently balanced forces and couples.

Vibrating machinery and equipment transmit this motion to structures, and to humans in contact with either of them. Building structures, particularly steelwork, are not rigid and act as sources of vibration. Building structures can be put into vibratory motion by equipment contained within, as well as external sources. The latter include vibration from subways and other underground railways, pile driving, and roadway compacting by vibratory rollers and tampers. These problems are particularly troublesome when the resonant frequency of the ground is similar. Sway motion induced by wind also can be involved.

Transmission of vibration through building structures occurs by longitudinal, transverse, torsional, and flexural waves (Bramer et al. 1977). Longitudinal waves act along the axis of the structural member (such as a beam). Transverse waves act perpendicularly. These can be induced by up and down movement of machinery and equipment. Torsional waves involve twist, shear and rotational forces. Flexural waves involve "rippling" of the surface of components. Flexural waves pose the greatest concern for building-borne vibration, as flexing is the easiest way for materials to move.

Resonance is an extremely important issue. Resonance is the large oscillation set up in a spring mass system at the natural frequency (Bramer et al. 1977). The natural frequency is the frequency at which a system vibrates when disturbed from rest and then released. That is, this is the frequency at which transfer of energy from the exciter to the receiver is most efficient. While an idealized system has one natural frequency, real-world systems have several. The number of natural frequencies depends on the complexity of the system. Resonance becomes a serious problem with equipment when the operating speed is close to the natural frequency (Jones 1994b). The solution to this problem is design, so that equipment does not operate within 20% of the natural frequency.

Bridges are especially prone to resonance phenomena. The classic destructive force on bridges was caused by soldiers marching synchronously across the structure. The tramping of feet onto the bridge surface occurred close to the natural frequency of the structure. The most famous recent collapse of a bridge structure involved the Tacoma Narrows Bridge. The deck twisted and turned during a high wind storm (Jones 1994b; Petroski 1992). The wind blew through the cables inducing a vibration. The design of the bridge was unconventional, employing stiffened girders rather than the deeper open trusses of conventional designs. This silhouette was dramatic and graceful, but very flexible in the wind. The roadbed flexed during construction and following its opening for service. The deck structure acted like an airplane wing subjected to uncontrolled turbulence.

Another example of the destructive outcome from resonance involved a turboprop aircraft (Jones 1994b). The speed of the engines was a forcing frequency that fed into to the roots of the wings. This caused a flexing action that led to severance of the wings.

Vibration causes numerous effects on machinery, equipment, and structures. These include mechanical failure resulting from excessive stress, fatigue, and destructive impacts. Other effects include excessive wear and noise, and inadequate performance. Humans and equipment, machinery, and structures have different tolerance to vibration (Peterson 1980). Vibration levels that are safe for a vehicle can be uncomfortable, annoying, or even dangerous for the occupant. Similarly, vibration in buildings that may have no impact on the structure can interfere with human performance and cause fear of collapse.

Humans are affected by vibration through standing, sitting, and lying on surfaces. Touching and gripping vibrating tools and equipment transmits vibration through the hands and arms. The human body is modeled by a mechanical system containing a number of mass-spring damper subsystems, each having characteristic resonance frequency (Table 3.8) (Guignard 1965).

Whole-body vibration is a generalized stressor that affects multiple organs simultaneously (Helmkamp et al. 1984). The focus of much of the concern is chronic exposure, as experienced by workers engaged in an occupation for long periods of time. This could amount to more than 40,000 hours over a period of 30 years. Acute studies show varying and inconsistent effects on blood pressure, pulse rate, and cardiac output. The direction and magnitude of these changes depend on frequency, amplitude, and direction of the vibration. These effects appear to be transient in nature. Epidemiological studies of long-term exposure indicate increases in disorders of the musculoskeletal, circulatory, and digestive systems. Driving vehicles as an occupation may be a risk factor in development of an acute lumbar herniated disc.

## Table 3.8
## Resonance Frequencies of Body Parts

| Body Part | Frequency Hz |
|---|---|
| Legs (knees flexed) | 2 |
| Thoracic-abdominal complex | 3 |
| Shoulder girdle | 4 to 5 |
| Abdominal mass | 4 to 8 |
| Upper arm-shoulder | 5 |
| Spinal column | 10 to 12 |
| Lower arm | 16 to 30 |
| Head | 20 |
| Legs (rigid posture) | >20 |
| Head | 25 |
| Eyeball | 30 to 80 |
| Hand grip | 50 to 200 |
| Chest wall | 60 |

Source: Guignard 1965.

Exposure to hand-arm vibration is associated with an irreversible circulatory condition known as Raynaud's disease, vibration white finger, or more recently, as hand-arm vibration syndrome (Wasserman 1996). The incidence of this effect is related to frequencies ranging from 40 to 125 Hz (Guignard 1965). Tingling and numbness in the fingers characterize the early stage and also acute exposure. This leads to blanching of the fingers on exposure to the cold, and impairment of touch, dexterity, and grip strength. Severe cases involve occlusion of blood vessels of the fingers and tissue necrosis and gangrene (Pelmear et al. 1992).

## Explosion

An explosion is an event characterized by the sudden release of energy (Cruice 1991).The violence of an explosion depends on the rate of release. Explosions result from chemical and physical processes. Chemical explosions usually are differentiated as deflagrations or detonations according to the rate of release of energy. Detonation of an explosive and rapid combustion of a cloud of flammable vapor or gas are examples of explosions resulting from chemical processes. Rupture of a container caused by overpressure or structural failure, and rapid vaporization and uncontrolled escape of a liquid or vapor or gas are examples of explosions resulting from physical processes. The effects produced by an explosion occur due to at least one of the following processes: production of gases, rapid expansion of gases, and motion of projectiles.

One of the main characteristics of an explosion is the shock wave or blast wave. The shock wave is characterized by the rapid change in atmospheric pressure caused by rapid expansion of gases. The shape of the pressure profile depends on the type of explosion. The shock wave produced during an explosion moves outward at subsonic or supersonic velocities. Again this depends on the type of explosion. A subsonic shock wave results from a deflagration, one type of explosion involving gas/vapor-air mixtures. During a deflagration, the pressure equalizes at the speed of sound following the overpressure. The pressure drop across the flame front is relatively small. Supersonic propagation of the flame from the source of ignition occurs during a detonation explosion. During a detonation, the rate of pressure equalization at the flame front is less than the rate of propagation. The result is considerable pressure drop across the flame front (Zabetakis 1965).

During an explosion atmospheric pressure rises almost instantaneously to a peak and gradually decreases to the starting level. As the shock wave travels outward, the height of the peak of the pressure wave decreases at the shock front with increasing distance. The shock wave in air usually is referred to as a blast wave because it may be accompanied by a strong wind. The peak wind velocity behind the shock front depends on the peak overpressure. Overpressure generated by the blast wave can injure or kill people and damage or destroy equipment and buildings. The overpressure in the shock wave is followed by a region of negative pressure, or underpressure. The underpressure usually is quite weak and usually does not exceed about 100 kPa (760 mm Hg). The negative pressure is caused by the inertia of air accelerated in the direction of the blast. When the positive pressure portion passes, this air mass is "pulled back" by the still air behind it. The high pressure components of the shock wave move outward at higher velocities. Initially, the shock wave from a detonation travels at supersonic speed. As the intensity of the wave subsides, it becomes sonic (Bodurtha 1980; Cooper 1996; Lees 2001).

The ears and lungs are the parts of the body most vulnerable to the shock produced by the air blast because of the presence of air in these structures (Bowen et al. 1968). Spalling and tearing occur at the air-tissue interface. Ear and lung responses depend on body orientation, as well as the impulse (the area under the overpressure curve) and pressure. Estimated overpressure (square wave) to produce 50% lethality in humans is 420 kPa (60 lb/in$^2$). In practical terms, this means that an individual in the open whose long axis of the body is perpendicular to the direction of propagation of the blast wave and is facing in any direction, standing or prone, is more susceptible top peak over pressure than when the long axis of the body is parallel to the direction of the blast wave(as when lying on the ground). Reflecting surfaces exacerbate the situation even further. The ears are more susceptible to injury than the lungs. Damage to the ears includes small tears or rupture of the eardrum. This is repairable. At higher levels of overpressure, damage to the eardrum is permanent. At higher overpressure than this, damage to the inner ear occurs.

Table 3.9 provides additional information about the effects of blast-produced overpressures (after Cruice 1991; Glasstone 1962; Robinson 1944); that is, pressures above atmospheric pressure (101 kPa at sea level). Slight overpressure can demolish ordinary buildings, as indicated in the table. Not surprisingly, these types of explosions have demolished even reinforced concrete buildings.

## Pressurized Fluids

Pressurized fluids have countless applications. Fluids in this context include both liquids and gases. At sufficient pressures, these fluids, both liquid and gaseous are injectable through the skin into the body. A practical application for the injection of liquids into the body is administration of vaccine during mass vaccinations. This eliminates the need for the hypodermic needle.

### Gases

Common pressurized gases encountered in workplaces include air, nitrogen, argon, and helium, and sometimes oxygen and carbon dioxide. Compressed air has many applications, including blowing off dust and debris from surfaces, and providing the motive force for spray coating and abrasive blasting equipment. Concerns about injection of air and particles contained in the air are reflected in regulations

## Table 3.9
## Effects of Overpressure

| Overpressure kPa | Structural Element | Failure Condition |
|---|---|---|
| **Physical Effects** | | |
| 3.4 to 6.9 | Glass windows | Shattering, occasional frame failure |
| 6.9 to 14 | Corrugated asbestos siding | Shattering |
| | Corrugated steel or aluminum siding | Buckling, attachment failure |
| | Wooden exterior siding panels; standard house construction | Failure at main connections panels blown in |
| 14 to 21 | Concrete or cinder block wall; 200 to 300 mm thick; not reinforced | Shattering of the wall |
| 48 to 55 | Brick wall; 200 to 300 mm thick; not reinforced | Shearing and flexure failure |
| >70 | Structure | Total destruction |
| 1930 | Ground | Crater formation |
| **Physiological Effects** | | |
| 15 | Knockdown | Knockdown threshold |
| 34 | Eardrum | Rupture threshold |
| 207 to 255 | Lung | Threshold of damage |
| 240 | Human body | Threshold of fatal injury |
| 345 | Human body | 50% Lethality |
| 450 | Human body | 99% Lethality |

Sources: Cruice 1991; Glasstone 1962; Robinson 1944.

controlling pressure used in blow-off guns. Air or other gases injected into the skin can enter the blood as a gas bubble. A gas bubbles can pass through the circulatory system to the brain or the heart where an embolism (blockage of an artery by the bubble of gas) can form. A cerebral air embolism is rapidly fatal if not treated by hyperbaric pressurization to compress the bubble followed by controlled decompression (NIOSH 1977).

Air embolisms are more commonly encountered during diving where ascent is too rapid and gas vaporizes too rapidly from the blood following the decrease in pressure. To illustrate, pressure at a depth of 40 m (130 ft) is 5 atmospheres (0.51 MPa or 74 lb/in$^2$). Pressures commonly provided by plant compressed air systems range from 0.55 to 0.83 MPa (80 to 120 lb/in$^2$). Hence, there is a possibility of injection through the skin into a blood vessel at these pressures. However, this must be tempered with the fact that gases are compressible and may deflect from the path of a barrier, such as the skin, more easily than a liquid which is incompressible.

### Liquids
Applications involving pressurized liquids include hydraulic systems which involve movement of liquids through closed circuits, and open circuit systems, such as spray systems used in water jetting and the application of coatings. Water jetting itself covers a wide variety of applications, starting with cleaning of surfaces and removal of coatings, and progressing to cutting of materials including steel and concrete.

Water jetting applications and ultimately the hazard posed by pressurized liquids depend on delivery pressure. Equipment operating at a wide range of pressures is readily available to consumers from big-box, as well as corner hardware stores. Hence, the risk of injury due to use of pressurized fluids is not restricted to workplaces. The same can be said for both air powered and airless spraying equipment. The Water Jet Technology Association defines the following ranges (WJTA, 1994). Pressure cleaning or cutting refers to pump pressures less than 35 MPa (5000 lb/in$^2$). High pressure ranges from 35 to 210 MPa (5000 lb/in$^2$ to 30,000 lb/in$^2$. Ultra high pressure includes pump pressures exceeding 210 MPa (30,000 lb/in$^2$).

Use of pressurized fluid technology poses serious safety risks at virtually all pressures. This caveat includes low pressures (Vijay 1993). Several papers in the literature have examined injuries (Figure 3.8) caused by fluid jets (Katakura and Tsuji, 1985, Summers and Viebrock, 1988, Vijay, 1989).

There are abundant anecdotes about 'old hands' running their fingers along a piece of pipe when trying to locate a fluid leak and shearing off a fingertip in the flow. This is one type of traumatic injury caused by pressurized fluids. Another type of injury results from direct injection of fluid into the skin.

Traumatic injury produced by injection of fluid jets differs from that produced by other types of cutting tools, such as a chainsaw. Injury or damage caused by the chainsaw usually is localized, whereas injury caused by a fluid jet can be widespread. Reports indicate that injection injury can occur at pressure as low as 0.55 MPa (80 lb/in$^2$). The severity of the injury depends on the pressure, duration of exposure, and site of the impact. The eye, obviously, is a critical target for low-pressure injection injury (Vijay 1989). Serious injury to the hand, the eye, and the abdomen has occurred due to injection by high-pressure jets. Jets carrying abrasive materials, as well as fluid, can amplify the severity of the injury. Injury in these applications has occurred at pressures as low as 0.35 MPa (50 lb/in$^2$) (Vijay 1993).

The site of the impact usually is small. Spread of the fluid in the tissue and the intensity of the ensuing injury is highly dependent on the kinetic energy of the jet (proportional to pressure and mass flow rate) and the duration of exposure. Studies of the cutting of seafood and meat products using high-pressure jets indicate that the fluid, in this case water, can penetrate extensively perpendicular to the plane of the cut (Vijay and Brierley 1983). This suggests, that once the fluid jet penetrates through the skin, there are no means to predict how far and how deep it will spread into the underlying tissue. The extent of spread can increase when the jet hits a resistant structure, such as bone. The bleeding or the sensation of pain that occurs when the jet penetrates through the epidermis and reaches the dermis (bottom layer of the skin) is only the obvious indicator of injury.

The amount of fluid entering the body depends on stand-off distance (distance of the jet from point of entry). Injury will vary according to the site of injection. Tissue necrosis (death) and breakdown of tissues which are surrounded by healthier ones result from deposition of bacteria and toxins contained in the injected fluid and accompanying material. Direct mechanical damage to tissue and blood vessels also can occur due to the abnormally high shear force induced by the jet. Release of gas (usually air) along soft tissue planes can cause severe emphysema. Gas-induced infections can be fatal if not treated early by adequate surgical debridement (cleansing of the wound by cutting away dead or infected tissue and foreign matter) (Vijay 1989).

Because, as mentioned, the superficial appearance of an injury from a pressure jet usually is small, this lack of size can induce a false sense of security leading the victim not to seek medical attention. Delay in seeking treatment can be extremely serious. Minor delay can drastically increase

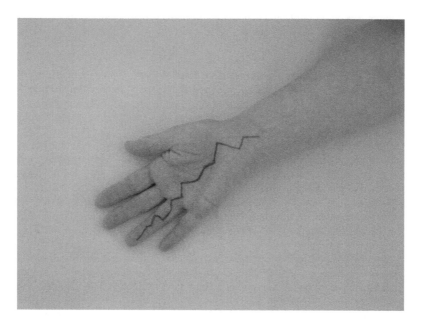

Figure 3.8. Fluid injection injury. The path taken by a fluid inside body structures during injection is unpredictable due to reflection on hard surfaces of tendons and bones. Gangrene and loss of the limb are serious potential consequences resulting from delay in obtaining treatment.

the probability of need to amputate the injured limb. When a person enters the hospital within a day of incidence of an injury of this type, there is a 20% chance that amputation of the affected limb will be required. When the patient delays 2 days, the probability of need to amputate increases to 60%. When the patient delays a week, loss of the limb involved in the injury is almost certain. Table 3.10 describes the three stages of injury due to fluid injection.

The best course of action is never to ignore any fluid jet injury, no matter how small it may first appear and to manage an injury caused by a fluid jet as a surgical emergency (Vijay 1989). The victim should report immediately to a hospital to seek attention from an experienced physician or surgeon, not the less experienced staff usually found in emergency departments.

### High Vacuum

Vacuum is the opposite condition to pressurization. In this case the walls of the container are subjected to negative pressure and inward-acting forces. Vacuum trucks operate at vacua reaching – 90 kPa (–27 in. Hg). This suction is considerably beyond the capability of the average person to overcome when a body part becomes attached to the open end of the hose and can cause severe traumatic injury, including evisceration and removal of limbs from sockets. Deadheading the vacuum by suction against a solid surface can cause a back and forth rocking motion of the chassis of the truck.

Rapid release of high vacuum from a container, such as the picture tube of a television set or the cathode ray tube of a computer monitor, produces an in-rush of air, followed by an overpressure and possibly explosion of the structure.

### Traumatic Injury

The one area not discussed thus far is the most consequential regarding accidents and contact with hazardous energy: the role of kinetic energy in the causation of traumatic injury. Most serious and fatal accidents involve rapid conversion of kinetic energy through rapid deceleration. This energy either is

## Table 3.10
## Stages of Fluid Injection Injury

| Stage | Description |
| --- | --- |
| Acute | Immediate symptoms of swelling and an increase in interstitial pressure, accompanied by edema. This can cause compression of the arteries or thrombosis making the tissues white and anaesthetic. |
| Intermediate | Presence of oleogranulomas (nodular tumors formed as a result of the reaction between the foreign fluid and the tissue). (Oleogranuloma formation may not occur with water jets.) These can remain within the body without change for a number of years. |
| Late | Breakdown of the oleogranulomas close to the surface of the skin is associated with the appearance of widespread lesions (tumors, etc). Malignancy also may occur. |

Source: Vijay 1989.

possessed by the human body when in motion, or by some object that strikes the body. In the former case, the body could acquire kinetic energy during a fall or as result of being inside or on a motorized vehicle. The injury occurs during the sudden stop in a collision by the body or the object that strikes it. In the latter case, the object striking the body could be a bullet, or a more mundane item, such as equipment or things put into motion mechanically, or by fluid systems or gravity.

Abrupt acceleration, deceleration, or impact are characterized by forces of very abrupt onset, short duration (less than 0.2 seconds), and high magnitude (Stapp 1961). The tolerances of humans in an abruptly decelerated system are related to the elastic and tensile limits of the tissues involved, as well as physiologic, psychologic, and metabolic effects (Snyder 1973).

Many factors complicate human tolerance to impact: gender, age, physical and mental condition, restraint systems, force altering systems, and the magnitude, direction, distribution, duration, and shape of the pulse of the force resulting from the impact. The force is expressed through a transient peak having a time-varying nature. As a result, the inertial reaction varies throughout the body during the impact (Snyder 1973). This is a very complex subject.

Workplace accidents are highly varied. Hence, reconstruction to determine forces and kinetic energy applied to the body is not usually undertaken. One area where systematic investigation has occurred is fall-related injury. These studies formed the basis for specification of fall protective equipment and strategies.

Information to answer these questions comes from sources such as fall effects research, studies of motor vehicle accidents, fall-related injuries during warfare, research into rapid deceleration from high speeds, parachute jumping, ejection of pilots from aircraft, projectile research, and possibly more recently, bungee jumping.

**Body in Motion**
Most injuries in falls, automobile crashes, and other abrupt deceleration result from the sudden transfer of mechanical energy as the moving body abruptly decelerates on contact with the immovable object (Gryfe 1991). The nature and extent of injury reflect direction of forces applied to the body, magnitude of the forces, rate of application of forces, and duration of the process.

Survival depends on extending the period of deceleration to the maximum possible. People have survived falls from great heights when deceleration extends over a prolonged period. This factor differentiates death due to a fall and a successful ski jump from the same height.

Gross movement of the body and its parts is describable in terms of velocity, acceleration, and jolt. Jolt has units of $m/s^3$ versus $m/s$ for velocity and $m/s^2$ for acceleration. Jolt is a strong limiting factor in tolerance. Acceleration is often compared to acceleration due to gravity ($9.8\ m/s^2$ or $32\ ft/s^2$).

Median survival (50%) for impaction against a hard surface occurs at 8 m/s (29 km/h or 18 mi/h) (White and Bowen 1959; Hammer 1993). This velocity corresponds to a calculated fall of 3.4 m (11 ft), kinetic energy at time of impact of 2240 J, and deceleration force of 2800 N for a 70 kg male and deceleration time of 0.2 s. Statistics on fall-related injuries in the construction industry (Table 3.11) compiled much later also reflect these data (Culver and Connolly (1994).

In a fall to the floor, there is sufficient potential energy in the standing position, which, if unabsorbed, could break any bone in the body. This is amply illustrated in hospital admissions due to falls on ice from standing height. Muscles play a key role in the absorption of energy, whether in falls in which the victim lands on the feet, or in compression of the chest following an impact (Gryfe 1991). Bone strength is a function of bone mass and integrity of bone architecture.

Snyder et al. (1977) examined free falls in a major report (Table 3.12). These authors concluded that virtually any fall greater than 3 m (10 ft) with head-first impact would produce skull fractures. Adults landing in a sitting position after falling more than 3 m (10 ft) would sustain fractures of the lumbar spine, especially the first lumbar vertebra. Feet-first falls greater than 4.5 m (15 ft) would result in leg and ankle fractures. Falls from 9 m (30 ft) and above would result in pelvic fractures. Absorption of 3400 J (2500 ft-lb) of energy in arrest from a free fall leads to uncertain survival.

The forces involved in the conversion of energy that occurs during arrest of a fall are proportional to the square of the change in velocity and inversely proportional to the stopping distance. Hence, doubling the stopping distance halves the resultant forces. Doubling the velocity quadruples the forces. Susceptibility to fatal injury in a fall also is a function of gender and age (Snyder 1963).

Studies of forces involved in fall-related research have shown that the human body can tolerate high deceleration forces (g-forces) for a brief period of time (Noel et al. 1991). Deceleration force of 45 g over the period of 0.1 s forms the limit of human tolerance (Damon et al. 1966; Haddon 1972). (g is the gravitational acceleration rate, $9.8\ m/s^2$ or $32\ ft/s^2$.) Humans can survive considerable deceleration stresses without serious injury, as indicated in research about falls involving considerable distances and rapid stoppage, for example, from 1017 km/h (632 mi/h) to 0 km/h (mi/h) in 1.4 s (DeHaven 1942; Stapp 1955). Deceleration time of 20 ms is common in automobile crashes and falls. Deceleration ranging from 200 to 300 ms produces hydrostatic effects and displacement of body fluids. This occurs without discomfort until the duration increases to 400 to 600 ms (Snyder 1963).

For impacts in the vertical direction, impact sensitivity is a function of duration (Snyder 1973). The body acts as a rigid mass for durations less than 70 ms. For durations greater than 70 ms, effects result from shifts in body tissues and fluids. A major change in the damage-no damage curve occurs at durations between 100 to 150 ms. The same applies to voluntary tolerance limits. That is, voluntary tolerance limits decrease as duration of peak acceleration increases. The critical entrance velocity for human survival of water impact is about 30 m/s (100 ft/s) (Snyder 1965).

Different segments and organs in the body have different inertial characteristics (Amphoux 1991). Deceleration due to force applied at one point in the body is only gradually obtained through the more or less rigid connections between the parts. Solid organs, such as the heart, liver, kidneys, and spleen, can be torn loose from connective tissue under external force. The brain and cerebral column can be crushed against the walls of the skull or edges of the occipital opening.

Injury to the head is the most frequent and severe result of impact (Snyder 1973). Usually, such trauma occurs due to contact with the structure or projections, rather than the action of acceleration forces on the head as a whole. The bones of the skull are uniquely sensitive to "punch-through" (bearing load) failures (SAE 1986). This is due to anatomical construction (sandwich), physical prominence, and lack of padding from overlaying soft tissue.

## Table 3.11
## Fall Distance and Lethality (Occupational Data)

| Fall Distance | Lethality | Cumulative Lethality |
|---|---|---|
| m | % | % |
| <2 | 2 | 2 |
| 2 to 3 | 8 | 10 |
| 4 to 6 | 22 | 32 |
| 7 to 9 | 20 | 52 |
| >9 | 36 | 88 |
| Not available | 12 | 100 |

Source: Culver and Connolly 1994.

Neck injury occurs due to hyper-extension rearward followed by abrupt flexion when the body is accelerated from back to front without head support. Wearing a helmet can aggravate neck injury due to increased mass.

Change in acceleration (jolt) is an important element in determining risk of injury (Amphoux 1991). Jolt measuring several thousands of $m/s^3$ is tolerable for a fraction of a millisecond. Maximum tolerable jolts are estimated to be 5 $km/s^3$ at 350 $m/s^2$ and 1.3 $km/s^3$ at 120 $m/s^2$. Duration of exposure is another important factor. The maximum tolerable deceleration is about 50 $m/s^2$ for the extended period of time needed to arrest a fall. Acceleration exceeding 300 $m/s^2$ and jolt of 10 $km/s^3$ to the head lasting 0.02 seconds have been recorded. Such situations are particularly threatening to the cervical spine.

Studies of parachutists indicate that 12,000 N (Newtons) is an upper limit for arresting force tolerated by the body (Amphoux 1991). Tolerating this level of force requires fitness training, as well as maintaining the body in a particularly favorable orientation with muscles tensed. Forces of this magnitude would injure normal individuals. Even in modern jumps involving forces around the bases of the transverse apophyses of 2250 N, the jumper must bend forward, and rest the chin on the chest at the moment of opening the chute to prepare for the shock. This is to protect the vulnerable cervical region in the neck, in particular the joint between the neck and head.

A limit of 6000 N proposed as the maximum tolerable during fall arrest (Amphoux 1991) was believed appropriate for the broader population and range of circumstances in which fall occurs. These include spinal conditions and the range of spinal mechanics, and the possibility of a head-first fall where the shock wave is damped by the shoulders and rib cage prior to reaching the vertebral column.

The limitation in considering the data in this manner appears from examination of the manner in which fall arrest occurs. Waist belts, for example, are forbidden as fall-arrest devices. Waist belts cause the body to jackknife during fall arrest (Amphoux 1991). Workers intuitively sensed that these belts would "cut them in two" during fall arrest. Actual testing indeed showed this perception to be correct. Falls arrested in this manner put the abdominal organs (liver and spleen) at considerable risk. Similarly, a fall arrested by an object striking the abdomen, or collision between the abdomen and an object is highly injurious to these organs.

Waist belts incorporating shoulder straps transfer force to the rib cage and cause fracture of the ribs following even short fall distances.

## Table 3.12
## Fall-Related Injury

| Injury | Fall Distance | Impact Force | Impact Energy |
| --- | --- | --- | --- |
| | m | N | J |
| Skull fracture (head-first impact) | 3 | 2695 | 2060 |
| Seated fall, lumbar fracture | 3 | 2695 | 2060 |
| Feet-first fall, leg/ ankle fracture | 4.5 | 3290 | 3090 |
| Pelvic fracture | 9 | 4550 | 6175 |
| Uncertain survival | 5 | 3465 | 3400 |

Basis of calculations: 70 kg male, u = 0, speed = final velocity (v), deceleration time = 0.2 seconds.

Source: After Snyder et al. 1977.

Actions that cause the head and shoulders to be thrown back violently can cause injury to the vertebrae of the spinal column and the neck. In the former case, this results from contact between the bases of the transverse apophyses of the vertebrae. In the latter case this can lead to fracture of the posterior arches of the cervical vertebrae, or injury to the medulla or nerves (Amphoux 1991). Vertical arrest of a fall can subject the vertebrae and intervertebral discs to compressive forces. These can cause fractures related to contact between transverse apophyses (projections from bone). The most sensitive region of the spine is the cervical region.

### Projectile Motion
The second area providing information on human injury due to interaction with kinetic energy is projectile research. This provides information about the effect of collisions between projectiles and the body. For example, the skull will fracture when hit by a 2.3 kg (5 lb) hammer falling through a distance of 3 m (10 ft). This corresponds to an energy contact of 68 J (50 ft-lb). Table 3.13 provides data on the effect of projectiles on limbs (after White and Bowen 1959).

### Hazardous Energy: Bioeffects in Context
This chapter has presented a wide-ranging discussion about the biological expression of the hazardous effects of some sources of energy. These types of energy are capable of and known to have caused harm to workers and other people, thereby making them hazardous.

Table 3.14 summarizes this information using the Pascal (Pa), the unit of pressure (1 Pa = 1 N/m²) and the Joule, the common unit of energy, as the basis for reporting. The range of expression of these interactions between energy and biological structures occurs at its most minute level in the breakage of the chemical bonds that form the chain of the helix of the molecules of DNA and RNA and some proteins. These expressions progress through cellular structures to larger structures including organs and the skeleton and the entire body. The key to the biological effect is the delivery of energy of the appropriate type at the location appropriate for participation in a mechanism capable of causing harm.

The feature that stands out in Table 3.14 is that the quantities of energy involved in causing harm range from the unimaginably small to the unimaginably large. This acknowledges the importance of

## Table 3.13
## Effect of Projectiles on Human Limbs (Cadavers)

| Effect | Energy J |
|---|---|
| Slight skin laceration | 15 |
| Penetrating wound | 21 |
| Abrasion, cracked tibia | 45 |
| Skull fracture | 68 |
| Travels through the thigh (spherical bullet) | 90 |
| Threshold for bone injury | 33 to 49 |
| Large bone fracture | 158 |

Source: After White and Bowen 1959.

proximity of the source to the target of the action and the mechanism by which the action of harm occurs.

## References

ACGIH: *A Guide for Control of Laser Hazards*, 4th Ed. Cincinnati, OH: American Conference of Governmental Industrial Hygienists, Inc., 1990.

ACGIH: *Documentation of the Threshold Limit Values and Biological Exposure Indices*, 6th Ed. Cincinnati, OH: American Conference of Governmental Industrial Hygienists, Inc., 1991.

ACGIH: TLVs and BEIs, Threshold Limit Values for Chemical Substances and Physical Agents & Biological Exposure Indices. Cincinnati, OH: American Conference of Governmental Industrial Hygienists, 2001a.

ACGIH: *Documentation of the Threshold Limit Values for Physical Agents*, 7th Ed. Cincinnati, OH: American Conference of Governmental Industrial Hygienists, Inc., 2001b.

AIHA: AIHA White Paper on Extremely Low Frequency (ELF) Fields. *Am. Ind. Hyg. Assoc. J. 63*: 679-684 (2002).

AIUM: Bioeffects Considerations for the Safety of Diagnostic Ultrasound. *J. Ultrasound Med. 7*: S1 (1988).

Allen, R.G.: Retinal Thermal Damage. In *Non-Ionizing Radiation. Proceedings of a Topical Symposium, November 26-28, 1979, Washington, DC*. Cincinnati, OH: American Conference of Governmental Industrial Hygienists, Inc., 1980. pp. 161-168.

Amphoux, M.: Physiopathological Aspects of Personal Equipment for Protection Against Falls. In *Fundamentals of Fall Protection*, Sulowski, A.C. (Ed.). Toronto, Canada: International Society for Fall Protection, 1991. pp. 33-48.

Andersen, F.A.: Sodium and Mercury Vapor Lamps. In *Non-Ionizing Radiation. Proceedings of a Topical Symposium*, November 26-28, 1979, Washington, DC. Cincinnati, OH: American Conference of Governmental Industrial Hygienists, Inc., 1980. pp. 239-243.

Barrow, G.M.: *Physical Chemistry*, 2nd Ed. New York: McGraw-Hill, Inc., 1966.

## Table 3.14
## Summary of Hazardous Energy Sources

| Example | Condition | Energy/Force |
|---|---|---|
| **Ionizing Radiation** | | |
| Production of ion pair | 30 to 35 eV | 4.8 to 5.6 x $10^{-18}$ J |
| LD50 | 4 to 6 Sv | 4 to 6 J/kg |
| **Ultra-violet Emission** | | |
| Welder's flash (UVC and UVB) | | 4 to 14 mJ/cm$^2$ |
| Corneal injury | | 10 J/cm$^2$ |
| Cataract (UVB) | | 0.15 to 13 J/cm$^2$ |
| Retinal damage, lens removed (UVB and UVA)) | | 5 J/cm$^2$ |
| Sunburn (UVC and UVB) | | 6 to 30 mJ/cm$^2$ |
| **Visible Emission** | | |
| Retinal burn | >45° C, internal temperature | 10 J/cm$^2$ |
| **Infra-red Emission** | | |
| Corneal burn (IRC and IRB) | >45° C, internal temperature | 10 J/cm$^2$ |
| Retinal burn (IRA) | >45° C, internal temperature | 10 J/cm$^2$ |
| Thermal burn (IRA, IRB and IRC) | >45° C, internal temperature | 10 J/cm$^2$ |
| Cataract (IRA and IRB), threshold | | 4 J/s/cm$^2$ |
| **Microwave Emission** | | |
| Thermal damage (0.3 to 3.0 MHz) | | 100 mJ/s/cm$^2$ |
| **Thermal Effects** | | |
| Thermal burn, contact | >45° C, internal temperature | 10 J/s/cm$^2$ |
| Heat stroke | 41° C, core temperature | |
| Contact frostbite | –7° C | |
| Frostbite on prolonged contact | –1° C | |
| **Electrical** | | |
| Ventricular fibrillation | 70 to 200 mA, average impedance of 600 ohms | 40 to 45 J/c |

## Table 3.14 (Continued)
## Summary of Hazardous Energy Sources

| Example | Condition | Energy/Force |
|---|---|---|
| **Sound** | | |
| Pain threshold | | 10 J/s |
| **Fluid Injection into the Skin** | | |
| Air jet | 800 kPa | 800 N/m$^2$ |
| Liquid jet | 350 to 600 kPa (50 to 85 lb/in$^2$) | 350 to 600 N/m$^2$ |
| **Fall** | | |
| LD$_{50}$ | | 2240 J (70 kg male) |
| **Overpressure** | | |
| Eardrum rupture | 34 kPa | 34 N/m$^2$ |
| Lung | 207 to 255 kPa | 207 to 255 N/m$^2$ |
| LD$_{50}$ | 345 kPa | 345 N/m$^2$ |
| **Projectiles** | | |
| Skull fracture | | 68 J |
| Large bone fracture | | 158 J |

Sources: ACGIH 1991; Cember 1996; Gandhi 1982; Hardy 1958; ICNIRP 1997; INIRC 1985a; INIRC 1988; Pizzarello and Witcofski 1967;Wertheimer and Ward 1952; Wilkening 1991b.

Berger, E.H., W.D. Ward, J.C Morrill and L.H. Royster (Eds.): *Noise and Hearing Conservation Manual*, 4th Ed. Akron, OH: American Industrial Hygiene Association, 1986.

Berger, E.H., L.H. Royster, J.D. Royster, D.P. Driscoll and M. Layne (Eds.): *The Noise Manual*, 5th. Ed. Fairfax, VA: American Industrial Hygiene Association, 2000.

Bernhardt, J.H.: Bioeffects of Radiofrequency Fields. In *Non-Ionizing Radiation. Proceedings of the 2nd International Non-Ionizing Radiation Workshop, Vancouver, British Columbia, Canada*, 1992 May 10-14. Greene, M.W. (Ed.). London: International Radiation Protection Association, The Institution of Nuclear Engineers, 1992. pp. 32-58.

Bernhardt, J.H. and R. Matthes: ELF and RF Electromagnetic Sources. In *Non-Ionizing Radiation. Proceedings of the 2nd International Non-Ionizing Radiation Workshop, Vancouver, British Columbia, Canada*, 1992 May 10-14. Greene, M.W. (Ed.). London: International Radiation Protection Association, The Institution of Nuclear Engineers, 1992. pp. 59-75.

Bernstein, T.: Physiological Effects of Electricity — Relationship to Electrical and Lightning Death and Injury. In *Electrical Hazards and Accidents, Their Causes and Prevention*, Greenwald, E.K. (Ed.). New York: Van Nostrand Reinhold, 1991. pp. 28-49.

Biegelmeier, G. and W.R. Lee: New Considerations on the Threshold of Ventricular Fibrillation for AC Shocks at 50-60 Hz. *Proc. IEEE 127(2)*: 103-110.

Birmingham, D.J.: Occupational Dermatoses. In *Patty's Industrial Hygiene and Toxicology*, 4th Ed., Vol. I, Part A, General Principles. Clayton, G.D. and F.E. Clayton (Eds.). New York: John Wiley & Sons, Inc., 1991. pp. 253-287.

BLS: National Census of Fatal Occupational Injuries, 1995 by G. Toscano and J. Windau. In Fatal Workplace Injuries in 1995: A Collection of Data and Analysis – April 1997 (Report 913). Washington, DC: U.S. Department of Labor, Bureau of Labor Statistics, Census of Fatal Occupational Injuries: Reports, (1997a). 12 pp. (Also published as Toscano, G. and J. Windau: National Census of Fatal Occupational Injuries, 1995. *Comp. Work. Cond.* September 1996. pp. 34-45.)

BLS: Lost-Worktime Injuries: Characteristics and Resulting Time Away from Work, 1995 (USDL 97-188). Washington: U.S. Department of Labor, Bureau of Labor Statistics, 1997.

Bly, S.H.P. and G.R. Harris: Ultrasound Sources and Human Exposure. In *Non-Ionizing Radiation. Proceedings of the 2nd International Non-Ionizing Radiation Workshop, Vancouver, British Columbia, Canada*, 1992 May 10-14. Greene, M.W. (Ed.). London: International Radiation Protection Association, The Institution of Nuclear Engineers, 1992. pp. 173-188.

Bodurtha, F.T.: *Industrial Explosion Prevention and Protection*. New York: McGraw-Hill Book Company, 1980.

Bowen, I.G., E.R. Fletcher, and D.R. Richmond: Estimate of Man's Tolerance to the Direct Effects of Air Blast (DA-49-146-XZ-372). Washington, DC: HQ NASA, 1968.

BPA: Electrical and Biological Effects of Transmission Lines: A Review (DOE/BP-524, Revised). Portland, OR: U.S. Department of Energy/Bonneville Power Administration, 1986.

Bramer, T.P.C., G.J. Cole, J.R. Cowell, A.T. Fry, N.A. Grundy, T.J.B. Smith, J.D. Webb, and D.R. Winterbottom: *Basic Vibration Control*. Sudbury, Suffolk, UK: Sound Research Laboratories Limited, 1977.

Bussmann: *Safety Basics, Handbook for Electrical Safety*. St. Louis, MO: Cooper Bussmann, Inc., 2000.

Cabanes, J.: Physiological Effects of Electric Currents on Living Organisms, More Particularly Humans. In *Electrical Shock Safety Criteria*. Bridges, J.E., G.L. Ford, I.A. Sherman and M. Vainberg (Eds.). New York: Pergamon Press, 1985. pp. 7-24.

Casini, V.: Overview of Electrical Hazards. In Worker Deaths by Electrocution, A Summary of NIOSH Surveillance and Investigative Findings (DHHS (NIOSH) Publication No. 98-131). Cincinnati, OH: U.S. Department of Health and Human Services/Public Health Service/Centers for Disease Control and Prevention/National Institute for Occupational Safety and Health, 1998.

CDC: Leading Work-Related Diseases and Injuries – United States. *Morbid. Mortal. Weekly Rep. 33*: 3-5 (1984).

Cember, H.: *Introduction to Health Physics*, 3rd Ed. New York: The McGraw-Hill Companies, Inc., 1996.

CIE: *International Lighting Vocabulary*, 3rd Ed., CIE Publication No. 17.4. Vienna: Commission Internationale de l'Eclairage, 1988.

Cleary, S.: Biological Effects of Low Intensity Microwave and Radiofrequency Radiation Exposure. In *Non-Ionizing Radiation. Proceedings of a Topical Symposium, November 26-28, 1979, Washington, DC*. Cincinnati, OH: American Conference of Governmental Industrial Hygienists, Inc., 1980. pp. 71-74.

COMAR: Medical Aspects of Radiofrequency Radiation Overexposure. *Health Phys. 82*: 387-391 (2002).

Cooper, P.W.: *Explosives Engineering*. New York: VCH Publishers, Inc., 1996.

Court, L. and D. Courant: Laser: Characteristics and Emissions. In *Non-Ionizing Radiation. Proceedings of the 2nd International Non-Ionizing Radiation Workshop, Vancouver, British Columbia, Canada*, 1992 May 10-14. Greene, M.W. (Ed.). London: International Radiation Protection Association, The Institution of Nuclear Engineers, 1992. pp. 289-306.

Cruice, W.: Explosions. In *Fire Protection Handbook*. 17th Ed. Cote, A.E. and J.L. Linville (Eds.) Quincy, MA: National Fire Protection Association, 1991. pp. 1-56 to 1-71.

Culver, C. and C. Connolly: Prevent Fatal Falls in Construction. *Safety. Health.* September, 1994. pp. 72-75.

Dalziel, C.F.: Deleterious Effect of Electric Shock. In *Handbook of Laboratory Safety*, 2nd Ed., Steere, N.V. (Ed.). Cleveland, OH: Chemical Rubber Co., 1971. pp. 521-527.

Dalziel, C.F. and F.P. Massoglia: Let-go Currents and Voltages. *AIEE Trans. 75 Part II*: 49-56 (1956).

Damon, A., H.W. Stoudt, and R.A. MacFarland: *The Human Body in Equipment Design.* Cambridge MA: Harvard University Press, 1966.

DeHaven, H.: Mechanical Analysis of Survival in Falls from Heights of Fifty to One Hundred and Fifty Feet. *War Med. 2*: 586-596 (1942).

DeVaul, R.: Just Shocking! Occup. Health Safety. Product Literature & Web Guide, Supplement to OH&S. Vol. 5, No. 2. September 2002. pp. 6-10.

Dunn, F. and F.J. Fry: Ultrasonic Threshold Dosages for the Mammalian Central Nervous System. *IEEE Trans. Biomed. Eng. 18*: 253-256 (1971).

Feldman, R.J. and H.I. Maibach: *J. Invest. Derm. 48*:181 (1967).

Floyd, H.L., J.J. Andrews, M. Capelli-Schellpfeffer, T.E. Neal, and L. F. Saunders: Electrical Safety: State of the Art in Technology, Work Practices and Management Systems (Session 712). Safety 2001. *Proceedings of the ASSE Professional Development Conference and Exposition, Anaheim California*, June 10–13, 2001. [CD-ROM].

Frey, A.H.: Human Auditory System Response to Modulated Electromagnetic Energy. *J. Appl. Physiol. 17*: 689-692 (1962).

Geddes, L.A. and L.E. Baker: The Specific Resistance of Biological Material — A Compendium of Data for the Biomedical Engineer and Physiologist. *Med. Biol. Eng. 5*: 271-293 (1967).

Gandhi, O.P.: Conditions of Strongest Electromagnetic Power Deposition in Man and Animals. *IEEE Trans. Microwave Theory Tech. 23*: 1021-1029 (1975).

Ghiassi-nejad, M., S.M.J. Mortazavi, J.R. Cameron, A. Niroomand-rad, and P.A. Karam: Very High Background Radiation Areas of Ramsar, Iran: Preliminary Biological Studies. *Health Phys. 82*: 87-93 (2002).

Glaser, Z.R.: Basis for the NIOSH Radiofrequency and Microwave Radiation Criteria Document. In *Nonionizing Radiation, Proceedings of a Topical Symposium*, November 26-28, 1979, Washington, DC. Cincinnati, OH: American Conference of Governmental Industrial Hygienists, Inc., 1980. pp. 103-116.

Glasstone, S.: *The Effects of Nuclear Weapons*, rev. Ed. Washington, DC: U.S. Atomic Energy Commission, 1962.

Grim, P.: When Lightning Strikes. *Discover 23*: August (2002). pp.46-51.

Gryfe, C.: Anatomical and Physiological Considerations in Design Criteria for Body Support Devices and Shock Absorbers for Women. In *Fundamentals of Fall Protection*, Sulowski, A.C. (Ed.). Toronto, Canada: International Society for Fall Protection, 1991. pp. 49-103.

Guignard, J.C.: Vibration. In *A Textbook of Aviation Physiology*, Gillies, J.A. (Ed.). Oxford: Pergamon Press, 1965. pp. 813-894.

Haddon, W.: A Logical Framework for Categorizing Highway Safety Phenomena and Activity. J. *Trauma 12*: 193-207 (1972).

Hammer, W.: *Product Safety Management and Engineering*, 2nd Ed.. Des Plaines, IL: American Society of Safety Engineers, 1993.

Hardy, J.D.: Thermal Radiation, Pain and Injury. In *Therapeutic Heat*. Vol. 2, Licht, S. Ed. New Haven, CT: 1958. pp. 157-178.

Helmkamp, J.C., E.O. Talbott, and G.M. Marsh: Whole Body Vibration – A Critical Review. *Am. Ind. Hyg. Assoc. J. 45*: 162-167 (1984).

Hétu, R. and J. Parrot: A Field Evaluation of Noise-Induced Temporary Threshold Shift. *Am. Ind. Hyg. Assoc. J. 39*: 301-311 (1978).

Hewson, G.S. and J.J. Fardy: Thorium Metabolism and Bioassay of Mineral Sands Workers. *Health Phys. 64*: 147-156 (1993).

Hinrichs, J.F.: A Bright Spot in Welding. In *Non-Ionizing Radiation. Proceedings of a Topical Symposium, November 26-28, 1979, Washington, DC.* Cincinnati, OH: American Conference of Governmental Industrial Hygienists, Inc., 1980. pp. 245-247.

Hodgkin, B., O. Langworthy, and W.B. Kouwenhoven: Effect on Breathing. *IEEE Trans. Power Appar. Sys. PAS-92(4)*: 1388-1391 (1973).

ICNIRP: Guidelines on Limits of Exposure to Laser Radiation of Wavelengths Between 180 nm and 1 000 μm. *Health Phys. 71*: 804-819 (1996).

ICNIRP: Guidelines on Limits of Exposure to Broad-Band Incoherent Optical Radiation (0.38 to 3 μm). *Health Phys. 73*: 539-554 (1997).

ICRP: *Recommendations of the International Commission on Radiation Protection*, Publication 60, Vol. 21 No. 1-3. Tarrytown, NY: Elsevier Science, 1990.

IEC: *Effects of Current Passing Through the Human Body, 2nd Ed.* Part 1, General Aspects (IEC 479-1). Geneva: International Electrotechnical Commission, 1984.

INIRC: Interim Guidelines on Limits of Human Exposure to Airborne Ultrasound. *Health Phys. 46*: 969-974 (1984).

INIRC: Guidelines on Limits of Exposure to Ultraviolet Radiation of Wavelengths between 180 nm and 400 nm (Incoherent Optical Radiation) *Health Phys. 49*: 331-340 (1985a).

INIRC: Guidelines on Limits of Exposure to Laser Radiation of Wavelengths between 180 nm and 1 mm. *Health Phys. 49*: 341-359 (1985b).

INIRC: Guidelines on Limits of Exposure to Radiofrequency Electromagnetic Fields in the Frequency Range from 100 kHz to 300 GHz. *Health Phys. 54*: 115-123 (1988).

INIRC: Interim Guidelines on Limits of Exposure to 50/60 Hz Electric and Magnetic Fields. *Health Phys. 58*: 113-122 (1990).

Johns, H.E. and J.R. Cunningham: *The Physics of Radiology*, 3rd Ed. Springfield, IL: C.C. Thomas Publisher, 1974.

Jones, R.M.: A Guide to the Interpretation of Machinery Vibration Measurements – Part I. *Sound Vib.* May (1994a). pp. 24-35.

Jones, R.M.: A Guide to the Interpretation of Machinery Vibration Measurements – Part II. *Sound Vib.* September (1994b). pp. 12-20.

Kahn, F. and L. Murray: Shock Free Electric Appliances. *IEEE Trans. Ind. Gen. Appl. IGA-2(4)*: 322-327 (1966).

Katakura, H. and S. Tsuji: A Study to Avoid the Dangers of High Speed Liquid Jets. *Bull. Japan Soc. Mech. Eng. 28*: 623-630 (1985).

Kavet, R., M.A. Stuchly, W.H. Bailey, and T.D. Bracken: Evaluation of Biological Effects, Dosimetric Models, and Exposure Assessment Related to ELF Electric- and Magnetic-Field Guidelines. *Appl. Occup. Environ. Hyg. 16*: 1118-1138 (2001).

KlimKova-Deutschova: Neurologic Findings in Persons Exposed to Microwaves. In *Biologic Effects and Health Hazards of Microwave Radiation.* Warsaw, Poland: Polish Medical Publishers, 1974. p. 268.

Kouwenhoven, W.B.: Effects of Electricity on the Human Body. *Electr. Eng. 68*:199-203 (1949).

Kryter, K.D.: Exposure to Steady-State Noise and Impairment to Hearing. *J. Acous. Soc. Am. 35*: 1515 (1963).

LANL: A Review of Criticality Accidents, 2000 Revision (LA-13638) by T.P. McLaughlin, S.P. Monahan, N.L. Pruvost, V.V. Frolov, B.G. Ryazanov, and V.I. Sviridov. Los Alamos, NM: U.S. Department of Energy, Los Alamos National Laboratory, 2000.

Lee, R.H.: Electrical Safety in Industrial Plants. *IEEE Spectrum*, June 1971.

Lee, R.: The Other Electrical Hazard: Electrical Arc Flash Burn. *IEEE Trans. Ind. Appl.* Vol. IA-18, No.3. May-June (1982). p. 246.

Lee, R.: Pressures Developed by Arcs. *IEEE Trans. Ind. Appl.* Vol. IA-23, (1987). pp. 760-764.

Lees, F.P.: *Loss Prevention in the Process Industries, 2nd.* Ed. Vol. 2. London: Butterworths-Heinemann, 2001.

Loral: *Non-Ionizing Radiation Handbook.* Hauppauge, NY: Loral Microwave-Narda, 1992.

Magrab, E.B.: *Environmental Noise Control.* New York: John Wiley & Sons, 1975.

McDonald, O.F.: Cold Stress: How to Deal With It. *Nat. Safety Health News*: February 1986. pp. 18-24.

McKinlay, A.: Optical Radiation. In *Non-Ionizing Radiation. Proceedings of the 2nd International Non-Ionizing Radiation Workshop, Vancouver, British Columbia, Canada*, 1992 May 10-14. Greene, M.W (Ed.). London, England SE6: International Radiation Protection Association, The Institution of Nuclear Engineers, 1992. pp. 227-251.

McManus, N. and G. Green: *The HazCom Training Program*. Boca Raton, FL: Lewis Publishers, 1999.

McManus, N. and G. Green: *The WHMIS Training Program*. North Vancouver, BC: Training by Design, Inc., 2001.

Metzger, R., J.W. McKlveen, R. Jenkins and W.J. McDowell: Specific Activity of Uranium and Thorium in Marketable Rock Phosphate as a Function of Particle Size. *Health Phys. 39*: 69-75 (1980).

Moore, W.J.: *Physical Chemistry*, 3rd Ed. Englewood Cliffs, NJ: Prentice-Hall, Inc., 1962.

Moran, J.M. and M.D. Morgan: *Meteorology: The Atmosphere and The Science of Weather*. 2nd Ed. New York: Macmillan Publishing Company, 1989. pp. 27-55.

Morgan, M.G.: Electric and Magnetic Fields from 60 Hertz Electric Power: What do we Know About Possible Health Risks? Pittsburgh, PA: Department of Electrical Engineering and Public Policy, Carnegie Mellon University, 1989. [Booklet].

Moss, C.E.: Radiation Hazards Associated with Ultraviolet Radiation Curing Processes. In *Non-Ionizing Radiation. Proceedings of a Topical Symposium, November 26-28, 1979, Washington, DC*. Cincinnati, OH: American Conference of Governmental Industrial Hygienists, Inc., 1980. pp. 203-209.

NCRP: Radiofrequency Electromagnetic Fields: Properties, Quantities and Units. Biophysical Interacting and Measurements, Report No. 67. Washington, DC: National Council on Radiation Protection and Measurements, 1981.

NFPA: *National Electrical Code* (NFPA 70). Quincy, MA: National Fire Protection Association, 2011

NIOSH: *Occupational Diseases, A Guide to Their Recognition*, Rev. Ed. (DHEW (NIOSH) Pub. No. 77-181). Key, M.M., A.F. Henschel, J. Butler, R.N. Ligo, and I.R. Tabershaw, Eds. Cincinnati, OH: U.S. Department of Health, Education and Welfare/Public Health Service/Center for Disease Control/National Institute for Occupational Safety and Health, 1977.

NIOSH: NIOSH Alert, Request for Assistance in Preventing Fatalities of Workers Who Contact Electrical Energy (DHHS(NIOSH) Pub. No. 87-103). Cincinnati, OH: DHHS/PHS/CDC/NIOSH, 1986a.

NIOSH: NIOSH Alert, Request for Assistance in Preventing Electrocutions Due to Damaged Receptacles and Connectors (DHHS(NIOSH) Pub. No. 87-100). Cincinnati, OH: DHHS/PHS/CDC/NIOSH, 1986b.

Noel, G., M.G. Ardouin, P. Archer, M. Amphoux, and A. Sevin: Some Aspects of Fall Protection Equipment Employed in Construction and Public Works Industries. In *Fundamentals of Fall Protection*, Sulowski, A.C. (Ed.). Toronto, Canada: International Society for Fall Protection, 1991. pp. 1-32.

O'Brien, Jr., W.D.: Introduction to Ultrasound. In *Non-Ionizing Radiation. Proceedings of the 2nd International Non-Ionizing Radiation Workshop, Vancouver, British Columbia, Canada*, 1992 May 10-14. Greene, M.W. (Ed.). London: International Radiation Protection Association, The Institution of Nuclear Engineers, 1992a. pp. 127-150.

O'Brien, Jr., W.D.: Ultrasound Dosimetry and Interaction Mechanisms. In *Non-Ionizing Radiation. Proceedings of the 2nd International Non-Ionizing Radiation Workshop, Vancouver, British Columbia, Canada*, 1992 May 10-14. Greene, M.W. (Ed.). London: International Radiation Protection Association, The Institution of Nuclear Engineers, 1992b. pp. 151-172.

Olishifski, J.B.: The Skin (rev. by J.S. Taylor). In *Fundamentals of Industrial Hygiene*, 3rd Ed. Chicago, IL: National Safety Council, 1988. pp. 47-58.

OSHA: Controlling Electrical Hazards (OSHA 3075, revised). Washington, DC: U.S. Department of Labor/Occupational Safety and Health Administration, 1986.

Parrack, H.O.: Effects of Airborne Ultrasound on Humans. *Int. Audiol. 5*: 294-308 (1966).

Pelmear, P., W. Taylor, and D. Wasserman (Eds.): *Hand-Arm Vibration: A Comprehensive Guide for Occupational Health Professionals.* New York: Van Nostrand-Reinhold, 1992.

Peterson, A.P.G.: *Handbook of Noise Measurement,* 9th ed. Concord, MA: GenRad, Inc. 1980.

Petroski, H.: *To Engineer is Human, The Role of Failure in Successful Design.* New York: Vintage Books, 1992.

Pitts, D.G.: A Position Paper on Ultraviolet Radiation. In *Hazards of Light, Myths and Realities, Eyes and Skin,* Cronly-Dillon, J., E.S. Rosen, and J. Marshall (Eds.). Oxford: Pergamon Press, 1986.

Pizzarello, D.J. and R.L. Witcofski: *Basic Radiation Biology.* Philadelphia: Lea & Febiger, 1967.

Rasmussen, A.T.: Outlines of Neuroanatomy, 2nd Ed. Dubuque IA: William C. Brown Company, 1943.

Rekus, J.F.: Shocking Experiences. *Occup. Hazards 59*(2): 23-26 (1997).

Robinson, C.S.: *Explosions: Their Anatomy and Destructiveness.* New York: McGraw-Hill Book Company, 1944.

SAE: *Human Tolerance to Impact Conditions as Related to Motor Vehicle Design* (SAE J885, Rev. Jul86, supercedes HSJ885 APR80). Warrendale, PA: Society of Automotive Engineers, Inc., 1986.

de Santis, V. and I. Longo: Coal Energy vs Nuclear Energy: A Comparison of the Radiological Risks. *Health Phys. 46*: 73-84 (1984).

Saunders, R.D. and J.G.R. Jefferys: Weak Electric Field Interactions in the Central Nervous System. *Health Phys. 83*: 366-375 (2002).

Sheppard, A.R., R. Kavet, and D.C. Renew: Exposure Guidelines for Low-Frequency Electric and Magnetic Fields: Report from the Brussels Workshop. *Health Phys. 83*: 324-332 (2002).

Sliney, D.H.: Biohazards of Ultraviolet, Visible and Infrared Radiation. *J. Occup. Med. 25*: 203-206 (1983).

Sliney, D.H.: Ultraviolet Radiation and the Eye. In *Light, Lasers, and Synchrotron Radiation,* Grandolfo, M. et al. (Eds.). New York: Plenum Press, 1990. pp. 237-245.

Sliney, D.H.: Ultraviolet Studies. In *Non-Ionizing Radiation. Proceedings of the 2nd International Non-Ionizing Radiation Workshop, Vancouver, British Columbia, Canada,* 1992 May 10-14. Greene, M.W. (Ed.). London: International Radiation Protection Association, The Institution of Nuclear Engineers, 1992a. pp. 268-288.

Sliney, D.H.: Measurements and Bioeffects of Infrared and Visible Light. In *Non-Ionizing Radiation. Proceedings of the 2nd International Non-Ionizing Radiation Workshop, Vancouver, British Columbia, Canada,* 1992 May 10-14. Greene, M.W. (Ed.). London: International Radiation Protection Association, The Institution of Nuclear Engineers, 1992b. pp. 252-267.

Smoot, A. and J. Stevenson: Report on Investigation of Reaction Currents (Subject 965-1). Melville, NY: Underwriters Laboratories Inc., 1968.

Stuchly, M.A.: Introduction to Bioelectromagnetics. In *Non-Ionizing Radiation. Proceedings of the 2nd International Non-Ionizing Radiation Workshop, Vancouver, British Columbia, Canada,* 1992 May 10-14. Greene, M.W. (Ed.). London: International Radiation Protection Association, The Institution of Nuclear Engineers, 1992. pp. 17-31.

Snyder, R.G.: Human Tolerances to Extreme Impacts in Free Fall. *Aerospace Med. 34*: 695-709 (1963).

Snyder, R.G.: Survival of High-Velocity Free Falls in Water (FAA-AM-65-12). Oklahoma City, OK: Federal Aviation Agency, April 1965.

Snyder, R.G.: Impact. In *Bioastronautics Data Book,* 2nd Ed. (NASA SP-3006). Parker, J.F. and V.R. West (Eds.). Washington, DC: Scientific and Technical Office, National Aeronautics and Space Administration, 1973. pp. 221–295.

Snyder, R.G., D.R. Foust, and B.M. Bowman: Study of Impact Tolerance through Free Fall Investigations (UM-HSRI-77-8). Ann Arbor, MI: University of Michigan, Highway Safety Research Institute, 1977.

Stapp, J.P.: Effects of Mechanical Force on Living Tissues. 1. Abrupt Deceleration and Windblast. J. *Aviat. Med. 26*: 268-288 (1955).

Stapp, J.P.: Human Tolerance to Severe, Abrupt Acceleration. In *Gravitational Stress in Aerospace Medicine*, Ganer, O.H.and G.D. Zuidema (Eds.). Boston: Little, Brown & Company, 1961.

Summers, D.A. and J. Viebrock: The Impact of Water Jets on Human Flesh. Paper H4. In *Proceedings of the 9th International Symposium on Jet Cutting Technology* (Sendai, Japan). Cranfield, Bedford, England: British Hydromechanics Research Association, 1988. pp. 423-433.

Summers, D.: Historical Perspective of Fluid Jet Technology. In *Fluid Jet Technology. Fundamentals and Applications*, 2nd. Ed. St. Louis: Water Jet Technology Association, 1993. pp. 1-21.

Tenforde, T.S. and W.T. Kaune: Interaction of Extremely Low Frequency Electric and Magnetic Fields with Humans. *Health Phys. 53*: 585-606 (1987).

Tengroth, B.M., E. Lydahl, and B.T. Philipson: Infrared Cataract in Furnacemen. In *Non-Ionizing Radiation. Proceedings of a Topical Symposium, November 26-28, 1979, Washington DC*. Cincinnati, OH: American Conference of Governmental Industrial Hygienists, 1980. pp. 169-170.

ter Haar, G.R. and C.R. Hill: Ultrasound. In *Non-ionizing Radiation Protection*, European Series 25. Suess, M.J. and D.A. Benwell-Morrison (Eds.). Geneva: World Health Organization, Regional Publications, 1989.

Tucker, S.B. and M.M. Key: Occupational Skin Disease. In *Environmental and Occupational Medicine*, Rom, W..N. (Ed.). Boston: Little, Brown and Company, 1982. pp. 301-311.

UL: Standard for Safety, Ground-Fault Circuit Interrupters (UL 943). Northbrook, IL: Underwriters Laboratories Inc., 2006.

UL: Standard for Safety, Transformer-Type Arc Welding Machines (UL 551). Northbrook, IL: Underwriters Laboratories Inc., 1987.

UNSCEAR: *Sources and Effects of Ionizing Radiation, Report to the General Assembly*. New York: United Nations Scientific Committee on Effects of Atomic Radiation, 2000.

Urbach, F.: Occupational Skin Hazards from Ultraviolet Exposure. In *Non-Ionizing Radiation. Proceedings of a Topical Symposium, November 26-28, 1979, Washington DC*. Cincinnati, OH: American Conference of Governmental Industrial Hygienists, 1980. pp. 145-156.

Vijay, M.M.: A Critical Examination of the Use of Water Jets for Medical Applications. Paper No. 42 In *Proceedings of the 5th American Water Jet Conference August 29-31, Toronto, Canada*. St. Louis, MO: Water Jet Technology Association, 1989. pp. 425-448.

Vijay, M.M.: High Pressure Safety. In *Fluid Jet Technology. Fundamentals and Applications*. 2nd. Ed. St. Louis, MO: Water Jet Technology Association, 1993. pp. 9.1-9.30.

Vijay, M.M. and W.H. Brierley: Feasibility Study of Cutting Some Materials of Industrial Interest with High Pressure Water Jets. In *Proceedings of the 2nd U.S. Water Jet Conference* (Rolla, Missouri). St. Louis, MO: Water Jet Technology Association, 1983. pp. 289-298.

Ward, W.D.: The Use of TTS in the Derivation of Damage Risk Criteria for Noise Exposure. *Internat. Audiol. 5*: 309-313 (1966).

Ward, W.D.: Effects of Noise on Hearing Thresholds. In *Noise as a Public Health Hazard*, Ward, W.D. and J.E. Fricke (Eds.). Washington, DC: American Speech and Hearing Association, 1969. pp. 40-48.

Ward, W.D., L.H. Royster and J.D. Royster: Anatomy and Physiology of the Ear: Normal and Damaged Hearing. In *The Noise Manual*, 5th Ed., Berger, E.H., L.H. Royster, J.D. Royster, D.P. Driscoll, and M. Layne (Eds.). Fairfax, VA: American Industrial Hygiene Association, 2000a. pp. 101-122.

Ward, W.D., J.D. Royster and L.H. Royster: Auditory and Nonauditory Effects of Noise. In *The Noise Manual*, 5th Ed. Berger, E.H., L.H. Royster, J.D. Royster, D.P. Driscoll, and M. Layne, Eds. Fairfax, VA: American Industrial Hygiene Association, 2000b. pp. 123-147.

Wasserman, D.E.: An Overview of Occupational Whole-Body and Hand-Arm Vibration. *Appl. Occup. Environ. Hyg. 11*: 266-270 (1996).

Wertheimer, M. and W.D. Ward: Influence of Skin Temperature upon Pain Threshold as Evoked by Thermal Radiation — A Confirmation. *Science 115*: 499-500 (1952).

White, C.S. and L.G. Bowen: *Comparative Effects Data of Biological Interest*. Albuquerque, NM: Lovelace Foundation for Medical Education and Research, 1959. p. 19.

WHO: *Lasers and Optical Radiation* (Environmental Health Criteria 23). Geneva, Switzerland: World Health Organization, 1982.

Wiegand, K. and S.P. Dunne: Radon in the Workplace—A Study of Occupational Exposure in BT (British Telecommunications) Underground Structures. *Ann. Occup. Hyg. 40*: 569-581 (1996).

Wiernicki, C. and W.J. Karoly: Ultrasound: Biological Effects and Industrial Hygiene Concerns. *Am. Ind. Hyg. Assoc. J. 46*: 488-496 (1985).

Wilkening, G.M.: Ionizing Radiation. In *Patty's Industrial Hygiene and Toxicology*, 4th Ed., Vol. I, Part B, General Principles, Clayton, G.D. and F.E. Clayton (Eds.). New York: John Wiley & Sons, Inc., 1991a. pp. 599-655.

Wilkening, G.M.: Nonionizing Radiation. In *Patty's Industrial Hygiene and Toxicology*, 4th Ed., Vol. I, Part B, General Principles, Clayton, G.D. and F.E. Clayton (Eds.). New York: John Wiley & Sons, Inc., 1991b. pp. 657-742.

Wilkerson, J.A. (Ed.): *Medicine for Mountaineering*, 3rd. Ed. Seattle, WA: The Mountaineers, 1985.

Wilkerson, J.A.: Frostbite. In *Hypothermia, Frostbite and Other Cold Injuries*, Wilkerson, J.A. , C.C. Bangs, and J.S. Hayward. Seattle, WA: The Mountaineers, 1986a.

Wilkerson, J.A.: Other Localized Cold Injuries. In *Hypothermia, Frostbite and Other Cold Injuries*, Wilkerson, J.A., C.C. Bangs, and J.S. Hayward. Seattle, WA: The Mountaineers, 1986b. pp. 96-101.

Wilson, A.J. and L.M. Scott: Characterization of Radioactive Petroleum Piping Scale With an Evaluation of Subsequent Land Contamination. *Health Phys. 63*: 681-685 (1992).

Wolff, J.-P.: Working Hot: How to Avoid Disaster. *Elect. Const. Maint. 98(5)*: 26-36 (1999).

Zabetakis, M.G.: Flammability Characteristics of Combustible Gases and Vapors (Bull. 627). Washington, DC: U.S. Department of the Interior/Bureau of Mines, 1965.

Zuclich, J.A. : Hazards to the Eye from UV. In *Non-Ionizing Radiation. Proceedings of a Topical Symposium, November 26-28, 1979, Washington, DC*. Cincinnati, OH: American Conference of Governmental Industrial Hygienists, 1980. pp. 129-144.

# 4 Human Factors and Hazardous Energy

## INTRODUCTION

One of the factors in fatal accidents detailed in Chapter 1 was failure of the interaction between individuals, between individuals and equipment, and between individuals and their environment. This interaction often was directed through procedures, written or unwritten, followed, not followed, or partly followed. Research presented in Chapter 1 indicated that some victims interacted with energy sources and materials containing energy when equipment and systems were live. This interaction sometimes occurred by choice, and sometimes through procedure. A deficiency in the interaction led to the fatal contact with the energy source.

Preparation of Chapter 1 involved review of hundreds of fatal accidents. Chapter 1 details the obvious elements that characterized the situations that occurred. Yet, in reading through descriptions of these accidents, one is left with the impression that the full extent of what occurred is not fully expressed or understood. The reason for this is that our daily lives are highly complex. Many factors, starting with those involving our home lives, affect the purpose and efficiency of our daily functioning. There are many detractors to performance, some of which are apparent from :reading between the lines" in the descriptions of the fatal accidents, while others are not.

Electrical contact accidents provide many examples of the problem. High-voltage lines and equipment are uninsulated. Protection of the general public against electrocution and other types of electrical injury relies on the distance between the lowest point on the conductor bundle and the highest point between the ground or protrusions above the ground. Normally, work does not occur at a position that puts the individual in proximity to high-voltage lines, or involve equipment that extends reach from the ground toward the lines. This, however, is not the always the case.

Urbanization of large expanses of land has meant that electrical services, formerly on agricultural or industrial land, now are in the middle of built-up, highly congested areas. This situation is even more extreme in urban areas containing above-ground wires. High-voltage electrical conductors usually are found on poles that line the sides of roadways. However, these are not the only places where high-voltage lines are encountered. High-voltage electrical conductors exist in close proximity to, and attached to many buildings, and sometimes are located close to, and sometimes above rooftops (Figure 4.1). The ever-escalating value of property has mandated use of land areas to the fullest extent possible. Buildings encroach onto minimum clearances mandated for preventing flash-over. This means that work at ground level or at heights near or on these structures increases the risk of contact or flash-over.

Work at ground level sometimes involve the handling of conductive structures that extend the reach of the individual, although in an unintended way (Figure 4.2). This includes, for example, the handling of aluminum ladders, metal poles, television towers, and aluminum siding. Any metal structure or object that extends human reach greatly increases the risk of electrical contact.

Work at heights can put workers into close proximity with bare electrical conductors located on utility poles or on the side of buildings.To illustrate, in one situation bricklayers on a scaffold could easily reach and touch energized conductors positioned on an adjacent building. One individual did just this to demonstrate to fellow workers that the lines posed no danger. This individual survived the encounter because the planks on which he was standing acted as insulators, and did not provide a path to the metal frame of the scaffold and to ground. Another individual who was touching the aluminum frame of the scaffold at the same time was not so lucky, and was electrocuted.

Electrical contact accidents also involve the use of equipment. Cranes used for unloading delivery trucks are involved in electrical contact accidents (Figure 4.3 and Figure 4.4). On many job sites workers routinely wear hardhats. The individual cannot see above brim of the hardhat, and since most of the work occurs at ground level, has no reason to look up. The focus is totally on the horizontal plane located at eye level and on the space below. Looking up requires the individual to tilt the head backward, an action that is short in duration due to the muscle strain involved. Truck drivers unloading

Figure 4.1. Congested wiring. With the development of industrial land into city environments, electrical contact with uninsulated above-ground wiring becomes increasingly likely during activities of normal commerce.

Figure 4.2. Fatal electrical contact accident. Electrical contact accidents often involve the handling of conductive structures that extend the reach of the individual, although in an unintended way. (Source: OSHA.)

Figure 4.3. Operating a truck unloading crane. The focus of the operator is the load. The top of the boom is above the plane of vision. This is especially the case when the operator wears a hardhat.

Figure 4.4. Fatal electrical contact accident involving a truck-mounted crane. (Source: OSHA.)

dump trucks focus on delivery of the load from the tailgate as the body goes up, not on the highest position reached by the front of the body. Crane operators focus on position of the load (usually close to ground level) and not the vertical position of the boom. This is a secondary concern. Individuals carrying long loads focus on the horizontal, usually the forward direction of progress. Individuals maneuvering vertically oriented equipment, such as ladders and extension poles, focus on the horizontal for a considerable portion of the time.

Many of the protrusions used at ground level are longer than the height of the overhead line. Some, like ladders and pole extensions to some equipment, are intended to be used in the vertical position.

The presence of uninsulated high-voltage lines in close proximity to the work zone introduces an additional complication in the performance of these tasks. The worker must divide attention between the task to be performed and the spatial relationship between the protruding device and the overhead line. The person naturally focuses attention on the task at hand, be it manipulating equipment or device, positioning the load, and preventing injury to other ground-level workers and property damage to the surroundings. Focusing on overhead power lines is secondary to this process, even when the person is aware of their presence.

The fatal electrical accidents described here serve as a surrogate for many other circumstances described as accidents waiting to happen. That is, the hazardous conditions await the performance of the task.

Among other considerations, this chapter explores the conflicting themes of task and conditions. These are parallel, mutually exclusive realities in workplaces and workspaces The existence of these realities and absence of interactions between them are crucial to understanding about the causation and occurrence of accidents involving hazardous energy. One cannot focus simultaneously on task and conditions during the performance of work. Rather, once can focus on one or the other of these realities.

The classic illustration of this concept is the expression about walking and chewing gum. These actions represent two different realities. Walking on dry pavement requires considerable concentration to ensure that the soles of shoes do not come into contact with unpleasant materials. Add ice and snow to that surface or other obstructions and the concentration needed to walk without slipping, tripping, and stumbling increases dramatically.

A further illustration of the concept is a well-known psychology experiment involving a basketball game (Curry et al. 2006; Simons and Chabris 1999). The instructor asks participants to focus on the ball as it moves from player to player and into the basket. Nobody notices the person wearing the gorilla suit who is readily obvious to less involved spectators when viewed on the video of the game. Again, people cannot focus simultaneously on task and conditions.

Our usual and natural focus is task. This observation perhaps offers an explanation about why seemingly well-experienced individuals, described by co-workers as being conscientious and highly safety conscious, are involved inexplicably in fatal accidents. This comment also applies to practitioners of occupational hygiene and safety. That is, as individuals, we also focus on task and not conditions during our activities at work. This focus on task is a natural human attribute. Focus on conditions requires thinking in the abstract about future events and the mechanisms by which they can occur, rather than the concrete level of the task, the level at which we normally operate.

## The Human Condition

The first considerations in a deeper exploration of these realities are the physical and physiological, and emotional and psychological status and capability of the worker and the demands imposed on these individuals. The influences on the well-being of the worker are not only those present and exerted at work. We spend almost 16 hours of every day away from work. This time extends to longer time during periods of rest and recreation on weekends and holidays.

Hence, first among these considerations are characteristics of the home environment (Table 4.1), about which nothing is revealed in the descriptions of accident situations. How many of these factors

## Table 4.1

## Detractors to Safe Performance of Work: Matters off the Job

| Factor | Destabilizing Influence on Performance of Work |
|---|---|
| Sleep | Insufficient sleep due to baby and child-rearing<br>Sleep disturbed by household and community noise |
| Diet | Quantity of food<br>Quality of food can affect level of blood sugar |
| Eating time | Eating time can affect performance at work |
| Health status of the worker | Chronic disease interferes with sense of well-being |
| Relationship between family members | Disharmony with spouse and/or family members<br>Separation<br>Divorce<br>Death of spouse, parent, or close relative |
| Financial matters | Worry about inability to pay bills |
| Health status | Sub-optimal health status due to disease<br>Worry about personal health status<br>Worry about the health status of a family member<br>Controlled mental disorder<br>Uncontrolled mental disorder |
| Substance abuse | Off-site intoxication may continue during work time<br>Intoxication not always detectable at work<br>Inability to perform at optimal level<br>Finances needed to support habit/addiction may lead to criminal behavior |
| Gambling addiction | Worry about inability to pay bills |
| Political matters | Worry/anger about domestic or foreign political situation |
| Interaction with strangers | Anger at actions of strangers with whom we have casual interaction (drivers, shoppers, etc.) |

detracted from the emotional and psychological, and the physical and physiological well-being of the victims whose stories were described will never be known to students of this subject.

Table 4.1 outlines factors that constitute some of the "baggage" that people bring to work and carry around with them all day. Some people discuss their situations with fellow workers to seek their support and advice, while others say nothing.

News reports in recent years have described many situations where distraught employees or former employees have "taken matters into their own hands" and entered and returned to workplaces to commit mass murder. Progressive organizations possessing sufficient resources operate Employee Assistance Programs to assist employees with such difficulties.

Additional influences on the performance of people are part of the workplace environment. While usually external to the work that people perform, these considerably influence the manner in which it is conducted (Table 4.2).

Elements mentioned in Table 4.1 and Table 4.2 influence the outcome of the actions of individuals from day to day and within any particular day. Taken together, the sum of these elements is human performance. Human performance also reflects the function of the biological structures within the body whose individual actions contribute to the outcomes of our daily lives.

Human factors are the behaviors, abilities, and limitations of humans that impact on the way that they perform. Human factors, therefore, influence the mode and success of interaction between humans and tools, machines, systems, tasks, jobs, and the environment (Sanders and McCormick 1993). This interaction can be positive, neutral, or detrimental. Human factors is a broad field devoted to scientific investigation to determine how humans interact with tools, machines, systems, and their environment, and to the design and engineering for the productive, safe, comfortable, and effective use of and interaction with these things. Human factors encompasses psychology, physiology, and several disciplines within engineering.

## The Senses

The fundamental components in human-machine/tool/system interaction are information and feedback passed through the sensing organs to the brain for interpretation and action. Humans normally employ sight, hearing, touch, smell, and sometimes taste in the conduct of work.

Sensory units contain three essential components: receptors, neural pathways, and interpreting centers in the brain (DeCoursey 1974). Although interpretation occurs in localized areas of the brain, sensations are projected to their source, such as a cut finger.

Sensory units are activated by interaction with energy sources in the environment. Receptors are grouped as enteroceptors, proprioceptors, and interoceptors. Sensors in the eye, ear, and skin that are sensitive to stimuli in the external environment are enteroceptors. Proprioceptors, located in the inner ear, in muscles, tendons, and around joints, sense position of parts of the body, and extent of stretch of muscles. Much of the action produced by the proprioceptors occurs below the level of consciousness. Interoceptors, located in the viscera, control autonomic functions. Receptors are tuned to change in the environment, rather than constancy.

Receptors have specialized to respond to mechanical, thermal, chemical, acoustical, and photonic energies. Special senses are those found only in the head: sight, taste, hearing, smell and head position and movement. General senses are those found throughout the body. Obviously, some of the senses are more relevant to a discussion of this type than are others. Hence, this discussion will focus on senses and receptors that could influence the outcome of situations involving contact with hazardous forms of energy and deterioration of human performance.

The simplest of the sensors structurally are the simple nerve endings (DeCoursey 1974). These are found throughout the body and are believed to be primarily pain receptors or concerned with tactile (touch) sensations. More complex receptors are specialized to pressure and touch. These receptors (and also those involved in sensing heat and cold) are encapsulated by connective tissue.

Pain is the most potent stimulus known to arouse and stimulate behavior (Leukel 1976). Pain sets off somatic and visceral withdrawal reflexes. Somatic withdrawal reflexes are the first line of defense against injury. Some areas are more sensitive to pain than others. These include the armpit and back of the knee, the neck, the bend of the elbow, and shoulder blade. The tip of the nose and soles of the feet are least sensitive to pain.

Pain can exert major influence on performance, judging by the emphasis on pain relief in-over-the-counter pharmaceuticals. Expressions of pain range from headaches to muscle aches to pain emanating from injury.

Meissner's corpuscles react to touch; that is, stretch or pull exerted on the skin (Leukel 1976). They are found throughout the body, but are most numerous in the fingertips, palm of the hand, and

## Table 4.2

## Detractors to Safe Performance of Work: The Work Environment

| Factor | Destabilizing Influence on Performance of Work |
|---|---|
| Hot temperature | Induces sweating, discomfort, and risk of heat stress<br>Major disincentive to wearing personal protective equipment that traps body heat and moisture |
| Cold temperature | Need to wear bulky outer clothing and gloves greatly reduces dexterity and adds weigh |
| Sun | Staring into the direct path of the sun considerably hampers visual function |
| Rain | Necessitates wearing of bulky clothing to keep dry and warm |
| Snow and ice | Items hidden under the snow<br>Slippery surfaces increase concentration, fatigue and risk of injury |
| Surface obstructions | Increase visual concentration and fatigue and risk of injury |
| Lightning | Necessitates ceasing work and sheltering |
| Loud noise | Impairs communication<br>Impairs ability to hear alarms<br>Causes fatigue |
| Lighting | Low levels impede ability to read critical information<br>Glare from high levels impedes vision |
| Solvent overexposure | Headache, nausea, drowsiness<br>Impairs concentration |
| Production conflict | Shortcuts in procedures taken to maintain or restore production |
| Workplace cultural conflict | Lack of support for routine procedures for safe performance of work because they are more complicated |
| Interruptions (breaks, end of workshift, reassignment by supervisor, telephone calls) | Interruption of procedure prior to completion at a critical step breaks concentration<br>Resumption may omit a critical step |
| Distractions (birds, voice calls, attractive members of opposite gender, alarms, use of entertainment devices) | Interruption of procedure prior to completion at a critical step breaks concentration<br>Resumption may omit a critical step |
| Interpersonal conflict | Breakdown of trust and cooperation<br>Possible inward and outward hostility |

**Table 4.2 (Continued)**

**Detractors to Safe Performance of Work: The Work Environment**

| Factor | Destabilizing Influence on Performance of Work |
|---|---|
| Recognition | Absence of recognition of ability or perception of low pay for quality of work can lead to lack of commitment to performance |
| Economic situation | Fear about layoff or termination of employment |

sole of the foot. They are dermal layer receptors. They are found in hairless areas of the body. The most touch-sensitive areas of the body are the tongue, and tips and backs of the fingers. The least touch-sensitive areas are the thick parts of the soles of the feet.

Pascinian corpuscles are sensitive to pressure (Leukel 1976). These are located deeply in subcutaneous tissue and throughout the body. They respond to deformation of the skin and are especially sensitive to vibration.

Among the more complicated of the sensory structures (Figure 4.5) are the semicircular canals located in the inner ear (DeCoursey 1974). They are deeply embedded in the bone and are oriented at 90° to each other. During normal orientation of the head, two of the three canals are nearly vertical. The third is nearly horizontal. The semicircular canals indicate motion of the head. This occurs when there

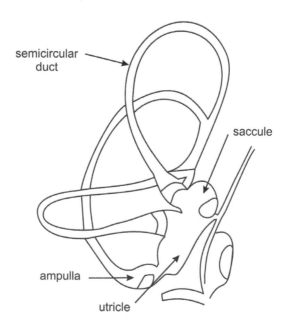

Figure 4.5. Human balance mechanism. Components of the balance mechanism sense static position and movement of the head. (Modified from Vander et al. 1990.)

is a sudden change of position, either in a direct line or rotation. Change refers to acceleration or deceleration, rather than constant motion. Whirling affects the horizontal canal much more so than the two semicircular canals in the vertical plane because they are perpendicular to each other. This motion can lead to nausea, sweating, and vomiting. Indication of motion occurs by activation of a sensory structure containing hair cells and a gelatinous coating that projects into the semicircular canal. Movement of the fluid induces movement of the hairs in the gelatinous structure.

The semicircular canals open into an oval membranous sac, the utricle. In the normal position of the head, the utricle is oriented horizontally. The utricle contains a gelatinous substance. The bottom surface contains hair cells. The latter project into the gelatinous substance. On the top surface of the gelatinous substance is a mat of minute calcium carbonate, stone-like objects called statoconia. These float on the surface of the gelatinous substance making it top heavy, and therefore more sensitive to movement (Leukel 1976). The structure is similar to molded gelatin with weights on top. Motion, especially wave-like, reciprocating motion, causes the structure to jiggle. Any position off the vertical causes displacement to the side. Either of these movements stimulates the hair cells which are embedded in the jelly.

The utricle indicates relative position of the head. During a change in position of the head to a new orientation, the relative position of the components of the utricle changes. This affects the position of the hairs, and the signal that they send to the brain. Continuous stimulation of the utricle is believed to be the cause of seasickness.

The utricle opens into a second similarly configured sac called the saccule. The saccule is oriented perpendicularly to the utricle. Hence the hair cells project horizontally.

The semicircular canals and the utricle play a role in maintaining equilibrium, postural reflexes, and righting reflexes. The role of the saccule is not known. Regular rhythmic motions that differ sufficiently from walking can cause motion illness. Motion-induced illnesses include carsickness, airsickness, and seasickness (Leukel 1976). The primary component is vertical motion, rather than front-to-back or side-to-side movements. The frequency of 20 Hz (waves per minute) and amplitude of 2.1 m (7 ft) seem to have the greatest impact. Adaptation does occur.

### Sensation and Perception

Sensation is the activity of receptors and the resulting activity leading to the sensory area of the cortex in the brain (Hebb 1972). Perception is the activity of mediating processes to which the sensation gives rise directly. Perceptive processes do not occur reflexively. They occur through mediated action and decision-making. Sensation is a one-stage process. A receptive surface is activated and sends a signal. Perception is sequential, starting with a sensation, leading to a motor reaction with feedback, and perhaps a series of exploratory reactions. Perception often involves input from more than one sensory agent and movement by the limbs. Perception is preparation for response. The appropriateness and adequacy of the response depend on circumstances.

Psychophysics is the science of functional relations, or said differently, relations of dependency between body and mind (Christman 1971). Psychophysical methods are used to study the response to signals and other inputs. Fechner developed the main methods in use (Boring 1950).

The strategy employed in the Method of Limits (or Method of Just Noticeable Differences) is to change the stimulus in discrete steps until the subject changes the response. In one variant, the tester increases signal strength. The subject responds on first reception and recognition of the signal. The signal may be undetectable when first presented and becomes detectable only on increased intensity. This is the technique used in audiometric testing. The tester increases signal intensity at a particular frequency until the subject first detects it. This establishes the audiometric threshold at that frequency.

The converse methodology is also utilized. In this case, the tester decreases signal strength until the response is no longer detectable. This methodology is used by optometrists when determining the appropriate strength of lenses for a patient. The optometrist decreases magnification from a stronger power to a weaker one until the patient just notices inability to read the eye chart. The optometrist then shifts the magnification back and forth several times to obtain the most precise judgement possible from the patient.

The approach in the Method of Right and Wrong Cases (Method of Constant or Constant-Stimuli) is to present the stimuli in random or quasi-random order. The subject indicates the stimulus as sensed or not sensed. A variant on this theme is the Method Constant Stimulus-Difference. In this case, the subject determines the detectability of differences between stimuli. Again the stimuli are presented in random order. Visual signals are presented together, while audible signals are presented sequentially.

The approach in the Method of Average Error (Method of Adjustment) is to require the subject to match a standard signal with a variable one. The variable signal is continuously variable, rather than varying in discrete steps. In one variation of this approach, the subject can overstep the end-point and modify the signal to home in on it. Common practical applications for this method include matching colors or intensity of light, matching the position of an indicated signal, such as the light meter reading in a camera, and matching readings of dials in control rooms.

Signal detection by humans is not precise, and does not produce go/no go, yes/no kinds of end-points (Christman 1971). Rather, there is a range that is subject to the vagaries of consciousness and acuity of the senses. The latter is affected by age, physiological and mental health, alertness, fatigue, and so on.

End-points are also subject to chance. Some signals are harder to discriminate than others. For example, color variations are harder to discriminate than presence of light itself. The preceding factors broaden the range between definite yes and definite no responses. Many individuals are deficient in color recognition (color blindness). In the case of sound detection, damage to the hearing detection apparatus and age alter the limit of detection.

Habituation is the expectation of a condition. Habituation is a source of error in these determinations. Habituation is an especial concern where the individual assesses unvarying signals for long periods, such as occurred in control rooms in process facilities and military surveillance operations during scans of long-range radar signals. This situation was an especial concern in the era prior to installation of computer monitoring systems which allow the operator to focus attention onto deviations in the signal once identified as such, rather than the search for the deviation.

The opposite of habituation is anticipation. This is the "jump the gun" or false-start that occurs during competitive sports events. The subject attempts to anticipate the arrival of the signal and wants to respond at the moment that this occurs, in this case for competitive advantage. These errors in response are controllable to a large extent through training, along with appropriate instruction. Identification of individuals prone to these characteristics is possible through testing.

A later development in psychological testing is the Method of Rank Order (Christman 1971). The strategy in this method is to require the subject to arrange a group of things into a successive order or ranking from least to most or vice versa. This approach is at best semi-quantitative, since success depends partly on the value system of the subject. There is no requirement for differences between members of the group to be uniform. This sorting out occurs in many aspects of social and industrial activity. In sport, this ordering occurs through performances measurable objectively by the stopwatch, and in other cases by a panel of judges.

A variation on the theme is the Method of Paired Comparisons. In this case, the subject determines the ranking between two items at a time. This method can be extremely time-consuming given the number of side-by-side comparisons between members of a group.

Rating scales are common in their application throughout industry in assessment of social interaction and perception of importance or concern in opinion surveys. The basis for these determinations includes quantitative methods of achievement or subjective perceptions. Scales rank between extremes that can have finite or infinite limits.

Signal detection (or rejection) can have critical connotations in many applications outside the lab. This is especially the case when the human observer is the processor of the signal. One application having life and death consequences was decision-making during the "Cold War" involving radar signals. A real signal intermingled with a background of electronic noise could have indicated the presence of foreign objects, the precursor indication of an attack. Operators were required to stare at radar screens for long hours looking and listening for something different beyond the background. Detection by a human observer of that "something" required training in recognition and decision-making skills. Normal practices in signal detection involving an analogue threshold were not considered satisfactory.

In fact, the decision was found to involve a statistical element concerned with the acceptance or rejection of the hypothesis concerning the correctness or inaccuracy of the observation (Peterson et al. 1954). This is in keeping with previous comments regarding the range of detectability intra- and interpersonally. This approach has since been applied to many other applications in decision-making (Swets 1961). What is critical in this approach is not detection of an absolute threshold, but rather integration of the context and meaning of information through the use of probability and intuition.

## Information Processing

Information processing is the sequence of steps related to receiving stimuli, attending to them selectively, storing information in memory, deciding how to respond, and assessing the adequacy of the response (Sheridan 1976). Response time is the time between presentation of the stimulus and the response to it.

Information processing occurs in stages. A stage is identifiable by characteristic functional properties. Stages vary in duration and from person to person, and depend on different features of the stimulus. Stages involved in information processing include receiving stimuli, deciding how to react, and checking on the adequacy of the reaction. Stages can occur sequentially or simultaneously.

Mental associations assist in processing sensory inputs. These associations include images evoked by some element of the sensory input. With verbal inputs, for example, words easy to pronounce evoke images. This phenomenon is used by students studying for exams. Reading the material out loud several times and concentrating on the words and the images that they evoke is a major aide to committing information to short-term memory. Linking the sound created by forming the words when reading them out loud to the image of the word on paper is critical to the process. Tracing images and formulas with the fingers further aids in the process.

Information processing is believed to occur in units or blocks (Christman 1971). One unit of information, the bit (binary digit), is the content needed to reduce uncertainty in the information by 50%. That is, where two events with equal probability can occur, there is one bit of uncertainty (Kantowitz and Sorkin 1983). When one of the two possible events occur, one bit of information is acquired. The information gain in going from 100 alternatives to 50 is the same as going from 50 to 25, or from 2 to 1. Information gain depends on ratios, not on absolute numbers. Hence, various types of information are measurable using the same unit, the bit.

A practical example will illustrate the concept. The question is, how many bits of information are needed to specify a particular location, randomly chosen, on a checkerboard containing 64 squares, without uncertainty and without guessing (Table 4.3). Each of the questions in Table 4.3 increases the information content by half (50%). A minimum of six questions is needed to delineate the marked square.

The number of bits of information needed to specify information unambiguously occurs according to the power of 2 law (equation 4-1).

$$N = 2^n \tag{4-1}$$

where:
N = the number of things to be differentiated.
 n = the number of bits of information needed to differentiate all of the information in an unambiguous manner.

That is, n is the minimum number of questions needed to differentiate the most difficult to resolve member of the group. Situations where the information gain is not equal to 50% introduce complexities into the mathematics.

One of the classic experiments in human information processing determined that humans could remember $7 \pm 2$ numbers. That is, depending on the stimulus, humans could reliably remember between five and nine items from a group of similar items (Miller 1956). This is one of the reasons for the choice of the seven-digit telephone number, and by extension, the difficulty that many people encounter when inserting the three-digit area code in front of existing numbers already committed to memory

## Table 4.3
## Specification of Information by Bits

| Question | Alternate Responses | Definitive Answer? | Bit Count |
|---|---|---|---|
| On which side of the board is the square? | Right or left | No | 1 |
| On which half of the board is the square? | Top or bottom | No | 2 |
| On which side of the section is the square? | Right or left | No | 3 |
| On which half of the section is the square? | Top or bottom | No | 4 |
| On which side of the segment is the square? | Right or left | No | 5 |
| On which half of the segment is the square? | Top or bottom | Yes | 6 |

when making local calls. As an added example of this phenomenon, Patterson (1982) found that subjects could memorize 7 of 10 auditory alarms, and recall them after a period of time. Again, the recurring number was seven.

The eye and the ear are capable of providing large quantities of information per second. The bandwidth of the ear is estimated to be 8000 to 10,000 bits per second, and the eye, about 1000 bits per second (Sanders and McCormick 1993). The human brain receives thousands of sensory inputs at any point in time, yet can discriminate only about 2.5 bits of information. The bit is independent of the particular dimension containing the information (Sanders and McCormick 1993; Sheridan 1976). People experience considerable difficulty when attempting to remember more than five to nine different standards against which to judge inputs.

The question of the capabilities and limitations of human performance in the face of ever increasing inputs of information is critical to ensuring health and safety. Despite many years of study, the limitations of information processing and the ability to process more than one input still are not known.

One model holds that humans can transmit information at a limited rate (Kantowitz and Sorkin 1983). According to this model, people only do tasks requiring transmission of a small number of bits of information each second. Exceeding that transmission rate leads to errors and reduced transmission of information per second. According to this logic, tasks requiring less processing than the maximum capacity leave open room for additional processing. Hence, multi-processing (multi-tasking) should be possible.

The multi-processor (multi-tasking) model holds that humans can do independent tasks simultaneously. One task inevitably becomes the primary task and the other, the secondary. Psychological studies have examined performance outcome as the primary task becomes more complex; that is, requires more information processing. Increased task complexity should lead to decrement in performance of the secondary task. The predicted outcome does not always occur. In some combinations of tasks, little loss in efficiency occurred. What was observed subsequently is that humans are efficient at acquiring information, yet are inefficient when many responses are required. The bottleneck occurs when outputs (responses) are required. The latter highlights the old adage about walking and chewing gum at the same time, or playing the drums with hands and feet seemingly doing different things, or playing the organ or piano with both hands on different keyboards or parts of the keyboard. Very few people can do these things.

To draw a parallel in the industrial world, work procedures in the electrical industry rely heavily on the use of personal protective equipment and temporary shielding to prevent contact with energized conductors. This approach relies on the ability to position the shielding in an exact configuration. This is not always possible. As a result, the worker must maintain constant vigilance about gaps in protection and how to avoid them, while performing other tasks, including the one that motivated the exposure to the energy source.

An additional factor in information processing is compatibility (Sanders and McCormick 1993). Compatibility refers to the relationship between stimulus and response to human expectations. Obviously, compatibility is one of the fundamentals of design of successful systems and equipment. Conceptual compatibility refers to the meaningfulness of codes, visual symbols, words, and acronyms to the people who must work with them.

Movement compatibility refers to movement of displays and controls and the response of the system affected by it. Clockwise rotation of a dial or movement to the right usually indicates an increase in some parameter.

Spatial compatibility refers to the physical arrangement in space of controls and displays. The preceding compatibilities are intrinsic in the situation or reflect cultural conditioning.

Modal compatibility acknowledges that some stimulus-response combinations are more compatible with some tasks than with others. For example, the most effective combination for a verbal task is auditory input and spoken output. The most effective combination for a spatial task is visual input (displayed on a video screen) and manual response (Wickens et al. 1983). This type of compatibility has its origins in the way in which the brain is wired.

Human memory is believed to contain three storage subsystems: sensory storage, working (short-term) memory, and long-term memory (Kantowitz and Sorkin 1983). Sensory storage holds information transmitted from the eyes and ears for a brief period. Working memory maintains information for the longer periods needed for perception. Rehearsal maintains information in short-term memory. Interrupting the process of rehearsal can cause memory loss. Long-term memory is permanent, although not readily accessible to the conscious mind.

Interferences can prevent the association of learning (Sheridan 1976). Proactive interference occurs when prior learning interferes with new learning. Retroactive interference occurs when newer learning displaces previous learning. This occurs when learning sessions occur for long periods and the instruction does not refer back to previous learning. An effective instructor uses "handles" in the instruction and refers back to these in later sessions. This back and forth motion of ideas keeps the older information in the foreground, so that retention will occur more easily.

An example of this problem confronts partygoers and teachers. Partygoers who experience a succession of introductions to strangers have a difficult time memorizing names for future reference during the evening. Similarly, high school teachers are required to learn the names and characteristics of 150 or more students. Associations between names and behavior are important cues. Boisterous students and troublemakers stand out quickly. Leaders in the student group and clever students also stand out quickly. Attractive faces stand out from the crowd. Quiet, unassuming students are among the last to be recognized. Name tags are an invaluable means for easing this burden. They also permit discrete checking on this information. This enables the observer to make the association with visual and other sensual cues, such as odor of perfume and aftershave worn by the person.

As important to learning information is forgetting it (Sheridan 1976). The image of a visual stimulus persists for a brief period following presentation. The information that it contains is available during this time. A large quantity of information can be lost in the brief time period between sensing the image and reporting about it, a period as short as one second and as long as five.

Forgetting information occurs due to disuse or interference from other learning or both (Hebb 1972). Disuse refers to regression of stored information. This is the popular notion of memory loss. Spontaneous forgetting is what affects the many thousands of inputs received by the brain in the course of a day.

As workers remain longer in the workforce, questions about age and information processing acquire considerable practical importance. The effects of aging on information processing are progressive throughout adulthood. Aging affects all processes discussed in this section. These changes usually become noticeable after the age of 65 (Czaja 1988; Welford 1981). Generally, older people require more time to retrieve information from long-term memory and experience disruption in working memory. Aging also leads to interference in coding of ambiguous signals. This is believed attributable to weaker signals from sensory inputs and increased processing "noise" in the brain.

There is a general decrease in efficiency of the sensory organs with aging (Singleton 1972). Older people rely more on visual cues than other senses, so there is suggestive evidence that these are profoundly affected. Sensorimotor performance declines about 60% from maximum capacity as we age from 20 to 60. There is little reduction in capacity for immediate memorization, but information retention decreases and is more easily disturbed through distraction. This decrease in information handling results in a narrowing of interests and restriction of activities and decreased versatility. Given these realities, older workers who attempt to "keep up" with the young run the risk of making errors, or they may reduce speed more than is required to compensate.

## Attention to Task

One of the critical factors in workplace activity is being alert for signals or being attentive to inputs that occur unpredictably (Sanders and McCormick 1993). The term attention has several subdivisions.

Selective attention refers to monitoring several sources of information to determine whether change has occurred. This could involve a scan of instruments, for example. During scans of multiple channels of information, humans tend to focus attention onto those where information changes most frequently, and hence, most predictably. As a result, an input that changes infrequently receives less attention and could pass unnoticed. Under conditions of high stress, humans sample fewer sources, and those that are sampled are the ones perceived to be most important and salient (Wickens 1984).

Patterson (1982) highlighted the fact that disruptive alarms divert attention from the task at hand, and possibly the emergency itself. This indicates that the disruption caused by the alarm itself can be a major contributor to the situation when it exceeds the seriousness of the problem being highlighted. A disruptive alarm monopolizes the attention of the operator. Silencing the alarm becomes the priority. Addressing the cause is secondary. This outcome, of course, has a cost in response time.

The solution to this situation is to design the signal provided by auditory alarms to match the urgency to be attached to them. This indeed is doable (Edworthy et al. 1991). Achieving this goal involves manipulation of fundamental frequency, amplitude envelope, harmonic delay, rhythm, speed, pitch range, and pitch contour. Perceived urgency increases and reaction time decreases as signal pulse increases and inter-pulse interval decreases (Haas and Casali 1995). Unfortunately, tailor-made solutions for particular situations and problems are not yet possible (Robinson and Casali 2000). The solution at this time requires thorough study to match the particulars of a given situation with products currently available in the marketplace.

Focused attention refers to addressing particular sources of information to the exclusion of all others (Sanders and McCormick 1993). Focusing attention is the mental process that occurs while reading a report in a busy airport lounge while waiting for a flight. This is exceedingly difficult if the distraction is within one degree of visual angle from the task under observation (Broadbent 1982). Distinguishing factors aid in focusing attention. These can include features, such as louder, brighter, larger and more centrally located features when displays are involved.

Divided attention (multi-tasking) occurs when two or more tasks must occur simultaneously and the observer must monitor them (Sanders and McCormick 1993). Examples include cell phone use while driving a vehicle, carrying on a conversation with another person while walking, and eating and reading. Some aspects of divided attention were considered in the previous section. Primary school teachers, receptionists in busy offices, and sales people in busy retail environments are examples of occupations involving many interactions with people or signals. Performance of the main task declines because of the interruptions. This is especially noticeable as complexity of the interrupted task increases.

Some combinations of tasks can coexist without problem (Wickens 1984). Tracking tasks disrupt each other, but not tasks, such as mental arithmetic, that require resources in central processing. Time sharing occurs better when tasks utilize different modes, for example, visual for the one and audible for the other (Isreal 1980). Time sharing occurs better when the tasks to be shared are spatial and verbal. Spatial tasks involve moving, positioning, or orienting objects in space. Verbal tasks involve words, language, or logical operations (Sanders and McCormick 1993). Time sharing also occurs better when the tasks to be shared are vocal and manual.

The last type of attention is sustained attention (Sanders and McCormick 1993). Sustained attention refers to monitoring or vigilance to detect infrequent and unpredictable signals. Examples include surveillance by radar operators for incoming missiles, visual inspection for defects or foreign objects in products moving along an assembly line, and inspection of baggage for weapons. This type of work demands considerable motivation. Accuracy and speed of detection decrease exponentially during the first 20 to 35 minutes of the task and then remain constant (Giambra and Quilter 1987). Motivation is a key component in sustaining the performance. Work-rest rotation and optimization of the signal both can aid performance.

## Motivation and Emotions

Motivation is the tendency of an organism to produce organized, effective behavior. Motivation differs from reflexive actions, since it involves information processing. Motivation is biologically primitive and occurs far down the evolutionary scale. This indicates that motivation has value in survival (Hebb 1972). Several motivational factors can confront the brain simultaneously. These can lead to conscious debate about the appropriate course of action to take. Ambivalence is the state resulting from challenge by a group of opposing motivations.

Emotion is not precisely definable. Emotion refers to state of awareness. Emotional states have both positive and negative impacts on human performance, and hence the manner in which it occurs. Positive emotional states include joy, love, pride, and fun. Negative emotional states include anger, jealousy and fear, grief, shame, and depression.

Emotional states have in common that they are special states of motivation, and closely related to arousal (Hebb 1972). Anger is a temporary heightening of arousal accompanied by a tendency to attack. Fear is a temporary heightening of arousal accompanied by the tendency to withdraw. Disgust or horror is arousal with no accompanying expectation of being injured, but the desire to avoid the thing that disgusts. Joy or love is arousal accompanied by desire to deepen immersion in, or continue contact with the activity or thing that produces the effect.

The capacity of sensory stimulation to guide behavior is poor when arousal is very low or very high (Figure 4.6). At low arousal, such as occurs during waking from sleep, sensory transmission to the cortex of the brain is low. Emotion is closely related to arousal (Lindsley 1951). As arousal increases, so also does emotion. Low levels of arousal are organizing. High levels of arousal, as occur during emotional disturbance, produce high levels of activity in the cortex leading to disorganized behavior. These states are identifiable through measurement of brain wave activity.

A simple, long-established habit should be less disrupted by anger than a complex response that depends on delicate interaction between processes. Highly creative, detail-oriented activity, such as writing this book, is adversely affected by heightened emotional status. Emotional arousal, both positive and negative, can impair the ability to concentrate and compose thoughts. Emotion is directly correlated with arousal. Emotion, or arousal, is motivating to the point where conflicting activities in the cortex begin to interfere with each other (Hebb 1972).

Motivation rises at first with arousal, and levels off and falls at high levels of arousal. Emotional excitement can impair behavior and the processes that control organized behavior, and reduce the effectiveness of response to zero. This occurs through a loss of motivation that involves abolishing the thought or intention of acting. This is the literal meaning of paralysis of terror. (Paralysis occurs in the figurative rather than the literal sense (Hebb 1972). This is the situation reported by Tyhurst (1951), for example, concerning people caught in disasters. Tyhurst reported that about 15% of victims showed organized and effective behavior; 70%, varying amounts of disorganization but ability to perform with some effectiveness; and the remainder, completely ineffective behavior. The latter percentage varies from 10% to 25%, depending on the event. Studies of infantrymen in battle indicate that only 15% to 25% reliably will fire their rifles in the presence of the enemy. Additional examples of motivational

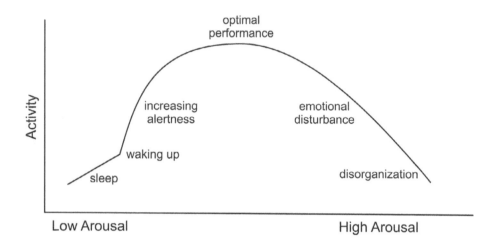

Figure 4.6.  Performance arousal curve. (Modified from Hebb 1972.)

impairment by a high level of arousal include stage fright, inability to perform in the face of great opportunity, and emotional attachment to a task, such as family ties to a patient requiring surgery (Hebb 1972).

One aspect of animal behavior that defies conventional logic is altruism (Hebb 1972). Altruism (good Samaritanism) is intrinsically motivated purposeful behavior whose function is to help another person or animal. The altruist expects to receive no immediate benefit from the action. Altruism is common in human society. It extends to the point of risk of life to the altruist. Altruism also exists in the animal world.

Another aspect of animal behavior that encompasses humans is the need to be physically and mentally active. This is expressed through play. Many animals play in the physical sense. Some also pursue mental challenges (Hebb 1972). Mental play involving the brain, as much or more than the muscles, becomes prominent in higher mammals. The need to avoid the conditions of boredom is fundamental to humans and animals. Boredom is the state in which the subject seeks a higher level of excitement, usually in the form of play. Avoidance of boredom is a crucial human pursuit.

A human lacking work often invents it. This need to invent situations can lead to risk-taking. Humans seek situations that produce fear, in the guise of thrill or adventure. During recreational time this is expressed through pursuits, such as rock and mountain climbing, whitewater rafting and kayaking, snowboarding and skiing, partaking of thrill rides in amusement parks, and auto racing. All of these depend on thrill for their attraction. This reality, if unrecognized, can have serious consequences in the underlying dynamics of workplace activity. Procedures, both verbal and written, intend to demonstrate the way to pursue work in the manner that poses the least risk. No doubt, there are circumstances that express themselves through accidents where individuals assumed undue risk. Altruism and risk-taking may not be as unrelated as first appears.

## Fatigue, Sleep Disorders, and Shiftwork

The comfort of workers at a worksite has a profound impact on productivity and their personal well-being and safety. Human factors considerations include the timing and length of workshifts, and provision of sanitary facilities and comfort stations.

An example (a composite situation) serves to illustrate just how wrong things can go, and why attention to these concerns is critical for safe conduct of work. The worksite was a field location about 100 km (60 mi) from the nearest town. The town has doctors and a small hospital. The worksite was located off-road about 5 km (3 mi) from the main highway. As conceived, the project was expected to be completed in 27 to 29 hours. Work was to proceed nonstop. Laborers were scheduled to work 12-hour shifts. Supervisory and technical personnel planned to work non-stop without a break. A contractor supplied heavy equipment and operators. The contractor planned to work non-stop until the work was completed. The owner provided a small trailer containing first aid equipment. This could not accommodate more than one person. Workers were expected to provide their own food and refreshments.

Due to an unforeseen problem in scheduling, the start of the project was delayed 3 hours. Instead of starting at 11:00, the work started at 14:00. Hence, the work did not start until late in the normal workday of the supervisors, technical personnel, and heavy equipment operators. All of these people were accustomed to working normal day-shift hours.

Daytime weather was hot and sunny with periods of rain. Nighttime weather was very cool, just above freezing, with periods of rain. Most of the workers were unable to take shelter from the rain during downtime or to keep active to conserve body heat. These conditions, compounded by the fatigue, were highly conducive to the onset of hypothermia. Several of the laborers were stricken and were provided with assistance, as best could be arranged, given the limited shelter available.

Partial flooding of the worksite led to an additional delay in advancing the project. The site was located in the path of seepage from a lake located at a higher elevation. The excavation created by the heavy equipment lowered the level of the grade below the water table. Pumps had to be located and a channel dug to remove the water.

The project was completed 42 hours after the start, almost 15 hours after the expected end-point. Some of the supervisory and technical staff collapsed from lack of sleep.

The work plan for this project contained no contingency for delay or emergency. As events unfolded, this decision led to imposition of unrealistic demands on capabilities and performance of supervisory and technical personnel, and the contractor's employees. This decision also had a direct impact on the safety of these individuals and the safety of others working at the site.

Most of the work involving operation of the heavy equipment (bulldozers and front-end loaders) occurred in close proximity to laborers and technical and supervisory personnel. This operation required intense concentration and high levels of diligence by the operators of the heavy equipment. Lighting was far from ideal, being provided by running lights on the equipment and portable generator-light sets. As well, pedestrians periodically passed close to the equipment. The latter activity necessitated gaining attention from the operator and waiting until the equipment had stopped operating. Long hours of this type of activity without breaks under these conditions is highly fatiguing and highly conducive to error. Safe operation in these conditions requires humans to perform within narrow tolerances of perfection.

People work best with a schedule that incorporates formal break periods, including time for sleep. Formal breaks for operators of heavy equipment are a legal requirement in many jurisdictions. There is a very real tendency under the pressure of time-critical operations, such as the one described here, to continue working, oblivious to the onset of fatigue. Contingencies to address fatigue and the outcomes of fatigue were not considered in this project.

An additional concern arising from the long work schedule mandated in this project was exposure assessment. Exposure considerations normally are based on shift lengths of 8 hours. Interpretation of results from exposure assessments during considerably extended workshifts is very difficult. Continuous long exposure to contaminants, even at very low levels, potentially was a factor in the fatigue experienced by the workforce.

In time-critical projects, the overall goal of timely completion always must remain clearly in focus. Not having replacements for supervisory and technical personnel put timely completion of this project at risk. Loss of one or more of these individuals from uncontrollable fatigue or injury could have increased project time beyond even that which was required and created other lasting consequences. Accidental damage to critical equipment resulting from substandard performance from overly fatigued operators could have produced the same result. The costs of failing to meet the goal because of unexpected events and failures can far outweigh any perceived savings through "short cuts".

Considerable useful information and even some wisdom can be gained from examining in some detail the processes and dynamics that occurred during this project. Considerable improvement in future projects can result by building on the lessons provided by this episode.

A major useful point to start is to acknowledge the limitations of human performance. By planning work within normally accepted work schedules, the project should progress more quickly since the minds and bodies of all participants will be rested and not overly stressed. This approach provides contingency and redundancy in the event of injury and other unforeseen circumstances.

Provision of food and hot and cold refreshments also deserves mention. This, however, becomes less of an issue when the first consideration (normal work schedules) is addressed. Weather is an important consideration. Susceptibility to cold stress especially is exacerbated by rain and long periods of inactivity in temperatures just above the freezing point. Having refreshments available to combat the effects of weather could be very important for maintaining well-being, and hence, productivity.

To illustrate, Wallin and Wright (1986) documented the impact of working conditions on performance in a methodological survey. This work illustrated the effects of situational aspects on the well-being of both white- and blue-collar workers. The results illustrated the impact of controllable and non-controllable conditions. Stress reactions can occur because of overload, lack of influence over working conditions, conflict, and low quality of work. Overload and pace were mentioned as the leading cause of stress reactions. The result was fatigue, listlessness, cardiovascular involvement, and tendency to depression. This also occurred in managers and others who held positions that provided varied work opportunities. The negative aspects of this work far outweighed the positive, according to this methodology.

The relationship between fatigue and accidents has been postulated since 1931 (Heinrich 1931). Heinrich postulated the role of fatigue in the sequence:

overtime → fatigue → incidents → injuries

Research has not proven that this sequence models reality. This model appears to be too simplistic and may need qualifiers. These include amount of overtime and recovery from fatigue, type of work, attitude toward overtime, pace of work and planning, and control and supervision (Gaunt 1980).

Many human physiological and psychological functions follow a 24-hour cycle (Figure 4.7). This is related to, but not necessarily controlled, by the diurnal activity pattern (Luce 1970). Diurnal activity patterns are evident in hundreds of physiological processes, including, respiration rate, urine excretion, mitotic activity (cell division), blood pressure, production of adrenal hormones, enzyme activity, body temperature, and reaction time. Kleitman (1963) showed that performance is related to change in metabolic rate of chemical processes in the brain. Reaction time is closely related to body temperature. This relationship also exists in the processing of simple information for which the speed of response is the major variable (Hockey and Colquhoun 1972). The diurnal temperature curve is an irregularly shaped sine wave that peaks at 18:00 and troughs at 4:00 (Aschoff 1981). More refined research indicates that there is a second, less severe trough of alertness that occurs between 14:00 and 16:00 (Walsleben et al. 1998).

The timing of the workshift or extended workday (as described in the example) can cause the demands of performance to be completely out of phase with diurnal rhythms. This is especially the case where individuals work considerably longer hours than normal (as in the example), or where the workshift changes. The cycle of body temperature remains unchanged by the shift in working hours (Colligan and Tepas 1986). Hence, the body temperature cycle could become dissynchronous with the activity cycle. During the night when body temperature drops, the individual attempts to remain awake and active. During the day when the individual needs to sleep, body temperature increases and passes through its maximum. Sleep deprivation can occur under these conditions. Not surprisingly, the biggest effect of shiftwork on individuals occurs on the quantity and quality of sleep.

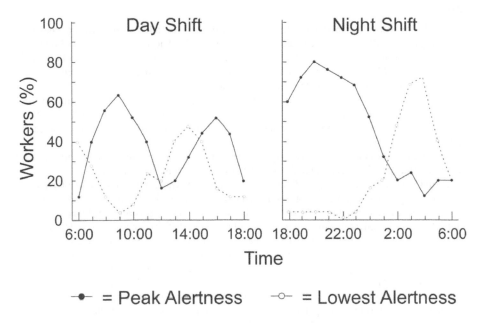

Figure 4.7. Circadian rhythms. (Modified from Colligan and Tepas 1986.)

Research indicates that night shift workers consistently sleep less than day and afternoon shift workers. Shift work produces qualitative and quantitative changes in sleep patterns (Walsh et al. 1981). There is evidence to suggest that shiftwork, in general, and night work, in particular, impairs normal sleep patterns, and has the potential to exacerbate pre-existing health problems, and to influence the action of prescribed medication (Colligan and Tepas 1986). Sleep loss makes people sleepier when awake, and less able to perform safely and efficiently on and off the job (Rosa and Colligan 1998). A worker who is sleep-deprived and whose diurnal temperature cycle is dis-synchronous is likely to be at higher risk for operational errors or accidents than a day-adapted worker working the day shift.

While there is no ideal candidate for shiftwork, the closest fit would be young workers whose diurnal temperature rhythms are relatively flat and who are introverted evening types. Older workers have more difficulty adapting to this lifestyle. Their temperature rhythms are more stable and more resistant to phase-shifting than those of younger workers (Reinberg et al. 1981). Similarly, individuals often describe their temperament as being "morning-type" or "night-type:. The latter are comfortable staying up late and sleeping during the morning.

In an idealized world, people would work according to their physiological preference. The real world, of course, is not amenable to such a process of selection. Many people work shifts with which they are not comfortable. Compounding matters is the fact that many two- and three-shift continuous operations rotate shifts on a regular basis. The former may include 12-hour shifts. The first night on a 12-hour shift is considered to be the most difficult, the least safe, and the least productive (Budnick et al. 1994).

The change from day shift acclimatization to night shift acclimatization occurs over the period of a week (Colquhoun 1971). The sine wave flattens somewhat (Colligan and Tepas 1986). This suggests that the activity cycle has an influence on body temperature but does not control or regulate it. Similarly, recovery from night work requires seven days. This is the same situation involved in recovery from "jet lag" following travel across multiple time zones.

## Aging

Until recently, the workplace model dictated that older workers would leave and retire and make room for the young and middle-aged somewhere beyond the age of 55. The stragglers and hold-outs would persist until age 65. Very rarely would individuals work beyond 65. Workplaces were not oriented toward the needs of older workers, the regulatory framework was hostile to their presence, and few workers wanted to continue or had the courage to challenge the system

This all changed with the arrival of the "baby boomers" on the threshold of the normal age of retirement of 65. Throughout their lives, the 'boomers' have had considerable impact on social themes and outlook. The arrival of the "boomers" rendered obsolete many social concepts. The concept of retirement is about to change also.

This situation does not exclusively reflect the collective attitude of the "boomers" toward aging. Three main influences that have suddenly converged will have long-term impact.

The first is the collective attitude of the "boomers" that government has no right to interfere in the lives of this group. That is, government has no mandate to dictate when retirement is to occur.

The second influence pertains to money and the expectation that government should provide sufficient to those of retirement age to ensure absence of hardship. Lifestyle choices have left many who are reaching 65 suddenly realizing that they have inadequate financial resources, meaning little money, with which to finance the coming years. This problem seems especially acute in the U.S. where "consumer spending" kept that economy moving for many years. Compounding the situation was the collapse of the global investment economy in which too many dollars and other currencies were seeking safe havens for continued economic growth through too few legitimate vehicles for investment. The suddenness and the depth and persistence of the economic collapse of 2008 have provided a profound lesson in economic reality to the boomers. A bout of inflation and overspending by government would rapidly destroy the value of fixed income government-provided pensions on which many people have planned their retirement and will be totally dependent.

The third influence pertains to healthcare and its cost. The elderly are major consumers of healthcare. The latter includes housing in seniors-oriented settings. Healthcare costs are increasing at 7% to 8% per year. The cost of elder care will be beyond the resources of government and will fall to the recipients. This means that the elderly must acquire considerably more money for these costs than they had realized.

These realities have forced many of the 'boomers' nearing the age of traditional retirement to reconsider the concept of "Freedom 55", to quote a slogan used in advertising copy several years ago. The new slogan is "Freedom 95." This situation is leading older workers to demand the opportunity to remain in their jobs well beyond the traditional age of retirement. This situation creates burdens and challenges and opportunities for employers

Legislation in recent years has attempted to protect workers and others whose capabilities are less than some ideal of perfection. One example is the Americans with Disabilities Act (EEOC 1992). The substance of this type of legislation is that workers who can perform the essential functions of a position (with or without accommodation) without substantial risk or threat to themselves or others should not be restricted in performing their duties or denied the opportunity.

In recent years, accommodation of the workplace to suit the needs of individuals has received considerable emphasis and effort. Confined spaces, for example, however conflict with this concept, since in many cases, the geometry of the space cannot be changed, and the individual must accommodate to the workspace. This can be very difficult, and potentially impossible for some individuals. For example, pressure vessels have small openings. These delineate who can and cannot enter. The size of openings is regulated by safety codes for pressure vessels. Changing the size to accommodate individuals who are large in stature or obese is not technically possible.

As discussed previously, most people are incapable of performing work that demands superhuman capabilities. Even those who can perform at this level lose this capability as they age. In order to enable individuals to continue to work as they age, or to permit wider access of work to both genders or to enable the physically and mentally disabled to work, accommodation is necessary. Accommodation to the needs of older workers has occurred for many years. For many industries, older workers are too valuable a resource to lose.

Yet, some activities require superior performance (Anderson, 1989). Superior performance requirements could lead to exclusion of many individuals and virtually all women. This would lead to disproportionate representation. Disproportionate representation is acceptable provided that it is job-related. Job-relatedness can be established through strength and endurance testing. Tests chosen for the protocol must bear high similarity to the job, be predictive of risk of injury and performance, have minimum bias, and be replicable. The critical problem for the physician and the organization is to define, as precisely as possible, the physical requirements of the work activity.

Aging, of course, is a natural and gradual phenomenon (Courau 1983). Aging and its effects on performance is a theme underlying various sections in this chapter. A number of recent articles in the safety literature have provided insight into the questions created by this shift in the paradigm (Drennan and Richey 2003; Findley and Bennett 2002; Haight 2003; Haight and Belwal 2006; Perry 2010; SOEH 2010). These articles focus on workplace realities rather than academic-oriented physiological research. Both approaches are essential to understanding and addressing this situation. Fortunately, these articles precede what could become unprecedented in its speed and breadth and depth of impact.

The effects of aging (Figure 4.8) usually become apparent between 40 to 45 as compensatory systems fail (Astrand and Rodahl 1986; Perry 2010). Degeneration has a major impact on older adults (Figure 4.9). Degeneration will have an impact on performance in the workplace. The effects of age are similar to what occurs due to oxygen deprivation. They also resemble the effects produced by fatigue (Singleton 1972).

Muscular strength reaches its maximum around the 25th year and declines thereafter (Figure 4.8). Yet, the impact of fatigue does not appear to be greater in the middle-aged than in the young. The ability to tolerate maximum effort declines with age. Older workers are less able to tolerate heat. Manual skill deteriorates after the 20th year. Physical agility and muscular coordination decline. Balance is less precise.

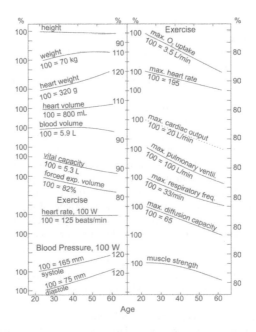

Figure 4.8. Aging and performance. (Modified from Astrand and Rodahl 1986.)

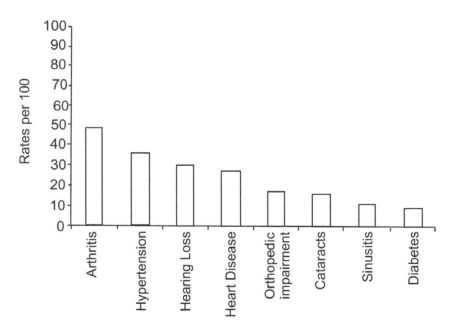

Figure 4.9. Degenerative influences of aging on older workers. (Source: Administration on Aging, Department of Health and Human Services.)

Cerebral functions peak between 25 to 30 years and then decline. These are a complex mix of crystallized functions involving the sum of personal knowledge and fluid functions, such as immediate memory, power of concentration and rapidity of comprehension and reasoning. Some capabilities improve with age, while others decline.

Work capacity remains relatively the same with increasing age. What declines is the ability to maintain the pace of work.

Reaction time needed to process information to make decisions increases with increasing age.

Increasing age of a stable workforce can have a significant impact on performance capabilities. Despite the limitations imposed by growing older, the demographics of fatal accidents involving confined spaces, for example, suggest that older workers are no more at risk than other groups (McManus 1999). Similarly, work in some workspaces requires individuals who are very strong and who can work in awkward postures. In neither case is the average person necessarily suitable to perform the task. Females generally are smaller and less strong than men, on average. Very tall or overweight individuals and older workers similarly can experience disadvantage.

Drennan and Richey (2003) comment on the role of obesity in degenerative disease. Obesity affects more than half of the population in the U.S. In older workers, obesity is a major factor in risk of degenerative disease and musculoskeletal injury because of lack of muscle tone, flexibility, and strength. These authors take the view that promoting fitness through exercise in the workplace would provide considerable benefit to both employers and workers. This would result from improving the health status of these workers, and by so doing, reduce the costs to employers from accidents related to degenerative conditions.

This generally is the case with occurrence of accidents. The young are considered especially vulnerable due to lack of knowledge and experience. This would strongly argue that the knowledge and experience of older workers are far too important to lose through lack of accommodation to their declining physical capabilities. However, older workers involved in accidents suffer more severe injuries (Findley and Bennett 2002).

Injury due to falls is a major concern in the workplace in elderly workers (Findley and Bennett 2002; Haight and Belwal 2006). Falls in the elderly result from loss of control of postural stability, loss of balance, and loss of strength.

In reviewing the literature, Haight (2003) concluded that errors committed by older workers were likely to be a serious concern, hut were amenable to improvement through change in workplace design. The data indicate that given sufficient time under conditions appropriate for information processing, the quality of decisions made by older workers was as good as those made by younger workers. The key then is to create the appropriate environment for information-processing and decision-making. This is the focus of recommendations for accommodation of older workers offered by Haight (2003) and Haight and Belwal (2006) in Table 4.4.

Specific measures for accommodation include installation of chain actuators for valve hand wheels, damper levers, and other control devices that otherwise tend to be located in areas of difficult accessibility (Haight and Belwal 2006). This measure brings the point of action to the normal zone of occupancy in a workspace. Eliminating ladder-type stairways and substituting more shallow-angle stairways where space permits provides benefits by minimizing the risk of fall for everyone.

## Substance Use and Abuse

Any discussion about the impact of human factors in the workplace would be incomplete without acknowledgment and consideration of the impact of substance use and abuse on human performance. Substance abuse usually encompasses unauthorized or inappropriate use of controlled substances, alcohol, prescription drugs, and tobacco (Peterson and Smith 1998).

Studies show continuing use of illicit drugs in the U.S. among people employed in the workforce (SAMHSA 1995). About 6% of the population over the age of 12 uses illicit drugs at least once per year. The frequency of use ranges to almost daily. Of these users, 74% are employed. This would suggest that intoxication and/or use of drugs on the job is a distinct possibility. Illicit drug use is highest in food preparation, food serving, and the wholesale and retail trades.

Over half of the U.S. population aged 12 and over use alcohol at least once per month (SAMHSA 1995). Of particular note in the survey are drinkers who described their use of alcohol as binge or heavy. Binge use is defined as five or more drinks consumed on the same occasion. Heavy use means consuming five or more drinks per occasion on five or more days per month. Of full-time employed persons 18 and over, 8% are heavy users of alcohol. As a result, these users could be legally impaired during work on the job following alcohol consumption during social activities or personal time. More than 20% of men employed full-time in construction, transportation, and wholesale trades are heavy drinkers.

Use of illicit drugs is associated with much higher rates of absenteeism, accidental injury, involuntary separation, and use of medical care (Sheridan and Winkler 1989; Zwerling et al. 1990). Alcohol abuse is correlated with higher incidence of workplace accidents. Tobacco smoking also is related to work injuries (Metts and Mill 1987). This is true even when smoking materials are not in use during work activity. Smokers appear to take more risk in performance of their jobs than nonsmokers. This may be due to personality and/or pharmacological factors.

Major substances of abuse are categorized as psychedelics, stimulants, opiates, and sedative-hypnotics (Seymour and Smith 1990).

Psychedelics produce an altered state of consciousness (Peterson and Smith 1998). They are also known as hallucinogens because they may produce sensory enhancements, illusions, and sometimes hallucinations. Intoxicating effects last from one to at least 24 hours after administration. These drugs increase awareness of sensory input, provide a feeling of enhanced mental activity and alter perception of body image, distort perception of environmental stimuli, and cause marked introspection.

Stimulants cause release of neurotransmitters from nerve cells (Peterson and Smith 1998). Some directly affect the nerve cell. Stimulants cause euphoria, decrease fatigue, increase feelings of sexuality, decrease appetite, increase energy, and interfere with normal sleep patterns. They also can cause impaired judgment and decision-making, and psychological changes.

## Table 4.4
## General Measures for Accommodating Older Workers

Improve illumination

Eliminate heavy lifting, overhead work elevated work from ladders , long reaches

Design floors and platforms with smooth and solid surfaces and some cushioning

Minimize clutter on control panels and computer screens

Minimize noise levels

Allow more time to complete steps in a task

Minimize the need to perform two or more steps in a task simultaneously

Minimize inconsistency between displays and controls

Maximize time allowed for making decisions

Consider the need for reaction time when assigning older workers to tasks requiring intense processing of information

Provide time for practice and developing familiarity with the steps in a task

Sources: Haight 2003; Haight and Belwal 2006

---

Opiates produce analgesia (pain relief), drowsiness, changes in mood, and clouding of mental function (described as a floating sensation) (Peterson and Smith 1998). Opiates retard psychomotor activity and cause confused mental state and sensory disturbances.

Sedative-hypnotics include the licit drug, alcohol. Alcohol disrupts tracking, visual vigilance, tasks involving divided attention, postural stability, and cancellation tasks (Peterson and Smith 1998). Visual and fine-motor coordination are disrupted at blood alcohol levels as low as 0.02%. For safety-sensitive positions, 0.04% is the cut-off for blood alcohol level. For transportation employment, the maximum permissible level in the U.S. is 0.02%. Hangover occurs after the body detoxifies alcohol. Manifestations of hangover include nausea, irritability, hypersensitivity to light, noise, and temperature, sluggish psychomotor responses, memory and cognitive disturbances, tremors, and generalized feelings of malaise.

## Hazardous Energy and Degradation of Human Performance

### Vision
Vision is the most important of the special senses used during the normal performance of worktasks. Vision is also the most vulnerable to damage induced by external sources of energy, as discussed in Chapter 3. The unprotected eye is also prone to injury from airborne particles and projectiles. These injuries happen despite protective measures, including the placement of the eye in the skull in a fat-cushioned pocket and presence of a protective bony ridge and eyebrows to ward off falling objects, and the curtain formed by the eyelids (DeCoursey 1974). Fluids from various sources also lubricate the eye.

Work in modern society involves extensive use of the eyes in visually oriented tasks. The eyes and the sensory images that they produce are central to performance of a task, not peripheral to it. Some examples include welding and machining. The ability to see small changes in appearance is central to determining and maintaining the quality of the work. At the same time, the process involved produces emissions and contaminants that are highly injurious to the eye. The worker must be able to see what is occurring. At the same time, the eyes require protection to prevent injury. This emphasis on use of the eyes makes them more prone to injury than demands placed on them in former times by hunter-gatherers and farmers. The evolution of protective structures has not progressed to match the conditions to which the eyes can be subjected during the course of work. Workplace injury caused by projectiles ranges from particles in the fluids of the eye to puncture by sharp-pointed objects. Injury caused by chemical and physical agents was discussed in Chapter 3.

Visual Function

The eye is a visual receptor. Its sensitivity ranges over almost a $10^{13}$ fold span of intensity (Kantowitz and Sorkin 1983). However, at any particular moment of adaptation, the actual operating range is about $10^2$-fold. The iris exerts some control (by a factor of 40) over the amount of light entering the eye (Konz 1990). The opening of the iris varies from 1.5 mm to 9 mm (a factor of six).

The normal eye is nearly spherical (DeCoursey 1974). The cornea is the outer surface through which light first enters the eye. The cornea normally is spherical with a smaller radius than the eye. Hence, a portion of a small sphere is imposed onto a larger one. The cornea plays a major role in focusing light. The cornea is highly enervated, and is, therefore, very sensitive. It receives its oxygen by diffusion from the atmosphere, and not from a blood supply.

The lens is a biconvex-shaped, colorless, transparent structure located behind the cornea. Together, the cornea and lens focus light onto the retina (DeCoursey 1974). In the normal eye, the relaxed lens will focus light from objects 6.1 m (20 ft) away. This is the reason for the 20 that appears in charts of visual acuity. The comparison is made against the ability to read letters at that distance.

Suspensory muscles change the shape of the curve of the lens. The action of maintaining correct focus for distance is accommodation. Accommodation is a reflex action. The closest distance on which the eye can focus is the accommodation distance. As aging occurs, flexibility of the lens decreases. The curve becomes more shallow. This decreases the ability to focus because curvature in the lens is needed for focusing on close objects. Contraction of muscles in the eye increases curvature in the surface of the lens, hence the use focusing muscles for close-up work. Looking at distant objects relaxes these muscles. As aging occurs, the focusing muscles are unable to maintain the curvature needed for close-up work. As a result, the accommodation distance increases. This condition is presbyopia.

The retina is the multilayered structure that lines most of the interior of the eye (DeCoursey 1974). Light-sensitive, sensory receptor cells located just inside the rear layer of the retina contain the black pigmented cells. These are the rods and cones, so named because of shape. There are about 18 times as many rods as cones in the human retina (125 million versus seven million) (Konz 1990). Rod cells are concentrated around the periphery of the retina. Cone cells are concentrated in the middle of the retina along the optical axis.

Rods are sensitive to all wavelengths of visible light (DeCoursey 1974; Leukel 1976). Sensivity peaks at 510 nm (Kantowitz and Sorkin 1983). They function in the dark and in low levels of light. Adaptation occurs when a person moves to a dark environment. This process requires about half an hour. During this transition, sensitivity of the rod cells increases by one million-fold. This is the reason that vision improves with time under these conditions. Night blindness is a visual impairment that is expressed through inability to see as well as normal people in dim light. This condition is related to a dietary deficiency of vitamin A. Vitamin A is directly involved in proper function of the rod cells.

Cone cells function in normal daylight (DeCoursey 1974). Cones are differentially sensitive to parts of the visible spectrum (blue, green, red). Collectively, the three types of cone cells cover the complete spectrum, and hence, provide the ability to capture the full range of visible light. The peak of the composite curve for sensitivity of the cones is 555 nm (Kantowitz and Sorkin 1983). The transition from dark-adapted conditions to daylight also includes a process of adaptation. In this process, sudden transfer from dark to light is accompanied by sensations of blinding brightness. After a few minutes, this sensation disappears. Full adaptation to daylight requires 20 to 30 minutes.

Cataract formation is a degenerative condition involving the lens. The lens becomes opaque or milky white. If left untreated, this condition can lead to blindness. Cataract formation is age related. Cataract formation also occurs due to exposure to ultra-violet and infra-red energy.

Glaucoma is a degenerative condition characterized by reduced rate of drainage of fluid from the body of the eye. This leads to build-up of internal pressure with resulting pain and possible damage to nerve fibers of the retina. This process is age related.

Visual Defects

Visual defects are caused by heredity, aging, and injury.

Hyperopia is the hereditary condition in which the refractive (light-bending) power of the eye is insufficient to focus the image onto the surface of the retina (Figure 4.10). Hyperopia can occur because the lens is too thin or the eyeball is too short (DeCoursey 1974; Leukel 1976). In this circumstance, the lens focuses the image behind the position of the retina. The result is a blurred image unless the focusing muscles in the eye can provide sufficient accommodation. Hence, accommodation occurs for much of the time. Individuals with hyperopic eyes at some point in their lives will not be able to see close objects or to read properly.

Myopia is the hereditary condition where the refractive power of the eye is too great. This occurs because the lens is too thick or the eye is too long (DeCoursey 1974; Leukel 1976). In this circumstance, the lens focuses the image in front of the retina. When projected to the retina, the image will be blurred. The individual with myopic eyes cannot see distant objects.

Another hereditary condition is astigmatism. The normal cause of astigmatism is variation in curvature of the cornea (DeCoursey 1974). This variation leads to more than one point of focus. In the normal eye, containing equal curvature in all cross-sections around the cornea, light passing through the cornea forms a cone that focuses at a single point. When curvature varies, light from the greater curvature focuses at a shorter distance than light from the lesser curvature. The focus in this case is not sharp

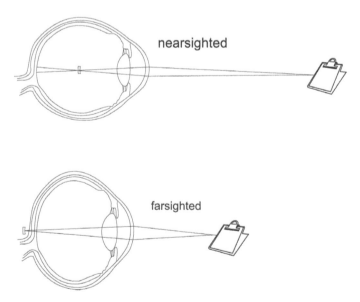

Figure 4.10. Hereditary visual defects. The nearsighted eye is elongated compared to normal and focuses distant images forward of the retina. The farsighted eye is compressed compared to normal and focuses close images behind the retina.

and images may appear blurred. Bright lights may appear to have a starred appearance, with rays extending outward from the central hub.

Color blindness (deficiency) is a hereditary condition caused by absence or deficiency of function of one or more types of cones (DeCoursey 1974). The most common deficiencies occur in the red or green cones. Deficiency in the blue cones is very rare. Individuals with color deficiency do see some color. Very few individuals are deficient in all three types of cones. In the latter case vision is similar to black and white images. As well, some individuals are color weak. Their eyes do not distinguish different hues as well as normal.

Visual Correction

Hyperopia, myopia, presbyopia, and astigmatism are correctable through eyeglasses containing prescriptive lenses. For the most part, the wearing of eyeglasses containing prescriptive lenses should not impair performance. In fact, many industrial operations require the wearing of safety glasses as a protective measure against injury from projectiles and other causes of eye injury (Figure 4.11). Hence, the wearing of safety glasses containing prescriptive lenses in these operations should be neutral, if not actually positive in benefit. At least these workers do not need coaxing, cajoling, or disciplinary measures to ensure the wearing of safety eyewear. They are very unlikely to experience an eye-related accident caused by failure to wear eye protection.

The need to wear prescriptive eyewear for proper vision can have negative implications. These relate to the necessity to use half- and full-facepiece respirator (McManus 1999). Wearing of prescription eyewear with the use of a half-facepiece respirator for long periods is barely tolerable. The lenses fog and smudge with skin oils carried on eyebrows and eyelashes. The situation with full-facepiece respirators is completely incompatible.

In order to receive a satisfactory seal from a full-facepiece respirator, protrusions such as the temple and earpieces on glasses are not considered acceptable. The only options available to individuals

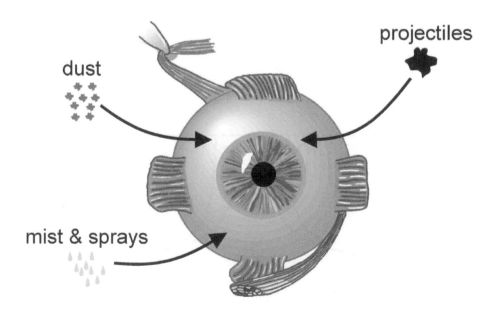

Figure 4.11. Causes of eye injury. There are many causes of eye injury, a major one being contact with airborne objects, particles, droplets, and gases and vapors.

needing visual correction are inserts for the facepiece that contain prescription lenses (Figure 4.12), or the wearing of contact lenses. Not everyone can obtain satisfactory performance from the use of a vision kit, as alignment is critical in some prescriptions (Cullen 1993).

Likewise, wearers of bifocals would experience considerable difficulty using a vision kit. Bifocals pose especial problems because the wearer must look downward to see through the magnifying section of the lens. This is compatible with reading tasks when the reading material is located below the face, as normally is the case. Situations where magnification is required to view targets on vertical surface pose considerable difficulty. When the target is above the head, the individual must tip the head backward and look up in order to obtain the necessary magnification.When the target is below the head, the individual must position the head in such a way as to capture the target in the magnifying part of the lens. This situation is very unsatisfactory.

Wearing contact lenses inside full-facepiece respirators has generated considerable controversy. On the one hand, they offer the wearer equal opportunity to use any full-facepiece respirator at a moment's notice. On the other hand, contact lenses are a barrier to the normal function of the cornea for which due consideration must occur. To illustrate, the very dry compressed breathing air used during winter conditions could dry out the lenses. Contaminated air that enters the facepiece could interact with the lenses.

DaRosa and Weaver surveyed U.S. firefighters about use of contact lenses (LLNL 1986). Despite the threat of legal sanctions against them and their employers, hundreds of firefighters (29% of respondents) admitted to wearing contact lenses during daily use of SCBAs. Of these, only six encountered problems serious enough to require removal of the facepiece. No firefighter was injured or died as a result of wearing contact lenses. These findings and follow-up on comments led the authors to conclude that contact lenses were not significantly more hazardous than eyeglass inserts. A survey performed by the U.S. Department of Energy obtained similar results (DOE 1986). OSHA subsequently took the position that the wearing of soft and gas permeable contact lenses would constitute only a de minimis vio-

Figure 4.12. Vision correction for full-facepiece respirators. The spectacle mounts into the facepiece using pressure exerted by the curved pieces. (Source: McManus 1999.)

lation of its standards (OSHA 1988). ANSI also adopted the position that contact lenses were no more hazardous than eyeglass inserts (ANSI 1984, ANSI 1992).

Wearers of contact lenses in normal industrial situations experience less-than-ideal conditions. Airborne particulates can cause major irritation and pain. This can lead to inability to function at a time of critical need. Vapors and gases in the atmosphere can permeate the polymers of breathable lenses, causing possible injury to the cornea.

Use of contact lenses by individuals in general industry has generated considerable controversy and inquiry (Nejmeh 1982; Stein and Slatt 1984; Rosenstock 1986; Randolph and Zavon 1987). Nejmeh (1982) reported on a detailed survey performed through the National Safety Council. This indicated highly variable policies and practices in participating companies throughout the chemical sector. In the three instances involving injury, the contact lenses were believed to have protected the eye. Stein and Slatt (1984) documented follow-up on reports of injury reportedly exacerbated by contact lenses and found no supporting evidence. These authors described measures for safe use of contact lenses. More recently, Randolph and Zavon (1987) reported on a survey of organizations. These authors found that the decision not to permit wearing of contact lenses appeared to be based on perception, rather than fact. Literature reviewed by these authors indicated that contact lenses were protective during injury situations and simulations in animal tests (Rengstorff and Black, 1974). In another part of the article, Randolph and Zavon provided guidelines for the use of contact lenses by workers in industry.

Cullen reviewed this question extensively (Cullen 1993, Cullen 1994). He noted numerous instances in which contact lenses provided protection and mitigated against more serious injury. The extent of protection depended on the nature of the hazard and the type of lens worn. Some types of contact lens provide added protection to wearers of eyeglasses against chemical splashes, dust, flying particles, and nonionizing radiation (arc flashes). As well, depending on type, contact lenses do not trap chemicals under or in them. Contact lenses offer added benefit over eyeglasses in many occupational circumstances (Cullen 1993).

Prohibition of use of contact lenses in the context of current knowledge about safety in their use puts the employer into conflict with legislation, such as the Americans with Disabilities Act (EEOC 1992; Blais 1994). Some people must wear contact lenses to obtain best visual performance or efficiency. An example is an individual with monocular aphakia (lens removed from the body of the eye) who must wear contact lenses to obtain binocular vision.

Environmental Impairment of Vision

Glare is one of the major factors influencing visual quality (Konz 1990). Glare is any brightness within the field of vision that causes discomfort, annoyance, interference with vision, or eye fatigue.

Disability glare directly interferes with visibility and ultimately visual performance. Disability glare occurs when viewing an object surrounded by bright backlighting, such as when looking at a person standing in front of a sunlit window (Sanders and McCormick 1993). Disability glare reduces the detail in the image of a person to that of a silhouette. Glare impairs the ability to resolve detail. Another example is the inability to see when driving at night and looking into bright headlights. Disability glare reduces the visibility of objects (Kantowitz and Sorkin 1983). Contrary to the perception that only discrete sources of light, such as oncoming headlights, are the cause of glare, every luminous point in space acts as a source of stray light. These sources can severely affect visibility of objects on or near the line of sight.

Disability glare results from the fact that the liquid contents of the eye and the lens and cornea are not a perfect ocular system (IES 1981). Inhomogeneities in the ocular media scatter incident light. Some of the light destined for one area of the retina impinges on another, and vice versa. This reduces contrast (definition) between details in the image.

Pulling et al. (1980) demonstrated an age-related decrease in threshold of disability glare caused by headlights. This could correlate to the accumulation of "junk" in ocular fluids with increasing age.

Disability glare also affects wearers of corrective lenses. Sensitivity to disability glare decreases in the following order: hard contact lenses > soft contact lenses > frame eyeglasses > no correction (Freivalds et al. 1983).

Discomfort glare is the sensation of annoyance or pain caused by high or nonuniform distribution of brightness in the field of view (IES 1981; Kantowitz and Sorkin 1983). A common example is performing a visual task, such as driving, while facing toward the sun. An overhead light shining on a computer screen containing a dark background creates the same problem. Discomfort glare is related to constriction of the pupil when exposed to bright light (Sanders and McCormick 1993). Assessments of discomfort glare usually consider the size, luminance, and number of sources of glare and the background luminance.

Sources of light in the field of view cause direct glare (Sanders and McCormick 1993). Windows are typical causes of the problem (Konz 1990). Another example is backlighting used to view transparencies or microfilm. Lights and incandescent objects in the visual field are additional examples.

Reflection of light sources on surfaces causes reflected glare (Kantowitz and Sorkin 1983). The reflection of the source of glare obscures or veils the object on which visual attention is focused.

## Noise

Chapter 3 described the function of the human hearing receptor and temporary damage caused by overloading. This section examines reception of sound as part of human-to-human and machine-to-human communications. Effective communication is critical for the transmission of dialogue and instructions between workers, warning signals between workers, and warning signals between monitoring instruments and workers. Impairment of communication can occur because of hearing loss and masking of sound signals.

### Permanent Hearing Loss

Chapter 3 outlined the mechanism for development of temporary hearing impairment (temporary threshold shift or TTS). Permanent loss is caused primarily by noise and is manifested through destruction of bands of hair cells on the basilar membrane in the cochlea. Even moderate exposure to noise causes subtle effects on the hair cells. Initially, these manifestations include twisting and swelling, disarray of the hairs, detachment of the hairs from the tectorial membrane (Figure 4.13), and reduction of enzymes and cochlear fluids (Ward 1986). As described, the system is fatigued. Additional energy is required to elicit the same level of response compared to the same system when refreshed. Increasing levels of damage are irreversible. Hairs fuse into giant cilia or disappear. Hair cells and supporting cells disintegrate. Ultimately, the nerve fibers disappear (Miller 1973).

The ear canal greatly amplifies sound in the 3000 Hz region. The net effect is that the head, earlobe, and ear canal amplify sound in the 2000 to 4000 Hz region by 10 to 15 dB. As a result, noise in this region is most hazardous to hearing (Ward 1986). This susceptibility to damage manifests as a notch in audiograms at 4000 to 6000 Hz (Figure 4.14).

Hearing loss due to noise energy also can occur due to single exposures at high levels, such as caused by an explosion. This results in disruption of attachments, such that hair cells are torn away from the basilar membrane. Tearing of the membranes allows intermixing of fluids. This, in turn, leads to poisoning of the hair cells (Ward 1986).

Sensory hearing loss also occurs due to aging (Spoor 1967). Loss due to aging becomes increasingly severe at higher frequencies. Note that the age-loss curves are corrected relative to 20-year-olds who have not suffered industrial noise exposure. These results thus apply only to a highly screened group of individuals. Finding individuals in today's society who have not experienced noise exposure is difficult. The upper limit of sensitivity (20,000 Hz) applies only to the young.

Hearing loss in the middle ear can occur due to a blow to the head, explosion, rapid decompression, or penetration of the eardrum (Ward 1986). This results in dislocation of the bones. This condition is treatable medically or surgically, whereas damage to the inner ear is considered irreversible.

### Speech Interference

Speech utilizes vowels and consonants. Vowels in the English language occur at frequencies less than 1000 Hz. Consonants occur above 1500 Hz and as high as 8000 Hz (Ward et al. 1961). However, 98% of speech energy (mostly vowel sounds) occurs at frequencies below 1900 Hz (Figure 4.15). Most occurs below 300 Hz (Loeb 1985). Hearing impairment to speech is not experienced until the average

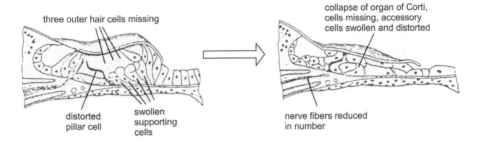

Figure 4.13. Destruction of the hearing apparatus by high levels of noise. (Source: USEPA.)

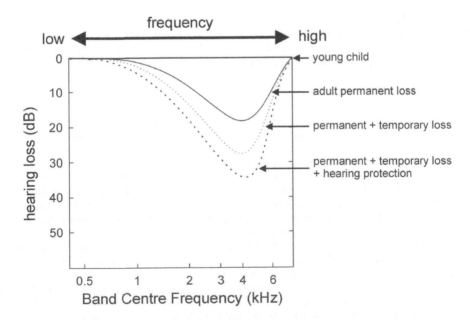

Figure 4.14. Audiometric spectrum showing hearing loss. Young children have a flat threshold of hearing. Age-related hearing loss and noise-induced hearing loss create permanent threshold shift. Noise-induced loss creates a characteristic dip in sensitivity in the region of 4000 to 6000 Hz. Temporary loss due to noise exposure and use of hearing protection further deepen the threshold of hearing.

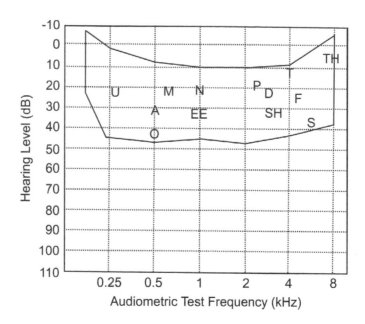

Figure 4.15. Speech sounds and frequency. (Modified from Ward et al. 2001.)

hearing threshold exceeds 25 dB at 500, 1000 and 2000 Hz (Table 4.5). Hearing loss at 2000 Hz can lead to misinterpretation of words. Misinterpretation of the letter "t" for the letter "p", for example, causes "tick" to become "pick". Losses in the 500 to 2000 Hz region are most assuredly accompanied by losses at higher frequencies especially, given the shape of the audiogram following noise-induced hearing loss.

Masking
Masking is the overwhelming of desired sound signals by other sounds. Other sounds include background noise or noise associated with a work process. A common example around the home is the inability to hear conversation from someone in another area when in the bathroom and the water is running in the sink or the shower. The sound of the running water "drowns out" the person's voice. The need to raise one's voice to be heard in a conversation in a noisy area at work is another common illustration of the problem. We need to hear specific sounds in an otherwise noisy environment (Robinson and Casali 2000). These include sound produced by machinery indicating need for adjustment or servicing. We also need to be able to hear fire and other alarms. Masking becomes a problem when sounds that must be heard and comprehended cannot be heard.

The need to provide audibly recognizable warning signals in these environments is a considerable challenge. Warning beepers and other sound signals from instruments can easily go unheeded. They become just another sound in the spectrum of sounds. That is, they can be heard, yet not recognized for what they are. In an attempt to overcome this problem, many construction sites use three long blasts on an air horn (an aerosol can containing compressed gas + the horn fitting) as a means of alerting people regarding emergency situations. Similarly, some industrial plants use dedicated horns or sirens to signal warnings. These are sounded infrequently so as to provide the necessary arousal of interest needed to elicit the appropriate response from recipients.

Seshagiri and Stewart (1992) produced one of the few reports in the industrial hygiene literature regarding this problem. These authors reported on an investigation of rundown accidents on railways

**Table 4.5**

**Effects of Hearing Loss on Interpretation of Speech**

| Loss dB | Effect |
|---------|--------|
| <25 | No significant difficulty in interpreting faint speech |
| 25 to 40 | Difficulty in interpreting faint speech |
| 40 to 55 | Difficulty in interpreting normal speech |
| 55 to 70 | Difficulty in interpreting loud speech |
| 70 to 90 | Only amplified speech can be understood |
| >90 | Amplified speech not understood |

resulting from the inability of workers to hear warnings from locomotive horns. The victims either wandered onto or remained on the track, oblivious to the approach of a piece of track equipment or a locomotive. The extreme of this situation occurred when a victim ignored warnings frantically shouted by coworkers. This equipment generates little noise in the forward direction and can easily come up on a person without obvious warning. Hence, the horn is an essential part of the warning system. In these and other similar situations, the workers were exposed to high levels of workplace noise.

Masking of desired sound occurs (Figure 4.16) when the frequency components of the background noise are louder than the sound we are attempting to hear (Hassall and Zaveri 1979). The masking noise can be narrow band, wide band or pure tone, steady or intermittent or impulsive. Generally masking is more of a problem in areas where the masking noise level is high relative to the signal to be heard (Robinson and Casali 2000).

Noise that causes masking induces vibration in a region of the basilar membrane (Meyer and Neumann 1972). Maximum vibration occurs at the frequency of the masking sound having maximum amplitude (assumed here to be the center frequency). At the same time, vibration occurs along the membrane in both direction. The brain will not recognize a second sound having a different frequency spectrum as separate from the first until it causes vibrations of the basilar membrane that are greater than those induced by the first sound. Masking noise exhibits a number of characteristics (Table 4.6).

Auditory Signals

Reception and interpretation of auditory signals is a logical corollary to the study of masking. In order to be heard above background noise, a real-world signal must exceed the background level by some threshold amount. Real-world signals usually are considerably more complex than the tones used in laboratory studies. Real-world signals possess multiple frequencies, periodic or aperiodic temporal pattern, and modulating frequencies and/or amplitudes (Robinson and Casali 2000). The multiplicity of these attributes increases the likelihood of hearing a real-world signal within a spectrum of noise. Complex signals obviously contain more information than simple tonal ones. This increases the likelihood that at least one component will exceed the masking threshold.

The most desirable means to improve communications is to reduce background noise (Robinson and Casali 2000). This reduces the potential for distortion in the cochlea of the signals from both the

## Table 4.6
## Characteristics of Masking Noise

**Characteristic**

A narrow band of noise causes more masking than a pure tone centered at the same frequency.

Masking occurs asymmetrically around the center frequency and produces greater effect at higher frequencies for the same level. That is, masking occurs over a narrow band below the center frequency and over a wide band above the center frequency.

At low levels, masking occurs in a narrow band around the center frequency of the masking noise.

Increasing the level of masking noise broadens the frequency range over which masking occurs.

Increasing bandwidth beyond a critical value does not increase masking of a pure tone at its center frequency. That is, sound energy near the frequency of the sound being masked contributes most to the masking effect.

The critical band over a wide range of frequencies above 500 Hz approximates the width of a 1/3 octave; hence, the use of 1/3 octave bands in noise measurement.

Sources: Hassall and Zaveri 1979; Ward et al. 2000.

---

background noise source and the desired signal. Another technique is to manipulate the spectrum of the desired signal so that it contains elements outside those of the background noise.

The brain estimates position in space from analyzing differences in phase and intensity detected by individual ears (McMurtry and Mershon 1985). As well, the brain uses perceived changes in position, level, and pitch to estimate speed and movement.

The threshold at which a signal is detectable above background noise is 0.5 dB for broadband sounds to 2 dB for pulsed tones (Small and Gales 1991). For reliable detection leading to action, 10 to 15 dB above threshold is considered adequate for most situations (ISO 1986). This assumes that hearing protection is not used and that the listener has normal hearing. Frequencies should range from 700 to 2800 Hz. Frequencies above 3000 Hz are potentially undetectable by individuals suffering from significant hearing loss at high frequencies. The level of the signal at the position of the listener, rather than that of the sender, is the critical concern. This is due to losses in the path caused by obstacles and absorption by surfaces.

As expected, the subject of warnings is complex. The reader should consult Robinson and Casali (2000) for further details.

Hearing Protection and Intelligibility of Signals

A critical issue in workplaces where noise levels above ambient are present is potential interference with messages when workers use hearing protection. Individuals wearing hearing protection typically can discriminate speech better than when unprotected in high noise areas (Berger 2000). This results from the fact that the hearing protection reduces the signal-to-noise ratio of the sound signal. This enables the signal to be received by the cochlea without distortion. This only occurs at levels below 90

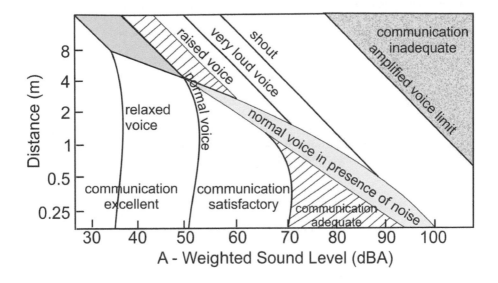

Figure 4.16. Masking of auditory signals. (Modified from Webster 1969.)

dBA (Lawrence and Yantis 1956). This effect is similar to the ability of the wearer of dark sunglasses on a bright sunny day to distinguish detail that otherwise would be obliterated by the glare.

For persons with normal hearing, hearing protection improves speech discrimination at noise levels above 80 dBA. Below this level, speech discrimination is reduced (Howell and Martin 1975). Hence, the protected noise level should be no less than 70 dBA (ISO 1996). For persons with hearing impairment, the situation is considerably more complex. Hearing protection decreases speech discrimination in low noise environments, and can prevent reception of higher frequencies rendered by the device below the threshold of audibility (Chung and Gannon 1979; Rink 1979). Regardless of hearing ability, visual cues are additional critical elements to speech recognition. The findings for warning signals are essentially the same as for speech discrimination (Wilkins 1980).

Nonauditory Effects
Levels of noise incapable of causing hearing loss can cause other serious impacts that affect human performance. These include masking of speech and sound signals, as already discussed, and impairment of sleep and performance of tasks (Ward et al. 2000). Sleep deprivation can considerably impair task performance. People differ considerably in susceptibility to annoyance. At levels of noise sufficiently high to be injurious to the hearing apparatus, other deleterious effects occur. These include increase in heart rate, blood pressure, and vasoconstriction in the extremities and dilation of the pupil. These effects result from increased production of catecholamine and adrenaline. Unexpected sounds produce the startle reaction.

Task Interference
Noise is perceived by the layman as deleterious to performance of tasks (Loeb 1985). Demonstration of this in lab studies is uncertain. This probably reflects that fact that performance involves the interac-

tion of many factors. These interactions occur in a complex, often unpredictable manner. Laboratory tests also use controlled conditions involving noise signals, lighting, temperature, and so on.

The real world in which we live, and in which noise can play a major role, is considerably more complex than what is achievable in the laboratory. Conditions other than noise usually are fixed, whereas noise is the variable. For example, one person can do homework or perform tasks, such as knitting, requiring little concentration, while music is playing. For other people, performance of complex tasks or tasks requiring intense concentration, such as writing this book, in the presence of music, or a radio or television program in which only background music is audible, or a car alarm sounding in the community, would become impossible. For still other people in industrial settings, background noise which tends to be constant has little impact on task performance. One consideration of potential concern occurs when the listener diverts attention to the sound signal at the expense of performing the task. This is especially important when the sound signal is music and the task requires intense concentration or involves safety-sensitive activities.

### Vibration

Closely allied to noise is vibration, since both are produced by surfaces that move with regular periodic motion. Among the more serious effects produced by whole-body vibration on humans are fatigue, annoyance, and interference with performance. These concepts are incorporated into standards (ISO 2631/1) on whole body vibration (ISO 1985). (Annoyance is not a consideration in standards addressing hand-arm vibration (Wasserman 1987).) ISO 2631/1 contains three sets of families of curves. The curves plot time dependence of acceleration in the vertical ($z$) direction) and lateral ($x$ and $y$) directions versus frequency. Families of curves in a particular coordinate direction have the same shape. The curves cover from one minute to 24 hours and frequencies from 1.0 to 80 Hz. The most restrictive of the families of curves depicts reduced comfort. Exposures at these levels are concerned with preservation of comfort during the work process. Curves 10 dB (decibels) higher express fatigue-decreased proficiency. Fatigue-decreased proficiency refers to the ability of a person to perform work without interference from the vibration. These curves intend to prevent the situation where an operator might lose control of the equipment or machine due to vibration. The highest group of curves (exposure limits) attempt to preserve the health and safety of the exposed individual.

Vibration affects the balance mechanism and other sensory actions that occur in the head (Figure 4.17). Visual performance is affected most by vibrational frequencies in the range of 10 to 25 Hz (Moseley and Griffin 1986). Amplitude is the key factor. The degradation in visual sensation is believed to be due to movement of the image on the retina and the perception of blurring.

### Cold and Cold Surfaces

Receptors for cold and cold surfaces are found in the outer portion of the dermis, the tip of the tongue, and the cornea of the eye (DeCoursey 1972). These are known as the end bulbs of Krause. These receptors normally respond to temperatures below normal (approximately 35° C). The stimulus for cold temperature receptors is 35° to 36° C or less. Sensation of cold causes constriction of blood vessels to preserve warmth in the core of the body.

Cold receptors are more abundant than heat receptors and lie closer to the surface (0.1 mm below the surface versus 0.3 mm). They are not evenly distributed across the surface of the skin. The face, for example, is highly sensitive to cold and has a high distribution of receptors (Leukel 1976). This provides some explanation for the delay in perceiving cold surfaces or cooling of skin and limbs of other parts of the body containing fewer receptors to temperatures capable of causing serious burns (frostbite). Thermal energy must move through the tissue to the cold surface. This decreases the temperature in the skin at the nearest cold receptor. The response of cold receptors is slightly faster than that of heat receptors.

Examples where the delay between exposure and sensation occurs include contact with cold surfaces outdoors or in deep freezes, or contact with cold running water. There is a definite delay between contact with the cold source and perception of cold and pain and the need for avoidance. The persistence of pain associated with exposure after withdrawal indicates that the warning is not instantaneous. One must learn this from experience.

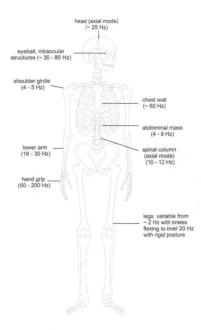

Figure 4.17. Resonance frequencies of parts of the body. (Modified from von Gierke et al. 1975; Rasmussen 1982.)

Work in winter conditions is difficult and potentially hazardous. Layers of clothing and hand coverings reduce mobility and grip strength. Icy, frozen surfaces become very slippery compared to temperate conditions (Figure 4.18a and 4.18b). Work must occur more slowly and deliberately. Under severe winter conditions, a warm refuge must be readily at hand to ensure safety and avoidance of injury from hypothermia and frostbite.

Statistics for the U.S. indicate that the major compensable cold-related injury is frostbite. Over 75% of the claims involved the fingers, foot, hand, toes, and nose (Jensen 1983). Occupational groups most affected were laborers, truck drivers, firefighters, police, and garbage collectors. These statistics do not capture less serious injury, and cardiac and traumatic injury resulting from less-than-optimal performance in which cold was a factor.

**Heat and Hot Surfaces**
Receptors for heat are located deep in the dermis (0.3 mm below the surface versus 0.1 mm for cold receptors) (DeCoursey 1974; Leukel 1976). These are the so-called end organs of Ruffini. The end organs of Ruffini are long, cylindrical capsules filled with fine nerve fibers. Receptors for heat are less plentiful than those for cold. They react to temperatures above normal (34° to 35° C or more). Sensation of warmth causes dilation of blood vessels near the surface to cool the warm area. The threshold of pain is 45° to 50° C.

Heat receptors are not evenly distributed over the surface of the body. The wrist provides a much better estimate of temperature than the fingertips. This provides some explanation to the delay in perceiving hot surfaces or heating of skin and limbs of parts of the body containing fewer receptors to temperatures capable of causing serious burns. Heat must move from the hot surface through tissue to the nearest heat receptor. Examples where the delay between exposure and sensation occurs include contact with hot surfaces of pots on a stove or contact with hot running water. There is a definite delay between contact with the heat source and perception of heat and pain and the need for avoidance. The

Figure 4.18a. Work in temperate conditions. Time of day and the weather complicate work.

Figure 4.18b. Winter conditions create an entirely different landscape and hazardous conditions, such as slip, trip, and fall from same elevation, that otherwise were controlled prior to covering existing surface features in an oblique layer of snow.

persistence of pain associated with the exposure after withdrawal indicates that the warning is not instantaneous. One must learn this from experience.

This absence of warning can have serious consequences during work. The first of these is a burn injury, discussed in the previous chapter. The second is an involuntary reflexive withdrawal from the exposure. This can lead to other consequences because control over actions of the affected limb is lost temporarily. The individual can no longer control placement of the hands and limbs in the precise manner that might be needed to avoid injury from other sources of energy. As well, the focus of the individual is directed toward the pain associated with the contact. Only when the pain subsides can the affected individual regain the composure needed to comply with requirements of a procedure.

Accident investigations documented in the first chapter indicated that profuse sweating has a role in some types of fatal accidents, especially those involving contact with energized electrical conductors. Working under hot conditions is extremely uncomfortable when the skin becomes wetted with running sweat. Unvented personal protective equipment traps perspiration and heat generated from activity, as well as that caused by radiant sources, such as the sun (Figure 4.19).

Statistics for the U.S. indicate that the major compensable heat-related injuries are circulatory system failure, followed by other body system failure, and entire body system failure. Hence, these are claims for heat stroke and heat exhaustion, rather than heat cramps. Occupational groups most affected were laborers and firefighters. These statistics do not capture disorders treated at the local level, which, like heat cramps, disable a worker for only a few hours, or which may have contributed to a cardiac incident or traumatic injury.

### Odor

Odor might seem an odd choice for inclusion in this type of discussion. However, odor and its detection can have profound influence on performance. Odoriferous compounds are vapors that have some water solubility (DeCoursey 1974). Water solubility is necessary for dissolution into the mucus that covers the olfactory sensors. Lipid solubility also is necessary for entry into the olfactory nerve endings. The olfactory center is located high in the nose. Some substances, such as ammonia, pepper, camphor, phenol, and ether produce irritation or a tactile effect, rather than an olfactory effect.

Odor is a complex subject. Since 1870, more than 30 theories have attempted, without success, to explain the functioning of the olfactory system and our response to odorants (substances to which the olfactory system responds) (Metcalf and Eddy 1991). We encounter odorants every day in the community. These range from cooking odors to commercial and industrial process emissions. Indeed, avoiding exposure to odorants would be difficult, since almost every airborne substance is detectable by nose, and therefore produces a sensation that we call an odor. The relatively few gases and vapors lacking an odor include oxygen, nitrogen, hydrogen, steam, hydrogen peroxide, carbon monoxide, carbon dioxide, methane, and the inert gases (helium, neon, argon, krypton, and xenon). Particles of volatile substances and particles to which volatile substances have absorbed or adsorbed can present an odor that can persist long after cessation of work. All that is necessary for sensing of odor is sufficient concentration of molecules for the olfactory receptors to send a signal of sufficient strength for the brain to respond.

Mercaptan (organic-substituted hydrogen sulfide) is a substance to which the olfactory center has an extreme sensitivity. Mercaptan is sensed at a concentration of $0.4 \, \mu g/m^3$. Since approximately 2% of inhaled molecules excite receptors, this leads to an estimate of 40 excited cells for detectability (Leukel 1976).

Researchers estimate that over 1000 genes (1% of the total genome) encode the olfactory receptors (McGinley and McGinley 1999). Humans have the ability to recognize and describe 10,000 different odors. Human response spans a spectrum of descriptors. To a carpet layer, habituated to the odor from solvents in the adhesive and emissions from the carpet, the odor is a nonissue. To an organic chemist, the odors from the adhesive indicate the presence of an aromatic solvent, toluene or xylene, and perhaps acetone. While the odor is strong, it is not objectionable, as this individual is interested in what it represents. To an office worker located in an adjacent area, the merest hint of a chemical odor can be highly offensive. No wonder this is such a complicated subject.

Figure 4.19. Entrapment of heat in impervious suits. The impervious material prevents radiation of heat and escape of moisture. As a result, the thermal condition inside the suit can differ dramatically from that outside.

Research has yet to show whether prolonged exposure to an odorous substance can produce loss of the sense of smell in a manner similar to loss of hearing following prolonged exposure to high levels of noise (Turk and Hyman 1991). What is known is that adaptation occurs, usually in one to 10 minutes, but always less than 30. Hence, many odorants, including perfumes, are effective at eliciting a response only very briefly. Recovery from habituation also occurs. Typically, 60% to 70% of recovery occurs within 2 minutes following cessation of exposure. The implication of this is that an individual exposed to an intermittent odorous plume would be able to detect episodes after a lapse of 5 or more minutes.

At low concentration, the compounds that together constitute an odorant pose a psychological rather than a toxicological problem to humans (Allen 1991). Objective symptoms of offensive odors include eye, nose, and throat irritation, shallow breathing, cough, headache, vomiting and nausea, decreased appetite and reduced consumption of water, emotional upset resulting in lack of sense of well-being, and irritability and depression. Physiological effects include decreased heart rate, constriction of blood vessels in the skin and muscles, release of epinephrine, and changes in the size and condition of cells in the olfactory bulbs in the brain. While these conditions may have little impact on health status, they can induce a serious decrease in performance.

## Cellular Telephones and Distracted Driving

Among other topics, this chapter has explored the ability of the human organism to process information resulting from input from the senses. Information processing is critical to understanding about the causality of accidents. Another theme highlighted in this chapter is the observation that humans cannot simultaneously focus on the task to be performed and the conditions of the surroundings. These are parallel, mutually exclusive realities in workplaces and workspaces.

The existence of these realities and absence of interactions reflect observation about the manner in which work actually occurs. Workplace accidents involving hazardous energy sometimes highlight these realities. These accidents occur very quickly. Oftentimes, the occurrence of the accident relieves the initiating stressor.

Our usual and natural focus during work is task. This observation offers a possible explanation about why seemingly well-experienced individuals, described by co-workers as being conscientious and highly safety conscious, are involved inexplicably in fatal accidents. This comment also applies to practitioners of occupational hygiene and safety. That is, as individuals, regardless of background education and experience, we also focus on task and not conditions during our activities at work. This focus on task is a natural human attribute. Focus on conditions requires thinking in the abstract about the likelihood of future events and the mechanisms by which they can occur, rather than the concrete level of the task, the level at which we normally operate.

Understanding the mutual exclusivity of task and conditions is crucial to understanding about the causation and occurrence of accidents involving hazardous energy. That is, one cannot focus simultaneously on task and conditions during the performance of work. Rather, one can focus on one or the other of these realities.

The recent intense attention and focus on driving distracted while using cellular telephones and other devices and pursuing activities unrelated to driving provide an opportunity for better understanding the situation highlighted above. Every observer in traffic can see the impact of these activities and actions on the performance of other drivers. They have difficulty maintaining speed and position in the allotted space of traffic lanes, stop abruptly, if successfully, in the stop and go of heavy traffic, and ignore traffic signals and other warning signs, all while looking straight ahead through the windshield in the direction of travel.

Curry (2002) and Moore and Moore (2001) reviewed the situation that had developed in the 1990s during widespread adoption of cellular telephones. This was the era in which governments world-wide were restricting or outright prohibiting the use of cellular telephones in vehicles. Curry (2002) commented that these actions reflected a response to demands from the public to do something to curb the on-going tragedy posed by accidents involving vehicles. At the time of writing, data on this situation were inconclusive. The crux of the concern in use of a cellular telephone while driving is distraction from the task of driving.

Peters and Peters (2002) concluded that distracted drivers pose a significant problem in public health. They comment that the recent and ready availability of cellular telephones at modest cost shifted use from the well-to-do and business users to individuals of all ways and means and capabilities and knowledge and experience. This shift greatly increased the risk associated with use of these products, especially while driving. These authors comment that dangerous events during driving can occur quickly, on the order of a 1 second time-frame. A sudden lane change followed by rapid braking by a vehicle ahead could occur within 2 to 3 seconds. Complex brain functions require up to 8 seconds.

The concern over use of cellular telephones in vehicles has evolved as has the product itself. What started out as a simple telephone is now a device on which one can read and reply to e-mail messages, play games, listen to music, watch video presentations, and conduct business and other activities. The cellular telephone of today has converged into a small, readily portable, multi-functional, and highly adaptable device from the single function device of the 1990s.

Historic research reported by Curry (2002) put the concern about these products into perspective, although unknowingly so at the time. This relates to distraction of various types of tasks undertaken by drivers. The U.S. Highway Traffic Safety Administration reported that distraction of various forms was a causative factor in 25% of traffic accidents (Stutts et al. 2001).

Considerable effort has been expended in the current decade to learn more about this situation. This followed many years of study of distracted driving and 30 years of study of use of mobile phones during driving (Peters and Peters 2002). As an example, a recent controlled study of driver performance during use of a cellular telephone found more errors maintaining position in the designated lane and during stopping, slower response time, and errors maintaining designated speed (Horrey et al. 2010). These observations, qualitatively at least, are readily apparent to anyone observing the driving behavior of an individual distracted by some activity. There was no difference between hand-held and hands-free performance. Drivers erred on average once in every five repetitions of a task. As well, drivers in the test were unaware of the impact of the distraction on deterioration of performance. The young seem especially prone to making these kinds of mistakes, possibly due to inexperience in driving.

The (US) National Safety Council has taken an active interest in the issue of distracted driving and its impact on safety. The Council recently published a "white paper" on this subject that advocates for prohibition of use of both hand-held and hands-free devices (NSC 2010). This document discusses the problem of distracted driving and provides a concise summary of published research from more than 30 studies and reports produced by scientists around the world. These studies used a variety of research methods, to compare driver performance with hand-held and hands-free devices. All of these studies show that hands-free devices offer no benefit to safety during driving.

One of the most important concepts highlighted in the "white paper" is inattention blindness. Inattention blindness describes the situation where a driver can look straight out the windshield in the direction of travel and fail to comprehend the significance of the situation ahead. Drivers impaired in this way look at objects but fail to see them.

Estimates indicate that drivers using cellular telephones fail to see up to 50% of the information in their driving environment (Strayer 2007). They do not process all of the information in the roadway environment necessary to effectively monitor the surroundings, seek and identify potential hazards, and respond to unexpected situations.

To explore how cell phone use can affect driver visual scanning, the Ergonomics Division of Transport Canada tracked eye movements of drivers during normal activity and use of hands-free cellular telephones. Drivers using hands-free cellular telephones looked less at the periphery and reduced visual monitoring of instruments and mirrors, and some drivers entirely abandoned those tasks. At intersections, these drivers glanced less frequently at traffic lights and at traffic on the right. Some drivers did not look at traffic signals (Harbluk et al. 2007).

Simultaneous use of a cellular telephone and driving a vehicle are examples of activities commonly called multitasking. Superficially, driving a vehicle involves visual and audible inputs and manual responses, while superficially, talking on the telephone involves auditory input and vocal response (Sanders and McCormick 1993). Processing code resources for the two tasks for the most part are separate on the superficial level. Driving involves spatial coding (location of other vehicles and fixed objects). Talking on the telephone involves verbal coding.

Both tasks are considerably more complex than this simplistic analysis would suggest. Use of a cellular telephone continually occupies one hand and one arm. Some driving tasks, such as turning corners, activating turn signals and windshield wipers and headlights, and shifting gears on a standard transmission require two hands. A driver must periodically focus attention on road and other signs (such as signs on businesses). In heavy traffic, for example, a driver must be prepared to stop quickly or to take evasive action when unexpected events occur. Telephone conversations are not always low key transfers of information. Some involve considerable arousal through expression of emotion. This makes use of verbal codes. This duality creates a competition for central processing resources.

Potential multitasking conflicts between driving and use of a cellular telephone were demonstrated in a study performed by Just et al. (2001). This study used functional magnetic resonance imaging (FMRI) to measure activity during high-level cognitive (conceptual) tasks that occur in different, unrelated parts of the brain. FMRI takes a three-dimensional image of the brain and subdivides it into subunits (voxels), each measuring about the size of a grain of rice. The number of voxels is a measure of the amount of "power" committed by the brain to completing a task. Each operation performed separately required 37 voxels. The number of voxels activated during combination of the tasks was only 42, instead of the expected 74. As well, the signal intensity of each voxel was lower than expected. A similar phenomenon occurred in sensory, noncognitive parts of the brain where auditory and visual processing occur. Performance decreased as well, when measured through response time.

This study implies that there is a limit to how well humans can perform concurrent tasks. That is, the brain gives each task less attention than when performed sequentially. Demanding driving cannot be time shared with other tasks. Every demand for attention, whether listening to the radio or talking to someone in the vehicle or on the cellular telephone, competes for resources otherwise used for driving.

There is a difference between talking to a passenger and a person on a cellular telephone. The passenger experiences the same environment as the driver and can pause at moments when concentration on driving is required.

The brain not only juggles tasks, but also juggles focus and attention (NSC 2010). When people attempt to perform two cognitively complex tasks such as driving and talking on a cellular telephone, the brain shifts focus. The brain drops and does not process important information. To illustrate, a driver may not "see" a red light. Very little information actually receives full analysis by our brains. Research shows we are blind to many changes that happen in scenery around us, unless we pay close and conscious attention to specific details, giving them full analysis to get transferred into our working memory (Trick et al. 2004). Other real-world examples of this problem are the collisions that occur at railway crossings despite the presence of operational warning signals. Drivers apparently failed to recognize the threat posed by the train, despite its large size (Green 2004).

An expression of the cost associated with attention switching is the slowing of driver reaction time when using a cell phone. For every input, the brain must make many decisions: whether to act on information processed, how to act, execute the action, and stop the action (NSC 2010). While this process may require only a fraction of a second, all of the time consumed by all of these steps is measurable and may be highly significant. When driving, fractions of seconds can be the time between a crash or no crash, injury or no injury, and life or death.

Simultaneous multitasking by the human brain is a myth. Human brains cannot perform two tasks at the same time. Instead, the brain handles tasks sequentially, switching between one task and the other. The brain can juggle tasks very rapidly. This leads to the perception that the brain is performing the two tasks simultaneously. Microprocessors in computers are designed to enable multitasking. The cost of the tasks is reflected in the heat generated and the operation of the cooling fan in portable machines. Especially intense processing operations cause the fan to operate at full speed. The human brain, by contrast, is not capable of such operations.

## A Metaphor for the Role of Distraction in Workplace Accidents

Distracted driving is only one of many situations in which humans perform many different tasks as part of their jobs. In each case, the person can perform only one task at any point in time (Peters and Peters 2006). This leads to the presumption that the brain performs best when tasks progress uninterrupted from beginning to completion. In this manner, the brain can devote its available resources and the conscious mind its fullest attention. If this type of work is considered to be the ideal from the perspective of least burden on the brain and its resources, more typical kinds of work illustrate the potential role of distraction in accident causation.

Very little work is structured to allow flow-through from beginning to completion. Distractors include telephone calls, coffee and bathroom breaks, breaks for smoking cigarettes, breaks to check the mail and e-mail, and breaks for other actions and activities. Many of these breaks are deliberate severances of the flow of work. That is, the person has had enough and needs a change, brief or prolonged, in order to allow the brain to process information. The person may experience fatigue that prevents clear thinking to progress to the end-point.

One extreme type of work involves many tasks, any one of which could occur and recur and start and stop and progress through part-way or to completion unpredictably at any point in time during the work day. One example of work of this type is that performed by receptionists in offices. Receptionists meet and greet visitors, answer the telephone and direct calls, and often perform additional tasks, such as word processing to prepare letters and reports, preparing and paying invoices, and possibly preparing coffee. The individual performs each task discretely, but may not be able to complete a task without interruption. The receptionist must stop one task, start another, and stop the second to start a third, then complete this task and return to and complete the second, and finally return to the first, and so on.

Distraction of the receptionist by continual shifting of priorities creates numerous problems. At minimum, the work is not satisfying because tasks may never reach completion or reach completion in reasonable time. Errors can arise in work involving information processing because of the break in continuity. These errors can have serious consequences for the organization.

Another extreme type of work involves long, complicated, consequential tasks. The task must progress sequentially in order to produce the desired end-result and/or to maintain safety for the individual performing it. Live-line electrical work is an example of work of this type. Maintaining safety is highly dependent on following the steps in the procedure in the exact sequence using designated tools and equipment and the use of protective equipment. The fatal accidents summarized in Chapter 1 provide many indications about the consequences of breaking the necessary sequence and restarting at some inappropriate point. These errors can have serious consequences for the individual.

Previous discussion in this chapter highlighted the many influences both on and off the job to which workers are subjected. Some are known to supervisors, while many others are potentially unknown. Figure 4.20 provides a model for visualizing the impact of distraction on safe performance of work. This model uses a tree as an analogy. The most effective, efficient, and safe procedure progresses from rootlet to root to trunk to branch to branchlet. Interconnecting rootlets, roots, branches, and branchlets constitute points of distraction to the sequential performance of the procedure. Each point of distraction can lead to secondary, tertiary, and quaternary distractions. For each distraction, the individual must retrace the steps back to the main path to completion of the task. The risk associated with the procedure depends on the number of disractors. The greater the number of distractions, the greater will be the risk of negative consequence.

## References

Allen, R.E.: Odor, A Legal Overview. In *Patty's Industrial Hygiene and Toxicology*, 4th Ed., Vol. I, Part B. Clayton, G.D. and F.E. Clayton, eds. New York: John Wiley & Sons, Inc, 1991.

Anderson, C.K.: Strength and Endurance Testing for Preemployment Placement. In *Manual Material Handling: Understanding and Preventing Back Trauma*. Akron, OH: American Industrial Hygiene Association, 1989.

Figure 4.20. Distraction and work. A tree is a metaphor for distraction and work. Distractions from a task divert us along branches. To return to the task, we must retrace the path of the distraction represented by the branch and move onward. This can occur many times before completion of the task is possible.

ANSI: American National Standard for Respiratory Protection — Respirator Use — Physical Qualifications for Personnel (ANSI Z88.6–1984). New York: American National Standards Institute, 1984.

ANSI: American National Standard for Respiratory Protection — Physical Qualifications for Respirator Use (ANSI Z80.2–1992). New York: American National Standards Institute, 1992.

Aschoff, J.: Circadian Rhythms: Interference with and Dependence on Work-Rest Cycles. In *The Twenty-Four Hour Workday: Proceedings of a Symposium on Variations in Work-Sleep Schedules*. Johnson, L.C., D.I. Tepas, W.C. Colquhoun, and M.J. Colligan (Eds.) (DHHS/NIOSH Pub. No. 81-127). Washington, DC: U.S. Government Printing Office, 1981.

Astrand, P. and K. Rodahl: *Textbook of Work Physiology*, 3rd Ed. New York: McGraw-Hill, 1986.

Blais, B.R.: Contact Lenses in the Chemical Industry. *J. Occup. Med. 36*: 706-707 (1994). [Letter to the Editor].

Boring, E.G.: Sensation and Perception in *A History of Experimental Psychology*, 2nd Ed.. New York: Appleton-Century-Crofts, 1950.

Broadbent, D.: Task Combination and Selective Intake of Information. *Acta Psychol. 50*: 253-290 (1982).

Budnick, L.D., S.E. Lerman, T.L. Baker, H. Jones, and C.A. Czeisler: Sleep and Alertness in a 12-Hour Rotating Shift Work Environment. *J. Occup. Med. 36*: 1295-1300 (1994).

Christman, R.J.: *Sensory Experience*. New York: International Textbook Company, 1971.

Chung, D.Y. and R.P. Gannon: The Effect of Ear Protectors on Word Discrimination in Subjects with Normal Hearing and Subjects with Noise-Induced Hearing Loss. *J. Am. Audiol. Soc. 5*: 11-16 (1979).

Colligan, M.J. and D.I. Tepas: The Stress of Hours of Work. *Am. Ind. Hyg. Assoc. J. 47*: 686-695 (1986).

Colquhoun, W.P. (Ed.): *Biological Rhythms and Human Performance*: New York: Academic Press, 1971.

Courau, P.J.: Older Workers. In *Encyclopaedia of Occupational Health and Safety*. 3rd rev. Ed., Vol. 2. Parmeggiani, L. (Ed.). Geneva: International Labour Organization, 1983. pp. 1565-1570.

Cullen, A.P.: Contact Lenses in the Work Environment. In *Environmental Vision: Interactions of the Eye, Vision and the Environment*. Pitts, D.G. and R.N. Kleinstein (Eds.). Boston: Butterworth-Heinemann, 1993.

Cullen, A.P.: Contact Lenses in the Workplace. In *Eye Injury Prevention in Industry*, 2nd Ed. McRae, E. and M. Grimm (Eds.). Edmonton, AB: Alberta Labour, Occupational Health and Safety Division, 1994.

Curry, D.G.: In-Vehicle Cell Phones: Fatal Distraction? *Prof. Safety 47(3)*: 28-33 (2002).

Curry, D.G., J.E. Meyer, and J.M. McKinney: Seeing versus Perceiving. *Prof. Safety 51:* 28-34 (2006).

Czaja, S.: Microcomputers and the Elderly. In *Handbook of Human-Computer Interaction*. Helander, M. (Ed.). Amsterdam: North-Holland, 1988.

DeCoursey, R.M.: *The Human Organism*, 4th Ed. New York: McGraw-Hill Book Company, 1974.

DOE: Amendment of the Occupational Safety and Health Administration (OSHA) Prohibition on Wearing Contact Lenses in Contaminated Atmospheres with Full Facepiece Respirators, Walker, M.L. Washington, DC: U.S. Department of Energy, September 23, 1986. [Memo].

Drennan, F.S. and D. Richey: Injury Prevention in an Aging Workforce. *Prof. Safety 48(9)*: 29-38 (2003).

Edworthy, J., S. Loxley, and I. Dennis: Improving Auditory Warning Design: Relationship Between Warning Sound Parameters and Perceived Urgency. *Human Factors 33*: 205-231 (1991).

EEOC: *A Technical Assistance Manual on the Employment Provisions (Title I) of The Americans With Disabilities Act 1990*. Washington, DC: U.S. Government Printing Office, 1992.

Findley, M.E. and J.O. Bennett: Safety and the Silver Collar Worker. *Prof. Safety. 47(5)*: 34-38 (2002).

Freivalds, A., J. Harpster, and L. Heckman: Glare and Night Vision Impairment in Corrective Lens Wearers. *Proceedings of the Human Factors Society 27th Annual Meeting*. Santa Monica, CA: Human Factors Society, 1983.

Gaunt, J.A.: Relationship between Overtime and Safety. *Prof. Safety 25(7)*: 11-15 (1980).

Giambra, L. and R. Quilter: A Two-Term Exponential Description of the Time Course of Sustained Attention. *Human Factors 29*: 635-644 (1987).

von Gierke, H.W., C.W. Nixon, and J.C. Guignard: Noise and Vibration. In *Ecological and Physiological Foundations of Space Biology and Medicine*, Vol. 2, Book 1. Calvin, M. and O.O. Gazenko (Eds.) .Washington DC: National Aeronautics and Space Administration, 1975.

Green, M.: Why Warnings Fail. *Occup. Health Safety 73(2)*: 30-36 (2004).

Haas, E.C. and J.G. Casali: Perceived Urgency of and Response Time to Multi-Tone and Frequency-Modulated Warning Signals in Broadband Noise. *Ergonomics 38*: 2313-2326 (1995).

Haight, J.: Human Error & The Challenges of an Aging Workforce. *Prof. Safety 48(12)*: 18- 24 ( 2003).

Haight, J.M. and U. Belwal: Designing for an Aging Workforce. *Prof. Safety 51(7)*: 20-33 (2006).

Harbluk, J. L., Y.I. Noy., P.L. Trbovich, and M. Eizenman: An On-Road Assessment of Cognitive Distraction: Impacts on Drivers' Visual Behavior and Braking Performance. *Accident Analysis Prev. 39 (2):* 372-378 (2007).

Hassall, J.R. and K. Zaveri: *Acoustic Noise Measurements*, 4th Ed. Naerum Denmark: Brüel and Kjær, 1979.

Hebb, D.O.: *Textbook of Psychology*, 3rd Ed. Philadelphia: W.B. Saunders Company, 1972.

Heinrich, H.W.: *Industrial Accident Prevention*. New York: McGraw-Hill, 1931.

Hockey, G.R.J. and W.P. Colquhoun: Diurnal Variations in Human Performance: A Review. In *Aspects of Human Efficiency: Diurnal Rhythm and Loss of Sleep*. Colquhoun, W.P. (Ed.). London: The English Universities Press, 1972.

Horrey, W.J., M. Lesch, and D.F. Melton: Distracted Driving. *Prof. Safety 55(1):* 34-39 (2010).

Howell, K. and A.M. Martin: An Investigation of the Effects of Hearing Protectors on Vocal Communication in Noise. *J. Sound Vib. 41:* 181-196 (1975).

IES: IES *Lighting Handbook, Reference Volume*. New York: Illuminating Engineering Society of North America, 1981.

ISO: Evaluation of Human Exposure to Whole-Body Vibration (ISO 2631/1). Geneva: International Standards Organization, 1985.

ISO: Danger Signals for Work Places — Auditory Danger Signals (ISO 7731-1986(E)). Geneva: International Standards Organization, 1986.

ISO: Hearing Protectors — Recommendations for Selection, Use, Care and Maintenance — Guidance Document (ISO/DIS 10452). Geneva: International Standards Organization, 1996.

Isreal, J.: *Structural Interference in Dual Task Performance: Behavioral and Electrophysiological Data*. Champaign, IL: University of Illinois, 1980. [Ph.D. dissertation].

Jensen, R.C.: Workers' Compensation Claims Relating to Heat and Cold Exposure. *Prof. Safety 28(9)*: 19-24 (1983).

Just, M. A., P.A. Carpenter, T.A. Keller, L. Emery, H. Zajac, and K.R. Thulborn: Interdependence of Nonoverlapping Cortical Systems in Dual Cognitive Tasks. *NeuroImage 14 (2):* 417-426 (2001).

Kantowitz, B.H. and R.D.Sorkin: *Human Factors: Understanding People-System Relationships*. New York: John Wiley & Sons, Inc., 1983.

Kleitman, N.: *Sleep and Wakefulness*. Chicago: University of Chicago Press, 1963.

Konz, S.: *Work Design: Industrial Ergonomics*, 3rd Ed. Scottsdale, AZ: Publishing Horizons, Inc., 1990.

Lawrence, M. and P.A. Yantis: Onset and Growth of Aural Harmonics in the Overloaded Ear. *J. Acoust. Soc. Am. 28*: 852-858 (1956).

Leukel, F.: *Introduction to Physiological Psychology*, 3rd Ed. St. Louis, MO: C.V. Moseby Company, 1976.

Lindsley, D.B.: Emotion. In *Handbook of Experimental Psychology*. Stevens, S.S. (Ed.). New York: John Wiley & Sons, 1951.

LLNL: Is it Safe to Wear Contact Lenses with a Full Facepiece Respirator? da Rosa, R.A. and C.S. Weaver (UCRL 53653). Livermore, CA: Lawrence Livermore National Laboratory, 1986.

Loeb, M.: *Noise and Human Efficiency*. New York: John Wiley & Sons, 1986.

Luce, G.G.: Biological Rhythms in Psychiatry and Medicine (PHS Pub. No. 2088). Washington, DC: U.S. Government Printing Office, 1970.

McGinley, M.A. and C.M. McGinley: The "Gray Line" between Odor Nuisance and Health Effects. *Proceedings of Air and Waste Management Association 92nd Annual Meeting and Exhibition St. Louis, MO: 20-24 June 1999.*

McManus, N.: *Safety and Health in Confined Spaces.* Boca Raton, FL: Lewis Publishers/CRC Press, 1999.

McMurtry, P.L. and D.H. Mershon: Auditory Distance Judgements in Noise, with and without Hearing Protection. In *Proceedings of the Human Factors Society 29th Annual Meeting.* Santa Monica, CA: Human Factors Society, 1985. pp. 811-813.

Metcalf & Eddy Inc.: *Wastewater Engineering, Treatment, Disposal and Reuse,* 3rd ed. Revised by Tchobanoglous, G. and F.L. Burton. New York: McGraw-Hill Inc., 1991.

Metts, A. and R.A. Mill: Work Injuries and Smoking Habits — Are They Related? *Prof. Safety 32(1):* 42-48 (1987).

Meyer, E. and E. Neumann: *Physical and Applied Acoustics.* New York: Academic Press, 1972.

Miller, G.A.: The Magical Number Seven, Plus or Minus Two: Some Limits on Our Capacity for Processing Information. *Psych. Revs. 63:* 81-97 (1956).

Miller, J.D.: Effects of Noise on People (Report No. NTID 300.7). Washington, DC: U.S. Environmental Protection Agency, 1971.

Moore, L.R. and G.S. Moore: The Impact of Cell Phones on Driver Safety. *Prof. Safety 46(6):* 30-32 (2001).

Moseley, M. and M. Griffin: Effects of Display Vibration and Whole-Body Vibration on Visual Performance. *Ergonomics 29:* 977-983 (1986).

Nejmeh, G.: Contact Lenses, To Keep Them in or to Keep Them out on the Job — That Is the Question! *Natl. Safety News:* June 1982. pp. 58-61.

NSC: *Understanding the Distracted Brain.* Itasca, IL: National Safety Council (2010). (www.distracteddriving,nsc.org).

OSHA: Contact Lenses Used With Respirators (29 CFR 1910.34 (e) (5) (ii), Stepich, T. through L. Carey. Washington, DC: Occupational Safety and Health Administration, February 8, 1988. [Memo].

Patterson, R.D.: *Guidelines for Auditory Warning Systems on Civil Aircraft* (Paper 82017). Cheltenham, UK: Civil Aviation Authority, Airworthiness Division, 1982.

Perry, L.S.: The Aging Workforce. *Prof. Safety 55(4):* 22-28 ( 2010).

Peters, G.A. and B.J. Peters: The Distracted Driver. *Prof. Safety. 47(3):* 34-40 (2002).

Peters, G.A. and B.J. Peters: *Human Error, Causes and Control.* Boca Raton, FL: CRC Press/Taylor & Francis Group, 2006.

Peterson, K.W. and D.R. Smith: Alcohol and Drug Abuse in Industry. In *Environmental and Occupational Medicine,* 3rd Ed. Rom, W.N. (Ed.). Philadelphia: Lippincott-Raven Publishers, 1998.

Peterson, W.W., T.G. Birdsall, and W.C. Fox: The Theory of Signal Detectability. *IRE Trans.* PGIT-4 (1954). pp. 171-212.

Randolph, S.A. and M.R. Zavon: Guidelines for Contact Use in Industry. *J. Occup. Med. 29:* 237-242 (1987).

Rasmussen, G.: Human Body Vibration Exposure and Its Measurement. *Technical Review I.* Naerum, Denmark: Brüel & Kjær, 1982.

Rink, T.L.: Hearing Protection and Speech Discrimination in Hearing-Impaired Persons. *Sound and Vibration 13(1):* 22-25 (1979).

Reinberg, A., P. Andlauer, and N. Vieux: Circadian Temperature Amplitude and Long-Term Tolerance of Shiftworking. In *The Twenty-Four Hour Workday: Proceedings of a Symposium on Variations in Work-Sleep Schedules.* Johnson, L.C., D.I. Tepas, W.C. Colquhoun, and M.J. Colligan (Eds.) (DHHS/NIOSH Pub. No. 81-127). Washington, DC: U.S. Government Printing Office, 1981.

Rengstorff, R.H. and C.J. Black: Eye Protection from Contact Lenses. *J. Am. Optom. Assoc. 45:* 270-276 (1974).

Robinson, G.S. and J.G. Casali: Speech Communications and Signal Detection in Noise. In *The Noise Manual,* 5th Ed. Berger, E.H., L.H. Royster, J.D. Royster, D.P. Driscoll, and M. Layne (Eds.). Fairfax, VA: American Industrial Hygiene Association, 2000. pp. 567-600.

Rosa, R.R. and M.J. Colligan: Shift Work: Health and Performance Effects. In *Environmental and Occupational Medicine*, 3rd Ed. Rom, W.N. (Ed.). Philadelphia: Lippincott-Raven Publishers, 1998.

Rosenstock, R.: Contact Lenses: Are They Safe in Industry? *Prof. Safety 31*(1): 18-21 (1986).

SAMHSA: *National Household Survey on Drug Abuse*. Rockville, MD: U.S. Department of Health and Human Services, Substance Abuse and Mental Health Services Administration, 1979-1995.

Sanders, M.S. and E.J. McCormick: *Human Factors in Engineering and Design*, 7th Ed. New York: McGraw-Hill, Inc., 1993.

Seshagiri, B. and B. Stewart: Investigation of the Audibility of Locomotive Horns. *Am. Ind. Hyg. Assoc. J. 53*: 726-735 (1992).

Seymour, R.B. and D.E. Smith: Identifying and Responding to Drug Abuse in the Workplace. *J. Psychoact. Drugs 22*: 383-405 (1990).

Sheridan, C.L.: *Fundamentals of Experimental Psychology*, 2nd Ed. New York: Holt, Rinehart and Winston, 1976.

Sheridan, J.R. and H. Winkler: An Evaluation of Drug Testing in the Workplace. In *Drugs in the Workplace: Research and Evaluation Data* (NIDA Research Monograph 91). Gust, S.W. and J. M. Walsh (Eds.). Washington DC: U.S. Department of Health and Human Services/Alcohol, Drug Abuse, and Mental Health Administration/National Institute on Drug Abuse, 1989.

Simons, D.J. and C.F. Chabris: Gorillas in our Midst: Sustained Inattentional Blindness for Dynamic Events. *Perception 28:* 1059-1074 (1999).

Singleton, W.T.: *Introduction to Ergonomics*. Geneva: World Health Organization, 1972.

Small, A.M. and R.S. Gales: Hearing Characteristics. In *Handbook of Acoustical Measurement and Noise Control*, 3rd Ed. Harris, C.M. (Ed.). New York: McGraw-Hill, 1991. pp. 17.1 to 17.25.

SOEH: *Healthy Aging for a Sustainable Workforce.*Report of the Conference on Aging, February 17-18, 2009, Silver Spring, MD. McLean, VA: Society for Occupational and Environmental Health.

Spoor, A.: Presbycusis Values in Relation to Noise Induced Hearing Loss. *Internat. Audiol. 6*: 48-57 (1967).

Stein, H.A. and B.J. Slatt: *Contact Lenses in Industry* (Ref. No. B01732). Toronto, ON: Industrial Accident Prevention Association of Ontario, 1984.

Strayer, D. L.: Presentation at Cell Phones and Driver Distraction. Traffic Safety Coalition, Washington DC (2007, February 28).

Stutts, J.C., D.W. Reinfurt, L. Staplin, and E.A. Rodgman: The Role of Driver Distraction in Traffic Crashes. Chapel Hill, NC: Highway Safety Research Center, University of North Carolina (May 2001). Prepared for AAA Foundation for Traffic Safety, Washington, DC.

Swets, J.A.: Is There a Sensory Threshold? *Science 134*: 168-177 (1961).

Trick, L.M., J.T. Enns, J. Mills, and J. Vavrik J.: Paying Attention behind the Wheel: A Framework for Studying the Role of Attention and Driving. *Theoret. Issues Ergonom. Sci. 5 (5):* 385-424 (2004).

Turk, A. and A.M. Hyman: Odor Measurement and Control. In *Patty's Industrial Hygiene and Toxicology,* 4th Ed., Vol. I, Part B. Clayton, G.D. and F.E. Clayton (Eds.). New York: John Wiley & Sons, Inc, 1991.

Tyhurst, J.S.: Individual Reactions to Community Disaster. *Am. J. Psychiat. 107*: 764-769 (1951).

Vander, A.J., J.H. Sherman, and D.S. Luciano: *Human Physiology*, 5th Ed. New York: McGraw-Hill Publishing Company, 1990.

Wallin, L. and I. Wright: Psychosocial Aspects of the Work Environment: A Group Approach. *J. Occup. Med. 28*: 384-393 (1986).

Walsh, J.K., D.I. Tepas, and P.D. Moss: The EEG Sleep of Night and Rotating Shiftworkers. In *The Twenty-Four Hour Workday: Proceedings of a Symposium on Variations in Work-Sleep Schedules*. Johnson, L.C., D.I. Tepas, W.C. Colquhoun, and M.J. Colligan (Eds.) (DHHS/NIOSH Pub. No. 81-127). Washington, DC: U.S. Government Printing Office, 1981.

Walsleben, J.A., E.B. O'Malley, and D.M. Rapoport: Sleep Disorders at Work. In *Environmental and Occupational Medicine*, 3rd Ed. Rom, W.N. (Ed.). Philadelphia: Lippincott-Raven Publishers, 1998.

Ward, W.D., R.E. Fleer and A. Glorig: Characteristics of Hearing Loss Produced by Gunfire and by Steady Noise. *J. Aud. Res. 1*: 325-356 (1961).

Ward, W.D.: Anatomy and Physiology of the Ear: Normal and Damaged Hearing. In *Noise and Hearing Conservation Manual*, 4th Ed. Berger, E.H., W.D. Ward, J.C. Morrill and L.H. Royster (Eds.) Akron, OH: American Industrial Hygiene Association, 1986. pp. 177-195.

Ward, W.D., J.D. Royster and L.H. Royster: Auditory and Nonauditory Effects of Noise. In *The Noise Manual*, 5th Ed. Berger, E.H., L.H. Royster, J.D. Royster, D.P. Driscoll, and M. Layne (Eds.). Fairfax, VA: American Industrial Hygiene Association, 2000. pp. 123-147.

Wasserman, D.E.: *Human Aspects of Occupational Vibration*. Amsterdam: Elsevier Science Publishers B.V., 1987.

Webster, J.C.: Effects of Noise on Speech Intelligibility. In *Proc. Conference Noise as a Public Health Hazard.* Washington DC. June 13-14, 1968. ASHA Reports 4, Washington, DC: American Speech and Hearing Association, 1969.

Welford, A.T.: Signal, Noise, Performance, and Age. *Human Factors 23*: 97-109 (1981).

Wickens, C., D. Sandry, and M. Vidulich: Compatibility and Resource Competition between Modalities of Input, Central Processing, and Output. *Human Factors 25*: 227-248 (1983).

Wickens, C.: *Engineering Psychology and Human Performance*. Columbus, OH: Merrill, 1984.

Wilkins, P.A.: A Field Study to Assess the Effects of Wearing Hearing Protectors on the Perception of Warning Sounds in an Industrial Environment (Research Contract Report No. 80/18). Southampton, UK: Institute of Sound and Vibration, 1980.

Zwerling, C., J. Ryan, and E.J. Orav: The Efficacy of Preemployment Drug Screening for Marijuana and Cocaine in Predicting Employment Outcome. *J. Am. Med. Assoc. 264*: 2639-2643 (1990).

# 5 Hazardous Energy and Operating Systems

## INTRODUCTION

Industrial activity covers a seemingly infinite spectrum of endeavor. At first glance, finding commonality in all of this is seemingly impossible. Yet, commonality is readily apparent when industrial activity, and indeed all human activity, is considered from the level of fundamental actions. Common to many activities is the harnessing, use, and application of energy sources and materials containing energy. Most of the time, the way in which energy sources and materials containing energy are used is so much an intrinsic part of the activity that we don't even notice or consider it.

Much of the activity of harnessing, use, and application of energy sources and materials containing energy occurs through the action of machines and equipment. Machines and equipment often are complex structures containing component parts or subsystems. Coordinated action of the subsystems produces the desired outcome, and sometimes, unexpected and unintended consequences.

The simplest device that can utilize energy is a subsystem. The subsystem can be a complete machine in and of itself, but in the context of complicated machines, is one part of a more complex whole. Subsystems need not be mechanical devices. They also can be electronic or electrical, or fluid-based. Subsystems and systems process energy in the following manner: input, transformation, storage, and output. Understanding the manner in which these processes occur in subsystems is critical to understanding the impact that they have on the system itself, for this is the level at which interaction with humans often occurs.

The automobile is a classic example of a complex system containing many subsystems that input energy, store, and transform energy and output energy, on which industrialized society is completely dependent.

The input for the automobile is fuel added to the storage tank at the filling station. The engine, through compression of a mixture of fuel vapor and air, aided by an electric arc from an electrode powered by conversion of chemical energy stored in the battery, combusts the fuel in a controlled detonation. Release of chemical energy through rapid combustion leads to formation of hot gases whose expansion pushes the piston. Movement of the pistons in a coordinated way rotates the crankshaft. Rotation of the crankshaft rotates a system of gears and rods or pipes or tubes. These, in turn, rotate the wheels. Rotation of the crankshaft also powers the electrical energy generating and storage subsystems. Electric motors power the fluid power subsystem. These subsystems operate so seamlessly that the operator is unaware of problems until a breakdown occurs.

## Mechanical Systems (Machines)

In the academic context, a "machine" is a mechanical device that harnesses and controls the direction of expression of forces in order to produce required work (Olivo and Olivo 1984). Mechanical devices change the magnitude, direction or intensity of forces, or the speed resulting from them. Mechanical devices can be as simple as the screwdriver, wrench or hammer, or as highly complex as a passenger aircraft. In the industrial context, a machine usually is an assembly of subsystems. At its simplest, a machine or subsystem contains one or more machine elements (Table 5.1).

Machine motion is either translational (straight line or curved path) or rotational (Raczkowski 1979). Machine elements must carry loads, for example, static and dynamic forces. Work performed by mechanical systems results from the action of forces or torques. (For information about torque, refer to the following section.) Resulting forces or torques balance input loads.

## Table 5.1
## Machine Elements

Lever

Inclined plane and wedge

Wheel and axle

Pulley

Gear

Screw thread

## The Lever

One of the simplest mechanical devices is the lever (Walton 1968). The lever is a rigid structure, capable of turning about a point or fulcrum. The tendency to move around a point is the torque or moment of force (Equation 5-1).

$$M = Fd \qquad\qquad (5\text{-}1)$$

where:
$M$ = moment of force or torque ($N{\bullet}m = J$).
$F$ = force applied on the lever (N).
$d$ = distance of the center of the force from the fulcrum (m).

Notice that the unit of torque, the Newton•meter is the same as the unit of energy, the Joule. (Refer to Chapter 2 for further information about energy and units.)

The lever comprises two segments that have the fulcrum as a common point, by which force applied at one point is transmitted or modified at a second. Equation 5-2 applies when the torques are balanced and in opposite directions. Under this condition, the lever does not move.

$$F_1d_1 = F_2d_2 \qquad\qquad (5\text{-}2)$$

where:
$F_1$ = force applied on the lever at position 1 (N).
$d_1$ = distance of the center of Force 1 from the fulcrum (m).
$F_2$ = force applied at the other point on the lever at position 2 (N).
$d_2$ = distance of the center of the Force 2 from the fulcrum (m).

The meaning of Equation 5-2 is that a small force applied through a large distance from the fulcrum produces the same effect as a large force applied through a small distance. That is, the lever amplifies force. This effect is known as mechanical advantage. Machine designers fully exploit mechanical advantage in machines. Mechanical advantage of force is the ratio of the forces balanced by the lever. Mechanical advantage of distance is the ratio of the distances at which the forces balance on the lever (Olivo and Olivo 1984).

The compound lever is a group of simple levers containing interlinked linkages. The compound lever provides the same mechanical advantage as the simple lever, but in a much shorter distance. Angled lever arms and complex designs can provide high mechanical advantage in a compact package. This means that a small force applied to one end of a compound lever can produce a very large output.

The weigh scale is an example of the application of complex levers in a compact package. In this application, a large mass, the vehicle, produces a small and highly predictable deflection of the scale measuring device. This application of levers is the inverse of amplification of force. However, when configured in the inverse, a small force applied to the pointer in the scale room could produce a change in the vertical position of the vehicle.

Unexpected or unintended movement in machinery and mechanical systems containing compound levers can cause serious traumatic injury during service work. Compound levers fit into a deceptively compact package. This combination can produce a high level of displacement for a minimal input of force and high level of force for seemingly minimal displacement. Most machines are designed to enable a small amount of force acting through a system of levers to produce a large effect. This is done in the name of energy efficiency. The paucity of effort needed to produce this level of output is highly deceptive.

Undoing the movement of a system of complex levers is extremely difficult because of the need to perform work against the mechanical advantage provided by the system. In other words, this is akin to using a vehicle to change the small deflection of the vehicle weigh scale.

### The Inclined Plane (and the Wedge)

The inclined plane is the second machine element (Olivo and Olivo 1984). The inclined plane enables the raising of a load to a higher level along a diagonal surface for less effort than would be expended in a vertical lift (Equation 5-3).

$$EL = Rh \tag{5-3}$$

where:
$E$ = effort force directed along the slope (N).
$L$ = length of the slope (m).
$R$ = resistance force (N).
$h$ = height of slope (m).

Note that Equations 5-2 and 5-3 are essentially the same. The mechanical advantage provided by the inclined place is the ratio of the length of the slope (hypotenuse of the right-angle triangle) to the vertical rise. This is the same as the ratio of the vertical force exerted by the load to the effort needed to pull it along the inclined plane. The inclined plane extends the distance needed to perform the vertical lift, thus reducing the magnitude of the effort.

A movable version of the inclined plane is the wedge. Movement of the load along the wedge forces movement of the load up the slope instead of vertically. Mechanical advantage provided by the wedge is the ratio of the length of the slope to the vertical rise.

Movable ramps and wedges require anchorage to prevent horizontal slippage and opening of a gap between the fixed upper surface and the edge of the ramp or wedge. A gap creates the conditions for an accident. Slippage also could lead to a shift in position of structural members dependent on the intact position of the wedge and creates the potential for collapse.

### The Wheel and the Axle

The wheel and the axle constitute the third machine element. The wheel is the ancestor of the pulley, the gear, and countless other applications (Walton 1968). The wheel is a rotary system of levers. Rotation of a wheel creates mechanical advantage along the axle in exchange for the distance created by the number of rotations. The converse is also true. The fulcrum is the center of the axle. Mechanical advantage is related to the distance from the center of the axle (fulcrum). An axle containing two (or more) wheels of the same or different radius acts as a common fulcrum. The wheel with the larger radius exerts mechanical advantage over the wheel having the smaller radius through the ratio of the radii. Again, this is an example of the application of distance.

The wheel and the axle transmit force or produce a change in speed (Olivo and Olivo 1984). To illustrate, rotation of one wheel causes rotation of another with which it is in contact, provided sufficient friction exists between them (Nelson 1983). While in contact, the surfaces travel equal distances at equal surface speeds. This also applies to power (energy) transmission devices, such as belts and chains, since they contact the moving surfaces. That is, the belt or chain travels the same distance at the same speed as the surface of the discs.

Placing an even number of wheels in contact changes the direction of rotation, clockwise to counterclockwise and vice versa, of the driven wheel. Placing an odd number of wheels in contact maintains the direction of rotation of the driven wheel. Placing wheels of different diameters in contact changes the speed of rotation of the driven wheel. Where the discs are unequal in size, in order to maintain continuity, the shaft of the smaller disc must rotate faster than that of the larger disc. Whenever the wheel or axle is used as a simple machine, only the force or the speed may increase, not both simultaneously. This increase comes at the expense of distance.

## The Pulley

The pulley is a logical application of the wheel and the axle. Pulleys extend the distance between the axles of wheels through interconnecting linkages, including ropes, smooth belts, toothed belts, and chains. The significance of these devices is that the interconnecting linkage can transmit force through distance (Walton 1968).

The pulley consists of one or more sheaves, an axle, and a housing. The sheave is a wheel grooved on its periphery to accommodate the rope, belt, or chain. The sprocket is a variation of the sheave. The sprocket contains teeth into which fit spaces in the links of the chain. Some types of sheave are hybrids. They contain teeth to accommodate toothed belts for more secure contact grip. Each tooth in a sprocket and each tooth in a toothed sheave acts as a lever and connects directly to the fulcrum at the axle through the material of the structure.

Like the wheel, the pulley is a round lever. In the case of a pulley fixed in position, the axle is the fulcrum. The load arm (side of the lever to which the load is attached) is the radius of the sheave. The effort arm (side of the lever through which effort to move the load is expended) is also the radius of the sheave. The two arms are equal in length. Hence, the rope on each side of a fixed pulley carries the load. All that changes is the direction.

In the case of a movable pulley, the fulcrum is the point where the rope contacts the sheave. The load arm is the distance to the fulcrum (the radius of the sheave). The effort arm is the distance to the point on the sheave opposite the fulcrum (the diameter of the sheave). Hence, the effort expended to lift the load is half the weight of the load.

Combining a fixed pulley and moveable pulley into a unit interconnected by rope (the block and tackle) considerably reduces the effort needed to lift a load. Application of this principle is visible in any crane that uses wire rope for lifting. This configuration is analogous to the compound lever.

## The Gear

The gear is a disc or wheel containing teeth around the periphery. A rack is a linear version of a gear. A gear is equivalent to a cluster of evenly spaced levers mounted on a common fulcrum (the axle) (Walton 1968). Gears transmit motion through extensive positive contact (meshing of teeth) between one another. That is, meshing of the teeth of gears is impingement of the ends of one cluster of levers into those of a second cluster. The positive contact of teeth when meshed together enables gears to drive parts or mechanisms without slippage and to increase or decrease force (Olivo and Olivo 1984).

A small gear mounted on the powered shaft meshing with a large gear on the driven shaft provides a considerable increase in torque. Increasing torque is a major application of gears and gear trains (three or more meshing gears). Gears provide an increase in speed at the expense of torque, and vice versa. Speed of the driven gear may be faster or slower than that of the driving gear.

Alignment is critical for proper performance. Properly aligned and fitting gear teeth transmit uniform velocity and work in part with rolling instead of sliding friction.

Lubrication and freedom from the presence of foreign materials are key determinants in the life of gears. For this reason, enclosures are used to house gears and to provide access to lubricants. At the same time, the housing prevents observation of the complexity of the gear train by anyone unfamiliar with the design.

Normally, the driving gear and driven gear (last gear) rotate in the same direction when an odd number of meshing gears is present. The converse is true when an even number of gears is present (Olivo and Olivo 1984). This is also true for gears whose teeth are on the outside of the circumference. When the teeth are on the inside of the circumference of one gear and the outside of the meshing gear, the gears move in the same direction. This arrangement is more compact, provides more contact between teeth, and is quieter than external gearing.

The speed of the driven gear is related to the ratio of the number of teeth to those in the driving gear. Hence, the number of teeth in a gear determines the speed.

A compound gear train results when a driving gear and a driven gear are affixed to the same stud or shaft and cause another gear or train of gears to turn. The compound gear train occupies less space than a simple gear train providing the same function and provides a greater range of speed.

### The Screw Thread
The screw thread is the last machine element. The screw thread applies principles operative in the inclined plane, the wedge, and the wheel and axle (Olivo and Olivo 1984). The screw thread is a ridge of uniform cross-section formed around a cylinder or cone. The distance between each thread (pitch) is constant. In machine applications, screw threads transmit motion, apply tremendous force with little effort, and hold parts together. The screw thread converts rotary motion to rectilinear (straight-line) motion. The distance traveled depends on the pitch of the thread and the amount the screw is turned. Screw threads act as wedges to provide holding action. The effort distance is the circumference of the circle of rotation of the turning device. Hence, a large circumference, as would be created using a socket wrench, applies considerable force to the thread.

## Mechanical Power Transmission

Power (energy) transmission equipment is the medium through which practically all motion occurs in industrial equipment (Nelson 1983). (Power is the rate of use of energy.) Power (energy) transmission equipment refers to specific parts of a machine assembly that function in the mechanical transmission. By far the most common form of power (energy) transmission is transfer of rotary motion from one shaft to another. This occurs axially (longitudinally), and from shaft to adjacent shaft. Axial transmission requires a clutch or coupling. Axial transmission is the means by which mechanical energy created in the engine is transferred in a controlled way to the drive train in a vehicle. Transfer of rotary motion between adjacent shafts occurs through physical contact. Machine elements also must carry loads, for example, static and dynamic forces. Resulting forces or torques balance input loads.

Belt drives provide quiet, compact and resilient power (energy) transmission (Nelson 1983). Belts are used singly or in multiples, depending on service requirements. Belt types include "V"-belts, flat belts and positive drive (gear) belts. In the case of the "V"-belt, transmission of power (energy) occurs at the plane of contact between the sides of the "V"-belt and the sloping walls of the sheave (pulley). Slippage of the belt during overload conditions protects vulnerable equipment from load surges.

Applications involving power (energy) transmission using flat belts are mostly historical. These were used in factories to drive equipment from centralized power sources. Flat belting is still widely used in conveying systems. Positive (synchronous) drive belts containing molded teeth interface with pulleys containing matching molded teeth. The interface between the teeth of the belt and those of the pulley is the plane of power (energy) transmission. Positive drive belts offer the quiet operation and flexibility of the belt with the power-transfer interface of gears and chains.

Chain drives consist of one or more driven sprockets and an endless chain (Nelson 1983). The links of the chain mesh with the teeth of the sprocket. This maintains a positive speed ratio between the powered and driven sprockets because they do not slip or creep. The principal advantages of chains are simplicity, economy, efficiency and adaptability. The type of chain most widely used for power (energy) transmission is the roller chain (bicycle chain). Roller chain is manufactured in single and multiple strands. The inverted tooth chain is similar in concept to the positive drive belt. The links contain inverted teeth. These engage the teeth in a cut tooth wheel. This design provides the flexibility and quiet operation of belts, the positive action and durability of gears and the convenience and efficiency of chain.

## Mechanical Equipment Hazards

### Catastrophic Failure

Rotating parts experience centrifugal force, which, at high speed, can cause catastrophic failure (Tarr 2002). Industry standards for allowable rim speed relate to the strength of the materials of construction. Maximum rim speed for pulleys made from cast iron is 6,500 ft/min (33 m/s), and 8,000 ft/min (41 m/s) for ductile iron. Steel and aluminum can operate at higher rim speeds. Belts, likewise, experience centrifugal force and are limited by strength in their applicability. Bendability and resistance to bending also limit the speed of application.

Catastrophic failure also relates to structural members in machines and equipment, including levers and structures, such as booms, that function as levers. Adherence to codes during design and manufacture and operation respectful to codes for maximum loading minimize the potential for catastrophic failure, such that this has become and remains a rare event.

### Electrostatic Hazards

Accumulation and discharge of electrostatic charge are major concerns in the operation of some mechanical equipment, especially that containing belt drives and moving expanses of material, such as paper and fabric. This concern is especially enhanced in climates experiencing low humidity, primarily in winter (Tarr 2002). All belt drives and moving expansive materials create electrostatic charge due to contact and separation between the belt(s) and the pulleys and the expansive material and surfaces of contact. This situation is most likely to occur when part of the system is not grounded. Charge generation increases with increasing speed. Conductive belts and charge dissipative devices provide a means to prevent accumulation of electrostatic charge.

### Traumatic Injury

Entanglement during operation is especially possible around exposed mechanical power (energy) transmission equipment where the protection afforded by the enclosure is not respected or is circumvented. This situation results in many fatal accidents. (Refer to Chapter 1 for further information.)

### Rotating Equipment Hazards

Rotation, meaning circular motion, is cyclic. At some point, every distinguishing characteristic of the rotating component returns to the same position in space. For this reason, all rotational motion is potentially dangerous. Even smooth, slowly rotating shafts can grip clothing and skin, and draw or force an arm or hand into a dangerous position. Examples of components that pose rotational hazards include horizontal and vertical shafts, especially those containing irregularities such as burrs or gouges, collars, couplings, cams, clutches, flywheels, shaft ends, spindles, and meshing gears. The danger increases with the presence of projections such as nicks, abrasions, and projecting keys, bolts, or set screws (Figure 5.1).

Rotating components of machinery also create hazards known as in-running nip points. These result when a rotating part is close in proximity to a rotating or stationary part.There are three main types of in-running nip points. The first is created by parts rotating in opposite directions along axes parallel to each other (Figure 5.2). Contact or close proximity between these parts produces the nip point. This danger is present in machines with intermeshing gears, rolling mills, and calenders. Stock fed into or between the rolls produces the nip point. The wringer section of the old washing machines used in the home is an example of equipment that contained in-running and tangential nip points.

The second type of nip point results from tangential contact with the roller, either moving or stationary . The narrowing gap between rotating and tangentially moving components creates a nip point. Examples include the point of contact between the power transmission belt and pulley, the chain and

Figure 5.1. Rotating equipment hazard. Any rotating component can entrain clothing and sometimes the hair. Any irregularity on the surface of the rotating part enhances the hazard.

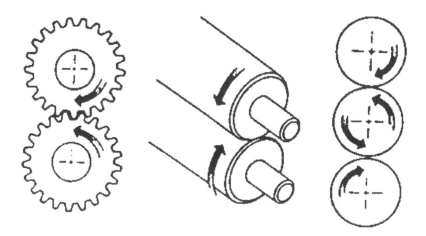

Figure 5.2. In-running nip point. The narrowing gap between rotating components moving in opposite directions on close parallel axes creates an in-running nip point. (Source: OSHA 1992.)

sprocket, and the rack and pinion (Figure 5.3). Contact between rotating and fixed parts resulting in shearing, crushing, or abrading action creates nip points. Examples include spoked hand wheels or fly-wheels, screw conveyors, or the periphery of an abrasive wheel and an incorrectly adjusted work rest (Figure 5.4).

### Transverse Motion and Reciprocating Equipment Hazards

Transverse and reciprocating motions occur in the horizontal and vertical planes, or diagonally with components in both directions. Transverse and reciprocating motions can cause injury from struck-by or kickback (Figure 5.5), caught-in, or caught-between a shear point by the moving part.

Reciprocating movement in machines produces actions such as cutting, punching, shearing, and bending. All of these pose risk of injury. These actions, singly or in combination, are integral to nearly all machinery.

Cutting can involve rotating, reciprocating, or transverse motion. The hazard of the cutting action exists at the point of operation. Injury results from inadvertent or inappropriate positioning of the finger, arm, and torso. Ejection of chips or scrap material can result in struck-by injury to the head, particularly in the area of the eyes and face. Such hazards are present at the point of operation when cutting wood, metal, or other materials. Examples of equipment posing cutting hazards include bandsaws, circular saws, boring or drilling machines, turning machines (lathes), or milling machines (Figure 5.6).

Punching action results when power is applied to a slide (ram) for the purpose of blanking, drawing, or stamping metal or other materials. The risk arising from this type of action occurs at the point of operation where stock is inserted, held, and withdrawn by hand. Typical machines used for punching operations include power presses and iron workers (Figure 5.7).

Shearing results from application of power to a slide or knife in order to trim or shear metal or other materials. A hazard occurs at the point of operation where stock is actually inserted, held, and

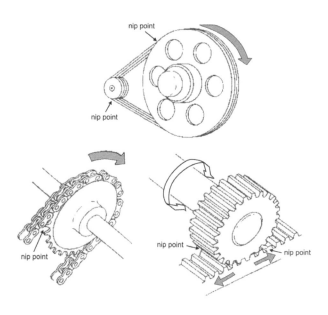

Figure 5.3. Tangential nip point between moving parts. Tangential contact creates a nip point. (Source: OSHA 1992.)

Figure 5.4. Tangential nip point formed between fixed and moving parts.

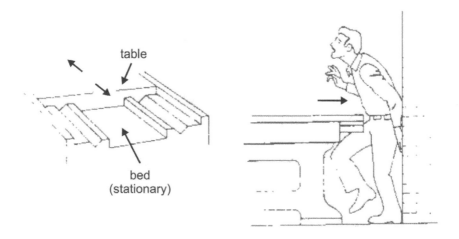

Figure 5.5. Hazardous transverse motion. Kickback of material in power tools is a significant cause of injury related to transverse motion. (Source: OSHA 1992.)

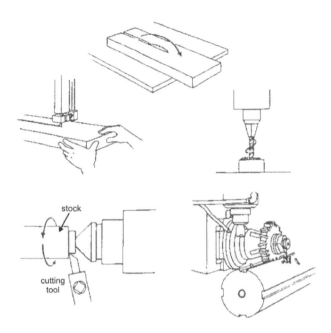

Figure 5.6. Cutting hazards. Many types of equipment pose hazards from cutting actions. These can involve traumatic injury from contact with moving parts, as well as contact with ejected parts and chips from cutting. (Source: OSHA 1992.)

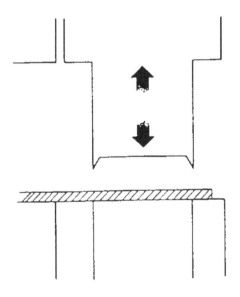

Figure 5.7. Punching. (Source: OSHA 1992.)

withdrawn. Machines used for shearing include mechanically, hydraulically, or pneumatically powered shears (Figure 5.8).

Bending results from application of force in order to draw or stamp metal or other materials. The hazard occurs at the point of operation where stock is inserted, held, and withdrawn. Equipment that uses bending action includes power presses, press brakes, and tubing benders (Figure 5.9).

## Mechanical Systems and Energy Storage

Power (energy) transmission equipment does not store mechanical energy. The energy content of the equipment dissipates rapidly following shutdown of the source. Energy storage can and does occur in equipment to which power (energy) transmission equipment is connected.

Energy storage in mechanical systems occurs in a number of ways. Gravity is the most widespread means. Parts in machines that move as part of normal operation, such as levers, knives, hammers, and many others, can stop in a position from which movement can still occur under appropriate conditions. As well, loss of anchorage during dismantlement can lead to pendulular motion of many types of functional and structural parts of machines. In a pendulum, energy transforms between potential energy (present solely at the top of vertical travel) and kinetic energy (present solely at bottom dead center). In all other positions, the energy partitions between potential and kinetic energy.

### Flywheels
A classic example of energy storage in machinery is the punch press (Juvinall and Marshek 1991). A punch press can operate with power from a motor or with power stored in a motor-flywheel combination. In the former case, the power must be available continuously. This design is wasteful and requires a large motor having the output to satisfy momentary demand (one sixth of a rotation) during which the punching action occurs. For a punch press that makes 60 punches per minute (1 punch/s) and a motor

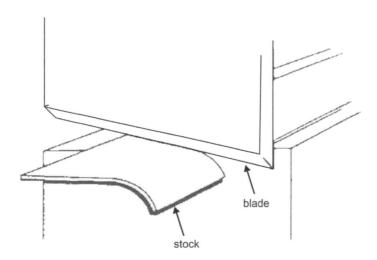

blade

stock

Figure 5.8. Shearing. (Source: OSHA 1992.)

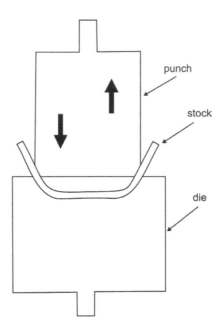

Figure 5.9.  Bending. (Source: OSHA 1992.)

rotating at 1,200 rev/min (20 rev/s), the demand occurs over a very small fraction of the operational cycle (0.17/20 revolutions) between punches.

In the case of the motor-flywheel combination, the punch is spread over one rotation. In addition, rotation of the flywheel accumulates and stores energy during the remainder of the operating cycle. Some of this energy is utilized during the punch stroke, thus slowing down the flywheel somewhat. Since the rotational movement between punch strokes is a large fraction of the operating cycle (19.83/20 revolutions), the flywheel recovers to its maximum energy between punches.

The dynamics of this system are amenable to modeling. Energy in rotational motion is given by (Juvinall and Marshek 1991):

$$E = \frac{I\omega^2}{2} \tag{5-4}$$

where:
E = energy (J).
I = the polar moment of inertia (kg•m$^2$).
$\omega$ = angular velocity (rad/s).

The moment of inertia for a hollow cylinder (a first approximation for the flywheel) is given by:

$$I = \frac{\pi(d_o^4 - d_i^4)L\rho}{32} \tag{5-5}$$

where:

$\pi$ = 3.14.

I = the polar moment of inertia (kg•m$^2$).

d$_o$ = outer diameter (m).

d$_i$ = inner diameter (m).

L = the length of the cylinder (m).

$\rho$ = mass density (kg/m$^3$).

Rotational inertia is an important mechanism for energy storage in some types of equipment. A rotating mass will continue to move after removal of the energy source. Flywheels have received consideration as energy-storage devices in trolley buses, reliable power systems and other applications. A less ambitious application is the reliable power supply present in or as an add-on for electronic equipment. In this application, the flywheel is part of an uninterruptible power supply (Lamendola 2001). Energy recovered during coast-down of the flywheel in one of these units permits orderly shutdown of the electronic equipment.

As evident in Equations 5-4 and 5-5, energy stored in a flywheel is proportional to the square of the rotational speed, the length and density of the material used in the flywheel, and the fourth power of the diameter. In the case of a ring-type flywheel, energy depends on the difference between the inner and outer diameters raised to the fourth power. The flywheel in some machines can store considerable residual energy.

## Springs
Machine parts, such as levers and flywheels, are deliberately constructed from rigid materials. Mechanical advantage derives from rigidity of components (Juvinall and Marshek 1991). Another type of energy storage in machine parts occurs because of elastic properties of materials. Elasticity can be a deleterious factor in performance of parts of machines and equipment. There are countless examples where restraint of elastic motion through the use of anchoring devices must occur. Inappropriate elastic motion can lead to vibration and production of noise, misalignment of critical interfaces, and wear on critical parts.

A serious concern with disassembly of structures maintained under such conditions is spring back of the part to its original position on removal of the restraint. This springing back to the unrestrained position occurs through conversion of potential to kinetic energy. The displacement of parts and structures that occurs during this process can cause damage and injury. In these cases, elasticity is neither intended nor considered desirable.

There are countless other applications where elasticity is considered highly desirable. Springs are a type of energy storage device (Figure 5.10). They exert forces or torques and absorb energy which is stored and released at a later time (Juvinall and Marshek 1991). Springs store energy by virtue of displacement from an equilibrium or resting position to one of disequilibrium or instability. That is, energy release occurs as the spring performs work in returning to its position of equilibrium.

Springs are classed according to function: controlled-action, variable-action or static (Jensen and Helsel 1996). Controlled-action springs have a constant range of action or well-defined function for each cycle of operation. Valve springs in engines, and springs in electrical switches are controlled-action springs. Variable-action springs have variable range of action. This reflects the varying conditions imposed on them. The leaf springs in the suspension of a vehicle, springs in the clutch and springs in seat cushions are examples of variable-action springs. Static springs exert comparatively constant pressure or tension between parts. Packing or bearing pressure, anti-rattle, and seal springs are examples of static springs.

Classification of springs also reflects the type of spring. Criteria that govern the type of spring include function, shape, application, or design. A compression spring is an open-coiled helix. The helix can contain cylindrical, or flat material. A compression spring offers resistance to compressive force. An extension spring offers resistance to pulling force. An extension spring is a close-coiled helical spring. Torsion springs exert force along a circular path. Torsion springs include helical and bar springs. A power spring is a flat metal coil spring that is wound on an arbor and usually confined to a

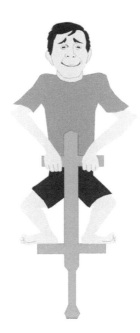

Figure 5.10.  Storage and conversion of energy. The "pogo" stick is a classic example of a device involved in the storage and conversion of energy. In this situation, energy converts seamlessly between potential energy (compression) and kinetic energy (expansion).

case or drum. The springs in a mechanical clock are power springs. Flat springs contain flat metal formed in a manner to oppose force in the desired direction when deflected in the opposite direction. The leaf spring is an example of a series of flat springs nested and held together and arranged to provide approximately uniform distribution of stress throughout the length of the assembly. Individual levers also can act as springs under certain circumstances.

A spring or a material or structure that exhibits elastic properties has a spring rate or spring constant k, which has units N/m (Newton/meter) for linear motion and N•m/rad (Newton • meter/radian) for circular motion (Juvinall and Marshek 1991). The spring constant for torsional motion is denoted as K.

For coil springs, the spring constant is:

$$k = \frac{F}{\delta} = \frac{d^4G}{8ND^3} \qquad (5\text{-}6)$$

where:
k = spring (rate) constant (N/m).
F = force (N).
δ = deflection (m).
d = diameter of the spring wire (m).
G = torsional modulus of elasticity (Pa = N/m$^2$).
N = number of coils in the spring.
D = diameter of the coil of the spring (m).

For a solid rod torsional spring, the equation for the spring constant is:

$$K = \frac{\pi d^4 G}{32L} \qquad (5-7)$$

where:
K = force constant (N•m/rad).
π = 3.14.
G = torsional modulus of elasticity (Pa = N/m²).
L = length of the rod (m).

For a beam spring (including leaf springs), the equation for the spring constant is:

$$k = \frac{F}{\delta} = \frac{Ebh^3}{6L^3} \qquad (5-8)$$

where:
F = force (N).
δ = deflection (m).
E = Young's modulus of elasticity (Pa = N/m²).
b = width of spring (m).
h = the thickness of each leaf (m)
L = length of the cantilever (half of the semi- and full elliptical spring) (m).

Mechanical systems containing the components mentioned here can convert, store, and transmit and translate energy within equipment. Energy release can occur slowly, as in the case of the flywheel or rapidly in the case of a compressed, stretched, flexed, or twisted spring, or member that falls from one height to another. Energy released from these sources in this manner, in conjunction with the compounding action of lever systems, pulleys and gears, can produce unexpected movement in machine parts. These unexpected movements have caused many fatal injuries to individuals working on this equipment.

Unstable Systems
The inclined conveyor and similar structures containing a load at the time of shut down represents an unstable system in which all of the energy is trapped as potential energy (Figure 5.11). Additional examples include raised dump bodies on dump trucks and loaded buckets on front-end loaders, and loaded vehicles parked on inclined roadways. Gravity attempts to pull the load to a lower position. This would result in motion and a conversion of potential to kinetic energy and the ability to cause serious damage and traumatic injury.

## Fluid Power Systems

A fluid power system transmits and controls energy through use of pressurized gas or liquid. Pressurized gases include compressed air, nitrogen and steam. Liquids include petroleum-based and synthetic oils, as well as water. Fluid power systems offer the option of separation between the power (energy) unit and the actuator(s); operation of multiple types of actuators by a single power (energy) unit; flexibility in speed; and precision of control of movement and motion. In addition, these systems contain few moving parts compared to equivalent mechanical systems (Johanson 2002).

A fundamental concept in fluid power is the relationship expressed in Pascal's law (Vickers 1992). Pascal's law states that a confined fluid (static system) against which a force is applied, exerts the same force perpendicularly on equal areas of the surface of the containment. That is, the force per unit area is constant on all surfaces. Force per unit area is pressure, a defined quantity. Pressure provides a practical means to integrate force, pressure, and surface area (Equation 5-9).

Figure 5.11. Unstable energy storage system. Material retained on an inclined conveyor at time of shutdown exists in an unstable status. The material attempts to move downward under the force of gravity.

$$P = \frac{F}{A} \tag{5-9}$$

where:
$P$ = pressure (Pa = N/m$^2$).
$F$ = force (N).
$A$ = area (m$^2$).

Equation 5-9 has many implications and useful applications. Equation 5-9 indicates that the fluid transmits pressure equally to all surfaces. As a result, a small pressure applied to a large area creates a large force. Similarly, a large pressure applied to a small area of surface creates a large force. Note that the fluid in the circuit is not flowing. It is trapped inside the confines of a circuit. Otherwise, the pressure would vary with distance from a source according to Bernoulli's theorem.

A fluid power system acts in the same manner as a lever. That is, a large displacement exerted over a small surface area is equal to a small displacement exerted over a large area. Again, the trade-off to obtain mechanical advantage comes at the cost of distance.

Fluid-powered machines are divided into two groups: velocity (dynamic) type and positive displacement (pressure) type (Stewart 1987). In the velocity (dynamic) fluid machine, interaction between a mechanical part and the fluid produces appreciable change in the velocity of the fluid. The centrifugal and axial flow compressor are examples of the velocity (dynamic) type of machine. In the positive displacement (pressure) type of machine, volumetric change or displacement occurs. Pressure is developed primarily by displacement action. Positive displacement fluid machines produce both reciprocating (back and forth) and rotary motion. Positive displacement pumps and hydraulic motors are the most prevalent examples (Vickers 1998).

The actuator is the component in a fluid power system that does the work. Cylinders containing pistons provide linear back and forth (reciprocating) motion. Vane or rack and pinion devices provide partial rotational motion. Motors provide continuous rotational motion.

Pneumatic and hydraulic cylinders are familiar applications of fluid power. Cylinders are designed to be single acting or double acting. The single-acting cylinder contains a spring that moves the piston against the pressure exerted by the fluid. In this manner, the piston returns to the neutral position after removal of pressure (Figure 5.12).

The double-acting cylinder receives fluid supply on both sides of the piston (Figure 5.13). This is the case in bulldozers, excavators, and other construction equipment. Fluid pressure is applied on both sides of the piston through a closed system. This is the mechanism used to control height of the blade when the machine is grading earth to achieve a level finish. The operator moves the control to raise and lower the blade to compensate for irregularities in the level. The control applies fluid pressure to raise or lower the blade through the same cylinder.on different sides of the piston. Disconnecting the fluid supply to one side of the piston does not inactivate the piston, however, since pressure exists or can exist on the other. This pressure, residual or operational, has caused considerable damage and traumatic injury during disassembly of loaded equipment.

**Pneumatic Systems**

Fluid power systems that utilize gas as the medium for energy transfer are referred to as pneumatic systems (SMC 2000). While the gas used in pneumatic systems usually is compressed air, other gases, such as nitrogen, are possible. Unlike liquids, gases decrease in volume during compression. Compressed gases, therefore, offer some "give" in positioning the load.While compressed air for use in such systems is obtainable from compressed gas cylinders, a dedicated or plant system compressor is the usual source. Typical pressures provided in plant systems range from 80 to 150 lb/in$^2$ (550 to 1,030 kPa). Despite the fact that these pressures are considerably lower than used in hydraulic systems, they

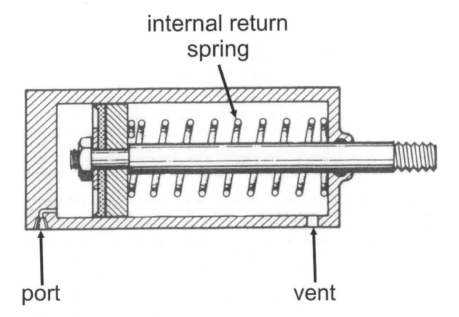

Figure 5.12. Single-acting cylinder. Force exerted by a spring located inside the cylinder moves the piston against the pressure exerted by the fluid. In this manner, the piston returns to the neutral position after removal of pressure. (Modified from Stewart 1987.)

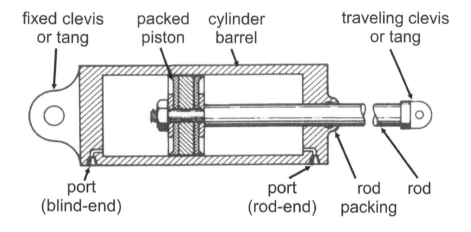

Figure 5.13. Double-acting cylinder. The double-acting cylinder receives fluid supply on both sides of the piston. Note the smaller area on the rod side of the piston. Retraction force is smaller than extension force for identical application of fluid pressure. (Modified from Stewart 1987.)

are sufficient to move the piston in a cylinder against a load or to rotate the rotor in a motor. Pneumatic devices present no spark hazard in an explosive atmosphere nor an electrical shock hazard. Compressed air is also more desirable in food processing or in applications in material transfer that benefit from the compressibility of air (Nachtwey 2002).

Pneumatic systems usually contain a compressor, receiver tank, air cooler/drier, piping, valves, line filter, lubricator, drain, speed controllers, and actuating devices (cylinder, motor, valves).

Various types of pressure regulation are used in these systems. One type maintains the intake valves in the compressor in the open position during the suction and compression strokes (Nelson 1983). Hence, no compression occurs in the cylinder. Another type uses a valve to close the intake line. This prevents air from entering the compressor. In the third type, the controller reduces the speed or shuts down the drive motor. Compressors controlled in the latter manner reactivate at any moment without warning. Guarding drive belts associated with this equipment, therefore, is critical to prevent injury.

Concerns associated with operation of a pneumatic system include the impact of compression on composition of the compressed gas. Compression reduces the volume of gases and increases the temperature. As a result, the concentration of contaminants in the air under compression also increases. A system operating at a gauge pressure of 90 lb/in$^2$ (620 kPa) has undergone a compression factor of six. An air intake inappropriately located could introduce air containing exhaust gases from vehicles, as well as fuel vapors and gases that could become ignitable following compression.

Compressed air systems normally exhaust to air, rather than to a reservoir for recycling. Hence, during use, the system is always consuming atmospheric air. The presence of exhaust gases from vehicles at elevated concentration is a serious concern where the air is to be used for breathing purposes. Potentially even more serious is the compression of gases and vapors into the ignitable range of concentration. This situation has led to explosions and fires involving compressed air systems. Vigilance to control composition of the air drawn into the intake of the compressor is the main defence to prevent contamination by ignitable vapors.

A parallel situation exists with transport of gas and aerosols. Aerosols include particulates and droplets of water and possibly oil. Particulates can result from dust in air entrained into the system, as well as materials shed from the walls of piping and other components. Particulates transported at high velocity in a gas stream can generate large quantities of electrostatic charge. This is an especial concern in systems containing isolated ungrounded metal components. This situation can result from use of components that are insulators. Hoses that lack metal reinforcing or continuous metal-to-metal bonding between metal components of the couplings and the reinforcing mesh can act as insulators.

Electrostatic discharge involving compressors and compressed gas systems has caused many fires and explosions. Bonding and grounding are essential to prevent electrostatic charge accumulation and discharge, especially in isolated systems.

Moisture is a serious concern in pneumatic systems (Heney 2003a). Compression of air increases the concentration of water vapor. At the elevated temperature created by compression, the hot air may still be able to accommodate the water vapor in gaseous form. However, much lower temperatures exist in components of the system, especially in winter. Condensation on metal surfaces is a serious issue. Lines can fill with condensed liquid at low points in plumbed-in systems. As well, freeze-up of critical components can occur. To illustrate the gravity of the concern, one specification for compressed breathing air requires removal of moisture to a dew point at least 5° C below the lowest anticipated temperature of the air (CSA 2000). (Dew point is the temperature at which moisture condensed on a surface exists in equilibrium with moisture in the air.)

Properly designed pneumatic systems contain several components for removal of moisture, starting with the dryer. Dryers utilize absorption (chemical reaction), adsorption (physical entrapment on a desiccant) or refrigeration as the means to remove moisture from compressed air (Anonymous 2000; Heney 2003a; SMC 2000). The second type of component for moisture removal is the filter. Filter units use centrifugal action to remove droplets of water and oil, and particulates from the air. The filter element removes finer particles of dirt, rust scale, and carbonized oil. Micro filtration units also are available. The third component for moisture removal is the drain unit. Drain units installed in low points remove and expel condensed water from the system. Points of installation include the aftercooler, receiver tank, and low points in piping. Expulsion of condensed water from the pneumatic system on locomotives gives rise to the characteristic "spitting" sound that is readily apparent when these units are stationary.

Removal of bulk quantities of accumulated water through drains in a system can require several minutes. Considerably more time is needed to achieve dry air in the piping of the system. Freeze-up in piping and critical components can lead to premature failure.

Removal of water from the system is critical for another reason: corrosion. Carbon steel used in piping in compressed air distribution systems, valve springs, and other components is subject to corrosive attack. While corrosion might not lead to leakage or even catastrophic failure, it could lead to production of rust and slag particles that can slough off and contaminate the air stream. Unless removed from the airstream by filtration, these particles could enter critical parts of control components and cause failure.

Almost every application involving pneumatic systems requires clean air (Schneider 2002a). Few applications can tolerate aerosols (liquid and solid particles) larger than 10 μm. As a result, there is a large market for filtration devices to remove aerosols from the airstream of compressed air systems. Standard filters act as strainers. Coalescing filters combine small particles into larger ones to effect the separation. Coalescing filters function into the sub-micrometer region.

Unconstrained components, such as hoses, pose especial risk in compressed gas systems in the event of unexpected coupling release or failure. Pressurized hoses and other unsecured components flail about in a completely unpredictable manner following disconnection and can cause considerable damage and traumatic injury. Secondary attachment is necessary to prevent whipping on failure or release of a coupling.

## Hydraulic Systems

Fluid power systems that utilize a liquid as the medium for energy transfer are known as hydraulic systems (Stewart 1987). Liquid fluids are incompressible. Incompressibility offers the advantage of precision in positioning and maintaining position. While this quality is important in some applications, especially construction equipment, such as excavators and hydraulic cranes, the usual situation for hydraulic cylinders in applications is either "fully open" or "fully closed." Leakage of hydraulic fluids can cause serious personal injury, fire, and environmental damage. Leakage of compressed air carries much less consequence (Johanson 2002).

The simplest hydraulic system consists of a tank or reservoir containing hydraulic oil or other hydraulic fluid, a pump, tubing or piping, control valving, and an actuator, such as a motor or cylinder (Stewart 1987). The pump exerts force on the oil. Pressure exerted by the oil against walls of the downstream containment causes the occurrence of some action. These actions include the longitudinal movement of a piston in a cylinder or rotation of the rotor in a motor. Depressurized oil then returns to the reservoir. A bypass valve controls the quantity of oil diverted from the load.

Hydraulic systems include both closed and open systems. Closed fluid systems are known as hydrostatic transmission circuits. The fluid returns from the actuator (hydraulic motor) directly to the pump. Closed systems primarily include circuits containing hydraulic motors. The fluid circulates in a closed loop.

Open systems are considerably more common (Vickers 1998). These include the system, as described, containing a reservoir at atmospheric pressure to collect and store the returning hydraulic fluid and to supply the fluid needs of the pump. In this type of system, the high pressure is present in parts of the circuit isolated by valves.

While almost all of the attention regarding hydraulic systems is directed to in-plant units and mobile equipment, small potable hydraulic equipment has widespread distribution, especially in construction applications. Some units require external power sources. Self-contained units incorporate the hydraulic pump into the hand tool.

Hydraulic jacks and cylinders provide force to lift, crimp or clamp, or spread objects and materials. These devices can generate pressures to 10,000 lb/in$^2$ (70 MPa) (Hohensee 1999). A useful piece of guidance is the 80% rule for both cylinder stroke and capacity. Extend the cylinder to only 80% of stroke and lift to only 80% of capacity to avoid damage to the equipment. Extensions (cheaters) on the handles of hand pumps create stability problems during operation. Some pumps are designed to operate vertically or horizontally, but not in both orientations. The hose end in pumps designed to operate vertically must remain in the down position in order not to admit air into the system.

The power source in hydraulic systems is the pump. Most systems use positive displacement pumps (Vickers 1998). The fluid-handling chambers in positive displacement pumps maintain metal-to-metal contact continuously during the operating cycle to prevent backward flow. Catastrophic failure of a pump could cause serious repercussions during operation of a hydraulic system. This potential is alleviated through careful design and use of check valves. However, faulty performance from a check valve and simultaneous catastrophic failure of the pump could lead to an accident situation.

About 80% of pump failures result from the erosive action of contaminants in the fluid. The other causes of pump failure include aeration and cavitation, 10% to 12%; overpressure, 5% to 6%; and other causes including poor fluid, defective parts, and accidental breakage. Contaminants in the fluid include metallic debris (burrs, chips, flash, weld spatter), and nonmetallic particles from dirt in the work environment, paint residue, pipe thread sealants, and sand used during casting of system components. Dirt can enter through a faulty seal, and the breather and fill cap on the oil reservoir (Vickers 1992).

Hydraulic fluid is the medium for energy transference in the system. The volume of a liquid fluid changes only slightly with applied pressure. Hence, hydraulic fluids are usually considered incompressible. On the other hand, compressed air undergoes compression until liquefaction of individual gases occurs. Liquid fluids also act as a lubricant of moving parts and as a heat exchange medium. When completely filled with liquid fluid, precise control of movement of a piston in a cylinder is possible. This is readily apparent from the precision of control demonstrated during operation of hydraulic excavators and cranes.

Most oil-based hydraulic fluids have flash point >150° C (>300° F). These products rarely pose significant operating, safety or maintenance problems, hence their popularity (Gere and Hazelton 1993; Zink 1992). When emitted from a circuit under high pressure, as occurs from a point of leakage, the fluid forms a jet containing tiny liquid aerosols. These droplets are ignitable and can travel considerable distance. Jets of hydraulic fluid emitted under such circumstances have caused fires on equipment and other disastrous fires (Kuchta 1985).

Fire-resistant hydraulic fluids greatly reduce this hazard (Gere and Hazelton 1993; Zink 1992). One approach to the formulation of fire-resistant products involved introduction of water into the fluid to act as a "snuffer" of ignition. These products are mixtures of water and glycols or emulsions and additives formulated to provide appropriate properties. The other approach was to create synthetic products whose chemistry resisted ignition or generated products of combustion that extinguished the flame. Older products contained phosphate esters, and more recently, polyol esters created from long-chain fatty acids and synthetic organic alcohols.

Hydraulic pumps can experience degrading conditions. Aeration and cavitation of hydraulic fluids cause serious erosive damage and possible internal failure (Totten et al. 2001; Vickers 1998). Aeration and cavitation produce noise, describable as the sound of marbles in the pump inlet. These conditions can cause rapid failure of the hydraulic pump.

Aeration arises from leakage of air into the hydraulic fluid. This can occur in the fluid reservoir or through leakage at fittings into lines near the inlet to the pump. Air bubbles can form in the fluid on the low-pressure side of the pump. On the high-pressure side, the bubbles collapse implosively and release considerable energy. This concentrated release of energy causes erosion of interior surfaces of the pump. Failure can occur rapidly.

Cavitation occurs when bubbles of fluid vapor form in the low-pressure side of the pump. This results from flow restrictions created by viscous fluid, undersized hoses, and operating the pump at too high a speed. Cavitation can occur in other components, such as valves and crimped lines where flow restrictions exist or have developed. The lower the pressure, the greater is the propensity for formation of vapor bubbles. Bubble formation occurs when the vacuum exceeds the vapor pressure needed for bubble formation. On the high-pressure side of the pump, as with aeration, bubbles collapse implosively.

Water can condense on metal surfaces in the reservoir and can enter the system through the breather or filler cap during careless replenishment of the fluid reservoir. Moisture transferred throughout the system can cause corrosion (rusting), sludge formation, and pitting.

Seals are integral to protecting hydraulic cylinders against premature failure (Vickers 1998). The most common failure in hydraulic cylinders is excessive leakage from one or more seals. Rod seals prevent entry of foreign material into the hydraulic fluid and prevent leakage. A seal is considered effective so long as no visible liquid passes by. Failure of rod seals is evident from visible leakage of fluid. Piston seals are not visible or readily accessible. Leakage and failure of the piston seal manifest through loss of performance. Downward drift of the piston due to leakage around a seal when the system is idle is not assured because the fluid must have an escape path beyond the cylinder. Valves may block this path. Fluid that leaks past a seal can increase pressure in this region of the system to destructive levels.

O-rings are among the most commonly sealing technologies used in hydraulic and pneumatic systems (Ashby 1999). The first O-rings appeared in 1939. O-rings are manufactured from elastomers. O-rings are unusual products because they are designed to deform in order to function properly (Messinger 2000). O-rings fail in service because of incorrect or inappropriate design, application, installation, insufficient lubrication, or incompatibility with the environment with which they make contact.

Hydraulic systems experience considerable opportunity for vibration and consequent production of noise (Heney 2003b). Fluid velocity, pressure, and line shock and pressure surge all contribute to the vibration and sound prevalent in hydraulic systems. Pumps and valves are the main causes of vibration and shock because of the occurrence of drastic changes in fluid pressure and velocity. These will cause flexing in unrestrained metal tubing. This flexing leads to eventual failure of the line. Flexing is especially a problem at connectors. Pumps are also the greatest source of noise. Fluid and surfaces transmit noise. Fluid reservoirs act as amplifiers.

Support for tubing, pipes and hoses is essential to minimize and to control this problem. Resilient materials containing wood, rubber, metal and thermoplastics are used in support devices. Plastics have largely replaced wood and less durable materials.

Hydraulic fuses in hydraulic circuits are the analogue of electrical fuses or circuit breakers in electrical circuits. They serve the same purpose: to regulate overflow of energy in a circuit (Hitchcox 2007). The hydraulic fuse is a type of check valve. It responds to occurrence of a predetermined signal, generally a sudden change in pressure or flow and stops fluid flow. Abrupt change in fluid velocity or pressure could result from catastrophic failure of a hose or tube. Hydraulic fuses include velocity fuses, quantity fuses, and pressure fuses.

Velocity fuses sense pressure differential inflow across a controlled orifice. The required pressure differential is normally at least 30% above rated flow. This design accommodates surges due to starting and stopping, valve actuations and motor reversals, thereby minimizing false trips. Time-delay features are available in these fuses as an additional measure to minimize false trips. Quantity fuses sense flow across two orifices sized to provide a specific volume. Pressure fuses sense low pressure needed for valve actuation.

### Pressure Transformation (Pressure Boosting)

The pressure booster (intensifier) is a component used to increase fluid pressure (Figure 5.14) (Stewart 1987). Designs exist for both hydraulic and pneumatic circuits.The pressure-boosting fluid and the pressure boosted fluid can be liquid or gas. The pressure booster is the fluid power analog of the electrical transformer. The pressure booster functions through the relationship between surface area and fluid pressure (Equation 5-10). That is, a low pressure exerted on a large surface area becomes a high pres-

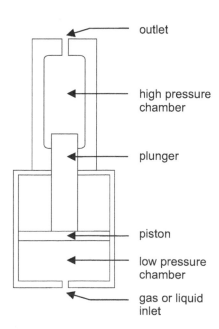

outlet

high pressure chamber

plunger

piston

low pressure chamber

gas or liquid inlet

Figure 5.14. Pressure booster (intensifier). The pressure booster (intensifier) in fluid power circuits is the analogue of the transformer in electrical circuits. The pressure booster applies the concept of equal force. Low pressure exerted over a large surface produces the same force as high pressure exerted over a small surface. (Modified from Stewart 1987.)

sure exerted through a small surface area. In the practical application of this principle, pressure applied by a low pressure fluid to a large piston becomes much higher pressure applied by a smaller piston.

$$P_1A_1 = P_2A_2 \qquad\qquad (5\text{-}10)$$

where:
$P_1$ = pressure applied to piston of surface area, $A_1$ (Pa = N/m$^2$).
$A_1$ = surface area of piston 1 (m$^2$).
$P_2$ = pressure applied by piston of surface area, $A_2$ (Pa = N/m$^2$).
$A_2$ = surface area of piston 2 (m$^2$).

Pressure increases of 50 to 1 and higher are achievable from this equipment. Pressure boosters can provide much higher pressure than high pressure pumps while producing less heat.

The pressure booster used in hydraulic circuits resembles a cylinder, except that the pressure-boosting device is a solid plunger, rather than a piston. Pumps used to boost pressure are present in water jetting equipment. This equipment has become common in residential applications. These products contain a plunger pump to boost pressure from residential pressure to discharge pressure. Pressure cleaning (powerwashing) or cutting refers to pump pressures less than 35 MPa (5,000 lb/in$^2$). These pressures are readily available from units sold for consumer use. Even inexpensive consumer units can provide pressures exceeding 10 MPa (1,500 lb/in$^2$). High pressure water jetting ranges from 35 to 210 MPa (5,000 lb/in$^2$ to 30,000 lb/in$^2$. Ultra high pressures exceed 210 MPa (30,000 lb/in$^2$). These pressures are sufficient to cut metals.

Pneumatic boosters are usable in explosive atmospheres. This permits use of hydraulic fluid at high pressure in an isolated circuit. Pneumatic devices present no spark hazard in an explosive atmosphere, nor an electrical shock hazard. This eliminates the need for the hydraulic power device. As well, low pressure valves can be used in the low pressure side of the system. The high pressure side has no valves. This means that equipment containing low pressure components can contain circuits that operate at very high pressure. The existence of a high pressure circuit in the system may not be evident from casual observation of low pressure components.

Another application of these principles occurs in the steam locomotive (Figure 5.16) and the marine steam engine. These units contain cylinders of increasing diameter to capture the remaining pressure contained in steam after partial conversion to mechanical movement. The steam locomotive contains a two-stage design. The marine steam engine contains up to four stages. The surface area of the pistons increases as the feed pressure decreases in order to maintain constant force in each of the cylinders.

Steam turbines and jet engines apply these principles in a similar manner. In the case of the steam turbine, as steam passes from stage to stage, pressure decreases and the surface area of the blades increases. The jet engine employs blades of varying surface area to act as a fan and in compression of incoming air and expansion of combustion gases.

### Fluid Power Systems and Energy Storage
Fluid power systems also can store energy. This discussion applies to both pneumatic and hydraulic systems.

### Check (Backflow Prevention) Valves
Valves in fluid power systems control the manner of use of energy transmitted through the fluid in a number of subsystems (Vickers 1998). One of the most important of the valve types is the check valve. Check valves allow one-way flow. Preventing backflow is essential to maintaining pressure in the system, and therefore, safety of operation. That is, the check valve can maintain fluid pressure in a circuit containing a cylinder even after reduction of the supply pressure to atmospheric pressure. Pressurization of isolated parts of a system is not apparent in the absence of a pressure-indicating device. This determination requires measurement and knowledge of circuit design and components. Cessation of

Figure 5.15. Compound steam engine. This design extracts energy remaining in steam as pressure decreases through cylinders of increasing cross-sectional area that maintain constant force. Marine steam engines can contain four cylinders. The pressure of the steam decreases through work performed but the force exerted in each cylinder remains the same.

rotation of a fluid-driven motor, for example, is a readily recognizable indicator that pressure in that part of the system is reduced below the threshold needed to start and/or maintain rotation.

Check valves contain springs to maintain the backflow preventer (a ball or poppet) in the closed position. Pressure against the backflow preventer compresses the spring to enable occurrence of flow. Springs lose strength over time. As well, being composed of steel, springs are subject to corrosion, and erosion due to cavitation or aeration. Inability of the backflow preventer to close the path means that backflow will occur. This could lead to system failure at a critical point in time. Some check valves are piloted. That is, they are operated by fluid pressure, electrical circuitry, or mechanical or other means. Piloted check valves can allow flow to occur in both directions, but only under deliberate control.

## Spool Valves

Spool valves provide directional control of fluid. The spool section moves directionally to expose flow channels. These valves incorporate check valves to prevent backflow due to leakage around seals and changes in position of the spool section. "O"-rings or seal plates form the seal between the spool section and the main body of the valve. Preventing leakage in valve assemblies is difficult. Leakage can occur due to destruction of the seals, and by stretching of bolts used to hold together the assembly, and by distortion of the valve casing. Spool valves can be manually operated or piloted. Pilot mechanisms for spool valves include fluid, electromagnetic, and mechanical piloting.

Hoses expand radially (circumference increases) when subjected to fluid pressure. The hose becomes rigid and repels external crushing when subjected to fluid pressure. This condition is especially apparent for hoses that handle hydraulic fluids.

## Loaded Cylinders

The external load on a cylinder creates and maintains pressure in hydraulic and pneumatic circuits (Figure 5.16). Ths pressure can be substantial compared to the pressure in the circuit created by the power source.This pressure acts against spool valves and check valves in isolated circuits.

## Pressure Gauge

Fluid power systems normally have a pressure gauge. The pressure gauge provides a window into the operation of the system, as well as an indication of malfunction. The window into which the gauge provides insight is the part of the system to which it is connected. The pressure gauge may not provide insight into the part of the system, isolated by valves, in which the pressure exists. Pressure indication by a gauge presumes that the gauge is functional and calibrated. This is not always the case (Figure 5.17).

## Energy Storage in Pneumatic Systems

### Air Receiver

The air receiver is the main energy storage device in pneumatic systems. The air receiver also dampens pulsations created by intermittent charging by the compressor and provides reserve capacity for peak demand. The receiver tank is sized according to the output of the compressor, size of the system and consideration about demand (constant versus intermittent) (SMC 2000). Sizing also reflects the strategy of avoiding overly frequent cycling of the compressor. For industrial plants, capacity of the air receiver should be at least equal to the output of compressed air per minute, with excess capacity factors ranging from 1.5 for large compressors to 3.0 for small systems. There is no limit to the size of a receiver tank. A receiver tank is a pressure vessel. A large receiver provides stability to a system, especially one requiring orderly shutdown in the event of power failure. Hence, a receiver tank can

Figure 5.16. External load on a cylinder. The external load on a cylinder continues to pressurize the fluid power system even when shut down, through prevention of fluid flow by backflow prevention valves (check valves).

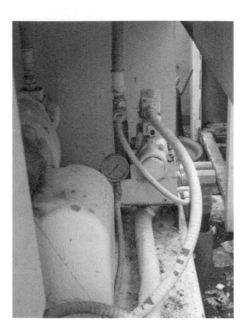

Figure 5.17. Pressure gauge. A pressure gauge provides a window into the pressure in a fluid power system. The utility of the reading depends on placement in the circuit. The accuracy of the reading depends on service and calibration of the gauge.

continue to supply air at or near system pressure for a period of time after shutdown of the compressor. The volume of the receiver determines the demand that can be met.

### Pulsation Dampener

The pulsation dampener is a two-chambered device used for removing irregularities in pressure in pneumatic systems (Figure 5.18). A bladder separates the two chambers. The top section contains compressed air or gas at a pressure lower than the discharge pressure of the pump. A pulse in the system created by the pump (positive displacement) or a valve pushes fluid into the liquid side of the dampener. This pushes against the bladder and the compressed gas on the other side. As the pressure in the system decreases, the force exerted by the compressed gas and bladder push against the fluid and force it into the system.

## Energy Storage in Hydraulic Systems

### Accumulators

The analogue of the receiver tank and pulsation dampeners in hydraulic systems is the accumulator. The accumulator is a fluid storage vessel that maintains pressure in the system through application of force. In parallel with the air receiver in pneumatic systems, the accumulator serves multiple functions (Table 5.2). An accumulator also can act as a surge or pulsation dampener (Anonymous 2003).

The accumulator maintains system pressure temporarily following failure or shutdown of the normal supply. This enables orderly completion of operation and shutdown of critical systems. Accumulators can easily range to 750 L (200 gal U.S.) in supply volume and system operating pressures to 140 MPa (20,000 lb/in$^2$).

Accumulators are used in critical circuits, such as brake and pilot control circuits. In a brake control circuit, the accumulator provides reserve power when the main pump fails or is shut down. An ac-

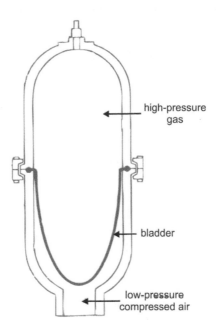

high-pressure
gas

bladder

low-pressure
compressed air

Figure 5.18. Pulsation dampener. A pulsation dampener removes irregularities from flow in pneumatic systems by reducing spikes in pressure.

cumulator in a pilot control circuit provides control when the main pump is shut down. This enables a raised bucket or boom to be lowered in a controlled way without operation of the engine and the hydraulic pump (Nelson 1983; Vickers 1992).

Accumulator designs include weight- and spring-loaded pistons, chambers containing hydraulic fluid, and pressurized gas or air. The weighted accumulator is used in large systems having a central supply and in equipment having long holding cycles under pressure (Figure 5.19). The pressurized fluid forces up the ram as long as the demand is less than the supply (Nelson 1983). When supply exceeds demand, the weight forces out the fluid from the body of the regulator into the circuit. The check valve prevents loss of fluid when the accumulator remains charged for a long period of time. Materials used for the dead weight include concrete, steel, water, or other heavy material. Force remains constant for almost the entire stroke of the piston (Stewart 1987; Vickers 1998).This results from the fact that the pressure applied depends only on the surface area of the piston and the dead weight applied to it. This is the only type of accumulator in which pressure remains constant during fluid delivery (Vickers 1992). A single accumulator can service a number of applications.

The spring-type accumulator uses pressure from a spring on a piston to maintain fluid pressure (Figure 5.20). The limitation of this design is the high spring rate (Vickers 1998). The high spring rate means that the force provided will vary considerably with distance. This, in turn, will increase and decrease the pressure on the fluid disproportionately. An added concern is weakening of the spring following long periods of constant compression. At this point, the spring cannot maintain the pressure against the fluid.

The third main type of accumulator uses gas to pressurize the interior of a container in which hydraulic fluid is also present (Stewart 1987). One design provides no separation between the fluid and the gas. Aeration or mixing of hydraulic fluid and gas is a possibility with this configuration. This would compromise performance of the unit through loss of pressure and the possibility of chemical reaction between the hydraulic fluid and the gas. Air and oxygen must never be used as the charge gas in

## Table 5.2

## Function of Accumulators in Hydraulic Systems

Smoothing pressure pulses developed in the hydraulic pump

Smoothing shock pressure (water hammer) due to rapid closing of valves

Providing additional pressurized flow to alleviate high transient demand

Providing for expansion due to increase in temperature

Noise control

Prevent pump cavitation

Sources: Accumulators 1998; Anonymous 2003; Bolton 1997; Nelson 1983.

this type of accumulator due to the potential chemical reaction under the high pressure of the system. This could lead to a diesel-type explosion (Vickers 1998).

The more prevalent type of gas-pressurized accumulator uses a barrier to separate the hydraulic fluid from the compressed gas (Figure 5.21). One design uses a piston to separate the hydraulic fluid from the gas. At equilibrium, the pressure of the gas and fluid in the piston is the same (Vickers 1998). This type of accumulator is double-walled. The gap between the inner and outer shell serves as a gas reservoir. Fluid leakage around the seals of the piston is a concern with this type of unit.

Most accumulators today use a bladder to separate the pressurizing gas from the hydraulic fluid (Bolton 1997). Some use a liquid, such as water, inside the bladder. At equilibrium, the pressure on both sides of the bladder is the same. Sizing the accumulator depends on the application. For accumulators that are filled and emptied gradually, where expansion and contraction of the gas occurs slowly enough for temperature to remain constant, Boyle's law (Equation 5-11) applies to the gas.

$$P_1 V_1 = P_2 V_2 = P_3 V_3 \tag{5-11}$$

where:
$P_1$ = pressure of the gas charge in the accumulator (Pa = N/m$^2$).
$V_1$ = the volume of the accumulator (L = liters).
$P_2$ = the pressure of the gas charge following liquid charging of the accumulator to operating pressure (Pa).
$V_2$ = the volume of gas following liquid charging of the accumulator to operating pressure (L).
$P_3$ = the pressure of the gas charge following partial removal of fluid from the accumulator (Pa).
$V_3$ = the volume of gas following partial fluid removal from the accumulator (L).

In situations where gas compression or expansion occurs rapidly due to rapid charging or discharging of the accumulator, thermal equilibration is incomplete. In these circumstances, the adiabatic equation for gas expansion is required (Equation 5-12).

$$P_1 (V_1)^\gamma = P_2 (V_2 g) = P_3 (V_3)^\gamma \tag{5-12}$$

where:
Pressures and volumes correspond to those mentioned in Equation 5.11.
$\gamma$ = the compressibility factor. $\gamma$ ranges from 1.0 to 2.0.

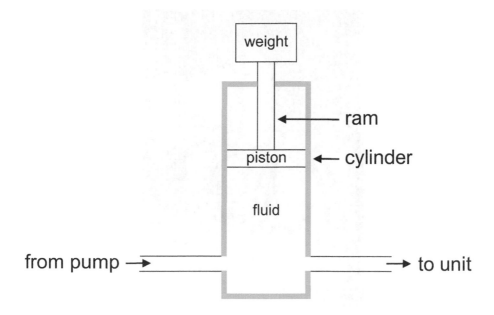

Figure 5.19. Weighted accumulator. The weight applies constant force on the piston in the accumulator. (Modified from Stewart 1987.)

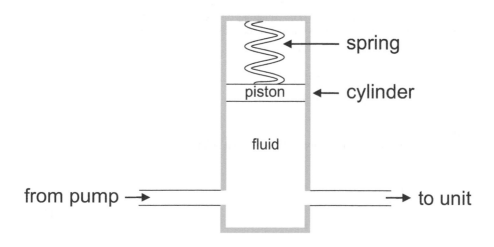

Figure 5.20. Spring-loaded accumulator. The force applied by the spring on the piston varies with the compression. (Modified from Stewart 1987.)

Figure 5.21. Gas-pressurized accumulator. The gas in the pressurized cylinder exerts force against the fluid usually through a bladder. The bladder prevents direct contact between the gas and the fluid.

The respective equations lead to equations for sizing the accumulator according to system requirements (Equations 5-13 and 5-14). The equations that follow are for the adiabatic case where $\gamma$ is required. The corresponding equations where heat equilibration occurs are the same, except for the absence of the $\gamma$ factor.

$$V_{outflow} = V_3 - V_2 = \frac{(P_1)^{\frac{1}{\gamma}} V_1}{(P_3)^{\frac{1}{\gamma}}} - \frac{(P_1)^{\frac{1}{\gamma}} V_1}{(P_2)^{\frac{1}{\gamma}}} \qquad (5\text{-}13)$$

where:

$V_{outflow}$ = the volume of fluid supplied by the accumulator under pressure (L).

$$V_1 = V_{outflow} \frac{\left(\dfrac{P_2}{P_1}\right)^{\frac{1}{\gamma}}}{\left(\dfrac{P_2}{P_3}\right)^{\frac{1}{\gamma}} - 1} \qquad (5\text{-}14)$$

Since the compressibility factor, $\gamma$, is larger than unity, the volume of the accumulator required under adiabatic conditions is larger than that for isothermal conditions. Note that during supply of fluid from the accumulator into the hydraulic system, the pressure in the air bladder decreases.

Charge pressure for the bladder used inside the shell of the accumulator reflects service pressure in the system (Bolton 1997). Dry nitrogen and not compressed air is required (Accumulators 1998).

This is necessary to avoid possible explosion due to compression of oxygen contained in air. Under high pressure conditions, the mixture of compressed air containing oxygen and hydraulic fluid encountered due to leakage can become explosive. Heat generated during compression could cause ignition.

### Pressurized Lines and Hoses

Energy in accumulators and pressurized lines has caused accidents in situations where this energy is presumed not to be present. As a result, activation of controls powered by energy stored in the accumulator can occur. One incident that highlights this situation involved a tugboat with steerable power units. During service in a shipyard, an electrical foreman was showing some visitors how the controls "worked," and moved the control unit. Unbeknownst to this individual, this caused the power unit to rotate under energy stored in the hydraulic system. This motion almost struck a surprised shipwright who was working under the stern of the vessel on the power unit. Neither had any idea about energy storage in this system.

Depending on type, hoses can elongate up to 2% when pressurized and can contract as much as 4% (Mueller 2002). Radial expansion, which also occurs, depends on the hose and the pressure. Hose installation is critical to protect against failure, and thereby to ensure safety. Twisting (bending in two planes) is highly detrimental to hose longevity. Flexing (bending in one plane) is an acceptable installation practice. Twisting causes strain on reinforcing wires, especially at the hose-to-coupling interface. Twisting a high pressure hose only 5° can reduce service life by 70%, and 7° up to 90%. Multi-plane bending of flexed hose has the same impact as twisting. About 80% of hose failures are attributable to external physical damage, abrasion being the main cause. Hose clamps and other supporting devices can reduce rubbing against surfaces.

The wire reinforcing in hydraulic hoses renders them electrically conductive. This situation can create an isolated conductor capable of accumulating electrostatic charge when isolated from other

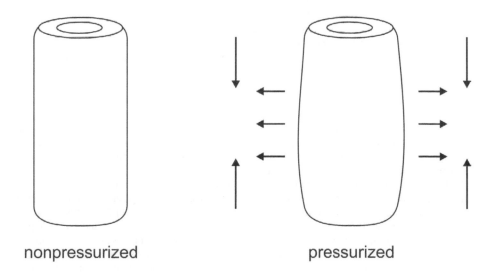

nonpressurized        pressurized

Figure 5.22. Pressurized hose. Hose expands radially when pressurized. The hose also becomes rigid, a sign that pressurization has occurred.

conductors by insulative fittings. This situation has led to electrostatic discharge through the wall to adjacent surfaces. This discharge can cause localized burning and development of a pinhole leak.

Piloted valves can remain active during maintenance when the system remains pressurized (Vickers 1992). The preceding example illustrates manual activation. Mechanical actuators rely on contact between an object and the actuator. This movement can happen through intended contact as would occur during normal operation. It also can happen through unintended contact involving tools or dismantled parts of the machine, as could occur during servicing and other maintenance. Electrical actuators can receive energy from energy stored in electrical and electronic systems. Pneumatic actuators receive their power from the compressed air system. The compressed air system can contain an accumulator or a large receiver tank that retains compressed air after shutdown of the compressor.

## Vacuum

Vacuum is the inverse of the norm in fluid power systems. Vacuum is the reduction of pressure below the benchmark of 101 kPa (30 inches) of mercury at sea level. Industrial applications of vacuum systems vary widely from large-scale applications, such as vacuum loader and vacuum pumper trucks, equipment used in brine treatment in food processing to remove moisture from meat , vacuum distillation in petroleum refining, controlled suction used in lifting objects, transport of fluidized solid materials, freeze-drying to remove moisture from frozen objects by sublimation, production of semiconductors, materials handling, and extrusion of plastics, among many others.

Vacuum applications typically fit into three classes. Low-level applications typically require high flow of air and low vacuum. Blowers typically are used in this application. Industrial applications require levels of vacuum ranging from −20 to −99 kPa (−6 to −29.5 inches of mercury). The most typical range is −40 to −70 kPa (−12 to −21 inches of mercury). Industrial vacuum systems can generate high vacuum. To illustrate, vacuum loading trucks operate at −90 kPa (−27 inches ) of vacuum. The highest level of vacuum is used in the scientific and process areas.

Design of vacuum systems that lift or hold using suction cups begins at the point of contact and works backwards to the source of vacuum (Schneider 2002b). A suction cup adheres to the surface of an object because the pressure under the cup is lower than atmospheric pressure. The lower internal pressure is achieved by evacuating most of the air under the cup into a vacuum system. Faster evacuation time results in less consumption of energy. The greater part of evacuation time occurs below −40 kPa (−12-inches of mercury). A larger cup operating at higher vacuum increases lifting and reduces force exerted on the cup. Energy consumption increases exponentially as vacuum increases. Increasing the vacuum level from −65 to −90 kPa (−18 to −27 inches of mercury) requires 10 times as much energy. As well, systems operating at higher vacuum are more susceptible to problems.

Air movers used in vacuum service include rotary vane, screw, rotary claw and rotary lobe, liquid-ring, scroll and turbo-molecular pumps, and side-channel blowers. Vacuum systems operating over a distance require pumps, piping and tubing, and valves, as do compressed air systems. Vacuum-on-demand systems use compressed air from a compressed air system as a power source and single or compound venturis to create the vacuum. Small-scale materials handling systems use suction cups and hoods to make the seal against a surface for lifting.

Operators of vacuum systems used for materials handling must exercise caution about working near and under loads. These systems require considered engineering in design and execution to ensure that loss of vacuum does not occur during use for lifting. This necessitates careful selection of components to minimize potential for in-leakage. Components incorporated into these systems include positive displacement pumps, check valves, and components to monitor and prevent loss of vacuum.

Systems that operate under vacuum generate design considerations that differ from those where pressure is present. Small levels of vacuum applied to structures designed to operate only under pressure can cause collapse (Figure 5.23). This highlights the relationship in Pascal's law of small vacuum exerted over large surface creating tremendous inward force. Structures normally are designed to resist pressure exerted from within. Vacuum creates different stresses on structures, and as a result, subjects them to forces which they were not designed to resist.

Figure 5.23. Vacuum-induced collapse. Slight vacuum applied to structures intended to operate under pressure can induce collapse.

This situation has caused disastrous consequences. Slight vacuum has collapsed structures that can resist considerable pressure. Condensation from gas or vapor inside the structure generates negative pressure. This usually occurs following decrease in temperature. This condition has resulted from use of steam for cleaning or inerting and failure to leave open a valve or vent to enable equalization of pressure following cool-down. Similar circumstances have arisen on sealing a tank containing residual warm or hot liquid and vapor, possibly also exposed to solar heating, followed by abnormal cooling. Abnormal cooling can occur following a sudden downward shift in temperature, after passage of extreme weather systems.

One example of such heating and cooling cycles occurs in western North America where warm air from the coast penetrates through mountain passes and heats areas in the lee side normally experiencing cold winter conditions. Shifts in temperature of 30° to 40° C (55° to 70° F) are not uncommon according to weather records. Rapid cooling to normal temperature can create vacuum inside structures initially heated to the higher temperature.

Additional situations have arisen during emptying of storage structures either by gravity or by active removal of contents by pumps or vacuum blowers. Failure to open or clogging of a vent can cause vacuum sufficient to collapse of the structure. Potential causes of vent failure include clogging by airborne dust and debris and formation of ice in the mechanism under cold weather conditions.

## Liquid Fluids in Process Systems

Process systems handle various types of fluids during operation. These fluids include gases, and liquids, and also solids transported in the fluid stream. Gases include fuel gases; shield gases for welding; process gases, such as hydrogen, carbon monoxide and nitrogen; compressed air (as already discussed); and steam. Liquids range from water to a myriad of process liquids. These fluids can be heated, as with hot water and heat transfer liquids, or they may be chilled, as with refrigerants, such as ammonia, propane, sulfur dioxide, and FREONS®. Solids transportable in process systems include powders, fibers, and pellets.

Process systems containing liquids operate at a range of pressures, sometimes under the influence of gravity. The liquids are contained in channels and piping.

## Pumps

Pumps provide the motive power needed to move the fluid or material in most industrial processes. A pump is a mechanical device that raises, transfers or compresses fluids. Pumps subdivide into two general classes: dynamic and positive displacement.

Dynamic pumps must rely on check (non-return or backflow prevention) valves to prevent backward flow. Internal components in positive displacement pumps make physical contact at all times during the pumping cycle. This prevents backflow of pressurized liquid through the pump from the positive pressure side.

Process systems employ pumps for circulation as well as filling storage containers. The pump is often positioned at the bottom of the circuit.

## Dynamic Pumps

Dynamic pumps further subdivide into centrifugal and axial flow pumps. Centrifugal pumps are the most common type of pump used to move liquids through piping systems (Lees 2001). Centrifugal pumps are typically used for large discharge through smaller heads.

### Centrifugal Pumps

The centrifugal pump converts rotational input from a power source to kinetic energy in the liquid through movement of a rotating device, the impeller. The most common type of centrifugal pump is the volute pump. Fluid enters the pump through an opening in the impeller along or near the rotating axis. The impeller rotates at high speed and accelerates radially outward from the impeller toward the casing. Expulsion of the fluid from the outlet creates a vacuum at the opening in the impeller that draws more fluid into the pump.

The energy transferred to the fluid depends on the velocity at the edge of the impeller. This, in turn, depends on rotational speed and diameter. That is, the faster the impeller revolves and the larger the diameter, the greater velocity and kinetic energy imparted to the fluid.

Directed upward, the discharge from a centrifugal pump will reach a particular height. This height is known as the head or shut-off head. The maximum head is determined by the diameter of the impeller and the speed of rotation. Note that when rotation of the impeller ceases, in the absence of backflow prevention (provided by valves), the fluid will flow backward through the pump until the pressures on both sides of the pump balance relative to the configuration of the system

Kinetic energy imparted to a liquid by the impeller in a centrifugal pump is subject to various resistances to flow. The first resistance occurs in the pump casing. The pump casing catches and guides and slows down the liquid ejected from the impeller. The loss of velocity converts to static pressure. The resistance to flow is the value read on a pressure gauge attached to the discharge line. A pump creates only flow. A pump does not create pressure. Pressure is a measurement of the resistance to flow.

In Newtonian fluids (non-viscous liquids like water or gasoline), head is a measure of the height of the liquid column the pump creates from the kinetic energy transferred to the liquid. The pressure from a pump will change when the specific gravity of the liquid changes. The head remains constant. This is the main reason for using the term head instead of pressure to measure the energy imparted by a centrifugal pump to a fluid.

Centrifugal pumps produce constant head. Note that the latter is not the same as constant pressure, since pressure is a function of head and density. A centrifugal pump will push all fluids to the same height when the impeller moves at the same rotational speed. The only difference between the fluids is the amount of power (energy) required to rotate the shaft connected to the impeller at the same rotational speed. The latter is related to the specific gravity of the fluid. The greater the specific gravity of the fluid, the greater the amount of power required from the power source.

## *Axial Flow Pumps*
Fluid enters and exits along the same direction parallel to the rotating shaft in an axial flow pump. The fluid is lifted, rather than accelerated by the action of the impeller. Axial flow pumps operate at much lower pressures and higher flow rates than centrifugal pumps.

## *Mixed Flow Pumps*
Mixed flow pumps combine features of both centrifugal and axial flow pumps. The fluid experiences both radial acceleration and lift and exits the impeller somewhere between $0°$ to $90°$ from the axial direction. As a consequence mixed flow pumps operate at higher pressures than axial flow pumps while delivering higher discharges than radial flow pumps. The exit angle of the flow dictates the pressure head-discharge characteristic in relation to radial and mixed flow.

## Positive Displacement Pumps
A positive displacement pump causes a fluid to move by trapping a fixed amount on the suction side and forcing (displacing) that volume through the outlet. Positive displacement pumps have an expanding cavity on the suction side and a contracting cavity on the discharge side. Liquid flows into the pumps as the cavity on the suction side expands and the liquid flows through the discharge as the cavity contracts. The volume delivered is constant given each cycle of operation. Positive displacement pumps produce the same flow at a given rotational speed regardless of discharge pressure. Internal components in positive displacement devices maintain continuous contact between moving parts. This prevents bi-directional flow-through when the device is not operating.

Positive displacement pumps are less common in process plants than are centrifugal pumps (Lees 2001). Positive displacement pumps are used mainly in high pressure and low flow applications. Positive displacement pumps subdivide two main groups: rotary pumps and reciprocating pumps.

Reciprocating pumps are particularly common in high pressure applications. Pressure relief on the discharge side is particularly important.

## *Rotary Pumps*
Rotary pumps include the lobed, external gear, internal gear, screw, shuttle block, flexible vane or sliding vane, helical twisted, and liquid ring vacuum pumps. Rotary pumps are very efficient because they naturally remove air from the lines, eliminating the need to bleed the air from the lines manually. Because of the nature of the pump, the clearance between the rotating pump and the outer edge must be very close, requiring rotation at a slow, steady speed. When operated at high speed, the fluid will cause erosion. Rotary pumps that experience erosion eventually exhibit enlarged clearance. This allows liquid to slip through and to detract from efficiency. Positive displacement rotary pumps can be grouped into three main types.

## *Reciprocating Pumps*
Reciprocating pumps include piston and plunger pumps and diaphragm pumps. Reciprocating pumps cause the fluid to move using one or more oscillating pistons, plungers or membranes (diaphragms). Reciprocating pumps require a system of suction and discharge valves to ensure that the fluid moves in one direction.

A plunger pump contains a cylinder and a reciprocating plunger. The suction and discharge valves are mounted in the head of the cylinder. In the suction stroke, the inlet valve opens, the outlet valve closes and the plunger retracts. This creates a vacuum that draws fluid into the cylinder. In the compression stroke, the inlet valve closes, the outlet valve opens and the plunger pushes the liquid out of the outlet.

In diaphragm pumps, the plunger pressurizes hydraulic oil which pushes against a diaphragm in the pumping cylinder.

The peristaltic pump contains the fluid in a flexible tube fitted inside a circular casing. A rotor containing a number of rollers compresses the flexible tube at several locations. As the rotor turns, the part of the tube under compression closes, thus forcing the fluid to move forward through the tube. Reopening of the tube to its natural state after passing of the compression device induces entry of fluid.

**Valves**

A valve is a mechanical device designed to direct, start, stop, mix, or regulate the flow, pressure, or temperature of a process fluid (Skousen 1998). Valves handle either liquid or gas applications. Humans have a long history of involvement with valves for isolation and control, starting with fallen trees that diverted water courses and stone and wooden barriers used to block or divert water in irrigation ditches. These approaches date to 5000 BC. Leonardo da Vinci analyzed pressure drop on either side of lock sluice gates and Newcomen in 1712 created a primitive valve from an iron plug to regulate steam in the steam engine.

Valves control the properties associated with the movement of fluids. All valves control flow (Frenck 2001). Some valves throttle flow, while others perform an all-or-nothing, on-off function. The main types of valves are gate, globe, quarter-turn (plug, ball, and butterfly), and check valves. All valves contain a body which contains the fluid and a moveable element inserted into the path of the flow. A stem which exits the valve body through a stuffing box moves the flow controlling element through either linear or rotary motion. Linear valves include gate and globe valves. Rotary valves include plug, ball, and butterfly valves.

Packing is the soft material used to seal the stem or shaft of the flow controlling device (Skousen 1998). Various components contain and compress the packing material. Compression is used to maintain a tight seal around the stem. Depending on material, packing experiences unique deformation under compression. Any irregularity in the surface of the stem is a potential path for leakage. Packing materials include asbestos, graphite, polytetrafluoroethylene, and perfluoroelastomers. Each of these materials hass advantages and disadvantages.

Lubricants are used to minimize friction between the packing and the stem. These products are most commonly silicone greases. The best products do not react with the process fluid or attract dirt or other particulate matter.

Gaskets are used to fill a predetermined space in a joint between two parts in order to prevent leakage (Skousen 1998). The space may be a counterbore, groove or retainer. Pressure applied to bolts or clamps holds the components and the gasket firmly in place. Gaskets generally are softer than the parts being held together.

In valves, gaskets perform several functions. A gasket forms the seal between the valve body and the seat in a linear valve to prevent leakage between the upstream and the downstream side. In split-body and top-entry valves, gaskets prevent leakage of fluid to the atmosphere. Gaskets prevent mixing in valves containing separate fluid chambers.

The choice of gasket material in a particular situation reflects function of the valve and temperature, pressure, and characteristics of the fluid transported. Gasket performance also depends on correct seating. The latter depends on compression applied to the gasket during assembly of the components. Incorrect compression (too much or too little) can create a leakage path. The most common gaskets are flat, spiral-wound, metal O-ring, metal C-ring, metal spring-energized, and metal U-ring. Materials of construction include plastics and elastomers, soft metals, asbestos, and graphite, mica, and ceramic materials.

A large number of types of valves are available in the marketplace. Understanding the capabilities of different types of valves compared to the needs and demands for performance inherent in the design of a circuit provides a basis for selection. A fundamental appreciation about the design of process circuits and selection of components, such as valves, is essential to understanding about the level of safety inherent in isolation of the circuit.

## Function-Based Classification

One basis for classification of valves is function (Skousen 1998). Functions performed by valves include blocking or allowing the passage of flow (on or off), passage of flow in one direction (non-return or check), and regulation of flow at any point between fully open and fully closed (throttling). The difficulty inherent in this approach toward classification is that specific types of valve can perform more than one function.

Block valves or on-off valves include gate valves, plug valves, ball valves, pressure-relief valves and tank-bottom valves (Skousen 1998). A majority are hand-operated, although automation through an actuator is possible. Ball valves used in high pressure situations can move suddenly to the on or off position rather than being open part-way. This abrupt action can cause damage and traumatic injury.

Non-return (check or backflow prevention) valves allow flow to occur only in one direction. During operation, the moveable element in these valves can accommodate any amount of flow from fully closed to fully open in the intended direction. Any flow or pressure in the opposite direction causes the moveable element to close. These valves protect pumps and compressors against backflow in the shut down condition.

Throttling valves regulate flow, temperature or pressure. The moveable element can maintain a range of flow from fully open to fully closed. Pressure-regulating valves contain an internal spring-loaded diaphragm that allows the flow to vary in order to maintain pressure. These valves ideally suit a powered actuator or actuation system.

Actuators used on throttling valves move the moveable element in response to signals concerning flow, pressure, or temperature. Power sources for actuators include electrical, pneumatic, and hydraulic units. The actuator contains return springs that oppose the motion of the power source.

The combination of the throttling valve plus the powered actuator creates the control valve. General service valves are usable in many applications without modification. Most control valves will function satisfactorily in most applications. Stated more quantitatively, more than 80% of control valves operate at pressure less than 100 lb/in$^2$ (700 kPa) and temperature less than 200° F (90° C) (Stepanek 2002). These operate between $-45°$ C to 345° C ($-50°$ F to 650° F) depending on rating. Special service valves are custom engineered for specific applications that are outside the range accommodated by general purpose valves. Severe service valves are equipped with special features to handle volatile products leading to high pressure drop and cavitation, flashing, choking or high level of noise. Severe service valves may require special actuation to overcome resistive forces.

## Movement-Based Classification – Linear Operation
Movement of the moveable element also can define the classification of valves. The main directions of movement are linear and radial. Linear operation of the moveable element occurs perpendicular to the direction of flow.

### Gate Valves
The gate valve comprises the body and a wedge-shaped moveable element (Figure 5.24) (Frenck 2001). When closed, the plate wedges between the sealing surfaces or seats, thereby blocking flow. In the open position, the plate is completely withdrawn from the flow. Achieving a bubble-tight seal is difficult for new valves, but even more so when parts are worn or corroded. As a result, gate valves are used exclusively for on-off operations where opening or closing is relatively infrequent. Repositioning the gate is a slow operation since a hand wheel is the likely actuator.

The gate traps liquid in a cavity bounded by the body and the sealing surface when in the closed position. An increase in temperature in a valve in the closed position causes an increase in volume and hence pressure exerted by the trapped fluid. In the absence of venting, this pressure can cause internal and external leakage. To avoid such a problem, manufacturers drill a small hole in the upstream seat. This approach creates a one-way valve, since it seals only on one side of the gate. This type of valve also requires indication of direction of flow to prevent backward installation.

The knife gate, a variation of the gate valve, employs a flat leading edge on the gate in place of a tapered one. These valves employ elastomeric or plastic materials to create the seal.

Gate valves are designed to operate fully opened or fully closed (Sahoo 2004). Fully opened, a gate valve creates little pressure drop and when fully closed provides a good seal against pressure. With proper mating of the disk to the seat ring, there is little or no leakage across the disk when the gate is closed. Some leakage can occur when the backpressure is very low (5 lb/in$^2$). The slow opening of gate valves prevents fluid hammer. Gate valves are not suitable for throttling applications because high

Figure 5.24. Gate valve. The moveable element slides across the path of flow.

fluid velocity near the seat and the leading edge of the gate leads to erosion. As well, in the partially open state, the valve is prone to vibration.

### Globe Valves

Globe valves achieve the same level of tightness as gate valves and can provide shut-off (Figure 5.25) (Frenck 2001). They are commonly used to throttle flow and are readily adaptable to automatic operation and control systems. Most control valves are globe valves.

The moveable element that changes flow has a circular cross-section and engages a circular seat. The principal operation of globe valves is the perpendicular movement of the moveable element toward or away from the valve seat (Sahoo 2004). The gradual closure of the annular space between the disk and the seat is the basis for the excellent throttling ability of globe valves. The disk to seat ring contact occurs at greater than 90°, hence the force of closure tightly seats the disk. The closed valve does not trap fluid. Hence, there is less leakage around the seat.

The main disadvantage of globe valves is the high pressure drop that occurs in the tortuous path through the valve. This occurs due to the presence of two or more 90° turns through which fluid must pass.

### Pinch Valves

Pinch valves use a pinching action of an external linear actuator to apply pressure to the liner through which fluid flows (Sahoo 2004). The liner defines a straight path through the body of the valve. As a result of this design, there are no crevices or moving parts in contact with the fluid. This type of valve can seal around solids and slurries. Pinch valves are suitable for on-off and throttling service, the latter only above the 50% pinched position. Near the closed position, flow is highly erosive. Natural rubber is a liner of choice in these products.

Figure 5.25. Globe valve. These globe valves provided long service in a residential plumbing system. Note the loss of the rubber seat from the moveable element due to the erosive action of water flow.

### Diaphragm Valves

Diaphragm valves contain a diaphragm that acts as a barrier between the moveable element and the fluid (Sahoo 2004). The moveable element pushes the diaphragm against the seat to effect the seal. Diaphragm valves are considered the cleanest valve and least likely to cause contamination. This design minimizes fugitive emissions (leakage to the atmosphere).

## Movement-Based Classification – Rotational Operation

The moveable element in these valves is able to start or stop flow following a quarter turn of rotation.

### Butterfly Valves

The butterfly valve is an alternative to the gate valve for on-off service (Figure 5.26) (Frenck 2001; Sahoo 2004). A butterfly valve contains a disk as a moveable element. The actuator rotates the disk in the vertical plane into the direction of flow. The butterfly valve opens and closes in a quarter turn instead of a long stroke. Butterfly valves can accommodate flow from either direction.

Generally, butterfly valves are unsuitable for control applications. The seals wear excessively and corrode when the valve is only partially open. As well, positioning the disk in the partly open position directs flow against the wall of the piping. This orientation potentially can accelerate wear and failure.

### Ball Valves

The moveable element has an opening through which flow occurs (Figure 5.27) (Sahoo 2004). A quarter turn blocks the fluid path. Ball valves are suitable for use with suspended non-abrasive solids. Ball valves use polytetrafluoroethylene and metal for seats. In flammable service, the ball and the valve

Figure 5.26.  Butterfly valve. The moveable element rotates across the direction of flow.

Figure 5.27.  Ball valve. The moveable element also rotates across the direction of flow.

body require bonding to prevent electrostatic accumulation and discharge where the seat is nonmetallic.

Ball valves also trap a volume of fluid when closed (Frenck 2001). A small vent on the upstream side of the closed port relieves the pressure. This creates a unidirectional shut-off. Again marking of fluid direction is essential to prevent backward installation.

### Plug Valves

Plug valves contain a tapered cylinder that rotates within the valve body (Figure 5.28) (Sahoo 2004). Plug valves are not normally used for throttling. Plug valves use polytetrafluoroethylene and metal for seats. In flammable service, the plug and the valve body require bonding to prevent electrostatic accumulation and discharge where the seat is nonmetallic.

Plug valves also trap a volume of fluid when closed (Frenck 2001). A small vent on the upstream side of the closed port relieves the pressure. This creates a unidirectional shut-off. Again marking of fluid direction is essential to prevent backward installation.

### Check Valves

Check (non-return) valves prevent backflow and maintain pressure (Figure 5.29) (Frenck 2001; Sahoo 2004; Skousen 1998). A properly functioning check valve closes as the pressure drops and in the absence of flow momentum or reversal of flow. When flow occurs, the moveable element swings completely out of the way of the flow path. This means that flow pressure is required to maintain the valve in the open position. As well, orientation of the valve in the piping is critical for proper function and performance. Internal springs aid in closing the moveable element.

The most common types of check valves are the swing, poppet, and ball type. Swing check valves are best suited in systems where flow velocity is low (Sahoo 2004). Swing check valves are best suited in systems containing low even flow and gate valves. They are not suitable in systems that experience pulsating flow because of damage caused by continual flapping or pounding against the seating surface. Lift check valves are commonly used in piping systems containing globe valves for flow control. The moveable element moves along a guide to ensure seating on closing. This type of check valve is subject to obstruction from dirt that enters the guide. Tilting check valves have straight-through geometry. The moveable element floats on the flow and closes as flow decreases. Corrosion and the presence of solids in the flow can cause sticking and malfunction.

The stop check valve used in boilers is a globe valve whose stem is not attached to the plug (Frenck 2001). When extended fully, the stem pushes against the plug to make the seat, thus creating a closed globe valve. When the stem is retracted, the plug rises out of the flow path. Backflow also pushes the plug against the seat.

### Pressure-Relief Devices

Safety devices prevent boilers and other pressure vessels from becoming potentially explosive devices when subjected to pressures greater than design (Garvin 1982).

A pressure-relief device opens to prevent a rise in internal fluid pressure from exceeding a specific value due to an emergency or abnormal condition. Pressure-relief devices include pressure-relief valves, non-reclosing pressure-relief devices and vacuum-relief valves. A pressure-relief valve is a relief device designed to close and prevent further venting of fluid after restoration of tolerated pressure. A safety valve is a pressure-relief valve actuated by inlet static pressure and characterized by rapid opening or popping action. These devices normally are used in steam or compressed air service. Non-reclosing pressure-relief devices remain open after occurrence of the release.

The main difference between a relief valve and a safety valve is the extent of opening at the pressure of the set-point (Sahoo 2004). A relief valve opens gradually above the set-point and only as necessary to relieve the overpressure. A safety valve opens rapidly at the set-point. A safety valve remains fully open until the pressure decreases below the reset. The reset pressure is lower than the pressure at the relief set-point. Relief valves are used for incompressible fluids, such as water or oil, and safety valves for compressible fluids, such as steam or gases.

Figure 5.28. Plug valve. The moveable element in the plug valve used to stop gas flow rotates across the direction of flow.

Figure 5.29. Check (backflow prevention) valve. The moveable element in these check valves drops into the path of flow following a decrease in pressure.

Process operations usually use pilot-operated pressure-relief valves in conjunction with globe valves (Garvin 1982). The relief valve has a gradual lift generally proportional to the increase in pressure above the set-point. The pilot valve is to be self-actuated and to fail in the open position.

A safety valve or safety relief valve subjected to pressure at or near the set point will tend to weep or simmer. This is the situation created by a pressure cooker on a stove. During this condition, deposition of contamination can occur in the area of the seat and disk. This situation could lead to failure to open at the set-point, at which time the valve no longer will perform as intended. The difference between routine operating pressure and the set-point of the valve should be sufficient to prevent the weeping or simmering condition. The pressure differential reflects the service in which the equipment operates.

## Valve Selection

Valve selection reflects a number of parameters already mentioned (Sahoo 2004). These include function, piping arrangement, material transported, temperature and pressure, sizing, pressure drop, shut-off requirement, and components.

Temperature governs the internal pressure at which iron and steel, the material of construction of most valves, can perform appropriately. ANSI/ASME B16.34. (2009) is the standard that specifies acceptable temperature. ANSI/ASME B16.34 specifies classes for nominal pressure designations for valves with various end configurations. These classes recognize the premise that the higher the pressure rating for a valve, the thicker must be the wall in order for the valve body to maintain its integrity. The temperature of the service affects the pressure rating. The net result is a pressure-temperature curve for each Class. Classes identified in ANSI/ASME B16.34 include Class 150, 300, 600, 900, 1500, and 2500 for carbon steel.

## Destructive Forces in Fluid Process Systems

Flow moves through valves and other components due to a difference in pressure between the upstream side and the downstream side (Skousen 1998). Fluid moves through a valve due to a difference between the upstream pressure and the downstream pressure. For piping of identical diameter on both sides and consistent velocity, the valve must reduce fluid pressure to create flow by frictional loss. Friction between the fluid and the walls of the structure is insufficient to create the pressure drop needed for sufficient flow.

The more effective means is to provide restrictions in the body of the valve. These include constrictions and abrupt changes in direction. These restrictions cause eddies in the flow of the fluid. Eddies are circular patterns that develop in response to the obstruction, both upstream and downstream.

The highest velocity and least pressure occur downstream from the narrowest constriction at the *vena contracta*. For this reason, the *vena contracta* is a measurable location, although not necessarily physically distinguishable from other locations in the structure. Downstream from the *vena contracta*, the fluid slows and the pressure rebuilds, although not to the original level of the upstream pressure. The decrease in pressure occurs due to frictional loss as the fluid passes through the structure.

High pressure drop during flow through structures creates flow-related problems, including cavitation, flashing, choked flow, and high levels of noise and vibration. Conditions that threaten the longevity of control valves include pressure drop exceeding 2/3 of the upstream pressure, hot process fluid (>300° C), or high velocity.(Miller 2002).

## Cavitation

Cavitation is the formation and collapse of microscopic bubbles of vapor in a flowing fluid (ISA 1998). In order for cavitation to occur. The fluid must pass from a region of low pressure to a region of higher pressure. In control valves, high local velocity in the *vena contracta* immediately downstream from the seating or port area causes the low pressure.

Cavitation is a physical process that occurs only in liquid service (Skousen 1998). Cavitation occurs when the atmospheric pressure equals the vapor pressure of the liquid at that temperature. Under this condition bubbles of vapor form in the liquid. This is observable in common experience as bubble formation when a liquid in a pot begins to boil. The same phenomenon can occur on reduction of atmospheric pressure when fluid accelerates to pass through the narrow restriction at the *vena contracta*. The pressure may decrease below the vapor pressure. This leads to formation of bubbles of vapor in the fluid.

As the vapor bubble formed in the vena contracta passes along the fluid during pressure recovery, the bubbles implode due to higher pressure in the fluid. The liquid-vapor-liquid transition occurs in a small path and within a timespan of microseconds. The cavitation phenomenon is known to involve propellers on ships, water turbines in hydroelectric operations, pumps, and valves.

Cavitation is a direct result from the manner in which valves and similar components function. Cavitation damage due to erosion can range from minor to severe and can affect valve seats, moveable elements and valve bodies, pressure vessels and piping, and pumps. The damage results from release of implosion energy against the surface. Cavitation damage creates an appearance on surfaces similar to abrasive blasting.

## Flashing
Flashing is the formation of pockets of gas in the fluid (Skousen 1998). This occurs when the pressure downstream from the *vena contracta* remains equal to or less than the vapor pressure. The result is that the bubbles formed in the *vena contracta* do not collapse. The fluid downstream is a mixture of vapor and liquid. Moving at high velocity, this can erode the valve and downstream piping.

## Choked Flow
Choked flow occurs when liquid flow is saturated by the fluid mixed with gas bubbles or pockets and can no longer be increased by decreasing downstream pressure (Skousen 1998). That is, formation of gas in the liquid crowds the *vena contracta*. This limits the amount of flow that can pass through the valve. Choked flow also occurs in valves handling gases as sonic velocity (speed of sound).

## Corrosion
Corrosion is the dissolution of a metal into an aqueous environment (Shackelford 1988). Metal atoms dissolve as ions.

Aqueous corrosion is a form of electrochemical attack. A variation in concentration of metal ions above two different regions of a metal surface results in flow of an electrical current through the metal. The region of low ionic concentration corrodes (loses material to the solution). Salt water from salted roads can increase the rate of the aqueous corrosion of steel used in vehicles. Carbon steel is at considerably higher risk in this environment than is stainless steel.

Galvanic corrosion occurs when two dissimilar metals, one more active chemically than the other, are placed in contact. The more active metal acts as an anode and is corroded (loses material to the solution). Iron and zinc will corrode when in contact with copper in the appropriate ionic environment, such as salt water. Galvanic corrosion can occur following substitution of inappropriate metals in fastenings and fittings.

Gaseous reduction involving dissolved oxygen is another corrosion process. In this case, the driving force is the difference in concentration of oxygen in different parts of the aqueous environment. Differences in oxygen concentration can develop in cracks and gaps in metal, in surfaces of the same piece of metal that are covered by dirt, grease, paint, or other impervious coating versus an exposed surface, and by the environment created by fastenings and fittings that hold together pieces of metal. Corrosion of this nature is readily obvious in the rust that forms at the joints of welded and non-welded steel structures.

Zinc in galvanized coatings can form a protective oxide layer on metal surfaces. Acidic rainwater (pH <7) can attack active metals, such as zinc, to form hydrogen gas. Acid attack on the galvanized coating, while removing the oxide layer and improving conductivity, eventually exposes the underlying steel to both oxidation and acid attack.

Mechanical stress enhances corrosion. Mechanical stress refers to both applied stress and internal stresses associated with microstructure (for example, grain boundaries). Inappropriate selection of washers can fail to maintain tension or over-apply tension in mechanically assembled components.

High stress regions (such as cold worked steel) act as anodes relative to low stressed regions. These regions can occur locally on the same piece of steel where flexing has occurred.


## Erosion

Erosion damage can occur from liquid droplets entrained in fluid vapor (Miller 2002). These are condensation droplets. Maintaining the temperature above the saturation curve avoids formation of liquid. This situation also is controllable through controlling velocity.

## Dead Legs

Dead legs are stagnant pockets where fluid collects (Schmidt 2001). Liquid can collect and remain in a dead leg even after cleaning because the cleaning fluid does not flow through this part of the circuit. Vertical dead legs that drain downward are preferable to horizontal dead legs, and horizontal dead legs are preferable to vertical dead legs that collect fluid by downward flow.

Maintaining dead legs as short as possible is critical to minimizing the concerns reflected by this issue. Length/diameter (L/D) ratio is a critical consideration in evaluating the significance of a dead leg. An L/D of 2 is a practical limit to removal of contamination. This reflects the reality of the occurrence of the concentration gradient and eddy and molecular diffusion in a dead leg.

Permanent dead legs result from design features and from operational modifications. These include instrument taps and pressure-relief devices. A permanent dead leg is also present in configurations containing the bleed valve in a double block and bleed circuit.

Configured dead legs result from closing valves. Manifolds, tees, and drains at low points become dead legs on closure of the appropriate valve. Locating a small valve close enough to a tee to respect the two-diameter limit may not be possible.

A parallel issue to dead legs is fluid retention in pockets in gate, ball, plug, globe, butterfly, and check valves.


## Water Hammer

One of the most serious outcomes from fluid flow in pipes is surge (or water hammer) (Debban and Eyre 1997; Faye and Loiterman 2001; Jennings and Boteler 2001). Water hammer is the banging noise heard in a water pipe following an abrupt alteration of flow with accompanying pressure surges (Debban and Eyre 1997). Water hammer also occurs in steam pipes when steam bubbles enter a cold pipe partly filled with condensate. Water hammer can lead to catastrophic failure of piping. This is especially problematic in utility piping systems that are old, and have endured high-temperature, high-pressure conditions for long periods and become embrittled. Investigation of the accident documented by Debban and Eyre (1997) demonstrated that steam and water cannot be mixed safely in a piping system without the risk of condensate-induced water hammer. Therefore, the operative principle is not to mix steam with water, either by injecting water into a steam piping system or steam into a piping system containing water. The only safe action is to depressurize and drain the system, and then to remove condensate as it forms.

In liquid applications, sudden cessation of flow causes shock waves of a large magnitude both upstream and downstream (Skousen 1998). Causes of the shock waves include sudden shut-off of a pump or valve. While these shocks produce noise, the real concern is structural and mechanical failure. Water hammer has caused pipe bursting, damage to pipe support and connections, and damage and failure of valve components.

Opening or closing a valve to change the flow rate changes the kinetic energy of the flowing column. This change introduces a transient change in static pressure (Zappe 1999). The transient change in pressure occurs progressively along the pipeline from the point of initiation. The kinetic energy stored in the liquid elements compresses and then expands the surrounding wall of the pipe. The remaining portion of the liquid column continues to flow at the original velocity until reaching the liquid column which is stationary. The velocity at which the compression zone extends toward the inlet end of the pipeline is uniform and equals the speed of sound in the liquid at that temperature. At the inlet end of the pipe, the liquid is stationary when the compression zone arrives. The static pressure is greater than the normal value. The unbalanced pressure now creates a flow in the opposite direction and relieves the rise in static pressure and expansion of the pipe wall. When the pressure drop has reached the valve, the whole liquid column is again under normal static pressure. Discharge continues to the inlet pipe, thus creating a zone of subnormal pressure starting at the valve. When this pressure wave completes the round trip, the normal pressure and original direction of flow are restored. This cycle repeats until the kinetic energy of the liquid column dissipates in friction and other losses. This situation leads to a back and forth motion of the pressure wave.

## Surge

Surge occurs when flow undergoes a step change in velocity (Jennings and Boteler 2001). One characteristic of surge is the metallic knocking heard in a piping system. Pressure surges in liquid systems occur during startup and shutdown, during valve opening and closing, and shutdown or startup of pumps. Shutdown and startup is an especial problem with positive displacement pumps. Water hammer can occur during every stroke of a piston or diaphragm pump. Additional situations that can result in momentary overpressurization, physical deformation and movement include the combination of a continuous gas phase (such as a steam system) with a discontinuous liquid phase (such as condensate), and localized occurrence of liquid flashing and condensation (such as leads to cavitation) (Fay and Loiterman 2001).

## Fuel Gases (Natural Gas and Propane)

Fuel gases are widely used for heating homes, and commercial, institutional and industrial buildings. Natural gas is transported by pipeline from point of origin, or more recently, following regassification from liquefied fuel. Propane is transported as a liquid, is stored on-site in above ground cylindrical tanks and vaporizes through external heating. The propane tank on the home barbeque is an indicator of the trust of regulators about the safety of these products in custody of nontechnical people under all conditions of use.

These gases are transported in underground pipelines that operate at various pressures. Gas pressure in residential applications is 0.5 lb/in$^2$. This value is set by use of pressure-reducing devices on delivery piping.The rarity of fires and explosions involving these products testifies to the engineering involved in control, but also to the training and practice of individuals who interact with this equipment.

The natural gas supply system uses plug valves and other valves to control distribution and for isolation of segments of the system in buildings. These segments use single valve isolations. This is also the case with propane tanks in home barbeques. The tank as provided for sale contains a single valve to contain the contents under normal temperatures of service.

## Steam Systems

One of the more common energy distribution systems is the steam system. Steam is used for heating and process applications. Steam systems differ from other fluid handling systems because heat provided to water in the boiler is the motive force. Steam is the gas produced when water boils. (Water vapor and steam both contain only free water molecules, but they originate at different temperatures. Water vapor is present in a mixture containing molecules of the other gases in air, whereas steam is pure gaseous water.)

Steam is classified according to quality and pressure (Graham 1983). Quality refers to moisture content. Wet steam contains steam plus moisture, mist, or spray. Dry steam, which includes saturated and superheated steam, contains no liquid water. Saturated steam is steam at a temperature according to its pressure. That is, saturated steam is present at the dividing point between water and steam. This strictly speaking is the only condition in which true steam can exist.

Superheated steam has a temperature greater than that corresponding to its pressure. That is, superheated steam is steam heated to a temperature beyond the evaporation point of the water contained in it . During boiling in a closed system, pressure and boiling temperature increase until all water has vaporized. At this point, further heating (that is, superheating) will not appreciably increase the pressure, but will increase the temperature. Superheat is the amount of heat to be lost by steam before condensation can begin. Superheating, therefore, reduces and potentially stops condensation in the system.

Steam pressure refers to pressure at or above atmospheric pressure. Vapor refers to water vapor. Atmospheric-pressure steam is pressure that is slightly below or above atmospheric pressure. This is the operating pressure of some heating systems. Low pressure steam refers to steam heating systems operating at 5 to 10 lb/in$^2$ (35 to 70 kPa) above atmospheric pressure (gauge pressure). Medium pressure systems operate up to 150 lb/in$^2$ (1,030 kPa). Standard working pressure in steam systems is 80 lb/in$^2$ (550 kPa) gauge pressure. High pressure systems operate from 150 to 300 lb/in$^2$ (1,030 to 2,070 kPa), and extra-high pressure systems from 300 to more than 2000 lb/in$^2$ (2,070 to 13,800 kPa). As boiling occurs in a closed system, pressure increases. Pressure increase suppresses vapor formation. As a result, the temperature at which boiling occurs also increases. Hence, this situation requires addition of more heat to water to cause boiling. At atmospheric pressure, the boiling point of water is 100° C (212° F). In a system operating at 80 lb/in$^2$ (550 kPa) gauge pressure, the boiling point of water is 162° C (324° F). The corresponding boiling point of water in a system operating at 600 lb/in$^2$ (4,100 kPa) gauge pressure is 254° C (489° F).

## Flowable Solid Materials

Solid materials are not generally considered in discussion about hazardous energy. Closer examination indicates that solid materials can exhibit many of the properties of fluids. Flowable solid materials are obvious solids whose energy content can transform almost instantaneously from potential to kinetic.

### Solid Materials in the Natural State

Materials in the natural state range from wind-blown and deposited sand to naturally exposed embankments of earth and sand and aggregate to solid rock. The state of aggregation of these materials ranges from little to semi-solid to full aggregation. These materials are stable only so long as conditions required for their stability remain satisfied. Major slides of materials in the natural state in mountainous regions into valleys have occurred over geological time, sometimes for no apparent reason. These slides sometimes block streams and result in formation of lakes. These occurrences provide evidence that what appears to be stable is not necessarily so.

Earthquakes are major promoters of changes to the status quo of materials in the natural state, although certainly not the only one. Slides in mountainous regions have occurred in the absence of earthquakes for unexplained reasons.

The most common promoter of instability of materials in the natural state is resource extraction and construction activity. Resource activity includes surface mining to remove sand, gravel, and other valuable materials. These operations use front-end loaders and excavators, sometimes assisted by blasting. The front-end loader and excavator remove material from the base of the exposed deposit. This action eventually steepens and undercuts the slope. The miner relies on instability caused by these actions to cause a controlled cave-in of material. Implicit in actions undertaken in this manner is that the cave-in occurs without risk to the operator of the equipment. This is not always the case, as numerous burials of equipment and operators have occurred following sudden slumpage of material. Blasting and other techniques aimed at controlled destabilization of the slope attempt to promote small-scale slumpage that minimizes the risk of burial.

Construction activity creates sloped embankments in many projects by recontouring existing slopes. Recontouring means removing existing material often to widen the base and to produce a new stable slope. The most obvious of these activities occur during road and rail construction. Creating the stable embankment requires controlled removal of material from bottom to top and contouring of the material remaining in place.

By far the most prevalent activities that expose materials in the natural state occur during trenching and excavation. Materials exposed during trenching and excavation include materials never previously exposed, materials exposed and replaced during previous activity, materials used as backfill, and a combination of these. Geotechnical engineers who assess stability of slopes exposed during trenching and excavation rely heavily on data generated during soil sampling and drilling. These samples provide evidence about the type of material underlying the surface. The reliability of the predictions depends heavily on the locations in which core sampling occurs.

Deposition of loose material onto existing material to create a new slope. This action can create instability through increasing the angle of the slope. The interface between the existing natural material and material deposited on top is a critical concern. This interface can create a slip plane that is subject to sliding. Movement of moisture along the interface between the two zones of material can provide the lubrication needed to enable the occurrence of a slide.

The tendency for flow of material in a natural formation or an intentionally deposited storage pile reflects competition between two sets of forces: those resisting failure of the slope and those tending to produce it (Terzaghi et al. 1996). Such a determination involves a stability calculation. A body of material is considered to be in a state of plastic equilibrium if every part is verges on failure.

The mass of material bordering a trench or excavation that is capable of slumping is a curved wedge, like the side of an axe. The vertical wall of the trench is equivalent of the centerline of the axe. During slumpage, the wedge transforms to a slab having a lower profile, and therefore, lower potential energy, and hence greater stability. The center of gravity of the material prior to slumpage is located at a higher position than that following slumpage. The (difference in height) × (mass) × (gravitational acceleration) is a first approximation of the energy liberated during slumpage. In doing so, some of the potential energy of position changes to kinetic energy of motion for a brief period.

Failure of a mass of material located beneath a slope constitutes a slide (Figure 5.30). A slide involves downward and outward movement of the entire mass of material that participates in the failure. Almost every slide of cohesive material is preceded by formation of tension cracks on the upper part of the slope or beyond its crest (Terzaghi et al. 1996). During the slide, the upper part of the slide area (root) subsides, while the lower part (tongue) bulges. The shape of the bulge depends on the nature of the material. The plane of the slide forms an S-shaped curve (Figure 5.31). This may be constrained when the flow of material reaches the opposite side of the structure.

Stability computations for slopes are based on moments of rotation about an imaginary point. Curvature of the surface of motion by the subsurface material resembles the arc of an ellipse. In stability computations, the arc of a circle substitutes as an approximation for the arc of the ellipse. Moments are calculated for the slice of material that tends to produce failure and the slice that tends to resist it. The moment is calculated using the center of mass for each slice and the distance along the perpendicular line drawn from the center of mass to the radius of the circle. In the simplest case, sliding resistance at equilibrium depends on the difference between the moments and the radius of the imaginary circle and the length of the surface of sliding.

In order to prevent slumpage and collapse, the wall of the trench or excavation requires sloping. Experience indicates that a slope ratio of 1:1.5 in soils (horizontal:vertical) is usually stable (Terzaghi et al. 1996). Certain types of soils are more stable and require shallower slopes. The ratio used in regulatory standards pertaining to trenches and excavations also could differ from these values. The variety of factors and processes that lead to slides are extraordinary and complex. As a result, excavations involving these materials require input from experienced geotechnical engineers.

A slope underlain by a dry cohesion-free material, such as clean dry sand, is stable regardless of height, provided that the angle between the slope and the horizontal is equal to or smaller than the angle of internal friction for the material in the loose state. The factor of safety is the tangent of the angle of friction divided by the tangent of the angle between the slope and the horizontal. Few natural soils and materials are cohesion-free.

## Processed Solid Materials

The processing, handling, and storage of flowable solid materials pose problems of energy storage and dissipation similar to those experienced with liquid fluids. Soils, and ground and pulverized rocks and minerals comprise a substantial portion of industrial flowable solid materials. Soil is an aggregate of mineral grains that can be separated by gentle mechanical action (Terzaghi et al. 1996). Soils are characterized by index properties. Index properties subdivide into classes: soil grain properties and soil aggregate properties. The principal grain properties are size and shape. The principal aggregate properties are cohesiveness, relative density, and consistency.

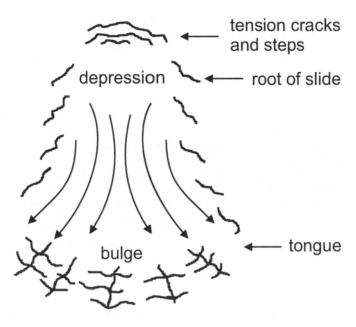

Figure 5.30. Slide of material. During the slide, the root (upper part of the slide area) subsides and the tongue (lower part of the slide area) bulges. (Adapted from Terzaghi et al. 1996.)

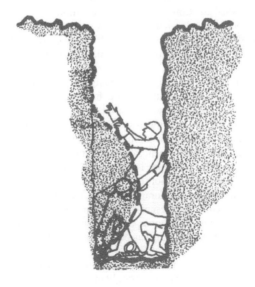

Source: OSHA

Figure 5.31 Cross-section through a typical slide of cohesive material. Note the "S"-shaped curve. (Source: OSHA.)

## Storage

An estimated 1,000 industrial and farm silos, bins, or hoppers fail each year in some manner in North America (Jenkyn and Goodwill 1987). Failure in this context ranges from cracks in a concrete wall or kinks in a steel wall to outright structural failure. Causes of failure include incorrect or incomplete considerations in design, poor construction practices, lack of maintenance, and change from intended use. In some cases, failure occurs by complete structural collapse. In other cases, distortion or deformation occurs (Carson and Holmes 2001). An improperly designed storage vessel is more likely to fail structurally than other plant equipment (Purutyan et al. 1999).

Out-of-round wall bending reflects application of unbalanced radial forces. These forces develop when flow does not occur along the centerline of the vessel. Off-center flow is one cause of silo failure. Nonuniform pressure causes horizontal and vertical bending moments (stresses at bend points). Internal components including support beams, inverted cone inserts, gravity-blending tubes, and other internal components can cause large concentrated loads or asymmetrical pressures.

Thermal ratcheting is another cause of structural failure (Carson and Holmes 2001). Thermal ratcheting occurs when the wall of an outdoor metal silo expands during the day and contracts at night when the temperature decreases. Free-flowing material inside the silo slumps during the day into the space created by the expansion of the metal. Contraction of the metal during cooling is unable to push the material upward into the decreased volume. This situation increases thermal stresses in the wall and repeats daily until failure occurs.

Silos experience vertical pressure and radial pressure (Carson and Jenkyn 1990). Vertical pressure acts downward and radial pressure acts against sidewalls. Loading conditions can contribute to silo failure (Table 5.3).

A bridge is a condensed layer of material that does not flow (Carson and Jenkyn 1990. The material under the bridge flows away from underneath, potentially leaving behind a hollow. The bridge may support the weight of an individual, but could collapse at any time (Figure 5.32). During failure, the bridge can fracture, in the case of solidified material, or can collapse as a cohesion-free mass in the case of flowing material. Broken pieces of solidified material intermingle with the victim and cause traumatic injury due to contact with blunt and sharp edges during the fall or following initial contact with the base prior to settling into final position. During collapse, settling of flowable solid material around the victim leads to engulfment and asphyxiation. The victim is unable to expand the chest against the radial pressure exerted by the flowing material, and suffocates as a result. Keeping up with flow by stepping above the collapsing bridges is almost impossible. This action explains the reason for burial in flowing materials.

The same considerations apply to hang-ups. These are masses of material that have solidified and adhered to walls. They can dislodge intact and fall to the base. Again, the broken pieces can intermingle with the victim and cause traumatic injury due to contact with blunt and sharp edges during the fall or following initial contact with the base prior to settling into final position.

Flowable solid materials contained in a sloping pile or on the walls of a silo have a tendency to move downward and outward under the influence of gravity (Terzhagi et al. 1996). If counteracted by the shearing resistance of the material, the slope is stable; otherwise, a slide occurs. Downward movement of material is highly desirable during normal operation, since this constitutes the normal operation of the silo in its function as a surge and storage structure. Under abnormal conditions, the material does not flow. Often this is related to the presence of abnormal levels of moisture. As fatal accidents that have occurred in these structures have shown, the moment in time of failure of the slope is not predictable (OSHA 1983, MSHA 1988). Slumping can occur slowly or suddenly, with or without apparent provocation.

A similar situation applies to material hung up on the walls in silos, except that the material that slumps is bordered by a curved wall rather than a straight one. Experience indicates that a slope ratio of 1:1.5 in soils (horizontal:vertical) is usually stable (Terzaghi et al. 1996). Slopes having smaller ratios than this occur in storage structures, such as silos, when the material hangs-up on the walls and refuses to flow. A slope underlain by a dry cohesion-free material, such as clean dry sand, is stable regardless of height, provided that the angle between the slope and the horizontal is equal to or smaller than the angle of internal friction for the material in the loose state. Few natural soils and materials are cohe-

## Table 5.3
## Loading Conditions Leading to Silo Failure

Large void (arch or rathole) that collapses and results in significant dynamic stresses on silo walls

Nonuniform radial pressure exerted on silo walls caused by an off-center channel in material adjacent to the wall

Local peak pressure where funnel flow channel intersects the silo wall

Mass flow in a silo designed for funnel flow

Asymmetric pressures caused by inserts (such as beams) across the cylindrical cross-section

Use of drastic flow promotion (explosives, excessive vibration, air injection

Migration of moisture from wet to dry particles within stored solids leading to expansion and radial pressure

Buckling of unsupported wall below an arch of stored solid

Source: Carson and Jenkyn 1990.

sion-free. In the case of flowable solid materials stored in silos, stable vertical slopes can exist, as demonstrated by the existence of hung-up material. Stability of a slope is a function of height.

## Storage
Design of storage vessels for flowable solid materials is critical to ensuring that flow occurs as intended and that bridges and hang-ups do not form. The critical point in design centers around properties of the flowable solid to be handled in the system (Craig and Hossfeld 2002). These properties govern behavior of the material when acting in bulk flow and in fluidized form during transport.

Flowable solid materials are the most difficult of the states of matter to handle (Wahl 1985). Failure of a material to flow by gravity frequently results from a combination of properties, such as density, compressibility and hygroscopicity. Compressibility is computed from the aerated density and the packed density of the material. The dividing line between free-flowing granular and nonfree-flowing powder is compressibility about 20%. A higher percent indicates a powder that is not free-flowing and is likely to bridge. Hygroscopicity refers to the ability of a substance to absorb moisture. Hygroscopic substances will flow erratically or not at all when exposed to high humidity.

Four angular indicators of the free flow potential of a bulk material are the angle of repose, the angle of fall, the angle of slide, and the angle of spatula (Wahl 1985). The angle of repose is the included angle between the edge of a cone-shaped pile formed by falling material and the horizontal. The smaller the angle of repose, the more flowable is the material. The angle of fall is the corresponding angle measured when the pile settles after a weight is dropped nearby on the same surface. The greater the difference between the two angles, the greater the free-flow potential. The angle of slide is the angle at which a material will slide down a surface under its own weight. To measure the angle of spatula, a spatula covered to the maximum is lifted from a pile of material. A weight is then dropped onto the same surface. The angle of spatula is the difference between the angle of the side of the pile before and after the dropping of the weight.

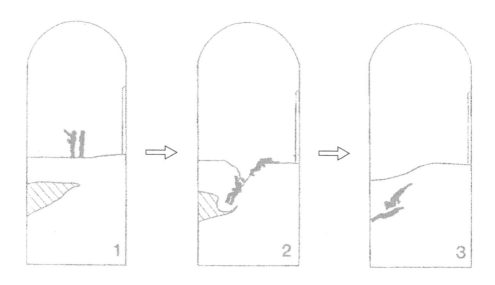

Figure 5.32. Bridge formation. A bridge is an arched structure that forms during storage of material. The bridge sometimes supports weight of added material, including occupants. The bridge can collapse into the hollow underneath without warning at any time. (Source: NIOSH.)

A classification scheme containing four divisions is used in the assessment of flowable solid materials (Wahl 1985). These classes range from granular, free flowing materials to fibrous or flaky materials of low bulk density that tend to interlock and absorb vibration. Flowability is a measure of the tendency of the material to form a bridge. A bridge is a condensed layer of material that does not flow. The material under the bridge may flow away, potentially leaving behind a hollow. The bridge may support the weight of an individual, but could collapse at any time. Bridging and ratholing (formation of flow channels through a condensed mass of material) in storage hoppers and silos is subject to mathematical treatment (Carson and Johanson 1985).

Hang-ups on walls and formation of bridges and other flow problems pose such problems, that a whole industry has developed to correct them. This industry manufactures equipment to dislodge adherent material, as well as to open channels to enable flow. Sending workers into storage structures to perform the same functions is a practice that need not occur. Dislodging hung-up material can require use of shovels, jackhammers, and even dynamite. In these situations, the purpose for the activity is to destabilize the structure of the material in order to induce flow, which, in turn, reduces the energy content.

Solids that can flow when constrained within a structure will flow in one or more modes: funnel flow, mass flow, or expanded flow. The type of flow in a particular situation is controllable by design and can happen in spite of attempts at control through design.

Funnel flow occurs due to friction between the wall and the material adjacent to it (Figure 5.33). The vessel experiencing funnel flow commonly has a shallow conical or pyramidal hopper constructed typically from carbon steel. The wall angle of the cone is typically 60° or less from the horizontal. This configuration lacks the steepness or the smoothness to force material to flow along its surfaces. The result is that the walls retain material while flow occurs toward the center of the structure. A flow channel develops typically directly over the outlet. Material adhering to the walls eventually falls from the top

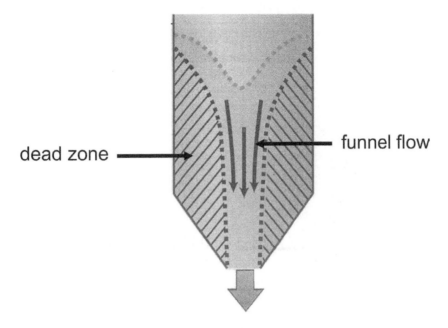

Figure 5.33. Funnel flow. During funnel flow, material flows from the walls of the vessel toward the flow channel centered above the outlet. Funnel flow produces a first in, last out flow pattern.

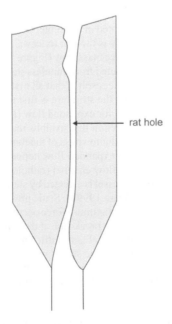

Figure 5.34. Flow channel (rathole) in material experiencing funnel flow.

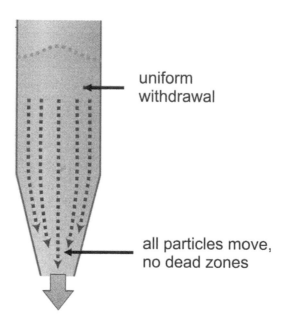

uniform
withdrawal

all particles move,
no dead zones

Figure 5.35. Material experiencing mass flow. Mass flow produces a first in, first out flow pattern.

of the retained zone into the flow channel located in the center (rathole) (Figure 5.34). As a result, material entering first into the structure typically is the last to leave.

The hopper in a vessel designed to produce mass flow (Figure 5.35) typically is higher and steeper than one exhibiting funnel flow and is constructed from stainless steel or coated metal. This design provides an interior surface of lower friction. The result is that all material in the structure moves downward at the same rate. Material first entering the structure is first to exit from it.

The hopper section of a vessel designed for expanded flow (Figure 5.36) contains an upper section that exhibits funnel flow and a lower section that exhibits mass flow. This design combines the benefits offered by both types of design: the shallowness of the funnel configuration and the even distribution of mass flow. A properly designed expanded flow hopper will exhibit even flow at the discharge and avoid development of the center flow channel (rathole).

Vertical flow of a material in a storage vessel is affected by static pressure, bridging, first-favored flow stream, and sloughing (Ware 1999a, Ware 1999b). Static pressure is the lateral pressure exerted against the walls of the vessel (Ware 1999a). Maximum force occurs at the point where the slope of the vertical walls of the vessel changes. This is the break line at the transition section of the hopper.

Bridging occurs when the outlet of the vessel isn't sufficiently large. The bridge forms from coalesced material and also reflects properties of the stored material. The bridge is a ring of particles compressed together. This provides strength to the formation. Downward pressure exerted by the material exerts lateral pressure against the wall to form the ring and the arch. The void space formed under the bridged material has an arched roof. The bridge must be structurally capable of supporting some of the material above it.

The first-favored flow stream results from the manner in which material flows downward. As material flows downward (dynamic motion), lateral forces against the walls increase. This is dynamic pressure. Dynamic pressure is about three times greater than static pressure. Again maximum lateral pressure occurs at the break line. First-favored flow creates either mass flow or funnel flow. During

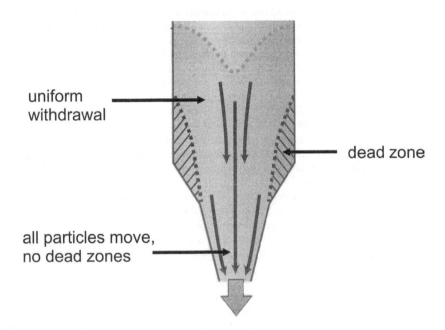

uniform withdrawal

dead zone

all particles move, no dead zones

Figure 5.36. Material experiencing expanded flow.. Expanded flow exhibits attributes of both funnel flow and mass flow to improve the flow pattern.

funnel flow, an active flow channel develops in the material having a diameter the same as the outlet diameter and located above it. Material on the walls flows into the flow channel in a process called sloughing. Sloughing involves shear forces that reflect failure of the horizontal stability of the material. Where stored material stabilizes on the walls around the active flow channel, this structure is called a rathole.

Freezing (ice formation) can occur in materials persistently exposed to temperatures below −4° C (25° F) (Ware 1999b). Frozen material forms a band around the inside wall of the storage vessel. The band of frozen material can easily become 0.6 m (2 ft) or more in thickness. Freezing is a greater problem in the transition section.

Spontaneous combustion is a concern with coal and similar materials. This is especially the case during long-term storage when air movement is minimal and heat can accumulate.

Fines separated from the main material during dust collection can exhibit considerably different flow properties than the bulk material (Wilkins 2004). Coarser particles remain with the bulk flow of material. Flow of fines is subject to slow flow, intermittent flow, and complete stoppage compared to the main material in storage. The function of a dust collecting system is to collect airborne dust generated at transfer points.

Part of the problem results from the design of the hopper in the dust collection system (Wilkins 2004). In order to minimize cost, these structures have small openings for discharge of dust and fines. This decision results from the choice of end fittings (rotary valves and airlocks) used to control flow and the small opening of the storage vessel. Adhesion to walls and formation of arches or bridges is a common problem with this type of storage.

The angle of taper is also a critical factor in mass flow of fines. A taper of 55° to the horizontal which is satisfactory to promote mass flow with coarse materials is insufficient for fines. Dust also can collect in seams and creases in corners of hoppers when the valley angle is insufficient. This can occur even when the vertical slope is sufficient to create mass flow. Designs that create mass flow in hoppers handling fines include slotted and round discharge openings. Establishing the flow properties of the material through quantitative testing is essential for establishing the parameters of the design.

The situation described for dust collection systems is similar to what occurs during disagglomeration. Disagglomeration occurs during handling of mixtures of materials containing diverse particle sizes (Pittenger and Carson 2008a, Pittenger and Carson 2008b). In this case, maintaining composition in a consistent manner is critical for product quality.

Segregation occurs through sifting, trajectory, fluidization, and particle entrainment in the airstream (Pittenger and Carson 2008a). Sifting is the most common mechanism. Sifting is the movement of smaller particles through larger ones. Sifting can only occur when the ratio of particle diameter is at least 1.3:1; mean particle size is less than 500 μm; the material is free-flowing; and particles move relative to other particles at different speeds. Sifting does not occur when particles adhere to each other. Sifting is likely to occur during filling of vessels. This can result in highly concentrated clouds of dust in the airspace above the material.

Fluidization refers to the tendency of lighter particles to remain airborne longer than heavier particles during free-fall through the air. Settlement by the heavier particles results in separation. Fluidization increases significantly for particles less than 100 μm.

Particle entrainment in an airstream occurs with finer, lighter-weight particles. Entrainment is significant for particles with sizes down to 50 μm and severe at 10 μm. Filling operations can cause major emission of dust due to entrainment. Large particles disengage from the airstream while small particles remain entrained. This condition is apparent in bins constantly evacuated by a dust collection system, vents or a significant difference in temperature from one location to another.

Mass flow eliminates flow problems, such as arching and rathole formation and reduces the severity of particle separation due to sifting and particle entrainment (Pittenger and Carson 2008b). The minimum size of outlet to prevent arching is directly related to the cohesive strength of the material (Purutyan et al. 1999).

Conveying

Flowable solid materials pass in bulk form through silos and hoppers, and in fluidized form through pneumatic and vacuum conveying systems. Pneumatic and vacuum conveying systems for flowable solid materials are similar in concept to those used for transport of other fluids. These systems contain valves and isolation devices as well as air movers that create the transfer medium.

Typical applications for valves in controlling flow of bulk solids include feeding and discharging mixers and process vessels, feeding from a non-pressurized hopper to a pressurized or vacuum conveying line, feeding from a hopper to a loss-in-weight feeder, and discharging a weigh hopper (Naberhaus 2004).

Many valves used in these applications were designed for handling liquids and gases (Naberhaus 2004). Inappropriate application can lead to leaking, jamming, or failing due to abrasion, cross-contamination and other problems. Abrasive solids can rapidly wear contact surfaces, seals, seats, and jam the moveable element or break internal elements.

Slide gate, butterfly and ball valves are liquid and gas valves. Small particles can leak between the sliding plate and the seal. This action can break large particles and create fines. As well, material can jam between these surfaces. The moveable element in the butterfly valve is subjected to wear from moving material. Bridging can occur when the valve isn't large enough to pass solid material. The moveable element can jam or cause breakage on closure. The ball valve is hard to clean. Cavities between the valve body and the moveable element can fill with material and cause eventual jamming

Iris, pinch and rotating disc valves are valve types intended for use with bulk solids. Iris valves and pinch valves use fabric or elastomer surfaces to control flow. These materials are subject to wear when abrasive materials are present. Elastomers are subject to degradation at high temperature and to chemical attack. The rotating disk valve contains a metal disk that rotates around a pivot point into the stream of material to start and stop flow. The internal pocket that contains the disk when rotated out of the path of flow can become clogged and contaminated.

Rotary valves (rotary airlocks and rotary feeders) located at the base of silos, hoppers and other storage vessels meter flow of material and regulate airflow (Figure 5.37). These devices receive flow

## Housing Vent

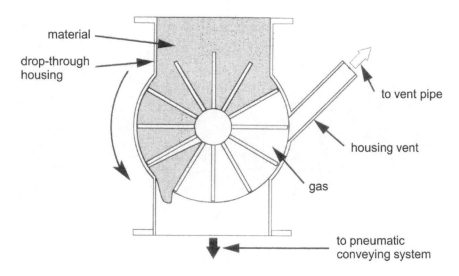

Figure 5.37. Rotary valve. The wedges in the "paddle wheel" contain flow and pass it by rotation from the inlet to the outlet.

by gravity from the interior of the storage structure. The movable element is an open- or closed-ended "paddle wheel." Solid material fills each wedge (rotor pocket) in the moveable element. Flow passes out the open bottom after moving with the paddle wheel through 180° of rotation. A motor rotates the moveable element. Designs include side entry and drop down types. Some rotary valves operate in pressurized systems. These designs act as an airlock. The feeder type operates under gravity. The feeder-airlock combines both attributes (Galuska 1999).

Dry bulk material doesn't flow like water. Hence, material flow will never be 100% efficient. No rotary valve is leakproof. Clearance leakage occurs between the rotor blades and the housing and the head plates (end plates). Displacement leakage is gas contained in the rotor pockets that passes through the outlet. The design containing the open-ended rotor forms the better seal because of multiple barriers to leakage. Some designs incorporate a venting system to vent displacement gas. Venting is especially important in pneumatic conveying systems (Pfeiffer 1999). Gas leakage flows upward, counter to the downward flow of material.

A variation on the rotary valve is the double flap airlock. This contains two chambers. Material flows into the first closed chamber from the source. Upon filling, the upper valve closes and the lower valve opens. Opening of the lower valve enables flow from the first into the second chamber and discharge into the outflow. These devices handle hot, abrasive materials.

Diverter valves (Figure 5.38) receive flow through an inlet port, alter the direction of flow and discharge through two or more outlet ports (Harkin 2004). These products are used in conveying systems. There are several types.

The moveable element in a rotary-plug diverter valves is similar in geometry to the ball in a ball valve (Harkin 2004). Rotation of the moveable element diverts flow to the alternate path. This type of product is best suited for handling pellets rather than powders. Powders can pack between the moveable element and the housing. The spherical dome valve is a variant of the ball valve. The moveable el-

Figure 5.38. Diverter valve. The moveable element diverts flow from the inlet to one or more outlet channels.

ement rotates across the path of flow of material in a quarter turn, similar to the ball valve and butterfly valve. This type of valve can provide a watertight seal (Naberhaus 1999).

The moveable element in a parallel-tunnel diverter valve contains two channels (Harkin 2004). Rotation of the moveable element switches flow to the alternate channel.

The moveable element in the rotary-blade diverter valve is similar in concept to that found in the butterfly valve. Rotation of the vane switches flow to the alternate path. This type of valve is used typically to handle granules and powders.

The flapper diverter valve contains a swinging metal flapper (gate) to divert flow from the upstream line to the downstream lines. Wear is a major concern with this type of valve.

The sliding-blade diverter valve contains a sliding metal plate to block and divert flow. This type of valve can handle both powders and pellets.

The flexible-tube diverter valve contains a flexible moveable element composed of hose material. Movement of a sliding plate on which the hose is mounted diverts the flow. The moveable element in this type of valve is susceptible to erosion from abrasive materials.

A variation of the diverter valve is the turnhead (Figure 5.39). This resembles a turntable in concept. Flow enters through the top and discharges through two or more outlet ports located in the bottom. These devices can operate in gravity feed and pneumatically or vacuum conveyed systems.

Conveying systems operate in either dilute phase or dense phase, and under pressure or vacuum (Harkin 2004). Pressure in dilute conveying systems ranges to 15 lb/in$^2$ (100 kPa), and vacuum down to $-30$ inches of mercury ($-100$ kPa). Pressure in a dense-phase system can range to 90 lb/in$^2$ (600 kPa).

Pneumatic conveying systems utilize fans, blowers and compressors for motive power (Solt 2001). In the absence of close clearances, a fan is not a positive displacement device. A positive displacement device displaces a fixed volume of air for each revolution. Fans are usually applied to dilute-phase (streamflow) conveying. In dilute-phase conveying, material typically remains suspended

Figure 5.39. Turnhead. The turnhead diverts flow to multiple outlet channels. These devices are common in agricultural applications.

in the airstream and does not deposit onto surfaces of the duct and other components. Fans can move large quantities with small pressure differentials.

A blower entrains a volume of air and discharges it to the other side.The blower has no influence on pressure and discharges a constant volume of air per revolution.. Resistance to flow downstream from the discharge of the blower controls pressure in the system. A blower is a positive displacement device. This results from continuous metal-to-metal contact between the lobes. Blowers are amenable to both dilute- and dense-phase conveying.

A compressor entrains air at a starting pressure, reduces the volume and discharges the air at a higher pressure. The compressor is a positive displacement air mover with internal compression. Compressors are best suited for applications to dense-phase conveying systems.

Safety Hazards
Safety hazards in pneumatic conveying systems include dust leaks in conveying lines, dust clouds in enclosed vessels and structures, electrostatic accumulation and discharge and hydrocarbon gases and vapors accompanying or emitted from solid materials (Agarwal 2002).

Dust leakage from enclosed lines can create a housekeeping and an exposure problem at the minimum, and in spaces having insufficient ventilation, an explosible dust hazard. Emissions of dust typically occur in pressurized conveying systems.

Enclosed vessels and structures in which explosible concentrations of dust can occur and can develop include dust collectors, bin vent filters, filter receivers, and bins and silos. Dust collectors, bin vent filters, and filter receivers are inherently more susceptible to developing explosible mixtures of dust because they collect dust and fines from the conveying gas.

Electrostatic charge generation and accumulation can occur on fibers of filters and on isolated metal structures. This can occur during normal transport in duct due to contact and separation between suspended particles and surfaces. This also can occur during pulse-jetting of the bags in a baghouse.

Electrostatic charge development and accumulation can occur during frictional contact between the bags and the wire support cages. Bonding and grounding of the cages are essential to prevent electro-static accumulation and discharge. A secondary strategy is installation of explosion-relief (blow-out) panels and explosion-suppression systems to reduce pressure in the contained structure.

Materials containing trapped hydrocarbon gases and vapors conveyed soon after production are considerably more prone to off-gassing, especially when fracturing and fine formation occurs due to rough handling. Evolution and accumulation also can occur downstream in bins and silos. Accumulation of ignitable gases and vapors and an electrostatically charged cloud of explosible dust considerably increases the explosion risk.

The use of nitrogen to inert the atmosphere in a closed-loop conveying system minimizes the ex-plosion risk (Crouch 2000; Tunnicluffe and Thomson 2004). A closed-loop system enables reuse of the inerting gas. This is especially important where the inerting gas is something other than nitrogen, such as one of the inert gases. Closed-loop systems include pressure and vacuum versions of dilute- and dense-phase conveying systems (Crouch 2000). These systems incorporate a gas analyzer to ensure that concentration of oxygen remains below the design limit, typically less than 5%. The most explo-sion-prone point in these systems is the airspace above the powder in the receiving hopper below the rotary metering valve. This vessel requires a continuous purge and explosion-prevention measures.

An explosible atmosphere can also develop inside milling equipment (Tunnicliffe and Thomson 2004). Milling can produce particle sizes in the range of 15 to 25 $\mu$m (micrometers). This size range be-haves similar to vapors and gases in explosible behavior. Containment of the explosion is another tech-nique in use in design of equipment, as are hybrids involving both strategies.

Many products shipped in bulk are fine and powdery (Hesketh 2001). Handling of granular and powdered materials poses issues caused by electrostatic accumulation and discharge. This concern pertains to all aspects of these operations and all components. Downstream from the producers of such products are the consumers. These products are shipped in bulk railcar and trailer quantities and in bulk bags, and bags and sacks small enough to be handled by a single individual. Dust collection and control and management of electrostatic accumulation and discharge are as critical on the small scale as they are on the large. Devices that contain dust emissions include spout closure bars, inflatable spout seals, and integrated dust collectors.

Flexible intermediate-bulk containers (FIBCs or bulk bags) are manufactured from woven syn-thetic fiber (typically polypropylene) (Gravell 2000). Electrostatic ignition sources can arise from the bag, the product contained in it, and a nearby ungrounded conductor. Electrostatic charges arise during transfer of nonconductive solids into or out of the bag, as well as rubbing surfaces of the bag against other surfaces.

Bags used in this service contain anti-static measures including conductive threads that form a network for connection to ground to dissipate electrostatic charge harmlessly (Gravell 2000; Hesketh 2001).

## Electrical Systems

Electrical equipment includes analogues to equipment described in pneumatic and hydraulic and me-chanical systems. Electrical equipment can modify voltage (analogous to pressure) and store energy. These processes occur through induction and capacitance.

### Induction

Induction is the process by which electrical energy is created in a conductor through the intercession of a magnetic field. Application of induction in systems that convert energy from one form to another de-rives from the work of Faraday and Henry (Halliday and Resnick 1962). Faraday showed that motion of a magnet toward a coil of wire oriented perpendicular to the direction of motion or vice versa in-duces current flow in the wire. Lenz further refined these observations by suggesting that the induced current will appear in the wire in the direction that opposes the change that produces it. That is, ap-proach of the north pole of a bar magnet induces a north pole in the magnetic field that surrounds the coiled wire. Current flows in the coil in the direction corresponding to the orientation of the magnetic field. Withdrawal of the north pole of a magnet from the coil induces a south pole in the magnetic field surrounding the wire.

Reversal of the poles of the magnet reverses the direction of flow of current. Current flows only as long as motion occurs in the magnet. This motion brings the magnet closer to the coil or vice versa. That is, the relative position between the magnet and the coil changes continuously. So also does the strength of the magnetic field. Hence, kinetic energy inherent in the motion of the magnet induces electrical energy in the wire. This also means that energy is converted in the process due to opposition of direction by the induced field.

This phenomenon is modeled through invisible lines of magnetic force that join the north to the south pole of a magnet and form concentric circles around a conductor through which current is flowing. Moving a conductor across the magnetic flux, and vice versa, cuts the lines of force, and induces an electromotive force (EMF) in the conductor. The magnitude of the induced EMF is proportional to the rate of cutting of the magnetic lines of force. The conductor can be a straight wire, a coil of wire or a block of metal in the shape of a vehicle (Palmquist 1988). Time varying the current flow and hence the magnetic field surrounding the conductor has the same effect as physical motion. That is, the lines of force become mobile.

Countless applications depend on the induction of current by movement of a magnet. Three of the best-known examples are the magneto, the generator, and the alternator.

The magneto is the subsystem in single-cylinder engines that generates the electrical energy to power the spark plug. The flywheel contains a small magnet. The body of the engine contains a coil of wire encased around a steel ring. Rotation of the flywheel causes the magnet to pass close by the steel ring. This induces a magnetic field in the steel which, in turn, induces a current in the wire. The time-varying change in position of the magnet and the steel ring is responsible for generation of electrical energy over a portion of the cycle of rotation.

The generator is the more advanced direct-current application of this principle. In a DC generator, the rotor (the moveable part) is the coil of wire in which induction of current occurs (Coates 2001). The coil must maintain physical contact with wiring in the stationary part of the generator (the stator) in order to transfer the current. This necessitates a transfer device (commutator) that is subject to wear. Overspeed rotation of the rotor can lead to arcing.

The alternator (AC generator) is the more advanced alternating current version of the DC generator. In an AC generator, the rotor is the magnet (Coates 2001). The rotor in an AC generator is lighter than the rotor in a DC generator. As a result, an AC generator weighs one-third as much as the corresponding DC unit. Since rotation of the rotor induces the magnetic field in the stator, there is no need for complicated energy transfer equipment. Output from a bank of batteries energizes the electromagnet in the rotor. AC generators can operate at high speed without concern about arcing.

In a second series of experiments, Faraday substituted a coil through which current was flowing from a battery for the magnet (Halliday and Resnick 1962). Opening or closing the on-off switch in the circuit induced current flow in the coil under study. Induction did not occur when current flow was constant. This study indicates that a time-varying magnetic field develops around a wire when current begins to flow and when flow is interrupted. The rate of change of current flow, and not the magnitude is the significant parameter. A variation of this concept involves circuits carrying time-varying electrical current. AC circuits and pulsed DC circuits are examples.

In the third set of experiments, Faraday observed that an induced EMF also occurs in a conductor located within a time-varying magnetic field (Palmquist 1988). The conductor may be a straight wire, coil of wire or a block of metal, such as an excavator. Maximum induction occurs when the lines of force are cut perpendicularly. The electrical transformer is a practical example of the application of induction caused by a time-varying magnetic field. Lines of magnetic flux vary with the change in current. The transformer is used as a voltage modifying device. Transformers function only in systems in which the voltage varies with time. Normally, these are AC systems, although transforming voltages should be possible in time-varying DC systems as well.

Transformers contain a square or rectangular ring of thin stacked insulated iron plates or strips (laminations) around which are wound isolated or non-isolated coils of wire (Palmquist 1988). The ring of laminations helps to concentrate the magnetic lines of force. The laminations of thin iron reduce to a minimum the induced current in the iron core. As a result of these energy conservation measures, conversion efficiency of transformers is as high as 99 %. Induced currents (eddy currents) are responsible for production of heat and loss of efficiency of conversion of voltages. Heating losses occur according to the power equation (Equation 5-15).

$$P = I^2 R \qquad (5\text{-}15)$$

where:
P = power (heating) loss (Watts = Joule/second).
I = current (Amperes = Coulomb/second).
R = resistance (Ohms = Joule•seconds/Coulomb$^2$).

According to Equation 5.15, a small resistance in the iron core could lead to substantive eddy currents even at low voltages.

The time varying voltage in the primary winding of a transformer induces a magnetic flux in the iron core. The lines of flux cut across the secondary winding and induce a time varying voltage that opposes them. Ideal transformers follow the relationship (Equation 5-16):

$$V_{secondary}N_{primary} = V_{primary}N_{secondary} \qquad (5\text{-}16)$$

where:
V = voltage (Volts).
N = the number of turns in the coil.
Primary refers to the input side.
Secondary refers to the output side.

In this equation the secondary voltage is controlled by the ratio of the secondary to primary number of turns in the coils. A corollary of Equation 5.16 is the relationship between current flow and the number of windings (Equation 5-17) (Middleton 1986).

$$I_{primary}N_{primary} = I_{secondary}N_{secondary} \qquad (5\text{-}17)$$

where:
I = current (Amperes).
N = the number of turns in the coil.
Primary refers to the input side.
Secondary refers to the output side.

In this equation, current in the secondary side is controlled by the ratio of the number of primary to secondary turns in the coils. Circuits containing transformers that convert current from inconsequential levels to levels above 15 to 20 mA pose particular concern because of exceeding the let-go level for a substantial proportion of the population and the potential to cause electrocution. The output of such a circuit would decrease voltage correspondingly as current increased.

The common factor in Equations 5-16 and 5-17 is the conservation of energy on both sides of the transformer. Energy in electrical systems usually is related through equations for power, in this case (Equation 5-18):

$$P = VI \qquad (5\text{-}18)$$

where:
P = power (Watts = Joule/s).
V = voltage (Volts = Joule/Coulomb).
I = current (Ampere = Coulomb/second).

Equation 5-18 indicates that the product of voltage and current in wires on both sides of the transformer is the same value. This means that the difference in current flow from one side to the other must occur through an increase in voltage.

Windings in transformers used in bathrooms provide the same voltage on the output side as is provided on the input side. This type of transformer isolates current flow in equipment used in the bathroom through that receptacle. This also enables control of power to the transformer by the position of the light switch.

Usually, input and output voltages differ. Step-up transformers increase voltage on the output side. These create the high voltages used in electrical transmission and distribution. High voltage, low current transmission minimize losses from heating according to Equation 5.15. Similarly, a step-down transformer reduces voltage at the distribution end of the transmission line. Associated with these lower voltages is the correspondingly higher current flow.

In isolation transformers, the primary windings and secondary windings are located at different positions on the core, and hence are independent of each other. In the case of autotransformers, the windings are not independent of each other (Palmquist 1988). That is, primary and secondary voltages derive from the same winding. One input terminal is common with one output terminal.

Transformers discussed thus far rely on a metal core to transfer energy from one set of wires to another. Another type of transformer is the current transformer. The current transformer uses a coil of wire or a core (loop) containing a coil of wire to intersect the path of a conductor or group of conductors. The time-varying magnetic field surrounding the conductor(s) induces a current in the coil or core containing the coil. The clamp-on ammeter uses a loop containing a retractable section for positioning around the conductor under study. The ground fault circuit interrupter (GFCI) likewise uses a current transformer to monitor current imbalances in groups of conductors. Imbalance leads to net current flow which, in turn, induces a current in the monitoring circuit of the GFCI that activates the electronics that open the circuit.

## Induction and Stored Energy

Induction usually is associated with transformation of energy from different forms into electrical energy, and transformation of voltage and current at constant level. Induction also is involved in energy storage through production of a magnetic field around the conductor. This energy is "given back" when current flow in the circuit ceases. At that moment, the lines of magnetic flux collapse into the coil or conductor. The movement occurring during collapse constitutes a magnetic field that cuts the coil turns and induces a voltage in the turns if the geometry is appropriate. This process is known as self-induction (Middleton 1986). This voltage can cause arc flash and arc blast at the switch contacts, depending on speed of opening.

Induction can occur at the most inopportune of times. These can have implications relating to shock and electrocution. Energy stored in a magnetic field created by an inductor or system of inductors is given by (Halliday and Resnick 1962):

$$E = \frac{LI^2}{2} \tag{5-19}$$

where:
E is energy (Joule).
L is inductance (Henry = Joule•second$^2$/Coulomb$^2$ ).
I is current (Ampere = Coulomb/second).

Inductance is influenced by the number of coils in a section of wire, length of the coil, and cross-sectional area of the coil. The number of coils is influenced by the length and the cross-sectional area of the wire. The magnetic field can store a much larger amount of energy than an electric field (Halliday and Resnick 1962).

Opening of the secondary circuit of a current transformer (the circuit containing the coil of wire or the loop and coil) under load must never occur because of the likelihood of arc flash/arc blast

(Palmquist 1988). (Load means passage of current through the wire surrounded by the coil of wire or the loop and coil.) Current in the primary line is present at all times and acts as the exciting current. This current can raise the flux and induced voltage in the secondary to a high value. This could lead to arc generation at the gap in contacts of the secondary circuit. High voltage capable of causing injury and damage to the iron in the loop due to flux saturation could develop in the secondary circuit.

A second example of unwanted induction can occur during shutdown of three phase, high-voltage systems. The conductors in the three phases often are positioned in parallel to each other in runs involving considerable distance. Opening the breaker to shut down an individual phase leaves that conductor potentially ungrounded and without a load. An ungrounded conductor in parallel with a conductor under load is susceptible to induction to the potential of the conductor under load. Opening the circuit in the last conductor can leave induced energy in the remaining two.

Similar circumstances have occurred during construction and maintenance of high-voltage power lines (NIOSH 2000). Ungrounded conductors, especially containing coils of wire, and guy wires and vehicles have become energized through induction. The magnitude of the induced voltage is related to the proximity to the energized conductor and the voltage carried by it. Induction can occur between power line conductors, and conductors and guy wires.

When conducting objects are immersed in an AC electrical/magnetic field, currents, and voltages are induced in them (DOE 1986). These induce current to ground (short-circuit current) (Table 5.4). Induced current increases linearly with increase in the strength of the electromagnetic field. That is, a field of 10 kV/m induces 10 times the induced current.

Table 5.4 indicates that induced current is related to physical dimensions of the object. That is, the induced current is larger in large objects. The key to understanding the implications of Table 5.4 is to determine under what conditions electromagnetic fields reach the magnitude of 1 kV/m. Large fields are expected intuitively to be present around high-voltage transmission lines. Under these conditions, the object is likely to be a considerable distance from the conductor because of distance above the ground. However, other conditions are possible where close proximity is possible to high-voltage insulated conductors in underground vaults (Table 5.5).

Results provided in Table 5.5 are based on area measurements. Actual estimates of personal exposure in these fields are somewhat lower (IERE 1988). Area measurements provide a measure of the environment experienced by inductors mentioned in Table 5.5 under the same conditions.

A third example of unwanted induction has resulted from use of portable electrical generators following outages involving downed power lines (NIOSH 2000). This has occurred when the main breaker at the boundary of the service has remained in the closed position. Backfeed into the service line has energized transformers and high-voltage circuits containing downed and ungrounded lines in the distribution system. Ungrounded lines in this configuration have caused electrocution when individuals handling the lines completed the path to ground.

A fourth example involved opening the main breaker upstream from transformers and switches in a distribution system. At the time of opening the breaker, the magnetic field remains around conductors in the distribution system. This energy must dissipate by contact with a ground strap. Similarly, opening a circuit breaker in the distribution system does not dissipate energy from conductors in the path between the opened main breaker and the opened circuit breaker.

## Capacity
Capacity is the process leading to the temporary storage of electrical energy in an electrical field. This energy is "given back" later, as in the case of inductance (Gibilisco 1997). Like inductance, capacitance can appear when not wanted. As well, the effect of capacitance in AC circuits becomes more important with frequency.

A capacitor is a device or structure that has the capacity to store electrical charge. A capacitor is constructed from two conductors positioned in close proximity, and separated by a nonconducting material (dielectric). The larger the surface area of the conductors, the greater the capacity to store electrical charge. Each conductor carries electrical charge opposite that of the other. The conductors can accumulate charge to the level at which the insulating property of the dielectric fails and charge transfer occurs. A capacitor charges to the same voltage as the battery or energy source powering the circuit

## Table 5.4
## Induced Current to Ground (60 Hz Field of 1 kV/m)

| Example | Induced Current mA |
|---|---|
| Person (1.75 m height) | 0.016 |
| Farm tractor | 0.1 |
| Station wagon | 0.11 |
| Camper truck | 0.28 |
| Farm combine | 0.3 |
| Large school bus | 0.41 |
| Large tractor trailer | 0.63 |

Source: DOE 1986.

(Middleton 1986). Hence, the critical issue in capacitor design is to optimize charge storage. Charge storage (capacitance) increases as the distance between the conductors decreases. Voltage on the conductors also decreases as the space decreases. Capacitance also depends on the dielectric and the shape of the conductor.

One of the earliest capacitors was the Leyden jar (Olivo and Olivo 1984). This consists of a stoppered glass jar whose inner and outer surfaces are covered with metal foil. A metal chain connects the stopper to the foil on the inside of the jar. More recent designs contain metal foils separated by a layer of oil-soaked paper and rolled into a cylindrical shape. Polymeric materials are now used as the dielectric (Gibilisco 1997).

Capacitance and Stored Energy
Electrical energy stored in a capacitor following the charging phase is released into the circuit once the circuit is opened (Palmquist 1988). This energy is capable of sustaining operation of equipment, as well as spark generation, depending on rate of discharge. Energy stored in a capacitor or charged system of capacitors is given by (Halliday and Resnick 1962):

$$E = \frac{CV^2}{2} \tag{5-20}$$

where:
E = energy (Joules).
C = capacitance (Farads = Coulomb$^2$/Joule).
V = voltage (Volt = Joule/Coulomb).

The electric field stores this energy. Capacitors are present in computer monitors and television sets containing cathode ray tubes, resistance welders, induction heaters, power stabilizers in electrical distribution systems, and conveyor-drive power supplies (UAW-GM National Joint Committee on Health and Safety 1985). Large power transformers can act as capacitors and store energy in some cases.

## Table 5.5
## Electric Fields from Various Sources

| Source | Electric Field kV/m |
|---|---|
| **Residential Exposure** | |
| Areas in the home (middle of rooms) | 0.008 to 0.013 |
| Electric broiler (30 cm from appliance) | 0.13 |
| Electric toaster (30 cm from appliance) | 0.04 |
| Electric blanket (30 cm from appliance) | 0.24 |
| Electric blanket (near heating wires) | 10 |
| | |
| **Workplace Exposure** | |
| 115-230 kV linemen | 1.6 |
| 115-230 kV substation | 1.5 |
| 400 kV substation | |
| • inspection tasks, time-weighted average | 4.7 |
| • everyday work, time-weighted average | 6.1 |
| • revision, time-weighted average | 6.7 |
| • testing, time-weighted average | 2.8 |
| 500 kV linemen | 6.3 to 7.5 |

Sources: DOE 1986; Knave et al. 1979; Stopps and Janischewskyj 1979; WHO 1984.

Capacitance also can develop in unusual situations. Capacitance exists between isolated, insulated conductors in a grounded metallic raceway and between individual conductors and the metal raceway (Palmquist 1988). The insulation on the conductor acts as the dielectric. The raceway must be grounded to prevent severe shock related to charge accumulation. Coaxial high-voltage cables are also subject to capacitive effects and high-voltage discharge.

Capacitance also can develop in non-electrical equipment. This includes metal structures containing dielectric materials in which electrostatic charge can accumulate (Jones and King 1991). A metal structure and a non-grounded metal surface in close proximity and separated from it by a dielectric material, such as a gasket can form a capacitor. Capacitive discharges are responsible for more than

90% of all dust and vapor ignitions due to electrostatic discharge. The minimum voltage needed to jump an air gap of 0.1 mm is 350 V. Minimum ignition energy for powders is about 5 mJ.

An unusual example of a capacitive system was a vacuum cleaning system constructed from plastic sewer pipe. The plastic pipe was held in position on a grounded steel frame by metal brackets. This system was used for cleaning metal shavings and fragments in an aluminum fabrication facility. Suction in the system created tremendous velocity and accompanying accumulation of electrostatic charge due to collisions with surfaces. Velocity in the system was insufficient to prevent deposition of the metal fragments onto the surfaces of the sewer pipe. This deposition led to formation of a capacitor containing isolated conductors. The charged metal debris formed one plate. The plastic pipe and nonconductive debris formed the dielectric, and the metal brackets holding up the pipe formed the opposite metal plate. Tremendous electrostatic discharges were observed in this system.

Electrostatic charge is generated in circumstances where contact and separation occur between dissimilar materials. This is especially evident during materials handling involving pneumatic and belt conveying systems. Generation of electrostatic charge also occurs in web-based operations including printing, and plastic and paper handling, especially where highly insulating materials and high-speed motion are present.

## Batteries

Some industrial facilities contain DC systems that utilize electrical energy from storage batteries. Batteries are usually part of backup systems. In some industrial facilities banks of batteries are integral to the process. Electrical generation systems utilize batteries to produce the magnetic field in the rotor. Other electrical installations also utilize DC current provided from batteries.

A major concern in battery installations is shorting between terminals and uninsulated buss bars. Shorting can occur following contact caused by dropped metal tools, such as screwdrivers and wrenches. In systems not protected by overcurrent devices, massive current flow can occur. These, in turn, can lead to explosion of the battery from rapid build-up of heat and gases.

Batteries contain a rating for delivery of current in units of ampere-hours or milliampere-hours. Performing the conversion indicates the ultimate unit is Coulombs, the unit of charge. The battery rating indicates at a superficial level that a battery will provide a current flow of some value for an hour. Within limits of delivery capability of the battery, the current will decrease over a longer period of time and increase over a shorter period. The question is what is the maximum current and the shortest period over which this can occur. The current deliverable over a very short period of time as would occur during the condition of a short is not immediately known.

A recent article highlights the concern about battery safety and rapid discharge (Reif et al. 2010). This article reflects concerns in the deepwater, oceanic environment. Energy density measured in Ampere-hours/gram of weight is an important consideration in choice of battery. Lithium batteries contain more energy per unit of weight than conventional batteries.

## References

Accumulators: Product information. Houston, TX: Accumulators, Inc, 1998. [Website].

Agarwal, A.: Preventing and Solving Some Common Pneumatic Conveying Problems. *Powder Bulk Eng. 16(3)*: 25-37 (2002).

Anonymous: Drying Compressed Air. *Indust. Process Prod. Technol.: 12(5)*: 44-45 (2000).

Anonymous: Smooth Operators. *Hydraul. Pneumat. 56(6)*: 38-43 (2003).

ANSI/ASME: Valves – Flanged, Threaded, and Welding End (ANSI/ASME B16.34). New York: American National Standards Institute/American Society of Mechanical Engineers, 2009.

Ashby, D.M. Recent Developments in O-Ring Sealing Technology. *Hydraul. Pneumat. 52(5)*: (1999). 2 pp.

Bolton, W.: *Pneumatic and Hydraulic Systems*. Oxford, UK: Butterworth-Heinemann, 1997.

Carson, J.W. and J.R. Johanson: Design of bins and hoppers. In *Materials Handling Handbook*, 2nd ed., Kulwiec, R.A. (Ed.). New York: John Wiley & Sons, 1985. pp. 901-940.

Carson, J.W. and R.T. Jenkyn: How to Prevent Silo Failure With Routine Inspections and Proper Repairs. *Powder Bulk Eng. 4(1)*: 18-24 (1990).

Carson, J.W. and T. Holmes: Why Silos Fail. *Powder Bulk Eng.. 15(11)*: 31-43 (2001).

Coates, D.: Alternator. In *Microsoft ® Encarta ® Reference Library 2002*. Redmond, WA: Microsoft Corporation, 2001. [CD-ROM].

Craig, D.A. and R.J. Hossfeld: Measuring Powder-Flow Properties. *Chem. Eng. 16(9)*: 41-46 (2002).

Crouch, C.: Safely Handling Your Powder with a Closed-Loop Pneumatic Conveying System. *Powder Bulk Eng. 14(3)*: 43-56 (2000).

Debban, H.L. and L.R. Eyre: Steam Distribution Systems: Managing and Preventing Condensate-Induced Water Hammer. *Prof. Safety 37(5)*: 34-37 (!997).

DOE: Electrical and Biological Effects of Transmission Lines: A Review (DOE/BPA-524), Revised. Portland, OR: U.S. Department of Energy/Bonneville Power Administration, 1986.

Faye, T. and H. Loiterman: Alleviating Unsteady-State Flow Headaches. *Chem. Eng. 108 (10)*: 96-103 October (2001).

Frenck, J.P.: Making the Most of Valves. *Chem. Eng. 107(5)*: 66-73 (2001).

Galuska. R.: Finding and Fixing Common Rotary Airlock Valve Problems. Valve Desktop Reference. *Supp. Powder Bulk Eng.* January, 1999. pp. 6–11.

Garvin, W.L.: Understanding Safety Valves. *Prof. Safety 27(4)*: 47-52.

Gere, R. and T. Hazelton: Rules for Choosing a Fire-Resistant Hydraulic Fluid. *Hydraul. Pneumat. 46(4)*: 1993. 5 pp.

Gibilisco, S.: *Teach Yourself Electronics*, 2nd Ed. New York: McGraw-Hill, 1997.

Graham, F.D.: *Power Plant Engineers Guide* (rev. by C. Buffington). Indianapolis, IN: The Bobbs-Merrill Co. Inc., 1983.

Gravell, B.: Use Flexible Bulk Containers Safely. *Chem. Eng. 107(9)*: 92-96 (2001).

Halliday, D. and R. Resnick: *Physics for Students of Science and Engineering*, Part II, 2nd Ed. New York: John Wiley & Sons, Inc., 1962.

Harkin, R.: Factors to Consider When Selecting a Diverter Valve. *Powder Bulk Eng. 18(3)*: 35-40 (2004).

Heney, P.J.: Drying Your Compressed Air Will Save Real Money. *Hydraul. Pneumat. 56(3)*: 32-41 (2003a).

Heney, P.J.: A Little Bit of Support Can Be a Good Thing. *Hydraul. Pneumat. 56(12)*: 39-41 (2003b).

Hesketh, D.: Bulk Bag Discharging: Tools For Dust and Static Control. *Powder Bulk Eng. 15(5)*: 95-103 (2001).

Hitchcox, A.L.: Hydraulic Fuses Add Safety and Control to Circuits. *Hydraul. Pneumat. 60(1)*: 12-14 (2007).

Hohensee, P.: Hydraulic Hand Tools: Some Safety Basics. *Heavy Equip. Guide 14(6)*: 15-16 (1999).

IERE: Epidemiological Studies Relating Human Health to Electric and Magnetic Fields: Criteria for Evaluation. Palo Alto, CA: International Electricity Research Exchange (Electric Power Research Institute), 1988.

ISA: Control Valves (Chapter 7). In *Practical Guides For Measurement and Control*. Research Triangle Park, NC: Instrumentation, Systems, and Automation Society, 1998.

Jenkyn, R.T. and D.J. Goodwill: Silo Failures: Lessons to Be Learned. *Eng. Digest.* September, 1987. pp. 17-22.

Jennings, P. and J. Boteler: Soak up Surges in Liquid Systems. *Chem. Eng. 108(7)*: 105-108 (2001).

Jensen, C. and J.D. Helsel: *Fundamentals of Engineering Drawing*, 4th Ed. Westerville, OH: Glencoe/McGraw-Hill, 1996.

Johanson, H.: Motion Control via Electro-Servo-Pneumatics. *Hydraul. Pneumat. 55(6)*: 20-25 (2002).

Jones, T.B. and J.L. King: *Powder Handling and Electrostatics*. Chelsea, MI: Lewis Publishers, Inc., 1991.

Juvinall, R.C. and K.M. Marshek: *Fundamentals of Machine Component Design*, 2nd Ed. New York: John Wiley & Sons, 1991.

Knave, B., et al.: Long-Term Exposure to Electric Fields: A Cross-Sectional Epidemiologic Investigation of Occupationally Exposed Workers in High-Voltage Substations. *Scand. J. Work Environ. Health 5*:115-125 (1979).

Kuchta, J.M.: Investigation of Fire and Explosion Accidents in the Chemical, Mining, and Fuel-Related Industries — A Manual (Bull. 680). Pittsburgh, PA: U.S. Department of the Interior, Bureau of Mines, 1985.

Lees, F.P.: *Loss Prevention in The Process Industries*, 2nd Ed. London: Butterworth-Heinemann, 2001.

Messinger, M.: How To Recognize and Avoid the Common Causes of O-Ring Failure. *Hydraul. Pneumat. 53(2)*: (2000). 6 pp.

Meuller, T.: Following Rules Maximizes Hose Life. *Hydraul. Pneumat. 55(9)*:39-42 (2002).

Middleton, R.G.: *Practical Electricity*, 4th Ed. Revised by L.D. Meyers. New York: Macmillan Publishing Company, 1986.

Miller, H.L.: Heavy-Duty Control Valves Live Long and Prosper. *Chem. Eng. 109(11)*: 78-80 (2002).

MSHA: Think "Quicksand": Accidents around Bins, Hoppers and Stockpiles, Slide and Accident Abstract Program. Arlington, VA: U.S. Department of Labor, Mine Safety and Health Administration, National Mine Health and Safety Academy, 1988.

Naberhaus, P.: Selecting a Dome Valve for Tough Bulk Solids Applications. *Powder Bulk Eng. 18(9)*: 23-30 (2004).

Nelson, C.A.: *Millwrights and Mechanics Guide*. New York, NY: The Bobbs-Merrill Company, Inc., 1983.

NIOSH: Worker Deaths by Electrocution (DHHS (NIOSH) Publication 2000-115. Cincinnati, OH: National Institute for Occupational Safety and Health, 2000. [CD-ROM].

Olivo, C.T. and T.P. Olivo: *Fundamentals of Applied Physics*, 3rd Ed. Albany, NY: Delmar Publishers Inc., 1984.

OSHA: Selected Occupational Fatalities Related to Grain Handling as Found in Reports of OSHA Fatality/Catastrophe Investigations. Washington, DC: U.S. Department of Labor, Occupational Safety and Health Administration (U.S. DOL/OSHA), 1983.

OSHA: Concepts and Techniques of Machine Safeguarding (OSHA 3067 Revised). Washington, DC: U.S. Department of Labor, Occupational Safety and Health Administration, 1992.

Palmquist, R.E.: *Electrical Course for Apprentices and Journeymen*. Revised by J.A. Tedesco. New York: Macmillan Publishing Company, 1988.

Pfeiffer, J.W.: Venting Your Way To better Rotary Valve Performance – Part I. Valve Desktop Reference. *Supp. Powder Bulk Eng.* January, 1999. pp 30-33..

Pittenger, B.H. and J.W. Carson: How to Minimize Feed Segregation to an Agglomerator – Part I. *Powder Bulk Eng. 22(2)*: 21-26 (2008a).

Pittenger, B.H. and J.W. Carson: How to Minimize Feed Segregation to an Agglomerator – Part II. *Powder Bulk Eng. 22(3)*: 55-59 (2008b).

Purutyan, H., B.H. Pittenger, and J.W. Carson: Six Steps to Designing a Storage Vessel That Really Works. *Powder Bulk Eng. 13(11)*: 56-67 (1999).

Raczkowski, G.: *Principles of Machine Dynamics*. Houston, TX: Gulf Publishing Company, 1979.

Reif, R.H., M. Liffers, N. Forrester, and K. Peal: Lithium Battery Safety. *Prof. Safety 55(2)*: 32-37 (2010).

Ruel, M: Control Valve Health Certificate. *Chem. Eng. 108(11)*: 62-65 (2001).

Sahoo, T.: Pick the Right Valve. *Chem. Eng. 111(8)*: 34-39 (2004).

Schmidt, M.: Selecting Clean Valves. *Chem. Eng. 108(6)*: 107-111 (2001).

Schneider, R.T.: Coalescing Filters Capture Oil Aerosols. *Hydraul. Pneumat. 55(3)*: 14-15 (2002a).

Schneider, R.T.: Designing With Vacuum and Suction Cups. *Hydraul. Pneumat. 55(7)*: 16-20 (2002b)

Shackelford, J.F.: *Introduction to Materials Science for Engineers*, 2nd Ed. New York: Macmillan Publishing Company, 1988.

Skousen, P.L.: *Valve Handbook*. New York: McGraw-Hill, 1998.

SMC: *Basic Pneumatics, A Manual for Fluid Power Components and Practical Applications*. Mississauga, ON: SMC Pneumatics (Canada) Ltd, 2000.

Solt. P.E.: Pneumatic Points to Ponder —Fans, Blowers, and Compressors in Your Pneumatic Convey-ing System. Pneumatic Conveying Components Desktop Reference, January, 2001. pp. 10-20.

Stepanek, D.: Control Valves for Real-World Service. *Chem. Eng. 109(3)*: 103-107 (2002).

Stewart, H.L.: *Pneumatics and Hydraulics*, 4th Ed. Revised by T. Philbin. New York: Macmillan Pub-lishing Company, 1987.

Stopps, G.J. and W. Janischewskyj: Epidemiological Study of Workers Maintaining HV Equipment and Transmission Lines in Ontario. Montreal QC, 1979. Canadian Electrical Association Re-search Report.

Tarr, W.: Consider Designs When Selecting Your Next High-Speed Belt Drive. *PEM Plant Eng. Maint. 26(4)* 17-18 (September 2002).

Terzaghi, K., R.B. Peck and G. Mesri: *Soil Mechanics in Engineering Practice*, 3rd Ed. New York: John Wiley & Sons, Inc., 1996.

Totten, G.E., R.J. Bishop Jr., R. Suzuki, and Y. Tanaka: Air Entrainment – How It Happens, How to Avoid It. *Hydraul. Pneumat. 54(7)*: 39-41 (2001).

Tunnicliffe, G. and M. Thomson: Explosion Protection for High-Containment Impact Milling Sys-tems. *Powder Bulk Eng. 18(6)*: 23-29 (2004).

UAW-GM National Joint Committee on Health and Safety: *Lockout*. Detroit, MI: UAW-GM Human Resources Center, 1985.

Vickers: *Vickers Industrial Hydraulics Manual*. Rochester Hills, MI: Vickers, Inc., 1992.

Vickers: *Vickers Mobile Hydraulics Manual*. Rochester Hills, MI: Vickers, Inc., 1998.

Wahl, W.D.: Properties of bulk solids. In *Materials Handling Handbook*, 2nd. ed., Kulwiec, R.A. (Ed.). New York: John Wiley & Sons, 1985. pp. 882–900.

Walton, H.: *The How and Why of Mechanical Movements*. New York: Popular Science Publishing Company, Inc., 1968.

Ware, B.: How Vertical Flow Effects and Material Characteristics Affect Discharge–Part I. *Powder Bulk Eng. 13(4)*: 17-24 (1999a).

Ware, B.: How Vertical Flow Effects and Material Characteristics Affect Discharge–Part II. *Powder Bulk Eng. 13(5)*: 27-36 (1999b).

WHO: Environmental Health Criteria 35: Extremely Low Frequency (ELF) Fields. Geneva: World Health Organization, 1984.

Wilkins, D.G.: Designing Your Dust Collector Hopper for Mass Flow. *Powder Bulk Eng. 18(10)*: 31-37 (2004).

Yokel, F.Y. and R.M. Chung: Proposed Standards for Construction Practice in Excavations. *Prof. Safety 28(9)*: 34–39 (1983).

Zappe, R.W.: *Valve Selection Handbook*, 4th Ed. Houston, TX: Gulf Publishing Company, 1999.

Zink, M.: Match Characteristics to Application Needs. *Hydraul. Pneumat. 55(5)*: 31-35 (2002).

# 6 Integration and Control in Systems

## INTRODUCTION

Machinery and equipment usually contain integrated subsystems. Today's vehicles provide excellent examples of the application of the concept. Some of the subsystems perform discrete functions, and some are involved in monitoring and controlling inputs, internal functions and outputs. Oftentimes the machine or equipment is greater than the sum of the contributions from the subsystems.

## Integration of Subsystems

Until recent years, the link between the subsystems was the human operator. The human operator would process information provided by sensory devices and act through actuators to effect change to the status quo. The subsystems operated discretely; that is, independently of each other. Now, machinery and equipment are greater than the sum of the parts contributed by subsystems. The reason is that the activity of the subsystems is coordinated by an on-board or remote microprocessor or installed computer system. The microprocessor or computer receives input from many more sensing devices and at a rate greater than the human brain can accommodate. The microprocessor processes this input, makes decisions based on instructions and algorithms programmed into memory and effects actions by actuators, all the while providing the human supervisor a report on its actions.

The automobile provides an excellent illustration of the positive and negative aspects of this development. Not long ago, the electrical system in the average vehicle consisted of a battery, alternator, a few motors, switches, relays, solenoids, fuses ,and lights (Freund et al. 1989). This era has long since passed. Next to fuel, electronic components and circuits have become the most important systems in the operation of a vehicle. The rate of development of electronic systems has far exceeded that of other automotive systems. This situation is parallel to what has occurred across industry.

Developments in automotive technology illustrate another trend. Electronic components are rapidly replacing mechanical ones in the automobile and many other types of equipment (Duffy 1995). Modern vehicles contain at least one computer. The computer monitors inputs concerning quality of fuel, combustion and engine performance, composition of exhaust gases, engine speed, vehicle speed, fuel mixture, and temperatures, among others. The computer, in turn, activates and controls the activity of actuators to which the operator normally has no access. Whereas previously automotive systems functioned almost independently from each other, various subsystems now operate in coordinated manner to increase efficiency and economy of operation.

These developments are happening due to incentives to contain cost, simplify operation, and optimize efficiency. At the same time, this change has introduced layers of complexity between the operator of the vehicle and the actuator. To illustrate, when automotive subsystems were isolated, discrete units, diagnosis and repair were well within the capabilities of a trained and qualified mechanic. With subsystems integrated through a computer, diagnosis and repair without connection to a diagnostic platform have become considerably more difficult.

The current situation affecting automotive manufacturers concerning inconsistent performance or accelerator and braking systems best illustrates the cost side of the situation. Prior to implementing electronic and computer technology on a vehicle-wide basis, some products had an unparalleled reputation for reliability. The current situation has highlighted problems that can occur with the changeover in engine control through the accelerator pedal leading from mechanical linkages to electronic sensors. This situation thus far has led to several years of problems concerning allegations of unpredictable and uncontrollable lurching and acceleration. Additional problems have surfaced regarding performance of the anti-lock braking system and the system that protects against roll-over on curves. Public discussion about these problems has suggested a link to the software that controls actions of the hardware.

A parallel example concerned the operation of therapeutic accelerator units that delivered high doses of ionizing radiation to cancer patients (Leveson 1995). These units seemingly and unpredictably went out of control and killed six patients through overdoses. Two of the most important causal factors identified in subsequent investigation and analysis were overconfidence and over-reliance on software to control this equipment. These factors were identified in subsequent accidents in other situations. Hence, this issue is not isolated, but is part of a theme involving computerized control of equipment. Another pertinent comment made by investigators and analysts regarding these accidents is the need for defensive design in the software. This has widespread applicability. Defensive design involves self-checks, and other forms of error detection and error-handling.

## Control Devices

Control devices activate and deactivate circuits involving industrial equipment. Today these circuits are mainly electrical, but also can involve hydraulic and pneumatic circuits and circuits containing process fluids and flowable solid materials. Electronic devices are increasingly used to control all of these circuits.

## Levers and Simple Machines

The first control devices were lever operated. The lever and the accompanying application of the principles applied in simple machines (outlined in Chapter 5) provided the force needed to move heavy structures.

One of the first industrial applications of the lever in control devices occurred in the mechanisms used to open and close the gates in locks (Figure 6.1). The gates are heavy and must move against the presence of water in the path of motion and form a tight seal at the point of contact. The gate arm is a lever that connects the lock gate to the mechanism that opens and closes it. The operating mechanism uti-

Figure 6.1. Use of levers provides mechanical advantage to control the gates in locks.

lizes rotary movement around a pivot and a gear mechanism to exert the force needed to open and close the lock gate.

Another familiar example is the mechanism used to open and close railway switches (Figure 6.2). Two basic manually operated designs are currently in use. The more widely recognized design contains a lever that the operator lifts, moves through 180° from the initiating to the final position, and then lowers. The switch mechanism rotates in the vertical orientation around the frame of the switch stand. The second device contains a weighted arm that the operator lifts from the initiating position and rotates through 180° to the final position. In this design, rotation occurs around a horizontal pivot. Ease of movement depends on lubrication of the moveable structure and the absence of ice, snow and debris between the track that could prevent full movement of the moveable structure.

### Hand-Operated Valves
Hand-operated valves (Figure 6.3) contain a wheel or lever that rotates around a pivot (Skousen 1998). Wheel-operated valves often require multiple turns in order to raise and lower the moveable element. The multiple turns needed to raise and lower the moveable element provide considerable mechanical advantage to the operator. The mechanical advantage in wheel-operated valves depends on the diameter of the wheel and the number of turns needed to move the moveable element from the open to the closed position. The mechanical advantage provided by the lever in quarter-turn valves is considerably smaller than that provided by wheel-operated valves. Mechanical advantage provided in operation of quarter-turn valves depends on the length of the handle. The wheel-operated moveable element is amenable to remote operation through the use of a continuous loop of chain.

### Wired Control Devices
Wired controls in electrical circuits (Figure 6.4) are the most commonly used and fundamental control devices worldwide (Rosenberg 1999a). They are highly visible features on control panels for industrial

Figure 6.2. Railway switch. The railway switching mechanism employs mechanical advantage to enable a person to move the heavy track to the desired position.

Figure 6.3. Hand-operated valves. Hand-operated valves employ a wheel or lever that rotates around a pivot to provide mechanical advantage.

Figure 6.4. Wired control circuit. Wired control circuits are common in industrial facilities. This photo illustrates the wired control in a lift station. Lift stations collect water and wastewater from the surrounding area and pump it to the gravity sewer at higher level.

machinery and equipment. They are also present in locations remote to control panels. These include the side of machines and equipment in locations beside or near the operating point chosen to enable rapid action, such as emergency shutdown (Figure 6.5). Casual observers may see these devices, yet not fully appreciate their significance in the overall operation. Operators and maintenance personnel who are highly familiar with the equipment and operate these devices by instinct can lose sight of their intended function.

Original devices used clock-like mechanisms (Figure 6.6) to monitor and track time, and switches and relays to establish and break electrical contact in control circuits. Classic examples of these devices include the timers on electric stoves and automatic washing machines. Electronic devices have largely supplanted the mechanical devices in this equipment.

Control Switches

Switches are among the most common of the fundamental control devices (Rosenberg 1999a). Switches used in electrical circuits experience the change in the status quo of the movement of electrical energy in the circuit. When moved to the closed from the open position, current rushes into the circuit beyond the switch. The in-rush current can be 10 to 15 times the current flowing during normal operation.

*Push-Button Switches*

Push-button switches are designed and manufactured in two main configurations (Rosenberg 1999a). These switches contain contacts and an operating mechanism. The contacts in these switches are either normally open (NO) or normally closed (NC). Normally refers to the condition of the contacts prior to imposing the action.

A spring-loaded return mechanism can return the contact to the pre-action position. This is the situation in ON switches. A brief push is all that is needed to activate the circuit. Some ON switches acti-

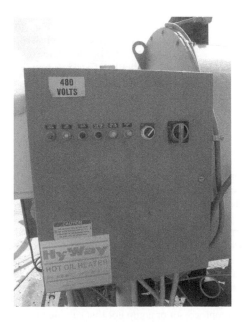

Figure 6.5. Local control panel. The local control panel provides the ability to effect control at the point of operation of machinery and equipment while also within the context of a central control panel.

Figure 6.6. Clockwork mechanisms were used in timers in control circuits. Electronic devices have supplanted mechanical controls.

vate the circuit only as long as depressed. This allows advancement of a machine (jog shuttle) but not operation. OFF switches often require manual resetting once depressed. That is, the mechanism remains in the depressed position (and the contacts opened), until pulled out to effect the reset manually.

A cylinder lock can immobilize these switches in a fixed position (Rosenberg 1999b). Immobilization prevents movement of the switch and hence the position of the contacts.

Industries use color coding (Table 6.1) to denote function of the control mechanism. Use of particular colors is not guaranteed across the spectrum of industry. Verification of intended meaning as exemplified by application of a particular switch in a circuit is essential in order to ensure safety during operation and maintenance. The machine tool industry assigns the colors mentioned in Table 6.1.

Selector Switches
Selector switches usually provide several options to the operator (Rosenberg 1999a). These switches typically contain two, three or four positions (OFF, ON, AUTO, MANUAL or HAND). Switching actions include manual and spring-loaded return to the starting position. The operator interface can include a knob or lever.

The ON position is a manual "on". The AUTO position implies an interface though a microprocessor or computerized control device, either locally in the machine or equipment or remote from it. The actions performed and actual control of a machine or equipment under AUTO control are not always known with certainty and become cause for concern because of the absence of certainty. The concern pertains to loss of control over the logic and actions programmed into the controlling device. This has occurred due to ad hoc, often unauthorized and often undocumented changes made through the interface to the controlling microprocessor or computer that suit the need of a particular situation.

A cylinder lock can immobilize the switch in a fixed position. Immobilization prevents movement of the selector and hence the position of the contacts.

## Table 6.1

## Colors Assigned to Control Switches (Machine Industry)

| Color | Function |
| --- | --- |
| Red | Stop or emergency stop |
| Yellow | Return or emergency return |
| Green/black | Start cycle |

Source: Rosenberg 1999a.

Solenoids
The solenoid is an essential device for the control of machines and equipment through electrical, and hydraulic and pneumatic circuits (Rosenberg 1999a). The solenoid utilizes electrical energy to cause mechanical movement. The solenoid consists of a frame of laminated steel, and a coil of wire inside which a moveable element (plunger) of iron or steel is positioned. Energizing the coil produces a magnetic field that attracts the plunger and causes movement into the coil. The force that produces this motion can be as high as 450 N (100 lb). The solenoid may contain a spring to return the plunger to the start position when current flow ceases. The device may also contain a second coil to produce movement in the reverse direction.

In-rush current is particularly high in large solenoids and can reach 15 times the current present during operation. Magnetic force created in the coil decreases rapidly with decreasing voltage. Low voltage in the control circuit and loss of pull are operating concerns with solenoids. Corrosion can cause the plunger to seize and fail to move.

Relays
The relay combines the solenoid with one or more electrical switches (Rosenberg 1999a). A control circuit connected to a relay opens or closes the contacts in a circuit of higher voltage. In-rush current is also a concern in the selection of relays.

Spring-loaded relays contain a return spring. The spring returns the plunger to the starting position following de-energization of the coil. These devices require continuous energization to maintain the plunger in the operating position.

Latching relays contain two control coils, one for movement in each direction. They also incorporate a mechanical latching device that holds the plunger in position. Some types lack the mechanical latch. In this case, movement of the plunger to a defined position requires selective energization of the coils. Under normal operation, energization of the coils occurs only during the time needed to move the plunger from one position to the other.

The contactor is a large relay (Figure 6.6). Contactors control high-energy circuits involved with lighting, heating elements, magnetic brakes, and with overload protection, motors and machinery. Contactors are electromechanical devices. They contain mechanisms to break the contact rapidly following de-energization of the electromagnet. Similarly, movement of the plunger in the relay is also rapid. The purpose for the rapid disconnection is to break the arc that establishes between the contacts during separation. This arc is responsible for the arc flash and arc blast that occur in electrical equipment. Note that these arcs can develop when the contactor opens under load. This is similar in concept to the arc created during welding during controlled separation of the electrode from the base metal. Some units contain arc-entrapment devices to minimize the impact of this situation.

Figure 6.7. Contactors. Contactors are large relays. The are used to establish electrical contact in circuits operating above household voltages (480 V and 600 V and higher). The contactor enables control of high-voltage circuits by low-voltage circuits, thereby increasing safety.

Status Annunciation

These devices often contain or are combined with indicator lights (Rosenberg 1999a). Indication of status is essential to ensure that the device is performing as intended. As with other mechanical devices, components in switches, solenoids, and relays are subject to weakening and failure of springs and corrosion of contacts.

Indication of status (annunciation) typically occurs through lights or more currently Light Emitting Diodes (LEDs). Indicator lights use color (Table 6.2) to communicate status to maximize the level of safety provided through the communication. Table 6.2 provides colors used in the machine tool industry.

Color-blindness is a major consideration in the use of these devices. Employers must ensure the ability of all individuals relying on these devices to be able to detect and interpret the information and warnings provided by them. Filaments in incandescent bulbs are subject to failure due to vibration. As a result, bulbs operating at 6 V to 8 V are the preferred type. Push-to-test devices provide the ability to confirm status of the indication, meaning that the bulb isn't energized or failure of the indication occurred due to failure of the bulb due to a failed filament or loose connection. Depressing the lens of the device passes a test voltage across the bulb.

## Servo Control Systems

The servo control system is a mechanism devised to ensure that a designated parameter in a machine or process remains constant (Stewart 1987). In simplest terms, a servo system acts in a loop containing the following principal sequential components: demand for change, action, feedback, compensation, feedback, compensation, and so on.

The simplest servo mechanism is a human-machine linkage. The human constantly monitors a sensing device or indicator, and makes corrections to maintain constancy in some parameter. This is what occurs when driving a vehicle not equipped with cruise control. The driver periodically gazes at

## Table 6.2
## Colors Used in Status Annunciation

| Color | Status |
|---|---|
| Red | Danger, abnormal operation or condition |
| Amber | Attention |
| Green | Safe condition, normal operation |
| White or clear | Normal condition |

Source: Rosenberg 1999a.

the speedometer and makes compensatory changes in foot pressure applied on the accelerator to maintain constant speed, or speed below a defined maximum, such as the speed limit. Moving up a steep hill at constant speed requires considerable increase in pressure on the accelerator, while moving downhill under the same constraints involves considerable decrease, and possible application of the brake, as well. Cruise control automates the process originally through mechanical and now through electronic devices.

The human-machine linkage in its various applications from controlling parameters in process operations to controlling the speed of an automobile to watching a pot on the stove, is prone to error. This results from the fact that the process is complex and subject to the vagaries and inefficiencies and ineffectiveness of human performance. Human performance is subject to many downgrading factors, as indicated in Chapter 4. There is much greater chance for overshoot or undershoot when a human operator is present, due to the considerably larger number of uncontrollable factors involved in dedication of the human organism to performance of repetitive tasks, especially monitoring ones.

What actually occurs in the human-machine linkage is more complex than indicated initially. An action (a deviation from the set-point for the system) results in a physically measurable quantity. The physically measurable quantity stimulates the human senses and initiates interpretation by the brain. Interpretation includes comparison of the deviation against the required value and a decision about the action needed to eliminate the deviation. Action leads to muscle contraction and movement of limbs which, in turn, lead to movement of an actuator in the machine. The latter directly affects the parameter of the machine under observation. Accuracy against the required outcome in order to minimize the need for further corrective action depends on the training, experience, skill, and motivation of the human operator. However, this situation is considerably more complex than reflected in the competence and motivation of the operator.

Accuracy of the action also depends on the human machine interface. This includes the manner in which the control provides tactile feedback and displays of monitoring instruments. Tactile feedback includes increasing resistance to produce greater effect. This feature is offered in some controllers for computer games. Controls and displays of information provided by monitoring equipment must reflect cultural conditioning. Cultural conditioning creates expectations in human operators. Cultural conditioning is reflected in design criteria presented in ergonomics textbooks.

A major consideration for the application of corrective action in a loop is the coarseness or fineness of the adjustment offered by the control system. A control system that allows overly coarse adjustment of the parameter relative to the deviation from the required value likely will not enable accurate correction on the first attempt. Undershoot or overshoot of the required value is the likely result, regardless of training, experience, and competence of the operator. Undershoot or overshoot leads to a new deviation from the required quantity and necessary repetition of the cycle. If the adjustment is too coarse for the deviation, the operator will not be able to match the required value. The system will require constant adjustment and the attained value will oscillate around the required value. At some point, the operator will become frustrated with the inability to control the system in a manner that provides success as a reward for the effort.

A control system that allows overly fine adjustment creates the opposite situation. The operator can position the adjustment accurately within the window of tolerance straddling the required value, but entering that window could require considerable time and effort in moving the controlling actuator. This again could necessitate repetitions of the cycle as the operator strives to make the appropriate adjustment, along with frustration with the inability to control the system in a manner that provides success as a reward for effort.

Setting an acceptable range of tolerance that straddles the set point within which the system is considered to be operating appropriately is critical for minimizing cycling and oscillation in a servo control system. This maximizes the achievability of the correction and reduces needless cycling due to undershoot or overshoot.

One feature offered by an intelligent system containing a human operator is the opportunity to change the set-point and the range of tolerance. A new set-point is also a deviation in condition from an existing set-point. The human operator then can decide the most effective strategy for attaining the new set-point through the minimum number of steps.

Mechanization of the process of control is readily possible. The simplest devices, such as the fly-ball governor, appeared in the earliest days of the steam engine (Stewart 1987). The fly-balls rotate around a shaft driven by the main shaft. As the speed of rotation increases, the balls move outward from the center of gravity of the rotation. This movement causes movement of the linkage that controls the position of the steam/fuel supply valve. The flow of steam or fuel, as the case may be, increases when the engine is under load, and decreases when the engine is coasting. Geared systems are available. These can sense the difference between a preset value and the actual value and send feedback to a correcting mechanism.

Small gasoline engines contain a vane attached by spring and linkage to the carburetor to control speed. Movement of the vane in response to airflow created by rotation of the flywheel controls tension on the spring, which in turn, increases or decreases the supply of fuel. As with the fly-ball governor, this system operates through mechanical linkages. As mentioned previously, electronic devices have superceded mechanical linkages.

The cruise-control system used in vehicles is another example of this evolution. The original system used mechanical components to monitor speed of movement. A vacuum system moved the accelerator linkage, and in turn, sped up or slowed down the engine. The driver overrode the system by exerting pressure on the accelerator or the brake. The modern iteration uses electronic devices to monitor speed and to control engine speed through the on-board computer.

An example of a servo control system that is left unattended for long periods is the temperature control system in home and office environments. This system uses a thermostat to control operation of the furnace or air-conditioner. The temperature control system operates at 24 V instead of the 110 V common in household circuits. This isolates the end-user from the higher voltage of the operational circuit. The 24 V circuit activates the solenoid that controls the flow of fuel to the combustion chamber or activates the refrigeration compressor of the airconditioner.

The original thermostat contained a sensor that used the movement of a strip composed of two pieces of dissimilar metals that are bonded together to open and close the electrical contacts. Metals expand and contract at different rates. Hence, the result of change in temperature on a strip of two metals that are bonded together is a slight curvature or change in curvature. The change in curvature opens or closes the electrical circuit. Newer versions of the thermostat utilize electronic sensing of current flow through a thermistor, an electronic component whose resistance changes with temperature. The electronics monitor current flow and activate the circuit at the set-point. The temperature control system operates through a predetermined increase or decrease in temperature. At the end of the range, the contacts open and disengage the solenoid controlling flow of fuel to the furnace or operation the refrigeration compressor.

Operation of the temperature control system under electrical or electronic control contrasts with the manually operated system in which the human operator controls startup and shutdown of the actuator. The electrical and electronic operators shut down the system at the predetermined setting at the end of the range of operation. The electronic controller is even more precise than the electrical one. Effectively, there is no overshoot or undershoot of the range. The frequency at which the system responds depends on the range permitted beyond the set-point and the rapidity of change of the set-point or deviations from the set-point.

Systems in which the set-point changes rapidly or in which deviations occur unpredictably are subject to overshoot or undershoot. A pH-maintaining system receiving inflow that is rapidly changing is an example of the latter. The servo control system must inject and blend stock solutions of acid or caustic or buffers rapidly in order to respond to the change in a flow-through process having little residence time.

In situations where small action of the control device produces large action of the actuator (that is, coarse control), overshoot or undershoot is also likely. In these situations, a back and forth movement of the signal is likely to occur as the controlling device locates the designated setting. There are circumstances where the designated setting lies between two operating points. In this case, the controlling device will cycle continuously. The latter situation will occur regardless of how coarse or how fine the adjustment becomes. There always will be a setting that will straddle two operating points. Setting the range appropriately is the key to minimizing cycling by the system as it seeks to achieve perfection (maintain control).

**Control Circuits**
Control circuits are the means to control the on-off, speed up-slow down, forward-reverse, in-out, pressurize-depressurize, open flow-close flow, and other actions of equipment and machines. Control circuits also control the movement of fluids in piping and other circuits. Understanding about control circuits is essential to understanding about the management and control of hazardous energy.

## Direct Actuation Control Circuits
Much of the preceding discussion concerned direct-actuation of control circuits. The actuator was the operator who acted through direct action or amplified control of equipment in the circuit (Figure 6.8). The operator exerted complete and direct control over the circuit. This concept has many applications in industry and is common in devices found in the home.

This type of circuit can contain safety devices, such as interlocks, fuses and circuit breakers. Interlocks interrupt the flow of energy (usually electrical current) when the circuit path is broken. Fuses and circuit breakers open the circuit based on designed-in parameters of performance. Refer to Chapter 7 for further discussion about these devices.

### Electrical Circuits
Electrical circuits provide many examples of simple, direct-actuation control circuits (Table 6.3). Control circuits also include the holding circuit and in the case of motors, overload protection. The holding circuit is a critical component in a control circuit (Figure 6.9) . The holding circuit may operate at the same or lower voltage than the device under its control. In the latter case a control transformer reduces the voltage.

The holding circuit contains a STOP device, a START device, and associated wiring. The STOP device is wired in series in the circuit and is normally closed (NC). All current in the circuit must pass through the STOP device. This configuration ensures that current flow in the circuit ceases on pushing the STOP button.

The holding circuit also contains a START device. This device is normally open (NO) and also wired in series. There is a normally open (NO) set of contacts wired in parallel with this device. These form the main component of the holding circuit.

Closing the START device energizes the contact device. This closes the contact in the contact device and enables current flow in the circuit regardless of the position of the contact in the START de-

Figure 6.8. Direct-actuation control circuit. The operator is the actuator and has direct control of the action of the circuit in this electric drill. Closing the switch activates the motor. Releasing the switch deactivates the motor.

vice. Pushing the START device is necessary to close the contacts that energize the device under its control.

The simplest control circuits operate at the same voltage as the device under control and involve a single phase. The next level of complexity involves three-phase conductors, and a control circuit operating at the same voltage and powered by two of the conductors. In more complex circuits, the holding circuit connects into the conductors feeding energy to the three phase device controlled by this circuit. Current flows in this circuit because the conductors are out of phase with respect to each other. In this configuration, the holding circuit operates at low voltage compared to the voltage used to energize the device.

The motor starter is similar to the contactor in design and operation (Rosenberg 1999b). The contacts are normally open (NO). Energization on starting the motor closes the contacts. The magnetic motor starter normally has three main contacts. The important difference between the motor starter and the contactor is the use of overload relays in the motor starter. These protect against overheating. These are located in the motor starter and remain normally closed (NC). These contacts will open in the event that one or more of the conductors overheats. Opening of one or more of these contacts will cause opening of the contacts in the motor control circuit and stopping of the motor.

Current draw is a reasonably accurate measure of the load on the motor and heat generation. Most overloads contain a thermally responsive element. Excess current flowing for a sufficient period causes the contact to open. High ambient temperature combined with heat caused by current flow also can cause opening of the contact. Some designs eliminate the role of ambient temperature in the role of the device as a monitor of high current flow.

Indirect Actuation Control Circuits
The direct-actuation approach to control limited the power of the actuation to the strength of the operator and to the force exerted by the amplifying system. A major consequence of this approach was expo-

## Table 6.3

## Simple Direct-Actuation Electrical Circuits

**Examples**

A push button switch used to energize a motor. The power supplied by the switch energizes a starter coil, closes the starter contacts and energizes the motor.

A thermostat contact that closes and energizes the coil of a contactor which closes contacts and energizes heating elements.

A limit switch contactor connected to energize the coil of a relay is held closed (operated) at the start of a cycle. A relay contact closes to energize the solenoid.

Source: Rosenberg 1999a.

sure of the operator to potential safety concerns because of potential close proximity to the operation produced by the actuation As well, this necessitated the presence of the operator at each location at which an actuator was present.

This approach was and remains entirely workable for simple configurations. As the need for more powerful actuation arose and the complexity of systems increased, the practicality of this approach became severely limited.

Another trend that limited the practicality of this approach was the centralizing of control. Operators became stationed in control rooms rather than roaming around the facility. Roaming by operators to complete a specific task in the days preceding availability of portable communication devices or wired communication devices at each location affected by the manual actuation created a break in communication between the lead operator in control of the system and the field operator who would execute the actuation at a location remote to the control room. This break in communication posed serious risks to process control and safe operation.

Coincident with these trends is the use of indirect-acting, low power subsystems to control the action of high-power (energy) ones (Figure 6.10). The high-power (energy) actuator can be a valve in the case of a fluid system, a switch, or voltage or current regulating device in the case of an electrical system. A small amount of work expended on the actuator in the low power (energy) produces a large change expressed through action in the high-power (energy) circuit. In all cases, the low-power (energy) circuit isolates the operator from the high-power (energy) circuit in space. This approach offers considerable benefit from the perspective of safety.

In some systems, the low power subsystem controls the high-power subsystem through the same medium, for example, a low-power fluid system controls a high-power fluid system. Hybrid control circuits are becoming increasingly common, as electronic circuits are integrated into controls of fluid and mechanical systems.

*Manual Subsystems*

The manual servo systems mentioned previously illustrate the concept of a low-power mechanical subsystem controlling a high-power subsystem, such as an engine or turbine. Human operators are unlikely to have the strength needed to effect change to the needed to operate the high-power (energy) actuator (Stewart 1987).

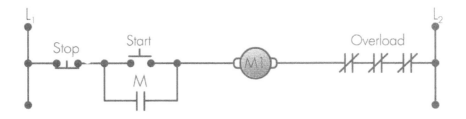

Figure 6.9. Holding circuit. The holding circuit is an essential component in the control of motors and machinery. In this configuration, the holding circuit is part of a direct-actuation control circuit. (Modified from Rosenberg 1999b.)

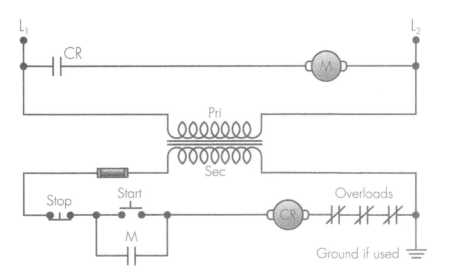

Figure 6.10. Indirect actuation control circuit. The low-power (energy) subsystem controls operation and change in the high-power (energy) actuator. (Modified from Rosenberg 1999b.)

*Mechanical Subsystems*
The mechanical servo systems mentioned previously illustrate the concept of a low-power (energy) mechanical subsystem controlling a high power (energy) system, such as an engine or turbine. Mechanical operators do have the strength to cause damage to the actuator. Overstressing the actuator must be a consideration during design of such subsystems (Stewart 1987).

A familiar example of a mechanically actuated high power subsystem is the "power steering" provided in many vehicles, especially large trucks. Finger effort is all that is needed to move the steering wheel when the vehicle is moving or even stationary. A hydraulic pump operated by a belt driven by the crankshaft powers a hydraulic motor in the car. This is the power in power steering. An axle containing a gear rotates along a rack and pinion device. Movement of the steering wheel actuates a control valve that regulates movement of the steering mechanism through the high pressure system. Attempts to move the actuator beyond the end of its range lead to slippage of the belt that powers the hydraulic pump or engine and a squealing sound as the system attempts to meet the demand placed upon it.

*Electrical Circuits*
The automobile provides a number of examples of low power (energy) subsystems that operate high power (energy) subsystems (Duffy 1995). The starting system for the engine uses a solenoid connected to the ignition switch to activate the starter motor. (A solenoid contains a soft iron plunger that moves following imposition of a magnetic field in a surrounding coil of wire.) The solenoid is a low current device. The starter motor is powerful enough to rotate the crankshaft of the engine at 200 revolutions/minute. This requires considerable power (energy) and use of heavy electrical cable in the connection to the battery. The starting system incorporates a low current circuit to activate a high current circuit. This design minimizes the use of heavy wiring, thus limiting the potential for heat losses, and short circuit and fire. At the same time, this isolates the operator from needless proximity to the heavy circuit.

High-voltage, three-phase systems contain a low-voltage holding circuit powered by a transformer (Figure 6.11). The transformer draws power across two of the phases of the high-voltage supply to the device controlled by the holding circuit. The voltage in the control circuit to which the operator may experience exposure is low compared to the voltage needed to operate the actuator. This design provides an external layer of safety to the operator of the equipment. As well, the high-voltage actuator is remote from the control device used to actuate the low-voltage circuit.

More powerful actuation creates its own safety risks. This is evident, for example, in the arc flash/arc blast accidents that have occurred in motor control centers and other high energy control circuits, about which much discussion has occurred in recent years.

*Fluid Power Circuits*
The pilot valve is a small valve whose operation requires low effort compared to that needed to operate large valves (Nelson 1983; Stewart 1987). Pilot valves activate the actuator of larger valve through movement of a plunger or exertion of fluid pressure (Figure 6.12). The piston makes mechanical contact with the actuator of the operated valve.

Output from pilot valves is used to control the position and hence the throughput of larger valves, and ultimately, operation of large-scale equipment. This means that the human operator can control the action of the equipment through relatively little effort, and at the same time can maintain precise control.

As well, the operated valve is located where required, as opposed to being near the human operator. This minimizes the need for high-pressure, large-diameter piping in the vicinity of the operator, since the pilot circuit requires only small-diameter piping. Excavators and other equipment of this type make extensive use of piloted circuits to operate high-pressure circuits. High-pressure circuits can operate well in excess of 34 MPa (5,000 lb/in$^2$). The operator of this equipment normally is not exposed to circuits operating at this pressure.

An important application of pilot valves is the servo control system. Servo control systems provide a large amount of force in the fluid power system for a small amount input of energy. Servo sys-

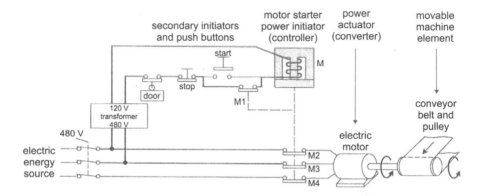

Figure 6.11. High-voltage electrical circuit containing low-voltage control circuit. The low-voltage control circuit controls actuation of the high-voltage circuit. This type of circuit is very common. (Modified from Hall 1992.)

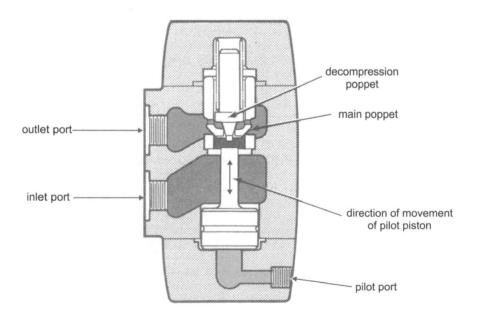

Figure 6.12. Piloted valve in a fluid power circuit. Valves in fluid power circuits contain spools, machined metal pieces that move back and forth. This movement controls fluid flow. The pilot valve provides the force to move the spool in the main valve. (Modified from Vickers 1992.)

tems also incorporate self-correction of overshoot or undershoot through a feedback mechanism (Vickers 1998). Feedback occurs on position of the valve spool, rather than the valve output.

The future for control devices for fluid power lies in hybrid subsystems. Most of these devices marry a mechanical component or an electrical component to a fluid power component.

Mechanical actuation of pilot valves or large-scale valves has existed for a considerable length of time. Mechanical actuation is often used for directional control (Stewart 1987). Directional controls operate through actuators such as roller cams, toggles, and trip pins. The latter are connected directly to the valve actuator (Vickers 1992).

Electrical actuators include motors and solenoids (Stewart 1987). Motors are used primarily on low-pressure valves, such as butterfly valves. Solenoids are more common. Solenoids can connect directly to the valve or to a mechanical operator, which, in turn, operates the valve mechanism. Solenoids normally are either on or off devices. Proportional solenoids contain solenoids that oppose each other. This enables precise positioning of spool movement, and hence control, in hydraulic valves. The relative energy passed through each solenoid controls the position of the spool. These units consume considerable electric power (energy).

Force motors are solenoid-like devices in which the plunger moves in one direction or the other, depending on direction of current flow. Pilot valves generally contain force motors in place of solenoids for bidirectional control. Hence, a small electrical signal from a controller can activate a force motor, that, in turn, actuates a pilot valve, that, in turn, actuates a large control valve. Piloting helps to minimize the size of the force motor and proportional solenoid. (Vickers 1998). These valves are used in hydraulic systems operating at pressures exceeding 20 MPa (3,000 lb/in$^2$).

A third variant on electromotive applications is the torque motor. The rotor of a torque motor is constrained to move through only a small fraction of the arc of a circle. Torque motors are reversible and consume low power. They are used with pilot valves in servo control systems (Vickers 1998).

## Process Systems

A manual operator is any device requiring the presence of a human to provide the energy to operate the valve as well as to determine position resulting from the action (Skousen 1998). Manual operators include devices, such as hand wheels and levers, to enable the action to occur under human capabilities of strength.

Generally manual operators are involved in on-off applications and simple throttling not requiring undue accuracy or immediate feedback. The advantage of manual operators is simplicity. The disadvantage is slow response and the need for considerable time to effect the change.

Actuators used on throttling valves move the moveable element in response to signals concerning flow, pressure, or temperature. Power sources for actuators include electrical, pneumatic, and hydraulic units. The actuator contains return springs that oppose the motion of the power source.

The combination of the throttling valve with an actuator creates the control valve (Figure 6.13) (Skousen 1998). An actuator is a device mounted on a valve that moves the valve to the required position in response to a signal using an outside power source. General service valves are usable in many applications without modification Most control valves will function satisfactorily in most applications. Stated more quantitatively, more than 80% of control valves operate at pressure less than 100 lb/in$^2$ (700 kPa) and temperature less than 200° F.(90° C) (Stepanek 2002). These operate between −45° C to 345° C (−50° F to 650° F) depending on rating.

A poorly functioning control valve can cause poor performance in a process loop (Ruel 2001). An estimated 25% of control valves in plants in North America perform below standards. Diagnostics performed on valves containing built-in sensors and microprocessors provide the basis for considerably improving performance by targeting service and maintenance only to valves requiring it. A healthy control valve reacts quickly to small changes in the output signal from the controller.

Actuators are critical elements in establishing and maintaining control in a process control loop (Skousen 1998). Actuators include pneumatically powered devices (diaphragm, piston, vane), electronic motor devices, and electrohydraulic devices. Process control loops include the actuator, a sensing device, and a controller. Sensing devices include temperature sensors, flow meters, pH sensors, pressure sensors, and other devices that detect parameters critical for operation of the loop.

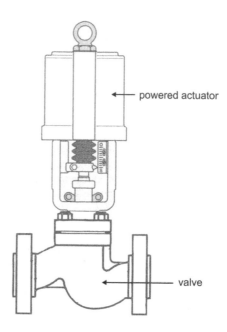

Figure 6.13. Control valve. A control valve combines a throttling valve with a powered actuator. (Modified from Skousen 1998.)

The controller compares the signal from the sensing device against preset parameters required by the process and determines the need for action by the control valve. The need for action leads to a signal sent to the actuator. These signals could indicate increase or decrease in pneumatic pressure, voltage, or hydraulic pressure. The positioner receives the electronic or pneumatic signal from a controller and compares that to the position of the actuator When a difference is present, the positioner sends the necessary power (energy) to move the actuator to the correct position. Depending on design, the positioner operates through electronic or pneumatic signals. The actuator uses some of the power (energy) to overcome the forces of inertia and the remainder to make the change to the status quo. These devices require precision to be able to make small adjustments to the setting, accuracy, and sometimes high speed (Skousen 1998).

Pneumatic actuators operate in one of two modes: air to extend or air to reverse. A disadvantage of pneumatic actuation is the compressibility of the air used to cause the action. The pressure applied must overcome the compressibility of the gas before the action can occur. Pneumatic systems offer a tremendous advantage in hazardous locations where ignition sources are not allowable.

Pneumatic lines are subject to pressure loss due to distance, leakage, and the effects of temperature on the compressed gas. Electrical lines are subject to stray voltages, surges, induction, capacitance, and ground loops. Hydraulic lines are subject to pressure loss due to distance, leakage, and the effects of temperature on the fluid. Almost all of the actuators currently in use are pneumatic and use compressed air, and sometimes compressed nitrogen as the pressure source.

Reliance on battery power is one of the drawbacks from electrical fail-safe actuators (Gibson 1999). Batteries can fail long before they are needed for a fail-safe situation. As a result, pneumatically powered fail-safe valves are the more popular choice in most situations. Electrically powered fail-safe valves become the better choice in remote locations where distance is a factor in the supply of compressed gas.

## Control Systems

In the 1940s, instrumentation used mainly pneumatic signals to transfer information from transmitters. Introduction of electrical signaling and clockwork mechanisms occurred in the 1960s, and in the 1970s, and then the computer arrived. The first digital smart transmitter appeared in the 1980s and the first field data bus in 1988 (daSilva Neto and Berrie 2004).

Motion control in industrial manufacturing testing or assembly is one of the most important aspects of process control. In most industrial processes, the position of an item dictates the operation.

Hybrid control subsystems containing solenoids and relays, and force and torque motors are but a small part of a much larger whole involving electronic sensors (transducers) and microelectronic process control. Terms such as magnetohydraulics and electrohydraulics, or perhaps even more descriptively, electromagneto-hydraulics (since an electrical current induces a magnetic field that produces hydraulic action) have appeared in an attempt to describe the fields of endeavor involving these devices. What has driven this activity is the ascendency of the microprocessor as decision-maker in process control.

Programmable logical controllers (PLCs) are solid-state devices containing one or more microprocessors that control machines or processes through input and output involving electronic devices. PLCs essentially are dedicated computers. The microprocessor of the PLC is programmed to control the system through decisions according to input from transducers. Transducers are specialized interfaces. Control in fluid power systems, for example, means control of pressure, flow, and direction of fluid (Vickers 1992, Vickers 1998). Control in process operations also using PLCs can also include level, position, density, temperature, pH, conductivity, and concentration of gases and solutions, among other parameters.

PLCs often became bottlenecks in process control due to the nature of the programming (ladder logic) inherent in these systems (Nachtwey 2002). Ladder logic programs are optimized for simple control functions, not for coordinating complex motion and human interaction. Microcomputers are taking on the role of human-machine interface with the PLC which continues to be used in the role of I/O (input/output) controller.

The trend today is to move away from centralized control to distributed control (Johnson 2002a; Vickers 1998). Distributed control occurs over a communication link known as a fieldbus. Each actuator contains a microprocessor and associated circuitry, forming a "smart" or "intelligent" actuator. These units contain feedback circuits to ensure correct performance of the task. Each actuator maintains contact with the central control, as do other elements of the circuit. Fieldbuses communicate using fiber optic or shielded twisted pair cable. Both of these should be resistant to electromagnetic noise.

This trend extends to vehicles and mobile equipment (Hitchcox 2003). Localized control by microprocessors simplifies wiring requirements, including protection and shielding of wiring harnesses and diagnostic troubleshooting.

The central controller is the workhorse of hydraulic systems containing these devices (Nachtwey 2002). These units are available as dedicated controllers, general-purpose programmable motion computers, and pre-programmed, general-purpose motion controllers. Each type offers advantages and disadvantages. Some units connect directly to transducers without the need for specific interfaces. This arrangement is more immune to noise in the signal. The best controllers generate analog signals, although analog signals are not as immune to noise as digital signals over long runs. As a result, controllers are best located close to the transducers and actuators.

A smart valve has a microprocessor-controlled positioner and is capable of bidirectional communication with external electronic devices, including computers and hand-held devices. The microprocessor is capable of receiving signals and implementing changes in both settings and diagnostic instructions and transmitting outputs and trends from testing data collected by the microprocessor (Simula 1999). The advent of microprocessor control of control valves has enabled delocalizing some of the control and diagnostics from the central computer to the valve. Another approach is to incorporate the microprocessor into the positioner.

A major benefit arising from an intelligent system is improved process control, especially through more precise control and assessment of performance of control valves (Fisher-Rosemount 1998). Imprecision and inaccuracy in performance of control valves is considered to be the leading cause of "off-spec" product, reduced throughput, and increased usage of raw materials, as well as other problems.

A complimentary advance in regard to intelligent systems is asset management (Gibson 1999; Masterson 1999). Smart devices are capable of performing diagnostics and calibrations, among their other functions, and can provide accurate feedback on the status and condition of equipment. This can reduce maintenance on equipment not requiring it, and can instead direct maintenance to where it is needed. To illustrate, until recently, common practice in the chemical process industries was to shut down a unit on a fixed frequency and to remove and service all of the valves, whether needed or not. With the availability of valves containing on-board microprocessors and diagnostic functions, service time has lengthened to once every four years (Gibson 1999). Functionality is coming so that users can analyze and troubleshoot from almost anywhere.

Intelligent systems offer opportunities for safer operation and higher productivity. Microprocessors can monitor thousands of inputs, whereas human operators are limited to $7 \pm 2$ (Miller 1956). (Chapter 4 provides more information about human factors and information processing.) The largest PLCs can handle 3,000 to 6,000 inputs. This is likely to increase as microprocessor technology and capability continue to advance. Intelligent systems can control all process activities normally performed by human operators, or by mechanical and electrical actuators. Additional functions include control of operation of pumps, agitators, heaters, chillers, vents, and so on.

Calibration of this equipment is still needed (Table 6.4) regardless of stability and the presence of on-board diagnostics (Anonymous 2000).

Process modeling is an advanced application of the data provided by intelligent systems (Richards et al. 2001). Process modeling is concerned with removing bottlenecks and optimization. A typical plant control system works through control loops. A complex operation can have hundreds to thousands of control loops. Control loops can experience many degrading factors (Table 6.5) revealed by process modeling.

"Tuning" reduces variability and improves product quality in the control loop. Yet, studies have shown that some plants use manual control in up to one third of their control loops. Loop tuning is based on trends in measurable parameters, such as temperature, pressure, pH, and conductivity versus the set-point. A loop containing a parameter that is off-target or oscillates excessively requires tuning. A loop containing a parameter whose amplitude is increasing requires more concerted action, as this represents an unstable condition.

Information provided by instruments for decision-making is only as good as the quality of operation of this equipment. Improperly selected, installed and maintained instruments can cause great harm to the operation of any equipment dependent on input of information all the way to process plants (McMillan et al. 1998). To illustrate, one study found that two out of three upsets in a plant were traceable to instrument faults (Sanders 1995). McMillan et al. (1998) contend that measurement errors and failures and instrument performance and reliability are largely established during design and construction (Table 6.6).

The biggest cause of measurement problems arises from the failure to anticipate and address adverse process conditions during design, and selection and positioning of measuring equipment (McMillan et al. 1998). Other sources of problems include electrical interference and grounding issues. Calibration of differential-pressure measurement devices depends, in part, on the density of the fluid. Density of the fluid necessitates constant composition.

Implementation of wireless transmission is occurring in process operations (Lewis 2004). One of the main arguments for implementing wireless communication is cost saving versus installation of wiring conduit and cable. Potential candidate situations for implementation of these options include operations that generate small quantities of data. These include manual reading of gauges and situations where installation of wiring would be very costly. Potential applications include monitoring pressure-relief valves, safety deluge showers and eyewash stations, on-off block valves, and temporary monitoring during engineering trials. Redundancy in such systems is essential. Strategies to provide redundancy include transmission at multiple frequencies and two-way acknowledgment of data. These strategies can provide highly reliable transmission of data.

## Table 6.4
## Benefits from Calibrating Equipment

**Benefit**

Compliance with health, safety, and environmental regulations

Compliance with quality assurance programs, such as ISO 9000

Compliance with regulations regarding weights and measures

Detection of long-term shift in performance of instruments or sensors due to environmental deterioration or poisoning by contaminants

Detection of compromised sensors (crimped lines, presence of condensate)

Detection of incorrect sensors

Source: Anonymous 2000.

Transmission requires line-of-sight conditions. Communication protocols used in consumer and commercial electronics, such as Bluetooth and the IEEE 802.11 family ("WiFi"), are unsuitable for application in process plants. (IEEE is the Institute of Electrical and Electronic Engineers.) They do not provide the redundancy of data transmission needed to ensure reliable communication. In addition, the equipment may not be adaptable to the requirements of hazardous locations. The range of equipment using industrial protocols is around 300 m.

Point-to-point radio transmission requires radio antennas. The current trend is toward wireless networks (Moss 2004). Wireless networks will reduce the problem created by the need for individual point-to-point radio transmission. Peer-to-peer networking allows individual nodes to communicate with each other.

Security has become a major issue for process operations, especially where hazardous materials and potentially explosive processes are present (GAO 2004; Rakaczky and Clark 2005). Threats to control systems used in industry can come from many sources (Table 6.7).

The most vulnerable of these systems are SCADA (Supervisory Control and Data Acquisition) systems. SCADA systems use radio frequencies to provide distributed and remotely controlled communication systems. SCADA systems are used in municipal water systems, hydroelectric power generation, and many other applications. These systems are especially vulnerable to unauthorized intrusion (Table 6.8), as they were never designed and engineered to withstand this threat.

The current focus on security against threats from sources external to countries ignores the reality of day-to-day threats to the integrity and description of these systems from sources within countries and inside organizations. Attacks on systems have occurred (GAO 2004). To date, these have involved domestic intruders. As software and inter-connectivity have progressed, the knowledge needed by intruders to enter a system and to cause damage has decreased. This lowers the level of sophistication needed for these exploits.

Fiber optic cable provides the best physical security for data signals (Rakaczky and Clark 2005). Typical connectors provide for immunity against vibration, dust, and moisture. Coaxial cable provides shielding against electrical interference. Unshielded twisted-pair cable is unsuitable for secure applications. It is susceptible to interference from magnetic fields, radio frequencies, extremes of temperature, vibration, moisture, and dust. The standard RJ-45 connector used in network connections (similar to a telephone connector) is neither water- nor dust-tight and will fail to provide reliable connection un-

## Table 6.5

## Bottlenecks and Inefficiencies Revealed by Process Modeling

| Parameter Affected | Bottleneck/Inefficiency |
| --- | --- |
| Instrument performance | Noise |
| | Calibration errors |
| | Instrument drift |
| Control valve performance | Sticking |
| | Hysteresis (deviation between the opening and closing sequence) |
| | Wear |
| | Poor accuracy due to improper sizing or lack of a positioner |
| Control tuning | Deviation from linearity in the process |
| | Process migration outside normal operating range |

Source: Richards et al. 2001.

der adverse conditions. The gold plating on the contacts will fail following exposure to vibration. The jacket material is so thin that the wire is subject to capacitance that will degrade performance when run inside a metal conduit.

The field of intelligent systems as applied to hydraulic systems (and for that matter, to equipment and machines, in general) is truly multi-disciplinary and demands the combined skills of several specialist practitioners: hydraulic system designer, electronic circuit designer, and controller programmer (Johnson 2002b).

The same and more can be said for process systems. Each participant in the design must communicate clearly and openly with the others. The designer of a hydraulic system must indicate fail-safe conditions for the operation to the designer of the electronic circuit. Together, they must indicate the mode for energizing and de-energizing critical solenoids to the programmer of the controller.

To illustrate, the system must fail in safe mode by dumping fluid through a solenoid-controlled valve. The controller programmer might become aware of some limitation of design in either the hydraulic circuit or the electronic circuit. The controller programmer must communicate this information to the other participants in a formal, organized, and managed manner that leads to acknowledgment, and as appropriate and necessary, to action. Coordination and cooperation among all participants is essential for the overall design and implementation to function correctly, predictably and safely. While this may sound obvious, miscommunication resulting from misperception has resulted in many failed start-ups. Misperception about direction of movement or rotation can lead to inappropriate sign conventions in programming logic.

Serious problems that affect performance of transducers, instruments, computers, PLCs, and controllers result from the quality of electrical power provided. Power quality is affected by ground loops, electrical surges, and harmonics (frequency multiples of 60 Hz). These affect voltage, current, and waveform (Lamendola 2000). An estimated 90% of these problems result from poor bonding and grounding.

Ground loops develop when instruments and controllers are connected to different grounds (DeDad 1999). The result of connection to different grounds is that the grounds can have different potentials (reference voltages) relative to each other, rather than the common potential that is required for

## Table 6.6
## Major Causes of Measurement Errors and Failures

Sensing lines that are plugged or contain liquid or solid foreign material when they should be dry and clean

Sensing lines lacking fluid due to vacuum condition

Sensing elements with excessive coating, fouling or abrasion

Excessive bubbles or solids in process fluid

Sensing elements containing cracks and holes

Stagnation due to insufficient turbulence and flow of process fluid that creates stagnation leading to coating, build-up, and freeze-up and eventual plugging

Leaking gaskets and O-rings

Inappropriate materials of construction

Sensing, pneumatic or electronic components affected by process or ambient temperature

Moisture on the sensing element or connections in signal transmission lines

Electrical interference

Ground loops

High resistance in connections or wiring

Non-representative sensing point

Inadequate run of straight pipe for the flow sensor

Plugged or fouled nozzle flappers

Feedback linkages that shift or contain excessive play

Incorrect or inappropriate calibration

Source: Modified from McMillan et al. 1998.

proper operation. Ground loops add or remove current or voltage from the process signal. As a result, the receiving device cannot differentiate between the correct and the supplemented signal. This problem is especially serious with distributed systems. In a facility that covers a large area and has many connections to ground, the probability of establishing a ground having more than one potential is high.

Eliminating ground loops may not be possible, since instruments, such as thermocouples and some analyzers, require grounding in order to make accurate measurements (DeDad 1999). In addition, analog control loops require grounding at one or more points. As well, grounding is required to maintain personal safety.

Protecting the integrity of instrument signals is possible through signal isolation. Signal isolators break the galvanic path (DC continuity) between all grounds, while permitting the analog signal to

## Table 6.7
## Sources of Threats to Plant Control Systems

| Agent | Source of Threat |
|---|---|
| Iinternal | Indifferent or incompetent recording of code changes |
| | Disaffected insiders including staff and contractors |
| | Failure to record changes to system logic by permitted users |
| | Disgruntled former employees |
| External | Hackers |
| | Common criminals |
| | General malicious code threat |
| | Illegal information brokers and freelance agents |
| | Intelligence/investigative companies |
| | Competitors |
| | Organized crime |
| | "Anti-corporate" political activists |
| | Nation states and governments |
| | Nonstate-sponsored terrorism |
| | Regional political activism |

Sources: GAO 2004; Rakaczky and Clark 2005.

continue through the loop. An isolator must provide input, output, and power isolation in order to prevent formation of an additional ground loop between the power supply of the isolating device and the signal from the process input and/or output. A dedicated, common ground bus is the preferred strategy to use for instrument grounding. All grounds connect to the bus.

Electrical surges of 20 V are sufficient to damage microelectronic circuits (Gorosito 2000). Surges result from lightning strikes, line disturbances, and electrostatic discharges (Gososito 2000; Turkel 2000). Equipment is designed to operate within an envelope of ±10 % of normal rated line voltage. Exceedences beyond this envelope can cause abrupt shifts in ground potential leading to current flow in a data line during an attempt to equalize the ground potential.

Problems involving electrical equipment also result from operation of too many units on one circuit, or insufficient isolation between circuits. The latter occurs during start-up of heavy consumers of electrical energy, such as large electric motors. Transients can persist from nanoseconds to milliseconds and can comprise hundreds of amperes at thousands of volts of potential. Transients can enter equipment through power and data lines. Transient-voltage surge suppressors grounded through a common bus will provide protection to both power and data lines. Use of a common ground again is imperative to prevent formation of ground loops.

The third detractor to power quality is the presence of harmonics in the 60 Hz electrical power supply. Harmonics are spikes that occur in the sine wave of AC voltage at multiples of 60 Hz. Such transients can affect uninterruptible power supplies. Uninterruptible power supplies (UPSs) protect only the load on the output side, not the system that supplies power to it (Hartfiel and Lamendola 2001). Uninterruptible power supplies do not necessarily stop the passage of harmonic transients. The ability to stop harmonics depends on the type of UPS, presence of isolation transformers, and harmonic

## Table 6.8
## Consequences of Unauthorized Entry Into SCADA Systems

### Concern

Disruption of operation by delaying or blocking the flow of information through control networks, thereby denying availability of networks to control system operators

Unauthorized changes to programmed instructions in PLCs (programmable logic controllers), RTUs (remote terminal units) or DCS (distributed system controllers), changing alarm thresholds, or issuing unauthorized commands to control equipment

Sending false information to control system operators either to disguise unauthorized changes or to initiate inappropriate actions

Modifying control system software producing unpredictable results

Interfering with operation of safety systems

Source: GAO 2004.

filters. Computers, variable speed drives, and UPSs all have nonlinear power supplies that introduce harmonics into the power system (Carr and Sitter 2001). These harmonics can create significant problems with the supply, especially if this is a portable generator (gen-set). High levels of harmonics can lead to an unstable voltage output. The UPS fed from the portable generator may not have harmonics filters. Small gen-sets (less than 200 kW) are more susceptible to problems with power quality, since the higher internal impedance (resistance) tends to amplify rather than absorb harmonics.

Power supplies used in computers and similar equipment operate in "switched mode" (CDA 1998). Switched mode produces high frequency pulses of DC, rather than AC, as would be provided by a transformer. Pulsed DC permits voltage conversion by transformer. These units can backfeed harmonics of 60 Hz into the supply system. The net result of backfeeding harmonics and transients into the system is possible malfunction of sensitive electronic equipment, and overheating of phase, and particularly neutral (white) conductors.

Intelligent systems offer too many advantages over previous methods of control not to proceed with their development and implementation. As with any advance, the change is not without risk and not without unforseen consequences. Newton's third law, roughly stated as, "for every action there is an equal and opposite reaction," is as applicable to change as it is to physics. In its colloquial corruption, Newton's third law is sometimes called the law of unintended consequences. Roughly stated, this means that for every benefit derived from a change there are unexpected or unintended consequences. The devil, as they say, is in the details. Table 6.9 summarizes some of the possible obstacles to successful implementation of intelligent systems in control loops. Note that many of the obstacles do not relate to the equipment itself, but rather, to the manner in which it is used and maintained.

Electronic equipment to be used in hazardous locations (areas) requires protection against potentially explosive gases and vapors (Anonymous 1989). Hazardous locations (areas) exist in many manufacturing industries.

Intrinsic safety is one of a number of techniques used to ensure protection of the hazardous area from operation of electrical and electronic equipment. Intrinsic safety uses the concept that electrical energy within the installation is insufficient to ignite the surrounding ignitable atmosphere, even under prescribed fault conditions. That is, energy that emerges as electrical arcs or hot surfaces is too weak to cause ignition, even in the most easily ignitable mixture of gas or vapor and air.

## Table 6.9
## Deleterious Conditions Associated with Process Control

| Deleterious Condition | Comment |
|---|---|
| **Design-Related** | |
| Miscommunication, misunderstanding | System design and implementation are multi-disciplinary. Each participant brings proprietary skills and knowledge. Lack of communication or understanding of seemingly insignificant elements can have deleterious consequences. |
| Software errors and omissions | These errors and omissions become evident only during operation of the equipment. |
| Unauthorized reprogramming of microprocessors or modification of software | This has already occurred, the result of actions of a disgruntled actions of a former employee who "hacked" into the system. |
| **Installation- and Operation-Related** | |
| Incompatible materials | Contact between incompatible materials leads to premature failure. Selection of appropriate materials for contact requires knowledge and experience. The knowledge and experience base in many organizations has been lost due to "rightsizing." |
| Signal noise | Digital systems are highly immune to noise over long distances, whereas analog systems are noise susceptible. Signal noise includes 60 Hz spikes from the electrical distribution system. |
| Ground loop | Components connected to different grounds and grounds having different potentials; balancing of current in the loop creates signal noise. |
| Electrical surges | Damage and destroy microelectronic circuits. |
| Harmonics | Voltage transients create signal noise. |
| High connection or wiring resistance | High noise and erratic reporting of signal. |
| Nonrepresentative sampling point | Nonrepresentative measurement |
| Inappropriate sensor location – inadequate straight pipe run | Error in measurement. |
| Plugged sensing lines | Long narrow lines filled with stagnant fluid are especially prone to coating, freezing or other modes of solids build-up. |

## Table 6.9 (Continued)
## Deleterious Conditions Associated with Process Control

| Deleterious Condition | Comment |
|---|---|
| **Installation- and Operation-Related (continued)** | |
| Sensing lines containing liquid when they should be dry | Source of error in measurement |
| Coating, fouling or abrasion of sensing elements | Meaningless measurements |
| Deformed, cracked or perforated sensing elements | Meaningless measurements |
| Process fluid with low Reynolds number | Error in measurement |
| Leaking gaskets or O-rings | Destruction of equipment |
| Temperature- or process-affected sensing, pneumatic or electronic components | Error in measurement |
| Moisture on sensing elements or signal connections | Error in measurement |
| Plugged or fouled nozzle flappers | Error in measurement |
| Shifting or excess play in feedback linkages | Excess cycling and oscillation In servo systems |
| Incorrect calibration | Error in measurement |
| Device failure | Can happen under any control scheme |
| Destruction of communications cable | Can occur in any system Redundancy critical |
| Controller destroyed | Can happen in any system Redundancy critical |

Sources: Carr and Sitter 2001; da Silva Neto and Berry 2004; DeDad 1999; Gorosito 2000; Hartfiel and Lamendola 2001; Johnson 2002b; Lamendola 2000; McMillan et al. 1998; Turkel 2000.

This concept derives from ignition curves that plot short-circuit current versus open-circuit voltage (Figure 6.14). This also considers energy stored in components capacitances and inductances (including interconnecting cables). Intrinsic safety is an inherently energy-limited, design concept. The usable region of current and voltage is bounded by limitations imposed by stored energy at higher lev-

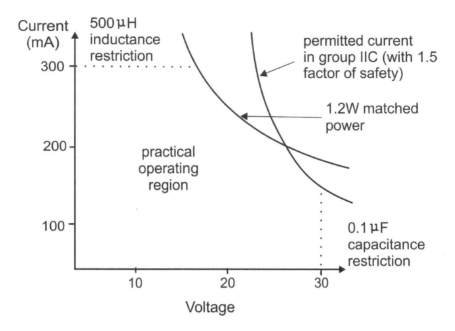

Figure 6.14. Intrinsic safety. Intrinsic safety utilizes the premise that the energy developed within an envelope defined by current and voltage is less than capable of acting as a source of ignition. The envelope encompasses ignition curves and restrictions on inductance and capacitance. (Modified from Anonymous 1989.)

els of voltage and current (300 mA and 30 V), the ignition curve and and a matched power restriction (typically 1.2 W at 60° C) to govern surface temperature.

## Human Interfaces to Control Systems
The interface between the human operator and the means of executing control is critical to the successful operation of equipment. Over the years tremendous progress has occurred as the interface has changed from a strictly mechanical one that often depended totally for success on the strength of the operator to an electronic device capable of being operated by anyone regardless of strength.

### Direct Controls
Hand- and pedal-operated levers connected to mechanical linkages and cables were the among the first of the control mechanisms (Figure 6.15). These provided to the operator direct input into the device subjected to control. These kinds of controls still exist in vehicles (gear shifters in manual transmission, clutches and brakes) and aircraft containing cable-operated controls.

Mechanical linkages are cumbersome and require a straight path (Heney 2002). Push-pull cables are more tolerant of changes of direction. Mechanical linkages require high actuating forces when mechanical advantage is not part of the design.

Operating some of these controls requires considerable strength because of the unassisted direct connection to the actuator under control. Cable controls were subject to failure under load due to corrosion and mechanical failure of the cable and the linkage.

### Assisted Controls
Hydraulic and pneumatic circuits were the basis for assisted controls in some applications. One example is the braking system used on trains. The braking system operates through pneumatic circuits to apply the brakes. The system on trains requires pressure to apply the brakes. Hence, a compressor

Figure 6.15. Cable control. Cable controls were used in early aircraft.


operating at all times is required to ensure braking capability on the train. Construction equipment had piloted and non-piloted valves connected to user controls. These controls are still to be seen on mobile equipment.

## Microprocessor-Based Controls
Microprocessor-based controls have largely replaced controls based on mechanical components and even pure hydraulic and pneumatic systems. Perhaps the greatest consequence from the move to microprocessor-based control and decision-making was the loss of contact, the touch and feel of the interface between the human operator and the system. Whereas a human could "poke and prod" at older systems, today's operator must rely on readings provided by instruments or on information in the form of graphics and numbers provided on a video display to investigate and make decisions about conditions deemed potential risks to personal safety. At some point, the operator or user must decide whether or not to trust the system.

Some of this situation may have resulted from difficulties in the reprogramming (training) of human operators in making the transition from the mechanical linkages and other control technologies to controls operating through electronic sensors. The transition for human operators and for manufacturers attempting to create an electronic analogue of a mechanical control system could be very problematic, and ultimately may not be possible. Humans trained on predictability of response from one type of system experience considerable difficulty making the transition to one that responds similarly, but not identically in all circumstances.

The most successful implementations of these changes have come from situations where the operator gains long-term, hands-on familiarity with the new equipment. Construction equipment, such as excavators, provide one such example. To a casual observer, the machine becomes an extension of the body of the operator. Controls in passenger aircraft provide another example.

Electronic or hydraulic circuits interfaced through a joystick have replaced the levers and mechanical linkages and clutches of the bygone era. Nostalgia about the previous design and implementation of this equipment fails to acknowledge reliability and predictability issues with previous implementations, and hence, safety of operation compared to modern control technology.

## Control Consoles

Control consoles were among the first centralized remote interfaces with equipment through control devices, such as solenoids and relays,.and hydraulic and pneumatic pilot circuits. The operator sat or stood at a desk and pushed button switches to activate or deactivate equipment (Figure 6.16). Some consoles included monitoring instruments that provided visual confirmation by the operator. The operator applied controls in a manner to produce a desired outcome and responded to feedback provided by the instruments.

Incorporation of Programmable Logic Controllers (PLCs) into control circuits enabled the operator to communicate with the uppermost level of control of the equipment. The operator needed only to push a single button or buttons in a particular sequence and the microprocessor would activate or to shut down all of the pieces of equipment. The switch would light when activated. At the same time, the PLC also controlled related equipment in a similar manner. That is, pushing the button to start up or shut down a particular piece of equipment in a sequence could start up or shut down related equipment or the entire sequence.

What occurred depended on the logic programmed into the PLC. The logic programmed into the PLC was never evident or available to the operator. This situation imposed a considerable burden onto the operator to learn and remember the outcome from every possible combination of button pushing.

## Joysticks

The "joystick" is an example of the change in interface to which humans must adapt. Among the first applications of the joystick as a control device in industry was aircraft control. The joystick replaced the central yoke used by pilots. Modern joysticks provide tactile feedback to the operator to assist in judging the extent of actuation (Heney 2002). To illustrate, the further the joystick moves in a particular direction, the greater is the response from the actuator.

Joysticks and similar devices are rapidly replacing levers and historic control devices (Heney 2002). Joysticks can function through direct connection to mechanical linkages and piloted circuits and electronic control. Piloted circuits simplify installation, require lower actuating force and simplify maintenance. Electronic joysticks operating through electronically controlled actuators on piloted circuits further simplify installation, operation and maintenance. Advanced joysticks provide tactile feel through use of springs or strain gauges.

The concerns with this equipment are failure due to vibration and electromagnetic and radiofrequency interference. Most joysticks spend the majority of time in the neutral position. In this orientation, low amplitude vibration from the equipment can cause dithering. Dithering can cause localized wear or a "dead" or "noisy" spot. Electromagnetic and radiofrequency interference are known to cause self-actuation output. These susceptibilities potentially can cause problems when work occurs near sources of electromagnetic fields, including 60 Hz transmission and distribution lines, and radio and cellular telephone broadcasting towers, among other sources.

## Mice and Touch Screens

Another direction taken in the development of interfaces for the automation of process control was the use of mice and touch screens. Mice and to a much lesser extent, touch screens are familiar to computer users. Their use as interfacing devices in the development of the Graphical User Interface (GUI) for process control was a natural migration from their use in home and office computers.

The GUI is a "shell" or layer situated above the programming language and logic that provides a familiar and non-threatening image to end-users. The end-user has no need to see or to be confronted with the behind-the-scenes operation of the control system. What is critical for the operator is to be presented with appropriate information on which to act. Typical of the GUI in current installations are

Figure 6.16.  Older console. Pushing a single control switch could activate or deactivate all of the components in a production sequence.

screens that indicate in simple graphics the equipment and its operation. The operator can monitor parameters of importance associated with operating equipment.

To change the status quo, the operator moves the mouse to a critical location and clicks on the appropriate button. In the case of a touch screen, the operator touches the graphic for the item with the tip of the finger. This creates the signal that causes occurrence of the desired action.

## References

Anonymous: *Introduction to Intrinsic Safety*. Annapolis, MD: Elcon Instruments, 1989.

Anonymous: HART Calibration Made Easy with New Documenting Process Tools. *Industr. Process Prod. Technol. 12*(2): 54 (2000).

Carr, M. and L. Sitter: Ensuring Power Quality and Reliability with Gen-Sets. *Elect. Contract. Maint. 100:* March (Suppl.) (2001). pp. 19-20.

CDA: Data Application Sheet, A Primer on Power Quality (A6018–98//98). New York: Copper Development Association Inc., 1998. [Booklet].

da Silva Neto, E.F. and P. Berrie: Who's Afraid of Control in The Field? *Petro Indust. News. 5*(5) 14-17 (2004).

DeDad, J.A.: *Elect. Contract. Maint. 98*(11): 14-20 (1999).

Duffy, J.E.: *Auto Electricity and Electronics*. South Holland, IL: The Goodheart-Willcox Company, Inc., 1995.

Fisher-Rosemount: Fieldvue Instrumentation. Marshalltown, IA: Fisher Controls International, Inc., 1998. [Brochure].

Freund, K., J. LaCourse, M. Stubblefield, B. Worthy, and J.H. Haynes: *The Haynes Automotive Electrical Manual*. Newbury Park, CA: Haynes North America, Inc., 1989.

GAO: Critical Infrastructure Protection, Challenges and Efforts to Secure Control Systems (GAO-04-354). Washington DC: General Accounting Office, 2004.

Gibson, W.D.: Valves Widen the Field of Operation: *Chem. Eng. 106*(2): 41-45 (1999).

Gorosito, A.: Simplifying Surge Protection. *Elect. Contract. Maint. 99*(11): 74-78 (2000).

Hall, F.B.: Safety Interlocks – The Dark Side. Niles, IL: Triodyne Inc., Safety Brief Vol. 7, No. 3. June 1992.

Hartfiel, M. and M. Lamendola: UPS Cures for Power Quality Problems. *Elect. Contract. Maint. 100*: March (Suppl.) (2001). pp. 10-12.

Heney, P.J.: Joysticks Are a Popular Control Option. *Hydraulics Pneumatics. 55*(10): 29-32 (2002).

Hitchcox, A.L.: On-Board Electronics Benefit Mobile Hydraulic Systems. *Hydraulics Pneumatics 56*(3): 20-24 (2003).

Johnson, J.L.: Local versus Centralized Buses for Pump Control. *Hydraulics Pneumatics 55*(4): 34-40, 74 (2002a).

Johnson, J.L.: Hydromechanical Aspects of the Millennium Pump. *Hydraulics Pneumatics 55*(3): March (2002b). pp. 20-24.

Lamendola, M.: Power Quality Planning: A "Big Picture" Job. *Elect. Contract. Maint. 99*: August (Suppl.) (2000). pp. 16-18.

Lamendola, M.: A Cure for the Battery Blues. Elect. *Contract. Maint. 100*: March (Suppl.) (2001). pp. 16-18.

Leveson, N.G.: *Safeware, System Safety and Computers*. Reading, MA: Addison-Wesley, 1995.

Lewis, C.W.: Industrial Wireless: Getting Your Feet Wet. *Chem Eng. 111*(7): 22-25 (2004).

McMillan, G.K., G.E. Mertz, and V.L. Trevathan: Trouble-free Instrumentation. *Chem. Eng. 105*(11): 82-94 (1998).

Menezes, M.: Improve Plant Safety through Advanced Measurement Diagnostics. *Chem. Eng. 107(10)*: 72-80 (2000).

Masterson, J.: Put a Smart Face on Asset Management. *Chem Eng. 106*(2): 108-111 (1999).

Miller, G.A.: The Magical Number Seven, Plus or Minus Two: Some Limits on Our Capacity for Processing Information. *Psych. Revs. 63*: 81-97 (1956).

Moss, G.: Industrial Wireless: What's Beyond the Wading Pool? *Chem. Eng. 111*(7): 26-28 (2004).

Nachtwey, P.: A Systems Approach to Electronic Control. *Hydraul. Pneumat. 55(3)*:50-52 (2002).

Nelson, C.A.: *Millwrights and Mechanics Guide*. New York, NY: The Bobbs-Merrill Company, Inc., 1983.

Rakaczky, E. and D. Clark: Thwarting Cyber Threats in Plant Control Systems. *Chem. Eng. 112*(4): 40-45 (2005).

Richards, M., R. Schmotzer, and G.E. Pogal: Modeling to Optimize Process Control. *Chem. Eng. 108*(8): 66-71 (2001).

Rosenberg, P. Correspondence Lesson 2: Basic Wired Control Devices. *Elect. Contract. Maint. 98*(3): 59-63 (1999a).

Rosenberg, P. Correspondence Lesson 7: Motor Controls. *Elect. Contract. Maint. 98*(8): 44-55 (1999b).

Ruel, M: Control Valve Health Certificate. *Chem. Eng. 108*(11): 62-65 (2001).

Sanders, F.F.: Watch Out for Instrument Errors. *Chem. Eng. Prog.* July 1995. pp. 62-66.

Skousen, P.L.: *Valve Handbook*. New York: McGraw-Hill, 1998.

Simula, M.: How to Choose a "Smart" Valve System. *Chem. Eng. 106*(9): 84-90 (1999).

Stepanek, D.: Control Valves For Real-World Service. *Chem. Eng. 109(3)*: 103-107 (2002).

Stewart, H.L.: *Pneumatics and Hydraulics*, 4th Ed. Revised by T. Philbin. New York: Macmillan Publishing Company, 1987.

Turkel, S.: Don't Let Surges And Sags Get You Down. *Elect. Const. Maint. 99*(11): 16-18 (2000).

Vickers: *Vickers Industrial Hydraulics Manual*. Rochester Hills, MI: Vickers, Inc., 1992.

Vickers: *Vickers Mobile Hydraulics Manual*. Rochester Hills, MI: Vickers, Inc., 1998.

# 7 Management of Hazardous Energy in Systems

## INTRODUCTION

Previous chapters have explored what is known about hazardous energy and our interaction with it. Hazardous energy is the level of energy capable of interacting with biological tissue to produce immediate and serious injury. Serious injury occurs immediately or following a short delay.

The mode of delivery is critical to the outcome. The outcome determines the quantity of energy needed to produce damage at a particular target. Hazardous energy comprises 26 of the 42 hazardous conditions readily identifiable in regulatory statutes. Accidents involving hazardous energy produce acute traumatic injury at levels ranging from nano-scale molecules, starting with damage to DNA, to the destruction of tissue and structural components, and at the macro level, injury of the whole organism.

Accidents involving hazardous energy occur because of direct contact with the energy source or energy released by it. Some examples include irradiation by nuclear sources, contact with electrical sources and strike by contact with moving components of equipment. Accidents also occur because of energy release during conversion from one form to another. This occurs in accidents involving components of machines and equipment and flowable solid materials that store energy by virtue of position and change position unexpectedly once operation stops or when disturbed. Rapid conversion from potential to kinetic energy occurs during these accidents.

This chapter acts as a bridge between what we know about hazardous energy and the unintended consequences that arise from contact with it and the means by which machines, equipment, and systems operate and are controlled. It is essential to understand what is possible and what is doable with regard to managing our contact with hazardous energy. Managing in this context intends to mean eliminating or at the minimum, controlling that contact to levels that produce no harm. What is possible with regard to managing contact with sources of hazardous energy unfortunately is not always doable without the imposition of considerable impracticability. Without a full understanding about what is to be managed and how management can occur practicably, the ability to manage is not possible to optimum capability. The result from this is that situations are not managed effectively, or worse, are allowed to happen.

The investigation of accidents involving contact with hazardous energy across a wide spectrum of industry indicates a number of causal factors (Table 7.1). The main themes summarized in Table 7.1 are knowledge, control, the limitations of information processing by the human organism, inconvenience or impracticability, and communication failure.

Knowledge about the environment in which work is to occur is critical for safety. Individuals who work long-term for an employer located in an unchanging location can gain considerable knowledge about the idiosyncrasies of that location and its equipment, processes, and hazardous conditions. This certainly does not apply to all employees, but the potential for acquisition of the critical knowledge needed to maximize safety is resident in these individuals.

This situation for long-term employees contrasts with that of short duration permanent and temporary employees and employees of contract employers. These individuals are at a distinct disadvantage regarding the knowledge needed to work safely in an unfamiliar premises. This situation is similar to learning a language. Short-duration permanent and temporary employees and employees of contract employers are operating at the level of beginners. Long-term employees have the potential to gain considerable finesse in the nuances of vocabulary and expressions of a language.

The second factor pertains to control. Control refers to the influence that an individual can exert over the conduct of activity in a workplace environment. Short-duration permanent and temporary employees and employees of contract employers also operate at a considerable disadvantage compared to long-term employees in this regard.

## Table 7.1

## Causal Factors in Fatal Accidents Involving Hazardous Energy

| Factor | Outcome |
| --- | --- |
| Knowledge | Inability to anticipate and recognize hazardous conditions |
| | Inability to assess potential for harm |
| | Ignorance of existing requirements and procedures |
| Control of conditions | Highest among long-term employees and least among short-duration and temporary employees and employees of contract employers |
| Information processing | Conflict for attention between task and conditions of work |
| | Distraction by personal circumstances |
| | Distraction by environmental factors |
| Inconvenience (impracticality) | Intangible quality of workplace culture and personal characteristics that sometimes leads to defiance of (failure to follow) existing procedures |
| Communication failure | Failure to appreciate or assess the impact on safety of an action within the progress of a task |

Long-term employees have the greatest potential through knowledge to recognize the hazard and the risk posed by a situation and also knowledge of the politics that exist within an organization. The latter are critically important for these underlie the decision about whether and how to exercise the right to refuse when appropriate. Short-duration permanent and temporary employees and employees of contract employers might recognize the hazard, but are much less likely to be able to judge correctly the severity of the risk of harm.

Exercise of the right to refuse work perceived to be unsafe leaves them extremely vulnerable to censure and expulsion in a situation that doesn't meet the perception of high enough risk in the minds of those in control or that exposes an obvious and embarrassing deficiency. This reality, coupled with the lack of knowledge of conditions at a worksite, leave short-duration permanent and temporary employees and employees of contract employers in a highly vulnerable position leading to reluctance to act or inability through lack of knowledge to act effectively.

The third factor pertains to the limitation of the human brain to process information. There are three main processes that compete simultaneously for our attention and the ability to process information. There is a natural competition for processing space that centers on the task to be performed, the conditions under which performance of the task occurs and distractions.

These accidents illustrated the concept of competition for attention between task and conditions. We cannot simultaneously focus on both, but rather one or the other. Task pertains to the actions that we take to reach an outcome. Conditions pertain to the safety hazards that surround us with which we interact as we pursue a task.

Painting the trim on the edge of the roof of a house from above without fall protection overstresses our ability to focus on application of the paint without making a mess or a spill and ensuring that we will not fall over the edge from lack of securement. Work on a ladder near the electrical power lines poses similar challenges in the competition between the task and the need to position ourselves appropriately to avoid contact with the wires.

Further complicating the situation is distraction from a black bear roaming in the neighborhood, from vehicles with noisy engines driving on the street and from people walking on the street. Distraction from environmental conditions is another element in the competition for information processing. Distraction also results from preoccupation with personal factors.

The fourth factor pertains to the inconvenience or impracticability of measures required, meaning imposed on workers, as a prerequisite for the satisfactory performance of work. This factor is intangible from the perspective of quantitation, but reflects the impatience of workers to perform tasks that require effort beyond that considered as tolerable or acceptable. This factor describes the situation when people who are knowledgeable about the hazards and risks implicit in a particular situation fail to follow procedures established to ensure safety during some task.

The fifth factor pertains to failure to assess the implication of the safety of an action versus the status of the work. Fatal accidents have occurred when individuals restarted deactivated and de-energized equipment, and in some situations, isolated and locked-out equipment, prior to the appropriate moment in a sequence of tasks. These situations have created fatal consequences to the person who performed the activation and to others following activation by a third party.

## Technical Strategies For Protecting Against Hazardous Energy

Strategies for protection against hazardous energy must reflect the mode of operation against which protection is required. The classic considerations include the concepts of time, distance, shielding, alternate path, and personal protective equipment.

### Exposure Limits

Time, distance, and shielding are logical outcomes from standards used to define acceptable levels of exposure to chemical and physical agents. In a book about hazardous energy, this discussion concerns physical agents almost exclusively. Some chemical reactions do produce radiant energy. However, the discussion about physical agents covers these emissions.

Discussion about the process of setting standards for permissible exposure is essential in order to understand the role of time, distance, and shielding in worker protection, and potential limitations in the application of these concepts.

The basis for many of the exposure standards is the observation that exposure as a surrogate measure of dose is an effective means to assess potential for harm. Dose is the total impact with which the body must cope following the exposure. Dose contains two elements: dose rate and duration of exposure.

Dose rate is the rate of delivery of the agent to the body in some measurable format related to units of (entity delivered)/(time). At some level, the entities that constitute dose rate describe individual units capable of causing harm.

Allowable dose may imply a level of damage, which requires repair and which may create no measurable impact on the body. On the other hand, allowable dose may lead to no damage. In the situation where the dose does lead to damage which is readily repairable, the manner in which one site of damage impacts onto another site of damage becomes the subject of discussion. On the one hand, the individual entities of damage may have no influence on each other. On the other hand, individual entities of damage occurring too closely within time and space may indeed exert an effect on each other. In the latter case, the rate of delivery of the entities (dose rate) becomes an important factor in the assessment of the harm caused by the exposure.

Duration of exposure is the timeframe within which the exposure must occur. The basis for setting the time duration of exposure limits is the response of the body to the mechanism of damage posed by a particular agent. This concept implies an assumed rate of delivery.

An implicit concern faces standard setters where duration of exposure is a consideration. This pertains to the question of difference of impact of delivery of the agent at a high rate for a short duration versus delivery of the same dose over a long duration at a low rate. Implicit in the acceptance of the duration is that there is no difference in harm between the two rates of delivery.

Given an exposure limit based on a time duration, employers and health and safety practitioners typically will divide the allowed exposure stated in the limit by the time duration to obtain the average rate below which overexposure would not occur. This then becomes the operative limit during work activity.

Some exposure limits carry an annual duration, as is the case with exposure limits for ionizing radiation These limits also included a quarterly duration. More typical are exposure limits based on a workshift of 8 hours. As well, there are limits based on shorter durations of 30 minutes and 15 minutes. Some limits are instantaneous ceiling limits not to be exceeded at any time.

Every one of these types of limits, including even ceiling limits, is affected by dose rate. In the latter case, this reflects the limitations of the equipment available for performing the measurement. Continuing improvements in measurement technology have considerably reduced the time needed to obtain a reliable and accurate measurement of conditions. Improvement in technology becomes most important in the relationship between the time needed for obtaining the measurement and the time frame stated in the exposure limit. The exposure limit can take on entirely new meaning as the time needed to obtain reliable measurements decreases.

To illustrate, whereas older technology required 15 minutes or more to obtain a measurement for a ceiling limit, current technology for some agents provides a continuous electronic signal for comparison against the exposure limit or guideline. This situation can create serious difficulty in the application of exposure limits or guidelines created to reflect the limitations of historic sampling technology. This is especially the case with ceiling exposure limits.

## Time

Time or the permissible duration of exposure limits the dose measured through exposure to a predetermined measurable quantity. Implicit in the use of time as a controlling factor in accumulating exposure is that the task is able to  be completed within the duration incorporated into the exposure limit.

In many circumstances, the time needed to perform a particular task considerably exceeds the time allowed within the context of the exposure limit. This situation poses no problem where a worker can complete the task over the period of several days. The situation becomes unmanageable when the time needed to work within the constraint of the exposure limit is considerably greater than justified by productivity. The sharing of a rate of exposure that exceeds the averaged exposure limit for a shorter duration limit within a group of workers generally is not condoned as a practice for administering the accumulation of exposure.

Circumstances have arisen following accident situations and other emergencies where the rate of exposure considerably exceeded the exposure rate averaged over the duration of exposure. In addition, these situations offered no alternative for providing worker protection. In these situations, the time needed to complete tasks considerably exceeded that available within the duration of exposure limits. This forced considerations about how to complete the tasks within the context of acceptable regulatory practice and to provide protection to workers in a situation where other options for protection were not available.

The only option that remained was to share the burden among a large numbers of worker for activity of very short duration compared to the time incorporated into the exposure limit. Each participant received the same total dose but at a much higher dose rate than would be acceptable in normal practice. This type of situation occurs very rarely.

Examples of this type of situation occurred decades ago, and could occur again. Technological advancements and innovations in robotics (Figure 7.1) would considerably reduce, but not eliminate, the need to subject humans to such conditions. A robot is expendable in hazardous circumstances, whereas a human is not replaceable.

In today's circumstances the use of time as a controlling factor in preventing overexposure is not practicable for protection against most sources. As mentioned, typical practice is to divide the allow-

Figure 7.1. The police robot (or remote mobile investigator) is one of the more familiar applications of robotics.These units contain cameras, claws for gripping, and climbing wheels. (Courtesy of Pedsco (Canada) Ltd., Toronto, ON).

able exposure by the number of working hours or minutes within the time duration to establish an operative exposure rate for purposes of administrative control.

## Distance

Distance of the individual from a source of radiant energy is another strategy for managing dose. This strategy is most effective with point sources. Energy from radiant point sources generally follows the inverse square law unless secondary mechanisms of interaction are occurring. The inverse square law indicates that exposure decreases with the square of the distance from a point source. Additional equations exist for line sources, disks, slabs, clouds, and other geometries and densities of materials.

The basis of the inverse square law is that all locations at the same distance in open space form a surface that receives the same level of energy. The surface area of the sphere is given by Equation 7-1.

$$SA = 4\pi r^2 \qquad\qquad (7\text{-}1)$$

where:
SA is surface area ($m^2$).
r is the radius of the sphere (m).

The energy density (intensity) measured at one distance provides the basis for predicting the energy density at a second distance based on the principle of conservation of energy. That is, the total energy is constant and distributes equally over the larger surface as the distance increases (Equation 7-2).

$$I_2 = I_1\left(\frac{r_2}{r_1}\right)^2 \qquad\qquad (7\text{-}2)$$

where:
$I_2$ is intensity at distance 2 ([energy unit]/m$^2$).
$r_2$ is distance 2 (m).
$I_1$ is intensity at distance 1 ([energy unit]/m$^2$).
$r_1$ is distance 1 (m).

The approach of using distance as a control of exposure works in some situations. This is the basis for establishing the zone of exclusion during nondestructive testing using sources of ionizing radiation. The source is a small piece of material (a point source). The exclusionary distance minimizes exposure of bystanders and prevents overexposure. Distance is also used as the basis for protection from emissions from communication towers and relay points located on existing infrastructure in the cellular telephone network.

With regard to preventing overexposure of workers, the use of distance as a protective strategy has limited utility. Sources of high strength and ever decreasing exposure limits mandate larger distances. Control of exposure of workers based on distance has limited practicality and is most suited for controlling exposure of bystanders.

Distance has great importance as a protective strategy for overhead electrical conductors in transmission and distribution systems. These conductors typically are not insulated. Protection against electrocution and flash-over is based solely on distance.

Distance of the conductors between each other and distance from the ground are the basis for the concept of limits of approach (Figure 7.2). Limits of approach are distances to which approach to the conductors can occur without the occurrence of electrocution and flash-over and without imposition of undue electrical effects, including induction and capacitance in conductive objects.

Figure 7.2. Limits of approach provide an effective means to control against electrocution and flash-over. Where any question exists about extensibility of equipment, measure the distance and contact the utility.

## Shielding

There are many situations where the worker must work in close proximity to the source of energy. Shielding provides the means to reduce the exposure rate to acceptable levels. This is especially the case for sources of radiant energy. Some of the materials used for shielding absorb energy from the source and possibly convert it to less harmful or harmless forms and levels of energy and re-radiate it. Other materials block transmission and reflect the energy.

### Ionizing Particulate and Electromagnetic Radiation

The techniques utilized in the nuclear power industry, medical practice, and industrial applications of ionizing radiation enable workers to work safely in close proximity to sources.

Alpha particles which densely ionize the materials through which they pass are easily stopped by common materials, such as pieces of paper and thin cardboard. The thickness required depends on the energy of the alpha particle (Cember 1996). The range of almost all alpha particles in air is less than 7 cm. The range in other materials is correspondingly less due to greater density. Use of materials of low atomic number minimizes or eliminates production of X-rays during interaction with the material of the absorber. This situation applies only to radioactive isotopes emitting only alpha particles.

Beta particles, atomic electrons emitted during nuclear processes and electrons accelerated in fields of high voltage, also dissipate their energy by ionization within materials. Stopping power depends on density of the absorber (Cember 1996). Bremsstrahlung, X-rays emitted during rapid deceleration of electrically charged particles traveling at high speed, is a major concern during consideration about protection from beta particles. Bremsstrahlung formation depends on the atomic number of the absorber and the maximum energy of the beta particle. For this reason, plastics and metals of low atomic number, such as aluminum, are used in shielding to protect against beta particles, atomic electrons emitted during nuclear processes and electrons accelerated in fields of high voltage.

Protons emitted during nuclear processes and protons accelerated in fields of high voltage dissipate their energy by ionization (Shapiro 1980). Protons behave like very slow moving alpha particles and densely ionize materials in their path.

Gamma rays and X-rays are electromagnetic waves that are distinguishable from each other only by wavelength. Gamma rays and X-rays dissipate energy by indirect ionization. The bases for attenuating gamma and X-ray emissions are the energy of the photon(s), and density and thickness of material in the path of the beam (Cember 1996; Shapiro 1980). The strategy for protection involves use of very dense materials, including lead and high density concrete in permanent installations. Emission of X-rays from secondary ionizations that occur within the shielding material complicates design of shielding for large-scale installations.

Neutrons are non-charged particles. They interact with matter without causing direct ionization. Attenuation of the energy carried by neutrons occurs only during collision with the nucleus of atoms in the absorbing medium. Maximum transfer of energy occurs during collision with nuclei of hydrogen atoms (protons) that have almost the same mass (Shapiro 1980). This transfer is similar to what occurs during collision of billiard balls. The collision creates an ion that moves at high speed. The ion transfers energy rapidly to the surroundings. Neutron emitters in the fuel in nuclear reactors are surrounded by water which is used as the medium for heat transfer, as well as shielding for neutron emissions. Where neutron emitters are present in other devices, water or other highly hydrogenous material (as present in plastics and some polymers) is a major component in the shielding strategy.

Very slowly moving neutrons (0.025 eV), known as thermal neutrons, pose an especial concern. Thermal neutrons have about the same energy as molecules of gas at the same temperature (Cember 1996). This corresponds to velocity of 2,200 m/s. Thermal neutrons are capable of continued interaction with biological tissue and the nuclei of susceptible isotopes. Absorption of a thermal neutron by the nucleus of a susceptible isotope can lead to production of energetic gamma radiation, generally around 7 MeV. This energy is highly penetrating and itself requires shielding. Neutron shields terminate in hydrogenous materials for this reason. Shielding for controlling exposure to neutron emissions is a complex problem.

## Nonionizing Electromagnetic Radiation

Electromagnetic radiation of wavelengths longer than X-rays generally has insufficient energy to cause ionization. Photons of nonionizing radiation interact with atoms and molecules in matter in various ways. The manner in which the interaction occurs depends on the energy in the photon. Protection by shielding against emissions of these sources utilizes strategies that exploit these interactions.

### Ultra-Violet Emissions

Shielding used against ultra-violet (UV) emissions employs the concept of absorption and reflection. Sunscreens used on the skin to reduce exposure to UV emissions from the sun employ these concepts. Molecular components absorb the energy with the promotion of electrons from ground state energies to excited states in quantum jumps. Return of excited electrons to ground state levels or transition of electrons from stable higher levels to the vacant positions leads to emission of photons of energy in the visible region (March 1985).

The energy needed to cause a particular transition is highly specific. A UV emission containing more or less energy per photon will not interact with the substance. Most situations encountered in industry, especially welding, involve UV emissions that cover a range of energies and wavelengths. Lamps containing mercury emit in the UV region of the spectrum. The spectrum of a UV emission is not "sharp" due to the existence of rotational and vibrational contributors to the absorption process.

Cleavage of molecules excited by UV emissions is a potential outcome from the excitation. Cleavage can result in formation of smaller molecules from the fragments, but also free radicals (highly reactive chemical species) (March 1985).

Structures in organic chemical molecules that absorb in the UV region of the spectrum are known as chromophores. Chromophores generally have double or triple chemical bonds involving carbon, oxygen, and/or nitrogen within the molecule. The presence of other chemical groups attached to the organic molecule also can act as a chromophore.

Glass is opaque to the passage of UV emissions. Quartz is transparent to UV emissions. For this reason, quartz is used in the envelope of lamps intentionally emitting UV. Glass is used in the envelope of lamps that do not emit UV. Glass is used as the outer envelope in lamps containing a quartz inner envelope in order to block UV emissions that pass through the quartz envelope. Failure of the glass envelope in this type of lamp has led on occasion to unintended exposure to high levels of UV energy.

UV-protective lenses in glasses and welding shields contain colorless coatings or formulations that block or absorb UV emissions. Products rated for use during welding require broad-spectrum capability due to the range of emissions contained in welding arcs.

Plastics and polycarbonate formulations used in UV-protective lenses in eyewear absorb a substantive portion of the UVA emissions and almost all of the UVB emissions. Chromophores incorporated into the formulation or applied as a coating further improve blockage of the UV of the lens material.

### Visible Emissions

Electromagnetic radiation in the visible region of the spectrum is readily absorbed by common and readily available materials and the pigment in coatings. Selective absorption of some wavelengths and reflection of others gives rise to the wavelengths (colors) detected (visible) to the eye.

### Infra-Red Emissions

Infra-red (IR) emissions cause internal heating of materials. Effective shielding against these emissions occurs by reflection of energy. Polished aluminum, along with other shiny metals, is a very effective reflector of wavelengths in the infra-red region of the spectrum. These are utilized to provide protection around furnaces and especially in rocketry and aerospace applications. Reflective materials used for shielding need not be substantive in mass or thickness. The reflective effect occurs at the surface. This is the basis for use of foils in fabrics of protective clothing worn around sources of radiant energy.

## Radiofrequency Emissions

Radiofrequency emissions occur across a broad spectrum of energies and wavelengths. The only distinguishing feature between any of the classes of signal is the wavelength. The means used to shield against exposure to these emissions is the Faraday cage. The Faraday cage is an enclosure formed by an electrical conductor or by a mesh of such material. Such an enclosure blocks the electric field component of the electromagnetic wave.

The Faraday cage acts as a hollow conductor. Mobile electrons that form electrical current position themselves along the surface. Hence, this positioning will occur whenever an electrical field is present, regardless of frequency. The important consideration is that openings in the enclosure, whether contiguous as in the case of the outer body of an automobile or the mesh in a screen are considerably smaller than the wavelength of the electromagnetic disturbance. Grounding the Faraday cage increases the capacity for acting as a shield The Faraday cage shields against time-varying electric fields generated both inside and outside the structure.

Faraday cages have formal application in lightning protection, protection against electrostatic discharges, testing of electronic components and circuits in an environment free from electromagnetic interference, and protection against emission of microwave energy through the door of microwave ovens (Figure 7.3). Rooms involved in secure communications employ Faraday cages to prevent emission of signals. Bags containing aluminized surfaces, otherwise intended to minimize entry of heat into frozen foods and chilled items act as Faraday cages and enable shoplifters to defeat security systems based on RFID (Radio Frequency Identification) technology.

## Sound

Sound is an energy disturbance that relies on the presence of a medium (air or other gas, liquid or solid) containing atoms and/or molecules for its transmission. Sound covers a wide range of frequency.

Figure 7.3. The mesh in the door of the microwave oven acts as a Faraday cage to prevent transmission of microwave energy.

Shielding against sound employs both reflection and absorption. Both phenomena occur when sound impinges on a surface. Low frequencies pose the greatest challenge for control regardless of the method employed.

Nonporous barriers of sufficient mass (at least 10 kg/m$^2$) can produce appreciable sound reduction when positioned between a source and a receiver (Magrab 1975). The effectiveness of noise barriers is readily apparent to residents who live adjacent to highways where installation of barriers has occurred. The challenge of shielding against low frequencies becomes apparent where a direct line of sight exists even at considerable distance. Slight changes in the contour of land can create noticeable differences in transmission of noise from a harbor or other industrial source up a mountain slope.

Absorption occurs during transmission of sound through materials. Sound energy dissipates by inducing vibration in the material, incoherent internal reflection off surfaces, and cancellation of waves of equal frequency and opposite phase (relation of the shape of the wave in time).

## Access Control

Some sources of energy are capable of causing immediate and serious traumatic injury up to and including death. This equipment requires a different strategy from that employed in control of energy sources with which the operator does not interact directly.

### Remote Manipulation

Remote operation allows the operator to perform routine, complex manipulations of objects that pose a life-threatening risk of exposure. The basis for this approach is to separate the operator from the environment in which the object is located and in which the operation occurs (Figure 7.4). The operation occurs through use of manipulators that act in place of the operator's hands. The operator works from an area totally protected from the harm posed by the source of energy. The operator operates the manipulator and once experienced sufficiently through practice in non-critical situations, then is able to per-

Figure 7.4. Remote manipulation from an area separated by shielding from highly hazardous items and materials enables operations to occur in safety. (Source: NASA.)

form complex manipulation of objects. This concept is widely employed in the nuclear industry. Similar applications occur in control of aircraft, construction machinery, and many other devices that perform complex actions, yet are operated remotely.

## Enclosures

Enclosure is the fundamental preferred approach taken with almost all equipment capable of causing immediate and serious traumatic injury. Enclosures are available in many configurations. They are also readily apparent in equipment found in the home. The washing machine and dishwasher provide examples of the use of enclosures to provide operator protection.

The enclosure that surrounds machines and equipment forms the boundary that prevents access to internal components and shields against interaction with moving parts and other sources of energy. Inside the enclosure, mechanical components and other sources of energy in this equipment may contain no shielding or guarding. This is most likely the case where entry into the enclosure during operation is considered to be an unrealistic expectation or prohibited or prevented by the action of interlock safety devices.

### Alternate Preferred Path

Another approach to worker protection from sources of hazardous energy, especially electrical, is to create a preferred path for the flow of energy. Energy will follow all available paths from a source to a sink, despite the existence of a preferred path.

This approach is widely used in protection against electrocution and is based on the concept of current flow in parallel circuits. The strategy is to create a path to ground having considerably lower resistance than other paths and large capacity for flow of current. Electrical current will take all paths available to it. One of these paths may include the human body. Creation of a path of considerably lower resistance and high capacity for current flow does not eliminate flow through other paths. Rather the flow of current through the body becomes so small as not to cause harm.

Another application of the concept is the diversionary channel created during construction of dams. Normally during construction, the space in front of the dam contains no water. Water flows through the diversionary channel. The alternate preferred path is the channel to be followed by existing flow of water. This path is deliberately constructed.

Sewer flows are often continuous. Work on existing sewer lines requires an alternate path. The alternate path often requires removal of water from the existing line by pumps and reintroduction into the line at a point downstream from the area in which work is to occur. Isolation of the existing sewer line within the zone of diversion is an essential part of this strategy.

### Personal Protective Equipment

Personal protective equipment is widely used as protection against sources of hazardous energy. Perhaps the first personal protective equipment used in this application was the shield and later the suit of armor used by warriors. The current iteration of this equipment is the body armor worn by police and the whole body protection worn by football and hockey players and mountain bike riders (Figure 7.5).

Personal protective equipment is not favored as the means of protection, as this usually imposes a burden on the wearer. Certainly this is true with the gloves worn by linemen as protection against electrocution. These gloves comprise an inner rubber glove tested to ensure protection against the voltage of concern and an outer leather glove. The outer leather glove protects the inner rubber glove against abrasion and puncturing. Together these gloves limit dexterity. Of greater concern is the sweat emits from the skin into the inner glove. This liquid wets the skin and causes considerable softening.

This system of protection is not always protective. Draining the sweat from the inner glove has led to electrocutions from the passage of a line of conductive droplets of water through the air (NIOSH 2000).

Protection of the eyes from emissions of welding arcs requires very dark lenses. The welder is unable to see through the lens prior to striking the arc. During welding, the welder can see the arc and the immediate surroundings, but can see nothing else in the further surroundings. The ability to see only a

Figure 7.5. Some modern-day personal protective equipment used against sources of hazardous energy is equivalent to the suit of armor. (Modified from Wiki Common, San Francisco.)

limited field of view imposes a considerable burden onto the welder and has led to accidents where awareness of conditions in the outer surroundings was essential to ensuring safety.

## Organization and Management of Industrial Activity

Much has been written about management. Usually the focus of the books and articles is management of the entity at the macro level with the emphasis on output and economic survival and prosperity. The books talk about managers and management styles. All too often managers are administrators and not leaders.

At this level, people who aspire to be managers are often trained through university business courses that lead ultimately to the famed MBA degree. The basis for the selection process for entry into these programs is academic performance. The programs offer the opportunity to participate in scenarios to analyze outcomes. This training provides the benefit of forcing consideration about alternative courses of action and the consequences of decisions and actions.

At the level at which hands-on activity occurs, the situation is considerably different. Training of the type mentioned above is a luxury and is certainly not focused on analyzing and solving problems that occur at this level of activity. Typically, the individual who becomes a supervisor is a worker of some experience who is promoted to the next higher level. Multiple promotions could considerably raise the status of this individual within the organization. This is especially the case with organizations that promote from within.

The person who is a worker on Friday and a supervisor on Monday is the same person. The change in status typically means only a change in the color of the employee's hardhat. That is, the only difference between workers and supervisors, and sometimes managers, literally is the color of the employee's hardhat. The newly appointed supervisor is expected intuitively to understand the difference between performing the work and the overseeing of others who perform in that role.

For most supervisors, the emphasis in the job is scheduling and ensuring productivity and production. Doing the job prepares the person for the outcomes that are required and expected. These are the familiar requirements of the job.

The job of supervisor is considerably different than that. In regulatory statutes, the supervisor is legally accountable for the safety of persons under supervision. This responsibility is often overlooked in the normal stresses of scheduling and ensuring productivity and production. Legal accountability is considerably more encompassing than knowing and memorizing the regulatory statutes that apply to work performed by subordinates.

An important part of supervision is to focus on conditions during the performance of work. Workers focus on performance of task. This approach is optimal where the supervisor and workers remain together in the same group during work activity. The situation, as described, is less and less common. More typical is the situation where the supervisor and workers are separated by time and distance. The supervisor cannot be present in more than one place at a time. This puts the emphasis on workers in these situations to focus on both task and conditions. As identified in previous discussion, the ability to focus simultaneously on both realities is impossible. In these situations, workers must look out for each other in a classic application of the buddy system.

What training that is available for the newly appointed supervisor tends to focus on interpersonal relationships and fails to cover the basics of managing activity and change, especially for purposes of safety and problem-solving. Activities and events are either managed or they are allowed to happen. The latter is readily apparent in many operations.

Management in the context of the workplace is the leadership that actively controls the direction of work activity with an emphasis on eliminating or at least minimizing and controlling the magnitude of downgrading factors. Control minimizes the impact and risk of harm to participants in the activity, meaning that control increases the level of safety. Absence of control or loss of control leads to higher than necessary risk and impact.

Management to prevent the occurrence of adverse conditions, events, and consequences at the hands-on level of an entity is at least as important as it is at the administrative level. At the administrative level, the consequences are financial. At the hands-on level, they are personal. Personal consequences can include death and loss of co-workers and friends; traumatic injury and loss of limbs; loss of trust; fear of working on equipment, machines and systems; and other factors. The impact is also financial. Financial impact occurs through loss of production and productivity, lawsuits and compensation costs from accidents, and loss of reputation of the entity.

## A Structured Approach for Management of Hazardous Energy

Management of hazardous energy is the same in concept as management of other activities. To rephrase an earlier statement, management is the activity of leadership that controls the direction of activity with an emphasis on eliminating or at least minimizing and controlling the magnitude of downgrading factors. Control minimizes the risk of harm to participants in the activity, meaning increases the level of safety. Absence of control or loss of control allows events to happen and leads to higher than necessary risk and impact.

The first consideration toward addressing the question of how to create a structured approach to the management of hazardous energy is to understand where the exposure occurs and can occur. The simple answer to the question is literally everywhere.

The approach proposed here can be visualized using two onions (Figure 7.6). One represents the universe of hazardous conditions posed by a workspace and the work to be performed in it. The other represents the universe of actions needed to render the workspace safe before and during work activity. The most hazardous of workspaces and the preventive actions that they require would be represented by the intact onions. Said another way, a life-threatening situation demands the highest level of action. On the other hand, a situation posing minimal hazards needs minimal response. The average situation is likely to be far less serious than the worst case, yet deserving more attention than minimal response.

The challenge is to create a systematic, defensible process for removing as many layers from the onions as justified by a particular situation. This approach conserves resources, so that they remain

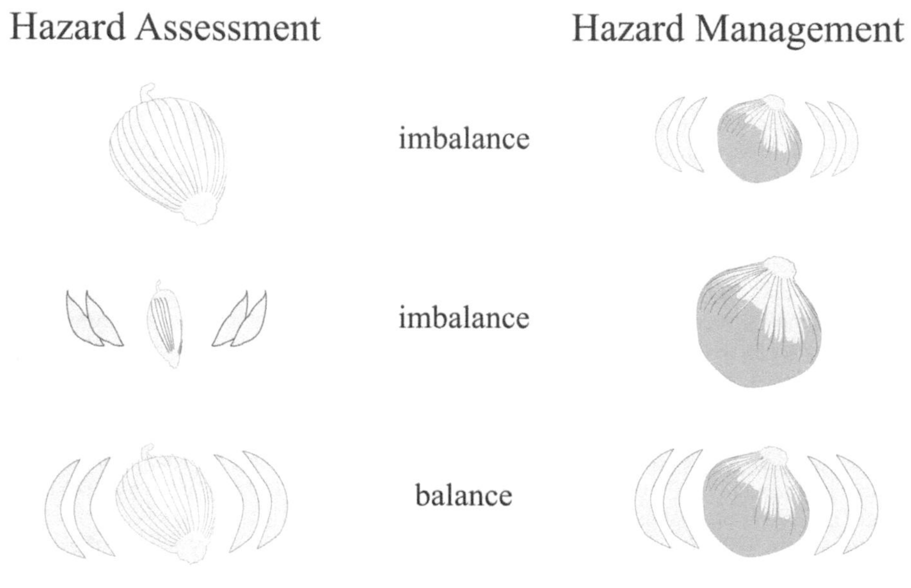

# Hazard Assessment                    Hazard Management

imbalance

imbalance

balance

Figure 7.6. Achieving balance in response to hazardous conditions in workplaces requires balance between the assessment of the level of risk of harm posed by hazardous conditions and the response. These concepts are achieved in a way analogous to peeling the layers from two onions.

available to be utilized where most appropriate. The caveat with this concept is to ensure that mismatch does not occur. Under-reacting to a hazardous situation could lead to a fatal accident. However, overre-acting to a nonhazardous situation also could produce deleterious consequences.

Anticipating, recognizing, and evaluating hazardous conditions created by energy sources and materials containing hazardous energy in equipment, machines, systems and structures is not intuitive. This is readily evident from the analysis of fatal accidents summarized in Chapter 1. Victims exhibited absence of the anticipation and recognition skills, and the knowledge needed to address hazardous conditions.

## The Uncharacterized Workspace
In the preceding examples, the external surroundings exerted no influence on the work to be per-formed. This is not always the case. In some cases, the exterior surroundings themselves pose the haz-ard. This situation has occurred, for example, during work on industrial robots. Industrial robots sometimes are configured in trains or groups. Work on any member of the train can expose the worker to harm from the actions of other members of the train.

Most industrial workspaces reflect some attempt at accommodating occupancy by people. They contain a roof; walls; windows; a smooth floor; lighting; tempered air provided by a ventilation sys-tem; a rest area containing seating; and washrooms containing toilets, sinks, and possibly showers.

Many other workspaces are outdoors. They subject workers to the ultra-violet emissions and heat from the sun; rain; wind; snow; ice; irregular surfaces on which to stand and sit; portable toilet facilities containing at best a toilet, hand sanitizer, and a sink; and rest areas that include the cabs of trucks, ser-vice trailers and vehicles, and construction trailers, and possibly the ground. The primitive nature of these conditions is widely recognized and acknowledged.

Both types of workspace can contain a third level of workspace. These workspaces occur both in-doors and outdoors. They are created from these workspaces by conditions requiring the shutdown and service of equipment. These workspaces often are unplanned and do not appear in drawings. They are sometimes created through dismantling equipment. As a result they are uncharacterized compared to workspaces in which work normally occurs.

Over the years, regulators recognized that certain workspaces posed higher risk of harm than normal (McManus 1999, McManus 2004). Some of these types of workspace have come to be known as confined spaces, while others possessing similar geometry were labeled as trenches and excavations, and still others that posed similar hazards received no distinction through a label or other direct recognition of concern.

Confined spaces do not distinguish themselves by size, shape, or nature of work activity. Some are large; some are small; some are partly open; others are completely enclosed. Some house equipment; others store liquid and solid bulk materials. Chemical processes occur in others. Still other confined spaces fit none of these descriptions. Some confined spaces are intuitively obvious, even to uninformed observers, while others gain recognition only because of events that occur in them. Still others defy visual recognition and require application of investigative questions.

Over the years, workers have routinely entered confined spaces without preparation, sometimes with fatal consequences. Workers have been severely injured or killed during work external to confined spaces while preparing them for entry. First contact with the environment in confined spaces has led to fatal injuries because of hazardous conditions overlooked during assessment, or not eliminated or properly controlled during preparatory activity. Lastly, the work activity can create the hazardous condition, while conditions in the space following preparation otherwise could be completely harmless.

The first thing that sets many confined spaces apart from "normal" workplaces is the function of the space. Confined spaces generally are not the type of place in which people work. While some people work in confined spaces every day, most people will never enter a confined space during their entire working lifetime.

Some workspaces are confined spaces because of design. They serve some function, such as containment of a chemical process or storage of bulk materials, or housing machinery. During normal conditions of operation, entry never occurs because entry cannot occur. There are various reasons why entry cannot occur. The contents could be pressurized and would spill out from the access opening. They could be highly toxic or could cause an immediate fire and explosion hazard. They could also be highly infectious. In other circumstances, while entry could occur by opening an access portal, the interior of the space could be so hot as to be lethal, or could pose an electrocution risk, or other situation that could result in death or serious traumatic injury. As well, some of these situations occur in workspaces that do not fit the legal definition for confined spaces.

Periodically, these structures and the equipment inside them require maintenance and replacement and improvement. Entry must occur in order to perform these activities. Preparation for entry leads to opening of spaces that meet the regulatory definition for confined spaces. Preparation for entry can create confined spaces from spaces that did not fit the legal definition. As well, preparation for work can create similar hazardous conditions and potential for harm in workspaces that do not fulfil the requirements of the definition.

Accidents that occur in confined spaces are very rare events. They are very difficult to predict and very expensive to prevent. Minor mistakes lead to consequences that far outweigh the significance of the error. These accidents often are more severe than those that occur in normal workspaces. Like a trap that is set and ready to spring, the hazardous condition that causes the accident acts rapidly, often without prior warning. Accidents involving confined spaces commonly injure or kill more than one person. After the accident has occurred, conditions often return rapidly to "normal," as if nothing has happened.

Statistics published in the U.S. by the National Institute for Occupational Safety and Health (NIOSH 1994) indicate that engulfment caused about 65% of the fatal accidents that occurred in workspaces having the characteristics of confined spaces. Atmospheric hazards caused about 30%. This situation appears to have shifted toward atmospheric hazards in the years following publication of the document following regulatory intervention and emphasis toward preventing cave-ins and collapse of earthen walls from slumpage by sloping and shoring.

A critical aspect that sets confined spaces apart from "normal" workspaces is the boundary surface. Boundary surfaces can prevent escape from a space. They can also amplify or magnify hazardous conditions in some circumstances. This is especially the case with noise and other forms of energy that reflect off the walls. The boundary surface can increase the risk of electrocution by acting as a path to ground. The relationship between the individual, the boundary surfaces and the source of energy or contamination or other hazardous condition constitutes the major factor in the onset and outcome of accidents that occur in confined spaces. The boundary surface need not be substantive or even contiguous. A chain-link fence or other enclosure can act as a boundary surface. The boundary surface can even be invisible, as is the case of the invisible envelope that surrounds robotic equipment and other machinery that operates in free space.

The term, confined space, is similar to a "zipped" computer file. These two words unzip to identify 42 hazardous conditions recognizable from regulatory statutes that can occur in these workspaces, and for that matter, in other workspaces as well. Of the 42 hazardous conditions, 26 pertain to energy, some of which can exist at hazardous levels (capable of causing immediate harm). Of course, this situation is not unique to confined spaces and can apply to any workspace. The boundaries for inclusion under the term, confined space, reflect the phrases used in regulatory definitions. The definition is the property of the authority having jurisdiction over the situation.

Attic spaces, crawlspaces, trenches, and excavations and other structures have the geometric characteristics to satisfy the requirements of the definition of confined space. There are circumstances where the definition definitely should apply to these structures because of the hazardous conditions that are present or can develop. Precautionary measures that follow solely from application of the definition of trench or excavation do not provide a satisfactory level of protection to workers in these environments under these circumstances.

Another disturbing situation is the number of accidents resembling those that occur in confined spaces that happen in workspaces that do not fit conventional definitions for confined spaces. The environments in which these accidents occurred do not fit any of the available positions in the matrix of workplace definitions yet posed hazards far beyond the level of harm normally encountered in workplaces (Table 7.2).

Additional structures that pose similar issues include the interior of duct connected to air-handling and air heating units. These ducts often are entered through doors of partial height. The doors open outward on the upstream side of the fan and inward on the downstream side. Inward pressure on the upstream side and outward pressure on the downstream side of the fan causes the doors to slam shut when opened during normal operation. Opening the doors requires considerable effort and is not easily achieved by an individual who is injured. The individual can remain entrapped in these structures until help arrives. Noise associated with ventilating equipment, coupled with isolation of the equipment, can severely impair the summon for help.

Other examples include satellite wastewater treatment systems and some types of wells. Satellite wastewater treatment systems can fit inside a shipping container. Entry occurs through a door of normal size from an above-ground structure. The interior of the structure is contiguous with the atmosphere above the wastewater. Depending on conditions and off-gassing from the influent, the atmosphere in these structures can be or can become life-threatening.

Some types of wells are known as "blowers and suckers" in the well-drilling business. These wells are known to exist throughout North America. They have the ability to inhale atmospheric air under high pressure conditions and to exhale gas under low pressure conditions. The exhaled gas can be highly oxygen-deficient. Some of these systems are used to air-condition homes, while others supply potable drinking water. Some of the drilled water wells are housed in buildings that contain pumps and other equipment. Exhalation in some locations is strong enough to force stones from the well.

The process of inhalation and exhalation is known to occur in mining. This involves rock containing sulfides and carbonates. The sulfide reacts with oxygen to form sulfates. This reaction removes oxygen from inhaled air. The sulfate formed in this reaction is acidic and can react with carbonates in the rock to form carbon dioxide. The processes are well known in mining and influences mine ventilation. These emissions can also create hazardous conditions during processing of water from mine drainage and tailings. Emissions can occur in buildings containing equipment that handles these waters.

## Table 7.2

## Accidents in Uncharacterized Workspaces

| Location | Description |
|---|---|
| Drum mold | The victim inserted head and shoulders into the mold (50 cm diameter by 80 cm deep) and was overcome by vapor from the perchloroethylene that he used to wipe the surfaces. |
| Waist-high paint mixing pot | The victim bent over into the pot while standing on the floor and was overcome by vapor from the methylene chloride that he used to clean the interior. |
| Shaft of a dumbwaiter | The victim opened the door to determine the location of the car and was struck by it. |
| Sand-mixing machine | The victim reached into a side hatch to repair a bearing and was struck by the blades when the drive motor started unexpectedly. |
| Empty 200 L (55 gal) drums | The hazardous (ignitable/explosive) atmosphere contained in these drums caused about 16% of fatal welding and cutting accidents. |
| Open-topped degreaser | The victim reached in over the top of the degreaser (2 m by 3 m by 1.5 m high) to recover a part that had fallen in and was overcome by the solvent-rich, oxygen-deficient atmosphere. |
| Enclosure of an abrasive blasting machine | The operator walked in through the access door to retrieve a part that had fallen and was asphyxiated by the nitrogen atmosphere used to inert the interior. |
| Enclosure of industrial robot | The maintenance worker was struck by a robot in the train unrelated to the robot on which he was performing instruction. |
| Sand quarry | Slumpage of sand at the working face of an embankment buried a front-end loader. |
| Mine tailings reclamation | Four people died from a hazardous atmosphere inside a sampling shed (2 m by 3 m by 2.5 m high) entered through a door of normal size. |
| Office | A carpet layer sealed-off the ventilation system, as well as gaps in walls and doors, to contain solvent vapors from adhesive used to anchor carpet tiles and was overcome by toluene vapor. |

Source: McManus 1999.

Traditional definitions for confined spaces have focused on the geometry of structures, and not on conditions. This then begs the questions, what exactly is a confined space and how should management of the hazardous conditions occur?

The test for effectiveness of a particular definition is simple: does it encompass all of the workspaces in which the hazardous conditions common to confined spaces exist or could develop? This is not always the case with definitions used in present regulatory and consensus standards. Definitions that do not encompass the unusual workspaces in which people work and are at risk of serious harm serve no one.

A major difficulty with the management of hazardous conditions in confined spaces is the fluid nature of the problem. A seemingly minor change or error or oversight in preparation of the space, selection or maintenance of equipment, or work activity, can change the status of conditions from innocuous to life-threatening. The work activity itself can create the problem. A space rendered innocuous by preparatory activity can become life-threatening because of hazardous conditions created during the work.

The reality is that people continue to die in similar numbers year after year in accidents involving confined spaces. They also continue to die in other workspaces of similar geometry covered in other regulations or that have similar hazardous conditions and are not directly covered in regulations. Many of these accidents are eerily repetitive. This begs the question about how to solve the problem.

The current approach to regulation forces workspaces into "pigeon holes" or a matrix based on definitions, for example, trenches and excavations. Expectations within the definitions fail to consider all work activities. Unconsidered work activities have resulted in development of hazardous conditions and unaddressed risk of harm.

The pertinent questions are what hazardous conditions does a particular workspace pose, what harm can arise, and what must we do to eliminate or to at least control them? The classification is irrelevant to safety. Provision of safety requires a practical mechanism to ensure that hazardous conditions existing or created in all workspaces receive due care and attention. That is, we need to establish "the right reasons to do the right things."

The critical reality in all accidents is the "golden hour," as taught in classes on emergency first aid. The "golden hour" reflects the observation that survival of victims of serious traumatic injury is considerably enhanced when they are moved from a workspace to the operating room in a hospital within one hour of the occurrence of the accident. Otherwise, the risk of death from shock increases significantly. Extrication from workspaces is often not considered in planning and can consume considerable time. Traffic on roadways delays response and increases transport time.

Viewed in the context of this discussion, confined spaces are a subset of a larger group of workspaces that fit the description of uncharacterized workspaces. This term also includes trenches and excavations, and workspaces that pose similar hazardous conditions and are not captured by the definition of confined space. This would include machinery spaces and enclosures that surround industrial robots and barriers that surround unguarded machinery that operates in free space. Industrial robots sometimes are configured in trains or groups. Work on any member of the train can expose the worker to harm from other members.

Uncharacterized workspaces are workspaces in which there is a reasonably foreseeable risk of harm from a hazardous condition that may exist or may develop. Hence, change the paradigm to require hazard assessment and work procedures for all workspaces that are uncharacterized. Under this scenario, the Qualified Person performs a hazard assessment and recommends measures to eliminate or control hazardous conditions. The Qualified Person prepares a written procedure for work to be performed. The assessment identifies probable accidents, proposes means to mitigate them, and focuses on the needs of the victim.

Given the nature of the work to be performed involving potential exposure to hazardous sources of energy, the surroundings in which this can occur are the uncharacterized workplaces discussed here. They could include confined spaces, but also could include workspaces in which similar conditions are present, even though they do not satisfy the requirements of the definition. In order to ensure safety, assessment of conditions in the surroundings in which work is to occur is essential. This is additional to assessment of the source(s) of hazardous energy.

**Assessing the Uncharacterized Workspace**
Hazard assessment is a process for identifying potential and actual hazardous conditions and assessing the level and acceptability of risk of harm. Hazard assessment is a difficult process, because many of the conditions that can produce acute or traumatic injury and are difficult to recognize and assess. Haz-

ard assessment is semi-quantitative at best and intuitive at worst. Yet, as indicated by real-world accidents, the tolerance for error in judgement is very small.

Hazard assessment equals hazard identification + detailed commentary. Hazard identification asks what hazardous conditions are encounterable during work activity. Hazardous condition is a situation capable of causing overexposure or impairment, injury, or death. Hazard assessment is a commentary based on experience and research. Hazard assessment leads to recommendations for control measures. The process must be repeatable and defensible. It builds on concepts of industrial hygiene: anticipate what could be present, recognize it when it is present, and predict the severity of the hazardous conditions created by its presence. This requires knowledge about processes and operation of equipment, machines and systems, potential magnitude of emissions, and measures to eliminate or control hazardous conditions, as well as personal and other protective equipment.

Determining the existence and extent of concern to apply to these conditions is the central role of hazard assessment. Hazard management focuses on work procedures and processes, and control of hazardous conditions associated with them. Hazard assessment and hazard management are co-operative and iterative processes.

A critical feature of the process is the occurrence of decision-making. Decision-making in a logical and defensible manner is a skill to be acquired. It is not intuitive. In 1958, social scientists C.H. Kepner and B.B. Tregoe founded a consulting practice dedicated to the field of decision-making (Kepner and Tregoe 1981). Kepner and Tregoe had observed that decisions were made in an irrational, often highly questionable manner. They concluded that the cause of this problem was deficiency in the process of gathering and organizing information. They developed a rational method for gathering and using information containing four key elements. These include organized enquiry to assess and clarify information; investigating the situation to determine cause and effect; applying logic to assemble facts from information and make choices; and anticipating the future and mitigating against its undesirable effects.

This system has received widespread application, and has stood the test of time, and is as valid and valuable in the present context as in any other. The Kepner-Tregoe model uses these patterns of thinking to varying extent in rational processes for decision-making (Table 7.3).

Several of the processes are directly applicable to the process of hazard assessment both for the uncharacterized workspace (Table 7.4) and also for sources of hazardous energy in equipment, machines and systems (Table 7.6). Note that Table 7.4 does not show the process of analysis for both the undisturbed or operating condition and the work activity(ies) to be performed. For assistance in completing Table 7.4, refer to McManus (1999) and McManus (2000).

Actions preceding preparatory activity could occur during operation. These could include photographing or recording a video of the action of equipment or establishing other physical parameters. This is especially helpful where operational drawings are not available. A video can establish the nature of movement of components. These measurements can establish, for example, storage and transformation patterns in mechanical linkages and complex levers.

A less comprehensive approach involves examination of drawings or plans of the workspace or pictures taken by a third party. However, the loss of comprehensiveness intrinsic to this approach requires acknowledgment. Experience acquired in assessing similar situations can compensate for this lack of presence to an extent. Lack of presence at the workspace by the Qualified Person represents potential loss of information. Plans or drawings of the equipment or structure and its surroundings and pictures should become part of any permanent record arising from this assessment.

Proactivity of this kind increases certainty and reduces uncertainty inherent in speculation about possibilities by establishing facts. The greater the extent of knowledge that can be established about conditions prior to pre-work preparation, the less is the unknown that remains to be discovered and addressed. Investigating conditions throughout the undisturbed or operational period is one strategy for reducing the level of concern during hazard assessment. The value of this initiative in the overall process of hazard assessment and hazard management cannot be overstressed.

Hazardous conditions identified during operation are addressed in subsequent segments of hazard assessment and hazard management. Most of the hazards identified in the operational phase likely will be controlled or eliminated during shutdown and pre-work preparation. It is possible that a

## Table 7.3

## The Kepner-Tregoe Model for Problem-Solving and Decision-Making

| Rational Process | Fundamental Questions |
| --- | --- |
| Situation appraisal | What is the deviation in the situation from the specification? |
| | Where is the deviation occurring? |
| | What is the magnitude of the deviation? |
| | What are possible/probable causes of the deviation? |
| | What option(s) are available for resolving the situation? |
| Problem analysis | What unexplainable deviation from what is expected is occurring? |
| | What is the nature of the deviation from the expected (identity, location, timing and magnitude)? |
| | What distinctions exist in the deviation? |
| | What distinctions reflect change in the *status quo*? |
| | What possible/probable causes can explain distinctions from the *status quo*? |
| | Does testing confirm the most probable cause to be the actual cause? |
| Decision analysis | What outcome(s) must occur as a result of the decision? |
| | What is/are the accompanying outcome(s) that are wanted as a result of the decision? |
| | What alternatives are available? |
| | What are the consequences of choosing a particular alternative? |
| | What actions about the potential problems are available now? |
| | What contingencies exist? |
| Potential problem analysis | What vulnerabilities exist in the proposed course of action? |
| | What could go wrong? |
| | What are the likely causes of these potential problems? |

Source: Kepner and Tregoe 1981.

hazardous condition is not controllable. This could persist through remaining segments of the hazard assessment. Continuous water flow is one such example. Reduction of the hazard represented by this condition may not be possible regardless of actions that are taken.

Hazardous conditions beyond elimination or control become the subject of study to determine requirements for emergency response and ultimately the decision about whether the proposed work can occur in safety.

## Table 7.4

## Assessing the Uncharacterized Workspace

| ABC Company | Uncharacterized Workspace Assessment | Work Activity |
|---|---|---|
| Owner: | Location: | Assessed by:<br>Tel:<br>Date: |
| **Confined Space Appraisal** | | |
| adequate size and configuration for employee entry? | limited means of access and egress? | intended for continuous occupancy? | **Confined Space?** ▲ |
| | | | » All entries **bold**? |
| Description: | Access/Egress: | Contents: |
| | Adjacent Spaces: | Equipment: |
| | Function/Use: | Process: |
| External Surroundings: | Downgrading Factors: | Potential Impact on Work Activity: |

**Hazard Assessment (Operational/Undisturbed Space) or (Work Activity)**

| | |
|---|---|
| **Hot Work** | **Yes** No **Possible** |
| **Atmospheric Hazards** | |
| **Oxygen Deficiency** | **Yes** No **Possible** |
| **Oxygen Enrichment** | **Yes** No **Possible** |
| **Bio/Chemical** | **Yes** No **Possible** |
| **Fire/Explosion** | **Yes** No **Possible** |
| **Micro/Biological** | **Yes** No **Possible** |
| **Ingestion/Skin & Eye Contact Hazard** | **Yes** No **Possible** |
| **Physical Agents** | |
| **Noise/Vibration** | **Yes** No **Possible** |

## Table 7.4 (Continued)
## Assessing the Uncharacterized Workspace

| | |
|---|---|
| Heat/Cold Stress | Yes No **Possible** |
| Non/Ionizing Radiation | Yes No **Possible** |
| Laser | Yes No **Possible** |
| **Personal Confinement** | Yes No **Possible** |
| **Bio/Mechanical Hazard** | Yes No **Possible** |
| **Hydraulic/Pneumatic/Vacuum Hazard** | Yes No **Possible** |
| **Process Hazard** | Yes No **Possible** |
| **Safety Hazards** | |
| **Rundown/Strikedown** | Yes No **Possible** |
| **Structural Hazard** | Yes No **Possible** |
| **Engulfment/Immersion** | Yes No **Possible** |
| **Entanglement** | Yes No **Possible** |
| **Electrical/Electrostatic** | Yes No **Possible** |
| **Fall** | Yes No **Possible** |
| **Slip/Trip** | Yes No **Possible** |
| **Visibility/Light Level** | Yes No **Possible** |
| **Explosive/Implosive** | Yes No **Possible** |
| **Hot/Cold** | Yes No **Possible** |

**Hazard Assessment (Operational/Undisturbed Space) or (Work Activity)**

| Hazardous Condition | Occurrence/Consequence | | |
|---|---|---|---|
| | **Low** | **Moderate** | **High** |
| | \| | \| | \| |
| hot work | | | |
| atmospheric hazards | | | |
| oxygen deficiency | | | |
| oxygen enrichment | | | |

## Table 7.4 (Continued)
## Assessing the Uncharacterized Workspace

bio/chemical

fire/explosion

micro/biological

ingestion/skin & eye contact

physical agents

noise/vibration

heat/cold stress

non/ionizing radiation

laser

personal confinement

bio/mechanical hazard

hydraulic/pneumatic/vacuum hazard

process hazard

safety hazards

rundown/strikedown

structural

engulfment/immersion

entanglement

electrical/electrostatic

fall

slip/trip

visibility/light level

explosive/implosive

hot/cold surfaces

## Table 7.4 (Continued)
## Assessing the Uncharacterized Workspace

In this table, **NA** means that the category does not apply in any normally foreseeable situation. **Low** means that exposure is readily identifiable but believed to be much less than applicable limits or that exposure to nonquantifiable hazardous conditions is unlikely to produce injury. **Low-Moderate** means that exceedence of regulatory limits is believed possible or that nonquantifiable exposure could produce minor injury requiring self-administered treatment. Control measures or protective equipment should be considered. **Moderate** means that exposure is believed capable of exceeding regulatory limits or causing traumatic injury requiring first aid treatment or attention by a physician. Protective equipment or other control measures are necessary. **Moderate-High** means that exposure is believed capable of considerable exceedence of regulatory limits or causing serious traumatic injury. Advanced control measures or protective equipment are required. **High** means that short-term exposure is believed capable of causing irreversible injury, including death. Advanced control measures or protective equipment are required.

**Action Required**

Source: Adapted from McManus 1999.

## Undisturbed or Operating Condition

This part of the assessment considers the work area prior to undertaking control measures. This establishes the status quo for the work area The status quo establishes hazardous conditions that exist or could develop and could affect the work to be performed. These conditions require attention prior to entering the workspace to perform the intended work.

Equipment, systems, and structures lie within the bounds of organizations. As such, they "belong" to some entity within the organization. This means that someone has responsibility for them. Active operations offer the opportunity for discussion about how the equipment or system functions, materials of construction, materials currently and historically handled, operating records and documentation, and the presence of operational utilities. Because work involving equipment, systems, or structures in a process operation can require extensive coordination with operations and maintenance personnel, this liaison is extremely important.

Operators are invaluable resources about the hands-on nature of the equipment, system, or structure. Maintenance personnel usually have valuable experience from performing hands-on work activities involving the equipment or structure. The experience, knowledge, and perspective of these individuals are invaluable to the process of hazard assessment. They can provide input regarding historical aspects about what was done, and why, and how successfully. How successfully is a relative term, especially in organizations where industrial hygiene surveillance was not performed during this work.

Downsizing through layoffs and early retirements similarly has created a loss in the information resources of every entity that has undergone these losses of people. With any of these circumstances, loss of the expertise and knowledge resident in former employees and operating and maintenance records can considerably complicate hazard assessment. The starting-point for hazard assessment in these situations is a thorough historical review of the site, its equipment, and structures.

Identification of materials that handled, stored, used, produced, or consumed in the equipment or structure defines the need for information about their properties. This information should be available from Material Safety Data Sheets (MSDSs) or resources that provide information about processes and individual chemical products and substances. Materials handled in the vicinity of the equipment or structure represent an additional question and a possible concern. These could cause exposure during the work activity or could react with substances removed from the equipment or structure. Subsidiary concerns with chemical products and substances include conditions of storage, since these could destabilize the product or lead to formation of new substances through reaction with moisture, oxygen in the air, or substances stored in proximity.

The status quo in these situations usually represents stability. That is, conditions have remained the same, or changed so gradually that nothing is likely to happen with passage of the several more days, weeks or even a month that are needed to decide how to evaluate address the potentially hazardous situations that could arise once the status quo is disturbed by opening the space.

Other essential inputs in the assessment process are detailed drawings of the equipment, system, or structure. Drawings should provide information about layout, including access points; piping, flanges, valves and vents; mechanical equipment and controls; hydraulic and pneumatic equipment and controls; electrical circuits; structural detail; and materials of construction. The information contained in these drawings is a vital resource for understanding about the workspace. This input may contain information that has been lost to memory over the passage of time and through changing of personnel.

The current physical condition of the workspace and the equipment and structures of interest are major concerns. An evaluation to determine the integrity of the equipment or structure should occur prior to starting any work activity. In cases where structural integrity is questionable, review of all structures by a qualified Professional Engineer is highly recommended. Under no circumstances should work occur on equipment or structures of questionable integrity.

Work Activity
The next segment of the hazard assessment addresses work activities to be performed in the workspace. Prior to the start of work activity, elimination or at least control of hazardous conditions identified in the first part of the document will have occurred. The work to be performed can create hazardous conditions that themselves require elimination or control. The work, as discussed here, does not pertain to the source of hazardous energy. This requires detailed consideration independent from this discussion.

Prediction of the impact of work activities is the most difficult task in the process of hazard assessment. This also is the aspect about which least is known. Work activities in maintenance have received little attention over the years. Unfortunately, there is little in the published literature to provide assistance. This puts reliance on the knowledge base and experience of the Qualified Person.

Emergency Preparedness and Response
Accidents occur in every workplace. They happen, despite the taking of all reasonable precautions and then some. There is no reason to expect that accidents occurring in workspaces affected by this hazard assessment should differ in type and severity from those that occur in ordinary workplaces when subjected to the same considerations. If preparatory and precautionary measures for work activity are adequate, the only remaining consideration should be unpreventable accidents.

## Identification and Description
This section identifies and describes the equipment, system, or structure under consideration. This consideration should lead to an electronic or paper-based file. The latter becomes the repository for all information gathered and derived about this situation. This become the basis for future enquiries.

**ABC Company** is the entity that performs the work involving the uncharacterized workspace. ABC Company may also be the Owner or host Employer.

**Work Activity** is the collection of tasks to be performed. Work Activity can group several or all tasks into a single document. This is possible for many tasks. This approach minimizes duplication of hazard assessments and eases the burden on the reader.

**Owner** is the entity on whose property the workspace is located. The Owner may not be the entity performing the hazard assessment for entry and work involving the uncharacterized workspace. The Owner may also be the host Employer.

**Location** is the recognized location of the uncharacterized workspace. At the time of entry for work activity, the location may appear considerably different from routine circumstances. This can occur because of disassembly of structures and reconfiguring the workspace. That is, the uncharacterized workspace may be a unique configuration of the normal workspace.

**Assessed by** establishes the name and credentials of the individual who performed the assessment and completed the document. The Qualified Person (identified in the regulatory requirements of some jurisdictions) should personally inspect (externally at the minimum) the equipment, system, or structure under consideration, unless the individual has extensive experience in assessing similar circumstances. Some jurisdictions specify professional designations that establish the capability of an individual to perform this work. Providing formal qualifications of the individual who performs this work "up front" is extremely important, since the outcomes from decisions taken during the assessment can have life or death implications.

Personal inspection is critical for making decisions about the status of the workspace. Personal inspection establishes first-hand the geometric configuration of the space, potential hazardous conditions, and its relationship to the external surroundings.

**Telephone Number** establishes a reliable means to contact the Qualified Person, should this necessity arise in the present at the time of the review that led to creation of the document or into the future.

**Date** establishes when the hazard assessment occurred. The date establishes the relative freshness and possible staleness of the document. Competent documents created for situations that remain static are useable long into the future.

**Confined Space Appraisal** indicates whether the workspace is a confined space. This document can also serve to assess hazardous conditions in the confined space (McManus 1999). This depends on the complexity of the relationship between the confined space and the external surroundings, for which completion of a separate document may be required. The questions used in the determination are unique to the jurisdiction. The questions used here derive from the definition found in 29CFR 1910.146 (OSHA 1993). This is the OSHA standard for permit-required confined spaces in general industry.

**Description** provides information about the physical configuration of the equipment, system, or structure. This section establishes the unique identity of the equipment, system, or structure (including a pile of material) under examination. This should include (approximate) dimensions, location relative to other equipment, structures, or configurations that could pose hazardous conditions and a risk of harm. Ideally, the hazard assessment documents should include plans showing the equipment or workspace and structures adjacent to it.

**Access/Egress** indicates the location of the entry and exit portal and the means by which workers are expected to gain access and egress to and from the workspace.

**Adjacent Spaces** indicates spaces that are located beside, above, below, and across diagonals from the workspace. This concept is especially important in ship environments where the adjacent space and its contents could pose a greater hazard than the space to be entered (NFPA 2009; OSHA 1993). This concept has universal applicability in assessment of the uncharacterized workspace. Workspaces usually are considered only in terms of the space enclosed by some defining geometry. This approach views the space in isolation from its surroundings. In reality, a space or structure is inseparably connected to its surroundings, just as visual inspection would indicate. What is misperceived is that conditions or activities that occur in the surroundings do not impact the workspace, equipment, or structure and vice versa.

NFPA 306 used by Marine Chemists in the practice of hazard assessment on ships and the OSHA standard on confined spaces in shipyard employment require evaluation of conditions across the boundaries (planar, curved, and angular) that enclose the space of immediate interest (NFPA 2009; OSHA 1994). Knowledge gained from tragic experience indicates that consideration must extend in all directions across the boundaries of the space. While this concept is foreign to land-based practitioners, it has considerable merit and deserves to be included in the hazard assessment. This can draw attention to areas around the workspace where activities unrelated to the work activity could have major influence on its safety.

**Function/Use** describes the function or use of the equipment, system, or structure. A workspace could have many applications. The level of hazard associated with conditions could vary considerably from application to application.

**Contents** identify the contents of the space at the time of the identification and hazard assessment. Listing of contents of historical importance should also occur. Previous contents are considered to be extremely important in some standards (NFPA 2009; OSHA 1994). Previous contents are a concern where residues either could react chemically with the later contents, or where these residues could cause exposure or a pose a risk of fire and explosion during subsequent work activity.

**Equipment** refers to equipment located within the structure. Operation of this equipment, expected or unexpected, could create conditions highly hazardous to occupants.

**Process** refers to the chemical, physical, or biological process that normally occurs in the space. Processes often involve intermediates, possibly uncharacterized. These processes can be the source of residual liquids and solids, and gases and vapors. Continuance of the process during work activity could create highly hazardous conditions.

**External Surroundings** identifies the area that surrounds the uncharacterized workspace. As discussed, the uncharacterized workspace could be the external surroundings to a confined space in which the work is to occur and for which a separate hazard assessment is required.

The external surroundings can pose more onerous hazardous conditions than the workspace itself.

**Downgrading Factors** identifies hazardous conditions that exist or can exist in the external surroundings to the uncharacterized workspace or to the confined space, whichever is the subject of the discussion.

**Potential Impact on Work Activity** indicates possible outcomes including overexposure and traumatic injury caused by exposure to the downgrading conditions.

## Hazardous Conditions

The first thing to note about the protocol is that it provides three options for decision-making: Yes, No, or Possible. Yes or No indicates that the assessor is confident about expressing an opinion about the situation posed by the condition. Possible indicates unsureness about the situation. Unsureness can occur for technical reasons and for "political" reasons. Following a question through to its logical end-point defuses any controversy surrounding an issue. A Yes or Possible response requires descriptive information about the situation, and will require remedies to be applied. A No response does not require additional information, although sometimes providing this for clarification benefits the situation.

This protocol is general in scope, and as a result, may have high applicability to some situations and low applicability to others. The protocol considers all conditions that may be present in the uncharacterized workspace. These include atmospheric hazards that could lead to overexposure to chemical and other substances, and other hazardous conditions not directly related to acute injury caused by rapid release of hazardous energy or materials containing hazardous levels of energy. These conditions require assessment as part of a comprehensive hazard evaluation. The latter could occur here or separately.

**Hot Work** is any type of activity or process that generates arcs, sparks, flames, heat, or other sources of ignition. Hot work can include welding, grinding, flame and mechanical cutting, gouging, soldering, brazing, burning, and drilling. Other sources of ignition include energetic chemical reactions, electrostatic discharge, abrasive blasting, space heating, electrical short-circuits, tobacco smoking, lightning, engines and compressors, and power tools, fixtures, electrical switches, and appliances that are not explosion-proof (ANSI 1995; API 2001a, API 2001b; NFPA 2009).

The indication of intent to pursue hot work during work activity externally on a structure containing or potentially containing ignitable substances either in the residual atmosphere or residual materials during pre-entry preparation is cause for major concern. Guidance for performing hot work is provided in a number of standards and regulations (API 2001a, API 2001b; Canada Gazette 1992;

Jones and King 1991; NFPA 1997, NFPA 2007, NFPA 2009; OSHA 1993, OSHA 1994, Canada Gazette 1992; Standards Australia 1995).

Hot work undertaken prior to opening a structure is sometimes perceived as not disturbing the status quo in any obvious way. However, hot work can act as a source of ignition of ignitable materials both inside and external to the equipment or structure. This situation again draws attention to the fact that equipment or structures usually are considered in isolation from their surroundings. The reality is that they are inseparably connected to their surroundings, just as visual inspection would indicate. Activities that occur in the space directly impact conditions in the surroundings and vice versa.

Surfaces in the area in which hot work is to occur may require cleaning to remove ignitable material, or shielding to prevent heating. These considerations apply to both sides of the surface affected by the heat, since hot work occurring on one side could ignite material on the other. Nothing happens in isolation. This is the basis for one of the key concepts found in some standards: concern for adjacent spaces (NFPA 2009; OSHA 1994). Hot work can act as a source of ignition of ignitible materials both inside and external to the structure, as amply demonstrated in fatal accidents discussed in Chapter 1.

The approach to hazard assessment utilized here is also to regard the hazardous condition represented by hot work as being greater than the sum of its contributing parts. This is the case because of the amplifying role of the boundary surfaces in creating the hazard in the workspace. A person working in the tight quarters posed by many workspaces would experience considerably greater difficulty in escaping from the danger zone during an accident caused by hot work. For this reason, plus the need to consider additional protective measures, hot work is considered as an independent category.

**Oxygen Deficiency** refers to the decrease in oxygen level below what is normally present at which a person will experience physiological impacts. This value is around 14% at sea level compared to the ambient level of 20.9%. The regulatory limit for oxygen deficiency in many jurisdictions is 19.5%. An oxygen-deficient atmosphere can develop in a stratified layer or throughout the space. The condition in a particular space depends on several factors: the process by which oxygen deficiency occurs, time, and turbulence.

Oxygen depletion occurs through a limited number of mechanisms. First is biological or microbiological action involving organisms, such as bacteria and fungi, resident on surfaces of the structure or in residual liquids. Oxygen depletion can also result from chemical action involving residual contents or the material of construction of the structure. Oxygen depletion can also occur through absorption by the walls of the concrete if left undisturbed for a prolonged period.

The key factor that affects air quality in these spaces is water. Water is essential for corrosion of metals and for the growth of microorganisms. The process of rusting involves chemical reaction of atmospheric oxygen with metal surfaces. The process continues deeper into the metal long after initial rusting of the surface. Microorganisms require oxygen for aerobic growth. Oxygen depletion through the action of microorganisms or oxidation of metals requires considerable time and quiescent conditions.

Ventilation during pre-entry preparation should eliminate the oxygen deficiency unless consumption of oxygen is rapidly and actively occurring.

**Oxygen Enrichment** refers to conditions where the level of oxygen is elevated compared to the ambient level of 20.9%. The regulatory limit for oxygen enrichment in many jurisdictions is 23.5%. Atmospheric enrichment by oxygen increases the ignitability of materials, including those not normally known to be combustible in air. Oxygen molecules adsorb onto the surface of the material.

The usual source of oxygen in these situations is leakage from fittings in oxyfuel systems or deliberate use of gaseous oxygen as a coolant during hot summer weather. Contact with the skin and clothing considerably enhances ignitability. An oxygen-enriched atmosphere can develop in a stratified layer or throughout the workspace. The condition in a particular space depends on the process by which oxygen enrichment occurs, time, and turbulence. Atmospheric testing must occur throughout the atmosphere of the space to ensure that all possibilities are examined where oxygen enrichment can occur.

**Bio/Chemical Hazard** refers to hazardous atmospheric conditions caused by biochemical substances (substances of biological origin) and chemical substances. Bio/chemical hazard indicates the presence or potential presence of a physical form of the substance (gas, vapor, mist, or particulate) in air. Contamination can develop in a stratified layer or throughout the space. The condition in a particular space depends on several factors: the process by which contamination occurs, time, and turbulence. Atmospheric testing must occur throughout the atmosphere of the space to ensure that all possibilities are examined.

The uncharacterized workspace can contain many sources and types of sources of airborne contaminants. The undisturbed or operational space can contain residues on surfaces as well as active sources of production of aerosols (liquid and solid particles). Ventilation during shutdown and pre-work preparation should eliminate or at least control any exposure hazard in the space. The only remaining potential sources of contamination would be residues. A hazardous condition arising from residual materials is considerably more likely where removal does not occur during shutdown and pre-work preparation. Removal of residual contents during shutdown and pre-work preparation is essential to eliminate the hazardous condition during work procedures.

Work activity can use a broad range of chemical products, including coatings, adhesives, grouts, mortars, and cements. Welding, cutting, and grinding are sources of many contaminants. Demolition can create airborne dust containing many substances from the materials of construction, including asbestos, lead, and silica (quartz).

Exhaust from vehicles and mobile equipment can enter the work space under conditions related to cold weather or inappropriate placement of equipment, and can cause needless exposure. This is especially probable where the geometry of structures prevents dispersion.

Exhaust from gasoline engines contains carbon monoxide (CO), carbon dioxide ($CO_2$), unburned fuel vapor, and particulates. Exhaust from vehicles is less of a problem at this time due to use of catalytic converters in the exhaust system. Small engines used in generator sets, pumps, and other small portable units are major sources of exposure to exhaust, especially where the geometry of structures prevents dispersion.

Exhaust from diesel engines contains nitric oxide (NO), nitrogen dioxide ($NO_2$, unburned fuel vapor, and particulates. Exhaust from the diesel engines of trucks is often directed horizontally at ground level. In cool or cold weather, vertically directed exhaust will cool rapidly and stratify in a layer just above the top of the truck. Descent to ground level is possible. In confining geometries, accumulation could pose a serious exposure risk.

Combustion gases discharged from propane-fueled air heaters used in cold weather conditions are sources of carbon dioxide and possibly carbon monoxide. Under cool or cold conditions, these gases do not disperse and can be entrained into the air provided by portable ventilation systems.

**Fire/Explosion Hazard** results from the presence or potential presence of an ignitable substance in the form of gas, vapor, mist, or particulate in the atmosphere of the workspace. Formation of an ignitable mixture also requires the presence of air or oxygen in the atmosphere of the space. Ignitible and explosible concentrations of many substances are much higher than those associated with the onset of toxicological effects.

Contamination can develop in a stratified layer or throughout the space. The condition in a particular space depends on several factors: the process by which contamination occurs, time, and turbulence. The possibility of an ignitable mixture indicates that testing must occur throughout the atmosphere of the space to ensure that all possibilities are examined. The action of testing should not itself create an ignition hazard. The latter has occurred during actions to open a space or in use of equipment that can generate electrostatic charge, a possible source of ignition.

Ventilation during shutdown and pre-work preparation should eliminate any fire or explosion hazard in the space. The only remaining potential sources of contamination would be residues or slag containing ignitible material.

A hazardous condition arising from residual materials is considerably more likely where removal does not occur during shutdown and pre-work preparation. Removal of residual contents during shutdown and pre-work preparation is essential to eliminate the hazardous condition during work procedures.

A cloud formed in the surroundings external to the space during removal of a vapor or gas could contain an ignitible/explosible mixture. Similarly, introduction of air into the space during forced ventilation could form an ignitible/explosible mixture by displacement. This mixture also could be present in the ventilation system. Bonding and grounding and use of explosion-proof equipment are essential practices under these circumstances. The cloud formed in the surroundings under these conditions could constitute a serious exposure hazard to persons working outside the space.

The process utilized during work activity could introduce fuel, as well as sources of ignition. A source of ignition used in a space containing residues of ignitible material could create an extremely hazardous situation. Some activities introduce ignitible materials into the space or generate them. Certain activities such as spraying or abrasive blasting, in addition, can generate electrostatic charge, a possible source of ignition.

Designation of an area as an NFPA/IEC (National Fire Protection Association/International Electrotechnical Commission) hazardous location requires use of electrical equipment rated for this service.

**Micro/Biological Hazard** refers to hazardous conditions caused by microorganisms and larger animals including insects, spider, and rodents. The usual concern relates to growth of bacteria and mold.

**Ingestion/Skin & Eye Contact Hazard** refers to substances that can cause body contamination by ingestion and adverse reactions on the skin and in the eye. Lead compounds are toxic by ingestion and reflect poor hygiene practices. Many substances can irritate the skin. Some are corrosive and some are allergic sensitizers.

**Noise/Vibration** can result from operation of equipment inside or adjacent to the workspace or from motion of internal machinery or contents. External sources of noise and vibration sometimes are used to promote flow and to prevent caking of flowable solids in silos and hoppers used in bulk material handling.

**Heat/Cold Stress** can result from thermal conditions in the space that differ from conditions of comfort. Shutdown and pre-work preparative activities attempt to moderate thermal conditions in the space to ambient. This may not be possible in all cases. Where this accommodation is not possible, these conditions require action in order to enable the worker to perform the work activity safely. This could require use of portable cooling or heating units, and personal protective equipment, as well as other supplies.

Weather conditions and location also can influence the onset of heat and cold stress. Radiant energy from the sun incident on a steel container considerably elevates the temperature inside. Damp, windy, wet, or winter conditions cause hypothermia in the unprotected.

Equipment and processes utilized during work activities can heat the interior of the space. Similarly, ventilation equipment used during winter can cause severe wind chill due to the velocity of air movement.

**Non/Ionizing Radiation** is potentially present from a number of sources. Sources of nonionizing radiation include intense electrical and magnetic fields from large electrical equipment, and ultraviolet, visible, infra-red, microwave, and radiofrequency emissions.

Potential sources of ionizing radiation include residues from naturally occurring and artificial radioactive materials deposited on walls of piping in upstream facilities that handle natural gas, materials of construction that are radioactive and radioactive sources that are present inside the space or external sources whose beams are directed through the space. Radioactive beams that are directed through the space are used in level gauges or for irradiating the contents.

Actions taken during shutdown and pre-work preparation normally are sufficient to render sources of nonionizing radiation inoperative, since these units require electrical power. Temporary shielding and other techniques also may be required for sources of ionizing radiation. Removal of deposits of naturally occurring radioactive material from interior walls of structures or in piping during shutdown and pre-work preparation may not be possible.

Work processes or equipment may produce or contain sources of nonionizing radiation including intense electrical and magnetic fields, and ultraviolet, visible, infra-red, microwave, and radiofrequency sources, or require work to occur during operation of existing equipment. Reflective surfaces present in many workspaces can exacerbate irradiation by these sources.

Some processes utilize sources of ionizing radiation from naturally occurring and artificially radioactive materials. The hazard arising from these materials depends on the isotope and quantity, shielding and mode of use. Conditions of use are provided in the licence for the source. The beam from sources used in industrial radiography can penetrate the wall of the structure. The beam could be directed through the wall of the space from the exterior or vice versa. In either case, occupancy must be prevented during this work.

**Laser** is an acronym for Light Amplification by Stimulated Emission of Radiation. Lasers have developed over a wide range of wavelengths and are found in many applications. Equipment that produces laser energy may be present in the space. Actions taken during shutdown and pre-work preparation are expected to render the laser source inoperative, since these units require electrical power. Work activity may introduce equipment containing a laser or utilizing laser emissions, or require work in the space during operation of the existing equipment. The hazard arising from this equipment depends on the type and strength of the laser, shielding, and mode of use. Reflective surfaces present in many workspaces can exacerbate irradiation by these sources.

**Personal Confinement** refers to the ability of the space, its portals and interior geometry to confine an entrant and hinder normal movement or emergency evasion. For some persons, claustrophobia is a major consequence of confinement. As well, individuals whose dimensions are large could experience considerable difficulty in passing through small openings in entry portals and interior walls, and available between piping and other obstructions. This could create considerable problems in situations where rapid evacuation is required. Complicated internal geometry can considerably complicate extrication of an injured individual following an accident.

**Bio/Mechanical Hazard** refers to biomechanical hazards and hazards due to moving components in equipment and machinery located in the space.

Start-up of equipment often occurs unexpectedly during normal operation. Mechanical equipment in enclosures sometimes has exposed moving parts. Motion can include rotation and reciprocation (back and forth and up and down). Unexpected motion of these parts can strike occupants and cause injury. Actions taken during shutdown and pre-work preparation attempt to render mechanical systems to the status of zero energy. This is not always possible. Mechanical hazard also includes discharge of stored energy through motion of parts. Effectiveness of the deactivation/de-energization analysis and the lockout procedure ultimately is confirmed during pre-work preparation or pre-work inspection.

Biomechanical hazards mainly occur during entry and work. Awkward postures required in many structures to gain access to components of interest increase the risk of musculoskeletal injury. Additional risk factors include stress and strain caused during the freeing of parts that have seized or "frozen" in position. Soft tissue injury and broken bones can result from parts that "let-go" from a previously immobile position and strike parts of the body. Disassembly and demolition pose significant risk of harm. Confinement, mentioned above, can exacerbate this situation.

Work activity can introduce mechanical equipment into the workspace. Operation of this equipment could pose considerable risk of harm because of the confines of the workspace. To illustrate, failure of a grinding wheel inside a confined workspace poses considerably greater risk of harm than a similar failure in a larger, less confined area. Rapidly moving objects reflect off walls, piping and other obstructions that form the boundaries of the space. Evasive movement to avoid a flying object is constrained by walls and other obstacles.

**Hydraulic/Pneumatic/Vacuum Hazard** pertains to systems containing pressurized fluids and systems that operate under reduced pressure.

Hydraulic systems routinely operate at pressures ranging as high as 205 MPa (30,000 lb/in$^2$). Storage can occur in dedicated devices (accumulators) and in isolated parts of the system. Pneumatic systems usually operate from plant compressed air systems containing pressures up to 1.03 MPa (150 lb/in$^2$). Pneumatic systems can contain one or more accumulators for storage of pressure. The air receiver tank also acts in the same function. Energy released from storage in hydraulic and pneumatic systems can cause unexpected movement of parts. Residual pressure retained in hydraulic circuits, including accumulators, valves and hoses is a form of energy that can release at unpredicted moments.

Stored energy normally is removed during shutdown and pre-work preparation. The verification step confirms the effectiveness of the deactivation/de-energization analysis and lockout procedure for mechanical equipment containing hydraulic, pneumatic and vacuum systems.

Work activity could introduce hydraulic and pneumatic equipment into the space. High pressure water jetting systems, for example, operate at 2000 to 3000 lb/in$^2$ and higher.

Vibration, ultra-violet energy from the sun (see above) and other sources, accidental contact with sharp-edged objects, and vandalism can result in failure of hoses and couplings. A jet of hydraulic fluid emitted from a hose leak or failed coupling has the ability to puncture the skin and to penetrate into the tissues. A jet of water from a high-pressure system can cause similar damage.

Injection injuries are particularly serious because the path of the fluid is unpredictable after entry. Injection of fluid can lead to gangrene and possible loss of the affected limb through amputation. Injection injuries require immediate attention from a well-experienced medical practitioner specializing in vascular injury.

Pinpoint leaks from pressurized hydraulic lines have also caused amputation of finger parts. Use of fingers to trace location of such leaks is highly hazardous and must not occur.

Some systems operate under vacuum. These can include structures into which entry can occur. The more common sources of hazard from vacuum systems are industrial vacuum trucks. The vacuum system on these trucks can generate ~27 inches of mercury vacuum (~90 kPa). Vacuum of this magnitude combined with high inward flow and relatively large diameter piping can entrain clothing and limbs during careless operation. Suction into the vacuum line can cause serious traumatic injury including evisceration and avulsion.

**Process Hazard** refers to hazardous conditions that arise from chemical, physical, and biological processes that normally occur in the space. These processes can include intermediates whose properties are not known.

Shutdown and pre-work preparation normally eliminate process hazards. Work activity often involves use of chemical products that ndergo reaction as part of their use. The quantities of these materials usually are small, and pose an exposure hazard to workers, rather than one involving energy release.

**Rundown/Strikedown** is the collision between humans and vehicles and mobile equipment. Mobile equipment can include front-end loaders, excavators, and other construction equipment as well as materials handling equipment, such as forklifts. Rundown can cause soft tissue and severe traumatic injury up to and including death. Rundown usually is preventable through traffic control measures.

**Structural Hazard** refers to internal instability in a structure in which or on which work is to occur. Structural instability can result from many causes, including overload and corrosive failure resulting from incompatibility between contents and materials of construction. The potential for harm results from unexpected or unpredicted change in the status quo that could include collapse of part or all of the structure.

Actions taken during shutdown and pre-work preparation could exacerbate an existing structural hazard or create one. Development of a hazardous condition under these circumstances may not be recognized at the time that the critical action occurs. Consideration about the impact of work procedures on structural stability and integrity is essential. Advice from a professional Structural Engineer about structural stability should be sought where this issue is a concern.

**Engulfment/Immersion** refers to the ability of flowable solids or liquids in the workspace to engulf (surround) or immerse a worker. Engulfment can result from failure of caked materials on walls or of bridged flowable solids or slumpage down a slope. Engulfment also can result from instability of walls and floors of trenches, excavations, shoring, and similar structures.

A prone individual can drown in 15 cm of water.

An engulfment or immersion hazard could exist at the time of work activity depending on the nature of preparation. Work procedures could exacerbate the hazard depending on circumstances.

**Entanglement** in this context pertains to the ability of immobile structural features of the workspace and its interior equipment to entangle clothing or other equipment such as lifelines. Entanglement impedes movement especially during action taken during avoidance. Entanglement hazards are an inherent property of the geometry of the space, its internal configuration, and that of equipment inside it. Few options may be available for ameliorating potential entanglement hazards.

**Electrical/Electrostatic** hazard pertains to AC and DC current electricity and to electrostatic accumulation and discharge.

Electrocution can result from exposure to or work on live electrical circuits in equipment that operates in the space. These can include cables and wires that pass through the space but have no role in work to be performed. Safe occupancy of the space depends partly on the integrity of the bonding and grounding network associated with electrical systems. Bonding and grounding networks are subject to corrosive attack and breakage during earthquake as well as other causes of damage. Effectiveness of the deactivation/de-energization analysis and lockout procedure require confirmation during shutdown and pre-work preparation or the pre-work inspection.

Work activity often introduces electrical equipment to the workspace. The safe use of this equipment is contingent on its integrity and performance. The integrity of electrical equipment depends on the quality and integrity of its internal and external wiring and components. Wiring used in portable equipment and extension cords is vulnerable and destroyed rapidly during accident situations. For this reason, pre-work inspection of this equipment and ground-fault protection are required on all electrical equipment and temporary supply lines. Circumstances could merit use of low voltage equipment or equipment rated for use in NFPA/IEC hazardous locations containing potential or actual ignitable or explosive atmospheres.

**Fall** hazard exists when vertical distances through which fall can occur could cause serious bodily injury through collision with equipment, structures, and people. Fall distance is defined in regulatory limits. Fall hazard is one of the first considerations following the entry into the workspace for the start of work activity. Installation of fall prevention and fall protection equipment is a prerequisite for any work activity where fall can produce serious injury or exceeds regulated limits.

**Slip/Trip** refers to conditions that can cause slipping and tripping accidents. Conditions conducive to occurrence of this type of problem include surfaces that are slippery due to residual contents or angle of slope. Tripping hazards result from geometric features of the interior structure.

Shutdown and pre-work preparation should address slip/trip hazards existing in the workspace. In spaces where thorough removal of residual contents has not occurred, slip/trip may pose a serious and underappreciated risk of harm.

Work activity can introduce slip and trip hazards. This is especially likely where hoses, lines, duct, pipe, and electrical cable introduced into the space during work activity are present on the floor. These objects create a considerable housekeeping problem.

**Visibility/Light Level** pertain to the ability to see in the workspace. Visibility hazard results from presence of aerosols (airborne particulates, mists or sprays) that obstruct visibility of equipment and signage and can pose explosibility hazards. Shutdown and pre-work preparation should eliminate the visibility hazard resulting from aerosols. Work activity can create the conditions where visibility is obstructed by the presence of aerosols. Air blowing to clean surfaces and activities that abrade and micronize, such as sanding and grinding can generate levels of airborne dust in the explosible range.

Light level refers to the illumination produced by presence of overly bright illumination or the absence of a source of illumination in the workspace. To illustrate, some workspaces are populated by ultra-bright luminaires used for lighting other areas, while others have no lighting. Light levels produced in the former situation could damage the retina. Low level of lighting could lead to trip and fall accidents.

Installation of lights should occur where illumination of the space is absent or substandard. Illumination level generally is sufficient when a person can read a newspaper unaided. Installation of lights in the space is one of the first work procedures to occur after entry occurs. Conditions could warrant use of explosion-proof or intrinsically safe equipment, since the lights could act as an ignition source. The potential for fire or explosion as an influencing factor deserves careful consideration.

**Explosive/Implosive** hazard refers to structures that are physically explosive or implosive. These conditions reflect internal atmospheric pressure that differs from ambient. A small difference in pressure impressed on a large surface can create considerable force. This situation is a serious concern where work occurs inside airhandling equipment during fan operation. Doors upstream from the fan open outward while those downstream from the fan open inward. An injured worker may not have the strength to open these doors under operating conditions. Equalization of pressure to atmospheric levels usually occurs during shutdown and pre-work preparation.

Explosive also refers to the presence of chemically explosive substances. Removal of these substances and products containing them must occur prior to the start of work.

Work activity that introduces explosive substances or devices into the workspace deserves exceedingly careful consideration because of the consequences of a deliberately planned or accidental explosion. Rapid pressurization and depressurization following an explosion could cause structural failure. Work activity is unlikely to involve equipment posing an implosive hazard.

**Hot/Cold Surfaces** refers to surface temperatures that differ markedly from ambient. These could cause burn injury, or if hot enough, act as a source of ignition. Burns on unprotected skin can occur following prolonged exposure to temperatures exceeding 45° C. Seriousness of the burn depends on temperature and duration of contact. Brief contact with surfaces below −10° C or prolonged contact for several hours with surfaces below −1° C causes cold burns (frostbite). Frostbite is akin to the white patches of "freezer burn" observed on the skin of a frozen turkey.

Shutdown and pre-work preparation should attempt to eliminate hot or cold surfaces. This action may not be possible in circumstances where maintaining a particular temperature is critical to the operation. Insulation applied to the surface or personal protective equipment could be the only means for assuring protection under these circumstances. Where this is not possible, this condition requires further action to enable the worker to perform the work activity safely.

Many processes used during work activity involve controlled generation and application of heat. These processes may create hot surfaces and point sources of heat. Hot surfaces created in this manner can pose considerable hazard due to proximity of the surface to the worker. Some processes apply cold to freeze contents in piping. The freezing agent poses a risk of harm from frostbite.

**Action Required** lists actions that are recommended or required for assessing and eliminating or at least controlling hazards identified during each stage of the hazard assessment. Actions required for hazardous conditions involving work activity could be considerably different from those required during pre-work preparation. This results from the changing nature of hazardous conditions.

## The Operational Cycle of Equipment, Machines and Systems

Industrial activity is cyclic, not haphazard in nature. Equipment and processes are characterizable by different modes in the activity cycle of an entity, namely production, standby, and shutdown. The operational cycle of equipment and systems provides an important basis for creating a structured approach to assessment and control of hazardous energy. The approaches needed and available to ensure safety differ according to the mode in which they are to be applied. Previous discussion about industrial activity alluded to this situation.

The operational cycle can be short or long, repetitive or non-repetitive, simple, or complex. Fatal hazardous conditions can exist or develop during any of these segments. This was amply illustrated in the fatal accidents summarized in the first chapter.

As appropriate, performing the hazard assessment for each segment can considerably clarify the process of achieving safe work conditions because the focus could change in each part of the cycle. This is the basis of the approach taken during machine safeguarding (refer to Appendix G). The level of concern about a particular hazardous condition could change drastically between any of the modes of the operational cycle. The only way to ensure that the level of concern and consequently the level of response for any segment are appropriate is to assess conditions for each. This need not occur literally every time in every situation, only where justified by the level of concern warranted by a particular circumstance.

### Production Mode

Production mode is intrinsic to the modus vivendi of an entity. The end-point of production is the creation of goods and services and the creation of wealth for the entity and employment for its workers.

One endeavor of production as an activity is to achieve the goals of wealth creation as efficiently as possible. Efficiency includes safety, because accidents and injuries are highly destructive to efficiency. Many studies have shown the destructive power of accidents and injuries on the individual(s) directly affected, but also on other people affected indirectly or even peripherally.

Production equipment operates during production mode. Production mode implies that the machine, equipment or system is fully energized and capable of performing or in the midst of performing its designed or intended function.

Production mode consists of two or more phases: action and standby. Action is the activity during which production machines and equipment create output.

The standby phase of production mode is the period during which output is not occurring. The machine or equipment remains deactivated and partially or fully energized awaiting the command to act.

The command to act can originate from a human operator or other source of authority, such as a computer. Machines and equipment of older vintage provide an illustration of these concepts. The human operator places items in a particular orientation into the point of operation, then commands action of the machine or equipment, and then removes the item once acted on by the machine or equipment. Following the action, the device remains deactivated and partially or fully energized in standby until commanded to act again by the operator. These actions are the basis for considerations about machine safeguarding (refer to Appendix G).

## Standby Mode

Standby mode is the state between production and shutdown. Standby can occur during production as mentioned above. Standby is a block to production mode and the ability to perform the designed or intended function imposed through action and application of control devices. These actions include deactivation and possibly de-energization of subsystems and components of machines and equipment and systems. Removal of deactivation and partial or full de-energization imposed by control devices enables production to reoccur. Imposition of actions additional to deactivation and de-energization (isolation, lockout, and verification of status) shut down the machine, equipment, or system.

## Shutdown Mode

Shutdown mode is the state of inability of a machine or equipment to operate or to release energy. Note that the system can retain energy during shutdown.

ANSI Z241.1-1975 produced by the American Foundrymen's Society introduced the concept of the Zero Mechanical State (ZMS) in a consensus standard (ANSI 1975). The concept had existed for a considerable period of time in the industry previous to this (AFS 1961). ZMS (Table 7.5) pertained to machines and equipment used in the foundry industry. ZMS was the state of the machine of minimum potential unexpected movement caused by stored energy.

Price (1975) advocated expansion of the concept of the Zero Mechanical State soon after publication of ANSI Z241.1-1975 to cover all forms of energy as the Zero Energy State (Table 7.5). The concept of the Zero Energy State attempts to ensure that the equipment or structure contains no source of energy that could expose an individual to harm during maintenance or disassembly. This means no introduction of energy through the input(s) or output(s) and no storage of energy in the equipment or structure.

The state of zero energy in the equipment, machine, or a system is not always achievable and often is not necessary for pursuing the work to be performed. Energy retention in equipment, machines, and systems occurs through deliberate storage and entrapment. In situations where energy can remain in equipment, a machine, or a system, an analysis to determine the potential for expression of the energy and procedures to prevent expression of this energy are required.

### Shutdown and Pre-work Preparation

Shutdown comprises the actions taken to shut down equipment, machines, or systems. Pre-work preparation comprises the actions taken to render the equipment, machine, system, or structure safe for work activity following shutdown. Safe means a state of zero energy or a state where energy retained in the equipment, machine, system, or structure cannot express itself during disassembly and work activity. Shutdown begins with any action that changes the operational status quo. Shutdown and pre-work preparation occur externally to the equipment, machine, system, or structure on which work is to occur.

## Table 7.5
## Zero Mechanical State and Zero Energy State

| Concept | Application |
| --- | --- |
| Zero mechanical state | External power sources |
| | Pressurized fluids (oil, air, other) |
| | Mechanical potential energy (springs, suspended parts) |
| | Kinetic energy (drop hammers, die tools) |
| | Workpiece or material supported, retained or controlled by the machine that could move or cause machine movement |
| Zero energy state | Equipment or structure contains no source of energy that could expose an individual to harm during maintenance or disassembly |
| | Introduction of no energy through the input(s) or output(s) |
| | No storage of energy in the equipment or structure |

Sources: ANSI 1975; Price 1975.

Shutdown and pre-work preparation can be the first instance of contact between a potentially hazardous internal condition and equipment introduced for the purposes of testing or for other purposes. The chemical, physical, and ignitability characteristics of the atmosphere and residual contents in the equipment, machine, or system, therefore, could influence selection of the type of equipment, mode of power, and materials of construction.

Establishing contact between conditions in the interior of the equipment, machine, or system and those outside changes the status quo immediately. Establishing contact literally could occur through insertion of a small diameter probe to perform atmospheric or other type of testing. Under some circumstances, insertion of a probe has generated a spark that exploded gases in the space and propelled manhole covers through the air. Explosions of similar type have also occurred in crankcases of engines and compressors (Lees 2001). Similarly, use of a lighter or match to provide visibility at the opening to an inspection port has caused tank explosions. Shutdown and pre-work preparation may be elementary or extensive. While the period occupied by shutdown and pre-work preparation usually is brief relative to the overall operational cycle, it can pose significant hazards to workers.

Actions preceding shutdown and preparatory activity could occur during operation mode once approval for entry of the the uncharacterized workspace that partly encompasses or surrounds the equipment, machine, system, or structure containing the source of hazardous energy has occurred. These actions could include photographing or recording a video of the action of the equipment, machine, or system or establishing other physical parameters. The inability to enter the uncharacterized workspace does not necessarily prevent the ability to capture this information, but this certainly can hinder it. The visual information captured photographically in pictures and in video is especially helpful where operating drawings are not available. A video can establish the nature of movement of parts. These measurements can establish, for example, storage and transformation patterns in mechanical linkages and complex levers.

Drawings or plans of the equipment, machine, or system and pictures or video taken by a third party are extremely important contributions to the knowledgebase. However, the loss of comprehensiveness intrinsic to this approach requires acknowledgment. Experience acquired in assessing similar situations can compensate for this lack of presence to an extent. Absence from the workspace by the Qualified Person represents potential loss of information.

Proactivity of this kind increases certainty and reduces uncertainty inherent in speculation about possibilities by establishing facts. The greater the extent of knowledge that can be established about conditions prior to shutdown and pre-work preparation, the less is unknown that remains to be discovered and addressed. Investigating conditions throughout the operational period is one strategy for reducing the level of concern during hazard assessment. The value of this initiative in the overall process of hazard assessment and hazard management cannot be overstressed.

Rendering equipment, machines, and systems safe for work requires a formal process that recognizes how energy passes from input to output. The process of rendering equipment machines and systems safe for work incorporates the following phases: deactivation, de-energization, isolation, lockout, and verification. These actions have stood the test of time and are incorporated directly or indirectly into consensus and regulatory standards and guidelines.

Deactivation and de-energization form the basis for conditions during the standby phase of production mode and the conditions of standby mode. In these circumstances, the equipment, machine, or system acts through control circuits to achieve this condition. The equipment machine or system is still energized through the main power circuit(s) and capable of reactivation to production mode, again through the control circuits.

Deactivation, de-energization, isolation, lockout and verification form the basis for achieving a state of zero energy or controlled energy. The following sections provide only a brief discussion of these concepts. This will serve as an introduction to the protocol for assessing energy in systems. Chapter 8 contains more detailed discussion.

## Deactivation

Previous discussion has indicated that some of the accidents involving hazardous energy occurred through contact with active or uncontrolled energy sources. Deactivation stops the action of the equipment, machine, or system. This occurs by blocking the flow of power (energy) to actuators that create the action. Deactivation of industrial equipment, machines and systems occurs through control circuits and can occur directly through blocking the flow of energy to the equipment, machine, or system.

## De-Energization

Previous discussion has indicated that sudden release of energy stored in equipment and materials has caused many of the accidents involving exposure to hazardous forms of energy. In order to create safe conditions of work around equipment and structures in which energy conversion, storage, or transfer occurs, de-energization must occur. De-energization comprises the actions needed to render the equipment, machine, or system in a state of zero energy or a state of known energy retained through deliberated storage or entrapment.

## Isolation

Isolation constitutes actions taken to prevent re-energization of the system or release of energy that remains stored in the system. Isolation occurs physically through severing or blocking pathways or connections between sources of energy (external or internal) or supply, or drain lines or utilities and the equipment or structure. A device that blocks the flow of energy is an energy-isolating device.

## Lockout

Lockout is the action of applying a lockout device and one or more keyed padlocks and tags to an energy-isolating device.

## Verification

Verification determines the effectiveness of the isolation procedure.

## Assessing the Flow of Energy in Equipment, Machines and Systems

Table 7.6 presents a generalized protocol for assessing hazardous energy in equipment, machines, and systems and for creating work procedures based on this information. The protocol incorporates and integrates the concepts of deactivation, de-energization, isolation, lockout and verification. The protocol responds to the need expressed by various authors and consensus and regulatory standards for analysis of proposed actions involving lockout in a defensible and repeatable manner.

The version of the protocol provided here is generalized to include terms common to all versions. Specialized versions apply to electrical systems, machines and equipment, fluid handling and fluid power systems, and to material flow systems, in particular flowable solid materials. Guide words change to suit the context. Examples of these applications appear in the Appendices. The protocol is applied to individual circuits in which hazardous forms of energy are present. This provides the means to understand clearly the situation in all parts of the system. The protocol provides the means to capture information pertaining to proposed actions and the consequences of actions that may, on reflection, prove to be inappropriate.

**ABC Company** is the entity that performs the work involving the equipment, machine or system. ABC Company may also be the Owner or host Employer.

**Work Activity** is the collection of tasks to be performed.

**Owner** is the entity whose property contains the workspace containing the equipment, machine or system. The Owner may not be the entity performing the hazard assessment for work involving the equipment, machine, or system. The Owner may also be the host Employer.

**Location** is the recognized location of the equipment, machine, or system. At the time of work activity, the location may appear considerably different from routine circumstances. This can occur because of disassembly of structures and reconfiguring the workspace to create an uncharacterized workspace, as discussed previously. That is, the uncharacterized workspace in which the work on the equipment, machine, or system occurs may be a unique configuration of the normal workspace.

**Assessed by** establishes the name and credentials of the individual who performed the assessment and completed the document. The Qualified Person (identified in the regulatory requirements of some jurisdictions) should personally inspect (externally at the minimum) the equipment, machine, or system under consideration unless this individual has extensive experience in assessing similar circumstances. Some jurisdictions specify professional designations to establish the capability of an individual to perform this work. Providing formal qualifications of the individual who performs this work "up-front" is extremely important, since the outcomes from decisions taken during the assessment can have life or death implications.

Personal inspection is critical for making decisions about the status of the equipment, machine or system. Personal inspection establishes first-hand the geometric configuration of the workspace and the equipment, machine, or system under study, potential hazards, and relationships relative to the surroundings.

**Telephone Number** establishes a reliable means to contact the Qualified Person, should this necessity arise in the present at the time of the review that led to creation of the document or into the future.

**Date** establishes when the hazard assessment occurred. The date establishes the relative freshness and possible staleness of the document. Competent documents created for situations that remain static are useable long into the future.

**Overview** describes the overall situation. The overall situation can encompass several circuits. Analysis and procedure creation occur for each circuit in which exposure to hazardous energy can occur.

**Hierarchy of Energy/Fluid Input** establishes energy relationships involving input sources of energy. This information can establish the hierarchy of energy sources from inputs to outputs.

**Conversion Energy Output** identifies sources of energy created in the equipment, machine, or system by conversion of the input energy. Conversion energy may lead directly to output or to storage or to storage and conversion to output. Other combinations are possible.

**Equipment, Machine, System Affected** identifies the destination of the output from the conversion. The destination can be internal, in which case this becomes a source of energy to be expressed in-

**Table 7.6**

**Assessing Hazardous Energy in Equipment, Machines, and Systems**

| ABC Company | Hazardous Energy Assessment | Work Activity |
|---|---|---|
| Location: | Equipment, Machine, System: | Assessed by:<br>Tel:<br>Date: |
| Overview: | | |
| Hierarchy of Energy/Fluid Input | Conversion Energy Output | Equipment, Machine, System Affected |
| | | |
| **Circuit Name** | | |
| Function/Description: | | |
| Energy, Fluid, Material Flow Isolation Strategy: | | |
| Failure/Consequence Analysis: | | |
| Input: | Storage: | Dissipation/Purge: |
| Output: | Conversion: | Immobilization/Isolation: |
| Primary Deactivation/Isolation: | Secondary Deactivation/Isolation: | Tertiary Deactivation/Isolation: |
| Location: | Location: | Location: |
| Action: | Action: | Action: |
| Energy, Fluid, Material Flow Isolation? | Energy, Fluid, Material Flow Isolation? | Energy, Fluid, Material Flow Isolation? |
| (Circuit Name) Energy, Fluid, Material Flow Isolation Procedure: | | |

## Table 7.6 (Continued)

## Assessing Hazardous Energy in Equipment, Machines and Systems

| Verification Test Procedure: |
| --- |
| Notes: |
| Overall Isolation Procedure: |
| Overall Verification Test Procedure: |

### Hazard Assessment (Operational/Undisturbed Space) or (Work Activity)

| Hazardous Condition | Occurrence/Consequence | | |
| --- | --- | --- | --- |
| | Low | Moderate | High |
| | &#124; | &#124; | &#124; |
| circuit name | | | |

In this table, **NA** means that the category does not apply in any normally foreseeable situation. **Low** means that exposure is readily identifiable but believed to be much less than applicable limits or that exposure to nonquantifiable hazardous conditions is unlikely to produce injury. **Low-Moderate** means that exceedence of regulatory limits is believed to be possible or that nonquantifiable exposure could produce minor injury requiring self-administered treatment. Control measures or protective equipment should be considered. **Moderate** means that exposure is believed to be capable of exceeding regulatory limits or causing traumatic injury requiring first aid treatment or attention by a physician. Protective equipment or other control measures are necessary. **Moderate-High** means that exposure is believed to be capable of considerable exceedence of regulatory limits or causing serious traumatic injury. Advanced control measures or protective equipment are required. **High** means that short-term exposure is believed to be capable of causing irreversible injury, including death. Advanced control measures or protective equipment are required.

---

ternally through further conversion, or external, in which case this becomes an output. These three boxes provide the means to trace movement of energy through the equipment, machine or system from input to output. This information becomes the basis for action to detail the flow of energy and to create procedures for deactivation, de-energization, isolation, lockout, and verification.

**Circuit Name** provides the means to identify individual paths for energy in the equipment, machine, or system.

**Function/Description** provides more specific information about the circuit for which analysis of hazardous energy and development of a procedure for deactivation, de-energization, isolation, lockout, and verification is to occur.

**Energy, Fluid, Material Flow Isolation Strategy** describes and discusses the approach proposed to deactivate, de-energize and isolate, and thereby to manage hazardous energy in the circuit. This provides an overview of how management of hazardous energy in the circuit is to occur and the rationale for the proposal. This section is intended to force intellectual enquiry in the writer to establish confidence that the proposed approach is reasonable in the circumstances and highly likely to provide the desired outcome leading to successful conduct of the work to be performed. Inability to describe the strategy in detail is a potential indicator of inadequate understanding about the situation.

**Failure/Consequence Analysis** forces consideration about the potential for failure of the strategy for isolation. This includes the opportunity to introduce contingencies to ensure minimization of risk during the proposed work.

**Input** provides a detailed description of the path of energy from the power source through the circuit to the energy isolating device. This information identifies the presence of components in control circuits.

**Output** indicates the outcome from operation of the circuit.

**Storage** indicates the means and location of energy storage in the circuit.

**Dissipation/Purge** describes the means by which to release energy stored in the circuit during de-energization.

**Immobilization/Isolation** describes how blockage of energy retained or stored within the circuit is to occur to prevent expression during work on the equipment, machine, or system. Energy retained or stored within a circuit must not express during this work.

**(Primary, Secondary, Tertiary) Deactivation/Isolation** indicates the means to isolate the circuit from the source of energy. Primary, secondary, and tertiary indicate the sequence or order in which this activity is to occur. Existence of a procedure for more than one of the three paths in the protocol indicates that isolation of the circuit has not occurred satisfactorily until completion of all levels of the procedure. The usual reason for this is that the primary isolation and sometimes also secondary isolation involves only control circuits. Complete isolation involves isolation of devices in the power circuit.

**Location** is the named or descriptive location of the energy isolating device.

**Action** describes the movement of the energy isolating device needed to isolate the circuit. Action states the movement in concise, command-style of language in statements such as, move to the OFF position; apply keyed padlock and tag.

**Energy, Fluid, Material Flow Isolation** indicates positively, positively with conditions, or negatively whether the isolation procedure leads to the locked-out condition. Again, to reiterate a previous statement, achieving a locked-out condition may require completion of the primary, primary and secondary and sometimes, primary, secondary, and tertiary procedures. While not at the level of lockable isolations, some of these procedures deactivate and de-energize the equipment, machine, or system so that the procedure that produces the locked-out condition occurs while the equipment, machine, or system is not under load.

**Circuit Name Energy, Fluid, Material Flow Isolation Procedure** is a synthesis of steps provided in the protocol for that particular circuit. The sequence of the steps is critical. The sequence may begin with some action unrelated to the primary, secondary, and tertiary isolation procedures. This action could be related to dissipating or purging, or immobilizing or isolating energy from the circuit or something not related even to that. Similarly, actions required to immobilize or to isolate stored or retained energy may occur between the circuit isolation procedures.

**Verification Test Procedure** is a critical component in the process of creating lockout procedures. Where practicable, this procedure can involve a quantitative test to prove that energy is not present in the equipment, machine, or system or to prove that retained or stored energy cannot express itself following measures for isolation or immobilization.

Verification must occur at each step during procedure development to ensure the appropriateness and sequencing of individual steps. During implementation, verification occurs at the level of each step and through the formal verification at the end of the procedure. Verification is critical during implementation of the procedure during routine work to ensure the truth of what people believe to be the reality of a situation.

**Overall Energy, Fluid, Material Flow Isolation Procedure** is the integration of individual procedures. The integration can be a complex process where several circuits are involved. This can result from unintended and unanticipated consequences of steps taken during procedures created for locking out individual circuits. These steps can produce impacts on other circuits. The critical issue during development of the overall procedure is the same as that during development of procedures for individual circuits: verification of appropriateness, efficacy, and safety.

**Overall Verification Test Procedure** is the final step in a complex sequence of activities prior to beginning work activity. Completion of the verification marks the point at which someone authorizes the beginning of the work. Acceptance of conditions based on the outcome from this verification represents a critical moment in the operational cycle of the equipment, machine, or system.

## Procedure Writing
Writing effective technical procedures that communicate effectively is a learned skill and an art. Table 7.7 provides information about the attributes of procedures for isolation and lockout. Some of these attributes pertain to outcomes to be achieved.

Language used in effective procedures is different from what people use during everyday speech. Everyday speech usually contains slang, colloquialisms, and soft forms of style. The latter remove some of the resistance to following instructions and commands issued in typical superior-subordinate relationships. People typically use soft persuasive and coercive language in order to move work forward in the chosen direction. This style minimizes resistance in communication between people.

Effective technical procedures require a focus on simplicity of language, conciseness in the number of words, and completeness of thought. Mastering this skill, like developing mastery in other skills, requires practice and focus on the outcome from the process.

Previous discussion has commented about the ability of the brain to process information. The limitations of the brain as an information processor were readily apparent in the descriptions of fatal accidents reviewed to create Chapter 1. Verbose, convoluted language as used in many procedures forces the brain of an otherwise overloaded individual to focus on interpretation of unnecessary quantities of information.

Procedural steps should contain command-style language. To illustrate, move X located in Y to the Z position; apply keyed padlock and tag. The vocabulary used in these statements requires some practice in order to remove uncertainty and ambiguity.

The Z position is a critical factor. If stated too restrictively, the appropriate word for Z could change following modification of control switches or a change to a computer system from a manual one. Words, such as OPEN, CLOSED, OFF, and ON are specific enough, and yet vague enough at the same time, to enable occurrence of a known action. Use of these words eliminates the need to revise the procedure when a push-button switch replaces a rotary switch and the wording of the action changes.

The concept being stressed here is that the words chosen for the command do not necessarily describe the exact action undertaken. People of normal levels of language comprehension should recognize from labels on control equipment the action required to move the control from the ON to the OFF position. This action could require rotation, push, toggle, or some combination of these individual actions.

## Multiple Energy Systems
Modern equipment often contains inputs from several sources of hazardous energy and can create additional sources through normal operation. While these systems could be independent of each other, the likelihood is that they are interdependent and/or interconnected. Interdependence and interconnection can considerably complicate the task of deactivation, de-energization, and isolation. In some cases, systems, and processes in equipment are controlled remotely as a result of activities occurring in other equipment. As a result, testing for deactivation is not as simple as pushing a START button or toggling controls. While this is one means of activation, activation by a signal from a microprocessor in a control circuit within the equipment or in a distant control room easily could be another.

To illustrate, electrical systems are used to operate valves in hydraulic, pneumatic, and process and utility systems (Stewart 1987). Unlike mechanical systems, electrical systems provide limited

## Table 7.7
## Critical Issues in Procedure Writing

| Critical Issue | Comment |
| --- | --- |
| Unambiguous labeling | Label equipment unambiguously with easy-to-read signage. Use the same name for all references to a component. Color-coding identifies components in the same circuit. |
| Secure deactivation | Lock out the primary (control) circuit during deactivation. |
|  | If not possible, devise means to assure deactivation of control circuit prior to isolating the power circuit. |
| Isolation under load | Deactivate the control circuit prior to isolating the power circuit. This will minimize the risk of arc flash and arc blast associated with deactivation under load. |
| Readability for comprehension | Use simple, concise words in instructions. |
|  | Use a single phrase in each repetition of the same concept to maintain consistency. |
| Style | Use direct action, command language. |
|  | Do this to that located in a particular place to produce a particular end-result. |

power for actuation. However, an electrically operated pilot valve can control the activity of a much larger fluid power valve. Electrically operated valves utilize motors and solenoids. Motors operate through gears and gear reducers. The solenoid produces straight-line motion of a plunger through electromagnetic action. The solenoid operates a mechanical operator that in turn operates the valve mechanism.

The only way to achieve safe conditions of work in equipment containing complex subsystems is to deactivate, de-energize, isolate, and lock out all energy sources. This must occur in a logical sequence that considers the implications of prior and subsequent actions as part of the overall process. Applying a logical sequence provides the benefit of consistency of approach. Table 7.8 provides the hierarchy recommended by the UAW-GM National Joint Committee on Health and Safety (UAW-GM 1985). Given the background of the authors, this protocol may best suit production machinery.

## References

AFS: What You Should Know about ZMS (Zero Mechanical State). Des Plaines, IL: American Foundrymen's Society, 1961.

ANSI: Safety Requirements for Sand Preparation, Molding, and Coremaking in the Sand Foundry Industry (ANSI Z241.1). New York: American National Standards Institute, 1975.

ANSI: Safety Requirements for Confined Spaces (ANSI Z117.1-1995). Des Plaines, IL: American Society of Safety Engineers/American National Standards Institute, Inc., 1995.

API: Requirements for Safe Entry and Cleaning of Petroleum Storage Tanks (ANSI/API Standard 2015-2001, 6th ed.). Washington, DC: American Petroleum Institute, 2001a.

API: Guidelines and Procedures for Entering and Cleaning Petroleum Storage Tanks (ANSI/API Recommended Practice 2016). Washington, DC: American Petroleum Institute, 2001b.

## Table 7.8
## Suggested Hierarchy For De-Energization, Isolation, and Lockout

| Energy Source | Comments |
|---|---|
| Momentum | During shutdown sequence, moving parts in machinery come to a stop. |
| Gravity | Immobilization through pinning and blocking should occur before the primary energy source is turned off. |
| Stored mechanical energy | While sources of stored mechanical energy should receive consideration at this point in the sequence, they could be neutralized later in the procedure; relieving residual and stored energy may require shutting down a primary energy source. |
| Electrical system | Normal operation of the equipment must cease prior to shutting down the electrical supply system; electrical energy may remain stored within electrical equipment, for example, capacitors and transformers. |
| Pneumatic system | Residual stored air pressure existent in accumulators must be released during de-energization. |
| Hydraulic system | residual stored hydraulic pressure in accumulators must be released during de-energization. |
| Process and utility systems | Deactivated, de-energized, and locked out as needed. |

Source: Adapted from UAW-GM 1985.

Canada Gazette: "Part XI, Confined Spaces," *Canada Gazette, Part II. 126*: (17 September 1992) pp. 3863-3876.

Cember, H.: *Introduction to Health Physics*, 3rd. Ed. New York: McGraw-Hill , 1996.

Jones, T.B. and J.L. King: *Powder Handling and Electrostatics: Understanding and Preventing Hazards*. Chelsea, MI: Lewis Publishers Inc., 1991.

Kepner, C.H. and B.B. Tregoe: *The New Rational Manager*. Princeton, NJ: Princeton Research Press, 1981.

Lees, F.P.: *Loss Prevention in the Process Industries*, 2nd Ed. London: Butterworth-Heinemann, 2001.

Magrab. E.B.: *Environmental Noise Control*. New York: John Wiley & Sons, 1975.

March, G.: *Advanced Organic Chemistry*, 3rd. Ed. New York: John Wiley & Sons, 1985.

McManus, N.: *Safety and Health in Confined Spaces*. Boca Raton, FL: CRC Press/Lewis Publishers, 1999.

McManus, N.: *Portable Ventilation Systems Handbook*. New York: Taylor & Francis, 2000.

McManus, N.: *The Confined Space Training Program*, Worker Handbook. North Vancouver, BC: Training by Design, Inc., 2004.

NFPA: Control of Gas Hazards on Vessels, NFPA 306 (2009 edition). Quincy, MA: National Fire Protection Association, 2009.

NFPA: NFPA 77: Recommended Practice on Static Electricity (2007 edition). Quincy, MA: National Fire Protection Association, 2007.

NFPA: NFPA 328: Control of Flammable and Combustible Liquids and Gases in Manholes, Sewers, and Similar Underground Structures (1997 edition). Quincy, MA. 02269: National Fire Protection Association, 1997.

NIOSH: Worker Deaths in Confined Spaces (DHHS/PHS/CDC/NIOSH Pub. No. 94-103). Cincinnati, OH: National Institute for Occupational Safety and Health, 1994.

NIOSH: Worker Deaths by Electrocution (DHHS (NIOSH) Publication 2000-115. Cincinnati, OH: National Institute for Occupational Safety and Health, 2000. [CD-ROM].

OSHA: "Control of Hazardous Energy Sources (Lockout/Tagout)"; Final Rule, *Fed. Regist. 54*: 169 (1 September 1989). pp. 36644-36696.

OSHA: "Permit-Required Confined Spaces for General Industry; Final Rule," *Fed. Regist. 58*: 9 (14 January 1993). pp. 4462-4563.

OSHA: "Confined and Enclosed Spaces and Other Dangerous Atmospheres in Shipyard Employment; Final Rule," *Fed. Regist. 59*: 141 (25 July 1994). pp. 37816-37863.

Price, C.: ZES–Zero Energy State. A Systems Approach to Guarding, Maintenance, Servicing Functions. *Nat. Safety News 112*(6): 56-57 (Dec. 1975).

Shapiro, J.: *Radiation Protection, A Guide for Scientists and Engineers*, 2nd. Ed. Cambridge, MA: Harvard University Press, 1981.

Standards Association of Australia: Safe Working in a Confined Space (AS 2865-1986). North Sydney, New South Wales: 1995.

Stewart, H.L.: *Pneumatics and Hydraulics*, 4th Ed.. Revised by T. Philbin. New York: Macmillan Publishing Company, 1987.

UAW-GM: Lockout. Detroit: United Auto Workers-General Motors Human Resource Center, 1985.

Wiggins, Jr., J.H.: Control of Hazardous Energy. *Compliance Magazine. 4(05)*: 14-17 (1997).

# 8 Technical Aspects and Issues

## INTRODUCTION

The previous chapter introduced protocols to assess hazardous conditions in the workspace in which work is to occur and for assessing energy flow through systems. This chapter will consider topics associated with energy flow in systems in more depth.

Understanding how the movement, storage, and transformation of energy occur in machines, systems, and equipment is essential for the safe conduct of work in and around them. In the simplest of machines, systems, and equipment, the workings are visible to the operator and maintenance person. Actions and consequences of actions that could produce injury are somewhat apparent. As a result, the measures needed to prevent these actions are perceivable, and able to be implemented and assessed at little risk of injury.

Many of the early applications of energy flow in systems and equipment centered around weaponry (Schwartz 1990). The war machine provided the means to deliver lethal amounts of energy in a controlled manner at a distance. The catapult is an early example of this type of system. Rotation of a rope winch flexes the arm and the hurling mechanism. A locking mechanism permits loading of the spoon with a rock and prevents release of the drum until the appropriate moment. At that point, releasing the trigger causes virtually instantaneous transformation of stored energy from potential to kinetic. The ability of this energy to hurl rocks considerable distance suggests the inherent ability of these units to injure or kill the individuals who operated and serviced them, as well as the intended recipients of the missile. This is especially the case when energy was present in the stored condition (tension in the spring).

Experience would have shown that the appropriate time to service these units was when the hurling arm was not under tension; that is, a state of zero stored energy. The rationale for this was that the unit was not storing energy that could release suddenly, and that securement of pieces could occur readily to prevent unexpected or unintended movement. Under tension, motion of the hurling arm is not completely predictable, even when secured. The ropes and the locking mechanism could fail. The timber in the hurling arm could split. Accidental release or failure of the trigger mechanism could occur.

Unexpected release of the tensioned hurling arm when an individual was working in the mechanism would cause serious or fatal injury. Action of the hurling arm would be even more aggressive than when used under battle conditions, since the spoon would have no weight to oppose movement. One hazardous aspect of this design is that the energy storage mechanism is fully exposed. Because the unit is fully open and operated manually, the sequence and consequence of storing energy in this manner are fully observable.

There are many additional examples of energy storage in systems in common experience. For example, capacitors in television sets and video monitors containing cathode ray tubes (CRTs) store electrical energy even after the unit is unplugged. This energy is released in an instant by bridging the contacts using a screwdriver.

Failure of a dam releases the stored water, and with it the energy involved in storage against the surface of the dam. Similarly, failure of a slope releases energy stored in material that is capable of downward movement. These events cause tremendous damage.

The cooling system in vehicles and other liquid-cooled equipment retains thermal energy for some time after the engine is shut down. Removing the cap from the radiator releases thermal energy and accompanying pressure. This process can occur despite prevention of start-up of the engine through removal of the key from the ignition and locking of the door. All that is needed is access to engine compartment. An additional complication is the remote starting system. Remote starting systems operate through wireless signals.

In more complex machines, systems, and equipment, the outer enclosure prevents observation of the actions performed by internal subsystems. Complicating matters further, sensors and feedback circuits feed information about operation of these subsystems to a computer that ultimately controls them. The user is several levels removed from the action and control of these machines systems and equipment. Without sophisticated analytical tools and procedures, the operator or maintenance person can do little to ascertain energy status.

Energy utilization in systems is not always consistent across different segments of industry. To illustrate, the air braking systems in trucks and trailers operate completely opposite to those in trains. In automotive systems, air pressure releases the brakes. This means that reduction of pressure allows springs to apply the brakes. In trains, air pressure applies the brakes. Hence, the brake releases when pressure decreases. This means that continuous operation of the engine in the locomotive and the air compressor are critical to safe operation of the train during transit from one location to another. The manually operated brakes on the locomotive and cars is insufficient to stop a moving train.

Working on or inside equipment under these conditions is akin to attempting to defuse a newly exposed, unexploded World War I bomb whose detonation mechanism is known, but whose internal condition is unknown, or an explosive device planted by terrorists about which nothing may be known. There is no assurance about the ability to perform this type of work in safety. As a result, munitions experts usually attempt to recover these devices using robotic equipment and to detonate them under controlled conditions. The robot is expendable in the event of premature detonation. However, the luxury afforded by this approach is not available to persons who must dismantle, or enter and service machines and industrial equipment.

## Describing the Flow of Energy in Systems

Much of the potential for contact with sources of hazardous energy occurs within the context of work around and inside equipment. Designers of equipment have exploited controlled storage of energy for many centuries.

Fundamental to gaining control over energy in actual systems is the ability to conceptualize what is occurring. Energy flow in systems occurs from input to conversion and/or storage to output. Many variations on this theme are possible. Actual systems can incorporate several inputs and outputs, as well as multiple modes of storage and transformation. Some of the energy outputs, like heat, often are peripheral to the intended action of the machine or equipment. Every one of these variations and others provides a pathway for expression that can lead to loss of control and the possibility of traumatic injury.

To this point, discussion about this subject, with the exception of the protocol introduced in Chapter 7, was discussion based. Formalizing consideration about these pathways and about expressing energy flow diagrammatically and quantitatively provides the basis for managing hazardous energy by establishing and maintaining control.

The simplest systems exhibit the sequence, input → conversion → output. An electrical circuit containing a simple fan is an example. Electrical energy converts to mechanical energy to cause rotation of the impeller. Note that such circuits contain an ON-OFF switch.

If the motor does not contain a flywheel, rotation ceases when the switch is opened following rundown. The duration of rundown depends on the speed of rotation, load on the motor, internal friction, and rotational momentum. So long as the energy of the device driven by the motor is present only during motion of the motor and is not stored (that is, converted to another form), the system will revert to zero energy when the motor stops rotating.

An electrical circuit containing a common light bulb provides an example of a double conversion of energy. A light bulb transforms energy from electrical (potential) to heat (thermal) and then emits light (visible) as the temperature of the filament increases. This transformation occurs almost instantaneously in the filament of the bulb, and only when the circuit is closed. While the outputs from the bulb are both heat and light, in most circumstances, light is the only desirable output. The heat is an undesirable end-product of the energy conversion and requires removal. There are some situations, of course, where thermal energy is desirable for its ability to heat the surroundings. An overly intense output of both light and heat creates conditions hazardous to workers. These result from the ability of wavelengths in the ultra-violet, visible and infra-red regions of the spectrum to damage the eye and possibly the skin. Similarly, the heat radiated from the surface of the bulb can cause deleterious consequences, including ignition, and the hot surface of the glass envelope poses a serious burn hazard.

The electrical circuit containing the light bulb or the motor contains a source of energy, a sink (the light bulb or the motor), and a switch. The switch provides control of operation. Electrical energy normally present in the circuit cannot pass the gap provided in the open switch.

Figure 8.1 shows a means of diagraming the flow of energy in this type of circuit. The circuit has two states, de-energized with the ON-OFF switch open in the OFF setting, and energized, with the switch closed in the ON position (Figure 8.2). Electrical energy can flow in the circuit only in the energized setting. In the OFF position, the switch blocks the flow of electrical energy, as everyday experience indicates. What is not normally obvious is the continuing presence of electrical energy in the wires upstream from the switch.

Energy conversion but not storage occurs in this type of circuit. Note that the quantity of energy flowing through this circuit is constant. That is, output and input are identical. The total amount of energy remains constant in order to respect the law of conservation of energy.

The graphical representations shown in Figure 8.1 and Figure 8.2 are quantitative. A number of choices for units are possible. Each contains the Joule, the unit of energy, either alone or multiplied or divided by other units. Hence, this aspect of the representation of energy is quantitative.

Energy storage (accumulation) occurs in the second type of circuit. Energy storage alone without conversion is not common in circuits. An example is an electrical circuit containing a large number of capacitors whose sole function is to accumulate charge for a controlled discharge. An equivalent analogy involving fluid power is pressurizing a circuit, such as an airbrake following discharge to set the brake. A bus driver pressurizes the brakes in order to move the bus forward following an interval of parking at a passenger terminal. If the rate of pressurization is slow or the volume of the system is large

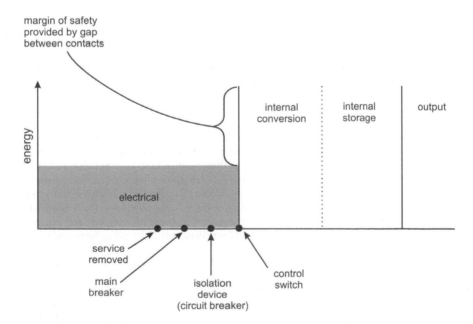

Figure 8.1. Flow of energy in an electrical circuit. The ON-OFF switch enables flow through the circuit. Until closure of the switch occurs, flow of energy through the circuit cannot occur.

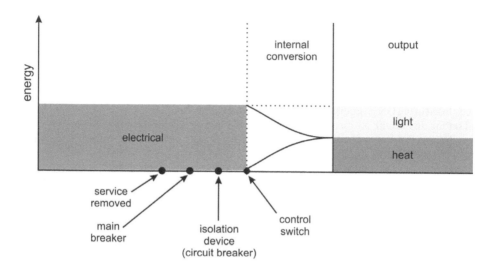

Figure 8.2. Energy flow and conversion in a simple system occurs on closing the switch to the ON position. The total energy remains constant in order to respect the law of conservation of energy and partitions between light and heat.

relative to the volume of pressurizing fluid, pressure in the circuit will increase slowly to the maximum attainable. In these cases, the amount of energy accumulated depends on the time during which accumulation occurs. A battery charging system is similar in concept.

Energy accumulation and storage in circuits are illustrated using a technique similar to that employed for energy conversion (Figure 8.3). In this case, the type of energy in the output is the same as that in the input unless or until conversion occurs. The visualization shown in Figure 8.3 presumes that losses in energy during conversion are small. This certainly is the case with transformers and other electronic equipment.

The graphical representation shown in Figure 8.3 is semi-quantitative. While energy levels are quantitative, the kinetics representing rate of change are only semi-quantitative until or unless additional information becomes available. In this type of discussion, rate of accumulation or rate of discharge usually are inconsequential. Where this information becomes important is likely to be rate of discharge pertaining to decrease in the level of energy stored in a circuit and gradually bled from it over the passage of time. The time profile of accumulation and discharge provides critical information in consideration about risk of traumatic injury during an accident involving release of this energy. This discussion applies to energy stored in capacitors and transformers in electrical systems and in isolated components in hydraulic and pneumatic systems.

The third type of circuit is a synthesis containing the energy conversion and accumulation and storage circuits. These circuits are very common in equipment. machines, and systems. In this case, the output could be much greater than the input, depending on the point in the accumulation-discharge cycle at which visualization occurs. This is especially the case with machines such as punch presses. The punch press uses constant rotation of a motor to move a heavy object into position for rapid delivery of energy stored in position.

Hydraulic and pneumatic systems are examples of conversion-storage systems. The pump or compressor driven by the motor pressurizes the fluid in the interior of the system. Some of the energy

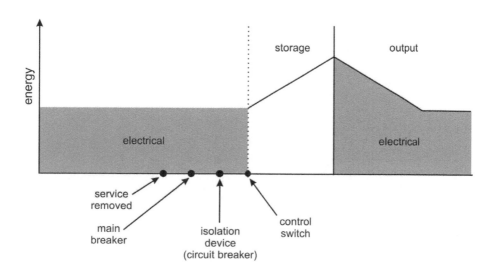

Figure 8.3. Accumulation and storage of energy. The output energy is the same type as the input energy until or unless conversion occurs.

acts against components in the accumulator and against bulk masses, such as the blade of a bulldozer or the boom of an excavator or hydraulically powered crane. These maintain pressure in the system (store energy) through position in space after shutdown of motive power. Fluid pressure dissipates only after these masses are supported by other means and prevented from moving or are lowered to the position of zero energy.

Diagraming energy flow in conversion-storage circuits is essentially the same as the other types (Figure 8.4). The magnitude of stored energy easily can be much greater than the rate of energy input, since storage implies accumulation over a period of time. The graphical representation shown in Figure 8.3 is semi-quantitative because of the lack of information about the kinetics of accumulation and discharge.

Considerably more complicated situations exist where the equipment receives energy from multiple inputs. These can include electrical, steam, compressed air and other gases, gaseous and liquid fuels, vacuum, chilled liquid, and hot and cold water. As well, equipment can receive energy input from process supply lines. Outputs can include condensate, hot and cold water, vented heated and cooled air and gases, and products from chemical processing. The diagrammatic format shown here can handle additional inputs and outputs.

While the diagrammatic techniques provided here may not be needed or appropriate for every situation and application, they provide a vastly expanded means to visualize and conceptualize what is occurring in these circuits.

## Deactivation

Deactivation is the action taken to stop the operation of the equipment, machine, or system. Deactivation occurs by blocking the flow of power (energy) to actuators that create the action. In simple circuits, these devices act directly (Figure 8.1). That is, there is a direct relationship between the action

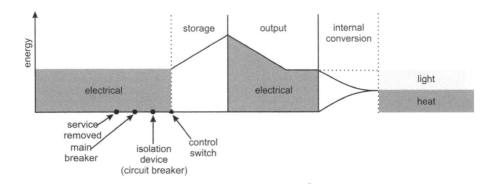

Figure 8.4. Circuit containing both storage and conversion of energy.

and the outcome. In more complex circuits, these devices act indirectly through control circuits. Deactivation occurs through control circuits and can occur directly through blocking the flow of energy to the equipment, machine or system or indirectly through controlling connection to high-energy circuits. Figure 8.5 shows a manual control switch in an electrical circuit. The simplest deactivation is directly observable by the operator.

Deactivation is not always what is perceived to occur in a device. To illustrate, the ON-OFF switch on a television set containing a cathode ray tube and other residential and industrial devices deactivates the external interface of the unit. Circuits in the television set remain energized even when the external control switch is turned off. That is, power circuits remain energized and control circuits are deactivated. Energization of some circuits enables the unit to start quickly when reactivated compared to the much longer time required for older television sets. Rapid start was an especially desirable feature in units containing cathode ray tubes (CRTs). Partial energization is common in electronic devices, such as fax machines, that restart rapidly when a message arrives.

The complexity of industrial operations led years ago to the centralizing of control functions using microprocessors in Programmable Logic Controllers (PLCs). With this innovation, the operator needed only to move a switch or push a button on a console to deactivate equipment in the process area (Figure 8.6). Again, these devices act through control circuits.

Development of advanced microprocessor technology led to dedicated control devices located in equipment and machines and distributed control systems. All of these devices operate through software. The most current generation of software provides a Graphical User Interface (GUI) between the operator and the functional control of this equipment (Figure 8.7). The software uses animation to indicate active operation of the component.

Part of the functionality is deactivation. The operator moves the cursor on the screen using the mouse or other pointing device to an icon that represents the component of interest. The icon is drawn to resemble the component. The component of interest can be a pump, valve, or other device, a subsys-

Figure 8.5. Moving the manual control switch to the OFF position provides a simple means to effect deactivation. Note that the MANUAL or HAND position or the AUTO position provides no assurance that deactivation will occur. Deactivation does not isolate the actuator from the main power circuit.

Figure.8.6. Control console. The operator needed only to move a single switch or push a button on a console to deactivate equipment in the process area. Note that this action involves only control circuits and does not reflect de-energization and isolation of the primary circuit.

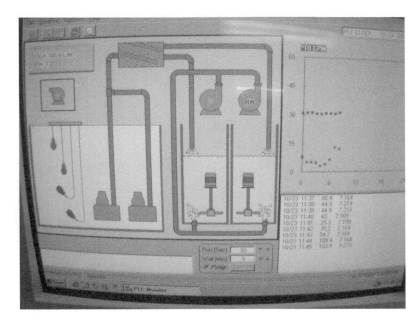

Figure 8.7. Microprocessor control and the Graphical User Interface (GUI) provide deactivation from the desktop.

tem or the entire system. The operator clicks on the icon or a box containing the name of the component. The software responds by opening a menu. The menu contains a command for shutting down the component, subsystem, and sometimes the entire system. A single click and action through the software is all that is needed to implement the command (Figure 8.8). Again, the action occurs through control circuits.

The functionality of the click in this software is the same as people experience in the home and in the office with other programs. In these environments, the click can scan and print documents, load and unload programs, activate music devices, and download pictures and video from cameras.

Computers, peripheral devices, and software used in the home and office have habituated people to certain expectations when utilizing computers as control devices when we click on icons on the computer screen. We are also habituated to the fact that software published for widespread use will contain defects. Where software is known to have errors in logic or function (bugs), the end-user is habituated to going onto the Internet to download an upgrade, service pack, or patch from the website of the publisher and then to install it on the affected computer. This is relatively straight forward and the consequences that arise from these errors are relatively minor.

The situation in the industrial setting is not as simple or as inconsequential. To begin, software used in workplace operations is either one-of-a-kind and started from the beginning, a one-of-a-kind customization of a base package, or at best a limited edition of a software package. This situation considerably increases the risk of error in logic and applications because of the limitation of resources available for creation and troubleshooting.

The issue of control over deactivation arose years ago when the operator lost direct control over the action of control devices and was compelled to rely on circuits wired into consoles. These circuits initially operated through clockwork mechanisms and relays and solenoids and later through microprocessors contained in PLCs, and later still in computers and advanced software. With each of these

Figure 8.8. Deactivated circuit on a computer screen. Note that this representation depicts only the action of control circuits and does not reflect de-energization and isolation of the primary circuit.

levels of complexity, reliance on equipment to do things in a predictable manner became more and more heightened and more and more of a concern.

Computerized control systems and the equipment with which they interact have a distinct life history. This is describable as design, commissioning, operation, modification and upgrading, and retirement.

Computerized control systems optimally receive several levels of input during design and commissioning. The Owner and the Designer must communicate so that the needs and expectations, and capabilities and limitations of what is do-able are clearly communicated and understood. The Designer of the system has expectations about how the equipment should operate and what units should operate together and what conditions should force shutdown of all units in a sequence. These concepts may or may not compliment expectations of the Owner through the consulting Operator. The consulting Operator functions in this discussion at the macro level. The Designer functions at a micro level. In many circumstances, the Designer presents a turnkey operation to the Operator without consultation and input.

The outcome from either of these directions is that the consulting Operator and subsequent system Operators likely have no knowledge about or access to the logic expressed through the design. They often learn about the logic through experience gained in operating the system. The logic underlying the design may not be readily apparent or documented or documented in a manner readily comprehended by those affected most by it.

Operating experience brings surprises. To illustrate, loss of personal control over action in a circuit whose control switch is set to AUTO is intuitive. Setting the control switch to HAND or MANUAL can also produce loss of of control, but provides no warning about how this could happen. To illustrate, the MANUAL setting on a switch located in the Lifeguard Office at an aquatic center controlled different equipment than the MANUAL setting on a switch controlling the same circuit located in the Pump Room.

Other situations showed that control switches operated circuits with no logical connection to the circuit of record. This example highlights total loss of knowledge about control of these circuits.

During creation and commissioning of a control system, the programmer has direct access to the logic that operates the equipment. The logic initially starts with the design. The logic intrinsic in the design usually is not communicated to the end-user or communicated in a comprehensible manner.

Operational modifications to these systems reflect different approaches. In one situation, electricians within the entity reprogram the PLC. They have the skill and the knowledge to make the changes and are given the privilege of doing so. Documentation of the change is not recorded for future reference. This approach creates the opportunity for complete loss of control over affected equipment, and ultimately the operation. This approach presents no barrier to the actions of a rogue individual directed to sabotage operation of equipment. Such action could occur, for example, because of acrimonious labor relations on the eve of a work disruption. This situation could have serious implications for the safety of other workers.

Another operation contracts a third party, a programmer, to make changes through a wireless connection or telephone modem. The system contains some security through password protection. The system affected is low profile and unlikely to attract attention. The programmer does not document the changes for the Owner. As a result, the changes and their possible unintended consequences are lost for future reference.

A third operation hires a local electrician to make the changes to the PLC in the control system. The electrician does not document the changes for the Owner. As a result, the changes and their possible unintended consequences are lost for future reference.

The situations documented here are a small number of the many possible examples in the universe of possibilities. There obviously are many other possibilities not documented here. The net outcome from these situations is the loss of knowledge about how a system is configured to operate through programmed instructions. Loss of knowledge is tantamount to loss of control over the actions of the equipment, machine, or system.

Loss of knowledge complicates the task of forensic investigators attempting to learn how the components are interconnected by wires and electrical components and controlled through instructions stored in memory. These systems are complicated (Figure 8.9) and represent the investment of considerable amounts of money. Forensic investigation to identify and to document interconnections between components to determine how an in-service system actually is controlled for the purpose of correcting errors, is a daunting exercise that can cost as much as the capital value of the equipment. An important caveat with such an exercise is that the status quo of knowledge beyond that moment in time depends totally on maintaining control and managing change through procedures and documentation of action

As a system matures through use and provides operating experience, the logic inherent in its design may no longer suit its current operation. At some point in the history of the equipment upgrading occurs. Upgrading reflects experience that the concept adopted in the original design did not work or did not work as expected, or contains bottlenecks. Upgrades may occur gradually or abruptly. Abrupt upgrades are subject to the sorts of disruptions already mentioned.

In each of these situations, Operators, whose safety directly depends on thorough understanding about how the system functions must subordinate their knowledge to a third party,.who is not affected by the operation. This situation creates little incentive for the third party to inquire about the global picture of the implications of change to the system. At the same time, this can create tremendous uncertainty in Operators about the level of safety inherent in the system on which their livelihood and safety depend.

Leveson has written a major part of a book on the subject of software control of systems (Leveson 1995). She comments about issues mentioned here and adds additional detail to the discussion.

She points out that changes to existing software do not occur in isolation and that change is likely to introduce errors and unforeseen consequences. This situation is evident in the current situation involving patches to computer operating systems and add-ons to correct vulnerabilities to attack and other deficiencies. Often these changes cause unforeseen conflicts with other existing software that require development of additional corrective measures. Change of this magnitude embrittles software

Figure 8.9. Control cabinet. Modern operations contain control cabinets such at this. Forensic investigation to determine and to document interconnections between components to establish how an in-service system actually is controlled is a daunting exercise that can cost as much as the capital value of the equipment.

and illustrates the limitations inherent in interconnectivity of components of the operating software and add-ons.

Errors in software ensure repeatability of the error because of consistency in expression of the software. This becomes a factor in system error. These errors are not random, like operator error or mechanical failure. Errors in software will express every time that that particular part of the software runs. An error in the software thus increases the risk of error in the system. Studies discussed in Leveson (1995) indicate that significant consequential errors occur in original material despite the extensiveness of testing prior to release.

Experience has shown that software testing is incapable of reliably detecting all of the possible errors contained in it. Experience with commercial software with widespread availability has proven this to be true numerous times. If large software publishers are unable to detect and correct all errors, publishers of one-of-a-kind or limited-edition software used in process control are considerably less able to follow this process. These companies lack the financial and personnel resources and the time needed to pursue these investigations during software development.

Errors in the underlying fundamentals in software, the mathematics and logic, are readily detectable and correctable. Experience has shown that many software-related accidents involved overload (Leveson 1995). Overload relates to the inability of the control system to receive and process the volume of input and thereby to provide the throughput needed in a particular situation. Lock-up in office software packages caused by overly rapid typing while the computer is running other programs is an example of this problem. These problems involve timing and are considerably more difficult to address than the correction of problems solely related to application software.

Redundancy is a well-accepted means for increasing reliability of hardware. Redundancy in software, however, has not shown effectiveness in reducing errors in system operations. Eliminating failures due to software errors is inherently more difficult than eliminating failures due to breakdown of hardware.

Reliability is a reflection of the ability of a system to do the same things consistently over a long period. Reliability does not guarantee freedom from error. Reliability means that an error will occur in a consistent manner. Increasing reliability through elimination of software error does not necessarily increase the safety of the system. Design error in hardware or software will re-express regardless of reliability. Safety is a property of the system and not the solely the software. Safety is application-, environment-, and system-specific.

Software control plays an obvious role in deactivation of equipment involved in processes. This discussion indicates that humans have a need to retain a level of suspicion about the reliability of decisions made by microprocessors acting through software and actions taken (Table 8.1). Safety demands certainty.

The questions raised in Table 8.1 suggest that operators must go beyond software deactivation as the sole means of preparing for work on equipment, machines, and systems where contact with hazardous energy can occur.

## De-Energization

De-energization describes the removal of energy from the equipment, machine, or system. De-energization occurs passively in many circuits following deactivation and requires active measures in other circuits. Experience has shown that obtaining verifiable and predictable de-energization of equipment, machinery, and systems can be considerably more complicated than simple disconnection from the source of energy/power because of the potential for retention of energy through storage or entrapment in circuits of the equipment, machine, or system. Deactivation stops the inflow through control circuits, but has no influence on storage and entrapment.

The logical starting point for determining how to de-energize equipment, machinery, or systems is to ascertain how energy enters, is retained, utilized, and converted and how this is controlled. The protocol introduced in Chapter 7 provides a means to guide this inquiry and capture of information for individual circuits.

In many real-world situations, the hazardous condition is apparent and so also is the strategy for elimination and control. This is the ideal situation at one end of the spectrum. Obviously, this is not always what confronts investigators.

The tiger in a cage at the zoo is a surrogate for an extremely hazardous source of energy. Luring the tiger into a holding pen, and closing and locking the door (a door that can withstand any level of physical attack) is an obvious strategy. Once behind the door, the tiger cannot re-enter the cage until permitted to do so by the keeper. This approach is satisfactory, so long as the door remains closed in some predictable and reliable manner.

Unfortunately, in many other real-world situations, the hazardous conditions inherent in a situation and the strategies needed for their elimination or control are not so obvious. Many machinery spaces receive inputs from utilities: electricity, water, steam, natural or fuel gas, compressed air, vacuum, and possibly gases, such as hydrogen, helium, argon, carbon dioxide, nitrogen, and oxygen. Process units can have any of these, as well as supply and drain lines and materials storage. The absence or lack of availability of design drawings for equipment, machines, and systems is not unusual. Even when design drawings are available, they provide little assistance in hazard assessment and control. Essentially what maintenance personnel confront in these circumstances is a black box that has some recognizable inputs and outputs. What occurs internally is not fully known from external examination of the situation. This lack of information originates in design.

Even today, current textbooks do not recognize or acknowledge the ramifications of design decisions on maintenance, modification, renovation, and rehabilitation. The text by Juvinall and Marshek (1991), for example, refers to safety primarily in product design. The concepts of deactivating, de-energizing, isolating, and lockout or immobilization for servicing are not mentioned.

The manufacturer is the logical resource for determining how to de-energize or immobilize equipment, machines, or systems. The manufacturer has designed the equipment, machine, or system to perform according to some specification. This should include calculation of forces, or at the very least, trial-and-error determination of performance of subsystems that contribute to desired output

## Table 8.1

## Questions about Deactivation

### Question

What outcome is the command to deactivate a component intended to produce?

What outcome does the command actually produce?

Can some other entity within the universe of the system override the command?

Can the override occur without communication that it will occur?

What means are available to prevent this occurrence?

Can an entity within the universe of the system override the the override?

What protection does the originator of the command have against override by another Operator?

from the entire unit. The manufacturer also knows precautionary measures needed for manipulating components during assembly. The same requirements logically should apply during disassembly.

Some manufacturers have addressed the question of safety during disassembly of their equipment, machines, and systems for servicing. They indicate points for securement and sometimes provide procedures for de-energization, isolation, and lockout.

The manufacturer may be unable to provide information, This can happen for several reasons. The focus of manufacturers is to create products that perform an intended function at a stated level of performance. There is no emphasis during design to enable disassembly in order to gain access for service. Many manufacturers lack the resources or the expertise to perform this type of analysis. All but the largest manufacturers buy off-the-shelf components manufactured by other companies and then assemble these into specialized configurations. Also, the manufacturer may no longer be in business or manufacturing the product in question.

Ultimately, the responsibility for resolving this issue rests with the owner and operator of the equipment, machine, or system. Depending on the complexity of the equipment, machine, or system, resolving this issue may require input from a group possessing different backgrounds. Potential participants include safety and industrial hygiene, engineering, and the hands-on personnel in operations and maintenance who work with and on the equipment, machine, or system. Collectively this group can provide the best knowledge-based and experiential expertise for addressing this problem. Gaps in representation could hinder the functioning of this group.

Safety and industrial hygiene practitioners can provide generalized knowledge about the anticipation, recognition, evaluation, and control of hazardous conditions. This knowledge also includes regulatory requirements and implications.

Engineering, operations, and maintenance personnel from the department that "owns" the equipment, machine, or system know what it does, how it functions, and problems that have occurred during operation and maintenance and disassembly. They have the knowledge about the electrical, hydraulic and pneumatic, mechanical, utility, and process and control systems that enable the equipment, machine, or system perform its function. Knowledge about control systems and feedback circuits is especially important where integrated systems and processes are involved. An action that occurs in one part of a piece of equipment or a machine or one process unit in an integrated chemical plant can instigate unpredicted and unanticipated effects in another. This knowledge provides the basis for discussion about energy flow and possible sources of energy retained through deliberate storage or entrapment.

Personnel from the department that "owns" the equipment also are potentially the most likely to work on the machine equipment or system to inspect or to perform work. Tradespeople acquire hands-on knowledge about the equipment and structure, its systems and subsystems, energy sources and control equipment (electrical breakers, for example). However, internal tradespeople are not always the individuals who acquire this knowledge. Many organizations contract out maintenance work to external service providers. As well, retirement, death, and downsizing take their toll on the knowledge resource. The problem (understanding, eliminating and/or controlling energy in equipment) persists. Failing to build a knowledge resource because of inefficient or insensitive use of people just compounds the problem.

One formalized technique for doing this is energy barrier and trace analysis (Wiggins 1997). Trace analysis in this context means locating sources of energy that are present in the equipment, machine, or system or can enter from the exterior. This technique also can involve tracing the path of energy through circuits to points of storage and/or conversion.

Once the pathway and control of flow of energy within the equipment, machine, or system are understood, the next step is to determine how to de-energize in a predictable, reliable, and verifiable manner. Determining the answer to this question could require experimentation with components within the equipment, machine, or system. These studies may indicate that while complete deactivation can occur, complete de-energization may not be possible. If this is the case, prevention of expression of energy retained through deliberate storage or entrapment must occur. Accomplishing this end-point could require immobilization of movable parts in a position from which damage cannot occur, or devising novel means to enable release of retained energy.

## Passive De-Energization

In many circuits, de-energization occurs as a consequence of deactivation. This is the case in circuits in which energy retention does not occur or where release occurs on deactivation. There are many examples that illustrate this concept. Rotation of a motor in a residential fan ceases on deactivation by pushing the OFF button. In this case, de-energization occurs rapidly. An electric heater provides a different way to consider the concept. Energy retention does occur, but the heat dissipates passively on removal of the input. The same is true for equipment and machines containing flywheels. Rotation slows gradually and stops without further input of energy. In none of these examples does the circuit retain energy following deactivation.

## Active De-Energization

Active de-energization describes measures that increase the rate of de-energization or enable removal of energy deliberately stored in components or entrapped in isolated circuits. Active cooling of hot objects or warming of cold objects to bring the temperature to ambient more quickly. are examples of active de-energization in situations in which passive de-energization occurs but at a slower rate. Application of the brakes in a coasting vehicle is another example. On a flat surface, the vehicle will stop following de-activation. Applying the brakes increases the rate at which stopping occurs.

Active de-energization is also utilized in circuits in which energy retention through storage or entrapment continues to occur following deactivation. Hydraulic and pneumatic circuits contain components that intentionally store energy following de-activation. This storage performs a function related to safe operation by maintaining pressure. Other circuits in the same equipment, machines and systems can retain energy unintentionally. Unintentional retention often occurs as a result of the geometry at the time of deactivation and occurs because of imposition of a load on the circuit.

## Practical Considerations

Wiggins (1997) commented that the most difficult sources of energy to identify and lock out are those found inside the equipment, machine, or system or are part of it. These sources often cannot be isolated from the equipment, machine, or system and must be positioned so that retention of energy occurs at the lowest level possible.

Sources of energy in equipment, machines, and systems that fit this description include batteries, capacitors, motors, transformer coils, flywheels; hydraulic accumulators, pressure vessels, pressurized piping systems; springs, mechanical arms and piping under load; parts at height; hot or cold parts; and reactive chemical substances. These sources are not exotic and the actions needed to de-energize them are well within the bounds of normal knowledge and experience both at home and at work (Table 8.2). Table 8.2 follows the sequence of actions listed in Table 7.8. Note that not all actions will apply in all situations.

Each action that disturbs the status quo can create hazardous conditions. These can result in harm for the worker performing the action or for someone else at a later time. Proposed actions deserve careful attention to ensure that the preceding activities fulfill their objectives without creating additional hazards. As there are many types of workspaces, pre-work preparation could involve additional actions not mentioned here.

Many of the actions mentioned above involve interaction with systems that retain, transport, and convert energy. Exposure to a burst of hazardous energy released during disassembly of confinement barriers could be fatal. While confined by the barrier surfaces of the equipment, machine, or system, this energy is controlled and isolated from contact with humans. However, once exposed to the surroundings during ill-conceived preparatory activity, fatal exposure could occur.

The state of zero energy is not always attainable in real-world situations, and may not be necessary or appropriate in many circumstances. This is often the case in equipment, machines, and systems containing many inputs. Deactivating and de-energizing all circuits prior to working on a single circuit unaffected by the others would provide no benefit. The protocol introduced in Chapter 7 provides a basis for making this decision.

## Identification of Affected Components

The next consideration is identification of affected components. Unambiguous identification is the only way to communicate the identity of relevant components. This is especially the case where identical components, such as breakers, fuses, or an array of valves, are present. An incorrect selection could lead to catastrophe.

An example of this problem readily familiar to do-it-yourself homeowners is the breaker or fuse panel. A completely unlabeled or incompletely labeled set of breakers or fuses is not unusual. The homeowner attempting to do electrical work in safety is then left to discover the breaker or fuse that affects the circuit in question. Discovering the answer to this question through trial-and-error disconnection and reconnection can lead to loss of settings in clocks, microwave ovens, DVD players and VCRs, and data in computer systems not connected to uninterruptible power supplies. Additionally, some equipment, such as compressors in refrigerators and freezers, if tripped during operation, may require resetting. Depending on the status of the relays involved, this may or may not occur satisfactorily, thus creating an additional problem. Electric dryers require restarting.

Adding to the confusion and unknowns is work performed through previous owners who tapped into circuits not expected to be carrying the load, but rather that reflected convenience of access. Once the sheathing is installed on walls, tracing of the path of wires to establish connections in a circuit requires considerable effort. The only readily obtainable source of information comes from selectively deactivating and de-energizing a circuit and determining using testing instruments circuits that remain live and those that are deactivated and de-energized. Such a situation leads to extreme frustration for the individual attempting to do a minor repair. Labeling or tracing of circuits is the obvious solution to this situation.

Preceding discussion provides a basis for identifying and characterizing the attributes required from signage used on equipment, machines, and systems involved in lockout and isolation.

### Identifiability
Signage is a prevalent feature in the production and process areas of many entities. Manufacturers and employers have affixed many more warning signs to equipment, machines, and systems than formerly

## Table 8.2
## De-Energization of Retained Energy

| Action | Potential Consequence |
| --- | --- |
| **Momentum** | |
| Allow moving parts to stop | Free-wheeling under the influence of gravity (loaded conveyor) |
| | Free rotation of untethered component(s) |
| **Gravity** | |
| Allow movable parts to settle to lowest possible position | Free movement of untethered component(s) |
| **Compression, Elongation, Torsion** | |
| Decompress springs and remove torsion and compression from structural members | Free movement of untethered component(s) |
| **Electrical** | |
| Open switch, open breaker, remove fuse | Arc flash, arc blast |
| **Pneumatic** | |
| Close valve to isolate line | Movement of untethered components caused by retained energy |
| Open bleed valve to relieve pressure | |
| **Hydraulic** | |
| Disassembly of components | Spray by pressurized liquid, vapor or mist |
| | Fire and explosion risk from discharge of liquid, vapor, or mist to atmosphere |
| | Movement by untethered components caused by retained energy |
| **Process and Utility** | |
| Close block valve(s) | Internal leakage at sealing surface |
| Open bleed valve(s) to drain residual contents | Fire and explosion risk from discharge of liquid, vapor or gas to atmosphere |
| | Inability to reseal bleed valve |
| Disconnect piping, install blanks, blinds, cups and plugs on disassembled piping | Spray by pressurized liquid, gas, vapor, or mist |
| | Fire and explosion hazard from discharge of liquid, vapor or gas to atmosphere |
| | Cracking of welds in piping during movement |
| | Leakage at sealing surfaces |

were present. As a result, signage to identify key components of equipment, machines, and systems affected by isolation and lockout can become easy to miss in the competition for visual attention.

Identifiability is an important consideration in signage. Component labels must be readily recognizable to everyone who is affected by them. This includes workers in the organization and personnel who work for external contractors. Identifiability of signage is often overlooked during site orientation.

## Simplicity

An identifier is a communication device. Workers use the names of components in communication. Simplicity in devising the name for identification of components is an art. Simplicity in naming is easily lost. Workers will not use descriptive codes that overwhelm the ability to use them in communication through complexity. Workers will soon find ways to abbreviate or replace overly long and convoluted names and identifiers.

Identification of components often occurs through use of simple number and letter codes. Letters, such as P for pump, V for valve, Tk for tank, B for breaker, and so on, are abbreviations that are readily identifiable with the affected component. These are examples of primary encodement. The receiver of the information makes the connection between the component and the code based on a single piece of information.

Choice of a particular name for a class of components is a critical decision. Avoiding conflict later in the process requires an appreciation for all components in all circuits that require labeling because of potential overlap between abbreviated names.

More complicated identifiers can require the reader to consult a secondary source of information in order to make the linkage between the component and the identifier. Requiring consultation with a secondary source of information can needlessly complicate the situation by adding an additional step. This additional step increases the inconvenience and motivates against using the labeling system already in place.

Signage includes signs, labels, and tags. Labels include commercial products that attach by adhesive or by physical attachment to the component. They also can be as simple as a marking made by permanent marker or etching by an engraving tool.

Labels from commercial sources have different colors, shapes, and sizes. Some are metal and metallic in color, while others are plastic materials. Use of devices obtained from a single manufacturer or vendor will help reduce the training and recognition burden associated with products having different colors and shapes. Color coding provides a way to identify components in different circuits, for example, yellow for ammonia and green for brine.

## Consistency

Industrial facilities are often in a seemingly continuous state of development. This means that as-built drawings and electronic files may never match the actual layout of the plant and its equipment, machines, and systems. An organization in this situation is always playing catch-up. Then, there is the reality that upwards of 75% to 80% of organizations have fewer than 20 employees. These organizations lack the resources to create as-built diagrams. The same can also be said for larger organizations that have downsized and have retained only a skeleton operating and maintenance staff. These organizations have lost the knowledge intrinsic to their most senior (and most vulnerable) employees. Instead of retaining elders as advisors, these organizations contract out engineering, design, and construction functions.

Hence, facilities where this style of operation is present become a hodge-podge of add-ons and modifications to the existing plant, as each addition leaves behind its mark. Complete integration of the add-ons to the existing plant never occurs. This situation has led to differences in nomenclature for each add-on. Instead of renaming components to maintain consistency, operators were required to make the link between different names for the same component in different parts of the operation. Consistency in nomenclature in the entire circuit is extremely important (Figure 8.10 and Figure 8.11).

The lack of consistency in nomenclature can pose considerable problems. In practical terms, long-term workers become familiar with differences in labeling and can make the translation between the old nomenclature on the breaker panel and new nomenclature on control devices on equipment. This might be marginally acceptable in a static situation. However, the reality is that new hires, temporary workers, summer students, and outside contractors have no way of reliably making the translation

Figure 8.10. Unambiguous labeling of components is essential for ensuring reliable communication. This example illustrates insufficiency of labeling on a control panel.

Figure 8.11. Unambiguous labeling should continue from the component to the point of isolation. This minimizes failures in procedures due to ambiguity. This example shows a well-labeled control panel.

when they enter this kind of operation. This sets the scene for miscommunication, selection of inappropriate components for shutdown and isolation, and the preconditions for an accident.

Another situation that illustrates the problem involves multiple locations containing similar equipment and circuits installed by different contractors and serviced by one entity. These situations include the lift stations or pump stations that collect wastewater in low-lying communities. When installed in different residential subdivisions and industrial areas in the absence of guidance from the Owner (the municipality that has ownership of the infrastructure), there is no consistency in naming of components in these systems. This situation creates needless confusion in service personnel.

Components in a circuit to be isolated and locked out require unambiguous signage. This includes, breakers, pumps, valves, instrumentation, tanks, process vessels, and so on. Only then can the procedure refer to specific actions to be taken on a specific component, and thereby provide the information needed by the least knowledgeable individual to perform them. Procedures created using the protocol introduced in Chapter 7 refer to components specifically by name.

**Durability and Substantiveness**
Signage must be durable and substantive. Durable means the product must be able to withstand the rigors of the environment to which it is exposed. There are many environmental factors that can degrade or destroy signage. Signage also must be substantive. Substantiveness means the inherent ability to withstand premature or unintended failure or removal due to disintegration or breakage and attempts by unauthorized individuals to destroy the product.

Markings applied using a water-soluble marker will readily wash off when exposed to bulk water sources, such as rain and water from hoses used for washing down equipment. This should be self-evident in outdoor environments and indoor ones where water is present. However, loss of identification created using water-soluble markers could also occur in environments not normally exposed to water during washing and less vigorous cleaning.

Accumulation of oil and grease on surfaces can release adhesives used to apply labels and dissolve inks used in permanent markers, as well as resins used in paint.

Exposure to the sun and other sources of ultra-violet energy causes fading with the passage of time of commercial labels and also those painted on or applied using a permanent marker. Some colors definitely are less resistant to fading than others. Determining the least susceptible color or product may require experiment. Another consequence of exposure to the sun and other sources of ultra-violet energy is embrittlement of plastic materials. Embrittled materials are subject to flaking, delamination, and breakage and crumbling, all of which can lead to destruction.

Hot temperatures are potentially capable of melting plastic labels and ties. Hot surfaces of signage can cause thermal burns on unprotected skin. Situations where hot surfaces are present include steam systems, heaters, and components that operate at temperatures above ambient. Cold surfaces of metal signage can cause frostbite. Cold temperatures embrittle plastic materials. Materials of construction must respect this limitation.

Chemically aggressive and erosive atmospheres can remove painted labels and those applied using markers, as well as more substantive products. The same situation can arise during repainting of equipment. Painters not instructed about the significance of the markings engraved, painted on or applied by marker onto the surface of a component easily could ignore and paint over them, rendering them impossible to read.

Another problem for readability is accumulation of material on surfaces. This can occur from sources, such as particulate and liquid aerosols. Settled material can bury small labels and markings.

# Isolation

Isolation constitutes actions taken to prevent re-energization of the equipment, machine, or system and/or to prevent expression of energy retained through deliberate storage or through entrapment. Isolation occurs physically through the severing or blocking pathways or connections between input sources of energy and the equipment, machine or system. A device that blocks the flow of energy is an energy isolating device.

One of the outcomes from analysis of energy flow, storage, and conversion in equipment, machines, and systems is an understanding about the components that participate in deactivation, de-energization and isolation. The article by Wiggins (1997), discussed earlier in this chapter, provides an extensive list of sources of hazardous energy internal to equipment, machines and systems. In practical terms, these components are deactivated, de-energized, and isolated through the action of electrical switches, electronic devices that operate through electrical switches, and valves and other components of fluid systems.

There are other devices that are less obvious. The latter are already installed in equipment, machinery, and systems or are installed by users for the purpose of preventing movement of parts and components. The action of some of these devices is similar in concept to the PARK setting on an automatic transmission in a vehicle. The PARK setting locks the transmission so that rotation of the wheels cannot occur, even though the engine can still operate. The PARK setting offers at least the same level of confidence, if not protection, against undesired motion on a slope, as is offered by the parking brake. Similarly, industrial equipment machines and systems that contain devices and configurations for isolating the flow of energy and materials containing energy in circuits and for preventing conversion of stored energy from one form to another considerably ease the burden on users and maintainers.

Isolation can mean different things between jurisdictions and different things in different regulations within a jurisdiction. Differences in concept between jurisdictions are to be expected, given differences in opinion between regulators.

However, isolation also can mean different things between regulations within a jurisdiction. To illustrate, the OSHA Standard on hazardous energy sources (OSHA 1989) considers isolation to be prevention of transmission or release of energy. The OSHA Standard on confined spaces in general industry (OSHA 1993) considers isolation to be the process by which a permit-required confined space is removed from service and completely protected against the release of energy and material into the space. According to OSHA this occurs by such means as blanking or blinding; misaligning or removing sections of lines, pipes, or ducts; a double block and bleed system; lockout or tagout of all sources of energy; or blocking or disconnecting all mechanical linkages.

Reliable isolation of circuits after dissipating internal energy in a predictable manner is critical to ensuring that work will occur at a state of zero energy. A format for diagramming isolation provides the means to explore implications about options for action and the risks that they entail. A diagrammatic process must contain a means to show the action of stopping energy flow. This can occur at single or multiple points of interruption. Ideally, the format integrates the depiction of energy flow, conversion, and storage, discussed earlier in this chapter.

## Visualizing Isolation

The simplest systems contain a source of energy, an output, and an isolating device (Figure 8.12). For the moment, this discussion will consider the OPEN position to represent a break in the circuit.This discussion, in particular, applies to electrical circuits. (Without getting into specifics at this point, some circuits operate in the opposite manner.)

Opening the switch is akin to raising the bridge on a highway. Energy, or in this analogy, vehicles, cannot jump the gap between sections of roadway. This, however, is not always the case. Numerous movies have shown stunt drivers who have driven vehicles at high speed onto the ramp of lift bridges to make the jump to the other side. Stunt driver Evel Knievel, for example, made a career from jumping even more extreme gaps on a motorcycle at high speed. These jumps included spanning long gaps of canyons. The point here is that given enough energy, as in a surge, the energy can jump the open gap.

One of the challenges in this type of analysis is to anticipate the cause and magnitude of the surges that can occur in various circuits and to ensure that the barrier to flow is wide enough, or that sufficient alternate paths are provided, so as to render them harmless. The latter comment is an important component of the analysis of energy in circuits and the effectiveness of the isolation.

Discussion about energy flow in circuits leads naturally to visualization of isolation. This has importance during later discussion about the consideration and strategy for isolation applied to real circuits, especially electrical ones. Isolation can occur at the ON-OFF switch in a circuit when the switch

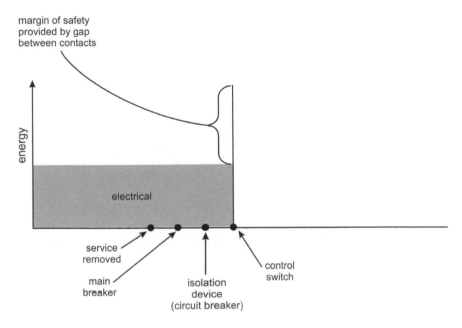

Figure 8.12. Simple electrical circuit. The open switch creates a gap across which electrical current cannot flow until and unless provided sufficient energy.

is in the OFF position. Provided that the switch is wired correctly, the energy is blocked at the switch, as described in Figure 8.12. This approach to isolation is strongly discouraged because of the many potential risks for error and exposure to electrical energy in the circuit.

The accepted practice for circuits at household voltages is to open the circuit breaker or to remove the fuse. This creates an additional block to the flow of current in the circuit and provides protection against exposure to electrical current in the event that the switch is closed or wired incorrectly. Circumstances in which work occurs on the panel require opening the main breaker in order to ensure safety. Given the disruption to operating buildings created when the main breaker is opened, work on the live panel is not an uncommon occurrence.

Severance of service at the service drop (Figure 8.13) is undertaken following major calamities involving the building. This ensures protection during work involving all wiring in the building. Major calamities usually involve fire and, increasingly, marijuana grow operations. Modification of wiring to bypass the meter to accommodate the heavy draw of the lights and other equipment used during these activities is characteristic of changes that occur. Ths only way to ensure safety during restoration of normal service and wiring is to sever the service at the drop and to rewire the affected areas.

### Isolation Strategy

Previous discussion focused on the actions needed to render equipment, machines, and systems to a state of zero energy. Properly implemented, isolation prevents normal as well as surge levels of energy from entering the equipment, machine, or system. Isolation creates a status quo for the equipment, machine or system. The weak point in relying on isolation as the end of the process is that it provides no means to maintain the security of the status quo, as established.

Small engines and the automobile, in particular, a complex integration of mechanical subsystems with which everyone has some level of familiarity, provide the basis for illustrating this situation. Stopping the engine of a lawnmower and preventing restart involves a series of progressively more

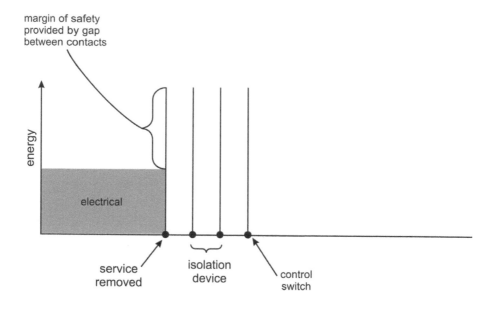

Figure 8.13. Isolation of electrical circuits at the service drop. Isolation in this manner is necessary in order to ensure protection against electrocution following fire and involving illicit activity, such as marijuana grow operations.

complicated steps. The simplest is to remove electrical energy from the sparkplug. This occurs by bonding the sparkplug wire to the engine. This floods the engine with fuel. Another approach is to run the engine to consume all the fuel in the tank. An alternative is to remove the sparkplug wire from the plug and lastly to remove the sparkplug. These steps make restart of the engine increasingly more difficult to the point of impossibility. Increasing the difficulty for an individual determined to restart the engine increases the challenge but provides little deterrent.

This example introduces and highlights the hierarchy of approach to isolation. That is, one approach to isolation does not apply in all situations. This is an important theme. It will recur in future discussion about strategy.

In the case of the automobile, a key deters, a number of undesirable activities. The word "prevents" is inapplicable in this context, since the ease with which a keyed system is defeated during theft, indicates that this will delay a determined individual only temporarily. A keyed door entry prevents unauthorized access to the fuel tank through the keyed cover panel and interior of the vehicle. Sometimes a keyed gas cap is also used. However, keyed locks are not successful at preventing auto theft or gaining access to the fuel tank. "Jimmying" the door lock or simply smashing the window is all that is needed to gain entry. Gaining access to the fuel tank is just as easy. A keyed ignition lock prevents unauthorized start-up and operation of the vehicle. Similarly, starting the engine requires only access to the starter wires or a forced ignition lock.

Forethought and simple tools are all that are needed in each of these cases to defeat the system of isolation and to gain access to the energy source. This discussion highlights the fact that there are very few systems resistant to the deliberate actions of a dedicated intruder. The software industry is replete with examples of successful attempts to "hack" into so-called secure systems. The result of this situation often is havoc. The motivation for such inventiveness can be deliberate for a particular outcome, but noncriminal, or even criminal.

The situation in industrial settings parallels that in the residential one. Motives for actions are the same, although criminal intent is considerably less obvious and considerably more difficult to prove. The situations are always called accidents, with the insinuation that the intent of the action that precipitated the outcome was not sinister.

Up to this point, this discussion has focused on gaining an understanding of equipment, machines, and systems, identifying components and the manner of eliminating and controlling energy flow. Equipment and machines are intuitive as self-contained entities of limited dimensions. Systems are only partly intuitive and encompass many possible variations. Since isolation and lockout also affect systems, some additional discussion is appropriate.

Systems can be small and very large, and open and closed. As introduced in Chapter 1, in large systems, the energy source can be unconfined, semi-unconfined, semi-confined, and confined. Unconfined energy sources are unconstrained by boundaries. These include weather systems and freely moving sources of energy such as large predators. Semi-unconfined sources include large piles of processed, loose material. They are confined by a single surface, such as the base on which the pile rests. Industrial materials include piles of ore, sand, gravel, crushed stone, road salt, coal, raw sugar, agricultural products, and other bulk materials. These piles are often found outdoors, but sometimes inside shelters, although the walls and roof do not influence energy flow, storage, or conversion by material in the pile.

Semi-confined sources are restricted by additional confinement. Snow and rock and mud confined within boundaries on mountain slopes are examples of naturally occurring materials that are capable of storage and rapid conversion of energy during avalanches and landslides.

The energy stored in materials in piles and on mountain slopes is not controllable through isolation and lockout. This requires measures reflective of the mechanics of bulk materials that relieve the instability that leads to the sudden cascade.

Semi-confined sources also include water that cascades down the spillway at a dam. This is a controlled release of surplus water and the energy contained within it. An open-topped furnace or ladle containing molten metal is another example of a semi-confined source of energy. Molten metal will rapidly emit from containment following addition of incompatible materials containing water. A similar but considerably less familiar example is the system that collects overflow at swimming pools and diverts this water to a surge tank.

Energy sources encountered in routine experience are mostly confined within closed systems. Closed systems confine energy flow, storage, and conversion within circuits, such as wiring and piping. Open systems are also present, but are less prevalent. Open systems primarily handle fluids and flowable solid materials where there is throughput, meaning input, processing, and storage and output. High-pressure water jetting systems are applications of open systems. The fluid or the flowable solid material experiences a discontinuity within the system (output), so that recirculation does not normally occur. Closed circuits recirculate the flow with or without the discontinuity. Confined systems have permanently connected circuits and sometimes disconnectable ones

Preventing energy flow is achievable in most circuits by opening switches in electrical systems, or opening or shutting valves (depending on context). These actions are process- or equipment-specific. They occur through energy-isolating devices (valves, switches, and so on). Energy-isolating devices alone provide an element of security. In some circuits three or even four energy-isolating devices are present. Conceivably, an individual could operate each of theses devices to isolate the circuit. How many are actually operated often depends on personal choice and comfort level. The question then is, how many energy-isolation steps are needed to provide security and therefore, safety; or stated another way, operation of how many energy-isolating devices is too many? Table 8.3 outlines some of the considerations and implications of using single and multiple energy-isolation devices.

## Single Isolation

Isolation provides the physical means to prevent the entry of energy and bulk materials containing energy into equipment and systems and movement from storage. What isolation does not offer is the ability to protect the status established by the energy-isolating device. Accident reconstruction reviewed in Chapter 1 shows that a single level of isolation often provides unreliable protection against the oc-

## Table 8.3

## Isolation Strategies

| Action | Foreseeable Advantages | Potential Disadvantages |
|---|---|---|
| Single isolation | Ease of reactivation; isolation device likely under direct control of user | Unintentional reactivation by user; unintentional reactivation has occurred in many accidents |
| | | Single layer of protection against surges |
| Double isolation | Prevent unintentional activation by user | Reactivation possible by third party; third party reactivation has occurred in many accidents |
| | Double layer of protection against surges | Possible inconvenience due to distance between work area and point of isolation |
| Triple isolation | Prevent unintentional activation by user | Increasing level of inconvenience for user, especially where isolation devices are remote from the work area |
| | Prevent activation by third party except in rare circumstances involving deliberate reactivation | Increasing potential for multiple circuits to be controlled by the more distal isolation device |
| | | Conflict or miscommunication likely |
| | Increased protection against surges | |

currence of accidents. This is especially true where the isolation device is a permanent part of a circuit, for example a switch or valve.

Fatal accidents have occurred because the operator or maintainer of a piece of equipment made unintentional contact with the start switch or an interlock device. At the time, the unit was in stand-by mode. Assistants have activated the inappropriate switch due to miscommunication in noisy environments, or even the inability to read the label. Computerized systems have activated isolated systems because of logic mistakenly programmed to restart equipment in stand-by mode. Similar situations have involved consoles in control rooms containing banks of switches. Control room personnel and visitors have inadvertently activated equipment by leaning on the console against switches, or by placing objects and clothing onto the bank of switches, and so on. Unprotected restart is a serious limitation of this approach. This approach works reliably only when the energy-isolating device is not readily accessible and is not activated, and remains under the exclusive control of the person who is affected by its status.

Double Isolation
At first glance, double isolation appears to offer an extra layer of protection to the operator or maintainer. Activation of the proximal energy-isolating device singly does not lead to activation of the equipment or system. Similarly, activation of the distal energy-isolating device singly does not lead to

activation of the equipment or system. This approach is potentially most successful where the individual affected by the isolation has exclusive control over the energy-isolating devices. Redundancy, however, can rapidly lead to complacency. Complacency results from the mistaken belief that there always is a second layer of protection. This is not always the case, since the distal energy-isolating device, unless highly visible to the operator or maintainer, can become active through miscommunication or misadventure. Again, this system is most effective when the status of the energy-isolating devices is obvious and not readily disturbed, and under the exclusive control of the person affected by the isolation.

Double isolation is least protective when the energy-isolating devices are part of a permanent circuit. Double isolation is considerably more substantive where the second energy-isolating device is easily separable from the circuit. This is the case with disconnectable items, such as electrical plugs. Isolating the first energy-isolating device provides access to the disconnectable device. In this case, the connection to the source is physically severed and must be reestablished, so that even the first level of isolation can be activated.

Double isolation can impose inconvenience where the distal energy-isolating device is some distance from the work area. The operator or maintainer must abandon the work area in order to tend to the distal energy-isolating device. Increasing distance between the machine or equipment and the distal energy-isolating device increases the inconvenience intrinsic to this approach. This especially becomes the case when the period of isolation is very brief or where many brief periods of isolation occur during the day. The perception that the hazardous condition is controllable by other means adds to the perception of inconvenience.

Double isolation can impose conflict where the distal energy-isolating device controls more than one circuit. Isolating more than one circuit through a common energy-isolating device means that the isolation can simultaneously affect the work of more than one individual. This can lead to conflict and miscommunication or misadventure about the availability of the circuit affected by the isolation.

Distal isolation also can create uncertainty about the isolation offered by the energy-isolating device. This can result from mislabeling or lack of labeling of the circuit(s) affected by the distal isolation. That is, the effect of the distal energy-isolating device is unknown unless tested against the ability to deactivate the affected circuit.

Triple Isolation

Triple or additional isolation adds additional protection against the impact of a surge in the circuit. At the same time, this increases the potential for conflict and miscommunication with users of equipment or systems affected by the same energy-isolating device. The greater the number of units affected by the same energy-isolating device, the greater becomes the potential for misadventure, deliberate or otherwise. Misadventure can occur within the same shift, but the potential increases considerably across shifts. This is readily evident in multi-shift environments. Things get moved, despite the presence of signage demanding the opposite. Miscommunication and conflict are exacerbated, especially among groups that have little contact with, and understanding, of each other.

Triple or additional isolation is effective only when the isolation remains under the control of the individual directly affected by it. Fatal accident reconstruction highlights situations where individuals have defeated multiple isolations in order to restart equipment. This occurred despite the seemingly obvious conclusion to be taken, that the isolation was present for a reason.

Defense in Depth

The concepts of multiple isolation raised here are consistent with an approach known as redundancy or defense in depth (Ortiz et al. 2000). Defense in depth reflects the premise that a single equipment fault or human error should not directly precipitate an accident. Multiple safety precautions act to protect the individual, especially when one protective measure fails. The likelihood of an accident arises when multiple layers of protection are not in place. This allows a single error or failure to develop into an accident situation.

Research presented in previous chapters, starting with accident analyses in Chapter 1, indicates that redundancy (defense in depth) is essential for accident prevention based on actions of individuals across the spectrum of industry in one dimension and technical failures producing exposure to energy sources along another. A single approach or protective measure is insufficient to ensure accident prevention at a high probability. Defense in depth (redundancy) has formed the approach in the nuclear and chemical process industries for many years.

The preceding discussion highlights a major shortcoming in the process of isolation: insecurity of the energy isolation in multi-employee and multi-shift workplaces. Unsecured energy-isolating devices work best and most successfully when under exclusive control of the individual affected by them. However, even exclusive control does not provide security in the face of memory lapses and inattention to detail and conditions, and uncontrolled and inadvertent body movements that change the status of energy isolation devices.

Another element highlighted in these discussions is the inconvenience factor. Inconvenience is a value that people correlate with the effort needed to perform some action. In effect, deciding whether an action is convenient or inconvenient involves a cost-benefit decision. In this circumstance, the cost is the effort (energy expended and time) needed to carry out some action. The benefit is an intangible quantity (that is, until recognized as needed) called safe performance of work. Safety is intangible and imperceptible until the opposite condition, absence of safety, demonstrated through high frequency of near misses and accidents, develops in a workplace. Many people refuse to take actions providing obvious benefits because the inconvenience factor exceeds some intrinsic threshold in their system of values. They would rather take the risk, especially one perceived to be small.

"Common sense" is often held high as the benchmark against which people make these value judgments. "Common sense", however, is relative to the level of knowledge and the value system intrinsic to an individual. Unfortunately, even highly knowledgeable individuals do not share the same value system with regard to risk-taking. Inconvenience and the accompanying risk-taking are related to other factors: haste to perform some action, physical well-being, and other factors related to the exertion required.

## Control Circuits and Energy Isolation

Most electrical equipment operating at voltages above 110/220 V contains control circuits. Control circuits have received discussion previously. At this point, the discussion focuses on the role of control circuits in isolation and lockout and their contribution to defense in depth (redundancy).

Control circuits energize and de-energize the device through action of the contactor. Control circuits, whether manually operated or software activated, are not considered energy isolating devices because the disconnect which is upstream can remain energized Figure 8.14). Hence, the control circuit does not provide the certainty of isolation that is available from isolation of the power circuit.

At the time of energization of the power circuit, the equipment, machine, or system is not under load when performed through the control circuit. Imposition of the load occurs at the time of closing the contactors in the power circuit. At the time of isolation of the power circuit, the control circuit disconnects the load by opening the contactors. The equipment containing the power circuits and control circuits is housed in Motor Control Centers (MCCs). MCCs are usually located in protected areas (electrical rooms) remote from the control switch.

Isolation of the power circuit occurs by opening the disconnect and removing fuses or opening the breaker. Performing the isolation and lockout when the equipment, machine, or system is not operating and the energy isolating device is not under load is critically important for minimizing the potential for arc flash and arc blast in electrical circuits. Arc flash and arc blast and pressurization accidents occur while power circuit is opened under load. The process is similar to generation of a welding arc when the welder slowly lifts the rod from the base metal.

Switches used in control circuits usually are not lockable. Relying on these circuits and the equipment, machine, or system to remain deactivated and de-energized during isolation requires careful coordination and communication where multiple operators and other personnel are present in an operation. This is necessary to ensure that reactivation of the control circuit does not occur during the interval between deactivation and de-energization of the control circuit and opening of the contactors and isolation of the power circuit. Switches in control circuits are often three- or four-position rotational switches or push-button devices. These must not move from the set position during the isolation

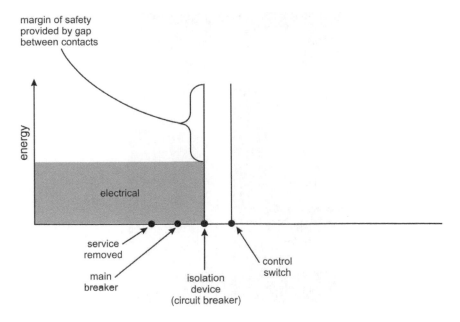

Figure 8.14. Isolation provided through a control circuit (Figure 8.1) compared to isolation through a circuit breaker (disconnect) or fuse. The barrier to passage of energy through the open power circuit is very high. The control circuit is de-energized following deactivation and de-energization of the power.

and lockout procedure. The same caveats apply to deactivation commanded through computer interfaces.

These situations represent examples of loss of control over conditions. One possible remedy for this situation is to retrofit control switches to render them lockable. This would provide contingency to the overall procedure through double lockout. Where the system operates through software control, the use of password-protected command status offers one means to minimize potential for override and reactivation of the circuit. Use of a spotter to observe operational status of the equipment, machine or system and two-way radio communication is a much less secure means to ensure that deactivation has occurred and remains effective at time of isolation of the power circuit.

MCC cabinets receive input power from the high-voltage supply. In the event of de-energization of the device, meaning opening the contactors that energize the device, the supply side of the circuit in the cabinet remains live, unless or until de-energized at the upstream disconnect or breaker.

Control cabinets contain safety devices that de-energize the load-side of the circuit on opening the door using the safety latching device (Figure 8.15). Bypassing the latching device while the circuits remain energized is easily accomplished (8.16). This action is familiar to operators and maintainers of this equipment. They make the argument that ensuring that the interior of the cabinet is electrically safe (de-energized) prior to making an adjustment to some component would require shutdown of the entire system dependent on the controls located inside.

Bypassing the safety latch leaves energized conductors on both the supply and load sides of devices located in the control cabinet. Partial bypass of the safety latching device by rotating the inside component to the OFF position de-energizes circuits on the load side in the cabinet. These practices negate the value of the safety locking device, but appear to be common, especially in small operations.

Figure 8.15. Typical cabinet containing control and power circuits. The device used to gain access to the interior of the cabinet deactivates and de-energizes the power circuit, but does not isolate it from the power supply. The supply side remains energized on opening the cabinet in the approved manner.

Figure 8.16. Defeat of the safety locking mechanism can occur using simple tools. In this configuration, all circuits inside the cabinet remain energized until or unless isolation occurs at a disconnect or breaker upstream from the cabinet.

Disconnection
One of the major strategies of isolation is disconnection. Disconnection entails physical separation of structures that confine energy, so that the path of flow contains a discontinuity.

## Disconnectable Circuits
One variation on disconnection is the disconnectable circuit. Disconnectable circuits occur in both electrical and piping circuits. Disconnectable circuits provide the means to sever the connection while the circuit is under load. While this is technically feasible based on design, as discussed in the previous sections, deactivation and de-energization prior to disconnection provide greater assurance of safety. This is an important consideration, given the fact that deterioration over time in hostile environments is a recognizable concern. Despite this caveat, these products provide valuable contributions to this area.

Electrical circuits provide the more readily recognizable examples. There are many examples in residential environments. Electrical equipment containing disconnectable circuits includes power tools and small appliances that operate on standard 110 Volt home wiring, and electric ranges and clothes dryers connected through plugs and receptacles into 220 Volt circuits. Plugs that operate at higher voltages are available (Figure 8.17). These units are designed to be disconnected under load. The cap of the receptacle (the energy-isolating device) is lockable.

Disconnectable couplings are also available for circuits containing pressurized fluids (gases and liquids). These offer the ability to connect and disconnect under pressure (Figure 8.18 and Figure 8.19). As well, hoses connected to wall-mounted valves from which they are readily disconnected can supply heating needs (steam) to small-scale reactors and tanks. The hoses are readily removable at the conclusion of use. Another example is dedicated geometry, disconnectable piping linkages used in some processes. These can handle liquids and gases. The piping system is discontinuous, and therefore unable to function, unless these linkages are installed.

Isolation in these examples is a manual two-step process. The first step is to close the appropriate valve or to move the control switch to the OFF position. This could occur, for example, when turning off a power tool and letting the rotation stop. In fluid systems, this could include shutting an upstream valve. Neither action is absolutely necessary. However, they are advisable, since removing plugs from receptacles while under load or disconnecting fluid lines while under pressure creates the potential for needless exposure to the energy source when the separation is not "clean." Parts of the body or jewelery or tools held in the same hand potentially can come into contact with live prongs in the partially re-moved plug. Likewise with disconnectable hose systems, disconnection under pressure increases the potential for emission of a high-pressure stream of fluid (liquid or gas).

Disconnection of this type of circuit means that the energy cannot flow past the point of isolation. This approach can provide satisfactory protection, provided that reconnection to the isolating device does not occur without the knowledge and approval of the individual affected by the action. In order for this approach to provide and maintain safe conditions, the disconnected plug or hose or pipe fitting must remain under the exclusive control of the individual who effected the disconnection or to be im-mobilized within a lockout device.

This approach alone is not satisfactory when the point of disconnection and the point of work are separated or when another individual can reconnect the components without the knowledge, consent, or direct control of the individual who made the disconnection. In fact, the only individual who should make the reconnection is the one who made the disconnection. This eliminates the potential for miscommunication and errors in selection where more than one choice is available. Lockout devices are available for enclosing plugs removed from receptacles to prevent reconnection (Figure 8.20). Large gauge sizes may also accommodate disconnectable hose and fittings. Devices are also available to immobilize the prongs of the plug (Figure 8.21).

## Lockout

The concepts of isolation and lockout bring specific language and meanings attached to words. One example is the language employed in the U.S. standard on controlling hazardous energy (OSHA 1989). Isolation is the act of preventing transmission or release of energy in a circuit. An energy-isolating de-

Figure 8.17. Disconnectable electric circuit. The plug in this product is designed for removal from the receptacle while under load. While this is technically feasible based on design, as discussed in the previous sections, deactivation and de-energization prior to disconnection provide greater assurance of safety. This is an important consideration, given the fact that deterioration over time in hostile environments is a recognized concern.

Figure 8.18. Disconnectable fluid circuit. While this is technically feasible based on design, as discussed in the previous sections, deactivation and de-energization prior to disconnection provide greater assurance of safety. This is an important consideration, given the fact that deterioration over time in hostile environments is a recognized concern.

Figure 8.19. Disconnectable fluid circuit. This design uses valves connected in series for isolation. Fluid trapped between the valves and released during disconnection is a small amount compared to fluid in the circuit. (Courtesy of Banjo Corporation, Crawfordsville, IN.)

Figure 8.20.  Lockout device to enclose a plug.

Figure 8.21. Built-in lockout device to prevent use of electrical plugs. (Courtesy of Meltric Corporation, Franklin, WI.)

vice is a mechanical device that physically prevents the transmission or release of energy. Lockout is the act of placing a lockout device on an energy-isolating device in accordance with an established procedure that ensures that operation of the energy-isolating device and the equipment being controlled cannot occur until removal of the lockout device. A lockout device uses a positive means to hold an energy-isolating device in a safe position, thereby preventing energization of affected equipment or machinery. Isolating and locking out a hazardous energy source prevents its activation (OSHA 1989).

The answer to the situation highlighted in the previous section regarding the insecurity inherent in the use of energy-isolation devices is lockout. In effect, lockout is the ritualized application of a security device to an energy-isolating device (Figure 8.22). The security device is a lockout device and lock or other tamper-proof device. Lockout requirements vary only slightly across jurisdictions. The functions of the lockout device are two-fold.

First, the lockout device and attached lock notify other workers and bystanders about the status of the energy-isolating device. This function is informational. The lockout device and lock indicate that someone is signaling that the position of the energy-isolating device is not to be changed. This can also occur through the use tamper-proof seals and other low-tech devices. An informational device does not confer protection.

The second purpose for the lockout device and attached lock is protection of the status quo. That is, the lockout device and attached lock are intended to prevent change in the position of the energy-isolation device. Attachment of a lock to the energy-isolation device sets a standard of protection. Defeating the lockout device requires deliberate and considerable application of force. Unlike the unprotected energy-isolation device or the tamper-proof seal, for all but the most determined, the lock and lockout device set a formidable barrier for physical tampering and interference that is not to be crossed.

This approach is successful, so long as everyone respects the sanctity of the lock and lockout device. Given sufficient time, appropriate tools, such as a hacksaw, bolt cutters, or a cutting torch, and de-

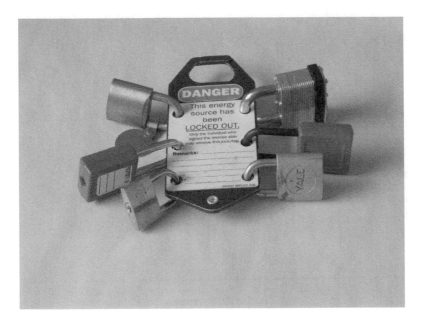

Figure 8.22. The lockout paradigm. Each person affected by an isolation applies a uniquely keyed personal pad lock to the lockout device prior to starting work on the affected equipment, machine, or system.

termination, every lock and lockout device will eventually yield to destructive force. Unfortunately, this situation has occurred and led to fatal accident. This, fortunately, is an isolated exception to the norm.

The paradigm that has evolved over the years is that each person affected by the status of the isolation attaches a unique lock to the energy-isolating device. Normally, the person who applies the lock also removes it.

### Lockout Devices
Locks alone are often insufficient to provide the coverage needed to prevent movement of energy-isolating devices from a set position. This results for several reasons, starting with the dimensions and geometry of the pieces requiring securement. Lockout devices attach to energy-isolating devices in a manner to prevent unauthorized movement. The lock(s) attach to these fittings to prevent their removal. Some energy-isolating devices are directly lockable. Lockable capability in energy-isolating devices sometimes becomes possible through a retrofit. Otherwise, the lockout device is an add-on. Considerable development of lockout devices has occurred during recent years. Some circumstances demand the use of a lockout device even though the energy-isolating device is also lockable.

Incorporation of the attachment for the lock on the energy-isolating device through design is the most desirable option. This means that the designer can incorporate a structurally sound attachment point for a lock in a compact, unobtrusive manner. The designer has the "big picture" concerning the equipment or component. Incorporating an extra feature at this stage is considerably easier than doing so as part of a retrofit.

Lockout devices include chains and cables, scissors, and hasps. Chains and cables need to be able to resist the environment to which they are subjected and attack by all but special tools, such as bolt cutters and other metal cutting tools (Figure 8.23). Some locks also incorporate chains and cables, so that no additional hardware is required. These products are often coated with polyvinylchloride (PVC) or

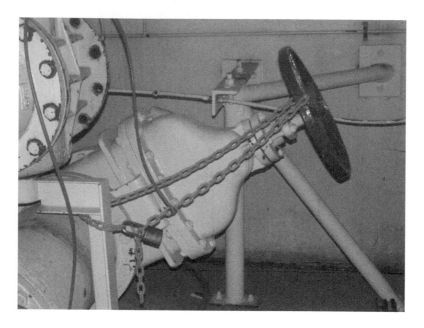

Figure 8.23. Coated chain used in a corrosive environment.

are fitted into PVC sleeves. The PVC provides protection against weather and hostile chemical environments. PVC also adds additional toughness to the product and increases cutting resistance. PVC-coated cables having diameters from 6 to 15 mm (0.2 to 0.6 in) are readily available from manufacturers.

Hasps are hinged devices that attach to doors and a fixed surface and provide attachment for the lock. The hinged closure prevents access to the mounting hardware. Some hasps incorporate padlocks. Brackets containing padlock eyes are also used. Roundheaded mounting hardware containing no surface machining is required for installations using these products. Plain roundheaded hardware prevents access to screwdrivers.

There is also the need for lockout devices that can accommodate multiple locks, since application of more than one lock may occur during a work activity. Scissor hasps can accommodate several locks. These devices and the locks themselves must be substantial, so that they cannot be pried apart and easily removed (Figure 8.24). Scissor hasps are available in PVC-coated metal, as well as anodized aluminum. The latter product incorporates a write-on tag. A nonmetallic nylon version is also available. This product is nonconductive and is promoted for use in electrical lockouts. Other versions are non-sparking and intended for use where ignitable atmosphere are present. The appendices to this book provide additional examples of lockout devices in specialized applications.

Lockout directives impose requirements on lockout hardware. These are intended to ensure that the hardware is readily identifiable, durable enough to withstand the environment to which it is subjected, and substantive enough not to be easily compromised.

Identifiability
The first requirement for locks and lockout devices is identifiability. Locks must clearly identify the individual who applies them. Identification can include a photograph of the person, engraving of the person's name and clock number, a color code, or other readily recognizable characteristic. The photo-

Figure 8.24. Hasps. Scissor hasps can accommodate several locks simultaneously. Each installed lock prevents opening of the hasp.  (Courtesy of Panduit Corporation, Tinley Park, IL.)

Figure 8.25. Keyed padlock containing the photograph of the individual applying it. A photograph rapidly communicates the identity of the individual. (Courtesy of Panduit Corporation, Tinley Park, IL.)

graph is probably the easiest identifier with which to work (Figure 8.25). It is large and shows the image of a person who is readily recognizable to others and the person's name and possibly department.

The photograph and name are examples of primary encodement. The receiver of the information makes the connection between the applier of the lock based on a single piece of information and does not need to investigate a secondary source in order to make the link.

The other identifiers, clock number, color, or other encoding, require the recipient to consult a secondary source of information in order to make the link to the person who applied the lock. That is, there must be a list containing the link to the individual. Requiring consultation with this list to establish identity needlessly complicates the situation by adding an additional step. This additional step increases the inconvenience associated with following the ritual.

Lockout devices generally are red in color, red normally being indicative of STOP and DANGER. The use of red as a color is not universal. Different colors are used on locks and lockout devices, one purpose being to comply with ASME A13.1 on pipe color schemes (ASME 1996). Some metal devices are metallic in color, while others are anodized or plastic coated.

Durability

The second requirement for locks and lockout devices is durability. Durable means they must be able to withstand the rigors of the environment to which they are or could be exposed. These can include contact with vapors, mist, and splashes of hydrocarbons and substituted hydrocarbons. These substances can dissolve into many plastics and compromise dimensional stability through swelling. Corrosive (acid and caustic) vapors, mists, and liquid splashes can attack and eventually destroy susceptible metal components. Erosive atmospheres, such as encountered during abrasive blasting, can destroy the structure of the device. Dust associated with these operations can compromise mechanical function of the device. Cold temperatures embrittle plastics. Embrittlement can lead to premature failure.

Durability should not be an issue with lockout devices incorporated into components as part of their design. Designed-in lockout devices are constructed from the same materials to the same standard of manufacture as the rest of the component.

Durability could be an issue with retrofits created by third-party manufacturers, and especially retrofits created by end-users. The manufacturer involved in original specification of equipment for a particular application gains knowledge about the rigors of the environment of the installation during tendering. Addition of a retrofit to an installed component likely occurs without interaction between the manufacturer and purchaser. Satisfaction with the installation is a function of the materials of construction and competence of the installer.

Add-on lockout devices face temporary exposure to the operating environment. This exposure could be as severe as that encountered by permanently installed devices, even though equipment is shut down to enable the work activity to occur. Circuits often operate in parallel in order to provide continuity of operation. Hence, the shut-down unit could be adjacent to fully operational equipment. Durability is not the issue with add-on lockout devices that exists with permanently installed devices, since the installation is temporary.

Hot temperatures of uninsulated valve stems and other components, as might be encountered in process operations involving heated liquids and steam systems, are potentially capable of causing severe burns during application and removal and melting plastic components.

Substantiveness

Lockout devices also require substantiveness. Substantiveness means the inherent ability to withstand premature or unintended failure or removal due to disintegration or breakage and attempts by unauthorized individuals to destroy the device. Breakage and destruction should require use of aggressive tools, such as hacksaws, cold chisels, bolt cutters, and other tools capable of exerting considerable force.

## Standardization

Regulators, such as OSHA, require standardization of lockout devices by shape, color or size (OSHA 1989). This requirement could cause conflict with the use of multi-colored devices, as described in a previous section. Certainly, use of devices sourced from a single manufacturer or vendor will help reduce the training and recognition burden associated with products having different colors and shapes, as is the intent of the regulatory requirement. This is especially the case in organizations that have a multiplicity of lockout devices. However, this can increase inconvenience where different groups within an organization or different organizations are working at the same worksite.

Differences in color and appearance can provide valuable cues to workers about which group is responsible for the placement of a lock, or lock and lockout device, or for that matter, the hazard of the circuit on which the device is placed. Color and shape cues can considerably reduce inconvenience. Otherwise, where homogeneity is enforced, individuals must read the lock or lock and tag to determine the identity and affiliation of the individual who installed the device.

Similarly, compliance with this requirement may not be possible in organizations containing a wide range of equipment having built-in, designed-in and retrofitted, and externally applied lockout devices. A broad range of equipment requires many different lockout devices.

## Retrofits

Retrofitting lockout devices is an acceptable alternative provided that the retrofit meets the criteria that apply to lockout devices. Retrofitting is the only option for older equipment for whose components isolation and lockout were not legal requirements at time of manufacture.

Manufacturers of some equipment and machines make available user-installable retrofit devices. Manufacturers are the logical suppliers of such components, since they know best the equipment or machine and have potentially the greatest means to establish contact with purchasers of older designs. The downside to this approach is the level of skill, finesse, knowledge, and experience of the individual who installs the retrofitted device. Installing a retrofit device onto an existing installation involves risk of injury to the installer, as well as risk of damage to the equipment or machine on which the installation occurs. Installation can require disassembly, drilling, cutting, filing, and other manual operations that can occur outside tolerances established by the manufacturer. Typical problems result from destruction and loss of parts involved in the installation and ill-considered improvisations.

The owner also can make the device used in the retrofit. This is best attempted only when retrofits for the equipment from the manufacturer or a third party are not available. The owner then undertakes the task of determining how to design the retrofit to prevent expression of energy isolated by the device, and then how to mount it onto the existing component in a manner capable of meeting requirements for performance of the equipment and security of energy isolation. This is especially critical where the component is designed for use in a hazardous or wet location, and where continuing electrical protection is required. Any modification could be construed as tampering with design and assembly features, and invalidating performance demonstrated through testing and certification.

Careful consideration about the internals of the component onto which the retrofit is to be mounted is necessary. This is especially important in electrical equipment where exposed conductors may be present and internal clearances are critical. As well, mounting hardware, such as bolts and screws must not permit disassembly from the exterior. This would defeat the security of the design.

The materials of construction of the retrofit must be compatible with those of the existing component. The retrofit, likely being external to the original component, must withstand the rigors of the environment to which it is exposed. A number of mechanisms can cause failure. These include oxidation, corrosion, physical disruption due to earthquake and excavation, use of incorrect or inappropriate components during assembly, mechanical stress, and possibly other mechanisms (Shackelford 1988).

Oxidation is the direct chemical reaction (attack) between metal and atmospheric oxygen. Metals affected by oxidation must have contact with atmospheric oxygen. Build-up of oxide scale on metals occurs by several mechanisms, each characterized by a specific type of diffusion through the scale. The zinc coating on galvanized metal is designed to oxidize (to form a protective layer of zinc oxide). The oxide layer can increase electrical resistance.

Direct chemical attack can occur through contact with other gases, including hydrogen sulfide and mercaptans which can be present in worksites due to anaerobic processes involving organic debris. Attack by hydrogen sulfide and mercaptans can lead to formation of a sulfide outer coating on metals.

Corrosion is the dissolution of a metal into an aqueous environment. Metal atoms dissolve as ions. Corrosion occurs by several mechanisms, each characteristic of the environment in which this occurs.

Aqueous corrosion is a form of electrochemical attack. A variation in concentration of metal ions above two different regions of a metal surface leads to an electrical current that moves through the metal. The region of low ionic concentration corrodes (loses material to the solution). Salt water draining into subsurface chambers from salted roads can increase the rate of aqueous corrosion. Both carbon steel and stainless steels are at risk in this environment.

Galvanic corrosion results when two dissimilar metals, one more active chemically than the other, are placed in contact. The more active metal acts as an anode and is corroded (loses material to the solution). Iron and zinc will corrode when in contact with copper in the appropriate ionic environment, such as the salt water that enters subsurface structures from salted streets and roadways. Galvanic corrosion can occur following substitution of inappropriate metals in fastenings and fittings.

Gaseous reduction involving dissolved oxygen is another corrosion process. In this case, the driving force is the difference in concentration of oxygen in different parts of the aqueous environment. Differences in oxygen concentration can develop in cracks and gaps in metal, in surfaces of the same piece of metal that are covered by dirt, grease, paint, or other impervious coating versus an exposed surface, and by the environment created by fastenings and fittings that hold together pieces of metal. Corrosion of this nature is readily obvious in the rust that forms at the joints of welded and non-welded steel structures.

Zinc in galvanized coatings can form a protective oxide layer on metal surfaces. Acidic rainwater (pH <7) can attack active metals, such as zinc, to form hydrogen gas. On occasion, accumulation of hydrogen in structures has led to fire and explosion. Acid attack on the galvanized coating, while removing the oxide layer and improving conductivity, eventually exposes the underlying steel to both oxidation and acid attack.

Mechanical stress enhances corrosion. Mechanical stress refers to both applied stress and internal stresses associated with microstructure (for example, grain boundaries). Inappropriate selection of washers can fail to maintain tension or can create inappropriately high mechanical stress in assembled components. High-stress regions (such as cold worked steel) act as anodes relative to low stressed regions. These regions can occur locally on the same piece of steel where flexing has occurred.

Add-on lockout devices span a wide range of design and functionality. Add-on lockout devices generally are temporary and do not remain in place following their use. Lockout devices, such as hasps and scissors, extend the scope of the lockout paradigm. Hasps and scissors enable several individuals to apply their locks to the lockout device.

## Locks

The lock specified for use in the lockout paradigm is a uniquely keyed padlock, rather than a combination type, under the control of a single individual. No one should have a key to the personal lock used by another individual except under highly controlled circumstances. This includes supervisory personnel. A lock whose keys are under the control of only one individual means that that person controls its application and removal. A common keyed lock allows access to the lock by many individuals. This means that an individual other than the one who applied the lock could remove it. This strategy could be beneficial where the individual in control of the lock was not available to remove it at a particular point in time. On the other hand, this defeats the purpose of maintaining the security of the isolation until removal is sanctioned by the affected individual.

Combination locks present problems, starting with loss of the combination. More serious than this is sharing the combination between individuals. This is the same as making duplicate copies of keys to a dedicated padlock and sharing them with other individuals. This situation completely defeats the intent and purpose of the security represented by the personal lock.

Attachment of a warning tag to accompany lockout provides visibility and additional information to spectators to the process. This heightens awareness and the need to respect the lockout.

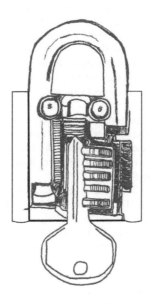

Figure 8.26. The humble keyed padlock is a complex mechanical structure containing a shackle, body, and locking mechanism.

Locks are ubiquitous, perhaps even humble. Despite their ubiquity and humble appearance, locks are complex manufactured items (Figure 8.26). Padlocks range in complexity from the lightweight products used on suitcases to provide token security to substantive pieces of hardware. They constitute a spectrum of complexity and secureness. Padlocks contain a number of components.

The shackle is the "U"-shaped piece of steel that opens when the key is turned in the lock. Shackle diameters range from 3 to 13 mm (0.1 to 0.5 in), although 7 to 10 mm are the most common. Internal clearance ranges from 13 mm to 145 mm (0.5 in to 5.75 in) in readily available commercial padlocks. Shackles are made from materials having varying strength, and therefore resistance to cutting. These materials range from chromium-plated steel to hardened steel and hardened alloy steel. Shackles made from hardened alloy steel resist cutting, prying, and sawing.

The locking mechanism holds the shackle in place and prevents opening of the lock. This occurs through movement of metal pieces that block the machined keyways in the shackle. Single and dual locking systems are available. Dual locking systems increase resistance to tampering by pulling and prying.

The body of padlocks includes laminations of chromium-plated steel, stainless steel, or brass, and solid aluminum, steel, brass, or polymeric resin. Some models of metal construction contain an external plastic coating to protect against harsh weather. The material of construction of the body reflects the service to which the lock is put. Laminations resist tampering and physical attack. Aluminum units are powder coated to provide visibility through color and to improve resistance to corrosion. Aluminum is also spark resistant. Brass provides extra corrosion resistance, and as a result, is suitable for use in harsh outdoor environments including maritime ones. Polymeric resins resist corrosion and other types of chemical attack.

Two types of locking mechanism are available: pin tumbler and warded mechanisms. Pin tumbler mechanisms are used in high security locks. Pin tumbler models usually contain four to seven pins, although four to six are most common. Five-, six-, and seven-pin tumblers offer high pick resistance and choice of thousands of key combinations (10,000 on five-pin systems and 12,000 on six-pin systems). Pin tumbler models offer keying flexibility. They can be keyed alike or differently and through coordinated keying schemes. (See the next section regarding keying systems.) Warded locks contain minimum moving parts and offer better service in environments where sand, dirt, or ice are present.

Some locks trap the key in the lock until insertion of the shackle. This prevents unnecessary removal and reduces potential for loss and external duplication.

The preceding discussion indicates that the subject of padlocks is considerably more complex than appears at first glance. Manufacturers produce many lines of keyed padlocks and variations on the theme.

The lockout paradigm imposes performance requirements on locks. These talk vaguely about withstanding the rigors of the environment to which they are exposed and preventing removal except by excessive force of special tools, such as bolt cutters or other metal-cutting tools. These directives provide no measurable quantitative requirements. Hence, these choices remain at the discretion of manufacturers and end-users (Figure 8.27).

For the purpose of the durability and tamper-resistance requirements of lockout, manufacturers suggest padlocks containing shackles of 5 to 10 mm (0.25 to 38 in) diameter (although 6 to 8 mm is most common), four- or five-pin tumblers and laminated steel construction. These units occupy the middle of the product line. Keyed padlocks are available from many sources. Provided that the locks available from these sources meet the general requirements, this provides flexibility to end-users.

Keying Systems
The lockout paradigm states that each person applies a personal lock to each energy-isolating device. This approach is functional where small numbers of individuals participate in a lockout process involving a small number of locks. Managing locks and keys becomes complicated when individuals re-

Figure 8.27. The marketplace offers many types of keyed padlocks. Most of these are substantive in construction, offer security against casual intrusion, and require aggressive tools to defeat the integrity of the product.

quire many locks and where large numbers of individuals are involved in the process (see group lockout and trapped key interlock systems).

Some of the keyed padlocks in the marketplace are rekeyable, meaning that the key cylinder is removable and the tumblers programmable. This design provides the opportunity to reconfigure locks so that an entire group is keyed alike. This means that one key opens all locks. This might suit the situation where one individual must apply multiple locks. There is no assurance the padlocks purchased off the shelf in retail stores are rekeyable or have different key combinations. Identical keying is most likely in the least expensive products.

Rekeying can occur on site, or through a locksmith or the manufacturer. The manufacturer restricts availability of key blanks and cut keys for high-security locks. Replacing key cylinders offers remedies for compromised security and employee turnover. Cylinder replacement is a simple procedure requiring only a hexagonal wrench to gain access to the interior of the lock body.

While this approach offers convenience, there is a potential loss of control in the sequence of application and removal. Lockout procedures require sequential application and removal of energy isolating devices and locks. Hence, this creates a rationale for maintaining each lock with a unique key even for use by one individual.

There are situations where keys are misplaced or where removal of locks must occur for other reasons. Normally, the only means to remove a lock for which there is no key is to destroy it through cutting or chiseling.

Keying systems offer the means to solve this problem (Figure 8.28). One key, the grand master, open all locks in a population. Subordinate to the grand master is the master key. This key can open all locks in a large group. Subordinate to the master is the sub-master. This opens all locks in a subgroup. Locks in a group or subgroup can be individually keyed or keyed alike. Several groups opened by individual master keys can subordinate to the grand master.

The grand master, master, and sub-master keys obviously provide tremendous power. Use must occur judiciously, under a rigorous system of control. Control could mean retaining these keys in a lock

## Keying Chart

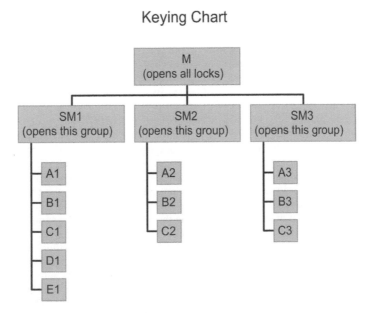

Figure 8.28. Keying systems with rekeyable locks provide the means to maintain security of lockout, and the means to open locks under tightly controlled situations. This system offers the flexibility for keying locks the same and differently within groups and between groups.

box held closed by two locks each unaffected by the grand master. In this way, the holders of these keys are the only individuals having access to the grand master, masters, and sub-masters. Both keyholders must agree to open the holding locks or else forcible means are required.

Group Lockout

The lockout paradigm holds that each person potentially exposed to a situation involving hazardous energy applies a lock, or lock and lockout device, to an energy-isolating device. In simple situations involving only one individual and one isolation device, one lock is required.

The process can rapidly become complicated and involve many hundreds or even thousands of locks when the lockout paradigm involves hundreds of energy sources and whole shifts of affected individuals. These can include tradespeople who work on the equipment, supervisors, technical people, and outside contractors and their employees. Obviously, there is an incentive to streamline the approach. Lockout directives provide no guidance about how to simplify the process, except the requirement to maintain protection for all equivalent to the application of an individual lockout device..

The minimum number of locks needed for a properly devised shutdown and energy isolation procedure is the same as the number of energy-isolating devices. One lock per energy-isolating device can provide as much protection as two, five, or ten. What matters is that the system is functional, has an analytical and logical basis, is correctly implemented, and is respected by all affected by it. The key issues are the appropriateness of the procedure for a particular application and respect for the presence of the lock.

Obviously, there is an incentive to streamline the approach. There are only limited ways to perform lockout. First, each affected individual can devise and apply a procedure each time that lockout occurs. This would adhere to the paradigm of one person applying a lock to each energy-isolating device. What is likely to result is a plethora of different procedures that reflect individual biases and levels of "common sense." There is, of course, no assurance that the procedure devised by any one individual will be effective or followed, or replicated in successive lockouts, especially when it is not recorded.

Another model involves input and consideration from many individuals. This leads to a recorded standard procedure. Independent audit following application of the locks to the lockout devices or energy-isolating devices prior to starting work provides back-up to ensure that the procedure is followed and followed correctly. This, surely, is a more appropriate model than the free-for-all where each individual acts out impulses for safety. The only limitation is that the sanctity of the lockout is respected by everyone. However, there is no more safety to be achieved for application of six locks than there is for application of a single one. An individual determined to overcome the lockout will destroy the weakest link, in this case, the hasp.

Once more than one lock is applied to a lockout device or the energy-isolating device, the weakest link in the system becomes the hasp. Cutting through a hasp is far easier than cutting through the shackle of a lock. Some hasps are constructed from materials such as nylon and aluminum which are far less resistant to destructive attack than steel. Nylon can be cut, sawn, or melted, the latter by way of a readily available lighter, propane torch, or even matches. Aluminum is easily sawn using a hacksaw.

One technique for ensuring sequencing in applying and removing locks is to color code the locks and keys in a sequence on a panel. That color code would match the color code and identifier at each energy-isolating device. That way, anyone following the procedure, from the summer student to the new hire to the 20-year veteran, can ensure correctness by matching the colors and numerical or word identifiers. Each lock and colored area on the panel corresponds to the application of a particular step in the shutdown sequence. Each lock uses a unique key. Absence of a particular lock from the panel and replacement in its position by its key indicates progress in applying the steps in the shutdown procedure. Similarly, replacement of the locks on the panel indicates progress in the start-up procedure. Locks are applied and removed only according to the sequence of the procedure. This ensures orderly application and reversal of the procedure, rather than something that suits the need of an individual in performing a particular task.

A variant on this technique is the group lockout. Group lockout operates on the premise that the keys for the locks used in the lockout are kept in a locked box. In one variant, the group leader or foreman locks the box and retains the key until completion of the work. At this point the reverse of the process then occurs. Criticism of this approach centers on the holder of the key. There is no control on use of the key by a person in authority to undo the lockout without acquiescence from those affected by it. A more secure system would result in the application of a minimum of two locks on the lock box, one held by the foreman and the other by a worker representative.

In another variant, each participant affected by the process puts a lock on the lockbox (Figure 8.29). The box remains locked until each participant has removed the individual lock. This prevents any participant from opening the box until all have acquiesced to this action. The problem intrinsic to this approach is that absence of a participant can halt the start-up process. This can occur for many reasons, legitimate and otherwise.

In these cases, the weak link in the system is the lockbox and the access that participants have to it. The walls of the box, hinges, and closure hardware are readily attacked by cutting tools such as hack saws and tin snips. The basis for success of this system is the same as all others: respect the sanctity of the process.

## Tagout

Tagout is the placement of a tagout device onto an energy-isolating device. Tagout without lockout is permitted in some jurisdictions in situations where lockout of the energy-isolating device is not possible (OSHA 1989). Tagout is intended as a bridging option between the status quo of impossibility and an eventuality where all equipment and machines handling hazardous energy will contain energy-isolating devices that are lockable. This transition is expected to occur as part of the equipment modification, repair, renovation, and replacement cycle.

Figure 8.29.  The lockbox provides a means to minimize application of keyed padlocks to lockout devices. In effect, the lockbox is the lockout device. The lockout box usually contains keys for padlocks, but can contain other devices, such as uniquely configured tools.

Tagout provides protection by identifying the energy-isolating device as a source of danger by indicating that the energy-isolating device and the equipment being controlled are not to be operated. That is, the status quo of the energy-isolating device is not to change.

Tagout on its own is not equivalent to lockout in security, and potentially, protection. A tagout device can be removed with considerably less effort than needed to remove a lock and lockout device. At best, tagout relies on the honor system, such that everyone respects the intent of the message on the tag. After all, the tag attachment can easily be removed and a new one reattached.

The attachment is the critical component in tagout. A unique attachment that is not easily obtainable can provide considerable control over unsanctioned removal. Unauthorized removal of the tagout device and reinstallation by unsanctioned means of attachment provides rapid confirmation of unsanctioned action.

**Tagout Devices**

A tagout device is a prominent warning device capable of being securely fastened to an energy-isolating device. Tagout devices include tags and their means of attachment. The tag warns against operating the equipment or machine to which it is attached until the tagout device is removed according to the energy control procedure (OSHA 1989).

Tagout devices must resist the rigors of the environment in which they are to be used. Tagout devices face temporary exposure to the operating environment, since the installation is temporary. Hot temperatures of uninsulated valve stems and other components, as might be encountered in process operations involving heated liquids and steam systems, are potentially capable of causing severe burns and melting plastic components. Cold temperatures embrittle plastic materials. Materials of construction must respect this limitation.

Tagout devices require substantiveness. Substantiveness minimizes premature or unintended failure or removal due to disintegration or breakage. Substantive means a minimum locking strength of 50 lb (220 N). Tagout devices must also be nonreusable, attachable by hand, self-locking, and nonreleasable.

Tagout devices must be standardized by color, shape, or size, as well as print and format. These requirements reduce visual confusion in the spectator by fostering recognition skills. Products standardized by appearance become engrained in recognition patterns and rapidly recognizable. Tagout devices also must identify the individual who installed them.

The tags require warnings about the hazardous conditions when the equipment or machine remains energized. Table 8.4 provides sample warning phrases.

**Interlocks**

An interlock is a device that interacts with another device or mechanism to govern succeeding operations (Krieger and Montgomery 1997). This is a concise description of interlocks and their function, but what it does not say about interlocks is much more important than what it does. Interlocks have wide application in industry, including deactivation, de-energization, isolation, and lockout.

**Interlocks and Lockout**

One of the premises of lockout is the action is taken as the final step in assuring a state of zero energy or a state of retention of energy through deliberate storage or entrapment . The intent of protection is to ensure that energy cannot flow through the circuit. Assurance of protection occurs only though the primary circuit. Protection through the primary circuit ensures that no energy enters any circuit. The role of secondary protection is to prevent signals from reaching the primary initiator.

Locking out the STOP switch on a piece of equipment provides protection only in the secondary circuit (Hall 1992). The STOP switch is part of the secondary initiator circuit. Locking out the STOP pushbutton does not prevent start-up of the equipment. The starter is manually closable using a screwdriver to move the position of the armature of the solenoid. This can happen for innocent or deliberate reasons.

**Table 8.4**

**Tagout Phrases**

DO NOT START

DO NOT STOP

DO NOT OPERATE

DO NOT ENERGIZE

DO NOT OPEN

DO NOT CLOSE

Nearly all initiators, whether in primary or secondary circuits, provide for manual actuation. That is, they provide for manual override. Manual actuation sometimes requires a tool; sometimes a button is present; and sometimes a detente is present. The position of the detente indicates that manual override can be a permanent condition. Hence, unless primary protection is in effect, there is no way to ensure that the secondary circuit is also open.

Secondary circuits containing interlocks receive energy from the primary circuit. Hence, in order for the secondary circuit to offer protection, the primary circuit must be energized. This means that upstream circuits are energized, even when secondary protection is in effect. Since, as mentioned, secondary initiators contain devices to enable manual control, they cannot provide assured protection.

OSHA in the U.S. addressed this situation partially by requiring disconnection of circuits and equipment from all electric energy sources prior to working on them where there is a chance of electrocution (OSHA 2011). Control circuit devices, such as push buttons, selector switches, and interlocks, are not to be used as the sole means for de-energizing circuits or equipment. Interlocks for electric equipment may not be used as a substitute for lockout and tagging procedures..

## Verification

Verification is the last step in the sequence, deactivation, de-energization, isolation, lockout, and verification. Verification receives the least amount of attention, being almost an afterthought. Yet, verification is the most important of all of the steps. Verification is the final action prior to commencing work on the equipment, machine, or system.

Verification is difficult to conceptualize. The normal focus and emphasis of effort involving equipment, machines, and systems is operation. Verifying that the equipment, machine, or system contains no energy or cannot express energy retained through deliberate storage or entrapment is almost a foreign concept.

The simplest starting point is visual. Verification may involve superficial inspection to confirm that nothing is moving and more detailed action to prove that nothing happens on activating control circuits that are not locked out. Pressure gauges may provide information about possible pressurization in circuits. Moving control levers is another possible action. This must occur under considerable care, as retained pressure in circuits may cause movement.

More in-depth investigation requires equipment, such as a voltmeter to test for the presence of electrical energy in circuits. This is a high-level activity requiring full precautionary measures against the possibility that the circuit is still live.

Regardless of the action chosen, the verification step must reflect knowledge of how the equipment or machine functions under normal conditions of operation and retains energy during shutdown through deliberate storage and entrapment.

## References

ASME: Scheme for the Identification of Piping Systems (ASME A13.1). New York: American Society of Mechanical Engineers, 1996.

Juvinall, R.C. and K.M. Marshek: *Fundamentals of Machine Component Design*, 2nd Ed. New York: John Wiley & Sons, 1991.

Hall, F.B.: Safety Interlocks – The Dark Side. Niles, IL:Triodyne Inc., Safety Brief Vol .7, No. 3. June 1992.

Krieger, G.R. and J.F. Montgomery (Eds.): *Accident Prevention Manual for Business and Industry, Engineering & Technology*, 11th Ed. Itasca, IL: National Safety Council, 1997.

Leveson, N.G.: *Safeware, System Safety and Computers*. Reading, MA: Addison-Wesley, 1995.

Ortiz, P., M. Oresegun and J. Wheatley: Lessons from Major Radiation Accidents. *Proc. 10th Congr. Internat. Radiat. Prot. Assoc. Hiroshima, Japan, May 2000*, 2000.

OSHA: "Control of Hazardous Energy Sources (Lockout/Tagout); Final Rule", *Fed. Regist. 54*: 169 (1 September 1989). pp. 36644-36696.

OSHA: "Permit-Required Confined Spaces for General Industry; Final Rule," *Fed. Regist. 58*: 9 (14 January 1993). pp. 4462-4563.

OSHA: Selection and Use of Work Practices, Electrical (Subpart S), 29CFR 1910.333. Washington DC: U.S. Department of Labor, Occupational Safety and Health Administration, 2011. [Website].

Schwartz, M.: *Machines, Buildings, Weaponry of Biblical Times*. Old Tappen NJ: Fleming H. Revell Company, 1990.

Shackelford, J.F.: *Introduction to Materials Science for Engineers*, 2nd Ed. New York: Macmillan Publishing Company, 1988.

# 9 Management and Administrative Issues

## INTRODUCTION

Previous discussion has focused on technical aspects of the subject of hazardous energy. Chapters have addressed topics concerned with energy as a physical property of matter, effects of hazardous energy on people, the conflict for attention between task and conditions, strategies for analyzing for the presence of energy in equipment, machines, and systems, and the means to deactivate, de-energize, isolate, and lock out circuits. At some point in the process, the decision is taken to perform the actual work. While the preceding discussion may seem prolonged and almost a slow-motion look at the situation, this, in effect, reflects the process that must occur during discussion and decision-making concerning technical problems.

This chapter will focus on the preparatory activities that are necessary to ensure that people are prepared for management of hazardous energy to provide and ensure safe conditions of work. As mentioned in previous discussion, events are either managed or they are allowed to happen. Thus, interaction can occur in a casual or haphazard way or through the structure provided by management.

## Management Issues in Isolation and Lockout

Every activity in industry, and indeed in human life, occurs through the actions of people. The vast majority of people will do what they understand to be the right thing concerning the well-being and safety of others. People are even more inclined to do the right thing when they appreciate the reason and necessity for performing the action. This means that people deserve to have the opportunity to participate in an outcome to determine what is the right thing to be done, and what is the right reason for doing it. Indeed, this is the fundamental captured in the approaches of consensus standards and regulations.

Someone must perform each step of the work, starting from analysis of the equipment, machine, or system to the hands-on component. That someone may be a single individual or it could be a group or groups of individuals. Each participant must possess the skills needed to perform the work with confidence, competence, and the ability to replicate it in a defensible manner. That is, the work must occur in an organized way that is repeatable and defensible. As a corollary, this work is easily visualized as the proverbial chain whose strength reflects that of its weakest link.

This work involves several groups of individuals whose efforts and identities may overlap. The first group performs the analysis of the equipment, machine, or system. This leads to a technical description of the movement of energy from input to output. The technical description is needed in order to determine the inner workings of the equipment, machine, or system. This leads to determination of circuits requiring deactivation, de-energization, isolation and lockout, circuits requiring immobilization to prevent expression of retained energy, and circuits not requiring attention. From there, this group must determine and confirm the means for achieving this end-point. Confirmation involves experimentation and testing to verify the effectiveness of the approach to be taken. In most circumstances, the process is quite simple and straight-forward. However, the process can become complex, sometimes unnecessarily so.

The second group implements the procedure for isolating and locking out and immobilizing circuits contained in the equipment, machine, or system. This activity requires knowledge about the process of isolation and lockout and immobilization, and the hands-on means to do the work and to verify the success or failure of the procedure. The manner in which the second group performs the work depends on recommendations from the first group. Participation in the two groups may overlap.

The third group performs the work on circuits in the equipment, machine, or system following the isolation and lockout. Participation in the three groups may overlap.

However, there is considerable variation from this model. One obvious deviation from the model involving the same people is use of outside contractors to perform this work. This is very common across the spectrum of industrial activity, ranging from new construction to upgrading, renovation, and restoration that interface with existing equipment, machines, and systems. In each of these situations workers from outside companies work on or in proximity to equipment, machines, and systems prepared through this process.

Sadly, the current situation indicates that de-energization, isolation and lockout and immobilization to prevent expression of hazardous energy retained in isolated circuits through deliberate storage or entrapment are less than perfect processes. Some workers are well aware of defensive measures to be taken during disassembly to protect their safety. This situation arises sometimes due to failure involving the analytical process and sometimes due to faulty implementation.

The fourth group affected by lockout and tagout includes people not involved in any of the other groups who may encounter the locked out or tagged out equipment, machine, or system as part of their daily activity, routine or otherwise. This group has the need to understand the meaning of the lockout and tagout system and the knowledge to know that interference with this protocol is not to occur.

The capability with which the groups perform their activities depends on knowledge from formal education, training in job-related skills, and experience gained in applying these skills.

### Establishing the Right Reasons to Do the Right Things

Knowing what to do in a particular circumstance is a critical skill that is not easily learned. This is almost forensic in scope because it is counterintuitive to normal practice with equipment, machines, and systems. Normal practice with equipment, machines, and systems focuses on operation and production. That is, using the entity in the manner in which it was designed. Disassembly and repair are counterintuitive to this perspective.

This fundamental principle is often lost when the formal organization of people into functional units occurs. The outcome of this situation can be failure to choose a particular course of action because, indeed, it is the right thing to do. This breakdown also can involve failure to understand what is the right action, because of failure to understand the rationale behind it or failure of the situation to alert to the need for an alternate approach to problem-solving.

Previous chapters have provided input for performing the analysis necessary for enquiring about and eventually making decisions about such situations. Chapter 7 mentioned the work of Kepner and Tregoe. In 1958, social scientists C.H. Kepner and B.B. Tregoe founded a consulting practice dedicated to the field of decision-making (Kepner and Tregoe 1981). Kepner and Tregoe had observed that decisions were made in an irrational, often highly questionable manner. They concluded that the cause of this problem was deficiency in the process of gathering and organizing information. They developed a rational method for gathering and using information containing four key elements. These include organized enquiry to assess and clarify information; investigating the situation to determine cause and effect; applying logic to assemble facts from information and make choices; and anticipating the future and mitigating against its undesirable effects.

Another approach directed toward technical problems is Failure Modes and Effects Analysis (FMEA). FMEA is an analytical technique used to identify, quantify, and prioritize risks (Wellborn and Boraiko 2009). In 1949, the U.S. military first used FMEA to improve the probability of successful strategic missions. Later, in the 1960s, NASA used FMEA to reduce the risks associated with manned space flight. In the 1970s and 1980s, the automotive industry used the technique to evaluate health risks to passengers in vehicles during operation, and later to minimize the occurrence and effect of quality defects.

A workplace accident results from failure of the work system to maintain safety. Failure modes and effects analysis provides a basis for managing risks associated with potential failures and root causes. As a proactive technique, FMEA can be used before the occurrence of work on the equipment, machine, or system. This is critically important because lockout and tagout and subsequent disassembly of the equipment, machine, or system are very similar in concept to the applications mentioned in this discussion. In this application, the emphasis is directed toward qualitation rather than quantitation, and elimination or at least control of hazardous conditions.

As mentioned in Chapter 7, actions used to establish background information could include photographing or recording a video of the action of the equipment, machine, or system or establishing other physical parameters. The visual information captured photographically in pictures and in video is especially helpful where operating drawings are not available. A video can establish the nature of movement of parts. These measurements can establish, for example, storage and transformation patterns in mechanical linkages and complex levers.

Equipment, systems, and structures lie within the bounds of organizations. As such, they "belong" to some entity within the organization. This means that someone has responsibility for them. Active operations offer the opportunity for discussion about how the equipment, machine, or system functions, materials of construction, materials currently and historically handled, operating records and documentation, and the presence of operational utilities. Because work involving equipment, systems, or structures in a process operation can require extensive coordination with operations and maintenance personnel, this liaison is extremely important.

Operators and maintenance personnel are invaluable resources about the hands-on nature of equipment, machines, and systems. The experience, knowledge, and perspective of these individuals are invaluable contributions to the process of hazard assessment. They can provide input regarding historical aspects about what was done, and why, and how successfully.

Downsizing through layoffs and early retirements similarly has created a loss in the information resources of every entity that has undergone the loss of these people. With any of these circumstances, loss of the expertise and knowledge resident in former employees and operating and maintenance records can considerably complicate hazard assessment. The starting-point for hazard assessment in these situations is a thorough historical review of the site, its equipment, and structures.

The following example will illustrate the skills and enquiry required. A large manhole in a live sanitary sewer system required entry and work. The contractor decided to block the flow using two inflatable plugs installed in series by a subcontractor. The wall formed by the leading face of the upstream plug would force the water to accumulate and to divert down an alternate sewer line. (This approach is a relatively common practice.) Workers from a second subcontractor would enter the manhole to powerwash the interior to remove residues resulting from the sanitary flow. Finally, workers from a third subcontractor would enter to perform work on the protective lining on the concrete surface of the manhole.

Additional background information about the strategy and conditions is appropriate to understand the concern. The geographic area in which this work is to occur receives considerable rainfall during much of the year. The drainage system usually can accommodate the peak flow. Overflow of manholes causing the cast steel covers to float on the water above the casting anchored in the ground at the top of the manhole is an infrequent but possible and known occurrence (Figure 9.1). As well, failure of the anchorage of the inflatable plugs has occurred with the result that manholes downstream from them have filled in seconds.

This situation should provoke questions in the minds of individuals charged with ensuring safety of the entrants (Table 9.1).

Companies that provide sewer isolation services approach this situation in two ways. One approach involves people who are well-experienced hands-on practitioners. They know how to choose and connect the hardware based on what works and what does not. Their approach is based on long-term operating experience. If something doesn't work correctly, they install additional or different equipment. When asked questions requesting more quantitative information toward substantiating their choices through documentation, they are perplexed and have difficulty providing answers.

Another company offering services in this arena approached this question from an engineering perspective. This perspective considered volumetric flows and fluid pressures and forces exerted on anchoring chains and strength of the anchorage to be used. This approach provides assumptions and employs some of the equations presented in Chapter 2. This company used the results from calculations to select equipment based on these and other considerations.

## Doing the Right Things for the Right Reasons

The right things often involve choices between alternate possibilities. Sometimes the choice is clearcut, and sometimes not. The following example illustrates the former situation.

Figure 9.1. Sewer overflow. High flow of water can occasionally overcome capacity of the sewer system. This situation can create life-threatening conditions for an entrant.

People occasionally have a reason to access the interior of a washing machine during operation. Typical reasons include adding extra items to the wash, adding cleaning products, untangling items wrapped around the agitator, and repositioning items to establish balance during the spin cycle. These tasks and those performed in operating equipment and machinery in workplace situations probably share common elements.

In order to gain access to the interior of a top-load washing machine, the operator can open the lid or deactivate the machine through the timer. We learn through experience often gained through impulsive action during a perceived crisis that opening the lid will stop the action of the agitator and the drum during a spin cycle involving a normally functioning machine. We also learn through observation that waiting for motion to stop is necessary in order to gain access to the clothing without being hurt. What we do not know from this situation is that there is an alternate means to shut down the machine, namely pulling out or pushing in the control unit.

Opening the lid on a clothes washer to stop the action of the machine is an almost instinctive action in a sequence of actions taken to solve the problem. The conditioning that we receive in working with this equipment is that we open the lid to insert and to remove the clothing. The actions to be undertaken to solve the problem are precisely the same. Hence, the logic of the action is entirely defensible. Breaking the conditioning intrinsic to the action is not easy.

On the other hand, the conditioning that we receive is that we operate the timer only to start the machine and to initiate a second rinse and spin cycle. The lid is always closed during actions involving the timer because we know by conditioning that the machine will not operate when the lid is open. We also know that there is no need to open the lid during actions involving the control unit. Another important consideration is that the opened lid blocks access to the control unit. The machine does not start when the lid is open even when activation of the timer occurs.

Opening the lid during operation fails to accord the level of respect needed in the circumstance. Opening the lid during operation exposes the user to needless risk of harm (the right reason for investi-

## Table 9.1

## Investigative Questions For Sewer Isolation

What is the diameter of the sewer?

What is the normal flow pattern during the day?

What is the normal flow pattern during the year?

Does the sewer receive combined flow?

What pressure would the flow exert against the plug?

What are the characteristics of the plug?

How will anchorage of the plug occur?

What reserve exists in the anchorage to ensure that failure will not occur?

What surveillance and security will occur on site before, during, and after the work?

Who is responsible for assessing this situation and what credentials does the person hold to ensure that this will occur in a competent manner?

gating the situation). The safer way to operate the machine to gain access to the contents during the spin or agitation cycles is to deactivate the unit through the timer/controller, rather than opening the lid (the right thing to do in the situation).

The effort and inconvenience involved in the two actions are essentially the same. Yet, one action exposes the operator to a serious risk of harm, while the other, for the same effort and inconvenience, provides a state of safety. The interlock in the lid is intended as a secondary protective device, rather than the primary one.

This example highlights the situation where the effort involved in both courses of action is approximately the same. What was absent was the knowledge that two options were available and that one provided a considerable increase in safety compared to the other.

An associated problem with this outcome is that the timer/controller is sometimes inconvenient to use. On older machines, this device is mechanical and operates through the seeming gnashing of gear teeth as rotation occurs. At the appropriate position, a pull-out or push-in must occur in order to activate or deactivate the circuit. What is more convenient: fighting with the timer/starter or lifting the door and waiting for the motion to stop?

This discussion highlights that the right thing to do for the right reason must occur in a manner that does not impose major inconvenience on the person performing the action. People perform actions that are easy to do and provide clear benefit. Voluntary and consistent performance of actions that are difficult or inconvenient, given the availability of simpler, although less appropriate or even inappropriate alternatives is much less predictable. When the right thing to do is the easier (easiest) thing to do, it will get done.

The preceding discussion highlights one possible reason for the failures (meaning accidents) that occur in situations where exposure to hazardous energy occurs. People fail to follow established procedures. They take "shortcuts" or devise unsanctioned alternate actions. Shortcuts arise from the perception of inconvenience and lack of understanding about why progression of an activity must occur in a particular way. They also reflect the desire for efficiency by the individuals performing the activity.

Developing the skill to acquire the awareness is one of the challenges involved in raising the level of capability of individuals who will participate in creation of procedures needed for deactivation, de-energization, isolation, lockout, and verification. An associated challenge is the means to instruct individuals affected by the procedure to follow this particular path rather than an alternate perceived to be more efficient that produces the same outcome.

## Programs for Managing Exposure to Hazardous Energy

Many jurisdictions require organizations to create a program to manage (meaning eliminate or at least control) exposure to hazardous energy. A program provides a framework for organizing measures and imposes structure to ensure that the right things occur for the right reasons. Table 9.2 provides elements present in a program for lockout/tagout. Table 9.2 provides the basis for ensuring identifying and training responsible persons and taking appropriate actions.

A program is considerably more than a collection of words created around a regulatory requirement. A program describes through policies and practices what people in an organization will do to respond to regulatory requirements. Since this document describes conduct and actions of individuals by job title, those individuals should have a role in its creation. All too often, someone either internal or external to an operation will describe in great detail actions to be taken, and then will have no role in their execution. Those affected by programs of this type usually have no role in their creation and are expected obediently to follow their dictates.

The finished documents may fill several binders and occupy space on a shelf in the office of a production or maintenance supervisor office, and then receive little use. This approach violates the intent and spirit of the regulatory requirement and produces no benefit for those affected by it. In order for a program to have the optimum chance for success, those affected by it must have the opportunity to participate actively in all aspects of its creation and implementation.

All too often, "programs" are merely restatements of regulatory requirements with the further statement that the affected party will do what is outlined in the restatement. A restatement of a regulation merely shows that the author can read and transcribe words. Anyone who is literate can do that.

This approach is ineffective because it fails to indicate how the organization will respond to the regulatory requirements. A program should act as an overview to indicate who will do what to whom within the context of an organization to ensure the effectiveness of the outcome of requirements required by regulation.

An audit of "programs" of this kind has a good chance of indicating that the organization is not complying with the spirit and intent of the regulation. To elaborate, a program is the framework on which rest the actions necessary to comply with the spirit and expectations of a regulation. Actions taken beyond the statements in a binder are the indicator of the existence and ultimate success of the program. These actions demonstrate how the organization responds to regulatory requirements possibly on a daily basis.

### Program Contents

Table 9.2 provided the basic organizational framework for a program on hazardous energy. As with other programs created within organizations, Table 9.2 indicates that involvement by a key person, a champion in senior management, is essential for success. Ideally, the program manager is experienced and knowledgeable about hazardous energy and its presence and occurrence in equipment and machines and systems and the means to eliminate or at least control exposure. A program of this type is on-going, and requires continuing maintenance and input, meaning the commitment of resources and people. Without these sustaining features, it will wither.

The term senior management implies a large organization. However, large organizations comprise a tiny proportion of industry in every jurisdiction. Hence, as with other regulatory statutes, the challenge is to assist medium and small business operations where the owner is the senior, and perhaps only, manager.

## Table 9.2
## Organizational Framework of a Program For Lockout/Tagout

| Element | Action |
| --- | --- |
| Management commitment | Employer |
| Roles and responsibilities | Employer → program manager |
| Inventory of equipment, machines and systems affected by lockout/tagout | Program manager |
| Hazard assessments and procedures | Program manager → qualified person(s) |
| Acquire equipment | Program manager |
| Training | Program manager |
| Recordkeeping | Program manager |
| Audit | Program manager |

The program itself requires additional detail. The classic model for hazard management (anticipate, recognize, evaluate, and control) provides the basis for creating the program for managing hazardous energy (Table 9.3).

Anticipation and recognition reflect background knowledge and experience. Without these, identification of sources of hazardous energy will be very difficult. Someone within or retained by the organization must possess these attributes. Anticipation and recognition imply a step involving the identification, description, and cataloging of all sources of hazardous energy within the confines of an organization.

Anticipation reflects the capability to know that a source of hazardous energy is or could be present or could develop. This capability is the product of experience, education, and training. The person possessing this skill must know about the operation of equipment and machines and systems.

Recognition implies the ability to recognize the type and potential magnitude of the energy source that is present or could develop. This ability also is the product of experience, education, and training. These capabilities are especially important where multiple sources of energy are present or where energy can be stored in equipment, bulk materials, or in structures. Anticipation and recognition of some, but not all, sources of energy is not satisfactory where a life depends on the appropriateness of the decision and follow-up actions.

Evaluation encompasses the skill of anticipating and determining, through measurement and experience, the potential magnitude of hazardous energy in a particular situation. Evaluation applies to potential and actual sources of hazardous energy identified within the confines of the organization to which exposure can occur. Identification and evaluation must occur in a repeatable and comprehensive manner in order to be defensible. Determining magnitude of hazardous energy where measurement is possible is necessary for specifying requirements for protection where the source must remain operational during work activity.

Control, or in this case managing hazardous conditions through elimination and control, describes all of the measures taken to ensure worker safety while working near sources of hazardous energy. These include procedures for deactivating, de-energizing, isolating, and locking out energy sources, as applicable, engineering controls, and personal protective equipment. Managing hazardous conditions through elimination and control, therefore, also requires education about concepts involved

## Table 9.3
## Program Elements

| Guide Word | Element | Subelement |
|---|---|---|
| Anticipate | Program manager recruited | Directive indicating resources to be made available to enable development and completion of program |
| | Experience and knowledge about hazardous energy and its presence and occurrence in equipment, machines, and systems | Training obtained for the program manager, as needed and appropriate |
| | Create statement of management commitment | |
| | Perform inventory of equipment, machines, and systems | |
| Recognize | Identify the occurrence of hazardous energy in equipment, machines, and systems requiring further attention | Identify qualified person(s) |
| | | Obtain training for qualified person(s) |
| | Identify equipment needed for in-depth assessment | Obtain equipment to be used by qualified persons |
| | | Obtain training for users of equipment |
| Evaluate | In-depth examination of equipment, machines, and systems that require lockout/tagout | Identify circuits containing hazardous energy and estimate and determine magnitude of sources |
| | | Determine flow of energy in circuits |
| | | Identify means to dissipate energy from circuits to attain zero energy |
| | | Identify circuits retaining energy due to deliberate storage or entrapment and means to prevent expression |
| | | Create procedure to lock out or tag out circuits in affected equipment, machines and systems and verify effectiveness |
| Control | Specify engineering controls to protect against exposure to sources of hazardous energy retained in the equipment, machine, or system | Fabricate or obtain, install, and verify effectiveness of these devices |

## Table 9.3 (Continued)
## Program Elements

| Guide Word | Element | Subelement |
| --- | --- | --- |
| | Retrofit isolation devices to accommodate lockout devices | Fabricate or obtain, install, and verify effectiveness of these devices |
| | Specify lockout devices for installation on isolation devices | Obtain these devices |
| | Specify personal protective equipment to protect against exposure to sources of hazardous energy where deactivation, de-energization, isolation and lockout are not possible | Obtain protective equipment required |
| | Train supervisors and affected workers | Create or obtain training material |
| | | Recruit instructor |
| | Recordkeeping | Maintain records of all activities pertaining to the program |
| | Audit effectiveness of the program | Audit actions of individuals |
| | | Review records for evidence of activity |
| | | Identify and correct deficiencies in the program |

in procedures, acquisition of equipment, recordkeeping, auditing effectiveness of these actions, and identifying and rectifying deficiencies and areas of weakness.

### Confounding Factors: Confined Spaces
Many of the workspaces in which exposure to sources of hazardous energy or materials containing hazardous energy can occur are also confined spaces (McManus 1999).

Confined spaces do not distinguish themselves by size, shape, or nature of work activity. Some are large; some are small; some are open; others are completely closed. Still other confined spaces fit none of these descriptions.

One feature that distinguishes many confined spaces from "normal" workspaces is the use of the space. Confined spaces generally are not the type of place in which people normally work. Some workspaces are confined because of design and size, or very limited access and egress. They are designed to serve some function, such as containment of a chemical process or storage of bulk materials or housing of machinery. Under normal conditions of operation, entry by people would never occur. However, periodically these structures and the equipment inside them require maintenance, improvement, and replacement. Entry must occur in order to do these things.

Another feature that distinguishes confined spaces from "normal" workspaces is the nature and severity of accidents that occur in them. Minor mistakes lead to consequences that far outweigh the significance of the error. Accidents that occur in confined spaces are more severe than those that occur in "normal" workspaces. Like a trap that is set and ready to spring, the hazardous condition that causes the accident acts rapidly, often without prior warning. After the accident has happened, conditions often return to "normal", as if nothing has happened.

The third feature that distinguishes confined spaces from "normal" workspaces is the boundary surface. The boundary surface magnifies or amplifies the severity of the hazardous condition. The boundary surface need not be substantive or even contiguous. A piece of paper or a thin layer of plastic can trap a hazardous atmosphere. The invisible envelope that defines the boundaries of the operating space used by a complex papermaking machine or robotic machinery determines the safe limits of approach to the space that surrounds this equipment. A chain-link fence surrounding an area containing a source of hazardous energy, such as rotating or reciprocating equipment, prevents entry. It also prevents escape.

Regulatory statutes and consensus standards contain many definitions to describe confined spaces. There is no universally accepted one. The definition in 29CFR 1910.146 (OSHA 1993) defines a confined space using the following terms: limited means of access or egress, (and) large enough for an employee to enter and perform assigned work, and not designed for continuous occupancy. Furthermore, a permit-required confined space meets the definition of a confined space and contains or has the potential to contain a hazardous atmosphere, (and/or) contains a material that has the potential for engulfing an entrant, (and/or) has an internal configuration that might cause an entrant to be trapped or asphyxiated by inwardly converging walls or by a floor that slopes downward and tapers to a smaller cross-section, and/or contains any other recognized serious safety or health hazard.

Permit-required confined spaces (under the context of 29CFR 1910.146) generally contain or have the potential to contain hazardous conditions that can cause acute injury and can impede egress in the event of an emergency situation. Entry and work involving permit-required spaces requires development and implementation of a written program and a permit-to-enter system.

Determination about the status of workspaces (confined space versus "normal," and permit-required versus nonpermit) and development of a program for work in permit-requried confined spaces should occur in parallel with the program to control exposure to hazardous energy. Many of the same aspects (such as knowledgeable personnel) are applicable to both regulatory requirements.

Significant differences exist between the requirements for isolation in 29CFR 1910.146 (Permit-Required Confined Spaces) compared to 29CFR 1910.147 (Control of Hazardous Energy Source, Lockout/Tagout). Readers affected by these statutes should check these differences carefully for possible application in their circumstances. Appendix A discusses consensus and regulatory standards applicable to lockout and tagout.

## Human Resources

The resource that is the limiting factor in facilitating this process is the human resource. The regulatory philosophy of the jurisdiction in which the organization is located can exacerbate this situation. Regulators approach the use of human resources in the workplace for achieving and maintaining health and safety in different ways.

One regulatory approach is to designate the employer as being solely responsible for all aspects of health and safety in the organization. This includes the hazardous energy (lockout/tagout) program as a particular example. This approach puts considerable stress on technical resources available within the management and administration of an organization. This ignores the reality that operators and maintenance workers know far more about equipment, machines, and systems and their idiosyncrasies than supervisors and higher level management and technical advisors. Yet, this regulatory approach to hazard management in workplaces deliberately excludes workers from the right and obligation to participate in the process of program development and implementation.

A way to approach this issue is to create a team composed of practitioners representing relevant disciplines within the organization: safety and industrial hygiene, and engineering, operations and maintenance. The team could then undertake the task, utilizing the strengths of each of the participants. This option offers the advantages of involvement of technical personnel in areas of endeavor beyond normal interests and responsibilities. Involving these groups provides the potential for major benefits, since they are involved in the design, operation, and maintenance of equipment and structures that ultimately become workspaces. To some members in this group, entry into these spaces for maintenance, repair, and modification may never previously have been a consideration. Involvement in the process can produce positive benefits in future endeavors through awareness-raising.

Safety and industrial hygiene professionals possess generalized knowledge about the anticipation, recognition, evaluation, and control of hazardous conditions. This group has specialized knowledge about requirements and implications of regulations and standards.

Engineering, operations, and maintenance personnel from the department that "owns" the equipment, machine, or system know what it does, how it functions, and problems that have occurred during its operation and maintenance. They have the knowledge about electrical, hydraulic and pneumatic, mechanical, utility and process and control systems. Knowledge about control and feedback circuits is especially important where integrated systems and processes are involved. An action that occurs in one circuit in a piece of equipment, machine, or process unit in an integrated chemical plant can sometimes cause unpredicted and unanticipated effects in another. This knowledge provides the basis for discussion about energy flow and possible sources of stored energy.

Another approach to workplace health and safety utilized by regulators is the internal responsibility system. The internal responsibility system holds employers and workers jointly responsible for achieving and maintaining health and safety. The internal responsibility system operates through the Health and Safety Committee. The Health and Safety Committee acts as a forum for raising and addressing concerns regarding the workplace. However, the Committee has a larger mandate and role, if given the opportunity. The employer is required to develop programs in consultation with the Health and Safety Committee. In its most superficial form, this process is merely a formality involving adoption of programs submitted by management for approval by the workforce through the committee. However, used in the manner envisioned by regulators, the Committee is a valuable resource for gaining the broad spectrum involvement needed to achieve compliance with the spirit and intent of regulatory requirements.

The downsizing of supervisory, management and senior technical ranks, and "re-engineering" of organizations has forced more effective utilization of the talent resident within and available to these Committees. This utilization can include extensive partnership in the development of programs, such as this. The fullest possible participation of all members can provide other benefits including the satisfaction that involvement and co-operation brings. Where workplace Health and Safety Committees exist, the expertise, talent, and viewpoint represented in the members is a resource that deserves to be utilized.

## Administrative Issues Involving Lockout and Tagout

At its most fundamental level, a program for managing hazardous energy interfaces with the performance of work. For this program to function in an optimum way, this interface should be as seamless as possible. This obviously will not be the case unless the interface between protocols demanded by the program and ultimately by regulatory requirement, and the way in which work actually occurs, receives considerable and appropriate attention.

This interface is not seamless because it forces considerable restriction on the free performance of activity as would formerly have occurred. Another way of expressing this is that the program becomes the guide to the way of performing work in an organization, and does not exist simply because demanded as a regulatory requirement. A number of key points of consideration follow.

### Shift Change

The normal approach for lockout involving overlapping shifts requires incoming workers to apply their locks to positions vacated by departing workers. The intent of this protocol is to prevent breakage of the continuity in the lockout cycle. This approach is workable only in small-scale situations.

In large projects involving many workers, this practice can cause serious operational problems as individuals crowd around the lockout device(s) containing the multiplicity of locks. This would result in considerable overtime cost, as well as confusion and frustration in change rooms because of the

presence of the extra people. This approach is wasteful of resources and introduces unnecessary inefficiency.

Departing workers usually have time allotted near the end of the workshift for clean-up prior to
leaving the work area. The choices for those involved in a lockout are to remove locks prematurely at
the end of the workshift prior to completion of the work or to leave them in place until the next return to
work.

Incoming workers often do not appear at the work area at the exact moment of the start of the shift,
for reasons such as changing into work area clothing, talking to supervisors, and so on. This gap in
time, which can reach as much as half an hour presents considerable problems to maintaining continuity of the lockout by involved workers.

This issue tends to argue for use of lockboxes (Figure 9.2) and perhaps multiple lockboxes to span
the different shifts. Solutions to this situation are not clear-cut.

One possibility is to use separate lockout devices for each shift. This would ensure continuity of
the lockout, since the locks of the leaving shift would remain in place until their next return to work.
This would mean that start-up of the equipment could not occur until their return even though the work
might be complete. This situation could create an intolerable delay in start-up of equipment especially
in long-shift operations (12 hour) where the group of workers might not return for several days.

A possible solution that builds on this approach functions through the supervisor, a worker representative, and a group lockout device. The supervisor and worker representative apply their locks first
to the device at the beginning of the shift followed by the affected workers. Affected workers would remove their locks at the end of the shift. The outgoing supervisor and worker representative would
remove their locks in the presence of the incoming supervisor and worker representative at the time of
the shift change. The incoming supervisor and worker representative would apply their locks followed
by application of locks by workers on that shift. In this scenario, the supervisor and worker representative retain ultimate control and responsibility for the lockout during the workshift and shift change.

Figure 9.2. Lockboxes populated by locks. One approach to the problem posed by inflow and outflow of
large numbers of workers is use of a lockbox dedicated to each group. Color-coding or other means can
provide unique identification for each lockbox.

## Forced Removal

Forced removal of a lock, lockout device, or tagout device to gain access to the energy-isolating device is an attack on the sanctity of the lockout paradigm (Figure 9.3). This situation, therefore, is very serious and requires an approach of considerable rigor. Removal of the lock, or lockout device, or tagout device and re-energization of the circuit can subject workers to hazardous conditions leading to injury and death.

The lockout paradigm indicates that defeat of a lock or lockout device requires considerable effort involving strengthened tools that provide enhanced mechanical strength (Figure 9.4). An alternate approach to destruction of the lock or lockout device is available through keying systems that use master and sub-master keying protocols. These protocols enable a key to open the lock under consideration. This approach, at least preserves the lock and lockout device for future use. This approach is available only where the Owner/Host Employer has control over all locks applied to strategic lockout points. Strategic lockout points refer to critical lockouts in the hierarchy of the lockout system.

Situations for forced removal should occur only under highly unusual circumstances. The presence of the lock indicates that the user is not present to be able to remove it. Destruction of the lock, lockout device, or tagout device to gain access to the energy-isolating device can occur for legitimate reasons. The most likely is that the applier of the lock has left the premises for a prolonged period or permanently, and has forgotten to remove it or chosen not to do so. Reasons for leaving the premises can include shift change, illness, injury, vacation, layoff, and termination of employment with the organization.

In the short term, this situation could mean that the individual was still working on equipment affected by the isolation and lockout, and had not sanctioned its termination. The individual could also have suffered an injury and was unable to be present at the time of the enquiry. The presence of the lock could indicate the occurrence of an unrecognized accident.

Figure 9.3. Defeat of a lock or lockout device represents an attack on the lockout paradigm. This situation requires an approach of considerable rigor.

Figure 9.4. Defeat of a lock or lockout device is intended to be a physically demanding process requiring strengthened tools that provide enhanced mechanical advantage.

Deliberate removal of the lock, lockout device, or tagout device also occurs for other reasons. The latter include accidental and willful negligence. Accidental negligence can include acts committed while under the influence of prescription and illicit drugs and legal intoxicants such as alcoholic beverages. Willful negligence can reflect noncriminal and criminal intent. Noncriminal intent can occur because of anger with the situation in the workplace and possibly a desire to cause inconvenience to an employer. Criminal intent is beyond normal comprehension for workplace conduct; yet, such actions can and do occur. Events described in this discussion pose the utmost seriousness and merit severe discipline or outright dismissal of the individual responsible.

Removal of the lock and lockout device or tagout device in the absence of the user must occur under carefully supervised and documented conditions (Table 9.4). A serious attempt to establish contact with the absent worker must occur. This means that contact telephone numbers for individual workers must be readily available, so that an attempt to make contact is technically possible. Individual workers, therefore, must provide answered telephone numbers as part of their contact information. This must be a condition for work on the site. This contact can establish the motive for failure to remove the lock, whether forgetful or willful. Contact with the user could establish the existence of information not available to the proponents of removal. This information could be very valuable.

Documents accompanying removal of a lock should include information about the whereabouts, well-being and approval of the owner/user of the lock. That is, the owner/user of the lock should know about and approve the intention to remove it from the lockout or isolating device. In this manner, the user can express objections to the action.

In the event of taking the decision to remove the lock, lockout device, or tagout device, assessment of the impact of re-energization on other work activity must occur. Re-energization of equipment, a machine, or system in one area of a complex could readily impact the energy status of equipment, machines, or systems in another. Similarly, re-energization of a subsystem in equipment or a machine could readily impact other subsystems in the same equipment or another machine. At the

**Table 9.4**

**Lockout Termination Protocol**

| ABC Company | | Lockout Termination | | Equipment, Machine or System | |
|---|---|---|---|---|---|
| Circuit: | | Lockout Point: | | Completed by:<br>Date:<br>Time: | |
| Reason forTermination: | | Urgency: | | Consequence of Delay: | |
| **Investigation** | | | | | |
| Employer: | | User: | | Supervisor: | |
| Telephone Number: | | Telephone Number: | | Telephone Number: | |
| Attempt | Outcome | Attempt | Outcome | Attempt | Outcome |
| First | | First | | First | |
| Second | | Second | | Second | |
| Third | | Third | | Third | |
| Status of Work: | | Status of Work: | | Status of Work: | |
| Consequence of Removal: | | Consequence of Removal: | | Consequence of Removal: | |
| **Resolution** | | | | | |
| Decision: | | Reason: | | Consequence: | |
| Authorization: | | Protective Action: | | Date: | Time: |

time of executing the removal, clearance of the work area affected by the isolation must occur and the whereabouts of all workers determined.

**Lockout Errors and Deficiencies**

Lockout errors and deficiencies are downgrading incidents that can lead to accidents. These incidents are ascribable to errors of omission and commission. These incidents also reflect deficiencies in the system created within the lockout program. On a practical level, lockout is one element in the comings and goings of people during the workday. The lockout paradigm of one person per lock can create tremendous operational and administrative difficulties. This situation leads to creative solutions (shortcuts) that attempt to address shortcomings and impracticalities of the status quo.

## Component Identification

Each component in a deactivation, de-energization, isolation, lockout, and verification protocol requires unambiguous identification. This is essential so that the person least familiar with the equipment can perform the steps in the protocol. Why is this important? Because at some point in time, that person might be required to do so. Who might that person be? That person could be an apprentice, a summer student, a new hire, a long service employee out on an emergency call at 3:30 a.m. during a blustery night, or a contractor.

Each of these individuals could be unfamiliar with the equipment involved in a particular protocol. Considerable turnover in industrial premises is occurring as a result of the tremendous economic adjustments that are happening. It is not unusual to be told that an operator is unfamiliar with the equipment and that the only knowledgeable individual in the group of operators is not present at a particular moment in time.

The only way to optimize the potential for safe conduct of work is to label each component in the sequence in an unambiguous manner.

## Lack of Supervision and Enforcement

In most circumstances, lockout imposes a burden on the individuals required to follow these protocols. Making a lockout program function beyond the paper stage requires co-operation and compliance by people on the shop floor. These include workers and their supervisors and managers.

In many of today's workplaces, the traditional model in which workers and supervision are present in the same space and time no longer applies. The supervisor is not present during the work and is therefore not available to enforce compliance internally with requirements in the lockout program. This situation severely strains the ability of an organization to ensure compliance at the workplace level in the manner specified in the regulatory framework.

Lockout often involves inconvenience because controls are not readily accessible. To illustrate, the electrical room containing the disconnects for equipment usually is located in protected areas remote from contaminants generated in a process. Primary shutdown of a piece of equipment or machinery may occur in a control room. The equipment or machine on the shop floor may have an energy-isolating device. Performing each of the steps associated with deactivation, de-energization, isolation, lockout, and verification requires time, effort, and travel.

In a situation where frequent repetition of these steps occurs during the day, the temptation to seek out a shortcut, an easier path, looms large. This is especially the case where production pressures intrude into these considerations and where authority figures are not routinely present. Individuals required to follow these practices could rebel because of the inconvenience and burden that they impose. Supervisors and managers could rebel because of the time cost imposed during critical operations.

In order for the program to function at the hands-on level, these circumstances must be anticipated and accommodated during design or retrofitted during modifications to minimize this type of inconvenience.

One approach to this problem is to perform a time-motion study to determine the most practical means for implementing the required actions. The time commitment required for the actions to occur then would be included as part of the cost of doing the work. This could establish an economic justification for implementing the technical means to eliminate the bottleneck created in a particular circumstance. Supervision and management must adopt the view that the procedures are to be followed regardless of time commitment. This then leads to the necessity to impose on workers the duty to follow the procedures. Failure to do so then would become a disciplinary matter.

An added outcome from this approach is an appreciation of the economic cost of the status quo. Appreciation of economic cost is one means to promote change to gain greater efficiency in operation of the equipment or machine.

Failure to Lock Out (Insignificance of Procedure)
Failure to lock out due to perception of the insignificance of a procedure on personal safety likely will continue to be a major reason for ongoing lockout accidents. This occurs because of the failure to make the link that seemingly insignificant mistakes during interactions between people and equipment or machines during start-up or continuous operation can cause serious traumatic injury.

These situations arise because of assumptions about predictability. These could involve machine operation (start-up, shutdown, cycle time, speed of rotation, speed of linear motion, speed of reciprocal motion), state of energization of a conductor or other conductive surface, pressure in a fluid circuit, and so on. The lesson to be learned from real-world accidents is that workers potentially in contact with sources of hazardous energy should assume nothing about operation of equipment and machines. This is becoming increasingly the case as computerization continues to advance.

## Loss of Control
Some jurisdictions allow the worker affected by an isolation and lockout not to apply a keyed padlock and tag in certain circumstances. These include situations where the individual retains exclusive control over the energy-isolating device (Figure 9.5). Exclusive control means that the individual who isolates the component will perform or directly supervise the work of others on the component and will remain constantly on the job site and in exclusive control of the component. Loss of exclusive control means that the individual who has control of the isolated component leaves the worksite.

This approach can work well in a home during removal and replacement of the furnace and water tank. Electrical panels in older homes are not lockable. Complying with the lockout paradigm would require installation of a lockout device to the circuit breaker or position of the vacant fuse in the panel. This requires the installer to maintain a stock of lockout devices for the many possible configurations of breaker.

Where the installer has exclusive control over the status quo, this approach can work well. Where this approach falters is the situation where the worker fails to consider the implications from a particular decision. A worker not locking out an isolated circuit need only to move the energy-isolating device to the active position. At that point, reactivation of the circuit has occurred. The action of removing a keyed padlock or even a tagout device forces a pause in the activity. This pause is highly significant because the symbolism involved in removal of either device forces the individual to consider whether the action is appropriate in the circumstances, or appropriate given the overall progress of a situation.

There are some additional situations where this approach is seriously flawed and adherence to the lockout paradigm must occur. To illustrate, many building operators hire contractors to service the infrastructure. This includes lights; heating, ventilating and air-conditioning (HVAC) equipment; plumbing; and drainage and sewerage equipment, to name the more prominent of the possibilities. Service to any of these systems can occur at any time. In order to do this, the Contractor must determine the location of isolation and lockout devices, and possibly prepare the equipment for this activity.

Sometimes more than one system will undergo service at the same time. This could occur without the knowledge and acquiescence of individual service providers. This is especially problematic for assuring the integrity of the lockout paradigm. This problem arises when breaker panels are remote from the equipment and individual circuits are not labeled clearly (Figure 9.5 and Figure 9.6).

## Key Left in Lock
This situation occurs when the applier of the lock onto the lockout device leaves the key in the lock. This can occur due to failure to remember to remove the key. More likely, this situation results when the applier of the lock intends for the key to be readily available for continued personal use or for use by another individual.

This situation could arise where equipment is undergoing frequent adjustment and many cycles of locking and unlocking are required. Another possibility is that lock is some distance from the work area and the applier signals to an assistant to apply and remove the lock prior to performing some procedure. This situation effectively defeats the purpose for the lockout paradigm of one key and one lock for maintaining personal control of the lockout since the applier no longer has direct control over the fate of the lockout.

Figure 9.5. Exclusive control. The individual working in this chamber containing pressure-reducing valves retains exclusive control of the isolation.

Figure 9.6. Absence of exclusive control. The worker servicing one piece of equipment or a machine in a complex environment loses exclusive control when other workers can interact with energy-isolating devices that are not locked out.

Surrogate Lockout (Lockout Performed by Others)
Asking another worker to perform a lockout is tantamount to giving away the psychological control conferred by the lockout paradigm. The individual applying the lock asserts control. Failing to apply the lock gives away that control to someone else. The person asking the second party to apply the lock has no assurance that this person will follow, as written, the procedure for deactivating, de-energizing, isolating, locking out, and verifying effectiveness. Any number of distractions can interfere with application of this procedure by even the most diligent of individuals. Fatal accidents researched for this book provide countless examples of the loss of control that occurs when the expected applier of a lock passes this authority to another party.

Locking through the Shackle of Another Lock
Situations where numerous locks are applied to a lockout device can lead to very crowded conditions. One possible approach to this problem is to position an incoming lock through the shackle of an existing lock. This situation puts the applier of the additional lock under the control of the applier of the existing lock. When the applier of the existing lock removes this lock, the second lock is also removed as there is no effective way to attach this lock to the lockout device.

The solution to this situation is to apply a second lockout device to the isolation device or to fit the second lockout device through a vacant position in the first lockout device. In this way the lock and the second lockout device are both secure and independent of other locks.

Failure to Use Tags or Other Identification
Tags provide identification and information and visual cues about the existence of the lockout or tagout. The visual sense is much more likely to be aroused by the presence of the tag and lock than by the lock alone. This is due to physical size and appearance of the tag. The lock has the appearance of belonging in the context of other equipment in the surroundings.

Failure to Verify Effectiveness of the Isolation
The confirmatory step is the last moment in time to determine the status of a circuit. The confirmatory step often seems trivial, yet testing prior to starting work has disclosed serious deficiencies in the deactivation, de-energization, and/or isolation steps. These arise because of unanticipated inputs and feedback loops in control circuits and sometimes errors in software algorithms.

## Interaction with Contractors

Contractors perform a large amount of the maintenance on existing equipment, machines, and systems, as well as modification and upgrading, and renovation and restoration. Each of these tasks involves isolation and lockout of existing equipment, machines, and systems. In order to honor the lockout paradigm, each worker potentially exposed to sources of hazardous energy is to perform the lockout.

This situation introduces a number of complications. Each worker employed by the contractor is expected to apply a lock to each lockout point potentially affecting the work of the contractor. In order to do this appropriately, the Owner/Host Employer must inform the Contractor about specific points in the lockout at which to apply padlocks. Even with the involvement of one worker, this situation could require many locks. With the involvement of many workers and several contractors, this situation rapidly becomes unmanageable.

One possible approach is to use separate lockout devices for each company. This would ensure continuity of the lockout since the locks of other workers and lockout devices of other companies would remain in place. Each company would use a readily distinguishable lockout device. This approach can work in small-scale situations.

A possible solution for larger-scale situations functions through the supervisor, a worker representative, Company lock(s), and a group lockout device. The Company locks apply to lockout devices involved with locks of the Owner/Host Employer.

The supervisor and worker representative apply their locks first to the device at the beginning of the shift followed by the affected workers. Affected workers would remove their locks at the end of the shift. The outgoing supervisor and worker representative would remove their locks in the presence of the incoming supervisor and worker representative at the time of the shift change. The incoming supervisor and worker representative would apply their locks followed by application of locks by workers on that shift.

In this scenario, the supervisor and worker representative retain ultimate control and responsibility for the lockout during the workshift and shift change. If failure of removal occurs, this approach localizes the problem to the Contractor.

## Education and Training

Adult education and training in the workplace are complex undertakings (McManus and Green 1999, McManus and Green 2001). The education and training of workers and front-line supervisors are just as important as the education and training of the technical people employed in an organization. Education and training of workers and front-line supervisors are all the more difficult because they are imposed by regulators as post employment requirements and are not part of a package offered by an individual as part of a hiring decision.

The role of educator and trainer is forced onto reluctant employers by regulatory requirements. Most employers, due to small size, are ill-suited to provide worker education and training. The main focus of employers, especially small businesses, is making widgets and survival. These employers lack the time and the skills needed for conducting training in the thorough manner ultimately required. This is evident from review of regulatory statutes habitually cited by regulators, such as the Occupational Safety and Health Administration (OSHA). For many years, hazard communication, an education-oriented standard, has consistently ranked at or near the top of standards cited in U.S. general industry. This is an indicator about the difficulty of providing education and training in the workplace.

The logical places in which to address education and training requirements about lockout fundamentals, as well as all other health and safety materials, are the trade and technical schools where students are captive. They must master the curriculum in order to obtain the certificate that becomes the key to obtaining a job.

This approach is fine for school-age and adult students who complete trades education and training through the high school and college system. This approach is not satisfactory for the many adults who left school prematurely to pursue employment and have no intent or ability to return. This means that in order to comply with regulatory requirements, employers must assume the role of educators and trainers.

Regardless of other factors, this role for employers is inescapable, at the least because of the specifics of each work environment. Few operations are carbon-copies of each other, such that learning gained in one is applicable to another. This is not likely to be the case even in tightly controlled franchises. Hence, the employer must become the educator and trainer for that specific workspace despite any other objections.

### Employer Perspective

Employers do not hesitate to send administrative and office personnel to training courses. These cover all manner of topics ranging from organizing the desktop and files and records, to use of software, to time management, to human relations and conflict resolution, and effective supervision. Training currently provided by training departments in large organizations concentrates on the development of management, supervisory, interpersonal, sales, clerical, computer, and other such skills. Yet, education and training of shop floor employees and first-level supervisors remains vastly underdeveloped, despite the existence of regulatory requirements to provide this to them and the observation that trained employees work more safely and productively.

There are a number of reasons for this situation. Some have appeared as themes in articles in the trade literature. Others have not received their due recognition.

The most important theme and the one that has received little, if any, attention, is economic. The worth of production and shop floor workers to an employer is many times the amount that they receive in wages and salaries. The wages and salaries of all other employees in an organization from the lowest- to the highest-paid derive from the economic value of work performed by production and shop-floor workers.

Hence, the net worth of the time of a production or shop-floor worker easily could be measured in thousands of dollars per hour, not the tens of dollars per hour that these individuals actually receive. Stopping production to perform training incurs a heavy cost in dollars and production hours. From this perspective, the disincentive to providing training to production and shop-floor workers is obvious. Training, that is neither efficient nor effective, wastes time and money and does not produce the intended outcome.

Instructors face additional realities in providing classroom training. In order to bring together a group of people, this often necessitates training after normal work hours and on weekends. Both of these options create learning problems. After normal work hours, people are tired and want to go home. Sitting in a classroom is not a pleasurable experience for them. Weekend training is only slightly less onerous. People are likely to be less fatigued unless they have attended a social function the night previously.

Another factor in this situation is the issue of compensation. Many employers cannot afford to pay workers to sit through training sessions conducted after hours or on weekends. The absence of pay for attending a training session creates resentment among workers and impairs the effectiveness of the training effort.

Another factor that complicates the delivery of technical information is the education level of employees. A workplace can have employees with elementary schooling as well as those with advanced degrees. What appeals to the interests and knowledge level of one individual may totally disinterest another.

The vast majority of organizations have fewer than 20 employees. As a result, the resources needed to deliver such technical information are unlikely ever to be available within most organizations.

**Worker Perspective**
The workplace is a close reminder of the formal institutional setting represented by the school. On-the-job training is an imposed form of information transfer and behavior specification or modification. Usually the supervisor, a person ill-equipped to carry out this function, performs the training. Many workers distrust and sometimes are outwardly hostile toward anything that originates from this source. An understanding of how people learn is fundamental for anyone wishing to train others.

Formal training situations can be a stressful experience for shop-floor workers and their supervisors. This occurs for several reasons. First, many members of the workforce are "doers." They are not accustomed to sitting for long periods. The instruction must involve these learners as actively as possible. A second possible reason is literacy level. For many immigrants, English is a second language. As well, many individuals for whom English is the mother tongue, have never adequately learned to read, write and perform arithmetic calculations.

Illiteracy is the inability to read, write, calculate, or solve problems efficiently (Goddard 1987). Illiteracy is either functional or borderline. Functional illiteracy refers to an inability to cope at the simplest level. Borderline illiteracy describes an ability to survive in society but with much difficulty. In many cases, the cause of illiteracy is lack of training in everyday English usage.

Another cause of adult illiteracy is learning disability. These disorders may arise at birth, or from brain damage or incorrect function. Often they are identified during early school years. Sometimes, however, they remain unidentified. These disorders can affect the ability to read, write, spell, listen, speak, or perform mathematical operations. When treated, these cases improve to varying degrees depending on the specifics of an individual case.

Workplace illiteracy is a recognized occupational problem. One study reported that illiteracy affects one in six employees in all fields. Employees who cannot read and comprehend the written word cannot recognize warnings and act in an appropriate manner. This puts themselves, and possibly others, at risk.

Some organizations have responded to the problem of illiteracy by setting up remedial programs to assist their employees. The immediate supervisor is the critical individual in addressing the literacy issue. An observant supervisor can recognize the outward manifestations of the problem in many ways. Potential indicators include the need for help in filling out complex forms or responding to written orders.

Trust between the supervisor and affected employee is absolutely essential for addressing the problem. Trust is an extremely important commodity, one that is difficult to establish and maintain and is easily lost. A person who feels secure will be more willing to admit to the problem and seek assistance. People who have literacy problems are reluctant to reveal them. Yet, once given the opportunity for help, they are often eager learners.

### External Resources

External resources are available to assist in the training burden. Progressive manufacturers already offer technical assistance to their customers under the guise of product stewardship. Some offer classroom training and/or videos in the use of their products. Trade associations sometimes offer training packages. Training packages produced through trade associations enable the pooling of resources. Safety associations and regulators also offer training materials.

Many jobs in industry involving electrical work, mechanical work, and welding, are similar regardless of where they occur. This situation provides a window of opportunity for vocational training through educational institutions. Training in hazard recognition and control could be added to existing programs and short courses directed to these trades.

Most external resources, such as accident prevention and safety associations, university health and safety centers, occupational health and safety consultants, and union resource centers, are located in large cities. This hinders the widespread delivery of person-to-person training to people in small centers and rural areas. This situation is considerably more serious given the view of regulators that the required knowledge embodies a skill to be practiced through hands-on activities. Skills training is more difficult and beyond the capabilities of many modes of delivery.

## Adult Learning

The place to start in this discussion is to consider adults as learners. Adults differ in their abilities and needs for learning from children. They are not, in effect, large versions of children in their learning styles.

Many adults regard adulthood as the period in their lives in which the need to receive formal education or training ceases. Many individuals prematurely left school because of lack of interest or fear of failure. Despite these views, adulthood is a period of learning. Knowles (1980) espouses four important attributes to adult learning (Table 9.5). The concepts espoused by Knowles provide a starting framework for understanding the nature of the problem, and ultimately, how to approach it.

Adult learning usually occurs through choice rather than obligation. The more obvious resources of information that contribute to adult learning include books, the media (newspapers, magazines, television, radio, movies), comments of peers at work and in social groups, and the lessons of life.

Information provided through these vehicles for the most part occurs in short bursts. Commercials in television programs repeat the message frequently in order to promote retention of key words. The process of learning is casual, subtle, and often long-term. There is no formal indication that it is happening. There are no exams or assignments and fear of failure. The learner can tune out or turn off the source and restart it at any time without penalty.

## Learning Theories

Learning theories attempt to model the reality of learning. No author in this field has yet devised a universal approach that incorporates all facets of human learning. Researchers have created several discrete models. Successful instruction incorporates different features from various approaches in the appropriate context at the appropriate moment. The following is a compilation of key points from four major theories of learning.

## Table 9.5
## Attributes of Adult Versus Child Learning

| Adults | Children |
|---|---|
| Though dependent in some situations, the adult learner strives to be self-directed. Dependence is only temporary. | Children depend on adults and other children. |
| Personal experience is a rich resource for learning. Learning from experience has personal meaning that can be shared with others. | Children expect their questions to be answered by outside sources. |
| Readiness to learn is correlated to the need to know in order to cope with a real-life task. | Children expect to be told what they need to learn. |
| Learning is a process to gain increased competence. Applying knowledge and skills to performance is necessary. | Learning is a process to gain the first level of competence. Knowledge comes with learning and is not present initially. |

Source: Knowles 1980.

**Sensory Stimulation Theory**
The underlying premise of sensory stimulation theory is that people learn through the senses. The most important of these in the context of work are sight and hearing (Laird 1985). We use sight for about 80% of sensory input, hearing for about 15% and the other senses for the remaining 5% (Figure 9.7). The greatest emphasis during learning is placed on sight (Figure 9.8) for the following reason. After 72 hours, people retain about 10% of what they have heard, about 30% of what they have seen, up to 70% of what they have simultaneously seen and heard and up to 90% of actively practiced activity (Dale and Nyland 1985). Note that retention when both senses are involved is far greater than simply the sum of the contributions from sight and hearing. This indicates a cooperative relationship between the senses and learning in the brain.

As an example of the effectiveness of sight and imagery in learning, compare the presentation of information through text with that provided by the images in Figure 9.7 and Figure 9.8. The adage about the picture being worth 10,000 words is highly applicable to education and training. Figure 9.9 applies the concept in regard to education about hazardous energy. This graphic shows that the expression of hazardous energy in some accidents is similar to the triggering of a loaded mousetrap. Once sprung, the conversion of energy has occurred. Relief of the unstable constrained system has occurred through the expression of hazard by conversion of the energy. Further expression of energy in this manner cannot happen until creation of the constrained system reoccurs.

Sensory stimulation lends itself to films, videos, and simple demonstrations. Long-term retention can occur following only a single exposure to these methods of presentation. This is especially true with movie productions shown in theaters. A single viewing can create life-long memories.

The negative aspect of this approach is that the learner plays a passive role. This method can transfer information but is inappropriate for development of hands-on skills. To be effective for development of skills, training must involve as many of the senses as possible and actively involve the learner in the process. The learner must gain tangible experience by handling equipment. The learner develops the skill through practice and numerous repetitions. A demonstration in the use of an instrument or device by an instructor at the front of a classroom will not lead to acquisition of the skill needed

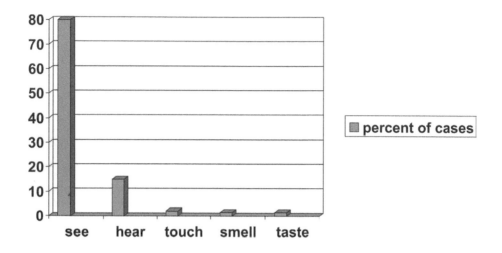

Figure 9.7. Senses and learning. Sight is by far the most important of the senses in learning. Compare the effectiveness in learning between this graphic and the listing of numbers in the text. One picture is definitely is worth 10,000 words.

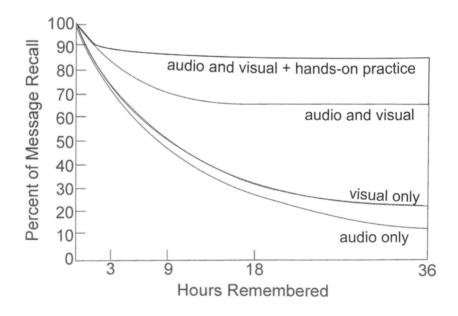

Figure 9.8. Learning retention depends on the mode of delivery. Learning involving the senses of sight and hearing is considerably more effective for long-term retention than learning involving only one sense. Practice further enhances the learning. (Modified from Dale and Nyland 1985.)

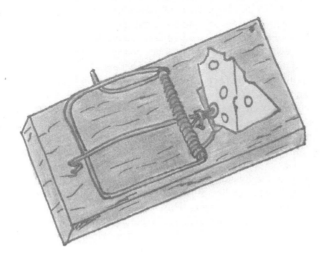

Figure 9.9. Powerful graphics create lasting memories and impressions following only a single viewing. In some situations, hazardous energy is constrained in an unstable system. Conversion during expression occurs rapidly and relieves the instability. Following expression, the system regains stability.

to operate it. The instructor, not participants in the demonstration, learns to operate the device in this situation.

**Reinforcement Theory**

Reinforcement Theory advances the view that animals, including people, perform tasks for rewards (Laird 1985). A reward is something proffered for performing an action Provision of the reward can occur prior to, during or after performance of the action. Reinforcement is the strategy employed in training animals, starting with pets. To an extent, this approach also works with people.

Reinforcement by reward is at best a communication tool. This approach works best with animals with whom we communicate at a primary level through this technique. We indicate to the animal through the reward that some action is desirable to us. Since the reward is a piece of food and since animals work for food, they will continue to repeat the action. As the action progresses to the next desired level of accomplishment, the animal receives another reward. Through progressive rewards, we communicate precisely what is required or desired.

Food is the only motivator of animals, such as pigeons and rats, used in laboratory experimentation. Higher animals, such as dogs and house cats, also respond to rewards other than food. These include verbal praise, play, grooming, and expressions of affection. These rewards provide reinforcement not only to the pet but also to the human with whom this interaction is occurring.

Humans reward working dogs for performing tasks such as detecting drugs and explosives in suitcases and packages, and now bed bugs, with food on successful detection. This approach corrupts the process since the diligence of the animal and dedication to task are at least as worthy of reward as the (hopefully infrequent) detection of the sought-out target.

Study of reward regimes has shown the key elements of reinforcement. Positive reinforcement (pleasant reward) following a desirable behavior reinforces and promotes the behavior (Laird 1985). Practice (repetition) of the desired behavior is necessary for learning to occur. To be effective, the reward must be individualized. That is, the person must value the reward. What appeals to one person will not necessarily appeal to another. Negative reinforcement (criticism, sarcasm, disapproving gestures) impedes the learning process. Negative reinforcement can lead to avoidance or escape.

Rewards must have strength. That is, a person must perceive the reward worthy enough to justify the effort. Continuous rewards are not necessary once mastery of the desirable behavior occurs. Rather, rewards provided unpredictably will maintain the behavior.

Using rewards as reinforcement of behavior with humans is complicated. Humans recognize application of the carrot and the stick (positive and negative reinforcement). They respond through language as a medium of communication and do not need reward for increments of progress toward some end-point because the reward is an acknowledgment of success and not a device of communication. The use of money and other devices of monetary value as rewards for anything less than special accomplishment rapidly corrupts the process and is viewed cynically for the manipulation that this represents.

A blatant example of the exploitation of Reinforcement Theory is its application to gaming and gambling. Some people will put an endless supply of coins into slot machines for a periodic payout when the machine"rewards" a lucky person. This situation reflects an obsession, an abnormal behavior. The decision to pursue the reward no longer is something over which the person has full control. Research into this situation has indicated that the reward is no longer simply the payout of money. The real reward occurs through the involvement of brain chemistry. These actions are considerably beyond the realm of simple learning.

### Social Learning Theory

People learn through observation or modeling (Bandura 1977). Observational learning must occur before the person actually performs a task.

Learning to dance is an example of this type of learning. The student follows the instructor from behind and attempts to place the feet in the exact location and sequence shown during the demonstration. Translating movements learned and practiced during the very controlled circumstances involved in modeling the actions of the instructor to the fluid movements needed during uncontrolled conditions of actual dancing is for many, a difficult transition. This transition is not always successful.

Modeling is an important component in development of hands-on skills. The art in instruction of this type is to break tasks into small components and then to focus on mastering each component. The next phase is to unify all of the individually learned components. This approach is similar in concept to learning to dance.

Reinforcement (reward), though not necessary, plays a facilitative role. That is, anticipation of reinforcement can be highly effective. The reward must have value in order to be effective. Informing a person about the benefits or rewards of doing something a particular way before observation can strengthen the learning. For example, anticipation of a valued monetary bonus for doing a task in a particular way likely will reinforce learning. This also could encourage the learner to be more attentive.

### Cognitive Learning Theory

Cognitive Learning Theory builds on sensory stimulation theory and reinforcement theory. Thinking, that is, information processing and memory, and synthesis of new ideas, supplement rewards and stimulation of the senses (Chambers and Sprecher 1983). Cognitive learning theory plays an important role in the system of formal education where practice exercises move students beyond the envelope created through formal instruction. Cognitive learning focuses on ideas and concepts and does not address the need for repetition and practice needed in development of hands-on skills.

Facilitated in part by memory, our minds are complex processors of information. Conscious memory of the here and now is a form of short-term memory. Unless a person deliberately tries to retain the information, short-term memory lasts for about 20 seconds (Gagné et al. 1992). Short-term memory holds only about four different items of information at a time. Hence, learning a complex task containing many bits of information by memorization is very difficult. Through repetition and practice, short-term memory develops into long-term memory.

Long-term memory is the information that we retain for many years. Recollection of a key item can trigger recall of a past event from long-term memory into short-term memory. The key item can be a taste, an odor, word, color, sound, or some other cue having no particular meaning in the present context. These memories can be pleasant, unpleasant, or neutral.

## Motivation

One of the most important elements in the learning process is motivation. Motivation is the spark that initiates the desire to achieve a result. Motivation can arise from an external stimulus, such as a paycheck, or from an undefined internal source. Motivation is an intrinsic factor. Without motivation, learning becomes very difficult. Many highly intelligent students quit school because of lack of interest or motivation. Motivation is the driving force that enables those with less than perfect potential to succeed.

Keller (1983) developed a model containing effective strategies for motivating learners. The strategies fall within four categories: interest, relevance, expectancy and satisfaction. Following are key points from Keller s strategies for motivation. Motivational strategies become embedded in the art of communication and instructional strategy.

### Interest – (Arouse and Sustain Curiosity)

Keller advises about the need to refer to new, unusual, out-of-place, conflicting, and strange events to generate curiosity and attention. These could include humorous anecdotes, challenges to common perception, extensions of behavior of individuals in the group, and items from the popular news media.

Instructors strive to create the "teachable moment." The teachable moment is a special time when a person senses the need to acquire additional knowledge and insight and provides the opportunity to fulfil it. Figure 9.10 and Figure 9.11 provide intellectual challenges intended to create teachable

Figure 9.10. Arouse curiosity and interest through the image of a tiger at the zoo. The tiger is a surrogate for a source of hazardous energy.

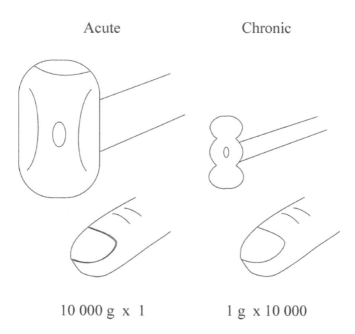

Figure 9.11. Arouse curiosity and interest. What would you prefer, one blow from a 10 kg hammer or 10,000 blows from a 1 g hammer. Both give the same total impact of 10 kg. The answer to the question is not intuitively obvious.

moments. They show images from common experience, a tiger in a cage at the zoo and hammers and a thumb or finger. There is nothing complex or complicated about the images, but in the hands of a skilled instructor, they provide a gateway to discovery and learning. The tiger is a surrogate for a source of hazardous energy. The picture of the hammers and the thumb convey information about dose. These pictures have elicited the same response from many people in different cities, regions and countries.

Another device is to describe personal experiences related to the situation of interest, perhaps the expression of energy under unexpected conditions. While using words to work around the situation, the instructor can appeal to the interest of the group and to provide an opportunity to learn more about a situation than people already know. The instructor should use familiar ideas to explain new concepts by building on previously acquired information. Common knowledge is an excellent source of previously acquired information as are the experiences of people in the group regarding a situation for creating the starting point for proceeding in a new direction. By asking the right questions, the instructor can stimulate the curiosity of participants.

**Relevance – (Appeal to the Needs of the Individual)**
One of the key approaches of the instructional strategy is to provide to the learner opportunities for success. Breaking the subject into small pieces is a technique for doing this This makes tasks more manageable. This strategy will appeal especially to people who are afraid of failure and those who strive to achieve high levels of success.

Another device is to provide to the learner opportunities to make choices, to assume responsibility, and to influence the direction of the training. This will appeal to people who are motivated by power.

To a greater or lesser extent, each person has a need to belong to the group. Therefore, avoiding situations that expose the individual to the possibility of failure, embarrassment, or humiliation is critically important in order to foster that sense of belonging. Within this context, participants need to understand the connection between personal effort and success.

### Expectancy – (Develop Confidence)

The preceding comments lead to expectancy. Establish a situation in which the learner expects to succeed and can succeed at all times. This optimizes learning potential. The oft-used expression, nothing succeeds like success, is no understatement when applied to training. In order to succeed, learners need to know about requirements and criteria for success. Inform them. Methods that allow the learner to define the completion of a step and when to move on provide some control over the learning process. This approach satisfies the need for expectancy.

### Satisfaction – (Manage Rewards)

Satisfaction with the process is the goal of both instructors and students. For instructors, this can affect both their own feelings of success and their livelihood because of the extensive use of evaluation techniques. This approach can be highly distasteful to instructors who have invested many hours of time to create the material. An instructor faces two challenges: the need to entertain and the need to convey information. Often the evaluation reflects the perception of the entertainment value of the presentation. These are not incompatible goals since effective learning occurs during entertainment.

From the perspective of the student, providing satisfaction is reflective of the art of instruction. Providing legitimate praise for accomplishment and contribution is highly effective in motivating participants. Praise given at unexpected moments in the learning process is likely to be more effective than that given in a predictable pattern.

Providing positive comments about progress is very helpful, especially where progress is difficult to attain. Even when little progress has occurred, accentuating the positive will encourage continuing effort. Instructors should avoid negative comments, half-hearted praise, threats, and so on. At the same time, the instructor should address evaluation in a low-key manner and stress the likelihood of success. Fear of evaluation can block progress. A person who fears evaluation is unlikely to be relaxed and may not perform to fullest potential.

Provision of positive feedback upon completion of a task is highly motivating. This is the case at sporting events when the crowd roars approval at a hockey game after the home team scores a goal. "Home-ice" advantage is a well-recognized morale booster.

Corrective feedback provided at critical points during learning is a necessary device. The manner in which the feedback is delivered is critically important and can affect the outcome of the instruction. The most useful time to provide corrective feedback usually is just before the learner repeats or practices the task. Provided in a low-key, sympathetic, and empathetic manner, this essential feedback can produce the necessary change without dampening enthusiasm.

## Instructional Strategies

Learning and motivation theory form the basis of instructional strategies. The purpose for using instructional strategies is to facilitate learning. Anyone can and will learn given suitable circumstances.

### Classify the Content

Training sessions often overload participants with information. This creates the situation where content delivery outweighs comprehension. This approach reflects the view that too much information is better than too little. This problem occurs especially during knowledge or fact-based training sessions. The outcome from this approach is that information may or may not be presented at an appropriate level of understanding. Sufficient time for practice and feedback, if any, is unavailable due to scheduling constraints.

Often the instructor attempts to convey far more information than the learner can assimilate during the allotted time. One strategy for addressing this problem is to separate content according to the hierarchy of need to know (Mills 1977). This concept classifies content into three groups (Figure 9.12)

## Content Selection

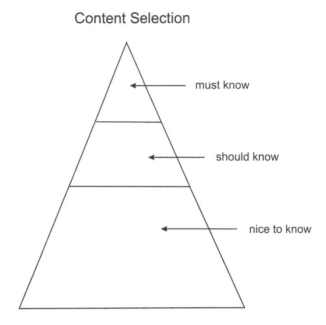

Figure 9.12. The hierarchy of content for instruction. Classifying content by importance to the outcome is essential to ensure emphasis during instruction on material that participants must retain and apply in the long term.

according to the overall goals of the training namely, the "must know," the "should know," and the "nice to know."

The instructor is the information manager and has the opportunity and obligation both to select and reject potential content for the instructional session. The need to perform the classification forces discipline into the process of content selection and presentation

The "must know" is the content essential to the instructional session. This is the information the participant must know and understand at the end of the session and be able to retain and apply in the long term. By focusing on the "must know," the instructor uses the usually limited time for training most economically. Any time left over becomes a bonus for reinforcing the "must know" of the information and possibly introducing the "should know" and the "nice to know".

As the words imply, the "should know" pertains to content that helps to explain and to supplement the essential information. The "should know" is the content that the instructor wants to present in order to gain satisfaction that the instruction is complete and must omit if time for presentation is not available. The "nice to know" is the content whose presence or absence will have no impact on the outcome of the instruction and its long-term application. Applying the discipline of selection provides a means for the instructor to gain the satisfaction of assuring completeness of the instructional package from what otherwise would be a large collection of interesting information.

### Events of Instruction
Gagné et al. (1981) devised a practical approach for applying cognitive learning theory to instruction. This approach contains nine steps known as the Events of Instruction. The events are designed to support internal learning processes. Most steps follow in sequence but these may vary somewhat depending on the situation. The following is a summary of the Events of Instruction. All of the events do not necessarily apply to all instruction.

## Gain Attention
Gaining the learner's attention at the beginning of a session, rather than at some later point, is very important. As demonstrated previously, powerful graphics grab attention and challenge the *status quo*. The challenge of maintaining attention then arises. Keller s motivational strategies (described previously) provide ways for maintaining interest (Keller 1983). Organizing information in a logical sequence facilitates sensory processing. This also involves breaking large amounts of information into smaller more manageable parts.

### Inform about the Purpose for the Lesson
Indicating clearly what the learner will be able to do as a result of the training brings the participant into a position of knowledge and removes some of the element of surprise and hopefully concern. Sometimes this type of statement appears in promotional material either as a money-back or no-risk guarantee. This step sets up internal expectancy by providing the learner with an end-point for completion.

## Review Prior Learning
New learning builds on previous learning. Mathematics is a good example. First a person learns to count, then to add and subtract, and later to perform higher-level operations. Reviewing previous work before proceeding with new concepts narrows and bridges the gap between the familiar and the new. The process of review facilitates the transition.

Workplace learners have different levels of knowledge about the subject of the instruction. Hence, before instruction involving new material can begin, build-up of knowledge is essential to ensure that everyone has the same starting-point. This then ensures that everyone has the ability to move forward simultaneously. An audience containing participants of varying levels of knowledge can create a conflict the classification concepts of Mills (Mills 1977). The instructor must have knowledge about the knowledge levels of participants in order to design the instruction to meet the needs of everyone. Content that otherwise would be nice to know for some participants is essential for others.

In situations where the background information is not needed, the instructor can skip this information and devote more time to other content. The critical issue is to recognize the potential need for additional content and to prepare the instructional package to reflect different possible realities.

## Present New Information Clearly
Wherever possible, use illustrations to explain concepts. This reinforces previous discussion about the use of powerful graphic images and the concept of the picture versus the 10,000 words. The corollary comment is always to present new information as clearly and concisely as possible. Clear presentation requires skill and simplicity. This is the basis for the use of pictures, namely, to avoid the need for the 10,000 words. There are many sources of illustrations: the internet, clip art provided in graphics packages, and photos and videos obtained on the premises of the employer.

While well-described verbal illustration is obvious, so also are well-chosen images. Visual material must support the message, and not detract from it or compete with it. A carefully chosen image and the appropriate words make the presentation memorable.

## Guide the Learner through the Instruction
People need clear visual cues to indicate the direction of progress. Websites involved in e-commerce provide an example of the concept. Some sites provide clear, easy to follow instructions about how to progress through the transaction. These sites typically use block arrows that point to the right to indicate the means to progress further. This device is a visual cue that is readily understandable to the viewer and minimizes the visual effort needed to find it.

Other websites use words. The words are often difficult to locate in a crowded visual field of information. The purchaser sometimes must expend considerable effort to find the word that enables progression. This effort is often needless compared to the economy intrinsic to the simpler system that uses the visual cues. Directional arrows have universal meaning. They need no explanation.

Similar concepts apply to instructional materials. Cues, such as color or arrows, provide direction. There is a need to avoid the under- or overuse of cues. Too few cues leads to ambiguity, while too many, distract attention. Providing the learner with clear instructions about how to proceed is critical. This is especially true with self-guided instruction. Providing clear instruction is partly the function of a live instructor during a classroom session.

Distance learning poses challenges to both instructors and learners. Instructors are not able to receive immediate visual and audible feedback as would occur during live sessions and has no way of knowing whether the group is ready to proceed. Members of the group must clearly comprehend the concept or there is nothing to be gained by proceeding and everything to lose when readiness is nonexistent.

In all situations of information delivery, the brain must process the information in order to make the transfer to short-term memory. The issue of timing in proceeding to new material partly reflects this reality. Since the brain can process only small amounts of information at a time, cuing the learner about the most important features of the material and aides memoires are essential.

## Involve the Learner Actively in the Process

The instructor needs feedback about progress as much as the learner. Asking participants how they are faring is a simple way to approach this important question. Practice exercises provide another device for making this enquiry.

Learning requires active repetition. However, the learner needs the practice, and not the instructor. This is a trap into which inexperienced instructors fall. They become very proficient at the skill while the audience watches passively. A demonstration has value only when the learner tries to replicate and repeat the activity on demand.

## Inform the Learner about Progress

Learning is most effective when the learner receives constructive feedback. People need feedback about their progress, and encouragement and help when experiencing difficulty. Keller (1983) provides strategies for motivation.

## Assess Performance

Testing and assigning a final project, or demonstration of competence provide a means of assessment. Assessment indicates how well the learner has assimilated the information and can apply new skills to meet some required level of performance or competence. Assessment also identifies areas needing corrective action.

## Apply Learning to New Situations

After practice using a simulation, assigning a real-work task, and monitoring the situation determines whether and how well the participant can perform the skill outside the well-defined boundaries of the simulation. The ability to apply knowledge and skills beyond the protected environment of the simulation and the classroom is critical point of departure in every training program.

This is especially the case with application of lockout concepts and procedures. Making this transition must be an integral requirement of the training program. After completing the section on concepts, the instructional package must provide the learner with realistic opportunities to apply the learning to on-the-job activity. The value in learning something new lies in the application toward solving actual problems. The training must include monitoring, feedback, and refreshment.Without these components, the learner rapidly loses the newly acquired skill. Refresher training is often necessary to ensure continued performance to the required level.

## Elaboration Model

Training often involves transfer of large amounts of information. This information could cover one large topic, several interrelated topics, or a large number of small related or unrelated topics. This situation poses potential problems because the information could rapidly overwhelm the learner. One approach for addressing this problem is the Elaboration Model (Reigeluth et al. 1980). The Elaboration Model provides a number of strategies (Table 9.6). These generally start from common experience and proceed to greater detail or to more advanced concepts (Mills 1977; Reigeluth et al. 1980).

## Table 9.6
## The Elaboration Model

**Device**

Sequence the information

Show the relationship between details

Present periodic summaries

Proceed from the simple to the complex

Proceed from the concrete to the abstract

Proceed from the particular to the general

Proceed from the known to the unknown

Proceed from the whole to the parts, and back to the whole

Proceed from observations to reasoning

Sources: Mills 1977; Reigeluth et al. 1980.

Movies and television dramas often employ these techniques. The camera first pans across a large expanse, then zooms onto a selected area and finally stops on the intended focus of interest. This technique provides the overall picture to the audience and then minute detail about a selected part or situation. It also enables one city to "stand in" for another. Applied to training, the learner can see the overall context of the instruction and hence its purpose.

One application of the elaboration model is induction training for new employees and site orientations for visiting contractors at worksites. The site may contain processes that require emergency action and possible evacuation. The new employee and visiting contractor must become aware about these processes and their interrelationships, and the meaning of alarms and evacuation routes, as quickly as possible. The first step in orientation might indicate the layout of the site, followed by areas of concern, the meaning of alarms and flashing lights, and routes of evacuation. A walking tour of the plant indicating key landmarks would reinforce what was seen on the diagram. This provides the overall picture to the trainee or visiting contractor. Provision of more in-depth information would reflect the need to know.

Training at the next level would include detail about specific operations. After receiving this information, the trainee would proceed to the work area, and as much as possible, apply information already learned about the plant and its layout to find the way.

### Performance-Based Training

The performance-based model (Blank 1982) is a carefully planned approach to designing a training program. This model is an especially useful for people lacking a formal background in education and training. Performance-based training consists of a series of steps that work together as a unit. The goal is to produce effective learning. Designing a performance-based training program involves a series of steps described in the following sections. Some or all of the sections apply to training provided for lockout and tagout.

### Identify and Analyze the Task

Job descriptions often contain many activities. Each activity contains at least one complete unit of work called a task. Each task contains procedures and each procedure contains steps. Performing a lockout is an example of a procedure (Table 9.7). A procedure has a starting point and ends in a specific result.

One approach to task analysis is to identify the critical steps of a procedure by observing and recording the ways in which a number of experienced people perform the same procedure. Since each person may perform the steps differently, the best procedure is the one that combines a logical sequence and all of the critical steps. Procedural problems refer to inefficiencies in the process. Examining operations requires a critical view. Often an outside observer can provide a more objective view.

Performance of the lockout can expose the individual to health and safety hazards. Health and safety hazards arise in several ways, including exposure to chemical and physical agents. The hazard may be present in the workplace itself. An example is the hazardous atmosphere present in a sewer. The task, for example, repairing a pump, may present no hazard in and of itself, once deactivation, de-energization, isolation, and lockout have occurred. In other cases, the work may cause the hazard. Discussion of hazard assessment and creation of procedures occurs in Chapter 7.

### Identify Procedural Problems and Evaluate and Revise the Task

Hazards associated with performing the task in the manner stated in the procedure are not always readily apparent. Evaluating hazards requires considerable formal education, experience, and background knowledge. This capability may not reside in an organization or may require a team effort. (See previous sections in this chapter for additional discussion.)

The person or group embodying this competence should examine each step in the procedure to determine the existence of hazardous conditions and to specify protective measures.

### Specify Preventive Measures

Preventive measures in these situations include engineering controls, work practices, and personal protective equipment. The choices of engineering controls in these situations are limited primarily to portable ventilation. Personal and other protective equipment are the likely choices available for use in these situations. Selecting a method that minimizes risk is another means of control.

### Write Standard Procedures

The preceding information forms the basis to create a standard procedure. A standard procedure is a synthesis of the information and ideas identified during the task analysis in conjunction with regulatory requirements. Discussion in Chapter 7 provides the basis for creating standard procedures for implementing isolation and lockout.

### Set Performance Objectives

A performance objective is similar to a road map. A performance objective provides a guide from a starting point to a destination. A standard procedure contains one all-encompassing performance objective that describes the result of training. A performance objective contains three parts. The first indicates what the learner will be able to do (always an observable action) at the conclusion of the training. The second part indicates conditions under which the learner will be able to do the action. The third part indicates how well the learner will be able to perform the action (must be measurable).

Following is a sample performance objective for a fictitious standard procedure involving lockout of Unit 31. "Using the information from the Lockout Procedure, the employee will perform the lockout on Unit 31 so that no hazardous energy remains." The "condition" in this objective is, "using information from the Lockout Procedure" The "what" is, "will perform the lockout of Unit 31" (observable). The "how well" is, "so that no hazardous energy remains" (measurable).

## Table 9.7
## Steps in Performing a Lockout

**Steps**

Obtain and read the standard procedure for the specific lockout to be performed.

Obtain the appropriate number of locks and lockout devices.

Inform appropriate individuals about implementation of the lockout.

Deactivate the equipment.

De-energize the equipment.

Isolate the equipment using the energy-isolating device(s).

Apply lock(s) to the energy-isolating device(s).

Verify the effectiveness of the isolation and lockout.

Report problem(s) in implementing the procedure; do not start the next task until resolution of the problem(s) has occurred.

Devise Test(s)

The standard procedure is a valuable tool for devising a test. Restated into question format, the steps become a performance checklist. Requiring the individual who performs the procedure to sign initials after each step creates an internal test and assertion of completion.

Assessment of knowledge or fact-based information, as provided during classroom-based sessions, requires a written or verbal test. The test should be objective and free from ambiguity. The best type is the short-answer test. In general, preparation of written tests requires considerable skill. For more information on this subject, refer to the text by Gronlund (1993).

Describe the Audience

The key to successful training is to know the audience in order to be able to adapt the training, as needed. Learners may differ considerably. What appeals to one person may completely repulse another. Factors contributing to individual differences include: age, experience, education, cultural background, literacy, personal problems, etc. (Refer to previous sections in this chapter for additional discussion.)

Before implementing the training program in a large organization, determining the entry point of each learner can provide valuable insight about content, comprehension, and level of approach (Dick and Carey 1985). This is usually not possible or feasible in small organizations. In these situations, discrete enquiry through management and supervision can produce invaluable insight about language difficulties, education, background knowledge, and other useful comments.

Customizing instruction to the needs of the individual is a difficult task requiring much knowledge and skill. This is the ideal situation. However, this usually is not the case and is not practical in real-world situations. At best, most instructors acquire only a general picture of their audiences. The sections on learning and motivation contain useful information to apply during training.

Select Training Methods and Materials

The goal of training is to ensure that the learner fulfills the performance objective. A number of training methods are available (Table 9.8). These range from passive methods, such as lectures and assigned readings without follow-up, to highly interactive methods, such as performance-based training.

Each method has strengths and limitations. The most suitable method also suits the reality of the situation. A method that on paper is among the least effective may in a particular reality be all that is available or practicable. Improvement is always possible for methods that have limitations. This can occur through use of multiple methods. Best success occurs through active involvement of the learner, regardless of the method.

Training materials (Table 9.9) support training methods. These products should reflect the needs of the group. Many ready-made, generic-type materials include videos, on-line and on CDs and DVDs, books, and computer software. Graphics packages provide line art and enable almost anyone to create materials. While these and other materials may be useful, they may not adequately reflect the information and skills needed in a particular situation. Regulators have commented about this situation and indicated that these materials, in and of themselves, do not provide sufficient detail about the specifics needed for application in a particular situation. Where possible, select a variety of materials, and use only parts of materials that contain "must know" information. Supplemental training can provide the specifics.

**Selecting the Training Environment**

In a perfect world, the training environment would accommodate the instructional methods and materials. In the real world, however, the area available for training often dictates the selection of instructional methods and materials. The environmental conditions of the training area (Table 9.10) are very important, since surroundings profoundly affect our moods and behavior, and ability to learn.

The reality of training is that this activity must occur in places that are far from ideal. Few entities have available a room dedicated to this activity. Training occurs in construction trailers located outdoors and sometimes inside buildings, lunchrooms, offices, and living rooms in homes, to name a few of the possibilities.

When training is performed in trailers, supply of outdoor air is a critical consideration. Construction trailers are generally well-sealed structures. The presence of a large number of people in a trailer can rapidly increase the level of carbon dioxide and induce drowsiness. Noise in industrial operations and construction sites conflicts with the need to open windows to provide outdoor air. Training performed in lunchrooms must occur around breaks and meals.

Effective training is possible in the many existing but far from ideal facilities. Analyze the situation, and wherever possible make improvements. Often significant changes are possible at minimal effort. Consider the characteristics compiled mentioned in Table 9.10 for optimizing a less than perfect environment. Where possible, apply those that will improve the training environment.

**Implementing the Program**

At some point, the decision to proceed with the training will occur. The decision to proceed is a commitment and a demonstration of action by the entity. This decision implies that follow-through will occur.

The decision to proceed will necessitate final preparations. Final preparations create the public face of the training. These are the aspects of the training that everyone will experience and will remember. These experiences will heavily influence the perception that will form in the minds of people. Following is a list of items to consider.

Follow-Through

Making the commitment to proceed and then making the necessary arrangements, but cancelling at the last moment creates a situation of great concern. The cancellation indicates a problem at the level of senior management. This situation creates a credibility gap within the organization. Follow-through is critically important for every organization

## Table 9.8
## Training Methods

| Advantages | Disadvantages |
|---|---|
| **Self Study** | |
| Minimal involvement by other individuals | No assurance that study of material will occur |
| Useful in situations involving isolated work locations | Relies on individual initiative and reading ability |
| Works best when information presented in easily digested quantities with accompanying practice exercises and assistance through availability of answers to questions | Without follow-through, there is little incentive to study the material |
| | Requires much effort to process and remember information without practice exercises |
| Study materials provide a resource for long-term reference and application | Obtaining assistance at the moment needed is not possible |
| **Socratic Method** | |
| Question and answer format | Very time-consuming |
| Either instructor or learner can ask a question | Best suited to small group of participants, awkward for one-on-one situation |
| Stimulates thinking | |
| Clarifies information | Feedback to incorrect answers requires sensitivity |
| Assistance readily available | Pace controlled by least knowledgeable participant |
| | Not suited for development of hands-on skill |
| **Lecture** | |
| Presents information quickly | Effective presentation requires expertise in public speaking and careful preparation |
| Accommodates large numbers of participants | |
| Audio-visuals reinforce verbal concepts | Obtaining assistance at the moment needed is unlikely |
| Handouts provide a means to follow presentation without need for note-taking and provide the basis for future reference | Not suited for acquisition of hands-on skill |
| Instructor-Led Discussion | |
| Requires clear focus on training objective | Negative viewpoints can interfere with flow |
| | Requires strong leadership to maintain direction and pace |
| | Easy to lose involvement by some members of the group |
| | Not suited for acquisition of hands-on skill |
| **Workshop** | |
| Suits discussion by small groups engaged in problem-solving | Not suited for acquisition of hands-on skill or transfer of large amount of information |
| Narrow theme possible | |

## Table 9.8 (Continued)
## Training Methods

| Advantages | Disadvantages |
|---|---|
| **Instructor-Centered Demonstration** | |
| Allows learner to see skill performed in linear sequence of steps | Suits small groups when performed in person |
| Instructor readily available | Staging is critical to ensure that information is shown clearly |
| Group can benefit from availability of scarce equipment | Without practice and feedback,skill will not be learned |
| Handout provides basis for future reference and reinforcement | Least able participant sets the pace of the group or is left behind |
| | Not suitable for transfer of large quantities of information |
| **Performance-Based Training** | |
| Suited for skills training | Expensive to set up and operate |
| Works best when each learner can try-out skill with adequate time available for practice and feedback | Each participant must have available equipment to practice the steps |
| Focus on individual steps to ensure mastery by all participants | Without adequate time for practice and feedback, acquisition of skill will not occur |
| Information presented by video demonstration Learner practices the skill until mastery achieved | |
| Each learner receives individual feedback | |
| Learning is self-paced; learner moves to next task only when ready | |
| **Simulation/Simulator** | |
| Uses models that represent or are exact replicas of work environments | Very expensive to create, maintain and upgrade |
| Used for training airline pilots, locomotive engineers, process operators | Limited applicability to the many possible situations |
| May include role-playing | Can accommodate only small number of participants at a time |
| Presents various conditions of situation | |
| Involves decision-making at critical points in operation | |
| Presents outcomes of choices | |
| Permits errors without consequences | |

## Table 9.9
## Training Materials

| Advantages | Disadvantages |
|---|---|
| **Books** | |
| Very effective when well designed and illustrated | Requires effective instructor |
| Can enable active participation | Difficult to provide detail relevant to specifics of an operation |
| Continuing resource | Expensive to produce in small quantities |
| Individual scheduling, on demand | |
| **Still Images** | |
| Show specifics of an operation | Passive format |
| Anyone can create using graphics packages and a camera (CAUTION: consider conditions of use of the equipment and inadvertent capture of safety infractions) | Only part of the complete package |
| | Requires effective script |
| | Requires darkened room for effective viewing |
| Clip art readily available | |
| **Video** | |
| Custom production possible using consumer equipment (CAUTION: consider conditions of use of the equipment and inadvertent capture of safety infractions) | Production values must reflect professional standards in order to command authority |
| | Passive format |
| | Requires effective script |
| Show specifics of the operation | Playback requires adequate television sets for audience; otherwise use a video projector |
| | Commercial production expensive |
| **Computer-Based Training** | |
| Highly interactive instruction and mastery learning possible | Program development and upgrading very expensive |
| Self-paced instruction | Requires graphics, animation, video, sound in order to be effective |
| Instruction available on demand | Human instructor necessary to respond to questions |
| High throughput for fact-based learning | |
| Delivery time compressed as much as 4:1 compared to traditional methods | Learner must be highly motivated |
| | Presentation of concepts must match knowledge base and values of the audience |
| | Difficult to provide detail relevant to specifics of an operation unless custom production |
| | Requires supplemental handout or book to enable long-term retention and application |

## Table 9.10
## Desirable Attributes in a Training Environment

**Attribute**

Adequately sized room to accommodate normal group and associated materials and equipment

Seating facing parallel to and facing away from windows

Drapes on windows, or windowless to minimize distraction

Isolation from noise and interruptions

Neatly organized area

No equipment, material or supporting columns to obstruct view of the front of the room

Comfortable chairs and nonreflective work surfaces

Round or square tables promote interaction between participants

Projection screen located at front of the room above heads

Adequate lighting level (at least 1000 lux)

Good ventilation (at least 0.6 $m^3$/min or 20 $ft^3$/min of outside air per person)

Prohibition of cell phone use and other distractions

Temperature controlled to comfortable levels

Source: Laird 1985.

Time Scheduling

While most instructors prefer to work regular hours, the reality is that education and training must accommodate to the operation of the host organization. Training during the daylight hours of Monday to Friday may not be possible. Saturday may be the best and only day available. Instruction that occurs on Saturday imposes the burden of finishing as early as possible in order to enable participants to recover some of their day for their own pursuits. Similarly, in multi-shift operations, the time available for education and training may be unconventional for people accustomed to working during the day.

The best time to train people is when they are fresh and alert. For day-shift workers, morning is most productive. Performance drops off just before or after lunch and toward the end of the day or workweek. People become irritable when hunger sets in before lunch and tired at the other times. Attracting and maintaining attention during these time frames can be quite challenging for the instructor. Schedule tests during time-frames when people are most alert and allow adequate time for practice and preparation. The goal is to enable learners to be successful.

Coffee Breaks and Meals

An old adage holds that the success of an army depends on its stomach. That is, a successful fighting force is well-fed. This applies also during education and training. Some organizations make no special preparation for education and training sessions. Some organizations provide coffee, and some also provide lunch. Anything that is unusual to the routine is a nice amenity both for the instructor and also for the participants.

What is most important is that participants know in advance about these arrangements in order to make preparations, as needed. In situations where lunch is not provided and prior announcement has not occurred and participants must travel off site to obtain food, the cost in instructional time can easily be half an hour to an hour. This situation can cause considerable difficulty when the training requires nearly the full shift for completion and considerable time is lost due to the extra-long break at lunch. The instructor must maintain a flexible outlook during these circumstances.

### Attention Span

Regardless of the time scheduling, the struggle for attention during training is continuous, from beginning to end. One rule of thumb is to change the instructional strategy about every 20 minutes. Experience indicates that using the same "prop" for this length of time is overly optimistic. People thrive on variety and the unpredictable and the unexpected. The latter are key strategies during delivery in the struggle to gain, keep, and sustain attention. Reference to previous topics is a way to "close the loop" and to reinforce those ideas and to link them with the one currently under discussion.

Workers are unaccustomed to sitting for long periods of time. Breaking for five minutes every hour can provide the change and refreshment needed to proceed. For some, the break provides the opportunity to smoke a desperately needed cigarette.

### Grouping

In large organizations, grouping of individuals for training is sometimes possible. With a small homogeneous group, the instructor is able to provide more attention to individuals and the group as a whole. Large heterogeneous groups make instruction more difficult, since the needs of the individuals can vary considerably. Assisting those individuals below the level of the group can be very time-consuming.

The unintended consequence from grouping is that some groups will gain the perception of inferiority or superiority in the minds of participants. This approach can leave a lasting impression about the perceived value of individuals within an entity.

### Availability of Training Area and Materials

Training on the premises of an entity is always an adventure and sometimes a misadventure for the instructor, especially one external to the entity. The instructor should always determine the nature of the area to be made available for the session and the number of participants. If a projector and screen are required, the instructor should determine their availability, and as necessary, to provide them. Indicate to the host about the need for paper and pens and markers and arrangements for preparing handouts.

Then there is the unexpected. This can start with forgetting to bring key items, such as the projector, or screen, or computer needed for the presentation, or handouts. There are several possibilities. The starting point for minimizing the occurrence of these situations is to assemble the required items in a central location that is isolated from other materials. In this way the items to be transported to the site are visible for loading into the vehicle. A check of the storage area prior to leaving will confirm the loading of the necessary items.

A USB drive can store the presentation for use in a loaned computer. Computers and presentation software are so prevalent now that this is a low-level issue. A wall or even a piece of cardboard or taped together paper can substitute for a screen. A spare bulb or even a spare projector can address the issue of failure of this equipment. Micro projectors and even a camera containing projection capability are available.

The preceding issues are controllable through planning for contingencies. There are other situations beyond control of the instructor, starting with lack of availability of the training area. To some extent, recovery from this situation is also possible. The key is to be both flexible and creative. Asking the group for assistance is an excellent way to gain rapport and cooperation.

Always be prepared! There is nothing worse than discovering that something is not ready or does not function as intended or as expected. Ensure that the training area, materials, and equipment are all available and ready before commencing.

## Evaluate, Revise, and Update the Program

Evaluation and revision of the training program prior to implementation are very important. Review prior to starting provides an opportunity to ensure that all of the parts are consistent, cohesive, and comprehensible. As well, actual errors that become apparent are better corrected in private than in public during the training session.

An effective strategy for creating a presentation or a training program is to create the material well in advance of the need. This approach enables the brain to purge the material from active memory. Approaching the material following such an interval will provide a clean and fresh view of the content and the manner in which it was prepared for presentation. Review in this manner will show the material from a totally different perspective than at the time of creation. What seemed inspired as an instructional device during preparation of the material may no longer seem so. Similarly, material that seemed marginal during preparation might take on a totally different perspective and considerable importance. Another advantage of this approach is that the instructor who prepared the material is the only person to detect the flaws and can correct them prior to exposure to other individuals.

External review also can provide valuable feedback. This is especially helpful when the time available for preparation is limited. People who are unfamiliar with the material can provide valuable feedback following exposure as first-time learners. Potential external reviewers include management, members of the health and safety committee or worker representatives and union representatives. These groups can provide valuable feedback of the proposed training package. Use this information to revise the program.

Feedback from participants after delivering a session is very important and valuable. After all, their need for education and skill development was the reason for developing the training program. Ask participants to complete a questionnaire that emphasizes their views on how to improve the material. Also consider results from tests and performance checklists. If the majority of participants perform poorly, reevaluate and revise the training material. Ask participants to indicate what they find to be difficult. Similarly, if most participants perform poorly on a written test, reassess the test items. Are the questions straightforward? Do they reflect the required knowledge as presented during the training session(s)? Focus the questions on the instruction and use this information to revise the program. As well, participants are more likely to feel at ease to express their opinions when the questionnaire is anonymous.

After each session, also jot down notes about what occurred, errors in the material corrected during the presentation, and new ideas and concepts that came forth. This approach is often overlooked and can provide valuable ideas and resources for revising and updating the program, if necessary, before repeating it. Evaluation is an ongoing process.

### Recordkeeping

Recordkeeping, though tedious, is an essential component of any program. Recordkeeping establishes the reality of the program to the auditor (the regulatory authority and other external parties). Records to keep include lists of attendance at training sessions, content of training materials, names of individuals holding various levels of training, and results of audits of performance and follow-up corrective actions.

## Workplace Learning and Lockout

Implementing adult learning in the workplace is considerably more complicated than application of learning theory would indicate. Previous discussion has identified many of the factors that will influence and control how this will occur. Reality, of course, always gets in the way of discussion in the abstract.

Reality has indicated that entities affected by regulatory requirements on isolation and lockout have had difficulty in implementation. This is to be expected because of economics and the struggle for survival, failing to appreciate the seriousness and importance of the consequences of these requirements, and sometimes, sadly, resistance to change in the status quo.

Education and training in lockout concepts and techniques need not be complex and time-consuming. What matters is that the training provides the desired outcomes in a sustainable way. Previous sections in this chapter identified the possible existence of four groups of workers needing education and training in the lockout paradigm.

The most important resources for education and training are to be found on the premises of the entity. These include people willing to assist in the process, and hazard assessments and procedures and equipment used to implement and ensure deactivation, de-energization, isolation and lockout of equipment, machines, and systems on the premises. The equipment, machine, or system involved becomes the model for pictures showing implementation of the procedure. The pictures and the accompanying words become a powerful instructional tool. This tool is personalized to the entity and can easily and sequentially show what to do and where to do it. Fortunately, lockout and isolation procedures are uncomplicated for most equipment, machines, and systems for which they are required

The lowest level of learner needs knowledge about the lockout process as it applies to the entity. This includes information about the application of locks, lockout devices, tags and tagout devices; the process involved in deactivation, de-energization, isolation , lockout, and verification; and the sanctity of the lockout paradigm. This presentation would fit at the level of a toolbox talk. Table 9.11 provides a document for recording the outcome from a toolbox talk. The spaces at the bottom of the form provide room for entering additional topics for these talks. The intent of the form is to record discussion about a different topic at each meeting in a rotation. The check box at the right edge of the topic list provides a way to mark the form to identify the next topic for discussion in the sequence of the rotation. In this way, the supervisor covers all topics in a recorded and consistent manner. The completed form provides a record that the talk occurred.

The next level of training applies to workers who are affected directly by the lockout or tagout. These individuals include maintenance, trades, and technical personnel employed by the entity and by visiting contractors. They require knowledge about the content at the lowest level plus additional information relevant to the lockout process within the entity. This includes the specifics of the lockouts that have occurred in preparation for the work that is to occur. The specifics include the hazard assessment and procedure, points of retention of energy in the equipment, machine(s), or system, and the means of immobilization and release of energy retained by the lockout.

The third level of training applies to workers who deactivate, de-energize, isolate, lockout, and verify the effectiveness of the procedure. These individuals would benefit from performance-based training. This type of training emphasizes the ability to understand and critique the hazard assessment document, as well as to perform all of the steps in the procedure. They must be able to demonstrate inquiry and judgement in these matters because the safety of personnel who work on the isolated and locked out equipment, machine, or system depends on the judgement and execution of procedures by these individuals. This is especially important in situations involving immobilization where energy retention occurs.

The fourth level of education and training pertains to individuals who assess circuits in the affected equipment, machines, and systems. This knowledge is harder to describe because it relates to the design of equipment and machines and the design of systems. The requirements of isolation and lockout apply to many types of equipment, machines, and systems.

One strategy in developing the training involves the use of photo images or line art to illustrate the premises and location of equipment, machines, and systems affected by the lockout. A further elaboration of this approach brings the level of detail down to the procedure used to deactivate, de-energize, isolate, and lock out and to verify the effectiveness of the lockout of individual circuits. Procedures containing pictures showing how to perform the steps employ the picture to replace the 10,000 words. This is extremely helpful to anyone performing these tasks on unfamiliar equipment or machines.

**Table 9.11**
**Toolbox Talk Record**

| ABC Company | Site Crew Meeting | Project | |
|---|---|---|---|
| Location: | | Date: | Time: |
| Supervisor: | | Signature: | |
| Attendees: | | Signature: | |
| Summary of Discussion: | | | |

| Possible Topics | | Possible Topics | | Possible Topics | |
|---|---|---|---|---|---|
| lockout/tagout concepts | | | | | |
| isolation devices | | | | | |
| lockout/tagout devices | | | | | |
| locks | | | | | |
| group lockout | | | | | |
| lockout termination | | | | | |
| | | | | | |
| | | | | | |
| | | | | | |
| | | | | | |

## What Adult Learners Need

Some principles (Table 9.12) are critical to beginning, maintaining, nurturing, and retaining dialogue with adults (Galbraith and Fouch 2007). Adults arrive at training with baggage when compared to children. They have many more life experiences, time demands, and psychological barriers (such as past negative experiences). The instructor should attract and maintain attention and evaluate the experiences that may hamper the learning process. Disengagement will occur when adults have had negative educational experiences they need to overcome before they can learn new skills. Adults also may be plagued by more incorrect information and knowledge than children, which impedes the learning process.

## References

Blank, W.E.: *Handbook for Developing Competency-Based Training Programs*. Englewood Cliffs, NJ: Prentice-Hall, Inc., 1982.

Chambers, J.A. and J.M. Sprecher: *Computer-Assisted Instruction: Its Use in the Classroom*. Englewood Cliffs: Prentice-Hall, 1983.

Dale, E. and B. Nyland: Cone of Learning. Eau Claire, WS: University of Wisconsin, 1985.

Dick, W. and L. Carey: *The Systematic Design of Instruction* (2nd ed.). Glenview, Illinois: Scott, Foresman and Company, 1985.

Gagné, R.M., L.J. Briggs, and W.W. Wager: *Principles of Instructional Design*, 4th ed. Orlando, FL: Harcourt Brace Jovanovich College Publisher, 1992.

Galbraith, D.D. and S.E. Fouch: Principles of Adult Learning, Application to Safety Training. *Prof. Safety 52(9):* 35-40 (2007).

Goddard, R.W. The Crisis in Workplace Literacy. *Personnel J.*, December 1987, pp. 73-81.

Gronlund, N.E.: *How to Make Achievement Tests and Assessments*, 5th ed. Boston: Allyn & Bacon, 1993.

Keller, J.M.: *Motivational Design of Instruction. In Instructional Design Theories and Models: An Overview of Their Current Status*, Reigeluth, C.M. (Ed.). Hillsdale, NJ: Lawrence Erlbaum Associates, 1983.

Kepner, C.H. and B.B. Tregoe: *The New Rational Manager.* Princeton, NJ: Princeton Research Press, 1981.

Knowles, M.S.: *The Modern Practice of Adult Education*, rev. Ed. New York: Cambridge, The Adult Education Company, 1980.

Laird, D. *Approaches to Training and Development* (2nd Ed.). Reading, Mass.: Addison-Wesley Publishing Company, Inc., 1986.

McManus, N.: *Safety and Health in Confined Spaces*. Boca Raton, FL: Lewis Publishers/CRC Press, 1999.

McManus, N. and G. Green: *The HazCom Training Program*. Boca Raton, FL: Lewis Publishers/CRC Press, 1999.

McManus, N. and G. Green: *The WHMIS Training Program*. North Vancouver, BC: Training by Design, Inc., 2001.

Mills, H.R.: *Teaching and Training. A Handbook for Instructors*, 3rd ed. New York: John Wiley & Sons, 1977.

OSHA: "Control of Hazardous Energy Sources (Lockout/Tagout); Final Rule," *Fed. Regist. 54*: 169 (1 September 1989). pp. 36644–36696.

OSHA: "Permit-Required Confined Spaces for General Industry; Final Rule," *Fed. Regist. 58*: 9 (14 January 1993). pp. 4462–4563.

Reigeluth, C.M., M.D. Merrill, B.G. Wilson and R.T. Spiller: The Elaboration Theory of Instruction: A Model for Sequencing and Synthesizing Instruction. *Instruc. Sci. 9*: 152-219 (1980).

## Table 9.12
## What Adult Learners Need

| Attribute | Comment |
|---|---|
| Needs assessment | Content of the material must reflect needs assessment. |
| | Instructor and learners should shape content based on relevance and applicability to adult needs and learning principles. |
| Safety and comfort | Respect for learners as decision-makers and an inviting environment for adults create safety and comfort. |
| | Safety reflects trust in the competence of the design and the instructor, relevance of the learning objectives, and an ability to express thoughts and logic in the sequence of activities. |
| Sound relationships | The initial meeting between the instructor and the student should establish a sense of inquiry and curiosity. |
| | Learning with the mind, emotions and actions can reduce the anxiety associated with starting a session containing unfamiliar material. |
| Action with reflection | Learning should occur through action with reflection. |
| | Inductive learning should invite reflection or action using new content. |
| | Deductive learning should encourage participants to consider ways to apply new content in different scenarios. |
| Small portions | Bite-sized chunks of information prevent overwhelming the learner and create opportunity for mastery. |
| | Adult learners prefer the "'whole-part-whole" strategy: demonstrate the new skill or use of knowledge, describe the details, and then apply and reinforce the concept using examples. |
| Teamwork | Teams simulate the participatory universe in which people live. |
| | Peers who can assist and mentor with clarity, sensitivity, empathy, and skill assist learning. |
| Engagement | Active involvement of everyone is essential to foster the desire and initiative to learn. |
| | Learning together can occur only when everyone is prepared to the same level of knowledge prior to proceeding to the unfamiliar. |
| Accountability | Meeting learning objectives is critical in learning events. |
| | Accountability ensures that what is proposed to be taught is taught; what is meant to be learned is learned; skills intended to be gained are demonstrated; and knowledge to be conveyed is evident in language and reasoning. |

Sources: Thoms 2001; Vella 2002.

Thoms, K.J.: They're Not Just Big Kids: Motivating Adult Learners. Paper presented at the Annual
  Mid-South Instructional Technology Conference, Murfreesboro, TN, 2001.
Vella, J.: *Learning to Listen, Learning to Teach: The Power of Dialogue in Teaching Adults*. San Fran-
  cisco: Jossey-Bass, 2002.
Wellborn, C. and Boraiko, C.: Proactive Safety. *Prof. Safety 54(10):* 37-41 (2009).

# A Consensus Standards and Regulations

## INTRODUCTION

Strategies for controlling exposure to hazardous energy are limited in number, and certainly not recent in origin (Grund 1995). They have existed since the earliest developments of devices that harnessed hazardous levels of energy.

The early 20th century saw increases in maiming injury and industrial accidents. Workers clearing jams or servicing equipment could not effectively eliminate the source of power because of continuous operation of the prime mover. Prime movers are large centralized power supplies that powered multiple machines through overhead driveshafts and belts and pulleys. Prime movers enabled consolidation of equipment and machines into factory buildings and creation of compact assembly lines.

The increase of risk of harm posed by the prime mover and increase in traumatic injury signaled the need for change. In the period from 1900 to 1925, progressive employers independently undertook initiatives to prevent such tragedies. These actions reflected both humanitarian and financial incentives. The workers' compensation system had not yet started and workers injured on the job were forced to seek recourse through the courts. These employers realized that an accident prevented meant money retained. At this point, there was no formal manner for interaction between employers who were competitors.

The implementation of the workers' compensation system of no-fault insurance and employer classes for assessment provided the incentive for collaboration and cooperation in the area of accident prevention. Safety organizations, such as the American Society of Safety Engineers and the National Safety Council in the U.S. and the Industrial Accident Prevention Association in Ontario, Canada undertook to promote these goals. At the same time, specific industry associations also formed with the goal of promoting safe work practices through specific committees. The concept of the Zero Mechanical State existed in the foundry industry for a considerable period of time previous to creation and implementation of ANSI Z241.1-1975 that introduced the concept for wider consideration (AFS 1961). Ultimately the implementation of these performance-based policies through specification-based procedures rested with the individual entities. (Refer to later sections for further discussion about these concepts.)

The biggest opportunity for improvement arose with decentralization of the power supply from a single prime mover to prime movers to power each piece of equipment and machine. This change reflected broad-spread electrification of cities and towns and the use of electric motors as prime movers. The change to prime movers for each piece of equipment and machine offered many advantages. This allowed continuation of the operation without interruption through parallel circuits and alternate circuits. At the same time, this innovation enabled service to occur on equipment and machinery that could undergo independent deactivation, de-energization, isolation, and lock out.

Procedures in this period tended to focus on specific energy sources, namely those of perceived or recognized greatest concern. In most cases, these sources were electrical or used electrical energy, since electrical energy was now the main source of power for equipment. Certainly this was the case until 1970 in the U.S. (Grund 1995).

Internationally, the greatest progress occurred after the end of World War II. During the period from 1945 to 1949, the International Labour Organization (ILO) based in Geneva developed a model code for use by government (ILO 1954). Parts of this code addressed safety measures for electrical repair and maintenance and repair of machines. These concepts obviously are not new and have a long history. Subsequent documents repeat these concepts and sometimes add additional nuances.

Efforts expressed through the advocacy of safety and industry associations eventually led to more more formal attempts to capture perceptions of best practices. These led to creation consensus standards for controlling exposure to hazardous energy. Generally these attempts occurred through industrial sectors by type of energy.

## Consensus Standards

Consensus standards are the output from a group of people who attempt to achieve common ground in pursuit of problem-solving. Producers of consensus standards include groups narrowly focused in industrial sectors. Each member receives one vote. These groups are often based in the private sector, having been set up by national partnerships of industry associations. On the other hand, many of these organizations are part of the governmental structure of their countries, or are mandated by their governments.

The process used by ANSI (American National Standards Institute) and other agencies that act in the role of coordinator and publisher is to set rules to be followed by Secretariats, groups that host and operate the standard-setting committee.

ANSI requires participation that reflects a balance between interests and interest groups. These can include manufacturers who have a self-interest in the outcome, other groups that have a self-interest in the outcome, end-users, individuals, and possibly other groups. The process intends to ensure balance in representation reflected in voting members. This is potentially the best way to ensure the expression of diverse ideas, points of view, and concerns. A document that reflects a single point of view, while not necessarily flawed, is open to criticism.

Standard-setting has occurred internationally within the European Community through the Comite Europeen de Normalisation. In this process, each country receives one vote toward adoption of a proposed document. That is, adoption is a country-to-country transaction. ISO, the International Standards Organization is the coordinating body in the broader international context. Again, adoption of a standard is a country-to country transaction. Each country receives only one ballot in the vote.

There are many consensus standards concerned with managing hazardous energy. (Managing in this context means eliminating or at least controlling to tolerable levels exposure to hazardous energy contained in equipment, machines, and systems or retained due to deliberate storage or entrapment in isolated circuits. The approach taken in this appendix is to provide information about key documents, especially those that provide high-level guidance to designers and other decision-makers.

An important consideration about consensus standards produced nationally in one country by different groups and internationally between countries is the overlap in coverage. Standard-setting organizations are publishers. In effect, they compete against each other for attention of governments and entities affected by the content of these documents. Reference to a consensus standard by a regulatory authority has a huge impact on the prestige and financial health of the publisher. This situation is not evident to the casual observer of the situation and reader of these documents. What superficially might appear to be orderly and not overlapping actually contains considerable overlap. The overlap is such that different standards that address the same topic can require different outcomes that ultimately lead to different costs. Entities affected by these documents naturally experience a temptation to adopt the one that better suits the philosophical and fiscal reality.

Standard-setters approach these situations from perspectives that occupy opposite ends of the spectrum: performance-based outcomes and specification-based outcomes and perhaps a blend of both (Table A.1).

### Performance-Based Standards

Performance-based standards specify end-results or outcomes to be achieved. This approach provides the flexibility needed to drive innovation and constant improvement. Another argument in favor of performance-based standards is that they allow entities to create solutions that uniquely suit their needs. The downside of this approach is that it provides few, if any, details about the process for attaining the stated outcomes.

A necessity with this kind of approach is a clearly stated expectation of outcome. To be fair to the process, the expectation must not bias the means to achieve it. Otherwise, the performance basis of the standard is lost because there is then only one or a limited number of means for attaining the outcome.

Intrinsic to implementing this type of standard is the need for knowledgeable, skilled, and creative resource persons. Those charged with implementing programs leading to these outcomes must be able to analyze the needs of an entity and to synthesize a strategy that best suits them. This requirement can be highly intimidating to individuals having low levels of education, experience, training, or skill.

## Table A.1
## Performance versus Specification Standards

| Attribute | Performance Standard | Specification Standard |
|---|---|---|
| Focus | Achieve outcome | Achieve details on path to outcome |
| Applicability to entity | High | Hit and miss |
| Innovation | End-user | Standard-setting committee |
| Modification | Not needed unless outcome changes | Needed when details change |
| Flexibility | Flexible within envelope defined by the outcome | Specific in detail |
| Knowledge resources | High dependency | Low to mid-level dependency |
| Error potential | Depends on competence of resources involved | Depends on competence of resources involved |
| Specificity | Limited to the outcome | Outcome + details reflecting influence on creation of the document |
| Cost of implementation | Likely higher because of dependency on knowledge resources | Likely lower |
| Cost of maintaining | Likely lower unless outcome changes | Likely higher because of changes in details |

There are few resources to assist in the process. Building a hazardous energy control program on a performance-based framework requires high-level, broadly-based expertise. This type of expertise is difficult to obtain; yet, the very real energy hazards present in workplace environments have no respect for this dearth.

Another challenge when working with performance-based standards is the need for real-world strategies that fit within the larger operational context of an entity. No single standard or regulation can be allowed to monopolize the attention of health and safety resources within an organization. This forces compromise. Compromise tends to limit coverage of the standard only to environments that contain major hazards.

### Specification-Based Standards
Specification-based standards contain detailed requirements for action to achieve an outcome. This detail can extend to choice of components and parts and specifics related to installation. While these details may address the characteristics of a particular hazard or reflect the needs of a specific industry, they may be completely inapplicable to others.

The requirements outlined in a specification-based standard provide comfort zones to less knowledgeable practitioners. However, the narrowness of tolerance in this type of standard can severely inhibit innovation and certainly cannot accommodate all combinations of hazardous conditions and situations. A comprehensive specification standard that would prescribe actions for all situations would contain intricate detail, and would be cumbersome in the extreme.

By necessity, because of all of the possible combinations, approaches for addressing hazards posed by sources of hazardous energy must be performance-based, rather than specification based, even within specific industrial sectors. Specification based standards are not practical for controlling exposure to hazardous energy given the broad-spectrum of equipment, machinery, and structures that exist.

## Association for Manufacturing Technology
Founded in 1902, the Association for Manufacturing Technology promotes technological advancements and improvements in the design and manufacture of products produced by members. In recent years, the scope of AMT has broadened to include all of the elements of manufacturing. These include design, automation, material removal, material forming, assembly, inspection and testing, and communications and control.

### ANSI/AMT B11 Series
The American National Standards Institute (ANSI) has published 24 safety standards for metal working machinery. Together, they comprise the ANSI/AMT B11 series, some of which have been available since the early 1900s. The ANSI/MT B11 standards are usually either updated or reaffirmed every five years. This means that the majority of the 24 standards were created, updated or reaffirmed after the year 2000. The newly created and updated standards are important because some of the technology has appeared only recently. On the other hand, OSHA machine guarding standards date to the 1940s and have not changed since 1975, and therefore do not reflect options currently available for machine safety.

The majority of the ANSI/AMT B11 standards are machine-specific. They offer best safety practices for one category of equipment only.

### ANSI/AMT B11.19
ANSI/AMT B11.19-2003 (R2009) serves as an umbrella standard for all equipment and machines addressed in the ANSI/AMT B11 series (ANSI/AMT 2009). ANSI/AMT B11.19-2003 (R2009) provides performance requirements for the design, construction, installation, operation, and maintenance of the safeguarding when applied to machine tools. These include guards, safeguarding devices, awareness devices, safeguarding methods, and safe work procedures. This standard does not provide the requirements for the selection of safeguarding for a particular application. ANSI/AMT B11.19-2003 (R2009) is not a total stand-alone safety standard and makes numerous references to other ANSI/AMT B11 standards.

The primary objective of ANSI/AMT B11.19-2003 (R2009) is to establish the requirements for the design, construction, installation, operation, and maintenance of the safeguarding used to eliminate or control hazards associated with machine tools. Safeguarding includes guards, safeguarding devices, awareness devices, safeguarding methods, and safe work procedures.

ANSI/AMT B11.19-2003 (R2009) relies on other standards to determine the safeguarding required or allowed to control identified hazards or hazardous situations, and is used with the standard in the ANSI B11 series for a given machine tool. ANSI/AMT B11.19-2003 (R2009) establishes responsibilities for the safeguarding supplier (manufacturer, rebuilder, installer, integrator and modifier), the user, and individuals in the working environment. The overall goal is to achieve safe work practices and a safe work environment.

In addition, ANSI/AMT B11.19-2003 (R2009) includes a comprehensive and informative appendix on safe distance. The appendix contains updated Liberty Mutual anthropometric data. The data used in the OSHA Standard (29 CFR 1910.217, Table 0-10) for safe distance were developed by Liberty Mutual in the 1940s. The data used in ANSI/AMT B11.19-2003 (R2009) were published in 1995. The updated data reflect use of larger anthropometric groups and are more inclusive of women and minorities. While the data sets are similar, several important modifications to the maximum gap size/minimum distance were suggested. These modifications appear in Appendix D (Table D.1 and Figure D.10).

ANSI/AMT B11.19-2003 (R2009) indicates that the words "safe" and "safety" are not absolute. Safety begins with good design. While the goal of ANSI/AMT B11.19-2003 (R2009) is to eliminate injuries, this standard recognizes that reduction of risk factors to zero in human activity is not practicable. ANSI/AMT B11.19-2003 (R2009) is not intended to replace good judgment and personal responsibility. The user also must consider operator skill, attitude, training, job monotony, fatigue, and experience.

## ANSI/AMT B11.TR3

ANSI/AMT B11.TR3-2000 is a technical report (ANSI/AMT 2000). It is part of the ANSI B11 series of technical reports and standards pertaining to the design, construction, care, and use of machine tools. This document is not an American National Standard and the material contained herein is not normative in nature.

ANSI/AMT B11.TR3-2000 defines a method for conducting risk assessment and risk reduction for machine tools, and provides some guidance in the selection of appropriate protective measures (safeguarding) to achieve tolerable risk. It also describes the risk assessment and risk reduction responsibilities of the machine tool supplier and user.

The method in ANSI/AMT B11.TR3-2000 requires the performance of a number of tasks. These include information gathering; determining the limits of the machine; identifying tasks and hazards over the life-cycle of the machine using a task-based approach; estimating risk associated with the task-hazard combinations; reducing risk according to a prioritized procedure; and documenting the results. The risk reduction process is not complete until achievement of tolerable risk. Flowcharts illustrate the process. The document includes examples of tasks and hazards.

This technical report explicitly recognizes that zero risk is virtually unattainable. It is intended for use on all new or modified machines and equipment designs and processes. The user may also utilize it to assist with risk assessment and risk reduction for existing tasks and hazards.

### American National Standards Institute

ANSI, the American National Standards Institute, is the original source of consensus standards on the management of hazardous energy in North America and possibly world-wide. In its earlier history, ANSI acted as the publisher of American National Standards, although ANSI itself did not develop them. In its more recent history, ANSI has shifted focus to the Secretariats that develop the standards. The Secretariat administers the ANSI process to ensure consistent development of consensus standards across the spectrum of industry.

## ANSI Z241.1

ANSI Z241.1-1975 arose from concerns in the foundry industry about accidents occurring during sand handling, preparing of molds that form the external surfaces of castings and coremaking (ANSI 1975). These tasks involve use of bulk storage hoppers, conveyors, mixers, and related machinery. ANSI Z241.1-1975 provided the first mention of the concept of the Zero Mechanical State (ZMS) outside the foundry industry. Zero Mechanical State is the condition that minimizes the possibility of unexpected movement of machinery or machine components because of the presence of incoming or residual energy (Hall 1975).

Zero Mechanical State was not a new concept to the American Foundrymen's Society, the Secretariat for ANSI Z241.1. This concept had appeared in AFS publications at least as early as 1961 (AFS 1961). ANSI Z241.1-1975 further extended the concept by requiring lock-out of every power source that could produce movement of a machine or machine component during servicing or maintenance.

The intent of ANSI Z244.1-1975 was to develop stronger safety measures for complex machines containing multiple sources of energy. This standard acknowledged that, while pulling a plug or flipping a switch might interrupt an energy source, this could not ensure safety due to the hazard posed by residual energy deliberately stored or entrapped in isolated circuits.

ANSI Z241.1-1975 promoted the application of discipline to maintenance by requiring formal analysis to identify and assess energy sources, and creation of procedures to enable work to occur with assurance of safety. Formal analysis of energy sources requires review of drawings of electrical, hydraulic, and pneumatic systems. Actions needed to deactivate, de-energize, and isolate equipment are the follow-through from this analysis (Table A.2).

ANSI Z241.1-1975 also acknowledged the problem posed by subsystems. Subsystems can transform and store energy in ways that are not readily apparent. Deactivating, de-energizing, and isolating the main system may be insufficient to prevent machine movement caused by unexpected energy release from a subsystem.

ANSI Z241.1-1975 advocated primary protection as the means to achieve the Zero Mechanical State. Primary protection referred to protective means that remained active despite malfunction or intentional bypassing. Primary protection also referred to prevention of machine movement by directly controlling the power source or using mechanical stops. The main electrical disconnect in a circuit is an example of primary protection.

Secondary protection refers to protective devices or structures such as control switches and circuits in a circuit in which the main disconnect remained closed. The circuit would remain energized in the event of bypass or malfunction of the control switch. These devices control a device that applies power but do not prevent the application of power. Secondary protective devices include limit switches and operator protective devices. Malfunction may not be noticeable until the protective function provided by the device is needed.

Zero Mechanical State had a limited life as a concept. As early as 1975, advocacy started to broaden the concept to include radiant energy, including ionizing and nonionizing forms, as well as chemical energy, and to rename it as the Zero Energy State (Price 1975).

ANSI Z241.1-1975 required lockout of multiple energy sources, blocking movable machine parts, and the use of multiple locks and tags.

ANSI Z241.1 was reissued in 1989. By this time, ANSI Z244.1 had taken the lead in control of hazardous energy in equipment, machines, and systems.

## American Petroleum Institute

The American Petroleum Institute is an important developer and publisher of consensus standards. Beginning with its first standards in 1924, API now maintains some 500 standards covering all segments of the oil and gas industry. Today, API standards have gone global through active involvement with the International Organization for Standardization (ISO) and other international bodies. API is an ANSI Secretariat and an accredited standards-developing organization.

API produces consensus standards, recommended practices, specifications, codes and technical publications, reports and studies that cover each segment of the industry. API standards promote the use of safe, interchangeable equipment and operations through the use of proven, sound engineering practices.

### API 598

API 598 covers inspection, supplementary examination, and pressure test requirements for both resilient-seated and metal-to-metal seated gate, globe, plug, ball, check, and butterfly valves (API 2009). Pressure ratings found in ANSI/ASME B16.34 determine pressure in the test for steel valves presented in API 598.

The allowable rate for leakage of test fluid past the seats for the duration of the tests varies according to the type of valve. For resilient-seated valves, zero leakage is allowable. For metal-seated valves in the liquid test, allowable leakage depends on pipe diameter. The allowable leakage rates are zero drops/min for valves smaller than 2 inches (50 mm) and 12 to 28 drops/min (0.75 to 1.75 mL/min) for valves ranging from 2 to 14 inches (50 to 350 mm) in diameter. For the liquid test, 1 mL is considered equivalent to 16 drops. For the liquid test, 0 drops means no visible leakage for the minimum specified duration of the test.

For metal-seated valves in the gas test, allowable leakage depends on pipe diameter. The allowable leakage rates are zero bubbles/min for valves smaller than 2 inches (50 mm) and 24 to 56

## Table A.2
## Actions Required by ANSI Z241-1975

### Action Required

Determine position of movable machine components to be maintained during maintenance or servicing and restraints or supports required to achieve this

Create methods and sequence for locking off each power source and reducing pressure in circuits, so that machine movement cannot occur

Create methods and sequence for stabilizing workpieces and flowable solid materials contained in hoppers

Create methods to test manual overrides and circuits to verify that power sources are inactive and that pressure remaining in circuits cannot cause unexpected machine movement

Source: ANSI Z244.1-1975.

bubbles/min for valves ranging from 2 to 14 inches (50 mm to 350 mm) in diameter. For the gas test, 0 bubbles means less then 1 bubble per minimum specified duration of the test.

For check valves, the maximum permissible leakage rate shall be 0.18 cubic inch (3 mL) per minute per inch of nominal pipe size in the liquid test and 1.5 standard $ft^3$ (0.042 $m^3$) of gas per hour per inch of nominal pipe size.

### API RP 2003

API Recommended Practice 2003 presents the current state of knowledge and technology in the fields of static electricity, lightning, and stray currents applicable to the prevention of hydrocarbon ignition in the petroleum industry (API 2008). API RP 2003 reflects scientific research and practical experience.

The principles discussed in this recommended practice are applicable to other operations where ignitable liquids and gases are handled. Their use should lead to improved safety practices and evaluations of existing installations and procedures.

### American Society of Mechanical Engineers

Founded in 1880 as the American Society of Mechanical Engineers, ASME is one of the oldest and most respected standards developing organizations in the world. ASME produces approximately 600 codes and standards, covering a multitude of technical areas including boiler components, elevators, hand tools, fasteners, and machine tools. ASME is an ANSI Secretariat.

### ANSI/ASME B16.34

ANSI/ASME B16.34 applies to new construction and covers pressure-temperature ratings, dimensions, tolerances, materials, nondestructive examination, testing, and marking for cast, forged, and fabricated flanged, threaded, and welding end and wafer or flangeless valves of steel, nickel-base alloys, and other alloys (ANSI/ASME 2009). ANSI/ASME B16.34 is the base document in which pressure/temperature ratings for steel valves are located.

## American Society of Safety Engineers

Founded in 1911, ASSE is the world's oldest and largest professional safety organization. ASSE is the Secretariat for several ANSI committees. ASSE organizes the committees that develop and maintain the standard(s), ensures that the revision process is timely and in accordance with ANSI procedures, and publishes the final product of the consensus process.

## ANSI/ASSE Z244.1

In 1973, Accredited Standards Committee Z244.1 held its first organizational meeting for the purpose of developing a standard on lockout/tagout (ANSI 2001). The National Safety Council functioned as the Secretariat. By the end of 1975, development work was complete and public review and balloting were finished. After considerable delay, ANSI published ANSI Z244.1-1982 on lockout/tagout of energy sources (ANSI 1982).

ANSI Z244.1-I982 outlined minimum requirements for lockout or tagout of energy sources enabling the safe conduct of work. ANSI Z244.1-1982 provided the framework for lockout and tagout used in numerous regulations and regulatory Standards.

ANSI Z244.1-1982 functions through the actions of the knowledgeable individual. The knowledgeable individual is someone who knows the effect of operating controls or equipment. The knowledgeable individual performs a survey to identify energy sources and related exposures to hazardous energy, and the ability to isolate equipment, machines, and systems. The survey includes physically locating and assessing the adequacy of energy-isolating devices. It can also include assessment through review of drawings, prints, and equipment manuals. This information provides the basis for creating procedures for deactivation, de-energization, isolation, lockout and verification of isolation, and clearance, release from lockout/tagout, and start-up.

The knowledgeable person assigns duties and responsibilities related to the lockout procedure to authorized individuals. Authorized individuals are deemed knowledgeable to perform specific assignments related to lockout and tagout and given responsibility to do so by the employer. Individuals assuming this responsibility require training in recognition of applicable hazardous energy sources and adequate methods and means for their isolation and start-up.

Other individuals whose work could interface with the lockout/tagout are called affected employees. Affected employees require training in the purpose and use of the lockout/ tagout procedure.

ANSI Z244.1-1982 operates through the deactivation and de-energization of the equipment or machine, followed by the application of a lockout and/or tagout device to an energy isolating device. The lockout/tagout forms at minimum a psychological barrier and at maximum an inconvenience requiring some effort to reactivation of the equipment or machine.

ANSI Z244.1-1982 specifies requirements for substantiveness, durability, and identifiability of locks, lockout devices, and tags. The standard also contains requirements for policies covering training, communication, and planning and implementing procedures for hazard assessment and control and work activity. ANSI Z244.1-1982 also includes considerations about the interface with contractors and outside organizations, interruption of the lockout/tagout for testing and adjustment, shift overlap, group lockout/tagout, and authorization for application and removal of locks and tags.

In 1987, 1992, and 1997, the ANSI Z244.1 Committee reaffirmed the standard without changes in content (ANSI 1987, ANSI 1992, ANSI/ASSE 2004).

During 1997, the ANSI 244.1 Committee voted to revise the existing standard after over 20 years without change (ANSI/ASSE 2004). These revisions are reflected through the modified title of the standard to acknowledge the broader universe of exposure to hazardous energy during operation of equipment and machines and development of control strategies. This change in direction also acknowledged the fact that lockout/tagout is but one facet in a much broader spectrum of strategies regarding safe conduct of work involving equipment and machines, energy sources, and bulk materials, all of which contain, transform, and store energy.

ANSI/ASSE Z244.1-2003 attempts to address the fact that the annual toll of injury and death in incidents related to release of hazardous energy remains unacceptable in spite of substantial efforts by employers, employees, unions, trade associations, and government during the past 25 years (ANSI/ASSE 2004). Review of accidents during preparation of the updated standard indicated that operational personnel are injured as often as maintenance workers and that often thermal, gravitational forces, and trapped materials under pressure are overlooked. These observations led to the conclusion that the lockout paradigm must explicitly address all forms of energy. This review also indicated that complex equipment and processes frequently demand unique approaches to energy isolation or control and that employers need to commit resources and substantial effort to planning, training, procedure, and infrastructure development before attempting to perform lockout or tagout.

ANSI/ASSE Z244.1-2003 addresses the need for greater flexibility through the use of alternative methods of control. Alternative methods, when used, are based upon risk assessment and application of the classic hazard control hierarchy. However, lockout/tagout continues to be emphasized as the primary approach for controlling hazardous energy. Alternate methods must provide protection equivalent to lockout and tagout. In addition, the standard emphasizes the responsibility of management for protection of personnel against the release of hazardous energy.

While total isolation is the primary method of energy control advocated in ANSI/ASSE Z244.1-2003, there are situations where this approach is cumbersome and potentially counterproductive (ANSI/ASSE 2004). Tasks that are routine, repetitive, and integral to production that exhibit most of the following characteristics are candidates for consideration. They are short in duration, relatively minor in nature, and occur frequently during the shift, day, or week. The are performed by operators or others functioning as operators, and do not involve extensive disassembly. They reflect predetermined cyclical activities, occur regularly, and minimally interrupt the production process. They occur even when optimal operating levels are achieved and require task-specific personnel training.

Operator intervention occurs to sustain the machine, equipment or process continuity within the nominal performance range and output quality. This usually occurs when the machine, equipment, or process is operating normally and the need for periodic service is predictable based on operating experience and product demands. Also, the tasks do not require that the machine, equipment, or process be taken out of the operational mode to accomplish them.

ANSI/ASSE Z244.1-2003 requires the organization to identify and examine all tasks deemed to be routine, repetitive, and integral to production, and to determine whether they possess the defining characteristics before proceeding in the development of alternative methods based on risk assessment. If the tasks do not substantially conform, then lockout or tagout is required.

ANSI/ASSE Z244.1-2003 puts the burden of assessing control of hazardous energy onto manufacturers, integrators of components, modifiers of components, and remanufacturers. This concept is similar to that expressed in the ANSI/AMT B11 series of standards discussed in a previous section. This will transfer the burden of investigation from the operator and maintainer of the equipment or machinery to the designer and manufacturer or remanufacturer integrator . Further to this, the standard calls for convenient positioning and clear labeling of energy isolating devices, as well as manuals that describe procedures for energy isolation. Where residual energy is to be present due to deliberate storage or entrapment in isolated circuits, ANSI/ASSE Z244.1-2003 requires incorporation of energy-isolating devices into the equipment or machine or reliable systems to prevent energy release.

When lockout or tagout is not used for tasks that are routine, repetitive, and integral to the production process, or traditional lockout/tagout prohibits completion of those tasks, the standard calls for use of an alternative method of control. Control shall occur in the following descending order of preference: elimination of the hazard through design; engineered safeguards; warning and alerting techniques; administrative controls including, safe work procedures, practices, and training; and personal protective equipment The objective of this process is to select the highest level of feasible control(s). In many cases, a combination of the methodologies is needed to provide individuals with protection equivalent or superior to lockout or tagout.

## ANSI/ASSE Z117.1

ANSI/ASSE Z117.1-2009) covers entry and work involving confined spaces (ANSI/ASSE 2009). A Qualified Person, knowledgeable in the operation to be performed and competent by virtue of training, education and experience to judge the hazards involved and specify controls and/or protective measures is the key to the process. The Qualified Person identifies confined spaces present in the operation, equipment, and premises and hazards associated with each confined space.

According to ANSI/ASSE 117.1-2009, hazards could include physical agents, mechanical hazards, and unexpected entry of liquids, gases, or solids during occupancy, as well as atmospheric contamination by toxic or flammable atmospheres or oxygen deficiency or enrichment, biological hazards, and isolation of occupants from external rescuers. In general, all energy sources which are potentially hazardous to entrants are to be secured, relieved, disconnected, and/or restrained prior to entry. ANSI/ASSE Z117.1 defers to ANSI/ASSE Z244.1 for lockout and tagout. ANSI Z117.1 prescribes that same measures for isolation of permit-required confined spaces as 29CFR 1910.146, the OSHA Standard (OSHA 1994).

The Qualified Person performs an assessment of the confined space as part of the process. Upon completion of the hazard assessment, the Qualified Person decides about the magnitude of actual and potential hazards. The Standard provides two choices, non-permit confined space and permit-required confined space, to determine the extent of the response.

For non-permit confined spaces the Standard requires development of written procedures that specify measures and precautions to ensure safe entry and work. These could include some of the measures needed for entry into a permit-required confined space. These procedures also identify conditions that would revoke the status as a non-permit confined space and require re-evaluation.

Persons working in non-permit confined spaces need training in entry procedures and conditions that prohibit entry. A condition for entry into a non-permit confined space would include successful control of atmospheric hazards within acceptable limits, as indicated by testing performed by the Qualified Person. The Qualified Person could waive testing if the space is ventilated properly before and during entry, provided that this means of control is shown to be satisfactory through formal hazard identification and evaluation.

A permit-required confined space contains actual or potential serious hazards. For permit-required confined spaces ANSI/ASSE Z117.1-2009 requires issuance of a permit prior to all entries. The intent of the permit is to ensure systematic review of hazards prior to each entry, communication of this information to potential occupants, and acceptance by them of requirements to be followed. The Qualified Person would complete the permit. Changes in conditions in the confined space or the nature of the work would terminate the permit.

## American Water Works Association

Founded in 1881, AWWA publishes a series of standards pertaining to valves and gates (Table A.3). Gates are plates that slide within a frame that mounts on the side of a structure. At the point of mounting, the structure has an opening in the wall to allow for flow of liquid when the gate is opened. These standards provide standards of performance for acceptable leakage.

AWWA is an ANSI-accredited standards developing organization for the water industry. AWWA regularly issues new standards and annually revises 20 to 25 standards.

## Canadian Standards Association

The Canadian Standards Association (CSA) was chartered in 1919 and accredited by the Standards Council of Canada to the National Standards system in 1973. CSA standards reflect a national consensus of producers and users, including manufacturers, consumers, retailers, unions and professional organizations, and governmental agencies. The standards are used widely by industry and commerce and often adopted by municipal, provincial, and federal governments in their regulations, particularly in the fields of health, safety, building and construction, and the environment.

A standard approved as a National Standard of Canada conforms to the criteria and procedures established by the Standards Council of Canada.

## CAN/CSA Z460

CAN/CSA Z460-05 is the first edition of CAN/CSA Z460 (CAN/CSA 2005). CAN/CSA Z460-05 specifies requirements for and provides guidance in a number of areas, including responsibilities of the principal parties involved in control of hazardous energy and design issues that influence the effective application of control methodology. CAN/CSA Z460-05 incorporates parts of ANSI/ASSE

## Table A.3

## Leakage Criteria and ANSI/AWWA Standards

| Leakage Criterion | Standard |
|---|---|
| Not to exceed 1 ounce/hour/inch (1.2 mL/hour/mm) of nominal valve size at rated working pressure differential | ANSI/AWWA C500-09 Metal-Seated Gate Valves for Water Supply Service |
| | ANSI/AWWA C507-05 Ball Valves, 6 Inch through 48 Inch (150 mm through 1200 mm) |
| | ANSI/AWWA C508-01 Swing-Check Valves for Waterworks Service; 2 inch Through 24 inch (50 mm through 600 mm) |
| | ANSI/AWWA C510-07 Double Check Valve Backflow Prevention Assembly |
| | ANSI/AWWA C511-07 Reduced Pressure Principle Backflow Prevention Assembly |
| Drip tight (zero leakage) after 500 operational cycles at the rated pressure differential | ANSI/AWWA C509-09 Resilient-Seated Gate Valves for Water Supply Service |
| | ANSI/AWWA C504-06 Rubber-Seated Butterfly Valves |
| 0.1 gal (US)/min/foot (1.24 L/min/m) on seating perimeter at the design head | ANSI/AWWA C513-05 Open-Channel, Fabricated Metal Slide Gates and Open-Channel, Fabricated Metal Weir Gates |
| | ANSI/AWWA C560-07 Cast Iron Slide Gates |
| | ANSI/AWWA C561-04 Fabricated Stainless Steel Slide Gates |
| | ANSI/AWWA C562-04 Fabricated Aluminum Slide Gates |
| | ANSI/AWWA C563-04 Fabricated Composite Slide Gates |
| Drip tight (zero leakage) at 250 lb/in$^2$ cold water working pressure and have a 500 lb/in$^2$ hydrostatic test for structural soundness for 4 inch through 12 inch (100 mm though 300 mm) | ANSI/AWWA C515-09 Reduced-Wall, Resilient-Seated Gate Valves for Water Supply Service |
| | ANSI/AWWA C517-05 Resilient-Seated, Cast Iron Eccentric Plug Valves |
| | ANSI/AWWA C518-08 Dual-Disc Swing-Check Valves for Waterworks Service |

Z244.1-2003, and ISO 14121:2004. (ISO 12100:2010 has superseded ISO 14121.) Hence, this standard is a hybrid of both.

CAN/CSA Z460-05 provides flexibility in methodology for controlling hazardous energy. However, lockout is emphasized as the primary approach to control of hazardous energy. Other methods are based on risk assessment and application of the classic hazard control hierarchy.

If the risk is deemed too high, risk reduction measures are required. Risk reduction is a hierarchical process that occurs through elimination through design, use of engineered safeguards, awareness means (including warning and alerting techniques), administrative controls (including safe work procedures and training), and use of personal protective equipment. Often, for a particular machine, piece of equipment, or process, the solution can include aspects of each of these elements.

Redesign to eliminate the hazard is the most desirable of the options in the risk reduction hierarchy. The risk is no longer present. With other options, the hazard remains but is subject to control of decreasing desirability and reliability.

Engineered safeguards or safety devices protect personnel from hazards that not reasonably eliminated or sufficiently reduced by design. Engineered safeguards include guards (both fixed and interlocked), trapped key devices, and trip devices (light curtains, laser scanners, pressure mats, and safety-rated switches. Safety devices include emergency stop buttons and enabling or hold-to-run devices. Most of these devices operate through electrical control circuits.

CAN/CSA Z460-05 requires safeguards or safety devices and the safety control system (electrical, pneumatic, hydraulic, steam, fuel gases, pressurized and gravity-flow liquids, flowable solids) to be suitably reliable for the required risk reduction. Suitable reliability imposes risk-based requirements for performance on components and the design of the safety system.

CAN/CSA Z460-05 requires use of warning and alerting techniques to protect from hazards not reasonably eliminated or sufficiently reduced by design, engineered safeguards, or both. Examples of warning and alerting techniques include attendants, audible and visual signals, barricades, signs, and tags.

Administrative controls, including safe work procedures, standard practices and checklists, and training can provide additional risk reduction after pursuing other options in the hierarchy. Training is an important adjunct to ensure correct and consistent implementation of all risk-reduction methods. Examples of safe work procedures, practices, and training include standard operating instructions, illumination, pre-job review, and establishing safe distances from a hazard.

Effective and consistent use of personal protective equipment sometimes is the only means available for achieving necessary additional risk reduction. Effective use of personal protective devices requires strong administrative procedures. Examples of personal protective equipment include safety eyewear and faceshields, protective footwear, protective gloves (insulating or cut-resistant), hearing protection, and protective headgear.

Following implementation of risk reduction measures the key questions that arise include the appropriateness of safety measures and compatibility with each other, and creation of new, unanticipated hazards or problems.

Appendix C provides examples of the application of the concepts developed in Appendix B. These examples are very important for illustrating the concepts mentioned here. Without having the opportunity to see and reflect on these concepts, they remain abstract and intimidating for the practitioner.

## European Standards
CEN, the Comite Europeen de Normalisation, creates and publishes almost all of the standards, European Norms or ENs, in Europe. CEN currently has 31 National Members and an outreach to over 480 million people. Hence, ENs truly are international documents.

## EN 954-1
EN 954-1:1996 was the internationally applicable standard in the area of safety-related control systems for machines, regardless of power type (electric, hydraulic, pneumatic or mechanical) (EN 1996).

EN 954-1:1996 established a procedure for the selection and design of safety systems for machinery. The procedure operates through five-steps: hazard analysis and risk assessment; establishing approved methods to reduce risk; detailing specific safety requirements of the control system; specifying the overall design and human interface needs; and providing methods to validate the system for safe operation. It also identified several safety-related functions and parameters such as stop, emergency stop, manual reset, start and restart, response time, local control, and fluctuation, loss and restoration of power.

EN 954-1:1996 defines categories for classifying different safety-related capabilities (Table A.4). The assignment of category is somewhat arbitrary. The phrases "well-tried," "reasonably practicable," and "suitable intervals" provide examples of the difficulty of expressing performance capabilities in words. In this context, "well-tried" has the meaning of widely used in the past in similar applications, or made and verified using principles that establish suitability and reliability in safety-related applications. Hence. this aspect of the rating system is experience-based. The decision ultimately depends on the application. There are no quantitative, probabilistic measures to quantify assignment to a particular class.

EN 954-1:1996 provides a means of describing fault tolerance of circuits. It does not provide a way of relating the extent of risk associated with the fault tolerance. One of the major drawbacks of EN 954-1:1996 is that safety-related control systems are restricted to the use of relay logic. EN 954-1:1996 also contains no special requirements for programmable electronic control systems. Complex logic is more suited to a configurable/programmable controller than by using relays. This approach also considerably reduces the potential for wiring errors. Furthermore, improved diagnostics can cut downtime and enable non-experts to troubleshoot machine stoppages.

EN 954-1:1996 does not address the handling of common-cause failures or the adequate frequency of diagnostic tests on category 2 systems.

## EN ISO 13849-1

EN ISO 13849-1:2006 is the standard designated to replace EN 954-1:1996 (EN ISO 2006). This occurred at the beginning of 2012 following several years of phase-in. EN ISO 13849-1:2006, received support internationally from all countries except the U.S., U.K., and Japan. This means that EN ISO 13849-1:2006 will influence the thinking about machine safety globally for years to come.

EN ISO 13849-1:2006 combines the complex probabilistic approach of IEC 61508 (following section) with the generally recognized concepts of the EN 954-1:1996. EN ISO 13849-1:2006 corrects the deficiencies of EN 954-1:1996 mentioned in previous discussion. As well, EN ISO 13849-1:2006 has striven to balance deterministic and probabilistic thinking, breaking down the new aspects into a requisite and practicable size for the average user.

This is achieved by enhancing reliability categories (Table A.4) to further contribute to risk reduction. In addition, there are new concepts, including, the reliability of parts, diagnostic coverage, and requirements for safety software. Requirements for safety software are critically important, because computer systems have taken over the tasks of safety functions.

In general terms, EN (ISO) 13949-1:2006 takes a four-step approach to the design of safety-related control systems. The first step is to perform a risk assessment. For the identified risks, the analyst allocates the Required Performance Level or $PL_r$.

Performance Level combines and renumbers and renames the Control Reliability Categories of EN 954-1:1996 and incorporates additional required information. Performance Levels range between a and e. Performance Level "a" corresponds to Category B in the ranking used in EN 954-1:1996. Similarly PL "e" corresponds to Category 4. Additional required information includes $MTTF_d$. DC and CCF.

$MTTF_d$ is Mean Time To Failure (dangerous). $MTTF_d$ is a measure of reliability of components. In this context, $MTTF_d$ is the average period before failure of components in the safety circuit causes harm to a worker. Ranking levels include Low, 3 years $MTTF_d$ <10 years; Medium, 10 years $\leq$ $MTTF_d$ < 30 years; High, 30 years $\leq MTTF_d$ < 100 years

DC is Diagnostic Coverage, a measure of fault detection in a safety-related system. Diagnostic Coverage indicates the ability of a component or circuit to detect and/or diagnose a fault within itself (a

## Table A.4

## Link between EN 954-1:1996 and EN ISO 13949-1: 2006

| | | EN 954-1:1996 | EN ISO 13949-1: 2006 Addition | | |
|---|---|---|---|---|---|
| Cat. | PL | Requirements | $MTTF_d$ | DC | CCF |
| B | a | Safety-related parts of control systems and/or safety devices and components designed, constructed, selected, assembled, and combined in accordance with relevant standards to withstand the expected influence. | low to medium | none | not relevant |
| 1 | b | Requirements of B + well-tried components and safety principles | high | none | not relevant |
| 2 | c | Requirements of B + well-tried safety principles; machine control system checks the safety function at suitable intervals. | low to high | low to medium | to be considered |
| 3 | d | Requirements of B + well-tried safety principles. Design of safety-related parts occurs such that: 1. A single fault in any part does not lead to loss of the safety function, and 2. Detection of the single fault occurs whenever reasonably practicable. | low to high | low to medium | to be considered |
| 4 | e | Requirements of B + well-tried safety principles. Design of safety-related parts occurs such that: 1. A single fault in any part does not lead to loss of the safety function, and 2. Detection of the single fault occurs during or prior to the next demand on the safety function, or, if not possible, an accumulation of faults should not as a result lead to the loss of the safety function. | high | high | to be considered |

Cat. is Category of Control Reliability. PL is Performance Level. $MTTF_d$ is Mean Time To Failure (dangerous). DC (Diagnostic Coverage) is a measure of the fault detection in a safety-related system. CCF is Common Cause Failure. Failures with a common cause naturally only occur in redundant systems.

short circuit, for example). The higher the Diagnostic Coverage, the lower is the probability of hazardous hardware failures.

CCF is Common Cause Failure. Failures with a common cause naturally only occur in redundant systems. CCF is the failure of different items, resulting from a single event, where these failures are not consequences of each other.

The next step is to devise a system architecture that is suitable for the Performance Level required and subsequently to validate the design to ensure that it meets the requirements of the initial risk assessment. The last step involves using data provided by manufacturers on the reliability of components and their configuration in the architecture. Charts in the Appendices provide data for producing the required parameters for cross-checking against the original assessment.

## Fluid Controls Institute

FCI was formed in 1921 when 17 manufacturers of steam specialty products organized an association known as the Steam Specialty Association. At this time, steam engines were prominent, and the manufacturers joined together to represent the interests of the industry.

### ANSI/FCI 70-2

The Fluid Controls Institute created one of the first standards on leakage, ANSI/FCI 70-2 (ANSI/FCI 2006). ANSI/FCI 70-2-2006 is an antecedent to standards created by other groups, such as the American Water Works Association (AWWA) and the American Petroleum Institute (API).

ANSI/FCI 70-2-2006 establishes six classes of leakage for seats of control valves. Classes of higher rank allow progressively less leakage. Also defined are specific test procedures to determine the appropriate class.

Class I is an open classification not requiring a test. Class I allows for specific agreement between the manufacturer and the customer regarding required leakage. Class I valves are identical in construction and design intent to Class II, III, and IV valves. Class I is also known as dust tight and can refer to metal or resilient seated valves.

Class II valves have a metal piston ring seal and metal-to-metal seats. Allowable leakage is 0.5% of full open capacity tested at maximum operating differential or pressure drop of 50 lb/in$^2$ (344 kPa), whichever is lower at 50° to 125° F (10° to 52° C) for water, or at maximum operating differential or 45 to 60 lb/in$^2$ (310 to 414 kPa) gauge pressure differential, whichever is lower for air at the same temperatures.

Acceptable rate of leakage for Class III valves is 0.1% and 0.01% for Class IV valves under identical test conditions.

For Class V valves, the acceptable rate of leakage is 0.0005 mL/min of water/inch of port diameter/lb/in$^2$ of pressure differential at the maximum service pressure differential and not to exceed the ANSI rating (ANSI/ASME B16.34) for the valve body and 50° to 125° F (10° to 52° C).

The acceptable rate of leakage for Class VI valves is 0.0005 mL/min of air or nitrogen/inch of port diameter/lb/in$^2$ of pressure differential at 50 lb/in$^2$ or maximum rated differential pressure across the valve plug, whichever is lower and 50° to 125° F (10° to 52° C).

### ANSI/FCI 70-3

ANSI/FCI 70-3-2004 is similar to ANSI/FCI 70-2-2004 (ANSI/FCI 2004). ANSI/FCI 70-3-2004 applies to pilot-operated and direct-acting, pressure-reducing, pressure-relieving (back pressure), differential pressure, and temperature regulators.

## International Electrotechnical Commission

The International Electrotechnical Commission (IEC) has published International Standards for all electrical, electronic, and related technologies for over 100 years, since its founding in 1906.

### IEC 61508

IEC 61508 is a series of seven documents (IEC 2010). IEC 61508-1:2010, the overview document, covers aspects to be considered when Electrical/Electronic/Programmable Electronic (E/E/PE) systems perform safety functions.

IEC 61508-1:2010 provides a generic approach for all safety life cycle activities for systems comprised of Electrical and/or Electronic and/or Programmable Electronic (E/E/PE) elements that perform safety functions. This unified approach provides a rational and consistent technical policy be developed for all electrically based, safety-related systems.

In most situations, safety is achieved by a number of systems which rely on many technologies (for example, mechanical, hydraulic, pneumatic, electrical, electronic, and programmable electronic). Any safety strategy must, therefore, consider not only all the elements within an individual system (for example, sensors, controlling devices, and actuators), but also all the safety-related subsystems comprising the safety-related system. Therefore, while this International Standard is concerned with E/E/PE safety-related systems, it also provides a framework for consideration of safety-related systems based on other technologies.

Use of systems comprised of electrical and/or electronic elements to perform safety functions has occurred for many years. For computer system technology to be effectively and safely exploited, those responsible for making decisions require guidance on safety to make these decisions. Hence, a major objective of this standard is to facilitate development of derivative standards that are applicable to different sectors.

A second objective of IEC 61508-1:2010 is to enable the development of E/E/PE safety-related systems where standards do not exist.

IEC 61508-1:2010 uses the concept of Safety Integrity Level (SIL) to indicate reliability of Safety Instrumented Systems when a process demand occurs. SIL is a statistical concept that correlates to the probability of failure of demand (PFD). PFD is equivalent to the unavailability of a system at the time of a process demand. SILs range from SIL 1 to SIL 4. The higher the SIL, the more reliable or effective is the SIS. A typical safety instrumented system includes devices from field sensors through logic solver to final elements (solenoids, valves, pumps, and compressors) to ensure a failsafe condition for the process. Such systems typically include interlocks, emergency-shutdown systems, and other safety-critical systems (Summers 2000).

Safety Integrity Levels directly correspond to Mean Time to Failure (dangerous) values incorporated into Performance Levels in EN ISO 13849-1:2006 (EN ISO 2006). PL "c" corresponds to SIL 1, PL "d" to SIL 2, and PL "e" to SIL 3, respectively.

The safety-instrumented system operates independently from process control. The safety-instrumented system operates beyond the passive mode when something goes wrong (Ondrey 2005). Most process control loops (about 60% to 70%) have no special safety requirements. The remaining 30% to 40% operate at a safety integrity level (SIL) of SIL1, SIL2, or SIL3. About 30% of the process control loops affected by safety requirements are SIL1 and the remainder SIL2 and SIL3 in a ratio of 2:1. There is considerable incentive to reduce the level of concern from SIL3 to SIL2, as the highest level of concern affects the overall rating required in the design.

Meeting requirements of IEC 61508-1:2010 and the derivative standards can involve design decisions concerning failure rate, redundancy of sensors and final elements, and "voting" to as the basis for making decisions regarding failsafe actions, functional testing and testing frequency, and other common factors, such as human error.

## IEC 61511

IEC 61511, Functional Safety Instrumented Systems for the Process Industry Sector is the implementation of the IEC 61508 series of standards for the process industries (IEC 2004). The basis for IEC 61511 is ANSI/ISA S84.01 (ANSI/ISA 1996) IEC 61511-1:2004 provides the overview to the series of documents that will form the standard. ANSI/ISA S84.01-1996 provides background about the eventual direction of IEC 61511.

ANSI/ISA S84.01-1996 specifies requirements for the assessment, design, installation, operation, and maintenance of safety-instrumented systems designed to prevent or mitigate potentially unsafe conditions (ANSI/ISA 1996). A typical safety instrumented system includes devices ranging from field sensors through logic solver to final elements (solenoids, valves, pumps, and compressors) needed to ensure a failsafe condition for a process. Such systems typically include interlocks, emergency-shutdown systems, and other safety-critical systems (Summers 2000).

ANSI/ISA S94.01-1996 requires documentation clearly stating what is required to minimize the risk, to demonstrate that the safety-instrumented system was designed to meet these requirements, and to show that the operation and maintenance practices will maintain the safety-instrumented system.

In March 2000, the Instrumentation, Systems, and Automation Society received confirmation from OSHA (Occupational Safety and Health Administration) that ANSI/ISA S84.01-1996 provides guidance for generally accepted good engineering practice for establishing safety-instrumented systems under Process Safety Management. Process Safety Management refers to process safety regarding highly hazardous chemicals, explosives, and blasting agents (OSHA 1992b).

ANSI S84.01-1996 establishes a cradle-to-grave process for managing safety-instrumented systems (Summers 2000). Complying with the standard requires three steps: decide the extent of risk reduction required, design systems to meet the requirements, and operate, test, and maintain the systems to ensure long-term risk reduction.

In general, risk reduction described in ANSI/ISA S84.01-1996 incorporates layers of protection. These normally operate independently from each other in order to ensure protection in the event of failure of one or more layers. These layers start with basic design of the plant and move outward through basic controls, critical alarms and manual intervention, physical protection (relief devices), physical protection (containment), plant emergency response and community emergency response.

Meeting these requirements can involve design decisions concerning failure rate, redundancy of sensors and final elements, and "voting" to as the basis for making decisions regarding failsafe actions, functional testing and testing frequency and other common factors, such as human error.

A Safety Instrumented System (SIS) is composed of a separate and independent combination of sensors, logic solvers, final elements, and support systems that are designed and managed to achieve a specified Safety Integrity Level (SIL). An SIS may implement one or more safety instrumented functions, which are designed and implemented to address a specific process hazard or hazardous event. The SIS management system should define how an owner/operator intends to assess, design, engineer, verify, install, commission, validate, operate, maintain, and continuously improve the system. This entails defining essential roles of the various personnel assigned responsibility for the SIS, and developing procedures, as necessary, to support the consistent execution of their responsibilities.

The safety-instrumented system operates independently from process control and operates beyond the passive mode when something goes wrong (Ondrey 2005). Most process control loops (about 60% to 70%) have no special safety requirements. The remaining 30% to 40% operate at a safety integrity level (SIL) of SIL1, SIL2, or SIL3. SIL1 is the lowest level of concern and SIL4, the highest. About 30% of the process control loops affected by safety requirements are SIL1 and the remainder SIL2 and SIL3 in a ratio of 2:1. There is considerable incentive to reduce the level of concern from SIL3 to SIL2, as the highest level of concern affects the overall rating required in the design.

## IEC/EN 62061:2005

IEC/EN 62061:2005 is the machine-specific implementation of the IEC 61508 series of standards (IEC/EN 2005). In contrast to EN ISO 13849-1:2006, IEC/EN 62061:2005 does not define any requirements for performance of non-electrical safety-related control elements for machines. Non-electrical elements addressed in EN ISO 13849-1:2006 include hydraulic, pneumatic, gravitational, and other sources of hazardous energy (EN ISO 2006).

IEC/EN 62061:2005 uses a six-stage process. The first steps are to identify the danger zones on the machine and to define risk parameters, Se, Fr, Pr, and Av, and then to determine the required Safety Integrity Level. Se is Severity of harm. Fr is Frequency and duration of exposure to the hazard. Pr is the Probability of occurrence of a hazardous event. Av is possibility to Avoid or limit the harm. IEC/EN 62061:2005 takes a quantitative approach. Hence, to an extent, all of the parameters are quantifiable.

For example, the Severity of harm (Se) carries 4 points for an irreversible injury (death, loss of eye or arm), down to 1 point for a reversible injury that requires on-site first aid. Similarly, points are scored for the other risk parameters, with the probability of occurrence of harm (Cl) being the sum of the points scored for $Fr + Pr + Av$. The standard contains a look-up table that shows the SIL required for a given combination of Se and Cl.

The remaining steps are to design and implement the necessary safety functions, to determine the the achieved SIL, and to compare this value with the required SIL.

## ISO Standards
ISO (International Organization for Standardization) is a network of the national standards institutes of 163 countries.

## ISO 12100
ISO 12100:2010 specifies basic terminology, principles, and a methodology for achieving safety in the design of machinery (ISO 2010). ISO 12100:2010 specifies principles of risk assessment and risk reduction to assist designers in achieving this objective. These principles are based on knowledge and experience of design, use, incidents, accidents, and risks associated with machinery. ISO 12100:2010 describes procedures for identifying hazards and estimating and evaluating risks during relevant phases of the machine life cycle, and for the elimination of hazards or sufficient risk reduction.

ISO 12100:2010 will enable designers to better identify risks during design of equipment and machines and hence to reduce the risk of future accidents during operation, servicing and maintenance. The risk assessment guidelines are presented as a series of logical steps to identify risks from crushing, cutting, electric shock or fatigue; and estimate potential dangers ranging from machine failure to human error.

Information obtained through this process will enable designers to determine whether a machine is sufficiently safe prior to bringing it to market. In the event that a machine is found to be unsafe, this information will provide the basis for subsequent redesign to reduce risk. Iterative repetition of the process would occur until achievement of risk at a level acceptably safe for use.

ISO 12100:2010 replaces ISO 14121-1:2007.

## ISO 14121
ISO 14121-1:2007 establishes general principles intended to be used to meet the risk reduction objectives established in ISO 12100-1:2003 (ISO 2007). These principles of risk assessment bring together knowledge and experience of the design, use, incidents, accidents, and harm related to machinery in order to assess the risks posed during the relevant phases of the life cycle of a machine.

ISO 14121-1:2007 provides guidance on the information that will be required to enable risk assessment to be carried out. Procedures are described for identifying hazards and estimating and evaluating risk. It also gives guidance on the making of decisions relating to the safety of machinery and on the type of documentation required to verify the soundness of the risk assessment carried out.

## Manufacturers Standardization Society
Officially founded in 1924, the Manufacturers Standardization Society (MSS) of the Valve and Fittings Industry develops and improves industry, national, and international codes and standards for valves, piping, and associated components. MSS valve standards have wide application in virtually every industry.

## ANSI/MSS SP-61
MSS SP-61was originally adopted in 1961 (ANSI/MSS 2009). MSS SP-61 was developed for the purpose of providing a uniform means of testing valves commonly used in fully open and fully closed service. MSS SP-61 is not intended for use with control valves. ANSI/MSS SP-61-2009 establishes requirements and acceptance criteria for shell and seat closure pressure testing of valves. This standard forms the basis for derivative standards created by organizations including the American Water Works Association and American Petroleum Institute.

The allowable leakage rates for valves tested to ANSI/MSS SP-61, 2009 are as follows: 10 mL/hr/inch of nominal pipe diameter for gate, globe, and ball valves; 40 mL/hr/inch of nominal pipe diameter, check valves.

The test for seat closure test must occur at a fluid (liquid or gas) pressure no less than 1.1 times the $1000°F$ ($380°C$) rating rounded to the next 5 lb/in$^2$ (30 kPa).

**National Fire Protection Association**
The National Fire Protection Association (NFPA) is an international nonprofit organization that was established in 1896. NFPA oversees the creation, updating, and publishing of 300 codes and standards.

## NFPA 70
NFPA 70 is the U.S. National Electrical Code (NFPA 2011a). NFPA 70 covers the installation of electrical conductors, equipment, and raceways; signaling and communications conductors, equipment, and raceways; and optical fiber cables and raceways in public and private premises, including buildings, structures, mobile homes, recreational vehicles, and floating buildings, and surroundings.

The National Electrical Code is a design, installation, and inspection code. NFPA 70 affects the outcome, but does not directly influence the manner in which electrical work is performed by employers and employees. NFPA 70 specifies requirements for performance of equipment and means of installation contingent on the hazards of the environment in which it is to be used. NFPA 70 is prescriptive. It describes what practitioners are to do to accomplish performance requirements. The prescriptive requirements in NFPA 70 result from a consensus process involving the collective wisdom of international experience (Paschal 1998).

## NFPA 70E
NFPA 70E is a maintenance and servicing standard (NFPA 2009a). NFPA 70E addresses electrical safety requirements for the practical safeguarding of employees during activities such as the installation, operation, maintenance, and demolition of electric conductors, electric equipment, signaling and communications conductors and equipment, and raceways. NFPA 70E applies to work performed within the envelope of NFPA 70.

NFPA 70E guides employers and employees in the manner in which electrical work is to be performed. One of the purposes for creation of the standard was to assist OSHA in setting safety requirements for electrical work. NFPA 70E-2009 contains four parts:1, safety-related work practices; 2, safety-related maintenance requirements; 3, safety requirements for special equipment; and 4, installation safety requirements. Part 1 on safety-related work practices, in part, covers lockout and tagout.

## NFPA 77
NFPA 77 is a recommended practice that covers static electricity (NFPA 2007). NFPA 77 applies to the identification, assessment, and control of static electricity for purposes of preventing fires and explosions in general applications.

Static electricity is important from the perspective of triggering fires and explosions of ignitable mixtures of gases and vapors and possibly aerosols. NFPA 77 discusses the origins and effects of static electricity and means for control.

## NFPA 79
NFPA 79 is the section of the National Electric Code (NEC) that focuses on the electrical wiring standards used with industrial machinery (NFPA 2011b). NFPA-79 applies to the electrical equipment used within a wide variety of machines, as well as groups of machines working together in a coordinated manner. Examples of industrial machinery include, among others machine tools, injection molding machines, woodworking equipment, assembling machinery, material handling machinery, and inspection and testing machines. NFPA 79 applies to all electrical and electronic elements of the machinery operating at 600 V or less commencing at the point of connection of the supply to the electrical equipment of the machine.

## NFPA 110
NFPA 110 covers performance requirements for emergency and standby power systems (ESPSs) that provide an alternate source of electrical power to loads in buildings and facilities in the event that the primary power source fails (NFPA 2010a). Power systems covered in this standard include power sources, transfer equipment, controls, supervisory equipment, and all related electrical and mechanical auxiliary and accessory equipment needed to supply electrical power to the load terminals of the

transfer equipment. NFPA 110 covers installation, maintenance, operation, and testing requirements as they pertain to the performance of the emergency power supply system.

## NFPA 111

NFPA 111 covers performance requirements for stored electrical energy systems that provide an alternate source of electrical power in buildings and facilities in the event that the normal electrical power source fails (NFPA 2010b). Systems covered in this standard include power sources, transfer equipment, controls, supervisory equipment, and accessory equipment, including integral accessory equipment needed to supply electrical power to the selected circuits. NFPA 111 covers installation, maintenance, operation, and testing requirements as they pertain to the performance of the stored emergency power supply system (SEPSSs).

## NFPA 306

NFPA 306 is a maintenance- and repair-oriented standard created for application on ships and other maritime environments. This standard applies to vessels that carry or burn as fuel, flammable, or combustible liquids (NFPA 2009b). It also applies to vessels that carry or have carried flammable compressed gases, flammable cryogenic liquids, chemicals in bulk, or other products capable of creating a hazardous condition.

NFPA 306 derived from a comprehensive series of standards intended to stop fires and explosions in the marine environment. These standards were originally adopted in 1922 (Keller 1982; NFPA 2009b). Current versions reflect concern about health hazards.

NFPA 306 applies specifically to those spaces on vessels that are subject to concentrations of combustible, flammable, and toxic liquids, vapors, gases, and chemicals capable of causing harm. This standard describes the conditions required before a space can be entered or work can be started, continued, or started and continued on any vessel under construction, alteration, or repair, or on any vessel awaiting shipbreaking. This standard applies to cold work, application or removal of protective coatings, and work involving riveting, welding, burning, or similar fire-producing operations.

NFPA 306 is also applicable to those spaces on vessels that might not contain sufficient oxygen to permit safe entry. This standard applies to land-side confined spaces, whether stationary or mobile, or other dangerous atmospheres located within the boundaries of a shipyard or ship repair facility as defined by 29 CFR 1915.11 (OSHA 1994). This standard applies to Marine Chemists performing activities related to inspection and certification procedures described in this standard and consulting services connected therewith on board any vessel.

NFPA 306 does not apply to physical hazards of tanks and confined or enclosed spaces on a vessel or vessel sections, or in the shipyard. For the purposes of this standard, physical hazards do not include fire and explosion hazards.

Preparation of ship spaces for work requires control of exposure to pressurized fluids (gases, hydraulic fluids and cargoes). This concern is addressed through inspection of pumps, valves, piping, and other components.

In following the requirements of NFPA 306, the Marine Chemist or other practitioner personally inspects the space(s) in which work is to occur and adjacent spaces on the other side of bulkheads and other boundary surfaces (Willwerth 1994). The Marine Chemist or other practitioner issues a certificate that describes conditions found in the space at the time of the inspection and precautions required for entry and work.

In determining the condition of the space, the Marine Chemist considers the three previous cargoes, nature and extent of the work to be performed, starting time and duration of the work, results of tests of compartment or spaces, cargo and vent lines at manifolds and accessible openings, and cargo heating coils, and tagging and securing of cargo valves in prescribed areas in a manner to prevent accidental opening and operation.

The certificate issued by the Marine Chemist carries several possible standard safety designations and outlines conditions necessary for complying with safe performance of the work to be undertaken.

## NFPA 780

NFPA 780 covers lightning protection systems (NFPA 2011c). NFPA 780 has its origin in a standard first issued in 1904. This document covers traditional lightning protection systems for ordinary structures, miscellaneous structures and special occupancies, heavy-duty stacks, watercraft, and structures containing flammable vapors, flammable gases, or liquids that give off flammable vapors.

NFPA 780 covers strategies for lightning protection of a wide range of industrial, commercial, residential, and recreational settings. Lightning as an energy source can injure and kill. It also can act as an ignition source for ignitable gases, vapors, and aerosols.

## Robotic Industries Association

The Robotic Industries Association is the champion in North America of robotic safety and an ANSI Secretariat for creating and publishing standards.

## ANSI/RIA/ISO 10218-1

ANSI/RIA/ISO 10218-1-2007 specifies requirements and guidelines for the inherent safe design, protective measures, and information for use of industrial robots (ANSI/RIA 2007). It also describes basic hazards associated with robots, and provides requirements to eliminate, or adequately reduce, the risks associated with these hazards.

ANSI/RIA/ISO 10218-1-2007 provides the basis for adopting emerging robot technologies, an important concern of the robotic industry in North America. One of the important innovations is the wireless teach pendant. Wireless teach pendants eliminate the need for cables. Cables pose trip hazards and can limit escape from the enclosure during emergency situations. ANSI/RIA/ISO 10218-1-2007 discusses safety aspects for wireless teach pendants.

Another feature of ANSI/RIA/ISO 10218-1-2007 is safety-rated axis limiting. Many of the new controllers include safety software to contain robot movement internally without any need for external safety sensors or other features. The new safety standard addresses issues that arise when robots and human workers share the same space and work together collaboratively. Lastly, the new standard provides guidelines for situations where simultaneous motion is occurring. This pertains to the movement of multiple manipulators coordinated by one controller.

## ANSI/RIA R15.06

ANSI/RIA/ISO 10218-1-2007 is a companion to ANSI/RIA R15.06-1999 (R2009) (ANSI/RIA 2009). ANSI/ RIA/ISO 10218-1-2007 pertains only to robot construction. It is not intended to replace ANSI/RIA R15.06-1999 (R2009). ANSI/RIA R15.06-1999 (R2009) provides requirements for industrial robot manufacture, remanufacture and rebuild; robot system integration/installation; and methods of safeguarding to enhance the safety of personnel associated with the use of robots and robot systems.

## RIA TR R15.206-2008

RIA TR R15.206 is a technical report. It provides guidelines for implementing ANSI/RIA/ISO 10218-1-2007 (ANSI/RIA 2008).

## Standards Australia

Standards Australia is an example of an organization outside North America that publishes consensus standards in occupational health and safety.

## AS2865-1995

AS 2865-1995 addresses confined spaces (Standards Australia 1995). While the focus of the standard is atmospheric hazards, the standard also addresses hazardous conditions arising from hazardous energy in its various forms. This document contains sections that address isolation from contaminants and isolation from moving parts. These sections comment about use of techniques of isolation for sources of hazardous energy.

Standards Australia has not published a comprehensive standard on controlling exposure to hazardous energy.

## Research Institutions

This section reports on research performed on hazardous energy with an emphasis on control.

### National Institute for Occupational Safety and Health (NIOSH)

In 1983, NIOSH published guidelines for controlling hazardous energy sources during maintenance and servicing (NIOSH 1983). These were developed under contract by Boeing Aerospace using methodology from systems analysis. The document provided results from a worldwide search of the literature on maintenance hazards and control techniques and review of 300 accidents. This document appeared eight years after publication of ANSI Z241.1-1975 on safety requirements for sand preparation, molding and coremaking in foundries, and a year after publication of ANSI Z244.1 on lockout/tagout of hazardous energy sources.

Analysis contained in the NIOSH report attributed the cause of accidents involving exposure to hazardous levels of energy to failure to control, inadequate isolation, failure to dissipate residual energy, and accidental reactivation of sources of hazardous energy.

NIOSH argued that accidents having these identifiable causes are preventable through employment of effective energy control techniques or procedures. NIOSH noted that work was usually performed without knowledge that hazardous levels of energy can exist in systems regardless of whether they were energized or de-energized. Unexpected and unrestricted release of hazardous energy can occur because of failure to identify all energy sources, failure to devise safe work procedures when energy is present, and unintended or accidental reactivation of energy sources without knowledge of the victim.

NIOSH argued that the means to control these situations are procedural in nature. In taking this position and offering a number of supporting guidelines, NIOSH supported these with comments obtained from its review of the literature and submissions made to the Occupational Safety and Health Administration (OSHA) following the Advance Notice of Proposed Rulemaking for the proposed OSHA standard on lockout and tagout. Most of the references cited by NIOSH in the report were, in fact, submissions to OSHA.

NIOSH analyzed the strategies used in the submissions for consistency with its advocacy of strategies of approach. There was no shortage of support for actions to identify and control sources of hazardous energy. The area of least support was the use of lockout in place of tagout. (Lockout requires the application of a lock and lockout device to an energy isolating device. Tagout requires the application of a tag and tagout device to an energy isolating device.) Many submissions to the OSHA rulemaking contended that tagout was as secure as lockout, and that lockout was not necessary.

According to NIOSH, control can occur through de-energization, or when sources of hazardous energy are present, through isolation. In both cases, identification of sources of hazardous energy must occur as a first step. During de-energization, control, and securement, verification of effectiveness of the procedure must occur for each source of hazardous energy. Control techniques include isolation, blockage, and dissipation. Securement means that accidental or deliberate re-energization cannot occur. Verification through testing is necessary to prove that energy flow cannot occur and to confirm the effectiveness of control techniques.

Documentation of procedures also must occur. Personnel who design and implement these measures must obtain qualifications through education, training, and experience.

The report also provides examples of methods for isolating or blocking energy sources and securing the points of control.

## Regulatory Standards and Regulations

Regulatory authorities worldwide set the regulatory framework in their respective jurisdictions. The outcome reflects their considered view about the most effective way to achieve safety. Regulators often borrow requirements from regulators in other jurisdictions. This is especially the case where particular regulators act as pioneers in regulatory action or can mobilize the considerable resources and dedication needed to create a Standard.

For more assistance with the regulatory situation in the U.S., the reader should direct attention to the comprehensive treatment, the reader should consult the monograph by Kelley (2001). The monograph by Ridley and Pearce (2002) provides a European perspective and the older monograph by Grund (1994) provides broad and specific coverage of regulatory concepts.

### U.S. Department of Labor (Occupational Safety and Health Administration)

Two federal government agencies in the U.S. Department of Labor, the Occupational Safety and Health Administration (OSHA) and the Mine Safety and Health Administration (MSHA), have produced regulations concerning lockout and tagout. The OSHA regulations are more general in nature, whereas those produced by MSHA are targeted to the mining environment.

### 29CFR 1910.147

OSHA's involvement in issuing a comprehensive regulatory Standard on lockout and tagout began in 1977 with a request for information on technical issues regarding machine guarding. This resulted in submissions commenting about maintenance and repair activities. OSHA published an Advance Notice of Proposed Rulemaking in 1980 soliciting comments about issuance of a generic regulatory Standard on lockout and tagout. In April 1988, the Occupational Safety and Health Administration (OSHA) proposed a rule on the control of hazardous energy sources.

In 1989, OSHA enacted 29CFR 1910.147 on control of hazardous energy sources (OSHA 1989). The Standard applies to general industry, but does not cover maritime operations, agriculture, or construction. The Standard also does not cover oil and gas well drilling; the generation, transmission, and distribution of electrical power by utilities; and electrical conductors and equipment. The long duration from proposal to final issue is a reflection of the complexity of the issues underlying the Standard and its antecedent references.

The purpose of 29CFR 1910.147 is to ensure uniformity of approach to controlling exposure to hazardous levels of energy across the broad spectrum of general industry. This Standard filled the gaps in coverage of general industry highlighted in the NIOSH review of U.S. regulatory standards (NIOSH 1983).

The preamble to OSHA Standards always provides useful insight into the process of rulemaking, in this case, lockout and tagout of hazardous energy sources. OSHA acknowledged the major contribution of ANSI Z244.1 (ANSI 1982), as the primary basis for development of the regulatory Standard. At the same time, OSHA did not adopt ANSI Z244.1-1982 by reference, and in some respects the agency deliberately departed from ANSI Z244.1-1982 in order to provide what it believed to provide a higher level of employee protection (OSHA 2004),

OSHA began the process of Standard creation with a review of accidents from various sources involving machinery and equipment. As a result of this analysis, OSHA believes that failure to control hazardous energy sources accounts for nearly 10% of serious accidents and 7% of fatal injuries (122 persons) in general industry in the U.S. (OSHA 1989).

Shutting down a machine or equipment is only a partial solution to the problem. Unanticipated movement of a component or of the material being handled can occur due to conversion of stored energy. OSHA believes that the most effective method to prevent injury from these sources is to dissipate or minimize remaining potential energy or to utilize restraining devices to prevent movement. Preventing inadvertent reactivation of deactivated and de-energized equipment also is essential.

OSHA also provided a summary of results from several reports on lockout-related accidents. OSHA concluded that the major factors in lockout-related accidents include servicing equipment while operating, failure to ensure de-energization of sources of power, inadvertent activation, lack of procedure, and lack of training.

The most contentious issue in the rulemaking process was the use of locks, locks and tags or tags alone as the means to control operation of the energy-isolating device. Regardless of what is used, the effect on workers is mainly psychological.

Determined individuals can defeat the hardware aspects of the process. Certainly substantive locks are available. At that point, the hasp becomes the limiting factor in the process. Some hasps are made from plastic materials. These are subject to destruction by melting using a cigarette lighter or match.

Successful control of the process comes through creating appropriate procedures, training all affected individuals about the meaning of the application of lockout devices, auditing actual performance, and applying discipline where adherence does not occur.

An important point made by OSHA and comments submitted in regard to the Standard is that locks and tags themselves do not control hazardous energy. Control is achieved by isolation of equipment from the energy source following procedures for deactivation and de-energization. Attachment of locks and tags should occur only after control has occurred.

The scope of 29CFR 1910.147 follows the general direction and strategy of ANSI Z244.1-1982 on which it was based. The Standard excludes certain practices for which it is not a practical requirement and incorporates regulatory rigor in other areas to ensure clarity of requirements to achieve compliance (Table A.5). For example, considerable effort is expended in the U.S. National Electrical Code administered by the National Fire Protection Association as NFPA 70 and NFPA 70E and regulatory requirements of 29 CFR 1910.269 covering electric power generation, transmission, and distribution and 29 CFR 1910.331-335 covering electrical safety-related work practices (NFPA 2009, NFPA 2011a; OSHA 1994a, OSHA 1991). Some of these contain considerable detail about requirements. The following sections contain an abbreviated look at standards and guidelines.

The Standard requires employers to create an energy control program prior to performing any work on equipment or machinery in which unexpected energizing, startup, or release of stored energy can occur. The program must contain energy control procedures and employee training.

To a considerable extent, 29CFR 1910.147 is a performance standard. It establishes general requirements, but provides latitude to employers to develop and implement specific methods for meeting those obligations. The Standard puts the onus on the employer to establish the relative risk of activities involved in the operation, service, and maintenance of equipment. These are difficult situations to assess. By following one set actions, a worker could remain entirely within the boundaries of the intent of the Standard. However, slight deviation from these actions borne from frustration of lack of results, or ignorance of outcomes, could produce an accident situation.

Not stated in the Standard is the means by which the procedures are to be created. However, implementation documentation indicates that procedure creation is to reflect a methodological approach including analysis and documentation of energy flow, storage, and dissipation in the machinery or equipment.

The Standard uses several terms to describe critical components. An energy-isolating device is a mechanical device that physically prevents the transmission or release of energy. Energy-isolating devices include manually operated circuit breakers, disconnect switches, manually operated switches that disconnect all poles from ungrounded supply connectors simultaneously, line valves, blocks, and similar devices used to block or isolate energy.

Lockout devices include mechanical devices that attach to energy-isolating devices into which the shackle of a padlock fits, as well as other mechanisms that provide an equivalent level of protection. These include bolted blanks and blinds in process piping systems which require at least as much effort to remove as locks, and which therefore are difficult to defeat intentionally or inadvertently. The Standard also provides detail about requirements for durability, standardization and substantiveness of lockout and tagout devices.

A periodic audit must occur at least annually to determine the effectiveness of the energy control process. The audit is to be performed by an individual capable of recognizing deficiencies who is not affected by the work under inspection. The intent of the audit is to identify deficiencies in training, procedures, and work habits.

## Table A.5
## 29CFR 1910.147: Inclusions and Exclusions

### Inclusions

Servicing and maintenance of machines and equipment in which unexpected energization or start-up or release of stored energy could cause injury.

Sources of energy covered in the Standard include electrical, mechanical, hydraulic, pneumatic, chemical, thermal, or other energy.

When the risk of injury is greater than present during normal production activity and unexpected activation or release of stored energy could occur.

### Exclusions

Situations where guarding provides protection against contact with energy sources and remains intact.

Repetitive minor adjustments and simple tool changes.

Cleaning and unjamming of equipment or machines where these activities do not expose the worker to risk of injury beyond that of normal production activity.

Servicing occurring in a way that prevents exposure to hazardous energy through use of special tools.

Alternate procedures that prevent contact of the body with energy sources.

Operational testing which must occur while machines and equipment remain energized.

Electrical equipment connected by cord and plug to the power source when removal of the plug from the receptacle eliminates unexpected energization or startup hazard, and the employee performing the servicing or maintenance retains exclusive control of the plug.

Hot tap operations involving transmission and distribution systems containing pressurized fluids, such as natural gas, steam, water or petroleum products when continuity of service is essential, shutdown of the system is impractical and the work occurs according to documented procedures using specialized equipment capable of providing proven effective protection. (A hot tap is a procedure used in repair, maintenance, and service activities involving welding on pressurized equipment in order to install connections or appurtenances, or addition and replacement of sections of pipeline.)

Source: OSHA 1989.

---

The employer must provide training to employees affected by the Standard. The level of training differs by level of involvement in the process of lockout/tagout. Affected employees operate or use machines or equipment on which servicing or maintenance is performed under lockout or tagout, or work in the area in which this work is occurring. Affected employees require instruction in the purpose and use of the energy control procedure. Authorized employees have the mandate to lock or implement a tagout system procedure on machines or equipment to perform service or maintenance on them. Authorized employees require instruction in recognition of sources of hazardous energy in the workplace, the type and magnitude, and methods and means necessary for isolation and control. Other employees require instruction in lockout/tagout procedures and the prohibition regarding attempts to restart or re-energize equipment that is locked or tagged out.

Retraining must occur when authorized and/or affected employees change jobs, or when a change occurs in machines, equipment, or processes that pose a new hazard, or when a change occurs in energy

## Table A.6
## 29CFR 1910.147 Sequence for Energy Isolation and Return To Service

**Step**

**Energy Isolation**

Notify affected employees of application of procedure

Prepare for shutdown

Shut down in orderly fashion

Isolate the machine or equipment by locating and activating energy-isolating devices

Lock or tag out energy-isolating devices

Relieve, disconnect, restrain, or otherwise render in a safe condition all sources of stored energy

Verify successful isolation and de-energization

**Return to Service**

Inspect the locked or tagged out equipment or machine to ensure removal of nonessential items and operational intactness

Ensure that affected employees are safely positioned or removed

Notify affected employees about removal of the lockout or tagout devices

Remove lockout or tagout devices (normally by the individual who applied them)

Source: OSHA 1989.

control procedures. Retraining must occur additionally when the audit indicates deviations or deficiencies in energy control procedures.

The Standard requires an orderly sequence of energy isolation and return to service (Table A.6). Normally, the worker who applies the lock removes the lock. In the event that the individual who applied the lockout or tagout device is not available to remove it, the employer may direct its removal under specific procedures developed, documented, and incorporated into the energy control program. This procedure must incorporate at minimum the following elements. The first is verification that the authorized employee who applied the lockout or tagout device is not present in the facility. The second element is proof of reasonable effort to contact the authorized employee to inform about the situation. The third element is documentation to confirm about informing the authorized employee about removal of the lockout or tagout device upon return of that individual to the facility.

Testing or positioning of equipment, machines, or components may necessitate removal of the lockout or tagout device and re-energization through the energy-isolating device. The Standard permits testing or positioning provided that the following steps occur. First, workers must clear materials and tools from the equipment or machine. This ensures that loose items do not become projectiles following sudden movement. Unaffected employees must move to a safe area prior to removing the lockout or tagout device and re-energizing the equipment or machine. This minimizes the number of individuals potentially exposed to the hazardous energy on re-energization. At this point, re-energization can occur. Lastly, de-energization and reapplication of energy control measures must occur.

Servicing or maintenance performed by a crew, craft, department, or other group introduces complexities regarding application of many locks to lockout devices. The Standard affords the opportunity for procedures to enable a group lockout or tagout to replace action by many individuals.

Specific requirements include vesting in an authorized employee responsibility for a set number of employees working under protection of the group lockout or tagout device. In following this approach, the authorized employee must ascertain the exposure status of individual members of the group regarding the equipment or machine. The employer must assign to one authorized employee overall responsibility for ensuring continuity of protection and co-ordination of work when more than one crew, craft, or work group is involved. The follow-up to this designation is to vest in the authorized employee the ability to affix one lockout or tagout device to the group lockout device, group lock box, or comparable device at the beginning of work and to remove it at the end of work.

The Standard requires the employer to devise procedures to ensure continuity of protection during shift changes through orderly removal and application of lockout devices.

The Standard also requires interaction between the employer and contractors affected by the lockout/tagout. Each party is mutually responsible for informing the other about lockout/tagout and for informing their personnel and ensuring compliance with restrictions and prohibitions outside the employer's energy control procedures.

## Conflict between Regulatory and Consensus Standards

This discussion has highlighted the existence of an ANSI consensus standard on lockout and tagout (Z244.1) and an OSHA regulatory Standard (29CFR 1910.147) on the same subject. This situation can function through collaboration or lead inevitably to conflict.

One approach to such situations is that the regulatory Standard references some version or the current version of the consensus standard. In this manner, the regulatory agency relies on the members of the consensus group to produce an outcome that provides for a superior level of worker protection through best practices and innovation at the same time. Operating rules and management of the committee that ensure exposure and amenable resolution of potentially conflicting points of view are critical to ensure that this process produces an outcome that is satisfactory to all participants.

The situation in the US is that OSHA is required to examine American National Standards for suitability for reference as regulatory Standards. The consensus standard undergoing examination is a version with a particular date. OSHA may choose to adopt the consensus standard in its rulemaking in whole or in part. As mentioned previously, the scope of 29CFR 1910.147 follows the general direction and strategy of ANSI Z244.1-1982 on which it was based. OSHA did not adopt the standard by reference, and in some respects the agency deliberately departed from the ANSI standard in order, in its view, to provide a higher level of employee protection (OSHA 2004). OSHA subsequently published the reason for its choice (see 58 FR 16617, March 30, 1993).

The regulatory Standard is the legal reference, hence the capital S. A consensus standard on the same subject no matter how rigorously created defers to the regulatory Standard. The regulatory Standard may have as its antecedent a particular version of a consensus standard in part or in whole. A subsequent version of a consensus standard has no standing as an advancement beyond the reference document.

While requiring employers to comply with OSHA standards, the OSH Act also authorizes OSHA to treat violations, which have no direct or immediate relationship to safety and health, as de minimis, requiring no penalty or abatement. OSHA's enforcement policy provides that a violation may be de minimis, when an employer complies with a proposed standard or amendment or a consensus standard rather than the regulatory Standard in effect at the time of inspection, provided that the employer's action clearly provides equal or greater employee protection. Hence, while a consensus standard may provide options for an alternate approach to a problem, these must provide worker protection equal or greater to that available through application of 29CFR 1910.147.

OSHA commented that in several important respects, ANSI Z244.1-2003 appears to sanction practices that may provide less employee protection than that provided by compliance with the relevant provisions of 29CFR 1910.147 (OSHA 2004). To illustrate, ANSI Z244.1-2003 employs a decision matrix that allows employers to use alternative protective methods in situations where the OSHA Standard requires the implementation of lockout/tagout or machine guarding.

In addition, ANSI Z244.1-2003 permits the use of tagout programs when they provide "effective" employee protection. The OSHA Standard allows the use of a tagout program only where the employer can demonstrate provision of full protection to employees. Full protection means a level of safety equivalent to that provided by a lockout program.

Further, the sections in ANSI Z244.1-2003 on Hazardous Energy Control Procedures, Communication and Training, and Program Review, while detailed and conceptually valuable, do not appear to mandate certain discrete practices that are prescribed in parallel sections of the OSHA Standard (OSHA 2008). When an OSHA standard prescribes a practice, design, or method that provides a requisite level of employee protection, employers may not adopt an alternative approach that provides a lesser level of employee protection.

## 29CFR 1910.146

In 1993, OSHA issued 29CFR 1910.146, the Standard covering permit-required confined spaces in general industry (OSHA 1993). This Standard is important to this discussion because some workspaces containing sources of hazardous energy are also confined spaces. As well, significant differences in requirements exist between the isolation and lockout requirements in 29CFR 1910.146 and 29CFR 1910.147.

The intent of 29CFR 1910.146 is to provide comprehensive protection from exposure to all types of hazardous conditions that can occur in confined spaces, including mechanical and physical hazards. OSHA intends to achieve this goal through direct linkage to 29CFR 1910.147 (reviewed in the previous section) (OSHA 1989). As a result, employers must evaluate hazardous conditions involving exposure to hazardous energy found in confined spaces and take all steps necessary to protect entrants.

The Standard applies to confined spaces in general industry. Major sectors that are affected include electric and gas utilities, manufacturing, transportation, agricultural services, and wholesale trade. Agriculture, construction, and shipyard employment industries are excluded.

A confined space has adequate size and configuration for employee entry, has limited means of access or egress, and is not designed for continuous employee occupancy. The preceding characteristics are common to all confined spaces.

OSHA took the view that segregating spaces having characteristics that are especially hazardous would make better use of available resources.. OSHA uses the terms, nonpermit confined space (nonpermit space) and permit-required confined space (permit space), to distinguish the level of hazard.

A nonpermit confined space (nonpermit space) does not contain or, with respect to atmospheric hazards, have the potential to contain any hazard capable of causing death or serious physical harm.

A permit-required confined space (permit space) has characteristics that make entry hazardous without special precautionary measures. A permit-required confined space (permit space) presents or has a potential to present an atmospheric hazard and/or an engulfment hazard and/or a configuration hazard, and/or any other recognized serious hazard. Recognized serious hazards include sources of hazardous energy.

Engulfment can include the surrounding and effective capture of a person by a liquid or finely divided (flowable) solid. This concept includes any solid or liquid that can flow into a confined space and can drown or suffocate a person.

The Standard provided no indication about what constitutes, any other recognized serious hazard. As a regulatory document, the meaning of this term will evolve through enquiry from interested persons through appropriate channels. OSHA stated this criterion in the broadest possible terms in order to embrace serious hazards and to afford protection to affected employees (OSHA 1996).

According to this view, these hazards could include conditions that necessitate preventive or protective actions, such as the wearing of personal protective equipment. On the other hand, they would not include conditions, such as poor communication. Other conditions would require case-by-case evaluation. These would include biological hazards, slippery surfaces and tripping hazards, noise and vibration, and illumination.

OSHA Standards, 29CFR 1910.146 and 29CFR 1910.147 are generic and interrelated. Both Standards may apply to the isolation of hazardous energy for a permit-required confined space, depending upon circumstances (OSHA 2008)). Application of the procedural and training provisions of 29CFR 1910.147 to permit spaces supplemented to ensure effective control of hazardous energy occurs when other standards require lockout and tagout.

Therefore, for any particular permit space, the question becomes, does 29CFR 1910.146 require lockout or tagout to isolate hazardous energy. The answer depends on the type(s) of hazardous energy to be isolated, whether lockout or tagout provides isolation (offering complete employee protection), and whether 29CFR 1910.146 requires the use of lockout or tagout.

Pursuant to 29CFR 1910.146 (including the preamble to the final rule), electro-mechanical hazards associated with a permit space require isolation in accordance with 29CFR 1910.147 (or guarded in accordance with requirements of Machine Guarding, Subpart O).

OSHA Standard 29CFR 1910.146 does not, however, allow lockout or tagout for isolating flowable material. The reason is that compliance with 29CFR 1910.147 does not in all cases adequately isolate hazards created by materials such as steam, flammable gases, flammable and combustible liquids. In a permit-required confined space, OSHA considers isolation of hazards associated with flowable materials to have occurred only by the use of blanking or blinding; misaligning or removing sections of line, pipe or duct; or use of a double block and bleed system. A double block and bleed isolation system usually utilizes the closure of two valves, the opening of a bleed valve, and the application of lockout or tagout devices. This approach offers complete employee protection; whereas an employer can comply with requirements of 29CFR 1910.147 simply by closing and locking out or tagging out a single valve. Closing and locking out a single valve could create atmospheric hazards in the space.

## 29CFR 1915

OSHA issued an updated version of the regulatory Standard addressing confined and enclosed spaces in shipyard employment in 1994 (OSHA 1994b). This update modernized existing statutes dating from 1971, and extended ship-based concepts to confined spaces in the remainder of the shipyard. This extension removed the overlap of jurisdiction with the standard for general industry, discussed in the previous section. Confined spaces in shipyards beyond ship structures can include tanks and spill containment structures, structures associated with welding and fuel gases, baghouses and cyclones associated with welding processes and coating application, stormwater and wastewater collection and treatment facilities, machinery spaces, attic spaces and crawlspaces, and possibly others.

OSHA Standard 29CFR 1915 is based on NFPA 306, a consensus standard (NFPA 1997). NFPA 306 undergoes progressive and routine updates. NFPA 306 and 29CFR 1915 utilize fundamentally different strategies for managing the hazards in confined spaces compared to ANSI Z117.1 and 29CFR 1910.146. These differences reflect the rapidly changing presence, configuration and design, and hazards of the confined spaces managed in the maritime industry, especially shipbuilding and shipbreaking (Willwerth, 1994). Encounters with confined spaces in the maritime industry occur daily rather than occasionally. In one extreme example, workers involved in construction of an aircraft carrier could enter 1,000 confined spaces per day.

OSHA Standard 29CFR 1915 is expert- and prevention-centered. An expert is readily at hand to address and assess concerns. As well, both management and the workforce are trained in recognition of hazards associated with work in confined spaces. Consultation by management and the workforce in hazard control with the expert is a routine part of hazard management in this environment. The expert routinely inspects, evaluates and certifies the status of confined spaces and specifies controls for the hazards of work to be performed. The emphasis in this approach is to discover and control hazards before work begins.

The focus of the shipyard employment Standard clearly is atmospheric conditions. The Standard relies on ventilation to prevent overexposure of employees to atmospheric hazards. This is accomplished through a built-in system of testing, visual inspection and application of ventilation principles. The intent of the standard is to prevent or eliminate hazardous conditions.

Critical to implementation of requirements of 29CFR 1915 is visual inspection and testing by the competent person. This activity occurs prior to entry by workers. A visual inspection is a physical survey of the space, its contents and surroundings. The purpose for the inspection is to identify such hazards as restricted accessibility, cargo residues, unguarded machinery, and piping or electrical hazards. Spaces that contain or have contained combustible or flammable liquids or gases and spaces adjacent to them specifically must be inspected. Visual inspection is a necessary adjunct to atmospheric testing. This is especially important where products, such as diesel fuel, may be present.

Pipelines that carry hazardous materials must be blocked or flushed and cleaned to prevent discharge into the space. Periodic testing must occur to ensure that safe working conditions are maintained in the space. Work must cease and the space vacated when conditions change and the space no longer meets the criteria specified in the certificate issued by the competent person.

**State Regulations**

State plan states have the right to enact their own lockout Standards. OSHA allows states with "approved plans" to issue and enforce comparable lockout/tagout requirements. Comparable does not necessarily mean identical. They must be as effective as U.S. federal requirements. Therefore, readers affected by state plan requirements must study these carefully to identify differences.

# References

AFS: What You Should Know about ZMS (Zero Mechanical State). Des Plaines, IL: American Foundrymen's Society, 1961.

ANSI: Safety Requirements for Sand Preparation, Molding and Coremaking in The Sand Foundry Industry (ANSI Z241.1-1975). New York: American National Standards Institute, 1975.

ANSI: Safety Requirements for Lockout/Tagout Of Energy Sources – Minimum Safety Requirements (ANSI Z244.1-1982). New York: American National Standards Institute, 1982.

ANSI: Safety Requirements for Lockout/Tagout of Energy Sources – Minimum Safety Requirements (ANSI Z244.1-1987(R)). New York: American National Standards Institute, 1987.

ANSI: Safety Requirements for Lockout/Tagout of Energy Sources – Minimum Safety Requirements (ANSI Z244.1-1992(R)). New York: American National Standards Institute, 1992.

ANSI/AMT: Technical Report for Machine Tools – Risk Assessment and Risk Reduction – A Guide to Estimate, Evaluate and Reduce Risks Associated With Machine Tools (ANSI/AMT B11.TR3-2000) McLean VA: Association For Manufacturing Technology, 2000.

ANSI/AMT:American National Standard For Machine Tools – Performance Criteria for Safeguarding (ANSI/AMT B11.19-2003 (R2009)). McLean, VA: Association For Manufacturing Technology, 2009.

ANSI/ASME: Valves Flanged, Threaded and Welding End (ANSI/ASME B16.34 - 2009). New York: American Society of Mechanical Engineers, 2009.

ANSI/ASSE: Control of Hazardous Energy – Lockout/Tagout And Alternate Methods (ANSI/ASSE Z244.1-2003). Des Plaines, IL: American Society of Safety Engineers, 2004.

ANSI/ASSE: Safety Requirements for Confined Spaces (ANSI/ASSE Z117.1-2009). Des Plaines, IL: American Society of Safety Engineers, 2009.

ANSI/FCI: Control Valve Seat Leakage (ANSI/FCI 70-2-2006). Cleveland, OH: Fluid Controls Institute, 2006.

ANSI/FCI: Regulator Seat Leakage (ANSI/FCI 70-3-2004). Cleveland, OH: Fluid Controls Institute, 2004.

ANSI/RIA/ISO: Robots for Industrial Environment - Safety Requirements Part 1–Robot (ANSI/RIA/ISO 10218-1-2007). Ann Arbor, MI: Robotic Industries Association, 2007.

ANSI/RIA: American National Standard for Industrial Robots and Robot Systems– Safety Requirements (ANSI/RIA R15.06-1999 (R2009). Ann Arbor, MI: Robotic Industries Association, 2007.

ANSU/RIA: Guidelines for Implementing ANSI/RIA/ISO 10218-1-2007 (ANSI/RIA TR R15.206-2008). Ann Arbor, MI: Robotic Industries Association, 2008.

API: Protection against Ignitions Arising out of Static, Lightning, and Stray Currents, 7th Ed. (API Recommended Practice 2003). Washington, DC: American Petroleum Institute, 2008.

API: Valve Inspection and Testing (API Standard 598, 9th Ed.). Washington, DC: American Petroleum Institute, 2009.

CAN/CSA: Control of Hazardous Energy —Lockout and Other Methods (Z460-05). Mississauga ON: Canadian Standards Association, 2005.

CEN: Safety of Machinery – Safety-Related Parts of Control Systems. Part 1: General Principles (EN 954-1:1996). Brussels: Comite Europeen de Normalisation, 1996.

EN ISO: Safety-Related Parts of Control Systems – Part 1: General Design Guidelines (EN ISO 13849-1:2006). Geneva: International Organization for Standards, 2006.

Grund, E.V.: *Lockout/Tagout, The Process of Controlling Hazardous Energy.* Itasca, IL: National Safety Council, 1995.

Hall, B.: Zero Energy State – Historical Background of the Concept. *Nat. Safety News 115(2):* 68-72 (1975).

IEC: Functional Safety of Electrical/Electronic/Programmable Electronic Safety-Related Systems – Part 1: General Requirements (IEC 61508-1 Edition 2.0 2010). Geneva: International Electrotechnical Commission, 2010.

IEC: Functional Safety – Safety Instrumented Systems for the Process Industry Sector – Part 1: Framework, Definitions, System, Hardware and Software Requirements (IEC 61511:2004). Geneva: International Electrotechnical Commission, 2004.

IEC/EN: Safety of Machinery – Functional Safety of Safety-Related Electrical, Electronic and Programmable Electronic Control Systems (IEC/EN 62061:2005). Geneva: International Electrotechnical Commission, 2005.

ILO: Model Code of Safety Regulations for Industrial Establishments for the Guidance of Government and Industry. Geneva: International Labour Organization, 1954.

ISO: Safety of Machinery — Risk Assessment — Part 1: Principles ( ISO 14121-1:2007). Geneva: International Standards Organization, 2007.

ISO: Safety Of Machinery – General Principles for Design – Risk Assessment and Risk Reduction (ISO 12100:2010). Geneva: International Standards Organization, 2010.

Keller, C.: Firesafety in the Shipyard. *Fire J.:* July (1982) pp. 60-66.

Kelley, S.M.: *Lockout Tagout, A Practical Approach.* Des Plaines, IL: American Society of Safety Engineers, 2001.

MSS: Pressure Testing of Valves (ANSI/MSS SP-61-2009). Vienna, VA: Manufacturers Standardization Society (MSS) of the Valve and Fittings Industry, 2009.

NFPA: Recommended Practice on Static Electricity, NFPA 77 (2007 Ed.). Quincy, MA: National Fire Protection Association, 2007.

NFPA: Standard for Electrical Safety Requirements for Employee Workplaces, NFPA 70E (2009 Ed.). Quincy, MA: National Fire Protection Association, 2009a.

NFPA: Standard for the Control of Gas Hazards on Vessels, NFPA 306 (2009 Ed.). Quincy, MA: National Fire Protection Association, 2009b.

NFPA 110: Standard for Emergency and Standby Power Systems, NFPA 110 (2010 Ed.). Quincy, MA: National Fire Protection Association, 2010a.

NFPA 111: Standard on Stored Electrical Energy Emergency and Standby Power Systems, NFPA 111(2010 Ed.). Quincy, MA: National Fire Protection Association, 2010b

NFPA: National Electrical Code, NFPA 70 (2011 Ed.). Quincy, MA: National Fire Protection Association, 2011a.

NFPA 79: Electrical Standard for Industrial Machinery, NFPA 79 (2011 Ed.). Quincy, MA: National Fire Protection Association, 2011b.

NFPA: Standard for the Installation of Lightning Protection Systems, NFPA 780 (2011 Ed.). Quincy, MA: National Fire Protection Association, 2011c.

NIOSH: Guidelines for Controlling Hazardous Energy during Maintenance and Servicing (DHHS (NIOSH) Pub. No. 83-125). Morgantown, WV: Department of Health and Human Services/Public Health Service/National Institute for Occupational Safety and Health, 1983.

OSHA: "Control of Hazardous Energy Sources (Lockout/Tagout); Final Rule," *Fed. Regist. 54*: 169 (1 September 1989). pp. 36644-36696.

OSHA: Electrical Safety-Related Work Practices (29 CFR 1910.331-335). Washington, DC: Occupational Safety and Health Administration, 1991. [OSHA Website].

OSHA: "Permit-Required Confined Spaces for General Industry; Final Rule," *Fed. Regist. 58*: 9 (14 January 1993). pp. 4462-4563.

OSHA: Electrical Power Transmission, Generation, and Distribution; Electrical Protective Equipment; Electrical Safety-Related Work Practices (29 CFR 1910.269). Washington, DC: Occupational Safety and Health Administration, 1994a. [OSHA Website].

OSHA: "Confined and Enclosed Spaces and Other Dangerous Atmospheres in Shipyard Employment; Final Rule," *Fed. Regist. 59*: 141 (25 July 1994b). pp. 37816-37863.

OSHA: Letter to Mary C. DeVany in Reply to Her Letter of Enquiry of June 13, 1996 by Richard S. Terrill through Ronald T. Tsunehara. Seattle, WA: Occupational Safety and Health Administration, July 1, 1996. [Letter].

OSHA: Letter to Edward V. Grund in Reply to His Letter of Enquiry of March 22, 2004 by Richard E. Fairfax. Washington, DC: Occupational Safety and Health Administration, November 10, 2004. [Letter].

OSHA: OSHA Instruction, The Control of Hazardous Energy – Enforcement Policy and Inspection Procedure (CPL 02-00-147). Washington: Occupational Safety and Health Administration, 2008.

Ondrey, G.: Process Control With Safety Built In. *Chem. Eng 112(5)*: 25-28 (2005).

Paschal, J.: *Understanding NE Code Rules on Grounding & Bonding*, 2nd Ed. (Edited by F. Hartwell). Overland Park, KS: EC&M Books, PRIMEDIA Intertec, 1998.

Price, C.: ZES – Zero Energy State: A Systems Approach to Guarding Maintenance Servicing Functions. *Nat. Safety News 112(12)*: 56-57 (1975).

Ridley, J. and D. Pearce: *Safety with Machinery*. Oxford: Butterworth-Heinemann, 2002.

Standards Australia: Safe Working in a Confined Space (AS 2865-1995). Homebrush, New South Wales: 1995.

Summers, A.E.: Setting the Standard for Safety-Instrumented Systems. *Chem. Eng. 107(12)*: 92-94 (2000).

Willwerth, E.: Maritime Confined Spaces. *Occup. Health Safety 63(1):* 39-44, 1994.

# B  Standards Issues

## INTRODUCTION

Appendix A and Appendix B are companion and interdependent documents. Appendix A focused on presenting and summarizing consensus standards and regulatory Standards of potential importance to management of hazardous energy in equipment, machines, and systems. Of necessity by the nature of its focus, Appendix A, while highlighting issues that have arisen and divergence in direction pursued by different agencies was unable to elaborate on these issues without submergence to the exclusion of the other information under discussion.

Appendix A introduced an important issue that affects all entities required to manage exposure to hazardous energy: the conflict between consensus standards and regulatory compliance,and the emergence of risk-based approaches to hazard management.

## Conflict between Consensus Standards and Regulatory Requirements

Consensus standards reflect the opinion of a group of people. The group is actually a collection of sub-groups, often structured and chosen to reflect diverse interests within the community. Diversity of opinion is an important consideration for ensuring balance in the advocating of interests that form the overall outcome of the process. Rare indeed would be the individual or entity represented on a standard-setting committee who acted selflessly and without motive for possible, potential, or actual gain. Standards organizations recognize these realities and consider them to be important to the overall direction and outcome of the standard-setting process.

Standards organizations seek balance in the output of standard-setting groups through the process of consensus. Consensus seeks resolution for differences that arise through the normal expression of opinions within a standard-setting group. The process of seeking consensus can be prolonged. Such is the value attached to resolution of these differences. Consensus-seeking can occur at multiple levels and can involve appeal to a higher authority for final judgement.

The hope is that the difference is resolvable through spirited attention to the detail of concern and the attempt to create compromise acceptable to both sides of the issue. Often this approach is acceptable to the parties involved in the dispute.

This, however is not always the case. Sometimes the matter proceeds to a vote of the committee involved in the standard-setting process and sometimes to resolution through appeal to a higher authority. At some point, someone or a group makes a decision in these unresolvable matters. That decision may not please all participants in the process.

The reader and follower of consensus standards must recognize the strengths and limitations of these documents. The strength is the involvement of the views of many different individuals and entities in the process of standard-setting. The limitation is the acknowledgment that the view presented in the final document may not reflect consensus of all participants identified in the introductory sections and that the document may not contain important input.

The nature of the relationship between consensus standards and the regulatory framework ensures that the former are always at a disadvantage. This occurs because the regulatory requirement has legal force. Requirements stated in consensus standards are voluntary and have no legal implications unless they counsel against following regulatory requirements.

Consensus standards often form the basis for regulatory requirements and sometimes are adopted as written. The latter situation provides tremendous economic advantage to standards-setting organizations. These organizations derive income from the sale of these documents. Hence, there are considerable economic implications for them following adoption by a regulator of one or more of their documents. This has led to competition for the attention of regulators when consensus standards effectively compete with each other by focusing on similar areas of endeavor. This situation creates considerable confusion in the minds of entities affected by regulatory requirements and solicited by the publishers of consensus standards.

Compounding this situation is the tendency of consensus standards to be interdependent and not self-contained. Thus, the standard of interest can refer to other standards and documents. Each reference widens the scope of documents required for purchase in order for an entity to be able to understand the full implication of a particular regulatory requirement.

The other approach taken by regulators is to create a regulatory Standard based in whole or in part on an existing consensus standard. This situation instantly renders obsolete the consensus standard. This means that the standard-setting committee no longer has a modus vivendi or must adopt a different approach. This situation effectively means that the consensus standard must "catch-up" to the regulatory Standard.

In either event, the regulators owe a tremendous debt to the committees that create consensus standards, for without them, regulation-setting would be considerably more difficult, if not impossible. The standard-setting committee created something from nothing, discussed alternatives, and resolved disputes regarding different points of view. The end-product is the most proactive document capable of survival and acceptance.

The rendering obsolete of the work of a standard-setting committee through the action of regulators poses a point of decision on these committees about which direction, if any, to pursue. One option is to explore areas outside the boundary of the regulatory Standard. This is the approach pursued by the ANSI/ASSE Z117.1 committee regarding the consensus standard on confined spaces (ANSI/ASSE 2009). The OSHA regulatory Standard applies only to permit-required confined spaces (OSHA 1993). Permit-required confined spaces pose a high level of risk of harm.

The ANSI/ASSE Z117.1 Committee has applied good practices to confined spaces that do not meet the permit-requirement (ANSI/ASSE 2009). OSHA indicated its belief that other regulatory Standards would suffice in the management of hazardous conditions in non-permit confined spaces. The OSHA viewpoint omitted consideration for the reality that hazardous conditions in confined spaces are not static in space and time. Extending confined space considerations to non-permit confined spaces establishes a seamless approach to hazard management in these workspaces.

Another option available to standard-setting committees is to propose approaches for hazard management not available in regulatory Standards. This was the approach taken by the ANSI/ASSE Z244.1 committee on control of hazardous energy (ANSIASSE 2004). ANSI/ASSE Z244.1 expressed considerable frustration with the lack of success of conventional approaches to managing hazardous energy during shutdown and servicing of equipment, machines, and systems. ANSI/ASSE Z244.1 decided to pursue alternate approaches based on risk assessment as alternatives to lockout and tagout.

OSHA (2004) pointed out the risks inherent in an approach that encroaches on the territory of a regulatory Standard. OSHA commented that the regulatory Standard (OSHA (1989) is fixed in time and space, unless or until amended. As a result, a more progressive consensus standard has no standing in consideration by OSHA compliance officers unless the protection offered as a result of following the recommendations provide equal or greater protection to workers. Even if the latter is true, not adhering to the requirements of the regulatory Standard still constitutes a violation, even though de minimis.

Important additional considerations for standard-setters are the compliance documents issued by regulators, such as OSHA. The compliance document for 29CFR 1910.147 on lockout and tagout has undergone revision and provides extensive guidance about application of the regulatory Standard (OSHA 2008). This document is the basis for actions taken by compliance officers.

NFPA 306 and 29CFR 1915 provide an example of peaceful co-existence between a consensus standard and a regulatory Standard (NFPA 2009; OSHA 1994). These documents apply to shipyards and shipbuilding and shipbreaking. Marine chemists adhere to the regulatory requirements of 29CFR 1915 and use NFPA 306 to provide recommendations for day-to-day upgrading and improvement to their practice. This is an example of the use of a consensus standard to govern the practice of a group of professionals given the reality of infrequent upgrading of the regulatory Standard.

This discussion highlights the fact that there is no parallel universe in the practice of occupational health and safety. There is no protection for entities to be had by adhering to the recommendations of consensus standards compared to regulatory requirements unless the former provide equal or greater protection to workers. This situation puts the onus on committees involved in standard-setting to be aware of and then to factor in regulatory requirements in order to protect the credibility and usability of the standard, and to ensure the well-being of entities that follow the recommendations provided in these documents. Requirements provided in consensus standards constitutes recommendations. Requirements provided in regulatory Standards are legally binding on entities. Regulatory Standards trump consensus standards at all times.

## Issues Concerning Standards on Hazardous Energy

Chapter 1 indicated that energy is everywhere in normal workplace surroundings and a major cause of accidents. The harnessing of energy from energy sources and materials containing energy is intrinsic to the activity that occurs in all societies, regardless of their state of development. The misuse and inappropriate use of energy sources and materials containing energy have caused countless injuries and deaths. Much of the modus vivendi of the safety profession focuses on minimizing the hazardous use of energy sources and materials containing energy.

Even though energy is always present, it is not always hazardous or dangerous. The danger from hazardous levels of energy occurs when the amount exceeds human tolerances. Yet, hazardous forms of energy are allowed to be present in our homes and places of work. Periodically we read about expression of hazardous energy through incidents that are rare events that occur at low levels of probability. These succinct phrases express the nature of the situation faced by standard-setters and regulators regarding the presence of hazardous forms and levels of energy and our interaction with them at home and at work.

A consensus standard, such as the current version of ANSI/ASSE Z244.1-2003 that attempts to provide a problem-solving model as an outcome, needs a clear statement of the problem as a starting-point (ANSI/ASSE 2004). The statement of the problem must reflect analysis to determine as much information as possible about the nature of the problem and the underlying contributing factors prior to proposing solutions.

Analysis of fatal accidents presented in Chapter 1 shows that workers either had the ability to influence the conditions of work or they did not, and that those who had no influence were involved more frequently in accidents involving hazardous energy. Some of the workers who had influence over the conditions of work took no precautions to prepare for the task that led to the fatal accident. They took these actions out of ignorance of hazardous conditions, and of existing requirements and procedures, and sometimes out of defiance (failure to follow existing procedures). In some circumstances, individuals restarted equipment and machines that were deactivated and de-energized through control switches, and sometimes removed locks and lockout devices placed by themselves and sometimes by others.

What is reported in Chapter 1 regarding the accidents is behavioral in nature and reflects the actions of people. These accidents are rare events, and as a result, are very difficult to predict and very expensive to prevent. The data show no indication of failure by equipment to provide protection. These observations suggest that the approaches that are necessary are to address and control the presence of energy in equipment, machines, and systems and the actions of people who interact with it.

### Hazardous Energy and Equipment and Machines

Recognition of the life cycle of equipment and machines is central to the challenge of formulating standards and regulations for managing exposure to hazardous energy. (Managing in this context means eliminating or at least controlling to safe levels.) Operation of equipment and machines usually requires levels of energy that are hazardous to humans. The whole modus vivendi and modus operandi of equipment and machines is to harness energy and forces in a manner that produces a desired outcome in the most efficient manner possible and to cause no harm to the installer and set-up person, operator, service person, and maintainer and upgrader in the process.

The life cycle of equipment and machines is complex and contains many identifiable events (Table B.1), as revealed through comments provided by interveners to Standard-setters attempting to create functional documents for guidance and enforcement (OSHA 1989). Each of these events contains

## Table B.1

## Events in the Life Cycle of Equipment and Machines

Erection

Installation

Commissioning

Normal operation

Lubricating

Cleaning

Unjamming

Minor adjustment

Simple tool change

Disassembly

Repair

Replacement of components

Upgrading

Decommissioning

Source: OSHA 1989.

sub-events. Each of the sub-events contains some characterized and likely many uncharacterized work activities, and hence, many possible routes of exposure to sources of energy and possibly hazardous energy. Compounding complexity of the situation is the broad spectrum of equipment and machinery used in industrial activity, the manner in which it is designed and intended to be used, and practices that have evolved around these uses in different industrial sectors.

As indicated in Appendix A, standard-setters and regulators have approached this problem by segregating operational issues from servicing and maintenance. As indicated in Chapter 7, this separation is difficult to achieve because there are elements characterized as operational that involve potential contact with sources of hazardous energy. The main difference between exposure to the sources of hazardous energy during operation versus servicing and maintenance, and the associated elements of erection, installation, commissioning, upgrading and decommissioning is that the exposure during the former is likely to occur more frequently.

The risk of harm introduced by the characterized and to some extent uncharacterized work activities that occur during operation is manageable through use of safeguards, as well as special tools and alternate procedures that prevent contact between the body and the source of energy. For the purposes of this discussion, these sources are related to the internally supplied energy sources used to power the equipment or machinery, and not to externally supplied energy used to move components for installation and removal. Internally-supplied energy sources include electricity (AC and DC), pressurized gases (air, nitrogen, steam), fuel gases (natural gas, propane), liquids (water, process fluids), and pressurized liquids (hydraulic fluids). Externally supplied energy is primarily related to lifting and movement of components and the intact equipment or machine.

Readers must consult the regulatory requirements applicable to their jurisdiction to ensure compliance with specific requirements. Subsequent sections examine recent trends in the development of consensus standards for managing exposure to hazardous energy.

**The Lockout Paradigm**

The lockout paradigm (Table B.2) is the approach taken during servicing and maintenance, and the associated elements of erection, installation, commissioning, upgrading and decommissioning. The lockout paradigm as expressed in consensus standards contains a number of elements. Everything in Table B.2 has appeared in previous discussion. This is to be expected because the lockout paradigm has a history that long precedes its expression in consensus standards and regulatory Standards. The various consensus standards add varying additional requirements to this framework.

As discussed in Chapter 7, the general starting-point for action is anticipation of the possible presence of sources of hazardous energy in the physical surroundings of an entity. Encoding the cue for the critical first elements, anticipation and recognition, in the process in workplace environments has proven elusive in consensus standards and regulatory Standards. Without anticipation and recognition of the presence of energy sources and the potential for development of hazardous conditions, satisfying the requirements of consensus standards and regulatory Standards is not possible.

The agent common to the implementation strategies of the approaches reviewed in Appendix A is the Qualified Person. The typical description for a Qualified Person is an individual, who by education, training and experience, is qualified to perform the required duties and actions. Rarely, however, are the qualities of education, training or experience, required in the Qualified Person outlined in a manner that enables a precise fit with content of programs available at educational institutions.

Qualified Persons perform duties requiring varying levels of knowledge and skill. Some jurisdictions vary the level of qualification according to the requirements of the task to be performed. This provides flexibility and spreads the duties of the Qualified Person among the skills acquired by many individuals. Thus, individuals perform tasks within the level of knowledge and expertise that they possess relative to the level demanded by a particular requirement. For example, one Qualified Person assesses conditions, while another effects the deactivation, de-energization, isolation, lockout, and confirmatory testing required prior to the start of work.

Another theme common to most approaches is top-down management style. Management, acting through the Qualified Person, identifies and assesses the hazard, and prescribes work practices and procedures and control and emergency measures. That is, management determines the conditions of work and specifies how the work is to be performed. The people who will perform the work and who are injured because of faulty decisionmaking at all levels often have no role in the process other than to follow instructions. Individual consensus standards often assign different names to the Qualified Person.

A variation on this approach mandates active involvement by the Occupational Health and Safety Committee in the process. In this approach, management engages a Qualified Person who assesses the situation, determines actual and potential hazards, and recommends actions to be taken and possibly generates work procedures. Management presents the hazard assessment to the Health and Safety Committee and then prepares work procedures, and control and emergency measures. Management then presents these to the Health and Safety Committee for review and discussion.

Consensus standards generally contain requirements for hazard identification and assessment, specification of work practices and procedures, control measures, training and education, recordkeeping, and communication and emergency measures.

Readers must consult the regulatory requirements applicable to their jurisdiction to ensure compliance with specific requirements. Subsequent sections examine recent highlights in the development of consensus standards for managing exposure to hazardous energy.

## Table B.2
## The Lockout Paradigm

**Element**

Identify sources of hazardous energy

Determine the manner of energy input, flow, storage, conversion, and output in equipment, machines, and structures

Devise means to deactivate, de-energize, and isolate sources of hazardous energy:
Shut down electrical circuits and apply grounds
Prevent inflow of compressed air, gas, steam by closing valves
Prevent inflow of water and other liquids by closing valves and/or breaking lines
Drain or bleed residual pressure in lines containing pressurized fluids
Block or immobilize components subject to action by gravity or free rotation or reciprocation

Apply keyed padlocks to lockout/energy isolating devices to prevent casual reactivation of sources of hazardous energy; each person affected applies a lock

Test to determine the efficacy of the lockout procedure to ensure that sources of hazardous energy are not present at the start of work

Reassemble and restart the machine, equipment, or structure according to a procedure or checklist and ensure that components are restored to appropriate position and setting

## Risk Control Strategies

Until recently, responsibility regarding control of exposure to hazardous energy by equipment, machines, and systems rested solely with the owner. The owner usually was an employer. Regulatory agencies directed responsibility solely in this direction. This meant the purchaser of equipment who was responsible for safety during its use had no influence upstream to other entities that had control over design, selection and integration of components, and assembly of the end-product, a piece of equipment, a machine, or system. That is, employers could exercise no influence on equipment, machines, and systems on whose intrinsic safety the lives of employees and the survival and prosperity of the entity were totally dependent.

This situation changed in 1995 with the imposition in Europe of the Machinery Directive (EC 1998). Under the meaning of the Machinery Directive, a machine is an assembly of linked parts or components, at least one of which moves. A machine contains the appropriate actuators, control and power circuits joined together for a specific application, in particular, the processing, treatment, moving or packaging of a material. The codified version of the Machinery Directive also covers components intended to fulfil a safety function, the malfunctioning or failure of which endangers the safety or health of exposed persons.

The Machine Directive obliges manufacturers of safety components and machines to perform an assessment to identify hazards intrinsic to operation of their products and to eliminate or at least to reduce them to an acceptable level. Hazard elimination or reduction is to occur through inherently safer design. The second requirement is provide protection against residual risks not possible to eliminate and the third requirement, to inform users about residual risks. The Machine Directive requires adherence to standards listed in the Official Journal of the European Union. Appendix A describes some of these standards.

The updated Machinery Directive added safety-related logic (safety-related parts of control systems) to existing requirements (EC 2006). The updated Machinery Directive Machinery increased the scope of coverage from machinery, safety components, interchangeable equipment, lifting accessories, and removable mechanical transmission devices to include chains, ropes and webbing, and partly completed machinery.

Regarding the safety and reliability of control systems, the updated Machinery Directive requires the parameters of the machinery not to change in an uncontrolled way, where such change may lead to hazardous situations. An automatic stop must activate in wireless control units when correct control signals are not received. This includes loss of communication.

Regarding control devices, the operator must be able to ensure in each control position that the danger zone is clear of people prior to restart. Alternately, the design and construction of the control system must prevent start-up while someone is present in the danger zone. Where there is more than one control position, use of one must preclude use of the others, except for stop controls and emergency stops. Emergency stop devices must back-up other safeguarding measures, not substitute for them.

The operating method must indicate the procedure to follow in the event of accident or breakdown and enable safe unblocking of equipment.

The updated Machinery Directive further emphasizes the importance of harmonization of safety standards applicable to equipment, machines, and systems. As a result, harmonized safety standards apply to components, subsystems, the assembled equipment, machine, or system and to the software that operates it.

The Machinery Directive provides an order of priority for the choice of methods to eliminate or to reduce risks through a five-step method. The challenge is to take actions that are effective and balanced to machine-friendly operation. The first steps are to eliminate or to reduce risks by design and construction and to move the work tasks outside the risk area. The next steps are to use guards and safety devices, to develop safe working procedures, information, and education, and lastly, to use warning devices, such as pictograms, lights, and sound sources. When one approach does not provide full protection, additional measures are required.

Safety standards created in the U.S. and Canada, and published by ANSI (the American National Standards Institute) and CSA (Canadian Standards Institute), respectively are attempting to harmonize with the European standards published by IEC (International Electrotechnical Commission) and ISO (International Standards Organization). The latter groups have shown aggressive leadership in this area.

## Deterministic Approaches

The first Machinery Directive operated through EN 954-1:1996 (CEN 1996). EN 954-1:1996 is a harmonized European standard, since superseded, concerned with safety-related parts of control systems. EN 954-1:1996 addressed architectural requirements in a deterministic way through five categories of increasing rigor. EN 954-1:1996 covers all technologies.

EN 954-1:1996 provided a means of describing the fault-tolerance of circuits. It did not provide a way of relating the extent of risk to the requirements of the fault tolerance. EN 954-1:1996 used phrases such as basic safety principles, well-tried safety principles and well-tried components and automatic checks at suitable intervals. Table B.3 provides information about design concepts that reflect basic safety principles (BGIA 2008). Table B.4 provides information about design concepts that reflect well-tried safety principles, and Table B.5, information about well-tried components.

The concepts reflected in "basic safety principles," "well-tried safety principles," and "well-tried components" have stood the test of time. The limitation intrinsic to these concepts is that they reflect a backward-looking dependence. Every additionally accepted inclusion to these groupings can occur only by reference to historically developed products and devices. By insinuation, inclusion cannot include current or future developments in the technology of control. This situation is highly restrictive and provides no avenue for the inclusion of improvement through change.

## Table B.3
## Basic Safety Principles

### General Basic Safety Principles for All Technologies

Materials and manufacturing processes selected in consideration of compatibility with anticipated operating conditions.

Dimensioning and geometry of components selected in consideration of compatibility with anticipated operating conditions. Further criteria include switching capacity and frequency, electric strength, static pressure, behavior of dynamic pressure, volumetric flow, temperature and viscosity of hydraulic fluid, and type and condition of hydraulic fluid or compressed air.

Components resistant to the environmental conditions and relevant external influences which are usual for the application. Important criteria include mechanical influences, climatic influences, the extent of sealing of the enclosure, and resistance to electromagnetic interference.

De-energization (closed-circuit current principle) as the safe state attained by removing the control signal (voltage, pressure). Important criteria include the safe state when the energy supply is interrupted, and effective spring resetting in valves used in fluid power technology.

Protection against unexpected start-up caused by stored energy or restoration of the power supply

### Pressurized Fluids

Pressure in a system or subsystem prevented from rising beyond a specified level by one or more pressure-relief valve(s). Pneumatic systems use pressure-control valves with secondary ventilation.

Avoidance of impurities in the pressure medium by filtration and dehumidification

### Electrical Technology

Connection of one side of the control circuit, one terminal of each electromagnetically actuated device or one terminal of other electrical devices to the protective grounding/earthing conductor. This side of the device is not used for deactivation of a hazardous movement. A short-circuit to ground cannot therefore result in undetected) failure of a de-energization path.

suppression of voltage spikes through RC element, diode, or varistor connected in parallel with the load (not in parallel with the contacts).

### Programmable Systems and Software

monitoring the program sequence/software modules for defective commands that occur despite the taking of all care during verification and validation. Monitoring generally occurs through an external, cyclically re-triggered watchdog capable of placing the equipment, machine, or system in a defined safe state in the event of detection of a defective program sequence.

Source: BGIA 2008.

---

EN 954-1:1996 was viewed as an oversimplification of safety concepts and very subjective or qualitative. The standard failed to force designers to assess the reliability of safety-related components. Standards currently in use (refer to subsequent discussion) have added quantitative calculations to the qualitative requirements of the previous standard by considering the likelihood of failure of components in the safety-related system.

## Table B.4
## Well-Tried Safety Principles

### General Well-tried Safety Principles for all Fechnologies

Over-dimensioning ensures use of equipment to loading below rated values to reduce the probability of failure.

Positive (reliable) actuation through contact by rigid mechanical parts rather than sprung connections. This provides reliable transmission of commands by the direct opening of a contact on actuation of a position switch, even when the contact is welded.

Limiting electrical and/or mechanical parameters (force, distance, time, and rotational and linear speeds) to permissible values in electrical, mechanical, or fluid power equipment to improve hazard control.

### Fluid Power Technology

Secure position maintains the moving element of a component mechanically in a secure position. Frictional restraint is not sufficient. Force is required to change the position.

Use only well-tried springs.

### Electrical Technology

Limit voltage, current, energ,y or frequency to prevent an unsafe state.

Avoid undefined states. The design of the control systemmust allow predetermination of status under all anticipated operating conditions. This is to be achieved through use of components having defined behavior (switching thresholds, hysteresis) and sequence of operation.

Separate non-safety functions from safety functions to prevent unanticipated influences upon safety functions.

### Programmable Systems and Software

Use self-tests (CPU function, memory) to detect faults in complex components such as microcontrollers.

Depending upon the application, fault detection by the process and fault detection by comparison between channels may be regarded as well-tried safety principles.

Source: BGIA 2008.

---

This is the situation expressed by OSHA in the letter concerning the review of the 2003 version of the ANSI/ASSE Z244.1 consensus standard on control of hazardous energy (ANSI/ASSE 2004; OSHA 2004). OSHA indicated that alternate means of approach were acceptable as de minimis violations provided that they provided protection to workers equal to or greater than that mandated in the OSHA Standard on lockout/tagout, 29CFR 1910147 (OSHA 1989). Neither OSHA nor any other interested party has provided any attempt at assessment to this point for determining the validity of a claim following publication of the letter.

### Risk-Based Approaches
The discussion in the previous section indicated the difficulty associated with approaches to safe operation of equipment and machines that depend on backward-looking approaches design of safety-related systems. The European Union approached this question in the 1990s and created a series of

## Table B.5
## Well-Tried Components

**Well-tried Components for all Technologies**

Well-tried components minimize or exclude critical faults or failures and thus reduce the probability of faults or failures which impact upon the safety function. General criteria for a well-tried component are wide use in the past with success in similar applications, or manufacture and verification through application of principles that indicate its suitability and reliability for safety-related applications. This consideration does not include complex electronic components, such as PLCs, microprocessors, and ASICs. Classification as a well-tried component in part, is application-specific. A component considered well-tried in one application is not acceptable in others owing to the environmental influences.

**Mechanical Technology**

Springs that meet rigorous performance criteria

**Fluid power Technology**

Directional control valves, stop valves, and pressure valves

**Electrical Technology**

Fuses and miniature circuit breakers used as intended and correctly rated are highly reliable.

Emergency off device/emergency stop devices with direct opening action for interruption of the energy supply. Type 1 switches contain only one contact element, in the form of a direct opening contact. Type 2: switches contain one or more break contacts and possibly one or more make contacts and/or one or more changeover contacts. All break contacts, including the contact-breaking parts of the changeover contacts, must feature direct opening contact elements.

Switches employing positive mode of actuation (direct-opening contacts) are available commercially as pushbutton, position switches, and selector switches with cam actuation for selection of operating mode. These switches have proved effective over many decades.

Non-complex and non-programmable passive components, such as resistors, diodes, transistors, thyristors, operational amplifiers and voltage regulators, having predictable failure modes.

Source: BGIA 2008.

harmonized consensus standards first published in the early 2000s and described in Appendix A. These standards have set the direction world-wide for forward-looking approaches to risk in the design, manufacture, operation, and maintenance of equipment and machines. These standards contain well-researched concepts relating to reliability and failure.

Prior to proceeding further, discussion about risk is essential. Risk-taking and risk-avoidance are parallel themes and intrinsic to life (Urquhart and Heilmann 1984). Fear drives people to avoid risk, yet risk is unavoidable in all endeavors. The balance between risk avoidance and risk acceptance reflects the age and knowledge of the individual.

Previous discussion has mentioned the OSHA requirement to provide a level of protection from hazardous energy. Accompanying attainment of that level of protection is the risk associated with the actions required to reach it. That is, attaining that level of protection is not free from risk. That risk currently is not quantified and not part of consideration in this type of discussion. The inconvenience factor reflects the acceptance of risk in taking an action in order to avoid the cost in time involved in a seemingly more onerous course of action.

The manner in which we respond to the underlying risk intrinsic to a situation is an important consideration in this discussion. Sudden, surprising news triggers shock, horror, and fear in people (Urquhart and Heilmann 1984). People crave stability in their lives and not destabilizing situations. Learning about risk through the element of surprise compromises our ability to balance risk and opportunity. This is the consequence arising from the manner in which information is transmitted by the news media. This situation has biased people toward overestimation of risk and underestimation of opportunity.

An accident that befalls a group has greater impact than loss of the same number of people in individual accidents. The larger the group, the worse seems the accident and the helplessness of the situation. This is especially true for traffic accidents where large numbers of people die in a single vehicle. The manner in which the news media operates is to focus on the victim. The plight of victims is emotionally gripping and impossible to ignore or to downplay. This situation with its micro focus on victims of a current disaster avoids focus on the macro situation concerning individuals not involved in the occurrence. This emphasis again creates a bias toward fear of the circumstance. The continual reportage of fear-inducing subjects and contradictions to long-standing "truths" leaves the public in the position of not knowing what to eat, what to drink, what to wear, where to live, and ultimately, what to think and whom to believe.

The word, risk, brings with it considerable confusion. If something can happen, it will happen. That something can be good, as in good fortune or good luck, or in varying degrees of bad, extending from minor to major inconvenience to minor to major calamity to death. The fact is that the something that can happen will happen is a given. The question remaining is when will this occur. The question of when introduces the element of probability into the discussion.

Discussion about the concept of probability can occur in different ways. When talking about defects in equipment, one defect occurs in 100,000 units, for example. The question that this information exposes is which of the 100,000 units is the defective one, one near the beginning of production, one in the middle of a production run or one near the end. This question then becomes a concern about reliability. The classic expression of reliability is the "bathtub" curve (Figure B.1). The "bathtub" curve is similar in shape to the transverse section of a bathtub. We learn from experience that items fail either near the beginning or near the end of their functional lives. The units of measurement of functional lifetime can include time units and repetition of cycles, as well as other possibilities. Determining a common unit for discussion becomes a challenge.

Reliability testing is tightly focused and controlled. When talking about risk involving events that occur in a distributed environment, such as a road network or a workplace or group of workplaces, the measure of risk is considerably less focused. Information about specific risks is considerably more difficult to obtain. To create an example in terms of time, let us say that one traffic accident occurs every 3 minutes, and one workplace accident, every 7 minutes. (These numbers are intended solely for illustration and are not based on specific information.) Each of these values provides a basis for calculating probability, but at a level of considerable vagueness.

Determining the values can occur through observation or testing, as occurs with components, and increasingly, with finished products and manufactured items. Testing can establish failure based on number of cycles of opening and closing a valve or some other critical component. Testing of components and products to establish reliability is becoming increasingly important in the quest for inherently safer design of safety-related components of equipment and machines.

Risk is voluntary or imposed. People accept voluntary risk whenever they undertake activity in life controlled by their decision-making. People tolerate these risks even when informed about their magnitude, as in the case of uses of vehicles for transportation and enjoyment-related activities, such as skiing and bungee-jumping. People readily accept these risks as a cost of being alive or avoid them by choice.

Imposed risk is the involuntary risk caused by other factors, including the actions of other people. These can include risk related to location of a business operation, power station, or other operation perceived as detrimental, or spillover resulting from the actions of other people, such as second-hand smoke. Imposed risk also results during and from workplace activity. These risks arise from the potential for disease and from traumatic injury

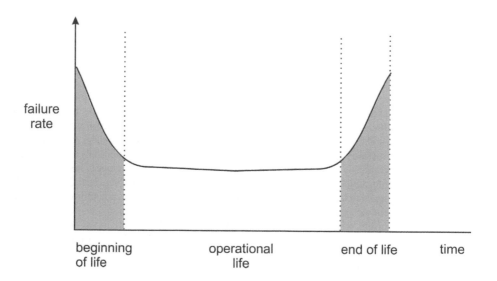

Figure B.1. The "bathtub" curve is a well-known visualization of the reliability of manufactured items. We learn by experience that products fail either early into their expected lives or near the end, but generally not in the middle.

People will accept levels of risk voluntarily that are 10 to 100 times greater than risks that are involuntarily imposed on them (Urquhart and Heilmann 1984). This factor is the cost of imposition of risk and is not something easily to be renegotiated as communicators have discovered over the years. Assumption of involuntary risk is not readily accepted by people. Their perception of safety is under challenge and reflects something about which people do not want to discuss. People locate in certain areas because of economic realities. They must accept the reality that this situation brings to their lives, but do not wish to engage in discussion about it or to be reminded about the implications of their circumstances. Discussion about imposed risk appears to reflect perceptual values of advantage versus disadvantage, as reflected in lifestyle forced on people by economic realities.

This situation and unnecessary concern brought about by perception has profound implications about regulatory decisions and future progress. Acceptance of risk by decision-makers who report through political masters instead of absolute safety and freedom from harm to all users is politically unacceptable. This absolute is reflected in decisions about the safety and marketability of prescription and nonprescription legal drugs, despite the fact that everyone knows that risk of negative consequences accompanies life and that legal drugs prolong and improve the quality of life (Urquhart and Heilmann 1984).

These considerations also apply to activities involving equipment and machines. Requirements for safety involve states of attainment. Safety is a relative and not an absolute term. Accident statistics show the decrease in the rate of injuries and fatalities over the years (Figure B.2). This decrease indicates that the risk of work, as shown in these statistics has also decreased considerably. The perception of risk of work is always relative to the time at which the view is solicited. In the absence of the statistical information shown in Figure B.2, coverage of workplace tragedies by the news media could easily create the impression that work activities pose high and unacceptably high levels of risk. Such is the relative nature of perception.

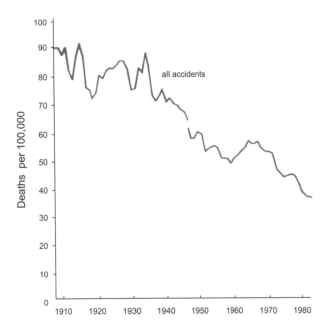

Figure B.2.  Safety is relative to the point in time following the keeping of statistical records. Maintaining the perspective of time is essential to understanding how concerns about risk have changed with time.

In the U.S., the appearance of consensus standards years ahead of regulatory requirements in the same area argues for the importance of these documents in improvements reflected in the trend of the graph in Figure B.2. The letter from OSHA to Edward Grund concerning the status of ANSI/ASSE Z244.1-2003 regarding 29CFR 1910.147 indicates the rigidity of the regulation-setting process. Regulatory Standards are fixed in time and space (ANSI/ASSE 2004; OSHA 2004).

The regulatory Standard on lockout/tagout set levels of performance based on the judgement of OSHA. Complying with the Standard to attain the level of performance involves risk not factored into the requirement. Depending on the level of risk, this can considerably detract from the achievement implicit in the level of attainment. Setting a regulatory Standard at too high a level introduces the issue of failure to comply willingly due to the inconvenience factor,. and the voluntary assumption of risk by individually affected employees.

To illustrate the nature of this concern, people detest impracticality and will often take the easiest path (the so-called path of least resistance). If that path involves recognized risk, people have accepted it as the price on one side of the balance against impracticality on the other side. So, the question then becomes how to minimize both the risk and the impracticality.

Any approach that improves on doing nothing without creating new and additional risk will provide benefit. The standard-setting group must decide at which point the benefit produced by an approach exceeds the threshold for inclusion in the group of acceptable approaches.

Lockout is easily defeated using readily available tools and considerable effort. This comment is even more applicable to tagout. Lockout and tagout rely on psychological mechanisms for their effectiveness. The psychology of respect for authority creates the barrier against unauthorized, deliberate defeat of the system. Hence, the threshold of effort needed to defeat lockout which is considered absolutely secure is very low. Hence, the benefit is potentially low also because of the ease of defeat.

Driving change in consensus standards at this time is the recognition that zero risk is an unattainable outcome. While standard-setters can acknowledge this reality, as discussed, Standard-setters cannot outwardly do so. This is the situation currently facing both groups. Review of the accident records in Chapter 1 indicates that failure of equipment and machinery was not an identifiable factor in these accidents. That is, the accidents reflect the actions of people.

An outcome that provides safety achievable by different means is not necessarily an absolute condition (Figure B.3). Different means can produce the same outcome (safety) with different levels of certainty. The level of certainty becomes a key determinant of the level of acceptability of a particular approach. Hence, the outcome becomes more of a relative than an absolute state of attainment. The key determinant in this approach is that the certainty of the attainment is acceptable to the group that sets the standard, and ultimately, the Standard.

Targets of this type are expressible in terms of probability (Table B.6). This provides a way to link between OSHA requirements expressed in 29CFR 1910.147 and the interpretive letter, and risk-based requirements expressed in ISO and IEC Standards (OSHA 1989, OSHA 2004).

Establishing a level of acceptable risk opens options for consideration. Some options likely offer higher certainty than others. Intrinsic to each option is a detrimental cost in risk of accidents of other types and economic cost in dollars. The risk of the option therefore incorporates the reduction of risk provided by the benefit plus the detriment imposed by the risk of performing the action to attain the benefit. Any approach that considers only the perception of benefit without considering the detriment incurred in its attainment is flawed.

Without formal analysis that explores all of the sources of risk intrinsic to this option, this will never be known with any certainty beyond perception and conjecture. Studies have shown that perception as a measure of risk is highly unreliable (Urquhart and Heilmann 1984). Establishing the level of risk of any approach requires information-gathering. This can occur though performance testing of components and circuits to establish reliability, accident statistics, and other means appropriate to increasing the level of quantitative knowledge.

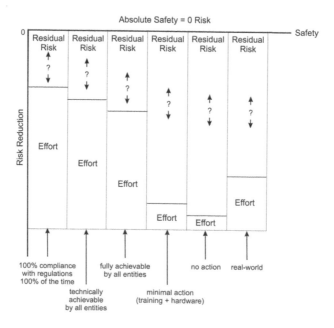

Figure B.3. Outcomes that provide safety are not absolute conditions.

## Table B.6
## Deterministic versus Risk-based Strategies

**No Action (no consensus standard or management action or regulatory Standard)**

What are the actions by which failure of safety (risk of harm) occurs?

What are the costs (risk of harm) imposed by each mode of failure of safety?

**29CFR 1910.147 (deterministic model)**

What are the outcomes specified by which safety is achieved?

What costs (risks of harm) do actions taken to achieve these outcomes impose?

By what mode(s) can the outcome fail to provide protection?

What is the risk of harm from failure of the outcome to achieve safety?

Is the level of protection at the required outcome fully needed to provide safety?

What reduction of risk of harm beyond what is needed does the outcome provide?

Do actions required to achieve the outcome deter routine implementation due to inconvenience?

What risk of harm does inconvenience impose on failure to implement actions to achieve the required outcome?

What is the net benefit achieved by implementing the required outcome?

**Risk-based Consensus Standards**

What strategy(ies) is(are) available to prevent injury for each action leading to loss of safety?

What reliability does each strategy provide?

By what mode(s) can failure occur?

What is the risk of harm from failure?

Do actions required to achieve the outcome deter routine implementation due to inconvenience?

What risk of harm does inconvenience impose on failure to implement actions to achieve the required outcome?

What is the net benefit achieved by implementing each strategy?

Does the net benefit achieved by implementing the strategy equal or exceed the benefit from implementing the required outcome in the deterministic approach?

## An Approach to Respond to the Requirements of OSHA

To the point of writing, no article has appeared in the public domain to address the requirements of OSHA regarding the value of alternate methods compared to isolation and lockout. The articles by Firenze (2005) and Piampiano and Rizzo (2006) appear to provide the most detailed and directed approach to this question.

Piampiano and Rizzo (2006) comment that OSHA has not formally recognized risk assessment as a means of evaluating and protecting against machinery hazards. Hence, the two situations are following parallel courses. These authors argue for the strong similarity between elements considered during a risk assessment and evaluation of compliance. One substantive difference is the absence of use of quantitative data on injury in the evaluation for compliance performed by OSHA. Use of quantitative data in the process of risk assessment considerably increases confidence in the judgement.

Piampiano and Rizzo (2006) comment that risk assessment techniques for safeguarding are qualitative and do not provide the certainty demanded by the rigor of comparison to OSHA criteria. The user must rely on judgement against words used to assess probability and severity. Correlating force to injury can considerably advance risk estimation by reducing subjectivity. This correlation enables more reliable prediction of worst credible harm that can occur.

In support of this concept, they cite data about contact temperatures needed to produce burn injury (Chengalur 2004) and pressures associated with crushing injury (Mewes and Mauser 2003). This information is critically important to a more quantitative approach to risk assessment in application of concepts contained in ANSI B11.TR3 (ANSI/AMT 2000).

The probability term is the most likely source of inconsistency in the risk assessment (Piampiano and Rizzo 2006). This assessment depends on evaluation of operator skill and behavior, characteristics that are not amenable to quantification. These authors argue that the strength of safeguards is a more reliable predictor of probability of accident. Redundant engineering controls are more reliable than single engineering controls, and redundant administrative controls are more reliable than single-level administrative controls. Single-level administrative controls (procedures, training, and certification) rely on human behavior and are intrinsically unpredictable. Redundant administrative controls (railings, presence-sensing devices, and integrated warning devices) still depend on behavior. Engineering controls reliant on hardware are more quantifiable because of testing to determine reliability.

A survey of outcomes of contested citations involving alleged safeguarding deficiencies involving machines and equipment performed by Piampiano and Rizzo (2006) indicated that OSHA must demonstrate not only the theoretical possibility of the exposure, but also reasonable foreseeability of entry into the danger zone. Foreseeability does not include misadventure by workers, according to this assessment.

Firenze (2005) wrote from the perspective of preparation for an appearance in court. In many respects, that is the nature of the challenge posed by OSHA in its letter to Grund (OSHA 2004). Firenze makes the point that design engineers do not fully assess the threat to people and property in their designs, the result being that the designs do not contain sufficient measures to ensure safety. Similarly, lawyers evaluating these situations fail to understand what the design engineer should have done to minimize the occurrence of the accident and its subsequent injuries. Firenze further comments that safety engineers provide the knowledge needed to bridge this gap.

Firenze (2005) injects the need for rigor in preparation for challenges of this type, and demonstrates the nature of argument and counter-argument related to issues involving risk. Compliance with a standard is not a substantive defense in an argument put forth to OSHA. This situation has received extensive discussion previously.

Firenze (2005) also introduces for consideration an index for calculating the probability of danger. Danger is the threat of harm to people and/or property resulting from the unreasonable or unacceptable combination of hazards, risks, and/or exposures coupled with failure to implement or use available safety measures.

## Functional Safety Standards

Standards, such as EN ISO 13849-1:2006 and IEC/EN 62061, and related standards focus on components, subsystems, and the overall machine or system (EN ISO 2006, IEC/EN 2005). EN ISO 13849-1:2006 is a machine-oriented and simplified version of IEC/EN 62061. These standards utilize reliability concepts.

EN ISO 13849-1:2006 and IEC/EN 62061 affect two types of users: designers of safety-related subsystems and designers of safety-related systems. Designers of safety-related subsystems face the burden of providing data ensuring that the design provides adequate integrity for the system. This requires testing, analysis ,and calculation. The designer of the safety system then takes that information and incorporates it into calculations to determine the overall performance level of the system.

These standards pertain to the functional safety of equipment and machines. Functional safety is the part of the overall safety that depends on correct operation of equipment or a system in response to inputs.

ISO 13849-1:2006 and IEC/EN 60261 require a multi-step approach to the design of the safety-related control system to ensure compliance with requirements of the Machinery Directive.

The first step is to identify safety-related functions that apply to the design. Safety-related functions carried out by safety-related parts of a control system reduce the associated risk to an acceptable level by bringing machinery to a safe stop, safe position, and/or safely limited speed (Table B.7). Each safety-related function requires a hazard assessment.

The hazard assessment begins with identification of hazard zones, and for each hazard zone, tasks and operations, and for each task and operation, accident scenarios. Accident scenarios indicate hazard, hazard situations and hazard events. IEC/EN 60261 provides a means for recording this information (Table B.8). (Rec. means Action Recommended.) For each hazard event, the analyst, has the choice of a simpler path using ISO 13849-1:2006 or a more complex path using IEC/EN 60261.

When using ISO 13849-1:2006, the analyst performs a risk assessment to define the required Performance level ($PL_r$) of the safety-related function(s). This follows the tree in a widely published graphic (Figure B.4). The first junction occurs with the decision about potential seriousness (S) of the injury. The choices are S1, slight (normally temporary or reversible) injury, or S2, severe (normally permanent or irreversible injury, including death). The next junction occurs at frequency (F) and/or duration of exposure to the hazard. The choices are F1, seldom to less often and/or short exposure time, or F2, frequent to continuous and/or long exposure time. The last junction occurs at possibility (P) to avoid the hazard or limit injury. The choices are P1, possible avoidance under special conditions, or P2, no realistic possibility of avoidance.

The path created by assigning the values of S, F , and P determine the level of acceptable risk in the performance of the safety-related function. The appropriateness of the assignment depends on the effort and skill expended on the choice of values for S, F and P. These selections require judgement and experience in evaluation.

The $PL_r$ needed for the individual safety function, defines performance into one of five levels, a, b, c, d or e. These levels form a hierarchy, with requirements for risk reduction increasing from a to e. The safety-related part that performs the entire safety function must maintain the required performance level. The designation from a to e describes the attributes of the safety-related system.

Systems generally contain three subsystems: input (a sensor or sensing element), logic and output (an actuator that does something). Dual parallel channel systems provide redundancy to reduce the potential for failure. That is, the probability of simultaneous failure of each channel of a dual channel system is extremely small and hence considerably increases reliability. A single logic processor can accommodate parallel channel input and output from each subsystem. Diagnostics that monitor for faults in each of the parallel channels and diagnostics incorporated into the logic subsystem further increase reliability. These features are the basis for satisfying the requirements for the designation from a to e.

The next steps in the sequence are to design the architecture of the safety-related system, to determine the actual Performance Level (PL) and to determine whether actual performance meets the requirements of the required Performance Level ($PL_r$). Validating the design to ensure that it meets the requirements of the initial risk assessment involves using data supplied by manufacturers on reliability of the components. These data include the Mean Time To dangerous Failure ($MTTF_d$), average Diagnostic Capability (DCavg) of the circuit and potential Common Cause Failures (CCF) of the components and circuits.

What remains is to determine what the safety function needs to do in order to fulfil its mandate. This decision requires knowledge and experience in the field of equipment and machine design. The user manual should indicate anticipated conditions of use. The user must confirm that these expectations match actual conditions of operation.

IEC/EN 60261 provides a more detailed path to analysis (IEC/EN 2005). This entails the table presented here as Table B.8 and an additional table (Table B.9) for determining the Safety Integrity Level required in electrical, electronic, and programmable control systems. Discussion in Appendix A indicated the relationship between required Performance Level ($PL_r$) and Safety Integrity Level (SIL).

On a practical level, danger points and hazardous areas are not accessible during normal production runs on machines containing proper safeguarding. However, tasks can necessitate opening of the

## Table B.7
## Safety-Related Functions

**Safety-Related Functions Performed by Safety-Related Parts of Control Systems**

Contactor monitoring with and without interlock/mechanical securing action

Permission switch

Safe timing relay

Standstill monitoring

Safe operation via two-handed control

Monitor open dangerous areas with contact mats

Highest speed monitoring

Safety-related stop function initiated by safeguard

Manual reset function

Start/reset function

Local control function

Muting function

Hold-to-run function

Enabling device function

Prevention of unexpected start-up

Escape and rescue of trapped persons

Isolation and energy dissipation function

Control modes and mode selection

Emergency stop function

Source: EN ISO 2006.

guards. In machinery containing multiple motor drives linked by material, such as paper or cloth, actions occurring in one drive must occur simultaneously in all in order to retain the relationship of forces applied to the transported material in all parts of the machine. These actions can include stop, start, slow to speed, speed up to speed, and so on.

Risk-based standards are intended to apply to the life cycle of equipment and machines. While this commitment sounds admirable and perhaps straightforward, this is a major undertaking. Design at some point must interface with commissioning, operations, and eventually, maintenance and modification and upgrade. The latter events in the life cycle of equipment and machines can severely

## Table B.8
## Hazard and Risk Assessment Table

**Risk Assessment**

| Consequence | Severity Se | Class (Cl) = Fr + Pr + Av | | | | | Frequency Fr | Probability Pr | Avoidance Av |
|---|---|---|---|---|---|---|---|---|---|
| | | 3-4 | 5-7 | 8-10 | 11-13 | 14-15 | | | |
| death, loss of an eye or arm | 4 | SIL 2 | SIL 2 | SIL 2 | SIL 3 | SIL 3 | $\leq$1 h = 5 | very high = 5 | |
| permanent injury | 3 | | Rec. | SIL 1 | SIL 2 | SIL 3 | >1 h to $\leq$24 h = 4 | likely = 4 | |
| reversible injury (medical) | 2 | | | Rec. | SIL 1 | SIL 2 | >24 h to $\leq$2 w =3 | possible = 3 | impossible = 5 |
| reversible injury (first aid) | 1 | | | | Rec. | SIL 1 | >2 w to $\leq$1 y = 2 | rarely = 2 | possible = 3 |
| | | | | | | | >1 y = 1 | negligible = 1 | likely = 1 |

**Hazard Assessment**

| Step | Hazard | Se | Fr | Pr | Av | Class | Protective Measure | Safe? |
|---|---|---|---|---|---|---|---|---|
| | | | | | | | | |
| | | | | | | | | |
| | | | | | | | | |
| Comments | | | | | | | | |
| | | | | | | | | |
| | | | | | | | | |

Source: Modified from IEC/EN 2005. **Rec.** means Action Recommended.

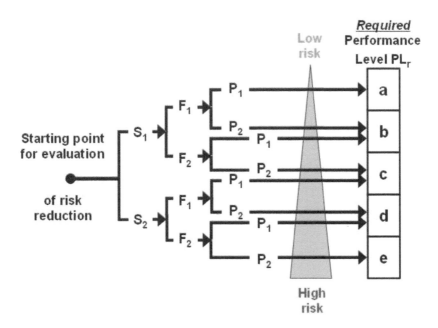

Figure B.4. Widely reproduced hazard analysis tree for EN ISO 13849.

challenge the capabilities of the safety-related system. For equipment and machines having a long life-cycle, the influence of such an approach is strongest at the beginning and weakest at the end.

In Europe, Machinery Directive (2006/42/EC) requires extensive technical documentation (EC 2006). This includes design drawings, risk assessments, testing and performance reports, health and safety requirements, explanations and instructions for use, protective measures implemented to eliminate identified hazards or to reduce risks and, when appropriate, the indication of the residual risks associated with the machinery. The documentation must include standards and other technical specifications used, indicating the essential health and safety requirements covered by these standards, and where appropriate, the declaration of incorporation for included partly completed machinery and the relevant assembly instructions for such machinery.

### System Safety Standards

Functional safety is only a part of overall safety, the determination of which occurs by considering the system as a whole and the environment with which it interacts. Machines do things to people, but people also do things to machines. The things that people do to machines are varied and reflect different motives and objectives. The things that people do to machines are deterministic. They reflect decision-making, however flawed. The actions taken by people against machines are not always predictable to designers. Accidents resulting from these actions are not random in this respect.

ISO 14121-1:2007 establishes general principles for assessing risk (ISO 2007). These principles consolidate knowledge and experience of the design, use, incidents, accidents, and harm related to machinery in order to assess the risks posed during the relevant phases of the life cycle. ISO 14121-1:2007 also provides procedures for identifying hazards and estimating and evaluating risk, and guidance on the type of documentation required to verify the soundness of the risk assessment.

ISO 14121-1:2007 uses a task-focused assessment. This reflects the interaction between the worker and the equipment or machine. Accidents involving the interaction of people with machines

**Table B.9**

**Relationship of Performance Level (PL) to Safety Integrity Level (SIL)**

| Performance Level (PL) | Safety Integrity Level | Average Probability of Dangerous Failure (1/hours) | Maximum Tolerated Dangerous Failure |
|---|---|---|---|
| a | no special safety requirement | $\geq 10^{-5}$ to $< 10^{-4}$ | 1/ 10,000 hours |
| b | SIL 1 | $\geq 3 \times 10^{-6}$ to $< 10^{-5}$ | 1/ 100,000 hours |
| c | SIL 1 | $\geq 10^{-6}$ to $< 3 \times 10^{-6}$ | 1/ 333,000 hours |
| d | SIL 2 | $\geq 10^{-7}$ to $< 10^{-8}$ | 1/ 1,000,000 hours |
| e | SIL 3 | $\geq 10^{-8}$ to $< 10^{-7}$ | 1/ 10,000,000 hours |

Source: EN ISO 2006; IEC/EN 2005.

are partly deterministic and partly stochastic. The overall rate is expressible in different ways. One way involves the overall operation in terms of incidents per number of hours or incidents per number of parts produced or incidents per number of operations. These statistics provide no information about the mechanisms by which the accidents occurred. These are expressible in terms of production or maintenance considerations. The two types of accidents can have distinctly different causes.

The first part of the process is to complete the hazard assessment. The hazard assessment form appears here as Table B.10 which is almost identical to Table B.8. For each machine, determine intended use and limitations. Then assess each machine for hazards and determine stages of the life and limitations of the machine. This occurs through listing the hazard zone and tasks and from them accident situations identifying the hazard, hazardous situations, and hazardous events. Tasks include, but are not limited to cleaning, setting, unblocking, inspection, maintenance, and repair. Associated with each task are many possible sub-tasks. Hazards include, but are not limited to, crushing, sheering, and impact.

For each hazard or hazardous situation (task), assess the probability of occurrence of harm (Table B.9). Table B.9 provides Safety Integrity Levels not found in the corresponding table in ISO 14121-1:2007. Brainstorming with knowledgeable people is advantageous because the process of selecting the probability of occurrence of an incident will be subjective because empirical data are unlikely to be available.

Risk estimation defines risk but not the severity of accidents and accounts for all foreseeable factors including misuse. The factors (frequency of exposure, probability of injury and avoidance of injury) are assigned a value and added together. This occurs regardless of the severity. That is, the severity is not included in the calculation of risk. The sum of the factors determines the class of the risk. The class of the risk compared with the severity of the injury determines the action required.

Additional factors to be considered include the number of people exposed, protracted time in the danger zone, level of skill and training of the operator, and length of time between access.

Analysis determines whether functional safety is necessary to ensure adequacy of protection against each significant hazard. Hazard elimination or risk reduction occurs through redesign, a safety related control system, work practices, information and training, and personal protective equipment.

IEC 62061 (discussed previously) uses an almost identical risk assessment and safety measures table as ISO 14121 (Table B.10). Note the assignment of Safety Integrity Levels to the boxes in the risk assessment table (Table B.8).

CAN/CSA Z460-05 on control of hazardous energy using lockout and other methods is another example of a risk-based consensus standard. CAN/CSA Z460-05 incorporated parts of ANSI/ASSE Z244.1-2003, and ISO 14121:2004 (ANSI/ASSE 2004; CAN/CSA 2005;.ISO 2004). (ISO 12100:2010 has superseded ISO 14121:2004.) Hence, this standard is a hybrid of both. Whereas ANSI/ASSE Z244.1 presents a list of tasks associated with performing risk assessment pertaining to lockout and isolation, CAN/CSA Z40-05 offers considerably more coverage to this subject. CAN/CSA Z460-05 is suitable for application at any time in the life-cycle of equipment or machines and provides a spot-check on conditions.

Appendix A provides examples of hazardous conditions and events associated with tasks. These provide a resource to the manufacturer and end-user to assist in identification. The lists of hazards, hazardous situations, and hazardous events are not exhaustive and are not presented in any order of priority. Therefore, the designer retains the obligation to identify and document all hazards, hazardous situations, or hazardous events that can occur in the machine, equipment, or process, regardless of occurrence in any of the lists provided in Appendix A.

Appendix B focuses on risk assessment and risk reduction. This process begins with identifying all tasks to be performed. These can occur during set-up, installation, removal, maintenance, operation, adjustment, cleaning, troubleshooting, and programming. The next step is to identify potential hazards associated with each task. These can include mechanical, electrical, thermal, pneumatic, hydraulic, radiation, and residual or stored energy hazards, motion, fuel, and human factors associated with each task. Human error, management system deficiencies, and foreseeable improper use of equipment also require consideration.

## Table B.10
## Hazard and Risk Assessment Table

### Risk Assessment

| Consequence | Severity Se | Class (Cl) = Fr + Pr + Av | | | | | Frequency Fr | Probability Pr | Avoidance Av |
|---|---|---|---|---|---|---|---|---|---|
| | | 3-4 | 5-7 | 8-10 | 11-13 | 14-15 | | | |
| death, loss of an eye or arm | 4 | Rec. | Action Required | | | | ≤1 h = 5 | very high = 5 | |
| permanent injury | 3 | | Rec. | Action Required | | | >1 h to ≤24 h = 4 | likely = 4 | impossible = 5 |
| reversible injury (medical) | 2 | | | Rec. | Action Req'd | | >24 h to ≤2 w = 3 | possible = 3 | possible = 3 |
| reversible injury (first aid) | 1 | | | | Rec. | AR | >2 w to ≤1 y = 2 | rarely = 2 | likely = 1 |
| | | | | | | | >1 y = 1 | negligible = 1 | |

### Hazard Assessment

| Step | Hazard | Se | Fr | Pr | Av | Class | Protective Measure | Safe? |
|---|---|---|---|---|---|---|---|---|
| | | | | | | | | |
| | | | | | | | | |
| | | | | | | | | |
| Comments | | | | | | | | |
| | | | | | | | | |
| | | | | | | | | |

Source: Modified from ISO 2007. **Rec.** means Action Recommended. **AR** means Action Required

Appendix C provides tables to assist in assessing the potential consequences for each task (Table B.11 and Table B.12). This assessment should consider the potential severity of injury to everyone who could experience harm. This consideration includes persons not involved in operating the machine, equipment, or process. The next step is to assess the potential exposure of all persons to the identified hazards. This assessment should consider the nature, duration, and frequency of exposure to the hazards.

Determining probability of occurrence is a critical factor in this assessment. This determination occurs through consideration of a number of factors. These include safeguards, safety devices, and safety systems either in use or to be used; reliability history and failure mode; operational or maintenance demands of the task; possibility of defeat or failure of safeguards; accident history relating to the particular task, activity, machine, equipment, or process, competence of everyone performing the task; and the working environment.

The final step is to evaluate the risk and to compare it against a benchmark deemed acceptable. This occurs through consideration of the information gathered for each identified hazard and task to determine the level of risk. The determination may involve risk comparisons, consensus appraisals, or informed value judgments. The key questions include adequacy of the safety level and the ability to perform the task without causing injury or damage to health.

If the risk is deemed too high, risk reduction measures are required. Risk reduction is a hierarchical process that occurs through elimination through redesign, use of engineered safeguards, awareness means (including warning and alerting techniques), administrative controls (including safe work procedures and training), and use of personal protective equipment. Often, for a particular machine, piece of equipment, or process, the solution can include more than one of these elements.

Redesign to eliminate the hazard is the most desirable of the options in the risk reduction hierarchy. Elimination ensures that the risk is no longer present. With other options, the hazard remains but is subject to control of decreasing desirability and reliability.

Engineered safeguards or safety devices protect personnel from hazards that not reasonably eliminated or sufficiently reduced by design. Engineered safeguards include guards (both fixed and interlocked), trapped key devices, and trip devices (light curtains, laser scanners, pressure mats, and safety-rated switches). Safety devices include emergency stop buttons and enabling or hold-to-run devices. Most of these devices operate through electrical control circuits.

The approach of risk-based standards is to put the major emphasis on design and development of equipment and machines. This shifts responsibility away from the end-user who has little control over conditions of operation back to the entities that envision and create these products. This shift should provide benefit to those most affected by the actions of equipment and machines.

## System Issues

The moment that a piece of equipment or a machine leaves the custody of the manufacturer it becomes subject to external influences. The life cycle of a vehicle provides an analogy familiar to people. Especially when the vehicle is new and while under warranty, the manufacturer through the dealer has strong influence on the operation and maintenance of this equipment. The owner returns to the dealer for free maintenance and improvements. The latter maintain the equipment to the level of expectation of the manufacturer. As the vehicle ages and especially following the demise of the warranty, the influence of the manufacturer wanes to the level of third-party repair manuals as third-party garages and eventually home hobbyists service and maintain this equipment. Near the end of serviceable lifetime, the manufacturer has no influence on the equipment unless it initiates a recall to address an issue of major importance. In some cases, the manufacturer may no longer exist.

In contrast to random failure of components, systematic failures occur through errors in some part of the system. They arise because of deficiencies in the design, specification of components, the manufacturing process, the operating methods, the operating environment, or the documentation. Once the manufacturer loses control over the servicing, maintaining, modifying, and upgrading of equipment and machines, third parties can this introduce additional possibilities for system issues, including errors.

## Table B.11
## Hazard and Risk Assessment Prior to Corrective Action

### Interpretation

| Severity | | Exposure | | Avoidance | | Risk Level | Urgency | Comments |
|---|---|---|---|---|---|---|---|---|
| S2 | serious injury, more than first aid | frequent | E2 | not likely | A2 | R1 | greatest | immediate hazard elimination or substitution |
| | | | | likely | A1 | R2A | ← | urgent engineering control required |
| | | infrequent | E1 | not likely | A2 | R2A | — | urgent engineering control required |
| | | | | likely | A1 | R2B | — | engineering control required |
| S1 | slight injury, first aid | frequent | E2 | not likely | A2 | R2C | — | engineering control required |
| | | | | likely | A1 | R3A | — | protective means required |
| | | infrequent | E1 | not likely | A2 | R3B | — | protective means required |
| | | | | likely | A1 | R4 | least → | awareness means recommended |

### Hazard and Risk Assessment Prior to Corrective Action

| Step Ref. | Task | Hazard | Severity S1 or S2 | Exposure E1 or E2 | Avoidance A1 or A2 | Risk Level |
|---|---|---|---|---|---|---|
| | | | | | | |
| | | | | | | |
| | | | | | | |
| | | | | | | |

Source: Modified from CAN/CSA 2005.

## Table B.12
## Hazard and Risk Assessment Subsequent to Corrective Action

### Interpretation

| Severity | | Exposure | | Avoidance | | Risk Level | Urgency | Comments |
|---|---|---|---|---|---|---|---|---|
| serious injury, more than first aid | S2 | frequent | E2 | not likely | A2 | R1 | greatest | immediate hazard elimination or substitution |
| | | | | likely | A1 | R2A | ↑ | urgent engineering control required |
| | | infrequent | E1 | not likely | A2 | R2A | | urgent engineering control required |
| | | | | likely | A1 | R2B | | engineering control required |
| slight injury, first aid | S1 | frequent | E2 | not likely | A2 | R2C | | engineering control required |
| | | | | likely | A1 | R3A | | protective means required |
| | | infrequent | E1 | not likely | A2 | R3B | ↓ | protective means required |
| | | | | likely | A1 | R4 | least | awareness means recommended |

### Hazard and Risk Assessment Subsequent to Corrective Action

| Step Ref. | Task | Hazard | Severity S1 or S2 | Exposure E1 or E2 | Avoidance A1 or A2 | Risk Level |
|---|---|---|---|---|---|---|
| | | | | | | |
| | | | | | | |
| | | | | | | |

Source: Modified from CAN/CSA 2005.

The ability of measures taken within expectations of EN ISO 13849-1:2006 during design to prevent adulteration of the system in this way is highly questionable. In order for this to occur, the equipment or machine would require a diagnostic system capable of detecting non-authorized work through changes in performance parameters. The diagnostic function would require the capability of accessing all performance parameters to compare them against default values and to report deviations.

This would be equivalent to the diagnostic system available for vehicles. Diagnostic testing for valves already is occurring as discussed in a previous chapter. This testing utilizes the embedded software and an external diagnostic device. A system for the global diagnosis of a complete system, such as equipment and machines, and ultimately a complex system comprising many components, equipment and machines is probably a concept for the future, given the small production runs compared to those involved in vehicle production.

## References

ANSI/AMT: Technical Report for Machine Tools – Risk Assessment and Risk Reduction – A Guide to Estimate, Evaluate and Reduce Risks Associated with Machine Tools (ANSI/AMT B11.TR3-2000) McLean, VA: Association for Manufacturing Technology, 2000.

ANSI/ASSE: Control of Hazardous Energy – Lockout/Tagout and Alternate Methods (ANSI/ASSE Z244.1-2003). Des Plaines, IL: American Society of Safety Engineers, 2004.

ANSI/ASSE: Safety Requirements for Confined Spaces (ANSI/ASSE Z117.1-2009). Des Plaines, IL: American Society of Safety Engineers, 2009.

BGIA: Functional Safety of Machine Controls – Application of EN ISO 13849 – (BGIA Report 2/2008e). Berlin: BGIA (Institute for Occupational Safety and Health of the German Social Accident Insurance), 2008.

CAN/CSA: Control of Hazardous Energy —Lockout and Other Methods (CAN/CSA Z460-05). Mississauga ON: Canadian Standards Association, 2005.

CEN: Safety of Machinery – Safety-Related Parts of Control Systems. Part 1: General Principles (EN 954-1:1996). Brussels: Comite Europeen de Normalisation, 1996.

EC: Directive 98/37/EC of the European Parliament and of the Council of 22 June 1998 on the Approximation of the Laws of the Member States Relating to Machinery. OJ EC L 207 (1998) p. 1; amended by OJ EC L 331 (1998) p. 1. 2007.

EC: Directive 2006/EC of the European Parliament and of the Council of 17 May 2006 on Machinery, and amending Directive 95/16/EC (recast). OJ EC L 157 (2006), p. 24; with corrigendum to Directive 2006/42/EC of the European Parliament and of the Council of 17 May 2006 on machinery, and amending Directive 95/16/EC of 9 June 2006. OJ EC L 76 (2007), p. 35. (2006).

EN ISO: Safety-Related Parts of Control Systems – Part 1: General Design Guidelines (EN ISO 13849-1:2006). Geneva: International Organization for Standards, 2006.

IEC/EN: Safety of Machinery – Functional Safety of Safety-Related Electrical, Electronic and Programmable Electronic Control Systems (IEC/EN 62061:2005). Geneva: International Electrotechnical Commission, 2005.

Firenze, R.J.: Establishing the Level of Danger. *Prof. Safety 50:* 28-35 (2005).

ISO: Safety of Machinery — Risk Assessment — Part 1: Principles ( ISO 14121-1:2007). Geneva: International Standards Organization, 2007.

ISO: Safety of Machinery – General Principles for Design – Risk Assessment and Risk Reduction (ISO 12100:2010). Geneva: International Standards Organization, 2010.

NFPA: Standard for the Control of Gas Hazards on Vessels, NFPA 306 (2009 Ed.). Quincy, MA: National Fire Protection Association, 2009.

OSHA: "Control of Hazardous Energy Sources (Lockout/Tagout); Final Rule," *Fed. Regist. 54*: 169 (1 September 1989). pp. 36644-36696..

OSHA: "Permit-Required Confined Spaces for General Industry; Final Rule," *Fed. Regist. 58*: 9 (14 January 1993). pp. 4462-4563.

OSHA: "Confined and Enclosed Spaces and Other Dangerous Atmospheres in Shipyard Employment; Final Rule," *Fed. Regist. 59*: 141 (25 July 1994). pp. 37816-37863.

OSHA: Letter to Edward V. Grund in Reply to His Letter of Enquiry of March 22, 2004 by Richard E. Fairfax. Washington: Occupational Safety and Health Administration, November 10, 2004. [Letter].

OSHA: OSHA Instruction, The Control of Hazardous Energy – Enforcement Policy and Inspection Procedure (CPL 02-00-147). Washington, DC: Occupational Safety and Health Administration, 2008.

Piampiano, J.M. and S.M. Rizzo: How Safe is Safe? *Prof. Safety 51*: 22-27 (2006).

Urquhart, J. and K. Heilmann: *Risk Watch, The Odds of Life*. New York: Facts on File Publications, 1984.

# C  Expression of Control

## INTRODUCTION

Previous discussion has highlighted concepts involved in control of equipment, machines, and systems. Control has different meanings in different contexts. Control can refer to end-points in a spectrum of options ranging from fully off or fully on or fully closed or fully open, and variable positions in between. This view of control focuses on outcomes. This is the view reflected in discussion about operation of equipment, machines, and systems during production activity, and actions taken during isolation and lockout.

Another view of control focuses on the journey involved in proceeding from A to B, rather than conditions at the origin and at the destination. The interest in this aspect of control beyond that expressed by specialists in design has arisen because of realization about the complexity of systems that enable the transition from A to B. This complexity has raised concerns regarding the ability to assure that A truly is A and that B truly is B. These concerns have arisen in operators and operations personnel, maintenance personnel, and regulators.

Change in the means to effect and express control has occurred rapidly and has become so complex that the ability of individuals to grasp the big picture of control of a particular system is rapidly disappearing. Coupled with this is the inability to communicate in a meaningful way with others about how the system actually works. This situation affects the very people who have a need in their daily activities to understand how control occurs and how misapplication of control and mistakes involving control have occurred and are occurring.

Those having the need for the global picture of control include creators or consensus and regulatory Standards, designers of equipment, machines and systems; installers and commissioning practitioners; operators; safety practitioners; and individuals involved in modification, upgrading, and decommissioning.

## The Evolution and Expression of Control

The earliest likely example to illustrate these concepts occurred when the fire started to cool and a guard or other vigilant individual added more wood. The fire provided heat and also warded off dangerous predators. The guard or other vigilant person sensed that the fire required additional fuel and took action to satisfy this requirement. At the same time, the guard or vigilant person ensured that the fire did not flare aggressively because of the addition of too much fuel.

The next example of control involves regulation of water flow in irrigation channels in ancient Egypt and other lands during the annual flood. The farmer opened up the channel to admit water and closed it at the appropriate moment to stop the flow. Stopping the flow at a particular moment was critical to preventing flooding in the fields and to avoid taking more water than allowed by quota.

Industrial development brought equipment and machines capable of performing actions involving expression of considerable energy. This applied to locomotives and other steam engines, machinery powered by movement of water, and machinery powered by pressure exerted by the wind.

The operator used control devices to activate and deactivate circuits involving the equipment and also to effect expression of energy intermediate to those two extremes. The first control devices were lever-operated. The lever and the accompanying application of the principles applied in simple machines provided the force needed to move heavy structures and to exert sufficient force to counter an action in progress.

One of the first industrial applications of the lever in control devices occurred in the mechanisms used to open and close the gates in locks. The gates are heavy and must move against the presence of water in the path of motion and form a tight seal at the point of contact. The gate arm is a lever that connects the lock gate to the mechanism that opens and closes it. The operating mechanism utilizes rotary movement around a pivot and a gear mechanism to exert the force needed to open and close the lock gate.

Another familiar example is the mechanism used to open and close railway switches. These utilize the mechanical advantage provided by a lever mechanism to move the heavy steel rails. More modern innovations include powered devices that provide the force to move the switch to the closed or open position.

Hand-operated valves are another example of devices that evolved to enable control. Depending on design, valves provide completely open or completely closed positioning, and in some cases intermediate positioning. Modern devices power the mechanism that operates the moveable element through pneumatic, hydraulic, or electrical actuators.

In each of these examples, the expression of control occurs in the same manner. A sensing device determines the state of a condition A processing device determines whether the state of the condition is acceptable or unacceptable. If unacceptable, the processing device activates a mechanism to return it to the starting position, to maintain it, or to change it to a state of acceptability.

In each of these examples, a human acted as the sensor or received information from a sensor, processed the information, made a decision concerning the information, and took action. The follow-up to the action was further sensing to assess whether the action was appropriate or sufficient, and to make a second decision, if necessary. In the event of inappropriateness of insufficiency of action, further action of the same or different nature would occur.

This is the basis for the action of control, whether by a human participant as originally occurred or by the action of a mechanical device, as occurred with the mechanization of operation of equipment and automation of control.

## Monitoring Devices
The development of monitoring devices enabled greater sophistication in the process under control. Original monitoring devices measured characteristics including temperature, pressure, weight, level or height, flow, and time. The absence of a means of transmission of information necessitated the presence of the operator in close proximity to the device and perhaps a routine on some recurring schedule to observe and to record the values, to process this information and to act on meaningful deviations from expectation or to impose some change at the appropriate moment in the process. The environment in which the instruments were located was often hostile to human tolerance. Outdoor locations necessitated work at all times of the day and in all weather.

## Information Transmission
Transmission of information between the monitor and the actuator is another critical component of the model for control. In the absence of communication technologies, the monitor of instrument readings and the person who acted upon this information were forced to use whatever technique was available. Early techniques included voice, flags and flares, mechanical signals, and more recently, two-way radios.

Development of data transmission devices and integration into monitoring and display instruments enabled the next step in the evolution of control: centralizing the monitoring function in a control area or control room. These areas contain remote monitoring and detection devices. Centralizing of monitoring offered many advantages, not the least of which was to provide a secure location for the operator from which to perform the monitoring function. At the least, this increased the safety of the job by minimizing the requirement to read gauges throughout the area of the process in all kinds of weather and at all times of the day and night.

In one iteration of the concept, the operator would determine from monitoring devices the need for an action such as opening or closing a valve. The operator would proceed to the location, undertake the action, and return to the control room. In another iteration, the operator and the control room occupied two separate locations. This situation necessitated a reliable means of communication between them. This was the situation on ships. The operator was located in the engine room. The bridge communicated orders for changing the status of the engines through a telegraph system.

## Control Devices

Electromechanical devices were the first of the fundamental control devices. Switches are among the most common of the fundamental control devices (Rosenberg 1999). Switches establish and break electrical contact for the start-up and shutdown of motors and lights. Switches coupled with sensing devices provided a means for creating a signal based on measurement of a physical quantity such as temperature, pressure, weight, height or depth, and flow. Switches coupled with timing devices provided the basis for start or stop at a particular moment in time.

The solenoid is an essential device for the control of machines and equipment through electrical, and hydraulic and pneumatic circuits (Rosenberg 1999). The solenoid utilizes electrical energy to cause mechanical movement. The solenoid consists of a frame of laminated steel and a coil of wire inside which a moveable element (plunger) of iron or steel is positioned. Energizing the coil produces a magnetic field that attracts the plunger and causes movement into the coil. The force that produces this motion can be as high as 450 N (100 lb). The solenoid may contain a spring to return the plunger to the start position when current ceases to flow. The device may also contain a second coil to produce movement in the reverse direction.

The relay combines the solenoid with one or more electrical switches (Rosenberg 1999). A control circuit connected to a relay that opens or closes the contacts in a circuit of higher voltage is the basis for the contactor.

## Control Mechanisms

The information processing device in the situations described previously was the human operator. The human operator responded in a manner to produce a desired or required outcome. This approach was satisfactory in the early level of industrialization but became less appropriate as industrial complexity increased through development of complex processes, the assembly line, and other manifestations of industrial progress. The burden on the human operator for reliability and repeatability and the need for functionality in often hostile environments in all kinds of weather at all times of day and night became too great.

Mechanical timing devices, such as spring-powered, clockwork mechanisms, provided the means of tracking time. These would alarm at preset moments, thus alerting the operator about the need to take some action. The operator would then take appropriate action based on the requirement of the procedure following sounding of the alarm.

When integrated together, the electromechanical components discussed: switches, relays and solenoids, and motorized clockwork mechanisms, have the ability to perform control functions. A clothes washer containing a motorized clockwork mechanism provides an excellent example of this functionality. The user pushes switches to set the water level for the size of the load, agitator speed, water temperature, and single use versus reuse of the wash water. The machine performs various actions including fill, agitate, pause to allow soaking, spin, spray, drain and pump out the water, and refill using recycled water. The rotary setting dial contains various settings that allow the end-user to control the sequence and duration of these actions.

The innards of the control mechanism perform the set actions through the selective contact and absence of contact of switches that occur in a specific sequence and for a specific duration according to progress of time. Contact of individual switches is sometimes linked with and sometimes independent of contact by other switches to the source of electrical energy. The sequence and duration of contact-making and contact-breaking produce the outcome observed by the end-user of this and other equipment. Electrical contact in these switches powers solenoids and relays and operates valves and motors subject to the opening and closing of other switches in the circuit that are connected to sensors.

An automatic washer operates through a linear sequence of events. The controlling factor is time as expressed through the position of rotation of the control setting. That is, the position of rotation is a measure of time. The position of the control setting determines the status of position of switches affected by the action. That is, activating the machine part-way through a cycle provides power to various circuits relative to the position of the setting. The control device does not revert to the starting point of the cycle and starts the machine at the particular time within the cycle.

Control of industrial equipment of the same vintage occurred in the same manner. Industrial equipment is considerably more complex than a clothes washer. One difference in execution of the control unit was the positioning of the control devices in large cabinets. Telephone exchanges formerly contained many relays as part of their design for routing telephone calls.

## The Logic of Control

Control mechanisms in home appliances and in much larger scale, industrial equipment, operate in a precise and predictable manner. This reflects deliberation and design. Achieving the desired control can occur through trial and error (tinkering) or through disciplined approaches. Trial and error can produce desired outcomes in equipment to a low level of sophistication. Beyond this level, an undisciplined approach is no longer satisfactory. This consideration becomes especially critical where control introduces issues related to safety, quality, and production optimization.

At the beginning of the process, involving a washing machine, the designer creates a time-line of actions to be performed by various components in the machine. The length of the time-line corresponds to the length of time of occurrence of the action. Within the sequence, various subsystems activate, as needed, and deactivate on fulfilling the need. The designer can visualize from the time-line exactly how the machine will operate at any moment in the cycle. This information includes how long subsystems will operate within the cycle and the duration of the cycle.

This drawing provides the basis for designing the mechanical analogue connected through the timing mechanism that will create selective energization and de-energization of electrical circuits based on the time within the cycle. Some of these circuits contain additional components including sensor-operated switches that respond to temperature and other parameters.

The designer of a process control system follows similar concepts. These require a visual presentation of time and conditions that are to exist at that point in time. The diagram would incorporate all relevant parameters based on time for batch processes such as clothes-washing. Capturing all of the elements required for describing the process from beginning to end provides the starting point for describing the control system and for determining components required and the manner in which they must function.

A continuous process implies a steady-state condition with inflow of material at one end of a structure and outflow at the other end. Conditions at the inflow, intermediate locations within the structure, and at the outflow are likely to be different from each other but constant in time. The visualization must accommodate to the change from time to distance.

Following description of the sequence and duration of control functions, subsequent steps include selection of components and determining the wiring patterns.

Wiring patterns express logic statements through various connections including series, parallel, series-parallel, and other combinations. Relay logic or ladder logic as it became known became the basis for expressing control of systems through electromechanical devices.

These actions also depend on electrical signals from sensors and set-points on these devices. The logic function is very simple. To illustrate, if X is true, then Y occurs, and conversely, if X is not true, then Y does not occur. The corollary is that when X becomes true, then Y will occur. The closing of a switch in a sensing device enables current flow in a circuit that, in turn, activates a relay. That, in turn, activates a motor or closes or opens a valve. This action continues so long as the signal is present.

Relay logic is a diagrammatic representation of the circuits. Diagrams used in relay logic resemble a ladder. They consists of power rails, hot on the left and neutral on the right, and individual circuits connecting the left to the right that form rungs. The logic of each circuit (or rung) flows from left to right. Power (logic) completes the circuit only when the circuit is complete, meaning that all switches are closed.

Relay logic contains many symbols. These include timers, counters, mathematics, and data moves. These allow representation of every logical condition or control loop in ladder logic. A small number of basic symbols, normally open contact, normally closed contact, normally open coil, normally closed coil, timer, and counter can represent most logical conditions.

## Servo Control Systems

The servo control system is a mechanism devised to ensure that a designated parameter in a machine or process remains constant (Stewart 1987). In simplest terms, a servo system acts in a loop containing the following principal sequential components: demand for change, action, feedback, compensation, feedback, compensation, and so on.

The simplest servo mechanism is a human-machine linkage. The human constantly monitors a sensing device or indicator, and makes corrections to maintain constancy in some parameter. This is what occurs when driving a vehicle not equipped with cruise control. The driver periodically gazes at the speedometer and makes compensatory changes in foot pressure applied on the accelerator to maintain constant speed, or speed below a defined maximum, such as the speed limit. Moving up a steep hill at constant speed requires considerable increase in pressure on the accelerator, while moving downhill under the same constraints involves considerable decrease, and possible application of the brake, as well. Cruise control automates the process originally through mechanical and now through electronic devices.

Mechanization of the process of control is readily possible. The original thermostat contained a sensor that used the movement of a strip composed of two pieces of dissimilar metals that are bonded together to open and close the electrical contacts. Metals expand and contract at different rates. Hence, the result of change in temperature on a strip of two metals that are bonded together is a slight curvature or change in curvature. The change in curvature opens or closes the electrical circuit. Newer versions of the thermostat utilize electronic sensing of current flow through a thermistor, an electronic component whose resistance changes with temperature. The electronics monitor current flow and activate the circuit at the set-point. The temperature control system operates through a predetermined increase or decrease in temperature. At the end of the range, the contacts open and disengage the solenoid controlling flow of fuel to the furnace or start and operation of the refrigeration compressor.

### Control Circuits

Control circuits are the means to control the on-off, speed up-slow down, forward-reverse, in-out, pressurize-depressurize, open flow-close flow, and other actions of equipment and machines. Control circuits also control the movement of fluids in piping and other circuits. Understanding about control circuits is essential to understanding about the management and control of hazardous energy.

## Direct Actuation Control Circuits

Much of the preceding discussion concerned direct-actuation of control circuits. The actuator was the operator who acted through direct action or amplified control of equipment in the circuit. The operator exerted complete and direct control over the circuit. This concept has many applications in industry and is common in devices found in the home.

This type of circuit can contain safety devices such as interlocks, fuses, and circuit breakers. Interlocks interrupt the flow of energy (usually electrical current) when the circuit path is broken.

### Electrical Circuits

Electrical circuits provide many examples of simple, direct-actuation control circuits. Control circuits also include the holding circuit and in the case of motors, overload protection. The holding circuit is a critical component in a control circuit. The holding circuit may operate at the same or lower voltage than the device under its control. In the latter case, a control transformer reduces the voltage.

The holding circuit contains a STOP device, a START device, and associated wiring. The STOP device is wired in series in the circuit and is normally closed (NC). All current in the circuit must pass through the STOP device. This configuration ensures that current flow in the circuit ceases on pushing the STOP button.

The holding circuit also contains a START device. This device is normally open (NO) and also wired in series. There is a normally open (NO) set of contacts wired in parallel with this device. These form the main component of the holding circuit.

Closing the START device energizes the contact device. This closes the contact in the contact device and enables current flow in the circuit regardless of the position of the contact in the START device. Pushing the START device is necessary to close the contacts that energize the device under its control.

The simplest control circuits operate at the same voltage as the device under control and involve a single phase. The next level of complexity involves three phase conductors and a control circuit operating at the same voltage and powered by two of the conductors. In more complex circuits, the holding circuit connects into the conductors feeding energy to the three phase device controlled by this circuit. Current flows in this circuit because the conductors are out of phase with respect to each other. In this configuration, the holding circuit operates at low voltage compared to the voltage used to energize the device.

### Radio-Controlled Equipment

Some equipment operates by radio control through hand-held units. This mode of control provides considerable safety to the operator, as in the case of radio-controlled self-unloading log barges, cranes on self unloading delivery trucks, and booms of concrete pumper trucks.

Self-unloading log barges tip to the side to unload logs (Figure C.1). This requires differential flooding of ballast tanks along one side. This tipping and partial righting after the logs have fallen off can produce violent oscillation and is only safely controlled from a distance.

Unexpected and unintended operation of this equipment is possible. This can occur during use of radio equipment that operates at the same frequency. This situation is known from experience to have occurred and should be expected, despite licensing of radio frequencies. There is no reason to presume that one user of a licensed frequency of short range is the only user. The problem develops on operation of equipment licensed for use in one geographic area in another where a legitimate licensee is also located. Other possible sources include overlap from close frequencies in use and hobbyists experimenting with radiofrequency generators.

Radio-controlled, self-unloading cranes on delivery trucks and boom positioners on concrete pumper trucks provide important safety measures for work occurring at construction sites where high-voltage overhead lines are present and contact is possible. Contact between booms and overhead lines is a major cause of electrocution during the unloading of construction materials at these sites because of problems of space and maneuverability. Similarly, the operator of a concrete boom must maneuver around form work in areas out of the direct line of sight. Eliminating the need to have direct contact with controls on the truck provides a considerable improvement to safety of the operator from flash-over and electrical contact.

### Indirect Actuation Control Circuits

The direct-actuation approach to control limited the power of the actuation to the strength of the operator and to the force exerted by the amplifying system. A major consequence of this approach was exposure of the operator to potential safety concerns because of potential close proximity to the operation produced by the actuation As well, this necessitated the presence of the operator at each location at which an actuator was present.

This approach was and remains entirely workable for simple configurations. As the need for more powerful actuation arose and the complexity of systems increased, the practicality of this approach became severely limited.

A major trend in control was the use of indirect-acting, low power subsystems to control the action of high power (energy) ones. The high power (energy) actuator can be a valve in the case of a fluid system, a switch, or voltage or current regulating device in the case of an electrical system. A small amount of work expended on the actuator in the low power (energy) produces a large change expressed through action in the high power (energy) circuit. In all cases, the low power (energy) circuit isolates the operator from the high power (energy) circuit in space. This approach offers considerable benefit from the perspective of safety.

In some systems, the low power subsystem controls the high power subsystem through the same medium, for example, a low power fluid system controls a high power fluid system. Hybrid control circuits are becoming increasingly common as electronic circuits are integrated into controls of fluid and mechanical systems.

Figure C.1.  Radio control of equipment, such as this self-tipping log barge, is essential for maintaining safety. (Courtesy of Seaspan Marine Corporation, North Vancouver, BC.)

*Manual Subsystems*

The manual servo systems mentioned previously illustrate the concept of a low-power mechanical subsystem controlling a high power subsystem such as an engine or turbine. Human operators are unlikely to have the strength needed to effect the change needed to operate the high power (energy) actuator (Stewart 1987).

*Mechanical Subsystems*

The mechanical servo systems mentioned previously illustrate the concept of a low-power (energy) mechanical subsystem controlling a high power (energy) system, such as an engine or turbine. Mechanical operators do have the strength to cause damage to the actuator. Overstressing the actuator must be a consideration during design of such subsystems (Stewart 1987).

A familiar example of a mechanically actuated high power subsystem is the "power steering" provided in many vehicles, especially large trucks. Finger effort is all that is needed to move the steering wheel when the vehicle is moving or even stationary. A hydraulic pump operated by a belt driven by the crankshaft powers a hydraulic motor in the car. This is the power in power steering. An axle containing a gear rotates along a rack and pinion device. Movement of the steering wheel actuates a control valve that regulates movement of the steering mechanism through the high pressure system. Attempts to move the actuator beyond the end of its range lead to slippage of the belt that powers the hydraulic pump or engine and a squealing sound as the system attempts to meet the demand placed upon it.

*Electrical Circuits*

High-voltage, three-phase systems contain a low-voltage holding circuit powered by a transformer. The transformer draws power across two of the phases of the high-voltage supply to the device controlled by the holding circuit. The voltage in the control circuit to which the operator may experience exposure is low compared to the voltage needed to operate the actuator. This design provides an external layer of safety to the operator of the equipment. As well, the high-voltage actuator is remote from the control device used to actuate the low-voltage circuit.

More powerful actuation creates its own safety risks. This is evident, for example, in the arc flash/arc blast accidents that have occurred in motor control centers and other high energy control circuits, about which much discussion has occurred in recent years.

*Fluid Power Circuits*
The pilot valve is a small valve whose operation requires low effort compared to that needed to operate large valves (Nelson 1983; Stewart 1987). Pilot valves activate the actuator of larger valve through movement of a plunger or exertion of fluid pressure. The piston makes mechanical contact with the actuator of the operated valve.

Output from pilot valves is used to control the position and hence the throughput of larger valves, and ultimately, operation of large-scale equipment. This means that the human operator can control the action of the equipment through relatively little effort, and at the same time can maintain precise control.

*Process Systems*
A manual operator is any device requiring the presence of a human to provide the energy to operate the valve as well as to determine position resulting from the action (Skousen 1998). Manual operators include devices such as handwheels and levers to enable the action to occur under human capabilities of strength.

The combination of the throttling valve with an actuator creates the control valve (Skousen 1998). An actuator is a device mounted on a valve that moves the valve to the required position in response to a signal using an outside power source.

Actuators are critical elements in establishing and maintaining control in a process control loop (Skousen 1998). Actuators include pneumatically powered devices (diaphragm, piston, vane) electronic motor devices and electrohydraulic devices. Process control loops include the actuator, a sensing device and a controller. Sensing devices include temperature sensors, flow meters, pH sensors, pressure sensors, and other devices that detect parameters critical for operation of the loop.

## Microprocessor-based Control Systems
In the 1940s, instrumentation used mainly pneumatic signals to transfer information from transmitters. Introduction of electrical signaling and clockwork mechanisms occurred in the 1960s and in the 1970s, the computer arrived. The first digital smart transmitter appeared in the 1980s and the first field data bus in 1988 (daSilva Neto and Berrie 2004).

Control systems based on relay logic were complex, often requiring hundreds, if not thousands, of components, all hard-wired to each other in various configurations. Relays are subject to failure. Failure could necessitate time-consuming diagnosis. The circuits are also subject to wiring errors. Diagnosis of the faulty logic expressed through a wiring error was also very time-consuming. Another major disadvantage of this technology was inflexibility to change. These systems reflected the physical expression of complex logical constructs. Change necessitated extensive rewiring of components. Change also introduced further potential for error and time-consuming diagnoses and commissioning of equipment and systems.

The answer to this problem was the Programmable Logic Controller (PLC). PLCs are industrial computers containing one or more microprocessors. They control machines or processes through input, application of logical processes, and output. Control in fluid power systems, for example, means control of pressure, flow, and direction of fluid (Vickers 1992, Vickers 1998). Control in process operations also using PLCs can also include level, position, density, temperature, pH, conductivity, and concentration of gases and solutions, among other parameters.

PLCs essentially are dedicated computers. The microprocessor of the PLC is programmed to control the system through decisions according to input from transducers. The purpose of the PLC was to replace electromechanical relays as logic elements, substituting instead a solid-state digital computer with a stored program able to emulate the interconnection of many relays to perform logical tasks. In an effort to make PLCs easy to program, the programming language was designed to resemble ladder logic. This eased the transition for industrial electricians and electrical engineers accustomed to reading ladder logic schematics. The ladder logic in the PLC is a computer program that the user can enter and change.

One advantage of PLCs that simply cannot be duplicated by electromechanical relays is the ability to communicate with other computers. This is the basis of modern process control.

PLCs often became bottlenecks in process control due to the nature of the programming (ladder logic) inherent in these systems (Nachtwey 2002). Ladder logic programs are optimized for simple control functions, not for coordinating complex motion and human interaction. Microcomputers are taking on the role of human-machine interface with the PLC which continues to be used in the role of I/O (input/output) controller.

The trend today is to move away from centralized control to distributed control (Johnson 2002; Vickers 1998). Distributed control occurs over a communication link known as a fieldbus. Each actuator contains a microprocessor and associated circuitry, forming a "smart" or "intelligent" actuator. These units contain feedback circuits to ensure correct performance of the task. Each actuator maintains contact with the central control as do other elements of the circuit. Fieldbuses communicate using fiber optic or shielded twisted pair cable. Both of these should be resistant to electromagnetic noise.

Security has become a major issue for process operations especially where hazardous materials and potentially explosive processes are present (GAO 2004; Rakaczky and Clark 2005). Threats to control systems used in industry can come from many sources.

The most vulnerable of these systems are SCADA (Supervisory Control and Data Acquisition) systems. SCADA systems use radio frequencies to provide distributed and remotely controlled communication systems. SCADA systems are used in municipal water systems, hydroelectric power generation, and many other applications. These systems are especially vulnerable to unauthorized intrusion as they were never designed and engineered to withstand this threat.

The current focus on security against threats from sources external to countries ignores the reality of day-to-day threats to the integrity and description of these systems from sources within countries and inside organizations. Attacks on systems have occurred (GAO 2004). To date, these have involved domestic intruders. As software and inter-connectivity have progressed, the knowledge needed by intruders to enter a system and to cause damage has decreased. This lowers the level of sophistication needed for these exploits.

## Software Control of Systems

The development of microprocessors and accompanying components and circuits that comprise the PLC and the microcomputer has revolutionized control and turned the focus onto the instructions that control the hardware. The software that controls these devices has become the component critical to safe operation of countless applications in process control. Rare today is the system that does not include a computer or microprocessor in its operation and control.

Computers used in safety-critical loops provide information to the human operator on request. They interpret data and issue commands for action directly without human input and also with human input. In each of these levels of control, there are consequences from software errors. In the lower levels of reliance on software, the human operator makes faulty decisions based on faulty information. In the higher levels of reliance, the software logic makes faulty decisions based on faulty information or because of faulty decision logic or both. The latter levels of sophistication considerably increase the difficulty of diagnosing faults due to programming errors. The presence of a human operator in the loop does not lessen the potential for error when reliance is put completely onto information provided through instrumentation and a programmed microprocessor.

Yet, few safety engineering techniques consider the potential for failure due to errors in computer software. Computers now control most safety-critical devices and often replace hardware safety interlocks and protection systems. Previous discussion has highlighted the comments of Leveson (1995) regarding software errors. Error-free software does not exist in the real world. In contrast to hardware faults, which occur as a result of random component failure, the causes of failure due to software error are systematic.

Software is generally complicated. This is the reason that the number of failures caused by software faults is on the increase, in contrast to the situation for hardware (BGIA 2008). As well, everyone who owns and operates computer equipment is familiar with the vulnerabilities exploited by creators of viruses, trojans, worms, bots, and of recent note, a worm that targeted one software application in one product that controlled the speed of one type of device.

Software control of equipment and processes differs fundamentally from electromechanical and human control (Leveson 1995). Random mechanical failures and human error are addressable through design and redundancy. Design errors in these systems are addressable through testing and reuse of proven designs. Software, on the other hand, has only design errors.

The primary approach to address reliability and safety is to eliminate these errors. The time and resources needed to build perfect software do not exist since much of the software is customized for a particular application. In addition, the experience base needed to identify and correct errors never develops because the software and the instrumentation with which it communicates are constantly changing.

Reliability of software differs from safety issues. Reliability is compliance with specifications. Errors in most safety-critical software arise from errors in the requirements for performance. That is, misunderstanding about what the software was intended to do. Software can be correct and reliable, and yet, still cause serious accidents. Safety and reliability overlap but are not the same. Hence, reliability does not assure an increase in safety.

This situation creates the dilemma of providing adequate and reasonable protection for workers and the public, while not impeding technological progress. The presence of a "bug" in a word processor is one thing, but the presence of a "bug" in software that controls the flight of airplanes is entirely another. As noted by Kletz (an internationally recognized expert on process safety), computers provide new and easier opportunities for making old errors rather than new ways of making errors (Kletz 1988). Computers allow more interactive, tightly coupled, and error-prone designs to be built. Computers increase the complexity of processes that can be controlled and the scope for introduction of conventional errors.

Deficiencies in safety-critical software applications indicate that divorcing human control from situations involving deactivation, de-energization, isolation, and lockout of sources of hazardous energy exposes workers to risks, at times completely unknown. One of the few studies on this subject found that up to 10% of program modules or individual functions in safety-critical systems deviated from the original specification in one or more modes of operation (Cullyer 1990). Discrepancies were present in software that had undergone extensive checking using sophisticated test platforms.

Normal software (simple software for simple functions) contains approximately 25 faults per 1,000 lines of code (Huckle 2003). Well-written software contains around two to three faults per 1,000 lines of code, and the software employed in the Space Shuttle has (according to NASA) fewer than one fault per 10,000 lines. What this means in practice is that the software that operates a cellular telephone contains up to 200,000 lines of code and up to 600 faults. These faults lie dormant in the program until, under certain conditions and in certain situations, they impact upon the function (BGIA 2008). Software and therefore also the programmers who create and maintain it assume a greater responsibility in the process of safe operation of equipment, machines, and systems than ever before.

Safety-related systems can contain Safety-Related Embedded Software and Safety-Related Application Software (BGIA 2008). Components contain embedded software and the system contains application software. Safety-Related Application Software used in safety-related systems parallels system-related application software and various types of application software that control other functions. The greatest separation between the latter two types of software occurs when written using separate programming systems (engineering suites) and run on separate PLCs. Economic factors usually dictate considerably less separation. Usually the entire suite of application software is prepared using a single programming system, possibly in the same engineering process.

A number of aspects require consideration whenever any interaction is acceptable (BGIA 2008). These include the requirement that non-safety-related parts of software, such as the program, function block, and function/instruction, must not overwrite safety-related variables, results, or outputs. Links between the two environments are permissible but only with the observance of specified conventions.

One such convention is retention of priority by safety-related signals and functions. Linking by means of an OR operation, for example, is not permitted under any circumstances. Current tools for software development tools support such control. Editors and compilers have implemented defined functions and rules with automatic checking. This functionality in the development software can prevent errors in logic operations not detectable with reasonable effort at the time of acceptance/commissioning which may have an effect only during unexpected operational situations. This functionality does not absolve the designer from the responsibility of performing a complete analysis to determine the influence of standard components of a control system on the safety-related parts and the influence of the safety-related functions on each other.

## Design, Verification and Validation of Safety-related Software

The historic approach to safety is to increase software reliability or to seek out bugs after damage through failure of performance has occurred. The current reality for software is that perfection is achievable only for the most elementary of systems. All computer users are well aware of the imperfections of software applications despite the development of repeated upgrades, patches, and other fixes. Even when software is written without errors, there is no way of knowing this with high confidence.

Methods to ensure the safety of computer-controlled systems have lagged behind their development (Leveson 1995). Adaptation of proven safety techniques to software will be difficult. This situation is compounded by communication difficulties between programmers and analysts, and designers and operating personnel. Superficially, the solution seems to be to detect and eliminate "teething problems." However, software errors have surfaced in some cases after many thousands of hours of use.

The V-model (Table C.1) generally forms the current basis for the design and validation of software used in safety-related systems (BGIA 2008). The V-model incorporates nested loops that address the modules that govern the safety-related functions and the subsystem that integrates the modules and possibly the system that integrates the subsystems. That is, the V-model is a series of nested loops each of which functions within the next more complex one.

Design followed by coding followed by testing for errors, defects, and deficiencies starts at the level of the modules. Detection of errors, defects, and deficiencies in a module leads to redesign and recoding and retesting, and as necessary, additional redesign, recoding, and retesting in a repetitive sequence. This proceeds to the subsystem containing the software in similar fashion, and finally to the system, if present. This model ensures detection and correction of errors, defects, and deficiencies at all levels of the software.

The V-model is generally used for very complex software. EN ISO 13849-1:2006 requires only a simplified version where there are only two levels to the software (EN ISO 2006). This form is considered to be appropriate for the practical conditions and the objectives for safety-related parts of control systems in the machinery sector and specifically for the development of safety-related application software. The objective is to create readable, comprehensible, testable, and maintainable software. Programmers who do not generally develop safety-related software are likely to consider these requirements tedious. However, they provide the certainty of software that meets the required standard for performance (BGIA 2008).

Developers of safety-related software employ a number of strategies to optimize performance (BGIA 2008). The first is defensive programming. Defensive programming assumes that internal or external errors may always be present, and that identification is critical. One means to do this is to create ranges of acceptability for strength of the signal in a measuring function. The program will not function and will indicate an error when the signal is outside the range. This is useful in calibration and zero-setting in instruments. This requires anticipation of the characteristic of the input signal over time.

Another technique is to analyze the code statically without execution, ideally using tools. These can determine whether the code is consistent with the previous software design and contains overrides

## Table C.1
## Software Design and Validation Model

| Sequential Step | Description /Comment |
| --- | --- |
| Specification | The required safety-related function governs the specification for capability and performance required from the safety-related software. The specification must anticipate environmental conditions, and normal and abnormal operational demands to be experienced by the component or the system for which the software must provide protective action. |
| System design | Reflects requirements stated in the specification and integrates the collection of software modules through coordinating software. System design is subject to change. |
| Module design | Reflects requirements for performance of hardware present as part of individual modules. Module design reflects requirements stated in the specification and is subject to change. |
| Coding | Code-writing for the individual modules and the integration software that forms the system. Coding is subject to change due to change in specifications for the system and the modules, and due to discovery of errors, defects and deficiencies during testing. |
| Module testing | Determines and ensures performance of the modules and identifies errors, defects and deficiencies. Correction of errors, defects and deficiencies occurs through redesign and recoding followed by additional testing, and as necessary, additional redesign and recoding in a continuous loop. |
| System testing | Determines and ensures performance of the integration of the integration software and identifies errors, defects and deficiencies. Correction of errors, defects and deficiencies occurs through redesign and recoding followed by additional testing, and as necessary, additional redesign and recoding in a continuous loop. |
| Validation | A concluding, special form of verification of the entire software to determine the nature of implementation of requirements of the software specification governing functionality, performance of the modules and identification of previously undetected errors, defects and deficiencies. Correction of errors, defects and deficiencies occurs through redesign and recoding followed by additional testing, and as necessary additional redesign and recoding in a continuous loop. |

Source: BGIA 2008.

that conflict with requirements of the design. Additional concerns include initialization and assignment of variables and conditional executions.

Developers of safety-related software employ software development tools, suitable and well-tried for the intended use. Certified tools for safety components are generally employed for development of safety-related application software. These tools, by ensuring avoidance and detection of semantic errors, observance of language subsets, and the monitoring of programming guidelines, enhance the quality of the software.

Libraries of software functions:are available. Some of these contain validated functions. The more the program contains functions already validated or certified, the fewer project-specific software components require validation prior to or during commissioning. These considerations also apply to system integrators. Modules developed to the requirements of EN ISO 13849-1:2006 are potentially reusable in subsequent applications.

### Application Software
Application software is the code that operates the system at the level of the PLC and provides the interface to the operator. The software that operates the PLC may be an updated version of the ladder logic software described previously.

The ability of the PLC to interface with a computer has provided major benefit in the advancement of control technology. Application software contains the code and creates the graphical user interface observable on displays in control rooms.

The graphical user interface provides the operator with a visualization of what is actually occurring in equipment, machines, and systems. This visualization includes depictions of processes, emptying and filling of storage structures, and operation of valves and rotating and other equipment. The animation used in these images can depict a wide range of situations.

### Management of Change
Experience has shown that safety-related application software that has received approval from testing will experience extension and adaptation during commissioning and subsequent operation of an installation or machine (BGIA 2008). This activity, following the development and extensive testing and validation that preceded it, is politely termed "modification." These changes are often so extensive that not only the code but also the original specifications are no longer appropriate and require revision. Change to a safety function in one position in an installation or machine can impact a safety function in another location. Alternatively, the modification may reveal gaps in the safety concept. This possibility requires examination and the relevant phases of the V-model repeated, as appropriate.

Practical experience also shows that in-service machines and installations sometimes require an additional modifications such as emergency stops or guards. The safety concept and existing software also requires upgrading following these modifications. This occurs because of the possible expression of dormant faults not previously expressed until implementation of the modifications. This can occur when the software was not adequately structured and individual modules or functions experience reciprocal impact.

In principle, modification to any part of the safety-related system requires resumption of the development process, meaning the V model at the point at which the change occurred, regardless of time into the life-cycle of a machine or installation. This action is necessary in order to maintain the integrity of the process required in EN ISO 13849:2006 (EN ISO 2006). This includes the specification and safety-related software. Change to the specification requires review to ensure that no fault has developed at a different location as a result of the change. Accordingly, repetition of all development and verification measures and also validation of the affected safety functions must occur.

This situation is not unique to safety-related software. During creation and commissioning of a control system, the programmer has direct access to the logic that operates the equipment. The logic initially starts with the design. The logic intrinsic in the design usually is not communicated to the end-user or communicated in a comprehensible manner. There are many examples to illustrate the loss of control over the destiny of software in computer-based, control systems. Operational modifications to these systems reflect different approaches.

In one situation, electricians within the entity reprogram the PLC. They have the skill and the knowledge to make the changes and are given the privilege of doing so. Documentation of the change is not recorded for future reference. This approach creates the opportunity for complete loss of control over affected equipment and ultimately, the operation. This approach presents no barrier to the actions of a rogue individual directed to sabotage operation of equipment. Such action could occur, for example, because of acrimonious labor relations on the eve of a work disruption. This situation could have serious implications on the safety of other workers.

Another operation contracts a third party, a programmer, to make changes through a wireless connection or telephone modem. The system contains some security through password protection. The system affected is low profile and unlikely to attract attention. The programmer does not document the changes for the owner. As a result, the changes and their possible unintended consequences are lost for future reference.

A third operation hires a local electrician to make the changes to the PLC in the control system. The electrician does not document the changes for the owner. As a result, the changes and their possible unintended consequences are lost for future reference.

The situations documented here are a small number of the many possible examples in the universe of possibilities. There obviously are many other possibilities not documented here. The net outcome from these situations is the loss of knowledge about how a system is configured to operate through programmed instructions. Loss of knowledge is tantamount to loss of control over the actions of the equipment, machine, or system.

As a system matures through use and provides operating experience, the logic expressed in its design may no longer suit the current operation. At some point in the history of the equipment, upgrading occurs. Upgrading reflects experience that the concept adopted in the original design did not work or did not work as expected or contains bottlenecks. Upgrades may occur gradually or abruptly. Abrupt upgrades are subject to the sorts of disruptions already mentioned.

Leveson has written a major part of a book on the subject of software control of systems (Leveson 1995). Leveson comments about issues mentioned here and adds additional detail to the discussion.

This author points out that changes to existing software do not occur in isolation and that the change is likely to introduce errors and unforeseen consequences. This situation is evident in the never-ending cycle involving patches to computer operating systems and add-ons to correct vulnerabilities to attack and other deficiencies. Often these changes cause unforeseen conflicts with other existing software that require development of additional corrective measures. Change of this magnitude embrittles software and illustrates the limitations inherent in interconnectivity of components of the operating software and add-ons.

Errors in software ensure repeatability of the error because of consistency in expression of the software. This becomes a factor in systematic error. These errors are not random, like operator error or mechanical failure. Errors in software will express every time that that particular part of the software runs. An error in the software thus increases the risk of error in the system. Studies discussed in Leveson (1995) indicate that significant consequential errors occur in original material despite the extensiveness of testing prior to release.

Experience has shown that software testing is incapable of reliably detecting all of the possible errors contained in it. Experience with commercial software with widespread availability has proven this to be true numerous times. If large software publishers are unable to detect and correct all errors, publishers of the one-of-a-kind or limited-edition software used in process control are considerably less able to follow this process. These companies lack the financial and personnel resources and the time needed to pursue these investigations during software development.

Errors in the underlying fundamentals in software, the mathematics and logic, are readily detectable and correctable. Experience has shown that many software-related accidents involved overload (Leveson 1995). Overload relates to the inability of the control system to receive and process the volume of input and thereby to provide the throughput needed in a particular situation. Lock-up in office software packages caused by overly rapid typing while the computer is running other programs is an example of this problem. These problems involve timing and are considerably more difficult to address than the correction of problems solely related to application software.

Redundancy is a well-accepted means for increasing reliability of hardware. Redundancy in software, however, has not shown effectiveness in reducing errors in system operations. Eliminating failures due to software errors is inherently more difficult than eliminating failures due to breakdown of hardware.

Reliability is a reflection of the ability of a system to do the same things consistently over a long period. Reliability does not guarantee freedom from error. Reliability means that an error will occur in a consistent manner. Increasing reliability through elimination of software errors does not necessarily increase the safety of the system. Design error in hardware or software will re-express regardless of re-liability. Safety is a property of the system and not the solely the software. Safety is application-, environment-, and system-specific.

In view of the potential consequences, the influence of a modification upon the safety functions requires study and systematic documentation. This demands creation of a process for management of change at an early stage. If appropriate, this should include appointment of the persons responsible.

Loss of knowledge complicates the task of forensic investigators attempting to learn how the components are interconnected by wires and electrical components and controlled through instructions stored in memory. Preparation of the software may have occurred years previously. The original programmer may have changed employers. The as-built drawings likely do not reflect the true status of the system if they are even available.

These systems are complicated and represent the investment of considerable amounts of money. Forensic investigation to identify and to document interconnections between components to determine how an in-service system actually is controlled for the purpose of correcting errors is a daunting exercise that can cost more than the capital value of the equipment.

This discussion highlights the need for a formal process and tools to enable reasonable control over the management of change involving computer-based control of equipment, machines and systems. This process is essential in order for change to occur in a controlled manner that provides benefit at minimum potential for risk of harm. Management of change is the process intending to maintain control over the actions involved in change. As mentioned previously, events are either managed or are allowed to happen. Nowhere is this statement more applicable than to this area. Table C.2 provides a basis for discussion of aspects of potential importance to management of change.

Owners and technical experts working on their behalf require diagnostic tools to determine the current status of programming of the PLC, the software that interfaces with the PLC software and provides the graphical user interface that provides the operator experience, and the hard wiring of the components of the system. These diagnostic tools may already exist, in which case they provide invaluable insight for diagnosis of errors, irregularities, defects, and deficiencies in the existing status quo of control. At the very least, these tools will provide the ability to greatly improve on the certainty of operation and maintenance of this equipment, these machines, and systems. If these diagnostic tools do not currently exist, their development is to be actively encouraged.

### Gaining Control through System Control

The reality today is that most, if not all process and much of operational control occurs under the guidance of a microprocessor and often a graphical user interface that is accessible to the operator. What is not currently under control of microprocessors and software inevitably will be in the near future.

Previous discussion has indicated some of the deficiencies that exist or can develop in these systems. The simplest circuits do not involve microprocessor technology (Figure C.2). The control switch controls the position of the contactor. The status of the contactor controls the movement of electrical energy through the device when the disconnect is closed.

The normal configuration in systems controlled by microprocessors is that the control circuit passes from the PLC through the control switch to the contactor for the component (Figure C.3). The operator can perform the primary deactivation through the computer interface. Moving the control switch to the OFF position isolates the contactor from the PLC. This is a hardware-based isolation.

Confounding this expectation is the situation where there is a second connection directly from the PLC to the contactor (Figure C.4 and Figure C.5). In these cases, the PLC can maintain the contactor in an energized condition, contrary to the operational expectation that deactivation and de-energization of the device occurs through the control switch.

In circumstances where the circuit between the PLC and the contactor remains live following isolation through the disconnect, signals from the PLC are evident. One means to remove this influence is to remove the fuse in the circuit board through which the signal passes. Since some of these boards can process four circuits, this decision requires thorough analysis and deliberation of consequences of leaving the fuse in place and allowing the control circuit to remain active. This situation emphasizes the diligence required in creating procedures for disconnection and the need for verification.

## Table C.2

## Requirements for Management of Change in System-Control

| Aspect | Requirement |
|---|---|
| **PLC Software** | |
| Code | Legibility, structure, intelligibility, conduciveness to straightforward, reliable modification, regardless of the programmer |
| Password protection | Use of passwords to establish who changed the code, how and when |
| Archive | Available for future examination and review |
| | changes to default code recorded at time of modification |
| Forensic analysis | Diagnostic tool to determine current logic used in control where historical changes not recorded |
| | Means to compare default code to current version to establish points of change and implications to operation and safety |
| **Graphical User Interface** | |
| Forensic analysis | Diagnostic tool to assess inconsistencies between information provided on the computer screen and actual status of affected equipment |
| Password protection | Use of password to enable user control over action taken through the graphical user interface |
| | Use of passwords to establish who changed the code, how and when |
| **Hardware** | |
| Connection | Diagnostic to identify components in the system and to map connection between them |

Information provided by the graphical user interface is not always reliable or true. In one example involving a relatively new system, the computer display following deactivation and de-energization at the control switch indicated that a pump was running, whereas the display for the variable speed drive indicated zero rotation by the motor connected to the pump, as verified by confirmation at the pump.

Deficiencies in control systems can occur in a number of ways. These include wiring used to connect control switches and PLCs to components, errors in programming of the PLC, and errors in the application software that links the PLC to the computer and provides the graphical user interface. Wiring patterns involving control switches and PLCs and components are not intuitively apparent to individuals not familiar with practices used in control. Both the PLC and the control switch in the control circuit can exercise control over components and in different ways. Operating experience brings surprises.

To illustrate, the MANUAL setting on a switch located in the Lifeguard Office at an aquatic center controls different equipment than the MANUAL setting on a switch controlling the same circuit located in the Pump Room. This occurred because, in one case, the PLC exercised control over devices through the MANUAL setting of the control switch and not in the other.

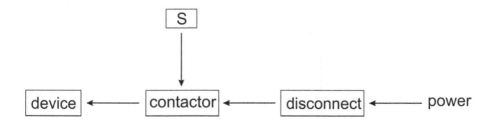

Figure C.2. A control circuit controlled by the human operator. Moving the control switch (S) to the OFF position deactivates the device and de-energizes the contactor. When the disconnect is closed, the upstream side of the contactor remains energized.

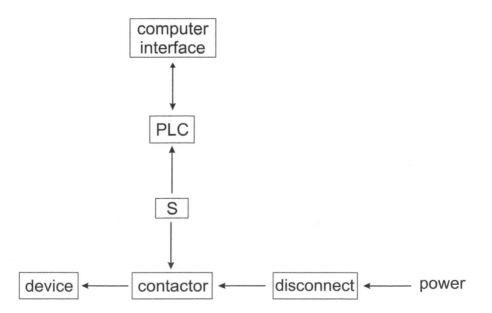

Figure C.3. A control circuit operated from the PLC normally passes through the control switch to the contactor. Moving the control switch to the OFF position de-energizes the contactor.

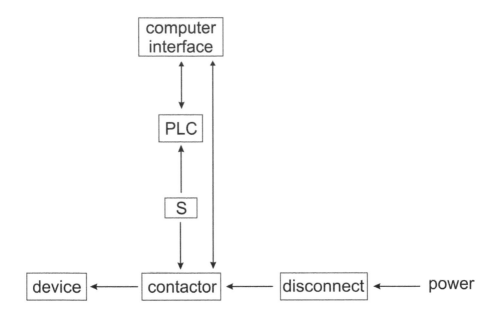

Figure C.4. Control circuit containing independent connections between the PLC and the contactor and the PLC and the control switch and the contactor. The PLC is able to send signals to the contactor at all times.

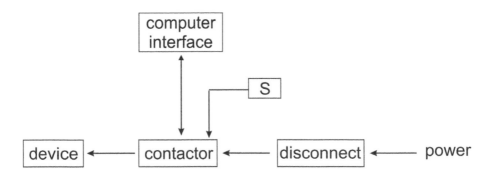

Figure C.5. Control circuit containing independent connections between the PLC and the contactor and the control switch and the contactor. The PLC is able to send signals to the contactor at all times.

Other situations have shown that control switches have operated circuits with no logical connection to the circuit of record. This example highlights total loss of knowledge about control of these circuits.

Software control plays an obvious role in deactivation of equipment involved in processes. This discussion indicates that humans have a need to retain a level of suspicion about the reliability of decisions made by microprocessors acting through software and actions taken. Safety demands certainty.

The discussion here indicates that operators must go beyond software deactivation as the sole means of preparing for work on equipment, machines, and systems where contact with hazardous energy can occur.

In each of these situations, operators, whose safety directly depends on thorough understanding about how the system functions, must subordinate their knowledge to a third party who is not affected by the operation. This situation creates little incentive for the third party to enquire about the global implications of change to the system. At the same time, this can create tremendous uncertainty in operators about the level of safety inherent in the system on which their livelihood and safety depend.

This discussion and these examples highlight only a few of the many possible situations that can occur. The operator must be aware about the four levels of control that now exist in computer-based control systems as compared to the two or three levels that existed in switch-based control systems operating through hard-wired relays. The operator should consider shutting down these systems in a sequence that involves all possible levels of control (Table C.3)

The situation posed by computer-based control highlights issues expressed in the requirements for isolation and lockout in 29CFR 1910.147 and in subsequent interpretive documents (OSHA 1989, OSHA 2004, OSHA 2008). This regulatory Standard contains a hardware basis for electrical isolation that requires severance of the energy source in the circuit. There is no allowance for circuit-powered devices to participate in the isolation and lockout. This approach contrasts with current philosophy expressed through risk-based standards. The latter standards are heavily Euro-centric in their philosophy and orientation.

The "local disconnect" switch is an example of this situation (Figure C.6). The current iteration of the hardware functions through a safety-related system The local disconnect provides a lockable disconnect switch close to equipment requiring service. The operator can initiate the shutdown through the computer or at the local disconnect. The local disconnect communicates a signal to deactivate, de-energize, and isolate the contactors. The safety-related system monitors the condition of the circuit following the signal to deactivate, de-energize and isolate, and communicates confirmation of zero energy status through an indicator located on the disconnect.

These units function through a power module and a control module, as well as one or more local disconnect switches for a particular piece of equipment or machine.

The power module contains redundant contactors. It receives the disconnect command from the local disconnect switch, then releases the contactors and confirms the status of zero-energy status to the local disconnect switch. In the normal context, this circuit would be viewed as a lockable control switch.

The control module acts as the interface between the power module and the local disconnect switch. The control module controls the status of inputs from local disconnect switches and the power monitoring relay.

This approach offers a number of potential opportunities and advantages from the perspective of safety of operation. The use of electronic, electrical, and electromechanical devices eliminates the need for direct human contact with the modules containing the contactors. This considerably reduces the potential for arc flash and arc blast (refer to Appendix G for further information) caused when the contactors open under load. This change in deactivation, de-energization, and isolation can considerably reduce the risk associated with this action since the equipment suffers the consequences of failure and not a human operator.

This approach also addresses the inconvenience factor associated with the effort needed to perform lockout. The latter relates to the time and effort expended in operating control switches and devices and moving to an (often distant) electrical room to move the disconnect to the OFF position and to apply the lockout device and keyed padlock. This approach also can reduce accident potential from

Figure C.6. Local disconnect switch. Local disconnect switches reflect the implementation of risk-based approaches to management of hazardous energy in electrical circuits. In some jurisdictions, these devices are regarded as lockable switches in control circuits.

unauthorized reactivation of a control switch during the sequence. This comment can also apply to failure to complete the lockout because of distraction that leads to forgetting to open and lock out the disconnect.

OSHA created a demonstration panel to illustrate the ways in which failure can occur in the sequence to lock out electrical equipment (OSHA 2008). This device provides an excellent counter to the argument proffered by proponents of risk assessment as the means to address this problem.

Energy isolating devices isolate machines and equipment from energy sources. The Standard prohibits the use of pushbuttons, selector switches, and other devices in control circuits as energy- isolating devices to control hazardous energy.

The electric circuit on the demonstration panel illustrates a motor control system The circuit consists of a power circuit which distributes electrical energy from the source (main disconnect) to the motor (connected load) and a control circuit to control the distribution of power. The control circuit contains a motor controller (contactor), system interlock device, on/off key switch, and start/stop push buttons.

The first situation of concern involves locking devices in a control circuit. The panel demonstrates several situations (Table C.4) where unexpected energization or start-up of the motor can occur following use of this methodology.

The second situation results from trusting the limit switch to stop a motor. These devices prevent energization of circuits by push buttons but will not prevent the start-up of a motor when voltage is present in the power circuit. A motor can start regardless of what occurs in the control circuit. Motor start can occur in several ways (Table C.4).

These situations identify some of the risks associated with reliance on control circuits as the primary method for controlling hazardous energy. A switch or other device in a control circuit is not an energy isolating device.

## Table C.3

## Sequence of Deactivation and De-Energization of Electrical Circuits

| Action | Rationale |
|---|---|
| Deactivate using control incorporated into the graphical user interface | Highest level of control<br>Affects action of the PLC<br>Provides a starting point for sequence of actions |
| Confirm visually the deactivation of the component | Visual confirmation indicates absence of motion or other indicator of use of electrical energy in the component at the time of moving the control switch |
| Move control switch to the OFF position | AUTO or MANUAL position of the control switch is subject to action by signals from the PLC |
| Apply lockout device and keyed padlock and tag if lockable | Prevents movement of the control switch enabling possible action by the PLC and possible reactivation of the component without the knowledge of the initiating individual |
| Confirm visually the deactivation of the component | Visual confirmation indicates absence of motion or other indicator of use of electrical energy in the component at the time of moving the disconnect |
| Move the disconnect to the OFF position | Prevent flow of electrical energy to the component from the power source |
| Apply lockout device and keyed padlock and tag | Prevent re-energization of the electrical circuit containing the component |
| Remove fuse from the board routing the signal from the PLC to the contactor | Prevent signals from reaching the contactor from the PLC following isolation. Note: one board may serve several circuits. |

OSHA has indicated the nature of the process to be followed in assessing the capability of alternate measures in a letter of interpretation to Procter & Gamble (January 5, 1998) which accepted their specific safety disconnect system (inherently fail-safe system) as equivalent to an energy isolating device (OSHA 2008). The equivalency determination was based upon specific facts concerning the process machine and a failure analysis. The failure analysis concluded that the inherently fail-safe system reliably prevented wired load circuits to (functionally interconnected) process machines from becoming energized when used in accordance with all design parameters, instructions, and limitations contained in the original report.

This example illustrates that possibilities exist for comparing alternative approaches against OSHA requirements through thorough design review and application for obtaining a variance described in 29 CFR 1905 (OSHA 2008).

## Table C.4

## Failure Modes in Electrical Control Circuits for Isolation

### Control Switches

Loss of control following manual reclosure of the relay in the motor controller (motor starter) enclosure by another employee.

Malfunction of the push button to the closed position.

Failure of a relay or motor controller (defective spring, welded contacts). This has occurred when a machine jam caused higher current in the motor circuit and freeze-up of the contacts in the controller relay. Arcing by the current welded shut the plunger-coil mechanism. This could be particularly hazardous to an employee relying on control circuits to clear jams because of possible start-up.

Contact between a loose wire and the conduit or enclosure.

Shorting out of two wires inside a damaged conduit. This can occur because of rubbing and wearing of insulation due to vibration. Rubbing and wearing of insulation under these circumstances has resulted in shorting and bridging of the control circuit.

bridging of contacts and circuit operation following entry of water, dirt, metal particles, or other conductive foreign debris into the control circuit enclosure.

sticking of a push-type control mechanism in the closed position due to ice, grease, dirt, wood, metal particles or other debris allowing flow of current.

### Interlocks

Closing the relay or motor controller (motor starter).

Shorting out the wiring in the conduit/enclosure.

Shorting out the wire against the conduit/enclosure.

Source: OSHA 2008.

Intrinsic to all approaches is an element of risk. This applies to the approach adopted by OSHA in 29CFR 1910.147 (OSHA 1989). The discussion in Appendix B and Appendix C provides a way to recognize that the use of software and computers for control of systems, including systems on which safety is totally dependent, is not itself without risk despite requirements for extensive testing and validation at the time of design. The risk intrinsic to this approach increases with age into the historical cycle of the equipment, machine, or system. Management of change in order to retain the validity of the software into the future of the application is the critical issue with this approach This issue is completely different from testing to assure performance of hardware.

## References

BGIA: Functional Safety of Machine Controls – Application of EN ISO 13849 – (BGIA Report 2/2008e). Berlin: BGIA (Institute for Occupational Safety and Health of the German Social Accident Insurance), 2008.

Cullyer, W.J.: Invited presentation to Compass '90. June 1990. Cited in Leveson, N.G.: *Safeware, System Safety and Computers*. Reading, MA: Addison-Wesley Publishing Company, 1995.

da Silva Neto, E.F. and P. Berrie: Who's Afraid of Control in the Field? *Petro Indust. News.* 5(5) 14-17 (2004).

EN ISO: Safety-Related Parts of Control Systems – Part 1: General Design Guidelines (EN ISO 13849-1:2006). Geneva: International Organization for Standards, 2006.

GAO: Critical Infrastructure Protection, Challenges and Efforts to Secure Control Systems (GAO-04-354). Washington, DC: General Accounting Office, 2004.

Huckle, T: Kleine BUGs, Grosse GAUs. Lecture entitled "Softwarefehler und Ihre Folgen." 2003. [www5.in.tum.de/~huckle/bugsn.pdf, Website].

Johnson, J.L.: Local versus Centralized Buses for Pump Control. *Hydraulics Pneumatics* 55(4): 34-40, 74 (2002).

Kletz, T.: Wise after the Event. *Control Instrument.* 20: October (1988). pp. 57-59

Leveson, N.G.: *Safeware, System Safety and Computers*. Reading, MA: Addison-Wesley, 1995.

Nachtwey, P.: A Systems Approach to Electronic Control. *Hydraul. Pneumat.* 55(3):50-52 (2002).

Nelson, C.A.: *Millwrights and Mechanics Guide*. New York: The Bobbs-Merrill Company, Inc., 1983.

OSHA: "Confined and Enclosed Spaces and Other Dangerous Atmospheres in Shipyard Employment; Final Rule," *Fed. Regist. 59*: 141 (25 July 1994). pp. 37816-37863.

OSHA: Letter to Edward V. Grund in Reply to His Letter of Enquiry of March 22, 2004 by Richard E. Fairfax. Washington: Occupational Safety and Health Administration, November 10, 2004. [letter]

OSHA: OSHA Instruction, The Control of Hazardous Energy – Enforcement Policy and Inspection Procedure (CPL 02-00-147). Washington, DC: Occupational Safety and Health Administration, 2008.

Rakaczky, E. and D. Clark: Thwarting Cyber Threats in Plant Control Systems. *Chem. Eng. 112*(4): 40-45 (2005).

Rosenberg, P. Correspondence Lesson 2: Basic Wired Control Devices. *Elect. Contract. Maint. 98*(3): 59-63 (1999).

Skousen, P.L.: *Valve Handbook*. New York: McGraw-Hill, 1998.

Stewart, H.L.: *Pneumatics and Hydraulics*, 4th Ed.. Revised by T. Philbin. New York: Macmillan Publishing Company, 1987.

Vickers: *Vickers Industrial Hydraulics Manual*. Rochester Hills, MI: Vickers, Inc., 1992.

Vickers: *Vickers Mobile Hydraulics Manual*. Rochester Hills, MI: Vickers, Inc., 1998.

# D The Qualified Person

## INTRODUCTION

The current approach of standard-setting organizations and legislators toward managing the conduct of work is performance-based rather than specification-based. A performance-based approach specifies end-results or key outcomes but provides few, if any, details about how to attain them. On the other hand, the specification-based approach dictates the action required in a particular situation. Of course, providing sufficient detail to cover all situations obviously is impossible.

The critical agent who is fundamental to the success of the performance-based approach is the Qualified Person. The Qualified Person is expected to have the knowledge, skills and experience needed in a particular situation to anticipate hazardous conditions that could be pres,ent, to recognize them when they are present, to assess their potential magnitude, and to advise about what must be done to eliminate or control them, and when and how. The Qualified Person thus personifies the encyclopedic information resource that would otherwise be needed to create the equivalent specification-based document.

While variously defined, a relatively comprehensive description for the Qualified Person is someone designated by the employer as capable, by virtue of education and/or specialized training and experience, of anticipating, recognizing, and evaluating the potential for exposure to sources of hazardous energy and specifying measures for elimination or control. As well, the Qualified Person must be capable of specifying action through procedures to ensure worker safety.

This definition along with the expectations of employers, employees, and regulators put an extraordinary technical and moral burden onto any individual who assumes the role of Qualified Person. Appreciation about the level of knowledge, skill, and performance required from the Qualified Person is extremely important to all concerned in order not to impose unrealistic expectations onto individuals.

The level of performance required from the Qualified Person leaves little room for error. As noted in Chapter 1, seemingly minor mistakes cause fatal outcomes when sources of hazardous energy are involved. This is especially the case where the source of hazardous energy is located in a confined space (McManus 1999). The Qualified Person must simultaneously focus an inordinate amount of knowledge, skill, and experience toward hazard assessment and decision-making in these situations compared to what is expected in the performance of normal work activity.

The aspects of performance from the Qualified Person that are critically important are the quality and level of decision-making. The safety of the Qualified Person and others could depend totally on these intangibles. Quality and level of decision-making both have similar characteristics. They depend not only on uncontrollable factors, such as the intrinsic capability of the individual, but also on controllable ones, such as knowledge, skill, and experience.

Performance-based standards and regulatory statutes on hazardous energy presently focus only on outcomes and provide no guidance on how to bridge the gap between beginning and outcome (ANSI 1992, ANSI/ASSE 2004). While, in theory, this approach provides maximum flexibility, this flexibility and the accompanying uncertainty really benefit only the true expert. The reality is that this approach provides no help to the average real-world decision-maker. The background knowledge, acquired from education, training, and experience needed to anticipate, recognize, assess, and devise measures to control potential and real hazards is extensive and unlikely to be part of the skill set of the average practitioner. This requirement should never be underestimated.

Other activities in human endeavor offer parallels to the challenges posed by this situation and remedies created to address them. In medical and firefighting situations, people totally depend on the knowledge, skill, and training of the care-provider. Society refuses to allow these individuals to operate without strict guidelines covering necessary education and skills development. In the case of medicine, these criteria have evolved over the centuries and are put into practice through medical schools and disciplinary colleges. This ensures development and provision of a consistent level of expertise. Such is not yet the case with the Qualified Person. While standards and legal statutes generally describe performance outcomes and actions required from the Qualified Person, they provide no guidance for practitioners on how to attain the level of proficiency implicit in them.

At the present time, the meaning of the term, Qualified Person, as used in context of managing and controlling hazardous energy, differs according to who provides the interpretation. To some, a supervisor having no training in the required areas could become qualified after attending a five-day, three-day or even a two-day course. To others, qualified means completing a rigorous process of certification in a professional discipline such as safety engineering. To even others, qualified means attainment of the preceding training plus additional study.

The concept of the Qualified Person and analysis of energy-related hazards is mentioned only obliquely in regulatory Standards, such as 29CFR 1910.147, on controlling sources of hazardous energy (OSHA 1989). Yet, compliance documents make clear that methodical analysis is required (OSHA 2008). The ephemeral nature of the reference to analysis of conditions performed by a qualified person in 29CFR 1910.147 has mitigated against attempts to establish a methodological curriculum for developing persons qualified in the anticipation, recognition, evaluation, and control of sources of hazardous energy.

As a starting-point for discussion, in the preamble to 29CFR 1910.146 (the OSHA Standard on permit-required confined spaces in general industry), the U.S. Department of Labor took the view that certification in safety, industrial hygiene, engineering and marine chemistry in and of itself did not qualify these professionals as Qualified Persons (OSHA 1993). This is among the clearest of expressions by a jurisdiction about the depth of knowledge needed to address the hazardous conditions posed by confined spaces, for example.

By extension, this also implies the extent of knowledge, skill, and experience needed in the Qualified Person. OSHA also argued that professional certification alone is not an automatic assurance of competence. On the other hand, absence of certification is not an automatic assurance of absence of competence. OSHA did acknowledge the extensive experience required by Certified Marine Chemists in attainment of their professional designation as an example of what is required. The understanding, knowledge, and skills of other professionals was recognized for its generalized nature.

Another view about the concept of qualified is expressed in terms of amateurs and experts (Wilson 1994). Wilson holds the view that the amateur has superficial knowledge or knowledge at the introductory level. The amateur has no idea that the subject is much more complex and/or isn't interested in learning more about it. The amateur operates at the level of instinct, impulse, and bureaucracy and upon a few basics or general knowledge of rules. More seriously, the amateur falsely thinks and represents that the superficial level of knowledge is adequate and does not realize or refuses to acknowledge the deficiency in this position.

The unprofessional or apprentice is the next level of attainment. What distinguishes the unprofessional from the amateur is recognition about the deficiency of knowledge and expression of desire to learn. During this phase, the individual gains competence in individual segments of the overall picture.

The professional has attained journeyman proficiency in most of the tasks. This is enough to sustain an operation. What separates the professional from the expert is the "big picture." The expert can handle almost any type of situation with confidence and competence.

The incongruities mentioned in the preceding discussion deserve clarification and resolution. Individuals who assume the burden of Qualified Person and their employers must know what is required from them. This is needed as soon as possible in the process in order for training needs to be defined and fulfilled.

In order for performance-based standards and statutes to produce the intended results and not to become well-intentioned failures, nurturing the Qualified Person must become a priority. The realities of the real-world workplace are very blunt and unforgiving.

A practical working definition for qualified as applied to hazardous energy control might well be the following:

**Qualified is a state of attainment such that no one will suffer accident, illness, injury, or death as a result of deficiency of judgement brought about by deficiency of knowledge.**

This is a bottom-line requirement. The knowledge, experience, assessment skills, and judgement of the Qualified Person will make the difference between success, and injury or loss of life. This is a sobering leveler of perceptions about qualifications and abilities. Responsible individuals will be humbled or perhaps even terrified by the fact that commitment of their words and signatures to paper could make this difference.

## Describing the Qualified Person

Without a starting description of academic or technical background or requirements, how can anyone describe a Qualified Person? Further, how can anyone decide what to do to become a Qualified Person? For that matter, how should organizations respond to the need to acquire or develop these highly knowledgeable, highly skilled, and highly creative resource persons?

A starting point in this process is recognition that there are two types of interests: personal and organizational.

Probably the most important starting point for developing individuals to become Qualified Persons is to determine what is needed to perform the role. That is, what knowledge is required in order to comprehend, assess, and make decisions about all of the elements that together form the scope of work involving the energy sources within an organization. For the individual, this means determining all that is possible to know about the equipment, machines, or systems containing sources of hazardous energy and the activities that occur within them. That is, this means ensuring that there are no unknowns. In the course of doing this, the individual hopefully can determine what is not known, and therefore, areas where additional information or study is needed. What one recognizes one does not know is by far the more important commodity.

The difficulty in determining what one does not know is comprehending how to identify the deficiency. If one does not understand the internal workings of the machinery, equipment, or system, and does not recognize the importance or consequence of some aspect, then one cannot identify deficiencies in knowledge or skill. Without knowing fully what knowledge and skills are required, one cannot identify what one does not know. This is a classic chicken and egg problem.

One approach for beginning this process is to identify the universe of performance outcomes and actions required from Qualified Persons. Standards and regulatory requirements on hazardous energy, and possibly confined spaces, are useful beginning sources of information about performance outcomes and actions.

Performance outcomes vary somewhat from document to document. What is specified in one in some detail may appear only as a word or phrase in another. Exclusion from one document or absence of detail does not necessarily mean lack of need for the performance outcome or action. A performance outcome or action absent from one document that appears in another may be essential to the process. The omission simply may reflect the skill of one group or the detail orientation of another. Once again, all that matters with performance-based standards and regulatory statutes is outcome. Lack of inclusion of a performance outcome or action merely may complicate the path for attaining the required end-result. This should not be interpreted as lack of need for the performance by the Qualified Person.

Table D.1 lists performance outcomes and actions identified in various sources as essential to the safe conduct of work involving sources of hazardous energy (ANSI 1992; OSHA 1989, OSHA 1993). Not all performance outcomes and actions necessarily apply to every situation that may impact a particular organization. Knowledgeable individuals should be able to recognize those that apply. For those who are less sure, starting from a broad, generic perspective offers greater assurance than underestimating the need.

Standards and regulatory statutes use two approaches for defining the role of the Qualified Person. The first specifies performance outcomes and actions that must occur. This approach does not identify specific job titles for the Qualified Person, nor does it imply that the same individual is required to perform all of the functions. This approach provides maximum flexibility for integrating the

## Table D.1
## Performance Outcomes and Actions

| Element | Performance Outcomes and Actions |
| --- | --- |
| Managerial | Assess impact of legislation on the organization |
| | Assess status of present program, compare against legal and other requirements and identify deficiencies |
| | Obtain support and commitment from senior management for necessary changes |
| | Express management support in a policy and generalized statement |
| | Assess capability of existing organizational structure to satisfy legal requirements and identify deficiencies |
| | Specify training needs |
| | Determine personnel needs |
| | Determine ability of internal/external resources to meet training needs |
| | Determine need for additional equipment and technical resources |
| | Devise recordkeeping systems |
| | Assess emergency response needs and capabilities |
| | Determine external interfaces and liaise with external resources |
| | Obtain funding |
| Technical | Identify sources of hazardous energy in machines, equipment, and systems |
| | Identify existing and potential hazards associated with each source |
| | Evaluate the impact of exposure to actual and potential hazards considering: |
| | - Number of employees affected by the exposure |
| | - Magnitude of energy expressed through the exposure |
| | - Likelihood of exposure |
| | - Seriousness of exposure |
| | - Likelihood and consequences of change in conditions or activities not considered initially |
| | - Strategies for control |
| | Assess severity of existing and potential hazards |
| | Assess capability of control and protective equipment presently installed and identify deficiencies |
| | Identify and assess resources available in the marketplace |
| | Recommend equipment for purchase |
| | Develop and review written procedures that specify measures and precautions needed for safe work activity |
| | Identify conditions that would change the hazard status and require re-evaluation |
| | Specify protective equipment required during work activity |

## Table D.1 (Continued)
## Performance Outcomes and Actions

| Element | Performance Outcomes and Actions |
|---|---|
| | Specify training requirements |
| | Assess specific needs for emergency response |
| | Develop emergency procedures |
| | Audit implementation of energy control procedures and revise, as necessary |
| OH&S Committee | Develop consultation process with the OH&S Committee on all aspects of the energy control program, including: |
| | - Hazard assessments by Qualified Person |
| | - Procedures developed for work activity |
| | - Results of compliance audit and inspections by Qualified Person |
| | - Selection of protective and other equipment |
| | - Selection of rescue equipment |
| All employees | Be informed about the mechanics of the lockout/tagout process |
| | Recognize locked out or tagged out energy-isolating devices through presence of hardware, tags, and other means of identification |
| | Understand that locked out/tagged out energy-isolating devices are not to be touched or disturbed in any way |
| | Understand that START switches or other start-up devices on locked out/tagged out equipment are not to be touched or disturbed in any way |
| Authorized employee | Understand nature of hazardous conditions associated with equipment, machinery, and systems due to energy sources |
| | Understand rationale for sequence of lockout/tagout procedures |
| | Demonstrate ability to implement lockout/tagout procedure (deactivation, de-energization, isolation, lockout/tagout, verification, prestart assessment, and verification of conditions, re-energization, start-up) in the appropriate manner |
| | Know hazardous conditions that may be encountered during implementation of lockout/tagout procedure |
| | Understand mode of exposure to hazardous energy and other hazardous conditions |
| | Recognize signs or symptoms or consequences of overexposure to sources of hazardous energy |
| | Recognize abnormal or prohibited conditions associated with implementation of lockout/tagout procedure |
| | Be aware of necessary protective equipment and its limitations |
| | Demonstrate proper use of equipment necessary for implementing the lockout/tagout procedure |
| | Communicate with supervision in a uniform manner to indicate status of implementation of procedure |

## Table D.1 (Continued)
## Performance Outcomes and Actions

| Element | Performance Outcomes and Actions |
| --- | --- |
| | Alert supervision about any indication of a dangerous or prohibited condition prior to or during implementation of lockout/tagout procedure |
| | Participate with incoming authorized employee during shift change in orderly transfer of lockout devices and account of progress and difficulties in performing the work |
| | Be familiar with emergency procedure |
| | Liaise with contractor(s) regarding implementation of lockout/tagout procedure by the host employer and proposals by the contractor |
| Affected employees | understand nature of hazardous conditions associated with equipment, machinery, and structures due to energy sources |
| | Understand the need to respect instruction from the authorized employee to move to safe areas of occupancy during testing and re-energization |
| | Warn other employees to stay away from the work area during testing and re-energization |
| | Inform other employees without valid reason to be present to leave the work area immediately during testing and re-energization |
| | Summon rescue services in the event of an emergency |
| | Know and follow work rules implemented by the employer |
| Operational/maintenance supervision | Understand nature of hazardous conditions associated with equipment, machinery, and structures due to energy sources |
| | Understand rationale for sequence of lockout/tagout procedures |
| | Know the hazards that may be present during work activity |
| | Understand mode of exposure to hazardous energy, and substances and agents |
| | Recognize signs or symptoms of overexposure to hazardous substances and agents |
| | Understand consequences of overexposure to the hazardous conditions |
| | Recognize possible behavioral effects resulting from overexposure |
| | Understand the need to co-operate with instruction from the authorized employee regarding the need to move other employees to safe areas of occupancy during testing and re-energization |
| | Warn other employees to stay away from the work area during testing and re-energization |
| | Inform other employees without valid reason to be present to leave the work area immediately during testing and re-energization |
| | Summon rescue services in the event of an emergency |
| | Know and enforce work rules implemented by the employer |

## Table D.1 (Continued)
## Performance Outcomes and Actions

| Element | Performance Outcomes and Actions |
| --- | --- |
|  | Recognize abnormal or prohibited conditions associated with implementation of lockout/tagout procedure |
|  | Be aware of necessary protective equipment and its limitations |
|  | Be familiar with emergency procedure |
|  | Participate with incoming supervisor during shift change in orderly transfer of lockout devices and account of progress and difficulties in performing the work |
|  | Liaise with contractor(s) regarding implementation of lockout/tagout procedure by the host employer and proposals by the contractor |
|  | Terminate work activity when deemed necessary |
| Testing/inspection | Be knowledgeable about use and calibration of testing instruments |
|  | Be knowledgeable about limitations of testing instruments and information provided |
|  | Demonstrate ability to interpret results |
|  | Just prior to start of work and during work, as appropriate, perform tests to verify conditions |
|  | Confirm exposure to chemical and physical agents singly or in combination in residual materials or leakage is less than regulatory limits, as appropriate |
|  | Confirm leakage of liquid, gases, or vapors of hazardous substance at secure points of disconnection or fitting of blank flanges is not occurring |
|  | Confirm successful de-energization of electrical and mechanical equipment and lines carrying pressurized fluids that could present a hazard due to release of stored energy |
|  | Confirm appropriateness of conditions during re-energization compared to procedural expectations |
|  | Confirm sufficiency of portable ventilation and other control systems |
|  | Prepare a signed and dated report outlining work performed and results obtained from tests |

energy control program within existing organizational structures. However, it does force organizational discipline since clear and unambiguous assignment of the requirements must occur in order to ensure that all are completed during the course of the energy control program.

The other approach involves mission-oriented job titles and specification of performances and actions required from them. This is the approach taken in 29 CFR 1910.147 (OSHA 1989). The focus of this approach is a hierarchical structure directed solely to performance of work involving equipment, machines, or systems containing sources of hazardous energy. The difficulty with the approach is that it can deviate markedly from the operational structure found in organizations. The latter structure is oriented toward the orderly completion of work tasks. Under the OSHA scheme, the person performing the lockout/tagout procedure may not supervise the work to be performed. This situation could create jurisdictional problems.

Table D.1 borrows heavily from ANSI Z244.1-1992 and 29CFR 1910.147 (ANSI 1992; OSHA 1989). These standards currently offer the most comprehensive list of performance outcomes and actions. ANSI/ASSE Z244.1-2003 could add additional and different performance outcomes and actions (ANSI/ASSE 2004). Table D.1 also contains performances not mentioned in these documents that are relevant in certain circumstances. Table D.1 partly organizes actions and performance outcomes according to the technical structure found in actual organizations rather than the mission-oriented approach used by OSHA. Performance of the actions outlined in Table D.1 could occur through one person or could involve a number of individuals. Overlap between actions provided by individuals also is possible.

As presented in Table D.1, the performance outcomes and actions fit a hierarchical structure. They also have temporal and sequential components. Once tasks required within an organizational context are identified and sorted, assignment of roles for available personnel can occur and the need for additional personnel identified.

In any organization there is a need for Qualified Person(s) to act at different levels. The role at the conceptual level is to assess organizational need and to create the program. At the managerial level, the role is to create policies and procedures within the program. The technical role is to assess energy sources in machines, equipment, and structures, to select instruments and equipment, and to train others. Lastly, the role in operations and maintenance is to implement the hands-on aspects of managing hazardous energy.

The reality in many organizations is that one individual will become the Qualified Person. This individual may need assistance to develop skills at all levels and be able to apply them on a continuing basis. This will be a difficult endeavor given the level of knowledge and skill that is required. Organizations caught in this situation may best be served through the services of consultants who can develop programs to fit their specific needs.

Larger organizations likely will have available several individuals nominated to become Qualified Persons. The availability of more than one individual offers the opportunity for specialization and "cross-pollination" to develop the skills of all using the strengths of each individual to remedy the weaknesses of the others. In this way no individual necessarily must become proficient in all of the skills.

## Qualifying the Qualified Person

The analysis provided in Table D.1 identifies an extensive number of performance outcomes and actions. The reader must exercise care not to underestimate the effort needed to master a performance outcome or an action . A single word or word phrase can signify a complex training program. Thus far, this discussion has attempted to separate consideration about the individual from the organization. In reality, both are intimately intertwined and separation should not occur. Hence, qualifying the individual also involves qualifying the organization.

Within any large organization, a hierarchy of individuals with capabilities applicable to this problem likely will develop. This is a natural outcome from the spectrum of skills, knowledge, and decision-making resident in the available technical resources.

Critical to the outcome of the process in the organization is the Key Individual, also called the Champion. The Key Individual or Champion accepts ownership for and organizes, implements and manages the issue, in this case the hazardous energy control program. The key individual must possess or develop a broader base of knowledge and skills than the other participants. More importantly, the Key Individual also must possess the analytical skill to recognize what is needed and the vision to determine how to achieve the performance outcome or action.

The critical starting point in the process for the Key Individual is to define competencies that are required in the program and to assess capabilities that already exist. The process starts with recognition of competencies, the knowledge and skills that are needed to satisfy the meaning of the term, Qualified Person, within the context of needs of a particular organization. The difficulty is to identify what participants should know but presently do not. This, presumably, was the basis for comments made by OSHA regarding the relevance of professional designations and the level of competence demanded in 29CFR 1910.146, the OSHA Standard on confined spaces in general industry (OSHA 1993).

Assessment of competencies is a difficult task. One possible approach is to prepare an inventory of needed skills to compare against those presently available. Actions and performance outcomes identified in Table D.1 provide a starting point. Another possible starting point for assessment is a detailed resume of relevant academic education and post academic courses and work experience. Again, the competencies outlined in this document provide the basis for comparison against performance outcomes and actions listed in Table D.1.

The need for certain types of competencies in high-level Qualified Persons is very clear-cut. Anticipatory knowledge and recognition skills are key attributes. The direction and development of the hazardous energy control program totally depend on the ability to anticipate and to recognize hazards against which action is required. Recognizing and addressing hazards that should have been self-evident after work activity is underway represents a loss of control in the process. This situation exposes maintenance workers to unnecessary risk. It also indicates a deficiency in knowledge or skill of the Qualified Person responsible for the hazard assessment. Further, this also could indicate failure of the program to ensure the ability of Qualified Person to anticipate and to recognize.

Table D.2 provides an outline of competencies potentially required in the Qualified Person. If required in their entirety, this would necessitate a very wide area of knowledge. The need exists, therefore, to identify only those competencies that are appropriate within the context of the organization. This approach would fulfil the need to upgrade skills and knowledge but would not overwhelm participants or the organization.

One means for obtaining the training in the competencies outlined in Table D.2 is specialized short courses or academic study. Table D.2 identifies possible areas of study. Individual situations could require knowledge at varying levels of proficiency in any or all of these topics. The list of competencies in Table D.2, while not expected to be all inclusive, provides a reasonable starting point for assessing the need. One of the most important questions facing individuals attempting to become Qualified Persons is determining what they should learn. The next is determining sources of training. Short courses are unavailable for many of the competencies listed in Table D.2.

Trade magazines are invaluable sources of information about areas such as hydraulic, pneumatic, and electrical safety. While these magazines are directed to practitioners within these functional interests, they can be invaluable resources to newcomers seeking to learn safety-related aspects. Publications from trade organizations can be vital resources of supplemental information. The American Petroleum Institute, for example, publishes standards on hydrocarbon safety. These identify hazards and preventive actions applicable to the handling of all quantities of these products. Similarly, the National Fire Protection Association publishes standards on fire and explosion protection and prevention. The information in these publications has widespread application across industry. Safety associations can provide assistance in specific sectors of industry. Governments and government agencies also are excellent sources of information.

Expecting a single individual within an organization to possess or develop the requisite working knowledge in all of the areas that could be needed is both unreasonable and unrealistic. Developing specific expertise by each member within a group is a realistic expectation and a possible solution to this dilemma. Each then could assist the others in specific aspects of a situation. Over time, the knowledgebase of the entire group would broaden and that of individuals would begin to merge and overlap.

## Organizational Aspects

To this point, discussion about the Qualified Person has focused on development of the skills and knowledge base of the individual. However, the need for involvement by the organization in this process should be apparent. A critical point in participation by the organization is recognition by management of the need to become involved in the process. Typically, this recognition occurs following emotional trauma induced by a near miss or fatal accident; action initiated by the Health and Safety

## Table D.2
## Possible Competencies Required in Qualified Persons

| Type of Hazard | Possible Competency |
| --- | --- |
| Process- and system-related | Hydraulic systems and control circuits |
| | Pneumatic systems and control circuits |
| | Mechanical systems and control circuits |
| | Process/machine control and feedback circuits |
| | Current electricity and electrical hazards |
| | Static electricity and electrostatic hazards |
| | Lightning hazards and protection |
| | Bulk liquid and solid materials handling |
| | Compressed and cryogenic gases |
| | Process operations and hazards |
| | Chemical incompatibility and instability |
| | Fire and explosion properties of matter |
| | Radiation sources and protection |
| | Structural soundness (integrity) |
| Work-related | Work processes (welding, spraypainting, etc.) |
| | Microbiological/biological agents |
| | Heat and cold stress |
| | Biomechanical/ergonomic hazards |
| | Noise |
| | Laser |
| | Chemical/material compatibility/incompatibility |
| Protective/preventive measures | Practices and procedures for lockout |
| | Crane and rigging safety |
| | Fall protection |
| | Extrication and retrieval techniques |
| | Fire prevention and protection |
| | Machine safeguarding |
| | Traffic control |
| | Noninvasive cleaning and servicing techniques |
| | Selection, calibration, operation and limitations of monitoring instruments |
| | Portable ventilation systems and fundamentals |
| | Selection and placement of protective equipment |
| | Isolation techniques |
| | Purging techniques |
| | Inerting techniques |

## Table D.2 (Continued)
## Possible Competencies Required in Qualified Persons

| Type of Hazard | Possible Competency |
|---|---|
| | Selection, use, maintenance, and limitations of personal protective equipment |
| | Decontamination techniques |
| | Contingency planning |
| | Communication methods and equipment |

Committee, by an individual, or the union; an order following inspection by a regulator; introduction of legislation; or perception of hazard. The preceding list is hierarchical. The least likely reasons for involvement in this pursuit by many, especially smaller organizations, are the last two.

The weakness in the process of organizational response is the total dependence on the actions of individuals. The ability of individuals to further the process on behalf of the organization depends on their level of awareness and training and personal initiative. As should be apparent, the actions of the organization closely intertwine with those taken by the key individual who undertakes to develop and implement the program.

The need to service a key piece of equipment within an organization or activation caused by one of the other factors mentioned above could set into motion an avalanche of activity. Once the avalanche begins, everything must happen yesterday. Fully Qualified Persons instantly must appear and address and fulfil legal requirements. That is, everything must happen at once.

Real-world realities, however, make things considerably different from the preceding situation. Time for making an organized approach to the problem does not exist. Individuals fully qualified to address the matter may not exist within the organization. The number of persons qualified to address requirements in any of the standards, but especially the OSHA Standard, is insufficient to fulfil the need. Skills development is a time-consuming, difficult exercise. This burden in many cases is added to the regular pressures of performing a day's work. Those charged with implementing the process must be able to analyze needs of the organization and to create a strategy that minimizes inefficiency. This process is likely to be intimidating and highly stressful for all concerned.

Addressing the need of an organization is best started through an audit. An audit could identify needed performance outcomes and actions against those indicated in Table D.1. A previously implemented program provides an invaluable starting point. A comparison of present operating practices with current requirements will indicate the extent of change that is needed. Previous work also could indicate reasons for service, nature of work performed, experience of workers, protective equipment, ventilating equipment, testing equipment, practices and procedures, emergency response procedures, training, recordkeeping, calibration and maintenance of equipment, and so on.

The present program may be structured and operate in a manner consistent with current requirements. All that may be needed to bring the program into line with present requirements is minor revision. Of course, the audit also could indicate that major improvement is needed. Even in the latter case, the existence of a structure, even with flaws, is a better position from which to build than no structure.

A recent article reported on the status of lockout programs in Quebec (Chinniah 2010). The audit determined that lockout programs from a sampling of small, medium, and large industrial establishments were incomplete and failed to meet regulatory requirements and requirements of consensus standards. This is not surprising given the onerous requirements for lockout.

Another possible benefit from an audit is an indication about how to develop a program that maximizes use of existing resources within the organization. A large organization, for example, may be able to create several programs, each containing a subset of the competencies listed in Table D.2. The ability to create subsets would simplify the need to train all Qualified Persons to the same level in all skills.

This approach is consistent with the wording in standards and legal statutes on controlling exposure to hazardous energy. These impose a variety of performance outcomes and actions required from a Qualified Person. Fulfilling these requires varying levels of knowledge and skill. These documents do not necessarily require the same individual to perform all actions. To a varying extent these documents permit approaches that utilize the capability of individuals. This means that the duties of the Qualified Person can be spread among many individuals, each performing at the level of a particular requirement.

The key is successful utilization of the knowledge and skills base of the senior Qualified Person. As well, the Key Individual, having created the big picture, then can focus on identifying and developing the skills and knowledge needed by the others.

The starting point for organizations needing but not already possessing a program for controlling exposure to hazardous energy is an inventory of skills and knowledge requisite in individuals participating in the future structure. The next step is to establish the level of skills, knowledge, and experience resident in individuals within the existing structure. This establishes the difference between what is needed and what is presently available. Matching the attributes of existing personnel to the demands of the future structure will minimize the learning curve.

## References

ANSI: Safety Requirements for Lockout/Tagout of Energy Sources (ANSI Z244.1-R1992). New York: American National Standards Institute, 1992.

ANSI/ASSE: Control of Hazardous Energy – Lockout/Tagout and Alternate Methods (ANSI/ASSE Z244.1-2003). Des Plaines, IL: American Society of Safety Engineers, 2004.

Chinniah, Y.: Equipment Lockout. *Prof. Safety* 55: 38-43 (2010)

McManus, N.: *Safety and Health in Confined Spaces*. Boca Raton, FL: Lewis Publishers/CRC Press, 1999.

OSHA: OSHA Instruction, The Control of Hazardous Energy – Enforcement Policy and Inspection Procedure (CPL 02-00-147). Washington: Occupational Safety and Health Administration, 2008.

OSHA: "Permit-Required Confined Spaces for General Industry; Final Rule," *Fed. Regist. 58*: 9 (14 January 1993). pp. 4462-4563.

Wilson, R.G.: Turn Your Servicemen into Experts. *The Cleaner.* June 1994.

# E  Model Hazardous Energy Control Program

## INTRODUCTION

This Appendix considers programs for work involving sources of hazardous energy in machines, equipment and systems. Many jurisdictions require creation of programs to satisfy regulatory requirements such as managing hazardous energy.

The performance outcomes and actions presented in Table D.1 provide a basis for determining what is required. Not all performance outcomes and actions necessarily apply to every situation that may impact a particular organization. Knowledgeable individuals should be able to recognize those that apply. For those who are less sure, starting from a broad, generic perspective offers greater assurance than underestimating the need.

A program provides a means to identify who will address these requirements and how and when this will occur. Regulators provide little guidance about what to include in a program. As a result, this is left to the imagination. ANSI (1992), ANSI/ASSE (2004) and CAN/CSA (2006) provide the view from the perspective of consensus standards. Grund (1995), McGuire Moran (1996) and Kelley (2002) provide the view of individual authors.

Programs are variable in time and space and size and needs of the entities to which they apply. As a result, no two programs should be expected to be the same. In order to function appropriately, the program must reflect the needs of the entity for which it was created.

This appendix provides an example of a program created for a large organization that has sufficient resources to employ an industrial hygienist as well as several safety officers. The organization is widely distributed across a large geographical area. Hence, the entity which is large overall is actually composed of many small sub-entities bonded together by a common name. The sub-entities contain several types of sources of hazardous energy. These include electrical, hydraulic, pneumatic, pressurized process fluids, thermal sources, and flowable solid materials. Not all types of hazardous energy are present at all locations.

This program assumes that the industrial hygienist and safety officers compliment other technical resources as advisors to on-site decision-makers. In this program, the high level advisors create the hazard assessment and energy control procedure following a visit to the site or other appropriate form of communication and discussion with operations and maintenance personnel. These inputs form the basis for preparing energy control procedures for use by the Qualified Person (authorized employee) at the site. To the extent possible, the procedures attempt to bridge situations involving the same piece of equipment throughout the entity. This obtains the greatest benefit possible for the entity from the investment of time by the resources involved.

On-site supervisors and the Qualified Person (authorized employee) are expected to contact the author(s) of the hazard assessment when the situation varies from what is described in the documents. This approach is consistent with general requirements in standards and legislation even though variations in terminology may exist. This also enables utilization of scarce corporate resources in an efficient manner.

## Model Program

### Introductory Remarks

Hazardous energy is a serious cause of workplace injury. Energy is used to power equipment, machines, and systems within this Organization. Some of the energy involved is capable of causing serious injury. Normally this energy is contained within the equipment, machine, or system and as a result poses no risk of harm. Opening the containment while the equipment, machine, or system is functioning can expose a person to hazardous levels of energy. Hence, maintenance and servicing, if done improperly because of failure to appreciate the level of energy involved and its capability for causing harm and to take steps to eliminate or control it, could lead to injury. Although individual operating departments have addressed this problem for some time, experience indicates the need for a consistent Organization-wide approach. The approach to be followed by this Organization will meet or exceed all regulatory requirements.

This document outlines the approach to be followed within this Organization for controlling exposure to hazardous energy and other agents that can arise during servicing and maintaining equipment, machines, and systems. It contains six main sections: education and training, recognition and identification, hazard assessment and energy control procedures, work activity and hazard control, emergency preparedness and response, and recordkeeping. This program identifies participants and assigns responsibility.

In carrying out this program, the Organization will take all reasonable measures to ensure competent evaluation and control of hazards arising from work involving sources of hazardous energy.

## Participants and Accountabilities

Participants involved in implementing and executing this program include:

• Health and Safety Manager
• Health and Safety Committee
• Industrial Hygienist and Safety Professional, other technical resources
• Operations and mechanical personnel
• Qualified Person (authorized employee)
• Affected employees
• Bystanders (other employees)

The Health and Safety Manager is responsible for the administration and operation of this program and is accountable to senior management.

The Health and Safety Committee assists in the identification and description of sources of hazardous energy at the site. The Health and Safety Committee reviews the program and offers input into its direction.

The Industrial Hygienist and Safety Professional and other professionals act as technical resources in the anticipation, recognition, assessment, and elimination or control of sources of hazardous energy in equipment, machines, and systems. They collaborate with operations and maintenance personnel to prepare hazard assessments and energy control procedures. Energy control procedures include lockout or tagout or other energy control activities. The Industrial Hygienist and Safety Professional are accountable to the Manager of the Health and Safety Department. Other professionals are accountable to the Managers of their respective departments. The Industrial Hygienist and Safety Professional both have certifications in their respective disciplines. These certifications are obtainable only by individuals who have suitable levels of education and professional experience. Both certifications contain requirements for adherence to strict codes of ethics.

Operations and mechanical personnel act as resources in operation of equipment, machinery, and systems containing energy sources and work activities that must occur. They prepare procedures for the hands-on people who perform the work after the energy-control procedure is put in place.

The Qualified Person (authorized employee) is responsible for implementing and sustaining the energy control procedure. The title, Qualified Person (authorized employee), does not exist as a regular workplace title. It obtains its existence from this program. This title applies in addition to regular job titles. The Qualified Person (authorized employee) provides expertise to site management and supervision. The Qualified Person (authorized employee) has education and training in the recognition of hazardous conditions associated with energy sources and the implementation of energy control procedures. The Qualified Person (authorized employee) may have other duties. The Qualified Person (authorized employee) is accountable to site management.

Affected employees, who may include the Qualified Person (authorized employee), perform hands-on work involving the energy source(s) and surroundings following implementation of the energy control procedure. Affected employees receive education and training about hazardous conditions caused by energy sources, the prohibition about modifying locked out or tagged out equipment, machinery or systems, and protective measures.

Other employees are individuals who have peripheral contact with sources of hazardous energy or work areas where maintenance or servicing are occurring. Other employees will receive education about hazardous energy sources and their effects, the lockout/tagout paradigm used during maintenance and servicing, and the requirement to stay clear of and not to touch or disturb equipment, machinery, and structures affected by lockout/tagout.

## Education and Training
The most important resources for assuring safe conditions for work involving sources of hazardous energy are well-informed and well-trained people.

The Organization will make general information about the hazards of energy sources available to all interested employees. This will occur primarily through safety meetings. Persons directly affected by energy sources will receive specialized training.

### Responsibilities for Education and Training
1. All persons working where sources of hazardous energy are present shall receive education and training to at least the level of bystander (other employee).
**Action:** Location Manager/Supervisor

2. Appropriate persons within this group shall be selected to receive education and training to become Qualified Persons (authorized employees).
**Action:** Location Manager/Supervisor

3. Persons nominated to become Qualified Persons (authorized employees) shall be provided the opportunity to receive this training.
**Action:** Location Manager/Supervisor

4. Other individuals who perform work on equipment, machines, or systems that undergo energy control procedures shall receive education and training to the status of affected worker.
**Action:** Industrial Hygienist/Safety Professional/Location Manager

5. Development of a curriculum for the education and training program shall occur using the performance-based model.
**Action:** Industrial Hygienist/Safety Professional/Maintenance Manager

6. Records of education and training for each individual shall be maintained.
**Action:** Location Manager/Supervisor

7. Education and training shall be valid for one year. An opportunity for retraining for maintaining this certification shall be made available.
**Action:** Location Manager/Supervisor

## Recognition and Identification
The Organization will create and maintain an inventory and other records pertaining to sources of hazardous energy. The inventory will be used for determining priorities in the program and for labeling of affected parts of equipment, machines, and systems.

### Responsibilities for Recognition and Identification

1. All machines, equipment, and structures containing sources of hazardous energy shall be identified and characterized. This information shall be maintained in a written (computerized) inventory and made available to operating departments.
**Action:** Health and Safety Committee/Industrial Hygienist/Safety Professional /Health and Safety Manager

2. The inventory shall be updated at least annually and shall include all current changes.
**Action:** Health and Safety Committee/Qualified Person (authorized employee)/Location Manager

3. All energy-isolating devices and other components affected by an energy control procedure shall carry a unique identifier (name, number or alpha-numeric) and be color-coded.
**Action:** Members of Health and Safety Committee/Qualified Person (authorized employee)/Location Manager/Supervisor

4. All locks and tags used for locking out or tagging out shall be uniform and carry the identification of the Qualified Person (authorized employee).
**Action:** Qualified Person (authorized employee)/Supervisor/Location Manager

### Hazard Assessment and Preparation of Energy Control Procedures

The purpose for the hazard assessment is to document and evaluate hazardous conditions that could be present or could develop in the machine, equipment, or system or the surroundings due to energy sources and residual materials and to act as background documentation in creation of the energy control procedure. Hazard assessment also considers work to be performed in the equipment, machine, or structure.

### Responsibilities for Hazard Assessment and Preparation of Energy Control Procedures

1. Hazard assessment and preparation of a lockout/tagout procedure shall precede work on equipment, machines, or systems containing sources of hazardous energy. Requests for hazard assessment shall be forwarded to the Industrial Hygienist and Safety Professional. These shall include a detailed description of the equipment, machine, or system containing the energy source(s) and work to be performed. Submission of the request shall occur at least one month prior to the start of the project.
**Action:** Location Manager/Supervisor

2. Hazard assessment and procedure writing shall proceed as expeditiously as possible. Hazardous conditions likely to be present shall be identified and evaluated. The assessment shall include review of specifications, operations manuals, drawings, and other relevant sources of information in order to determine the means to eliminate or control sources of hazardous energy and to ensure the safety of the proposed work.
**Action:** Industrial Hygienist/Safety Professional/Operations and Maintenance Personnel

3. Equipment, machines, or systems containing sources of hazardous energy shall not be opened for maintenance or servicing prior to completion and implementation of the entry-control procedure. The basis for the entry-control procedure is the hazard assessment.
**Action:** Location Manager/Supervisor/Qualified Person (authorized employee)

4. Instruments and methodologies capable of assessing conditions that may be encountered in the dismantled equipment or machine shall be investigated and recommended to the Organization. Where reasonable and practicable, emphasis shall be placed on monitoring instruments that provide continuous, real-time output.
**Action:** Industrial Hygienist/Safety Professional/Operations and Maintenance Personnel

5. Instruments suitable for assessing conditions shall be purchased in numbers sufficient to satisfy the need.
**Action:** Location Manager/Supervisor

6. A facility for calibration and repair of monitoring instruments shall be established and sustained.
**Action:** Location Manager/Supervisor

7. All testing equipment shall be maintained and calibrated according to manufacturers' recommendations.
**Action:** Location Manager/Supervisor/Qualified Person (authorized employee)

8. A written log of performance, calibration, and service shall be maintained for each piece of testing equipment.
**Action:** Location Manager/Supervisor/Qualified Person (authorized employee)

9. Equipment suitable for controlling exposure to hazardous conditions shall be investigated and recommended to the Organization. This shall include ventilation equipment and personal protective equipment.
**Action:** Industrial Hygienist/Safety Professional

## Work Activity and Hazard Control
Control can be achieved in a number of different ways as outlined in this section. Engineering control occurs through use of equipment, such as portable ventilation systems, blocking and positioning devices, bleed valves, and machine safeguards to control or eliminate hazards. Procedural control utilizes procedures to control hazardous conditions by regulating the potential for exposure to them.

Personal protective equipment is a barrier worn by the individual and includes hardhats, faceshields, safety glasses and goggles, respirators, protective clothing, gloves, and footwear. Personal protective equipment provides the only feasible means of control of exposure in many situations. Personal protective equipment must be chosen carefully since this can impose considerable burden on the wearer. The burden imposed by personal protective equipment itself may constitute a safety hazard.

Administrative control occurs through implementation of energy control procedures by an on-site Qualified Person (authorized employee). The Qualified Person (authorized employee) assesses conditions at time of start of work and authorizes work to occur under specified environmental conditions.

## Responsibilities for Work Activity and Hazard Control
1. Persons working on equipment, machines or systems affected by a lockout or tagout procedure shall have training to at least the level of "affected employee."
**Action:** Location Manager/Supervisor

2. The Qualified Person (authorized employee) shall define the conditions under which work may proceed pursuant to the energy control procedure. This includes inspection of control measures. The Qualified Person (authorized employee) also shall determine conditions under which cessation of the work must occur.
**Action:** Location Manager/Supervisor

3. Testing, as described in the energy control procedure, shall occur prior to starting and during work activity.
**Action:** Qualified Person (authorized employee)

4. If specifications for acceptable conditions stated in the energy control procedure cannot be met, the Qualified Person (authorized employee) shall notify the Supervisor and Location Manager immedi-

ately, and attempt to contact the Industrial Hygienist and Safety Professional. Entry and work shall not proceed until clarification has been obtained from the Industrial Hygienist and Safety Professional.
**Action:** Qualified Person (authorized employee)/Supervisor/Location Manager/Industrial Hygienist/Safety Professional

5. If requirements set forth in the energy control procedure are not followed, work shall be stopped and the Supervisor notified about the noncompliance.
**Action:** Qualified Person (authorized employee)

6. If corrective actions required for compliance with conditions stated in the procedure are not taken prior to continuing the work or the deficient action occurs again, the work shall be stopped and the Location Manager notified verbally and in writing.
**Action:** Qualified Person (authorized employee)

7. Following evaluation by the Qualified Person (authorized employee) of potential atmospheric hazards, ventilating, and personal protective equipment shall be obtained in sufficient quantity to address the need.
**Action:** Location Manager/Supervisor

8. Liaison between the host employer and contractor(s) working on equipment, machines, and systems affected by an energy control procedure shall be established.
**Action:** Local Manager/Supervisor/Qualified Person (authorized employee)

9. In the event that the contractor(s) fails to respect requirements of the host employer, the work of the contractor shall be subject to stoppage imposed by the host employer.
**Action:** Qualified Person (authorized employee)/Supervisor/Location Manager

10. An audit of performance of energy control procedures shall occur annually. The audit shall involve a third party not involved in performance of the procedure.
**Action:** Industrial Hygienist/Safety Professional

## Emergency Preparedness and Response
Preceding sections of this document address routine situations. Emergency preparedness is an important action for lessening the consequences of accidents that can occur despite competent hazard assessment and implementation of energy control procedures. Planning for emergency response involves anticipating situations reasonably likely to occur and taking the steps needed to lessen their impact.

### Responsibilities for Emergency Preparedness and Response
1. Accidents still considered possible after taking control measures shall be identified during hazard assessment. Specialized equipment needed for accident situations shall be investigated and recommended to the Organization.
**Action:** Industrial Hygienist/Safety Professional

2. Equipment needed to address accident situations shall be identified as part of implementation of energy control procedures.
**Action:** Qualified Person (authorized employee)

3. Equipment needed for emergency response shall be acquired in sufficient quantity to satisfy the anticipated need.
Action: Location Manager/Supervisor

4. Equipment needed for emergency response shall be readily available on the site at the time of work activity.
**Action:** Location Manager/Supervisor

5. A written plan for emergency response shall be prepared for the operating unit.
**Action:** Qualified Person (authorized employee)

6. Emergency response personnel shall be appointed and trained. (These persons also may function in other voluntary emergency response activities such as the fire crew.)
**Action:** Location Manager/Supervisor

7. Where emergency response personnel are not available within the Organization, suitable external sources of assistance shall be identified and contacted. The external response group shall be informed fully about possible accident situations and assistance that might be required.
**Action:** Location Manager/Supervisor/Qualified Person (authorized employee)

8. Persons working in remote work locations shall be provided with the means to communicate with emergency response providers.
**Action:** Location Manager/Supervisor

9. The emergency response plan shall be tested through drills. Testing shall occur at least once per year.
**Action:** Location Manager/Supervisor/Qualified Person (authorized employee)

### Recordkeeping
Records ensure continuity as functional staff changes. Records also provide invaluable links to the lessons of the past. They help to reduce the probability that something will go wrong in the present and future. Recordkeeping establishes the credibility of the energy control program by proving performance, competence and compliance. The main areas in which records are kept include:

- Inventory of equipment, machines, and systems containing sources of hazardous energy
- Hazard assessments including drawings for equipment, machines, and systems
- Work procedures
- Performance, maintenance, and calibration of instruments
- Monitoring results
- Training records
- Records of qualifications of individual participants
- Records of inspection of safety equipment including harnesses, fall protection, extrication equipment, and so on

### Responsibilities for Recordkeeping
1. Written records shall be maintained for all aspects of the hazardous energy control program.
**Action:** Location Manager/Supervisor

2. Records of hazard assessments and energy control procedures shall be maintained.
**Action:** Industrial Hygienist/Safety Professional/Qualified Person (authorized employee)

## References
ANSI: Safety Requirements for Lockout/Tagout of Energy Sources (ANSI Z244.1-R1992). New York: American National Standards Institute, 1992.

ANSI/ASSE: Control of Hazardous Energy – Lockout/Tagout and Alternate Methods (ANSI/ASSE Z244.1-2003). Des Plaines, IL: American Society of Safety Engineers, 2004.

CAN/CSA: Control of Hazardous Energy —Lockout and Other Methods (Z460-05). Mississauga, ON: Canadian Standards Association, 2005.

Grund, E.V.: *Lockout/Tagout, The Process of Controlling Hazardous Energy*. Itasca, IL: National Safety Council, 1995.

Kelley, S.M.: *Lockout Tagout, A Practical Approach*. Des Plaines IL: American Society of Safety Engineers, 2001.

McGuire Moran, M.: *OSHA's Electrical Safety and Lockout/Tagout Standards, Proven Written Programs for Compliance*. Rockville MD: Government Institutes, Inc., 1996.

# F  Electrical Systems

## INTRODUCTION

Previous discussion about electrical equipment has focused on isolated components and how these can store and transform electrical energy. Real-world applications of electrical energy involve systems often containing many components. This Appendix focuses on management of exposure to hazardous electrical energy in the electrical system.

## Electrical Accidents

Data presented in Chapter 1 indicate that in about 93% of electrical accidents, the electrical hazard existed prior to the start of work. The remainder occurred because of self- or other activation or reactivation of circuits.

Most of the victims in accidents involving high-voltage were journeymen linemen or electricians performing similar tasks. Victims also included other members of crews performing this type of work. These included apprentices, support workers, and first and second level supervisors. Victims also included individuals employed by contractors (painters and construction laborers) performing work on utility property unrelated to line installation and maintenance. Most victims were performing what appears to be normal work.

Of the nonelectrical workers who were victims in accidents involving high voltage, the data strongly suggest that they had no control over the conditions under which the activity was to be performed. Coupled with this was the complexity of the task being performed.

Most of the victims of accidents involving low voltage were hands-on workers in the trades (laborers, electricians, and others). The predominant trade was electrical. A large proportion of the victims were performing work that would be normal to them. This included primarily the installation, testing, and maintenance and repair of electrical equipment.

Victims performing unrelated work often were directly affected by the quality of work performed by electrical workers. These individuals, laborers, machine operators, cleaners, supervisors and managers, whose daily activity ordinarily would not lead to contact with electrical equipment, were affected by wiring errors and sloppy work practices including failure to close electrical junction and switch boxes. In some instances, machine operators, and in others, senior individuals, including supervisors and managers, attempted to troubleshoot faulty equipment by gaining access to control cabinets containing energized unshielded conductors.

Testing to determine the presence of energized circuits was not performed in these accidents. Testing would have alerted many of the victims to the presence of live circuits or faulted circuits containing leakage paths. This presumably would have warned them not to venture forward.

Wiring errors occurring during initial installation and during renovations were responsible for numerous electrocutions. A subsidiary contributor was the removal of ground prongs on plugs. This led to loss of ground protection. Perhaps even more horrifying were substitutions of improper plugs in order to fit certain receptacles. This led to mismatching of conductors and prongs and subsequent energization of outer cases of equipment. Use of plugs containing damaged fittings and equipment containing exposed conductors also contributed to the electrocutions. Owners of equipment failed to conduct periodic inspection to determine correctness and integrity of wiring.

Many of the victims involved in testing, installation, and renovation accidents worked "hot" as a matter of routine. Some individuals resorted to the use of nonconductive fibreglass ladders in the belief that these would isolate them and prevent establishing a path to ground. This strategy failed to account for the presence of conductive paths in metal structures (structural beams and the frames of suspended ceilings) with which they would come into contact.

Several individuals performing unfamiliar work at elevation contacted bare conductors used to supply power to overhead cranes. These individuals were unfamiliar with the means of powering the crane and either stepped on or grasped the conductors as a means of support while performing unrelated tasks.

These observations concerning electrical accidents highlight two main themes concerning human interaction with their environment: knowledge and control. Knowledge describes the sum of all formal training and experience applied to a particular concept. Individuals confronted by a particular hazard are either knowledgeable within the context of anticipating it when it could be present, recognizing it when it is present, and knowing what to do to prevent exposure.

Control refers to the ability to influence actions that occur or do not occur in workplace situations. Exercising control can occur through direct power granted by an employer or indirectly through the right-to-refuse granted to workers by regulators. The ability to exercise the right to refuse is certainly removed from the action of actual exercise of the right in real-world workplaces. Employees of a host employer have considerably greater ability to exercise this right than employees of a visiting employer (contractor). This inability reflects fear that arises through difficulty with language and perception of status.

Electrical accidents illustrate the application of these concepts. Many of the accidents highlighted lack of knowledge (unknowing) by the victim and persons with whom this individual interacted on the worksite and real or perceived inability to do anything feasible about the hazard.

The hazard of the energy source was not perceivable through the senses or recognizable by other common means, or the person lacked the knowledge needed to recognize its presence. Electrocution that occurs during performance of unrelated work around live electrical wires or wiring also highlights this concept.

Some of the accidents highlighted knowledgeable helplessness. The victim recognized the presence of the energy sourc, but performed the task regardless. The victim had no training in control of the energy source or perceived the ability to perform the task without undue risk or perceived the lack of availability of protective measures. Protective measures include shutdown of the energy source, installation of shielding between the source and the person, and use of personal protective equipment to prevent contact with it. The victim was helpless through absence of ability to influence or to exercise control over the operation.

Some of the accidents reflected knowledgeable defiance. In these situations, the victim attempted to perform the task without adherence to procedures or use or proper use of protective measures, despite the possibility of contact and having full knowledge that the energy source was active. Full knowledge can arise from visual or other cues from the senses, as well as knowledge gained through training. This classification also covers actions taken in defiance of procedures or direct orders, or requests from supervisors and others. An example that illustrates this classification is knowingly working without protection on electrical circuits or uninsulated, energized electrical conductors.

## The Electrical System

The electrical system is part of the infrastructure that is familiar to everyone in industrialized countries. The electrical system is essential to normal function of society. The transmission system guides the energy from the source to the point of distribution. The distribution system subdivides into many branches and eventually connects to end-users.

### The Above Ground Electrical System

Electrical energy often travels long distances from source to consumer. The physics of energy loss due to conversion to heat during transmission mandate for higher voltages and smaller currents. Transmission occurs in both AC (alternating current) and DC (direct current) modes. Voltage of transmission lines range typically from 230 kV to 1,100 kV although older lines range down to 69 kV (BPA 1986). Transmission lines directly affect the safety of few people. That is the intent of their design and positioning on the landscape.

The distribution system is readily obvious to all observers. This forms the seeming tangle of wires that provides the service to customers. The distribution system covers the range of voltage upward from 110 V that supplies homes in different phases to industrial voltages of 600 V to supply voltages ranging commonly from 4 kV to 25 kV. Distribution in industrial buildings for power uses typically ranges from 110 V to 600 V. Most industrial equipment appears to operate at 600 V.

## The Underground Electrical System

Underground wiring has become considerably more common in recent times, especially in cities. This has resulted from concerns about appearance as well as practical realities related to routing of high-voltage lines in the middle of major cities.

Historically, as part of rural electrification and development of housing subdivisions, cables were initially laid unprotected in trenches, the separation between conductors being controlled by the width of the excavation. Multi-phase secondary lines containing individual conductors were laid in parallel, 6 inches to 1 foot (15 cm to 30 cm) apart, or sometimes intertwined, with the ground cable being positioned somewhat haphazardly on top. Splices and T-taps received no special attention during installation. Underground residential cable and streetlight cable, in use since the late 1950s, constitute a very high proportion of the underground system.

The current underground electrical system consists of cast-in-place or precast concrete vaults interconnected by ducts (Figure F.1). The vaults hold splices and provide access to the cable between electrical substations. The ducts contain the cables. Early duct was manufactured from asbestos-containing cement. Current duct is manufactured from PVC (polyvinyl chloride) plastic.

Transmission cable ranges to 250 kV. Distribution cable ranges from 4 kV to 25 kV. Low-voltage cable ranges from 110 V to 600 V. Communication cable includes fiber optic cable, as well as cable for telephone and cable television. A particular vault can carry all types of cable. High-voltage cables are mounted on walls using spacers and carry labels for identification. The type of cable in a particular vault reflects the evolution of the underground electrical and communication systems in a particular area or at a particular site. Underground electrical systems are common in industrial, commercial, and institutional installations.

A critical aspect in creating underground electrical systems at user facilities is labeling. The identity of cables, otherwise identical in appearance, is easily forgotten. The Owner should maintain current operating line diagrams for the underground system. These diagrams should indicate circuits that pass through individual vaults, and location of individual conductors on cable trays. These documents become invaluable when mud and other foreign material that have entered the vault or soot from fire obliterates labels. Unhindered access to these documents by service and maintenance companies is essential in order to enable orderly and safe work practices especially during emergency situations.

Underground electrical vaults in outdoor locations are subject to the environment in which they are situated. These structures can contain saltwater resulting from use of salt on roads for deicing and mud from drainage (Figure F.2). Salt and mud can obliterate labels on cables. These structures also can contain hypodermic needles and other sharps, solvents, and chemical products poured into them for disposal and human and animal waste from surface run-off.

## Cable Used in Underground Electrical Service

Cable used in overhead lines, especially high voltage, usually has no insulation. The insulating properties of air and the concept of "limits of approach" form the strategy to protect against unintended electrical contact and resulting electrocution. Electrical protection of workers who work in underground electrical vaults requires a different strategy because of close proximity to these cables and the potential for contact.

Secondary underground lines, described above as being positioned in parallel lines or intertwined with the ground cable laid haphazardly on top, provide only a small margin of safety in the event of physical damage and disruption. Damage to a single conductor or subsequent contact by a metallic object provides a possible alternate path to ground. Depending on the magnitude of current leakage and tolerance of protective devices, the electrical system may experience no trip.

Figure F.1. Installing a 250 kV underground transmission line along a residential street. The ducts are encased in concrete and electrical cables pulled into them. Underground electrical vaults contain splices and related equipment.

Figure F.2. In-service underground electrical vault. Flooding by mud and other debris creates a considerably different environment compared to that at the time of installation.

Cables in current use contain a single hot conductor surrounded by a concentric layer of insulation and a concentric neutral conductor. This geometry shields the hot conductor from external physical damage and provides a direct return path in the event of puncture and arcing. This arrangement provides an almost instantaneous rerouting of current with the consequent trip of the circuit by protective devices.

The first cables in some areas contained a copper conductor surrounded by paper insulation impregnated with a non-draining, wax-like compound and an outer lead sheath. The outer lead sheath connected to ground through clamps to provide electrical protection. These cables date from pre-1970 to the mid-1970s.

Paper-insulated, lead-sheathed cables leak at joints and splices The non-draining compound melts and drips due to overheating. Cables overheat in service for several reasons, starting with exceedence of design current. Maximum design temperature for paper insulated cables is 80° C. Mutual heating results from radiation of heat from the surface of one cable to another due to proximity on the cable tray.

Deposition of material onto the surface of the cable as a result of flooding of the manhole impedes radiation of heat from the surface. Additional causes of overheating include high-voltage contact with the conductor or a short circuit. The fault current causes a trip of the circuit breaker. Trips of this nature occurring within design tolerances create no resulting impact provided that protection equipment also functions within design tolerances. However, protection equipment can trip sluggishly due to weakening of springs because of aging and other problems.

Entry of water leads to bulging and explosion inside the lead sheath of the cable. Water is present in vaults due to accumulation of rainwater, salty runoff from roads, groundwater, and condensation. The lead sheath is also known to be fragile and susceptible to failure on disturbance.

Cable of more recent manufacture encased the hot conductor inside extruded polyethylene along with an outer grounded metal sheath and an outer protective layer of PVC (polyvinyl chloride) or medium density polyethylene. Polyethylene is an effective insulating material. The first generation (1960s and 1970s) of cable (15 to 25 kV) experienced failure due to formation of tree-like channels in the polyethylene.

This problem was traced to entrainment of water used for cooling during extrusion of the polyethylene. The result was arcing from the conductor into the polyethylene to the ground sheath and subsequent explosion. Failure occurred after about 10 years of service. These cables are highly susceptible to water damage resulting from long-term migration of moisture into the insulation through the outer jacket. Exposure of this cable to high-pressure water jetting while energized renders it highly susceptible to failure. Removal of much, if not all of this cable from service, has occurred as a result of these failures.

The most current iteration of cable contains cross-linked polyethylene (XLPE) as the insulator that surrounds the conductor. Cross-linking protects the polyethylene against the arcing that forms the tree-like structures. This design also incorporates layers of semiconductive material to minimize the spikes in the electromagnetic field that led to the arcs. Maximum design temperature for XLPE-insulated cable is 90° C.

Cables that have overheated pose a high risk of failure. Overheating is not readily evident from visual inspection unless charring is present. Charring occurs as a result of arcing between the conductor and the grounded metallic sheath. Charring is the only visual evidence of failure. Charring would not necessarily be evident in cable covered by mud resulting from a flooding episode.

Another cause of cable failure is super-curvature. Super-curvature occurs when the tight geometry of the vault forces the cable to bend into a smaller radius of curvature than allowed by design specifications. As installed, the cable is positioned in a particular orientation between two vaults. On sloped terrain, the mass of the cable is affected by gravity. As a result, migration due to sliding can occur. The result is that cable stretches in the upgrade vault and compresses in the downgrade vault. Accumulation of excess cable in the downgrade vault can lead to displacement from the rack and super-curvature. Over-compression during bending can fracture the internal structure of the cable.

Another location of cable failure is the splice. The splice is the location of connection between two lengths of cable. Splicing is a potentially complicated practice that by necessity must establish and maintain close contact and continuity between the two sections of conductor, the insulation, insulant or insulating oil, layers of semiconductive material, grounded metal sheathing, and the outer protective and waterproof cover. Establishing a waterproof seal at the cover is critical to ensuring that water will not enter the splice.

The splice is a source of leakage of cable insulant as well as inward migration of water. Splices are sometimes encased in bituminous compounds. Heating of the splice beyond specifications due to over-load can melt this material leading to deposition on the floor and reduced insulating value and waterproofing.

Failure of cable and splices can also occur due to aging and attack by transient spikes, attack by rodents, deterioration caused by chemical attack, entry of water, and poor workmanship during instal-lation. Poor workmanship during installation is regarded as a major cause of failure.

Cable failure can lead to internal arcing and explosive failure of the sheathing and cover of the ca-ble or the splice. The overpressure created by the explosion could expel the manhole cover from the top of the underground electrical vault and cause considerable injury and damage. Failure leads to cessa-tion of transmission in the affected cable and possible damage to adjacent cables on a tray or on the floor of the vault, to splices and equipment, and the interior surroundings of the vault. As mentioned, arcing is identifiable visually only by the presence of blackened cable that shows visible damage. Visi-ble damage on a particular cable could be difficult to spot when it occurs on the side facing away from the observer or when the cable is coated in mud. Internal arcing that has not penetrated the cable sheath is not visible.

Additional obvious evidence of cable or splice failure can include emission of smoke, hot air and odorous compounds through the lifting holes in the manhole cover. More subtly, the air inside the vault might be warm or hot. Heated air alone could indicate overheating and not outright failure of a cable or splice.

## Reliable Electrical Supply

Reliability of supply is a normal presumption of consumers of energy from the electrical system. How-ever, some installations require greater, and in some cases, considerably greater than normal reliability in order to enable orderly shutdown. Some industrial operations demand reliable power. Some exam-ples include chemical plants, nuclear and other thermal power stations, steel mills and other facilities heavily reliant on uninterruptible sources of electrical energy. There are a number of means to achieve this end-point.

Unreliability during peacetime situations in external supply circuits results from collisions be-tween vehicles and power poles, lightning strikes, contact accidents involving mobile equipment capa-ble of causing a trip in the circuit, blowdown of trees during storms, landslides and avalanches, and rare events of very low probability such as earthquakes. Circuit overload due to insufficient capacity in the infrastructure is also a possibility. Failure of components in the external infrastructure due to the stress of overuse beyond capacity and old age are increasingly likely with increasing demand on the electri-cal system to power electric vehicles and other new applications is increasingly likely.

Unreliability in internal supply circuits within a facility results from insufficient capacity in the internal infrastructure, accidents that sever supply cables, wiring errors during modifications to the ex-isting system, and failure of components in the internal infrastructure due to the stress of overuse beyond capacity, and old age.

Reliability is achievable through the electrical supply system to several levels. The simplest ap-proach is to provide power from two circuits within a facility. The two circuits receive power from a common source that supplies the facility. This approach offers some relief from overload of individual circuits within the facility. This approach can provide relief from the heavy draw of electric current by motors during start-up. This approach, however, leaves open the possibility of failure of one or both of the circuits.

The more advanced approach for continued operation is to establish reliability from the in-feed source of power. At the lowest level, this can occur through in-feed from two circuits originating from the same substation in the distribution network. Provision of electrical energy from two different cir-cuits in the distribution system increases reliability by eliminating many of the causes of unreliability in both the internal and external infrastructure. This approach does not address the potential for issue of loss of power due to failure of the source or the transmission line that links the source to the distribution system.

Obtaining in-feed from two separate substations considerably increases reliability, but this approach still leaves open the concern about reliance on and failure of a single source of electrical power.

The most reliable situation for consumers of large quantities of electrical energy is to receive in-feed from two independent sources, namely two generating stations, through separate transmission systems. That is, the two separate generating stations feed the facility and possibly equipment and machines within it by combining right at the disconnect. Failure of both sources is extremely small and would reflect failure of the entire electrical grid.

Modern electrical systems maximize reliability by interlinking feeds from many independent sources hence the electrical grid. The limitation of the grid is the loss of parallelism and independence in supply to individual consumers and the ability to switch from one supply to another.

The component that switches circuits, the transfer switch (Figure F.3), is a key component in the circuits that provide reliable power. The transfer switch can influence safety during isolation and lockout of electrical systems.

### Emergency Power Systems

In many operations, reliability of power (energy) sources is a critical component for maintaining base levels of some functions and providing time necessary for orderly shutdown of other functions. Consumers of power provided by these systems include evacuation lights, computer memory dependent on continuous supply of power, and other equipment requiring time for orderly shutdown.

The presence of sources of emergency power is not always apparent in an operation, or considered during preparation for deactivation, de-energization, isolation, and lockout of the active circuit.

Figure F.3. Manual transfer switch. Some facilities have an urgent need for reliable and emergency power. Circuits affected by the system that provides emergency power are potential sources of exposure to electrical energy when otherwise believed to be de-energized.

Some circuits contain internal components that store energy as part of design for such situations, while other components retain energy as an intrinsic characteristic. In other cases, the source of energy used to maintain emergency levels of operation is external to the circuit. Both sources of energy can exert profound and consequential impact on the safety of work on circuits believed to have been deactivated, de-energized, isolated, and locked out.

### Internal Sources

Previous discussion has mentioned about retention of energy that is deliberately stored in circuits. Energy deliberately stored in circuits is intended by design to be available for use without intervention by the operator, human or otherwise.

Energy storage in this manner can occur in electrical, fluid power (hydraulic and pneumatic), and mechanical circuits. The intent of the source of energy is to enable operation of the circuit for a brief period in the event of failure of the normal source. In electrical circuits, capacitors and sometimes transformers can perform this function. This energy can provide life-saving capability in an otherwise disabled system.

Discharge of energy from these sources during emergency operation or deactivation and de-energization does not necessarily produce a zero energy state in these circuits or even a harmless level of energy. The circuit will operate until the energy decreases to the threshold required to operate the consumption device. Beyond this level, the circuit will retain energy.

### External Sources

External sources of energy provide supply of longer duration. Sources of emergency power range from the invisible to the inconspicuous. An installation can contain several sources of emergency power of different types. As a result, circuits provided by energy from these sources are unlikely to be widely known. This situation has the potential for causing exposure in circuits not known to be energized. The best way to inform workers affected by this situation is to post conspicuous diagrams of electrical circuits (Figure F.4).

#### Diesel-Powered Fixed Installations

Diesel-powered generators in fixed installations are the most conspicuous of the sources of emergency power (Figure F.5). This results from the location on an outside wall of the building, the radiator grille and the exhaust stack located in the wall of buildings, and sometimes the noise from operation of the engine and odor of the exhaust.

#### Battery Systems

Electrical generation plants contain banks of batteries to energize windings in the rotor of the generator. Municipal transit systems operating DC powered trolley buses and electric railway systems also have rooms containing banks of batteries to provide emergency power.

Batteries (Figure F.6) can supply direct current (DC) power without modification and alternating current (AC) power through an inverter. Battery power is temporary and allows for implementation of other arrangements. A battery room in a facility is unlikely to be widely recognized and appreciated because of its inconspicuous nature.

#### Mechanical Momentum Sources

Mechanical momentum sources containing heavy flywheels are an example of one type of product (Lamendola 2001a). Rotation of the flywheel/motor rotor in these units after loss of power generates electrical energy.

Wired in series upstream, these units can supplement batteries for short-term situations thereby prolonging service life. Service life of batteries depends on the number of uses, no matter how small. These units can supplement batteries for starting generator sets as well as emergency power systems that rely on batteries.

Figure F.4.  Label indicating flow of electrical energy through circuits. Labeling is critically important to maintain the flow of information about energy flow in circuits. All workers need ongoing reminding no matter how familiar with the equipment.

Figure F.5.  Generators powered by diesel engines are the most common type of of reliable and emergency power source installed in buildings and other installations. These units are conspicuous by the radiator grille and exhaust stack.

Figure F.6. Battery rooms are unobtrusive in facilities as part of the uninterruptible power system. Electrical energy originating from these locations can cause exposure to energized circuits.

### Steam Systems

Steam generating systems must also operate reliably and shut down in a controlled manner following loss of electrical power. Circulation of cooling water and movement of combustion gases are two possible examples of circuits that must perform reliably.

Steam plants can contain steam powered pumps, fans, and generators to aid in orderly shutdown in the event of power failure (Figure F.7). These maintain water circulation, airflow, and electrical power in circuits in parallel to those normally powered by electric motors.

### Portable Electrical Generators

Portable electrical generators powered by small gasoline or diesel engines are ubiquitous by their availability and potential for use. They are especially valuable to homeowners living in areas subject to severe tropical storms and hurricanes. These units are readily connected to the power supply system in a home through minor modification of the wiring at the panel.

### Backfeed

Backfeed refers to the energizing of the power supply system by reverse flow of electrical energy from a consumer. Backfeed occurs intentionally when small producers of electrical energy feed into the grid. This aspect of backfeed occurs in a recognized manner. Of considerable concern is backfeed that is not authorized or recognized.

All sources of emergency power are potentially capable of backfeeding into the electrical grid. Output from a 110 V generator in a garage or carport at a home can feed back into the electrical supply system when the main breaker is closed as would be the case in normal operation.

Backfeed from small generators operated in this manner in areas where power lines had fallen and no longer had normal ground protection has caused electrocution of linemen and other electrical workers.

Figure F.7. Steam-powered pump. The steam system in a steam plant provides reliable power to this pump to maintain water circulation to enable orderly shutdown in the event of electrical failure.

The unit that can prevent this situation is the manual transfer switch (Figure F.8). Both manual and automatic versions are available. The transfer switch routes electrical energy either from the grid or the temporary source, but not both, into the consumer. This prevents inadvertent connection of the temporary source into the electrical grid.

## Power Quality

Power quality is a generic term used by electrical engineers and the electrical industry to compare actual performance of an electrical system in a facility to its design or needed capabilities. Said concisely, power quality is the right power at the right place at the right time (Lamendola 2000). Right power means tailoring voltages and current to the power needs of the load. Some loads require clean power; others do not. Right place means keeping noise from electrical signals that must remain noise-free. Right time means consideration about issues of timing (voltages and currents in phase) and duration of performance of uninterruptible power supplies prior to cut-in by the standby generator set.

Power quality is a recurrent theme throughout this book. Power quality covers bonding and grounding and the presence of unexpected noise in signals transmitted by instruments. A thorough understanding about power quality is as essential for anyone involved in control of exposure to sources of hazardous energy as it is to electricians and electrical engineers responsible for maintaining operation of electrical equipment in a plant.

This understanding is necessary because computers are increasingly taking decision-making in equipment, machines, systems, and plants away from human decision-makers. If we decide to give this ability to electronic decision-makers, then we must have absolute confidence that they will make consistent decisions all of the time. (We control through logic and programming whether the decisions are correct or incorrect.) If microprocessors cannot make consistent decisions because of imperfect input and information processing, then we need to know this and act accordingly.

Figure F.8. Manual transfer switch in a home. Electrical generators in homes can backfeed into the electrical grid in the absence of a transfer switch. This situation has led to electrocutions from contact with downed energized conductors. Automatic transfer switches in circuits containing reliable power also can lead to unexpected energization of circuits.

A second consideration about power quality is that electrical energy is the source for creating almost all other forms. That is, almost all other forms of energy result from the transformation of electrical energy in equipment, machines, and systems. Electrical power of poor quality at the macro level can lead to unpredictable and deleterious performance of equipment and machines. Power of poor quality at the micro level can lead to unpredictable and deleterious performance by the electronics used to regulate and control operation of large sources of energy in equipment and machines.

Safe systems of work depend on reliability and certainty. Reliability and certainty breed confidence. Confidence in performance of equipment and machinery is absolutely essential for the placement of trust by human operators and maintainers. If trust in performance by equipment, including safety-related systems that operate through electrical power does not exist, it will not be used. Hence, the investment will be lost.

About 90% of problems with power quality develop because of inattention to issues of bonding and grounding (Lamendola 2000). These can arise during construction from inappropriate practices and during operation from poor maintenance practices. Poor grounding leads to noise in electronic circuits and data transmission networks. Noise leads to inappropriate signals and uncertain operation. Ground loops lead to electric shocks and outright electrocution from equipment that supposedly is grounded.

Large electric motors cause voltage sags when starting and surges when stopping and affect power factor when running (Hartfiel and Lamendola 2000). Imbalance of more than 2% in voltage is considered a death sentence for a motor (Lamendola 2000). Large heaters, such as process heaters in molding and extrusion, cause severe voltage sags when starting (Hartfiel and Lamendola 2000). A combination of motors and heaters drawing from the same circuits can cause serious problems. Voltage sags caused by start-up of the heaters can cause premature failure of the motors.

The situation regarding voltage surges in electronic circuits is even more serious. An electrical surge as low as 20 V can severely damage or destroy electronic components (Gorosito 2000). Transients can carry several thousand volts and several hundred amperes of current lasting from nanoseconds to several milliseconds. These events can lock-up electronic equipment and cause loss of memory.

Protecting sensitive electronic equipment requires protection of the AC power line and the DC communications line. DC communications lines include twisted pair (for example, telephone lines) and coaxial cable. Protection of twisted pair lines occurs by diverting energy beyond a set level to a common ground. Usually the set point is 1.4 times the nominal operating voltage (similar to overcurrent devices, such as fuses and circuit breakers).

An additional factor in power quality is the role of lightning strikes on electronics and data networks used especially in process industries (Chowdhury 1999). During stormy weather, transients less than 3 V peak or energy levels of $10^{-7}$ J (Joules) can damage or "confuse" electronic components leading to shutdowns and process upsets rarely attributed to lightning.

Uninterruptible Power Supplies (UPS) condition power and supply stored energy by providing continuous, transient-free, sinusoidal power (Stamper 2001). These units offer protection from voltage swings, sags, surges, and loss of power in critical circuits (Hartfiel and Lamendola 2001). While UPSs inject harmonics into supply circuits, they protect circuits from harmonics on the load side.

Applications of UPSs can contribute to problems with power quality (Stamper 2001). To illustrate, the rectifier section of the UPS can inject harmonics into incoming power lines. (Harmonics are signals superimposed onto the 60 Hz sine wave that occur at multiples of 60 Hz.) Harmonics are sources of electronic noise for some equipment. Some equipment is not affected by harmonics in the incoming power.

Harmonics interfere with operation of clocks that receive their syncing signal from the 60 Hz sine wave. Harmonics also have caused unpredictable operation of microprocessor-controlled equipment such as vending machines and production equipment.

Additional injectors of harmonics into the power system include switched mode power supplies used in personal computers, variable speed drives, battery chargers, large electric motors containing brushes (start-up mode), electronic dimming systems, ballasts used in lighting systems (especially fluorescent), arc welders and other arc devices, and medical equipment such as magnetic resonance imaging (MRI) and X-ray units (CDA 1998a).

Double conversion UPSs (AC to DC and DC to AC) do not work well with generator sets (Carr and Sitter 2001). Portable engine-generator sets can contribute significantly to the harmonic problem, much more so than the power grid. The voltage regulator in the gen-set cannot accurately read voltage when total harmonic distortion exceeds 20%. Thus, high levels of harmonics can lead to unstable voltage output from the gen-set. This can render the gen-set nonfunctional.

## Operating Characteristics of the Electrical System

In order to gain an appreciation of events that can occur during the interaction of workers with the electrical system, information is required regarding the characteristics of system operation. The electrical system is the sum of its components and the manner in which they are connected together and interconnected.

### Step and Touch Potential

Electrical current flowing to ground as a result of a fault condition can subject the ground rod or grounded object, and surrounding earth to very high voltage while current flows (BC Hydro 1998). The grounded object could be a vehicle such as a mobile crane containing extended stabilizers that makes contact with an energized line. Voltage decreases rapidly with distance from the position of the ground rod or grounded object (Figure 3.4a, Figure 3.4b, and Figure 3.4c). A conductor making contact with the earth at two points at different distances from the ground rod or other grounded object provides a parallel path through which current can flow. This phenomenon can have fatal consequences when the conductor is the human body.

Step potential is the condition where the feet are positioned at different distances (one foot closer than the other) from the ground rod. The current follows the path from the foot to the leg to the abdomen to the other leg to the other foot. This path is least consequential to the body because of bypassing the heart.

Touch potential results when the body contacts the ground rod or grounded object, perhaps through the arms and hands while the feet remain at the same distance from the ground rod or grounded object. Touch potential is potentially more hazardous than step potential because of the large decrease of voltage with distance close to the ground rod.

Touch-step potential results from contact between the foot and hand on surfaces at different potentials. The hazard from touch-step potential depends on the path of the current through the body.

Electrocution due to touch potential has caused many fatal accidents (NIOSH 2000). Victims touched or sat on objects that became energized. Touch potential exceeding 30 V is potentially hazardous because of the current that can flow through the body (NFPA 2004).

## Flashover

Flashover occurs when a conductive object, including the human body, approaches too close to an energized conductor for the insulation protection offered by the air. Flashover is an electric arc resulting from ionization of molecules in the air in the path between the source and the grounded object. (See the section on Arc Flash Protection.)

Flashover is short-circuit, fault process. This can involve transmission of very high current to the earth through conductive and semi-conductive paths including the human body. Flashover is one of the major causes of the fatal electrical accidents reported by NIOSH (NIOSH 2000).

## Contact

Contact refers to the physical contact established between the conductive object and the energized conductor. Contact is a short-circuit, fault process. This can involve transmission of very high current to the earth through conductive and semi-conductive paths, including the human body.

Contact is a major cause of the fatal electrical accidents reported by NIOSH (NIOSH 2000). Contact is an especial concern on construction sites where established buildings carrying uninsulated conductors and existing transmission lines are located in close proximity to cranes, concrete pumper trucks, excavators, and other mobile equipment containing long booms. Precautionary measures include establishing limits of approach and use of temporary insulation on uninsulated conductors where de-energization is not feasible.

High-voltage lines and equipment isolated from contact are uninsulated. Protection of the general public against electrocution and other types of electrical injury relies on the distance between the lowest point on the conductor bundle and the highest point between the ground or some protrusion above the ground. Work on and around lines and equipment often occurs while they are energized. This is the means by which electrical utilities maintain the high level of reliability of service to which we are accustomed and that we expect. Otherwise, service outages would be common rather than highly infrequent.

Work with high-voltage equipment involves concentration and discipline and precise control over position. Failure to exercise any of these is amply illustrated in the NIOSH document (NIOSH 2000). Even momentary contact with an energized conductor is fatal.

Concentration entails attention to the detail of the task to be performed. Many factors affect concentration: personal affairs, events of the moment, fatigue, the weather, and so on. The role of these factors in these accidents is not assessable from the data available.

Discipline is a personal characteristic. Failure to exercise discipline is amply illustrated in these accidents. Electrical workers employed by utilities and their contractors undergo extensive training, either in-house or through union halls. Many of the organizations described in the NIOSH document conduct regular safety meetings and perform on-site inspections with emphasis on ensuring adherence to procedures. The fatal accidents illustrate lack of adherence to procedures. Factors include lack of knowledge, momentary lapse in concentration, and possibly defiance.

Many contact accidents have occurred on construction sites and in the industrial setting. Contact with overhead electrical utilities is an especial concern during work involving use of the excavator and front-end loader and concrete pumper during construction on residential streets. Contact with overhead wires can occur during raising and extending the bucket or the boom. Similarly a potential for contact exists when the unit is driven with the boom or bucket in the raised position. Even low-voltage wires contribute to the risk, as insulation often separates from older lines leaving the conductors exposed (Figure F.9). A small amount of current (70 to 100 mA) passing through an individual in the appropriate path is sufficient to cause electrocution.

The operator often cannot see the relationship between the lines and the bucket or boom due to obstruction from the body of the vehicle and lack of spatial perspective. As a result, the operator tends to have a two-dimensional view of a three dimensional situation (Figure F.10, Figure F.11 and Figure F.12). An additional contributor to this problem is the focus of the operator on the position of the bucket or the boom, and not on the surroundings, including the position of the highest point of the boom. Contributing to this lack of visibility is loss of peripheral vision caused by the brim of the hardhat. The brim of the hardhat restricts upward peripheral vision such that overhead wiring is not visible. Positioning of the boom in a smooth, controlled manner depends on the dexterity of the operator and precision of control of position offered by the hydraulic system. The only way to be assured of clearance is to measure the highest point of the boom.

Another complication is the overhead wire itself. Overhead wires are part of the congested visual field. They appear to be completely innocuous since they provide no warning to the senses of the danger that they pose other than the passive one of their presence in the visual field and occasionally, buzzing. Warning flags strung near these lines greatly improve visibility but do nothing to reduce the consequences of contact (Figure F.13). Shielding devices hung on the line provide some protection against the consequences of accidental contact (Figure F.14).

The urban milieu has grown around the existing infrastructure. This is readily apparent in inner cities in areas containing old buildings where above ground electrical lines are found in alleys. There is little room to maneuver in these areas. Use of materials such as aluminum siding and long pieces of conductive material compound the probability of contact. This comment applies equally well to industrial facilities where the presence of uninsulated conductors is not always identified.

Contractors are especially at risk in these situations because of unfamiliarity with the operation and its electrical infrastructure.

Homeowners are accustomed to seeing insulation on wiring on street poles and cables that supply electrical energy to the building. An intact layer of insulation is absolutely essential for protection against electrocution during the many activities undertaken by unaware homeowners not specifically warned by the electrical utility against such practices. These include use of ladders for painting and pruning trees and bushes, poles for pruning trees and bushes, installation of aluminum siding, hanging and removing Christmas lights and decorations, and other activities. The presence of protective insulation on supply lines is not a given quantity.

Contact is not restricted to overhead electrical lines. Contact with underground lines occurs in several ways. First is exposure by the bucket of an excavator. This could easily happen with lines that service homes and streetlights where tracing by a locate service has not occurred.

Underground transmission lines of recent vintage use plastic pipe (duct) to hold the individual cables. Spacers hold the duct in position. Concrete surrounds the structure. Embedded in the top layer of concrete is a warning banner. This structure is hard to violate under the controlled conditions that characterize construction sites of today. A reasonable expectation is that the location of the structure would be well known.

This is not necessarily the case for work occurring decades into the future. Experience has shown that the general location of underground structures is retained while the specifics become lost. The specifics can have grave consequences to the safety of construction.

Hydroexcavating technology using heated water enables work on faulted or damaged lines to occur in the coldest part of the winter. This offers major advantages in speed and reduced risk to workers since direct contact with tools, such as picks and shovels, does not occur. In situations involving unprotected cable, the water jet exposes the cable free from adhering soil. Following repair, these cables reseat and refreeze. In warmer times, rain penetrates the soil and, in some cases, penetrates the insulation, causing a new failure. Some cables are so prone to damage of this nature that hydroexcavating is contraindicated as an approach to expose them.

Figure F.9.  Deteriorated insulation on residential utility lines indicates that protection by a protective covering is often more potential than real.

F.10.    The visual field containing electrical lines is a two-dimensional representation of a three-dimensional world. The inability to perceive depth and relative position of objects in the three-dimensional world is a cause for the occurrence of many contact accidents.

Figure F.11. Viewed from the side, the obstructions become considerably more intimidating to avoidance of contact.

Figure F.12. Electrical contact accident involving congested work area and conductive structures. Electrical contact accidents often involve unknowing and knowledgeable helplessness. Many circumstances have involved conductive extensions of the reach of people when handling long pieces of material. (Modified from OSHA.)

Figure F.13.  Flags and other visual devices  increase visibility of overhead electrical lines.

Figure F.14.  Shielding devices reduce the consequence of electrical contact.

The pressure at which the water jetting equipment operates (1,500 to 3,000 lb/in$^2$) is sufficient to cut or damage insulation on wires and cables when used in a narrow jet on prolonged contact. This damage manifests as tearing of the cover.

Change in grade compared to depth of overcover indicated on as-built drawings can mislead about the actual versus indicated depth profile.

## Current Leakage

Current leakage refers to electron flow or transfer of electrical energy outside the intended path. Current leakage occurs from a region of high electrical potential (voltage) to one of low potential. The zone of low potential most often is ground or earth. Low potential does not necessarily mean zero potential. Current leakage that establishes a path through the human body can produce fatal consequences.

Current leakage can occur through insulation applied to the surface of wiring. There is no perfect insulator. This includes even new insulation and systems (Palmquist 1988). The current is attempting to return to source and not to ground (Holt 2001a).

The most likely cause of current leakage through intact insulation is overstressing caused by the level of energy transmitted. This can occur during testing designed to determine functionality as well as overheating due to high load and in-service fault conditions. Current leakage also occurs as insulation ages through corrosive attack, embrittlement from loss of plasticizers, or degradation due to exposure to ultra-violet radiation from the sun or from man-made sources such as welding arcs. In these cases, the insulation cracks and establishes a pathway through which conductive materials in rainwater or other water contact, such as mud, can enter.

Current leakage also can occur through the plastic materials used in the casings of portable electric tools. This situation begins when a crack develops. Cracks in tool housings arise from stress as well as aging, embrittlement, and degradation. Conductive material transported in liquids, such as perspiration, can deposit in the crack over time thereby creating a conductive pathway. All that is needed for development of current leakage is an internal breakdown in the insulation that leads to linking of the two pathways.

Stray electrical currents and water are potent factors in electrocutions. The NIOSH and OSHA reports on fatal accidents illustrated this situation numerous times (NIOSH 2000; OSHA 1988, OSHA 1990). An energized conductor that is exposed to water can produce currents in the water (Smoot and Bentel 1964). The magnitude of the current will depend upon the shape and the size of the conductor-water contact surface, conductivity of the water, and the resistance in the current path to ground (Novotny and Priegel 1974). The water source in these situations can include sweat on the skin, as well as water or other conductive liquid, such as mud, that is present in workspaces. As well, metals, such as steel and aluminum, which are used as materials of construction in many structures, are excellent conductors.

## Electromagnetic Induction

Electromagnetic induction refers to the creation of voltage and current flow in a non-energized conductor by an energized one not physically in contact with it. Associated with the electromagnetic energy present in energized conductive materials are electrical and magnetic fields. In an AC system these fields display the characteristic sine wave pattern. The electric and magnetic components are oriented at 90° to each other. The strength of the electric field component of the energy field is measured in units of V/m (volts per meter). The strength of the magnetic component of the energy field is measured as magnetic flux density in units of A/m$^2$ (amperes per square meter).

Electromagnetic induction is a serious concern where ungrounded conductors and large metal objects such as excavators, cranes, concrete pumpers, other large construction equipment, farm equipment, and a myriad of other types of equipment are operated near high-voltage transmission and distribution lines. All that is needed for current flow is completion of the path to ground.

Energy transfer from an energized line to one that is isolated and grounded occurs because the latter cuts the lines of the electromagnetic field set up around the former (BC Hydro 1998). Current will flow when two grounds are present to complete the secondary circuit. One ground could be a system ground in a substation while the other could result from accidental contact with the line by a piece of equipment known or unknown to the service crew.

## Electrostatic Induction

Electrostatic discharges are most commonly associated with sources of direct current or materials that contact and separate. (See the section on Electrostatic Discharges.) A less commonly recognized cause of electrostatic energy is electrostatic induction (BC Hydro 1998). Electrostatic induction occurs between power lines and ungrounded conductors, especially large metal structures such as excavators, cranes, concrete pumpers, other large construction equipment, and metal objects such as steel water pipe. The magnitude of the induced current varies with field strength, frequency of the field, size and shape of the object, and the object-to-ground resistance (BPA 1986).

The power lines and the metal objects constitute plates of a capacitor. Air is the dielectric. Capacitance develops between phases of an energized line and an ungrounded isolated line or any large metal object in the vicinity. The alternating voltage on the energized line creates an alternating electrostatic charge on the isolated line or object. The electrostatic charge on isolated conductive lines is greatest when the lines run close together and parallel for long distances as occurs with transmission lines. Electrostatic current created in this way can be fatal. Electrostatic induction is discharged to ground through one connection. For grounded objects, the induced alternating current to ground is the short-circuit current (Table F.1).

Induced current increases linearly with increase in the magnitude of the electrical field. To illustrate, a 10-fold increase in field strength produces a 10-fold increase in induced current. To put these values into perspective, 1 kV/m is a benchmark being considered or enacted for electric fields at the edge of rights of way for transmission lines (Morgan 1989). Fields of this magnitude exist near the surface of electric blankets and some household appliances. ANSI/UL (1992) permits current leakage of 0.5 mA from portable household appliances and 0.75 mA from fixed appliances. Currents of this magnitude are barely detectable as shocks.

## RF Induction and Ignition

To this point, discussion has focused on 60 Hz, line sources. Broadcast towers are point sources of electromagnetic emissions (Denton 2002). Broadcast emissions are AC with frequencies in the high kHz and low MHz range. The entire AM radio tower acts as the antenna. Elongated metal structures, such as the lattice booms of cranes and the hoisting cable, are susceptible to induction in the presence of these emissions. Resonance of objects occurs at full, half, and quarter wavelengths. Accumulation of current can lead to shock.

## Inrush Current (Inductive Reactance)

Closing a circuit allows entry of the disturbance that constitutes electrical energy. The inward rush of the disturbance is somewhat akin to the rapid rush of water into a pipe when a valve opens rapidly to full bore.

Circuits containing coils of wire behave in an unusual manner. When voltage is suddenly applied to a coil, magnetic lines of flux develop and expand in and around the coil (Middleton 1986). As the lines of magnetic flux enlarge, they cut the turns of the coil and induce an electromotive force (emf). The induced emf (voltage) is not the same as the voltage applied to the circuit that created the lines of magnetic flux in the coil. The induced voltage opposes the applied voltage. This opposition is inductive reactance or inductive opposition to sudden change in current in the circuit. The opposition to flow of current increases the time needed for the current in the circuit to build to maximum level.The inrush current is the waveform of the input current measured immediately after first turning on an electrical device (Anonymous 2011).

To illustrate, incandescent light bulbs and other heaters containing coils of wire experience high inrush current until the filament heats up and increases the resistance. Alternating current electric motors and transformers draw several times their normal full-load current when first energized for several cycles of the input waveform. Power converters also experience high inrush relative to current draw in steady-state. This is typically the charging current of the input capacitance.

## Table F.1
## Short Circuit Current (60 Hz Electrical Field, 1 kV/m)

| Object | Induced Current mA |
|---|---|
| Person (1.75 m) | 0.016 |
| Farm tractor | 0.10 |
| Station wagon (car) | 0.11 |
| Camper truck | 0.28 |
| School bus | 0.41 |
| Large semi-trailer + rig | 0.63 |

Source: BPA 1986.

When a transformer first energizes, a transient current 10 to 50 times greater than the rated current can flow for several cycles. This happens when the primary winding is connected around the zero-crossing of the primary voltage. For large transformers, inrush current can last for several seconds. Toroid transformers can have inrush up to 80 times the operating current.

Selection of overcurrent protection devices is complicated by the need to tolerate high inrush currents. The overcurrent protection must react quickly to overload or short circuit but must not interrupt the circuit when the (usually harmless) inrush current flows.

### Outflow Current (Self Induction or Switching Surge)
At the time prior to opening a switch, a magnetic field exists in the space surrounding the coils in the circuit containing coiled structures in components including electromagnets, transformers, and motors. Opening the contacts of switches in these circuits often produces an arc (Middleton 1986). This situation can have highly deleterious consequences including arc flash, arc blast, and related effects. This outflow of current is known as a switching surge.

Opening a switch in circuit containing coils allows the field maintained by the applied voltage to collapse inward, again cutting the turns of the coil (Palmquist 1988). Cutting the turns of the coil by the collapsing electromagnetic field induces an emf. The emf induced (generated) in the circuit opposes any change in current flow. A magnetic field contains energy which is transformed during this process. The generation of the emf is called self-induction and occurs in circuits containing coiled conductors. Inductance is the inherent property of coiled conductors that opposes any change to direction and magnitude of flow of current at time of opening the switch.

### Protection Strategies

Traditional approaches to protection involving energy sources are broadly classed under time, distance, and shielding. More current approaches are based on risk assessment. If the risk is deemed too high, risk reduction measures are required. Risk reduction is a hierarchical process that occurs through elimination through design, use of engineered safeguards, awareness means (including warning and alerting techniques), administrative controls (including safe work procedures and training), and use of personal protective equipment. Often, the solution can include aspects of each of these elements.

Electrical issues typically have relied on code-based specifications. These can become highly prescriptive and lead to challenges for annual review of electrical standards, such as NFPA 70, the US National Electrical Code® (NEC) (NFPA 2011a). Electrical trade publications and consultants are gainfully occupied communicating about changes in the NEC.

Themes of major concern that affect nearly all systems include protection from electrocution and overcurrent, means of disconnection, and selection of components and cable, wire and enclosures appropriate to the service. Themes that affect some systems include protection against ignition of ignitable atmospheres, protection against deleterious effects of moisture and liquid water, and protection against chemical attack of conductors and components. Critical to ensuring protection is identification of critical factors at the beginning of the process of designing an electrical installation.

## Design

Design is a critical part of creating a system composed of more and more complex components interconnected by wire and cable that poses minimum risk to workers who have frequent or infrequent encounters with it. Encounters include at-a-distance and direct contact. The approach of the NEC to design initially is to require classification of the application of the service. This then leads to sections in the NEC and possibly to subsidiary standards that focus on the particular application (NFPA 2011a). Classification governs the type of electrical equipment to be used in the installation.

Redesign to eliminate the hazard is the most desirable of the options in the risk reduction hierarchy. The risk is no longer present. With other options, the hazard remains but is subject to control of decreasing desirability and reliability.

## Hazardous Locations

Preventing ignition of ignitable atmospheres through unexpected discharge of electrical energy (arcing) or heating of surfaces during normal operation can be paramount to protecting people and facilities. This is especially the case where the atmosphere in a workspace or work area containing equipment is ignitable or can become ignitable. Table F.2 assigns a level of hazard to the space according to the type of substance or material that can be present and its probability of occurrence (NFPA 2011a).

Class I contains four groups (A, B, C, and D). Groups list chemical substances having similar ignition temperatures. Class II contains three groups (E, F, and G). Division 2 locations normally are not hazardous. They apply to areas that become hazardous in the event of accidental discharge of ignitable materials from confined systems.

An explosive mixture of gas, vapor, dust, fibers, or flyings can exist under normal operating conditions in a Division 1 location. An explosive mixture can exist only under abnormal conditions in a Division 2 location. A Division 2 location normally surrounds a Division 1 location in physical space.

The International Electrotechnical Commission (IEC) created the system (Table F.3) used for classifying hazardous locations outside of North America and parts of Central and South America (IEC 2008, IEC 2009).

The IEC classification system bears similarities to that contained in the (U.S.) National Electrical Code®. Presently there is a move to harmonize the system for classifying hazardous locations in the National Electrical Code® and Canadian Electrical Code® with the IEC concept.

The maximum surface temperature of electrical or mechanical equipment must always be lower than the ignition temperature of the mixture of the surrounding gases or vapors with air at normal pressure. This is to ensure that the surface cannot act as a source of ignition. The ignition temperature of a flammable substance is the minimum temperature at which the material will ignite and sustain combustion (autoignition temperature). The autoignition temperature of different gases varies considerably. To illustrate, a mixture of hydrogen in air will ignite at 560 °C but a mixture of gasoline in air will ignite at 280 °C.

To assist manufacturers in the design of equipment for intended service, standard setters segregate autoignition temperatures into six classes ranging from 85 °C (T6) to 450 °C (T1) (Table F.4). Test-

## Table F.2

## Hazardous Locations (U.S. National Electrical Code)

### Classification   Description

Class I: A location in which flammable gases or vapors are or may be present in the air in quantities sufficient to produce explosive or ignitable mixtures.

Division 1      Location in which flammable gases or vapors are or may be present in the air in quantities sufficient to produce explosive or ignitable mixtures,

or

Ignitable concentrations of such gases or vapors may exist frequently because of repair or maintenance operations or because of leakage,

or

Breakdown or faulty operation of equipment or processes might release ignitable concentrations of flammable gases or vapors, and might also cause simultaneous failure of electrical equipment that could act as a source of ignition.

Division 2      Location in which volatile flammable liquids or flammable gases are handled, processed, or used, but in which the liquids, vapors, or gases will normally be confined within closed containers or closed systems from which they can escape only in case of accidental rupture or breakdown of such containers or systems, or in case of abnormal operation of equipment,

or

Ignitable concentrations of gases or vapors are normally prevented by positive mechanical ventilation and might become hazardous through failure or abnormal operation of the ventilating equipment,

or

Ignitable concentrations of gases or vapors might occasionally migrate to a class I, division 1 unless such migration is prevented by adequate positive-pressure ventilation from a source of clean air and effective safeguards against ventilation failure are provided.

Class II: A location that is hazardous because of the presence of combustible dust.

Division 1      Location in which combustible dust is in the air under normal operating conditions in quantities sufficient to produce explosive or ignitable mixtures,

or

Mechanical failure or abnormal operation of machinery or equipment might cause such explosive or ignitable mixtures to be produced and might also provide a source of ignition through simultaneous failure of electrical equipment, operation of protection devices, or from other causes,

or

Combustible dusts of an electrically conductive nature may be present in hazardous quantities.

## Table F.2 (Continued)
## Hazardous Locations (U.S. National Electrical Code)

| | |
|---|---|
| Division 2 | Location in which combustible dust is not normally in the air in quantities sufficient to produce explosive or ignitable mixtures and dust accumulations are normally insufficient to interfere with the normal operation of electrical equipment or other apparatus, however combustible dust may be in suspension in the air as a result of infrequent malfunctioning of handling or processing equipment and combustible dust accumulations on, in, or in the vicinity of the electrical equipment may be sufficient to interfere with the safe dissipation of heat from electrical equipment or may be ignitable by abnormal operation or failure of electrical equipment. |

Class III: A location that is hazardous because of the presence of easily ignitable fibers or flyings but in which such fibers or flyings are not likely to be in suspension in the air in quantities sufficient to produce ignitable mixtures.

| | |
|---|---|
| Division 1 | A location in which easily ignitable fibers or materials producing combustible flyings are handled, manufactured, or used. |
| Division 2 | A location in which easily ignitable fibers are stored or handled (except in the process of manufacture). |

Source: NFPA 2011a.

ing determines to which class a particular piece of equipment belongs. The manufacturer then labels the equipment accordingly based on the maximum surface temperature of any relevant part that might make contact with the flammable gas that has the same or lower T value. For flameproof and pressurized equipment, the maximum surface temperature refers to the outside of the enclosure. For increased safety, the hottest point is inside.

The surface temperature of any parts of the electrical equipment potentially exposed to the hazardous atmosphere should not exceed 80% of the autoignition temperature of the specific gas or vapor in the area where the equipment is to be used.

Several techniques (Table F.5) are available for protection when electrical and electronic equipment is used in hazardous locations (Paschal 1998a).

Explosion-proof equipment generally is required for Class I Division 1 locations. Explosion-proof electrical equipment is not gas-tight. In fact, threads in connecting piping, other passageways in the enclosure, and beveled surfaces vent the gases to atmosphere after sufficient cooling. Five full threads are required in threaded connections (NFPA 2011a). In order to rank as explosion-proof, the cast-metal electrical enclosure must prevent escape of flame and operate below the ignition temperature of the ignitable material in the ambient environment. Essentially, the purpose for an explosion-proof enclosure is to prevent initiation of a fire or explosion in the ambient atmosphere.

Nonincendive refers to techniques that prevent arcing or thermal conditions from becoming ignition-capable under normal operating conditions. These techniques are allowed only in Division 2 locations. Normally, nonsparking equipment or apparatus that has make-or-break contacts immersed in oil or is hermetically sealed is used in Division 2 areas. This equipment can cause ignition only when malfunction occurs at the same time as development of an ignitable concentration. This event is unlikely, considering the tiny probability of simultaneous occurrence of electrical failure and release of ignitable materials. When nonsparking equipment is not available, use of explosion-proof equipment or electrical apparatus contained in explosion-proof housings must occur.

## Table F.3
## IEC Classifications for Hazardous Locations

**Classification  Description**

**Gases and Vapors**

| Zone 0 | A place in which an explosive atmosphere is present continuously, for long periods or frequently. |
|---|---|
| Zone 1 | A place in which an explosive atmosphere is likely to occur in normal operation, occasionally. |
| Zone 2 | A place in which an explosive atmosphere is not likely to occur in normal operation but, if it does occur, will persist for a short period only. |

**Combustible Dust**

| Zone 20 | A place in which an explosive atmosphere in the form of a cloud of combustible dust in air is present continuously for long periods or frequently |
|---|---|
| Zone 21 | A place in which an explosive atmosphere in the form of a cloud of combustible dust in air is likely to occur occasionally in normal operation. |
| Zone 22 | A place in which an explosive atmosphere in the form of a cloud of combustible dust in air is not likely to occur in normal operation but, if it does occur, will persist for a short period only. |

Sources: IEC 2008, IEC 2009.

Purged and pressurized systems can contain uncontaminated air or an unreactive gas. This system must be fail-safe. A pressurized control room is an example of a space that meets these requirements.

Intrinsic safety means that the electrical equipment is designed to release insufficient electrical or thermal energy under normal or abnormal conditions to ignite a specific hazardous atmosphere. Abnormal conditions include accidental damage to field-installed wiring, failure of electrical components, over-voltage, adjustment and maintenance operations, and other similar conditions. The low-energy requirements limit the use of intrinsically safe equipment to low-power devices such as process-control instrumentation and communication equipment (Elcon 1989).

Classification of location has important implications in the specification of portable ventilation equipment. Users of portable ventilating equipment must realize that the classification of the location in the vicinity of proposed use governs the rating required. The area of proposed use may never have been classified. In this case, this could be an essential prerequisite before specification decisions occur.

## Damp Location
Using the terminology in NFPA 70, a damp location is an outdoor location partially protected under canopies, marquees, roofed open porches, and like locations, and interior locations subject to moderate amounts of moisture. These include some basements, some barns, and some cold-storage warehouses.

A dry location is not normally subject to dampness or wetness. A location classified as dry may be temporarily subject to dampness or wetness as in the case of a building under construction. A wet loca-

## Table F.4

## Surface Temperature Classes for Electrical Equipment (NFPA 70)

| Maximum Surface Temperature °C | Class I | Class II | Class III |
|---|---|---|---|
| 450 | T1 | | |
| 300 | T2 | | |
| 280 | T2A | | |
| 260 | T2B | | |
| 230 | T2C | | |
| 215 | T2D | | |
| 200 | T3 | Group E: all conditions; Group F: equipment not subject to overload; abnormal operation of overloaded equipment | |
| 180 | T3A | | |
| 165 | T3B | Group G: equipment not subject to overload; abnormal operation of equipment subject to overload | equipment not subject to overload |
| 160 | T3C | | |
| 150 | | Group F: normal operation of equipment subject to overload | |
| 135 | T4 | | |
| 120 | T4A | Group G: normal operation of equipment subject to overload | normal operation of equipment subject to overload |
| 100 | T5 | | |
| 85 | T6 | | |

Source: NFPA 2011a.

tion includes installations underground or in concrete slabs or masonry in direct contact with the earth, and locations subject to saturation with water or other liquids, and locations exposed to weather and unprotected.

Chemically Incompatible Location
Chemical incompatibility between materials of construction and materials present or produced in the operating environment is an important consideration in design. This concern is easily overlooked by designers unfamiliar with incompatibility. Incompatibility includes chemical incompatibility, exposure to sources of ultraviolet emissions starting with the sun in outdoor locations, and including process emissions from welding arcs and other sources.

## Table F.5

## Measures for Protection of Equipment in Hazardous Locations

**Protective Measure**

Explosion-proof equipment (Class I locations)

Dust-proof equipment (Class II locations)

Dust-tight equipment (certain Class II and Class III locations)

Purged and pressurized equipment

Intrinsically safe systems

Oil immersion

Hermetically sealed equipment and contacts

Nonincendive circuits and equipment

Source: Paschal 1998a.

### Bonding and Grounding

Bonding is the establishment of an electrically conductive path between conducting objects that has the capacity to conduct safely current likely to be imposed. Grounding is the routing of current from a conducting object to ground or earth (NFPA 2011a). Bonding and grounding are fundamental to establishing and maintaining the safety of electrical equipment and systems.

Failure to maintain an effective bonding and grounding system is a factor in many fatal electrocutions (Cantwell 1981). Bonding and grounding provide protection from overvoltage from inadvertent crossing of primary and secondary leads to transformers or high-voltage and low-voltage lines. Bonding and grounding dissipates lightning or electrostatic charges or other types of surge voltages and maintains noncurrent-carrying parts of an electrical system at zero potential relative to ground.

Fundamental to understanding the role of bonding and grounding is to recognize the behavior and action of electrical current that has leaked from containment at a fault in the electrical system. Electrical current attempts to return to source and not to ground (Holt 2001a).

Bonding of components of the electrical service (inside a building), through low-resistance connections back to the neutral of the electrical system, provides the necessary return path. These connections involve the grounding connector to earth and ground wire and include metal surfaces at loads and components of the electrical service (Figure F.15). The latter include enclosures of outlets, panels, main breaker, and the meter. The neutral of the service and the neutral of the utility transformer are connected to ground. Hence, the neutral is connected to ground throughout the electrical system.

The resistance of copper conductors used throughout the electrical system is about one billionth of the resistance of ground. Hence, ground is not the preferred path for currents resulting from faults in the electrical system. The preferred path is to source through the neutral conductor.

The earth and ground are not the same entity (Lamendola 2001b). Airplanes and spacecraft that have no contact with the earth during flight have grounding systems. Ground is a common reference plane of electrical potential. Ground may or may not be absolute zero of electrical potential. The earth is not even a good conductor and has high resistance compared to an equivalent length of copper wire. Hence, use of the earth, in place of an electrical conducting cable, as a bonding jumper between two independently grounded systems is potentially hazardous because of the requirement for development of considerable voltage in order to enable current to pass.

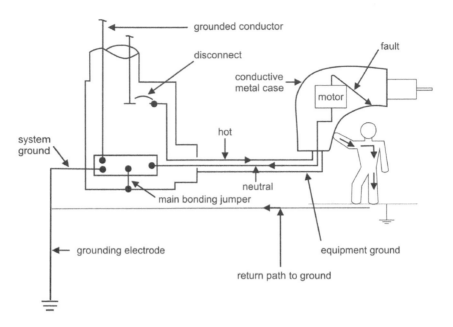

Figure F.15. Bonding is a series of interconnections between components and the neutral conductor in an electrical service. Bonding provides a return path for electrical current.

Bonding metal parts of an electrical system and connecting them to ground limits voltage to ground (Holt 2001b). This prevents destruction of electrical components as well as electric shock that can occur from superimposition of voltages from lightning and voltage transients. Resistance of the ground determines the effectiveness of the grounding system in dissipating surges of high voltage into the earth. Grounding metal parts to earth does not assist in removing dangerous voltage from line-to-ground faults by opening the overcurrent protection device for systems that operate at less than 600 V.

Ungrounded systems were popular in the early half of the 20th century (CDA 1998a). The reason for this popularity, aside from saving tremendous quantities of copper in wire, was to ensure that motors would not stop simply because of a short-circuit. The downside from this approach was that the frame of the equipment could become energized and pose a shock hazard when an individual simultaneously touched it and another conductive object. As well, a grounding system provides a low impedance (low resistance) path to ground for currents produced by lightning and other transients.

Bonding jumpers tie the grounding system together through a reliable low resistance path of electrical cable. Bonding ties together not only components in the electrical system but also nonelectrical components including structural steel in buildings, metal cladding, water piping, and metal gas piping that are capable of becoming energized. A completely bonded system ensures a low potential difference exists between conductive items located adjacent to each other. This will prevent a person from acting as a conductive path between conductive objects at different potentials and receiving a shock when touching the two items simultaneously (Hartwell 1999).

Traditional bonding protocols involving water and gas piping are undergoing review because of the increased use of nonconductive plastic piping in both applications (Paschal 1998b). Previously extensive networks of metal components underground now are discontinuous and fragmented due to the use of nonconductive plastics. This severely limits the continuity of the grounding network provided by these structures. Compounding this situation is the retrofitting of installations using nonconductive plastic materials that invalidate assumptions used in design calculations. This situation can considerably alter the effectiveness of a grounding system.

Bonding occurs through use of cable connected through fittings to various conducting materials including structural steel and aluminum. Bonding to structural steel is one of the most difficult tasks to undertake (Switzer 2000). Mechanical lugs, crimp or compression lugs or exothermically welded lugs are used to establish the bond. Mechanical and compression lugs attach to the structure using nuts and bolts. As well, attachment of the bonding cable to the structure can occur through clamps or direct welding to the steel. Problems with creating the bond include contact surface area and corrosion.

Mechanically bonding using nuts and bolts and clamps and fittings make true contact only at peaks in the surfaces. The surface area of these peaks is much smaller than overall surface area.

Exothermically welded bonds do not suffer from these problems because there is no mechanical interface in the electrical path.

Effectiveness of grounding systems depends on minimizing impedence (resistance) of the path to ground (Hartfiel and Lamendola 2000). The end load can be considerable distance from the neutral ground bond. The extra wire length can contribute to ground problems. One solution is to locate distribution transformers close to the end load.

The resistance of the ground path is contingent on establishing optimum metal-to-metal contact in order to dissipate high-voltage surges to ground (Holt 2001b). Deficiencies in the system tend to become apparent with age. Improper or inadequate grounding causes many of the problems affecting electrical systems. These range from electrical noise affecting data signals to shocks to electrocution from supposedly grounded systems.

Grounding is effected through direct connection of a circuit of low resistance between the conductive object and the ground. Effective grounding means that undesired current from lightning strikes and electrical surges will flow preferentially to ground rather than through alternate pathways such as conductive paths and electrical circuits. This could necessitate resistance in the bonding and grounding system as low as 1 Ω or 2 Ω, although specification of 10 Ω is more common and well below the 25 Ω recommended in electrical codes (CDA 1998b, CDA 2002; NFPA 2011a). (The ohm is the unit of resistance to flow of electrical current in conductors.)

Ground loops develop when the earth is used as a jumper between independent grounding systems (Hartfiel and Lamendola 2000). A ground loop forms when two or more points in an electrical system normally at ground potential are connected by a conducting path such that either or both points are not at the same potential. Current flows between grounds by way of the process loop (DeDad 1999). Rather than discharging noise, ground loops act like antennas to collect it.

Ground rods are the physical entities used to make contact with the soil (Paschal 2000a). Ground rods inject current into the surrounding volume of soil. Grounding electrodes do not make positive connection with soil. What occurs is surface-to-surface contact with the edges of soil particles. Hence, anything that can increase electrode-to-soil contact will reduce resistance in the grounding system. Current flow actually occurs through electrolytes present in the soil. Development of electrolytes depends heavily on the presence of water. Resistance of the ground is highly variable.

Rod materials (in decreasing order of conductivity) include copper, copper-clad steel, zinc-coated steel, and stainless steel. Corrosion and other considerations (soil resistivity and rod diameter) govern the type of rod that is chosen for installation since conductivity has little impact on earth electrode resistance. Aluminum is not permitted in grounding systems due to its high chemical activity (Table F.6).

Ground rods longer than 3 m (10 ft) generally provide little added benefit to the grounding system except in highly unusual circumstances. These include the need to protect sensitive equipment on mountaintops where the base is rock and lightning strikes occur (CDA 1998b). In one example at the peak of Mt. Washington in New Hampshire, ground rods 180 m (600 ft) were installed into the bedrock. Improvement in earth electrode resistance can occur with ground rods extended below the moisture line in extremely dry soils and below the frost line in areas with deep frost or permafrost. Soil resistivity heavily depends on soil moisture (Paschal 2000a).

Installation of ground rods demands care since failure of the grounding system can occur over time due to mechanical failure or corrosion. During installation of ground rods, the soil or material surrounding the rod should have low resistivity. Resistivity is the measure of conduction of soil. Resistivity of soil depends on composition, moisture content, salts and mineral content and temperature (Palmquist 1988). Dry, washed sand and gravel is a very poor conductor and would be a poor choice to use as fill next to a grounding conductor. Wet, salt-containing fills, ashes, cinders and brine wastes should be excellent conductors. Bentonite (a clay material) and concrete are also preferred choices. Concrete and bentonite are hygroscopic, and as a result, retain water. Conductive salts added to the mix further decrease resistivity.

Corrosion of the ground rod is an additional concern (Paschal 2000a). Soils containing sulfides or hydrogen sulfide ($H_2S$) will form copper sulfide on the surface of the rod. This situation is also a concern with airborne $H_2S$ and exposed copper in uninsulated cables in wastewater treatment facilities. Airborne $H_2S$ will turn the surface of exposed copper black due to formation of a sulfide layer. This layer is likely to affect the resistance of the cable since electron flow occurs on the surface.

Oxidation and sulfide formation will dramatically increase resistance in the bonding and grounding system and will accelerate failure. Stainless steel is required in these circumstances. Damage to the copper cladding of steel rods during installation will promote rapid corrosion through galvanic reaction between the steel and electrolytes in the soil (Table F.6). The steel sacrifices to protect the copper. Similarly, in the case of zinc galvanized rods, the zinc protects the steel (Shackelford 1988). When the zinc coating has disappeared due to corrosion, the steel corrodes.

Mechanical connections between the ground rod and cables tend to loosen over time. These require periodic tightening. Failure of the bonding and grounding system can occur due to inadvertent physical damage or even theft of components such as copper bars and cable for perceived value as scrap.

Use of longer electrodes, use of additional parallel electrodes, or addition of chemical additives to the soil near the electrode can improve grounding in unsatisfactory situations. Reduction of ground resistance through use of chemical treatment is not considered a permanent solution since surface and groundwater can dissolve away the salts. Of the three techniques for improving the system, driving additional ground rods is the preferred route.

Bonding and grounding are especially important in hazardous locations (Hartwell and Cannatelli 1999). Conductive structures must be reliably safe to touch even under fault conditions. This requires high levels of craftsmanship in all parts of the system. Connections used in bonding and grounding under these circumstances must be reliable. Reliability extends in time far beyond the initial connection. Intrinsically safe systems require low impedance (resistance) paths to ground. Wrench-tight connections are required to prevent sparking and to maintain the explosion-proof capability of some equipment.

## System Grounding

System grounding refers to intentional connection of a current-carrying conductor of the electrical system (neutral conductor) to earth (or some conducting body in place of the earth) (Hartwell 1999). The purpose for system grounding is to limit voltages imposed by lightning, line surges, or unintentional contact with lines of higher voltage and to stabilize voltage with respect to earth during normal operations (NFPA 2011a). The intent of system grounding is to trigger overcurrent devices (circuit breakers and fuses).

System grounding depends on the presence of an intact neutral conductor from the electrical utility to the service in a building (Holt 2002). In the event that the grounded service conductor (neutral) is missing or open due to a break in the line, neutral current will flow through metal parts of the electrical system. The grounding function of the grounded neutral shunts potentially dangerous energy from the system into the earth.

An indication that the neutral is open is that electrical equipment and switches do not function in the accustomed manner. As well, protective devices cannot clear a ground fault. Metal structures and piping could become energized to line voltage. A low resistance ground is not equivalent to the grounded neutral.

## Table F.6
## Activity of Metals Used in Electrical Components

| Activity | Metal |
|---|---|
| most | aluminum |
| | zinc |
| | chromium |
| | iron |
| | cadmium |
| | nickel |
| | copper |
| | mercury |
| | platinum |
| least | gold |

Source: Shackelford 1988.

## Equipment Grounding

Equipment grounding occurs through bonding all noncurrent-carrying metal components and connecting to the system grounded conductor (neutral), the grounding electrode or both, at the service equipment or at the source of a separate system (Lamendola 2001b). Noncurrent-carrying metal components include cabinets, conduits, support brackets and fittings, and raceways and cable trays. This maintains all exposed metal surfaces at ground potential even during a fault condition (Bernstein 1991a).

A low-resistance, internal fault in contact with an exposed metal surface probably will trip the overcurrent protection device if the equipment is properly grounded. This disconnects power from the equipment and removes the hazard. The overcurrent device operates because of the low-resistance path provided by the series combination of the internal fault and the equipment ground.

A properly installed equipment ground in a fault condition involving high internal resistance will tend to maintain exposed metal surfaces at or near ground potential. In the latter case, the high internal resistance limits flow of the fault current to a level below that necessary to activate the overcurrent device. Thus, even though disconnection of the power does not occur, surfaces of the equipment still are safe to touch.

Without equipment grounding, a fault that enables an energized conductor to contact exposed metal parts may not trip the overcurrent device (Bernstein 1991a). Activation of the overcurrent device depends on the existence of a path to ground of sufficiently low-resistance from the exposed metal parts.

A situation involving a faulted home appliance, such as a toaster oven, resting on a dry wooden table likely would not satisfy this requirement. The path to ground from the exposed metal parts likely would have too great a resistance to activate the overcurrent protection device. The equipment could continue to function normally with its exposed metal parts at a potential of 120 V. Simultaneously touching the exposed metal and a grounded object could provide the current pathway leading to a lethal shock.

An example of a potentially lethal situation involving isolated grounding systems is the two wire system found in many metal lighting standards (Holt 2001b). These systems use a ground rod at the light standard in an attempt to drain fault currents to ground. The earth is an inefficient path for current attempting to return to source and only a small current can pass. This current is insufficient to trip the overcurrent protective device. Hence, the metal of the light standard can pose a serious electrical hazard through touch potential. In order for an overcurrent device to protect humans, it must trip in less than one second. (See the section on overcurrent devices.)

## Bonding and Grounding in Underground Electrical Vaults

Bonding and grounding are part of the system for electrical protection in the underground electrical transmission and distribution system. Bonding establishes pathways of low electrical resistance through permanent and reliable metal-to-metal connection to provide a return path to the system during a fault. Grounding connects the bonding network through a conductive path of low resistance to earth.

The bonding network in each vault contains reliable, low resistance, metal-to-metal connections between the galvanized steel used in racking and support structures, the track molded into the concrete of the structure, and possibly the reinforcing steel. In newer structures, the bonding network utilizes both galvanized steel (zinc-coated steel) and copper connections, and stainless steel fastenings and fittings. Older structures utilized only galvanized steel. Welded bonds in which the components become fused together are favored for longevity over mechanical bonds.

The manhole rim and cover which sit on the top of the concrete of the vault or are extended to grade using concrete rings or cemented bricks normally are not tied into the bonding network. This situation requires assessment and consideration because of the presence of isolated conductors.

The grounding network usually includes one or more ground rods driven at each vault. Interconnection between vaults occurs in some situations. Where interconnection between vaults is present, current flow between substations is possible and can set up a network of circulating currents.

The bonding and grounding network on which the safety of entrants is highly dependent is subject to failure through a number of mechanisms. These can include oxidation, corrosion, physical disruption due to earthquake and excavation, use of incorrect or inappropriate components during assembly, and possibly other mechanisms.

Oxidation is the direct chemical reaction (attack) between the metal and atmospheric oxygen (Shackelford 1988). Metals affected by oxidation must have contact with atmospheric oxygen. Build-up of oxide scale on metals occurs by several mechanisms, each characterized by a specific type of diffusion through the scale. The zinc coating in galvanized metal is designed to oxidize (to form a protective oxide layer). The oxide layer can increase electrical resistance.

Direct chemical attack can occur through contact with other gases including hydrogen sulfide and mercaptans which can be present in vaults due to anaerobic processes involving organic debris. Attack by hydrogen sulfide and mercaptans can lead to formation of a sulfide outer coating on metals.

Corrosion is the dissolution of a metal into an aqueous environment (Shackelford 1988). Metal atoms dissolve as ions. Aqueous corrosion is a form of electrochemical attack. A variation in concentration of metal ions above two different regions of a metal surface leads to an electrical current through the metal. The region of low ionic concentration corrodes (loses material to the solution). Salt water draining into the vault in surface water from salted roads can increase the rate of the aqueous corrosion process. Carbon steel is at considerably higher risk in this environment than is stainless steel.

Corrosion results when dissimilar metals (copper and steel) are held in contact in the presence of moisture and oxygen. Steel will corrode in the presence of moisture to "protect" the copper (Table F.6). Table F.6 applies to pure metals. Differences may apply to alloys containing these metals.

Galvanic corrosion results when two dissimilar metals, one more active chemically than the other are placed in contact. The more active metal acts as an anode and is corroded (loses material to the solution). Iron and zinc will corrode when in contact with copper in the appropriate ionic environment such as the salt water that enters the vault from salted streets and roadways. Galvanic corrosion can occur following substitution of inappropriate metals in fastenings and fittings.

Gaseous reduction involving dissolved oxygen is another corrosion process (Shackelford 1988). In this case, the driving force is the difference in concentration of oxygen in different parts of the aqueous environment. Differences in oxygen concentration can develop in cracks and gaps in surfaces of the same piece of metal that are covered by dirt, grease, paint, or other impervious coating versus an exposed surface, and by the environment created by fastenings and fittings that hold together pieces of metal. Corrosion of this nature is readily obvious in the rust that forms at the joints of welded and non-welded steel structures.

Zinc in galvanized coatings can form a protective oxide layer on metal surfaces. Acidic rainwater (pH < 7) can attack active metals, such as zinc, to form hydrogen gas. Acid attack on the galvanized coating, while removing the oxide layer and improving conductivity, eventually exposes the underlying steel to both oxidation and acid attack.

Mechanical stress enhances corrosion (Shackelford 1988). Mechanical stress refers to both applied stress and internal stresses associated with microstructure (for example, grain boundaries). Inappropriate selection of washers can fail to maintain tension or over-apply tension in mechanically assembled components. High stress regions (such as cold worked steel) act as anodes relative to low stressed regions. These regions can occur locally on the same piece of steel where flexing has occurred.

External excavation around underground electrical vaults for installation of additional ducts or for concrete repair or other modifications can sever the bonding connection to the ground rod or network.

Design specifications for the bonding and grounding network set resistance as low as possible or technically achievable. The specification was created to ensure protection of workers potentially in contact with live circuits in these vaults. Performance of a bonding and grounding network to this level was relative to the time of installation. At that point in time, metal surfaces were not oxidized, not corroded and not contaminated by deposition of mud and other foreign material. The resistance of the bonding and grounding network in vaults experiencing real-world conditions is likely higher than the design specification.

The status of the protective bonding and grounding network in an underground electrical vault whose contents are submerged in water or covered in mud is uncertain as this cannot be assessed visually. Unless the status of the bonding and grounding network in the vault is assessed prior to entry, the benefit to the safety of entrants is not predictable.

## Hydroexcavating and the Underground Electrical System

Hydroexcavating utilizes a jet of water at high pressure and a vacuum system, both contained on a large truck. The high-pressure water jet breaks up the soil and debris and the vacuum system removes it. Hydroexcavating is used in many applications, some of which involve exposing conductors of the underground electrical system.

Exposure during high pressure water jetting of conductors carrying large quantities of current at low voltage and contact with an ungrounded system (the wand or boom tubes + the truck) pose an electrocution risk of unknown magnitude to the operators who may simultaneously touch something to complete the path to ground. Current flow to the truck and its components is limited only by the tolerance of the circuit protective device. While the hazard of circuits operating at 480 V or 600 V is intuitively obvious, the hazard posed by circuits operating at 110 V or 220 V is as real but not as obvious. Many fatal accidents have occurred due to exposure to circuits operating at 110 V. Procedural controls fail to offer the level of protection that is needed during this work. Protection is achievable only through use of insulating devices or by the technique of equipotential bonding and grounding.

Insulating devices prevent transmission between the exposed conductor and workers touching the boom tube or the wand and water hose. Removal of the boom tube and the wand from the excavation removes the potential for contact with the exposed conductor. Insulated or nonconductive boom tubes and wands must be able to withstand transmission when new and clean, but also when aged and wetted on the inside by conductive substances, such as salty water and mud.

Equipotential bonding and grounding utilizes the concept of rendering all components at the same potential (voltage) through interconnection to ground. This concept is used in servicing microcomputers and electronic components. In these applications, the worker stands on a conductive mat or rests the feet on the conductive mat while sitting and wears conductive wristlets. The electronic component rests on a conductive surface. All of the conductive components are connected together through cabling to a common ground. Each component has the same (equi)potential. This concept works satisfactorily only where the work to be performed occurs within a short definable distance, usually a table or desktop, from the connecting point of the ground.

In the case of hydroexcavating, equipotential bonding is achieved by connecting all components in the system through cables of sufficient gauge to the truck, and the truck to ground through a cable and ground rod. The ground rod must establish a predictable and reliable level of charge conduction. This can be difficult to achieve in soils encountered in prairie regions, especially during summer weather, and very difficult in rocky terrain. Insufficient grounding renders the level of protection insufficient. When equipotential bonding is employed, the worker stands on a mat containing a woven interior of conductive wires. The mat is connected to the system.

Realties of this work considerably complicate application of this technique in real-world circumstances. This work is wet and dirty. Mats, clothing, and other equipment become wetted and covered in mud and possibly earth spoil. This can create a conductive path between the mat and surrounding ground. This system is effective only so long as workers remain on the mat.

Stepping from the mat to unprotected ground can create a step potential. That is, the voltage experienced by each foot is different. This situation is counterproductive to the realities of this work. Workers must have flexibility to move around the work area to move and fetch tools such as pan shovels for lifting stones and debris from the hole, bars for breaking obstructions, and so on. They must also be free to move around the site to tend to the truck, traffic cones and signage, and interact with bystanders. The equipotential concept requires careful planning and supervision.

## Ground Resistance Testing

Earth electrode resistance is the resistance measured between the ground rod and a distant point on the earth (remote earth). At this distance, typically 7.5 m (25 ft) for a 3 m (10 ft) ground rod, the earth electrode resistance no longer increases appreciably when the distance from the grounding electrode increases. Earth electrode resistance effectively is the resistance of the soil surrounding the ground rod.

Testing is the only means to establish the efficacy of a given ground (Palmquist 1988). The three-point method uses a DC generator to generate current and two electrodes driven into the ground, one at a distance D from the ground electrode under test, and the second at a distance, 62% of D. Suggested distance for D is 30 m. This provides sufficient distance, such that the electrical fields surrounding each electrode do not overlap. The distance, 62% of D, is an inflection point on the resistance versus distance curve.

The current best method for testing soil resistivity is the four point method of Wenner (Lyncole 2005). This method uses a 4-pole digital ground resistance meter, probes installed in a straight line and equally spaced through the test area (Figure F.16). The probes establish an electrical contact with the earth. The 4-pole test meter injects constant current through the ground from the tester and the outer two probes. The current flowing through the earth (a resistive material) develops a voltage (potential difference) measured between the two inner probes. The instrument determines the amount of current that is flowing through the earth and the voltage drop across the two inner probes. With this information the meter uses Ohm's law ($R=E/I$) to calculate and display the resistance in ohms and soil resistivity measured in ohm-meter. The ohm-meter expresses the resistance of 1 $m^3$ of earth.

The quality and effectiveness of a ground are profoundly affected by weather and seasons. Since a ground fault, potential fire or accident, can happen at any time, the protection offered by the bonding and grounding system is only as good as the ground condition during the worst time of year.

## Dedicated Power Circuits

Numerous lockout/tagout incidents have occurred in circuits containing a common power source and shut-off. That is, one power source energizes several circuits. This is the case in the home where one fuse or circuit breaker controls the activity in several circuits.

This situation could occur in several ways in industrial, commercial, and institutional settings. The most obvious is that a single line splits to feed several circuits. This strategy normally conserves wire and components, but considerably increases the difficulty of isolation and lockout especially where the remaining circuits must remain active. This could mean that the only isolation device on the equipment or machine is the START-STOP switch in a control circuit.

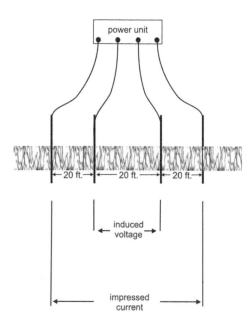

Figure F.16. The method of Wenner for establishing soil resistivity. The outer probes create a circuit and measure current flow. The inner probes measure potential difference (voltage).

More preferably, each circuit should contain a breaker or fuse in addition to START-STOP switches. This would enable complete isolation of each piece of equipment or machine.

## Limits of Approach

Distance of the conductors from each other and distance from the ground are the basis for the concept of limits of approach. Limits of approach refer to distances intended to protect against electrocution by uninsulated conductors and beyond which flashover from an energized to a nonenergized conductor cannot occur. A glance at any "hydro" pole on any urban street will illustrate the application of the concept (Figure F.17). The high-voltage supply line is positioned at the top of the pole. Well beneath that are the lines that provide electrical energy to the houses, and below them, communication cables for telephone and cable television.

The basis for protection against shock hazard in NFPA 70E for point sources is a series of concentric spheres and the regions of space between the surfaces (Figure F.18) (NFPA 2004). The innermost sphere, the prohibited approach boundary, encloses the energized equipment and a region of space called the prohibited space. The next sphere, the restricted approach boundary, encloses the restricted space. The outermost sphere, the limited approach boundary, encloses the limited approach space. These boundaries define the type of protective equipment required to minimize the possibility of electric shock to personnel.

The basis for protection against line hazards involving line sources is a corresponding series of nested cylinders that define limits of approach.

The prohibited approach boundary defines the outer limit within which the work is considered the same as making contact with the live part. Crossing this boundary requires the same protection as making contact with a live part. Only properly equipped body parts of qualified persons may cross this boundary.

Figure F.17. Utility pole with lines at different heights. The distribution line occupies the top position on the pole and feeds the step-down transformer that produces residential voltages. Residential supply lines occupy the next level. Telephone and cable lines occupy the lowest level.

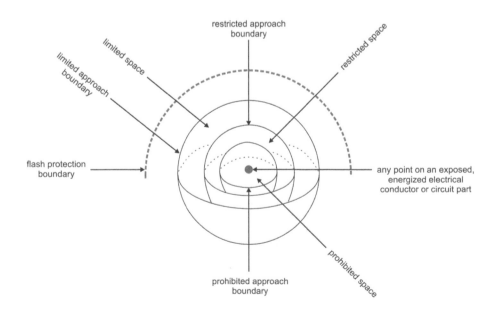

Figure F.18. The concept of limits of approach reflect a group of nested spheres for point sources and nested cylinders for line sources. (Modified from NFPA 2011.)

The restricted approach boundary is the limit of approach within which there is an increased risk of shock due to electrical arc combined with inadvertent movement for personnel working in close proximity to the live part. Only qualified persons working at a distance from the live part and employing shock protection techniques and equipment may cross this boundary.

Qualified persons may bring conductive objects to or within the restricted approach boundary only under special conditions. First, the qualified person is insulated or guarded from the live parts and that no uninsulated part of the body enters the prohibited space. The second condition is that the live part is insulated from other conductive objects at a different potential. Lastly, the qualified person is insulated from other conductive objects such as during live-line, bare-hand work.

The limited approach boundary is the limit of approach at a distance from an exposed live part within which a shock hazard exists. Where unqualified persons are working close to the limited approach boundary, the person in charge of the work space shall advise about the electrical hazard and warn about remaining outside the limited approach boundary. Advisement could occur through a safety meeting or crew talk. Warning indicators include signs, barrier tape and flag tape in temporary situations such as construction and fences in permanent installations.

Where there is a need for an unqualified person to enter the limited approach boundary, a qualified person shall advise about the possible hazards and provide a continuous escort during the entry. The qualified person shall prohibit access of the unqualified person beyond the restricted approach boundary.

The numerical values of the limits of approach depend on the nominal voltage in the exposed parts. That is, these represent a steady-state condition. These limits appear to be conservative and reflective of real-world experience.

In the case of electrocution prevention, 3 ft (approximately 1 m) is the recommended safe working distance up to 600 V for electrical equipment, depending on clearances (NFPA 2004). The actual distance at which flashover can occur is much shorter than this. Hence, the intent of this approach is to prevent unexpected contact due to forgetfulness induced by inattention or determination to complete a task that requires a closer approach.

The distances provide a buffer for controlled movement with a focus on the location of the exposed live part. This approach functions for the most part when the exposed live part is the focus of attention and movement and actions of people. This approach works best for electrical workers who receive training and reinforcement about these concepts as a routine part of their working experience. They are aware of the danger posed by uninsulated conductors and of the concepts of the protective boundaries and of protective equipment required to prevent contact.

Difficulty arises because the boundaries are invisible and require perception of depth that is not part of the human visual sense and constant provision of warning about penetration of the space within the boundaries. This situation is impossible to confront without shielding of all exposed live parts within the work area so that the focus becomes solely the live part of interest. This concept is similar to the approach taken by dentists by installing devices that shield the interior of the mouth and provide access to the tooth of interest. Increasingly, electrical workers, especially during live-line work, are shielding live parts unaffected by the work.

The only practical way to ensure compliance with this application of a limit of approach is to place barrier tape or a barrier of sheet plastic across the path of access to the energized conductor at or beyond the recommended distance. This approach puts the onus on the individual worker and supervision. NFPA 70E contains similar recommendations for clearances for conductors in electrical equipment operating at higher voltages.

For people who are unaware of the danger posed by uninsulated conductors or who are handling materials capable by length of making physical contact with the conductor during deliberate or accidental movements, this approach is not likely to succeed. The continuing and on-going contacts with energized conductors illustrate this concern.

An added concern is the occurrence of contact by high-voltage conductors onto conductors of lower voltage and other causes of electrical surge. This situation invalidates the boundaries presumed for a particular energized conductor based on nominal voltage. This situation reflects the rare occurrence of these events and the intercession of protective devices in the electrical circuit.

NFPA 70E also lists limits of approach for overhead powerlines (NFPA 2004). These limits reflect experience about what has occurred in previous situations involving contact between energized overhead power lines and conductive structures. Primary agents of contact include booms of mobile equipment such as cranes, excavators, concrete pumper trucks, and grain augers. Contact between this equipment and uninsulated power lines offers some possibilities for prevention, since the motion generally is controlled, rather than haphazard. Distance calculations performed prior to starting the work can determine the appropriate location to avoid contact should the boom positioning mechanism fail. An additional possibility in situations where maintaining distance is not feasible is the use of spotters to alert the operator about position.

Another major cause of fatal electrical contact accidents involves movement of conductive objects including ladders, poles, pipes, reinforcing rods, metal siding, and tree pruner poles through areas in which contact could occur because of restrictive geometry. These objects considerably extended the contact zone in a hemisphere surrounding the victim. This situation negates the use of distance alone as a means of protection since there is little to stop the extension of an individual in space though an objec, such as a light piece of pipe, from striking an uninsulated conductor on a transformer, or for that matter, an uninsulated high-voltage line.

The control that people exert over objects being lifted or carried is not as precise as that provided by mechanical devices. As a result, movement can be jerky and lacking in precise control of position. This lack of precision of position leads to uncontrolled entry into distance protective zones surrounding unshielded energized conductors. Limits of approach fail to protect in this type of circumstance.

## Assured Equipment Grounding Conductor Program

An assured equipment grounding conductor program is one technique sanctioned in regulatory statutes on electrical safety. In the U.S., see for example, 29 CFR 1926.404 on wiring design and protection and 1926.405 on wiring methods, components and equipment for general use, for specific guidelines on ground fault protection (OSHA 1998). In this statute, assured grounding is called the Assured Equipment Grounding Conductor Program.

The concept behind assured equipment grounding is a program of continuous inspection, testing and recordkeeping performed by a competent person. The intent of the program is to ensure that the grounding conductor in temporary wiring and cables, receptacles, and extension cords remains intact and functional. Inspections are to occur daily prior to use to determine the presence of external defects such as deformed or missing pins, damage to insulation, and potential internal damage.

OSHA requires removal from service and repair of damaged or defective equipment. Testing includes continuity of ground conductors and correct attachment of the grounding connector to each receptacle, attachment plug, and cord set. Tests must occur before first use, before return to service after repairs, and before reuse following any incident that could have caused damage. Routine tests of equipment exposed to possible damage must occur at intervals not exceeding three months, and six months for equipment not exposed to damage. One or more individuals trained to recognize safety hazards and have the authority to take corrective action must constantly monitor the program.

Manufacturers of electrical equipment have devised ways to simplify the monitoring process. One example is the ground continuity monitor (Ericson 1995). This device monitors the condition of the grounding conductor while in use. Glow from an LED (Light Emitting Diode) in the device (Figure F.19) indicates continuity. Failure of the LED to glow indicates that one or more of the following conditions has developed: reversed polarity, open hot, hot on neutral, hot unwired, open neutral, open ground, or hot and ground reversed. The condition has occurred within the cord, jobsite receptacle, or branch wiring.

## Double Insulation

Double insulation is a design feature that has increased the safety of electrical tools and equipment (Bernstein 1991a). The term, double insulation, means that insulation between internal energized conductors and any possible point of external contact consists of two insulating systems that are physically separated (UL 2004).

Figure F.19. The ground continuity monitor indicates, through the illumination of an LED, the integrity of the ground and measure of other functions.

Functional insulation is the insulation necessary for normal operation of the unit. Functional insulation is the insulation on windings of a motor or a transformer or internal wiring in the product. Protective insulation provides protection against electric shock following failure of functional insulation. Protective and functional insulation are independent of each other. A portable tool containing a protective enclosure manufactured from an insulating material is an example of protective insulation (Bernstein 1991a). Exposed metal parts in such tools, such as the exposed metal chuck on a drill, have internal protective insulation. Double-insulated appliances are required to have only a two-bladed plug.

Double insulation is not sanctioned as a protective measure in regulatory statutes on electrical safety. There is no method to assure that the integrity, and hence performance of the insulation, remains satisfactory throughout the life of the product (Carbone 1994).

Wetting of the insulation and introduction of foreign substances can deteriorate performance. Wetting, for example, can create a continuous conductive path between live conductors inside the tool and the outside surface.

Plastics are subject to stress cracking caused by hard use or loss of plasticizers due to aging, corrosive attack, or exposure to sources of ultra-violet energy. As well, double insulation of the interior of the tool does not protect against damage to cords and plugs or against entry of moisture and foreign substances through vents in tool cases. Moisture and foreign substances can establish a conductive path in the interior of tool cases.

### Overload Current (Overcurrent) Devices (Fuses and Circuit Breakers)

Overload current or overcurrent devices play an essential role in maintaining the health and safety of individuals who work on electrical equipment and electrical components of other equipment during normal operation and maintenance. They also protect the integrity of the equipment itself (Cooper Bussmann 2000a). Overcurrents are classed as overload currents or short-circuit currents.

Electrical circuits do not normally handle constant levels of current (Rosenberg 1999). Overcurrent devices must accommodate these fluctuations. To illustrate, when a circuit is first closed, current rushes in. Inrush current is highest when inductive reactance is least.

Fuses and circuit breakers are the overcurrent devices used in electrical systems. Fuses date from 1880 and circuit breakers from the 1920s (Roberts 1996). Overcurrent devices protect the wire insulation of the ungrounded conductor from overheating and damage because of overcurrent flow (Bernstein 1991a). They offer no protection from fires caused by the heat of an arc, overheating of a small conductor or component, or the heat generated by a loose terminal or connection.

Currents less than 15 A can start fires. Hence, the function of an overcurrent device is to protect wiring and electrical equipment, nothing more. Overcurrent devices also offer no protection from electrical shock. Leakage current above which humans cannot "let-go" of an energized circuit is about 0.05 A (50 mA). This level is well below the usual activation level of overcurrent devices.

Overcurrent devices act in two ways. The first is to stop the flow of electrical current in a circuit when it exceeds some preset value. Overcurrent devices function through an inverse current-time relationship. That is, the higher the current, the faster the device acts (Figure F.20). Fuses are set to open in 1,000 s at 120% to 150% of the rating (Littelfuse 1993). Circuit breakers perform in similar fashion.

The second way in which overcurrent devices act is to block high levels of current (short circuit or fault current) from passing through the device.

Overload currents are excessive currents relative to normal operating current (Cooper Bussmann 2000a). They remain confined to the conductive path provided by conductors and other components and loads of the distribution system. Overload currents by definition range from 100% to 600% of normal current level. Usually these are caused by temporary surges that occur on energizing heavy consumers of electrical energy (large motors, heaters, or transformers) during inrush.

Overload currents are regarded as harmless under most circumstances since they are short in duration and cause minimal rise in temperature. In such circumstances, reaction by circuit protective devices would be counterproductive. However, sustained continuous overload caused by defective or

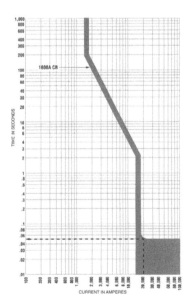

Figure F.20. The inverse relationship between time and current in overcurrent devices such as fuses and circuit breakers.

overloaded equipment or operation of too many devices in the circuit can cause damage. In these circumstances, the circuit protective device must act before damage occurs. Usually, removal of the overload current within a few seconds will prevent damage. Sustained overload will cause deterioration of insulation and components through heating.

Short circuits (or fault currents) flow outside the normal conducting path. They can be considerably larger than overload currents (starting at 600% of normal current level) and can easily range above the 50,000 A available in the distribution system. If not terminated in a few milliseconds, current of this magnitude can severely damage or destroy insulation, melt and vaporize metal conductors, ionize gases in air, develop arcs, and cause fires. Similarly, high current creates huge magnetic field stresses between buss bars and other conductors. The latter can lead to warping and distortion of these components.

Fault currents passing through overcurrent devices can destroy the overcurrent device. This destruction can occur with explosive force depending on the overcurrent device and its ability to handle high currents.

The measure of capacity of a protective device to maintain its integrity when challenged by fault currents is the interrupting rating. Most circuit breakers used in branch circuits have interrupting ratings of 10,000 A to 15,000 A and possibly higher, depending on cost of the product. Most modern fuses have interrupting ratings of 200,000 A or 300,000 A. Current-limiting fuses (Figure F.21) can cut-off a fault current in less than half a cycle (half a 60 Hz wave) or less than 8 ms (0.008 seconds). Circuit breakers require about 1.5 cycles (24 ms) to open under the same conditions (Cooper Bussmann 2000b). Because of the difference in time required for the trip, considerable damage could occur because of the passage of grossly excessive levels of current.

Circuit breakers function through thermal and magnetic mechanisms. The thermal mechanism protects against overload by means of a bimetallic element. Heat causes differential expansion of the two metals. This leads to bending that results in breaking of the contact. The bimetallic element is part of the current path (Roberts 1996).

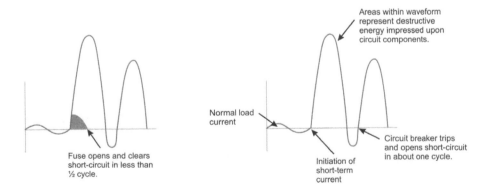

Figure F.21. Current-limiting fuses can open the circuit in less than half a cycle (0.008 s) and thereby prevent damage to the circuit from high fault current. (Modified from Cooper Bussman 2000b.)

The magnetic mechanism protects against short circuits (fault currents). At these levels of current, the breaker can trip in less than half a cycle; that is, in the same time-frame as a fuse, although this depends on the product. The circuit breaker uses a molded plastic case to contain the protective components. The interior of the case is air-filled. The plastic case and interior components are subject to rapid destruction by melting and vaporization following passage of fault current that exceeds the interrupting rating.

Fuses act as conductors up to their rated capacity. Beyond this point, the fuse element heats up. If the overcurrent persists, heating may be sufficient to melt the weak point in the fuse element (the internal conductor ribbon). Multiple weak points in the fuse element protect against high fault current (Littelfuse 1993).

In the event of high fault current, the weak points in the fuse element melt simultaneously within 2 to 3 ms (0.002 to 0.003 seconds). In this application, circuit breakers can require up to 17 cycles (0.3 seconds). Silica sand or other filler material in the fuse acts to quench the arc and to contain it within the body of the fuse. This is the technical basis for the superior ability of fuses to withstand high fault currents prior to failure compared to circuit breakers. For this reason, current-limiting fuses are used to protect circuits containing circuit breakers from damage from high fault currents.

Fault current capable of bridging the gap created during opening of a circuit breaker or melting of the linkages in a fuse can re-energize a de-energized circuit. This could lead to electrocution during work on electrical equipment and components contained in that circuit or an explosion that can affect work on adjacent components.

A circuit deactivated by opening the circuit breaker but not pulling the protective fuse still remains protected. The opened breaker will protect against normal current and low fault current and the fuse will protect against abnormally high fault current.

This then is the essence of the answer to the question about whether opening a circuit breaker or removing a fuse provides sufficient protection as part of a lockout procedure versus severing the service at the service drop. The only way to increase protection within the circuit is to remove the protective fuse as well as opening the circuit breaker. This then imposes the gap of the fuse block onto the gap of the circuit breaker onto the fault current. Severing and reconnecting the service imposes risk onto the lineman on the line side for what appears to be a very small increase in benefit to workers on the load side.

## Leakage Current Protection Devices

Leakage current protection devices are a class of devices that detect leakages in electric circuits (Schiff 1999). These devices monitor imbalances in current flow. These devices incorporate current transformers or zero sequence transformers (Palmquist 1988). Voltage produced in the sense coil wound on a magnetic core that encircles all conductors in the circuit is proportional to the difference in current in the conductors. That is, the magnetic flux is zero when the current is balanced.

## Ground Fault Circuit Interrupter (GFCI)

A ground fault occurs when an unintended path to ground develops. This could occur from a hot conductor or an ungrounded surface energized by a hot conductor through a person to ground.

The GFCI (Figure F.22) detects low-level ground faults and opens the circuit at currents equal to or exceeding the preset value of 6 mA (UL 2006). This is considerably less than current flow needed to trigger a conventional household circuit breaker or fuse opening at 15 to 20 A. The rate of interruption of flow of current in the circuit is inversely proportional to the imbalance in current. At 6 mA, the GFCI will de-energize the circuit within 0.025 second (1.5 cycles at 60 Hz). The trip times for the GFCI are designed to be in the safe range for human response (Bernstein 1991a). By tripping at 6 mA, an individual will be able to let go of the circuit and not receive a lethal shock. The shock may be painful but is not lethal.

What GFCIs do not do is to protect against a short-circuit condition, as occurs when a person contacts both the live and grounded conductors simultaneously (Roberts 1996). The individual acts as another load in contact with energized conductors.

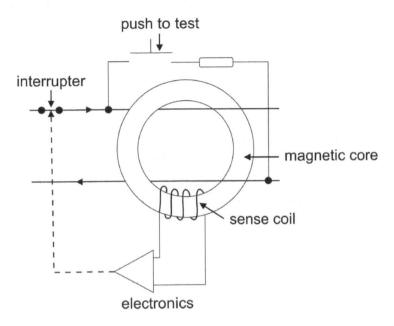

Figure F.22. The Ground Fault Circuit Interrupter (GFCI) senses imbalance in current between the supply side and the return path. An imbalance activates the electronics and opens the contact.

GFCIs also monitor the load side of grounded circuit conductor to ensure that grounding of this conductor on the load side of the GFCI does not occur (Roberts 1996). GFCIs are to be used only on systems containing a ground at the service drop. The concern is that a second ground on the load side would enable current to pass through a person, return to the circuit through a second ground, and through the sensing transformer in the GFCI. Under these conditions, there would be no current difference and the GFCI would not respond. GFCIs use a second current transformer and related circuitry to monitor the load-side, grounded conductor (neutral). A ground in this conductor on the load side of the GFCI trips the GFCI. This innovation ended the practice of using the grounded neutral conductor as an equipment ground.

GFCIs are available only for household voltages (110 and 220 V). GFCIs are sanctioned in regulatory statutes as a protective device in electrical circuits. (In the U.S., see for example, 29 CFR 1926.404 on wiring design and protection for specific requirements on ground fault protection (OSHA 2011b).

Requirements for GFCIs mandated for use in construction applications differ from those approved for use in the home (Ericson 1995). The reason for this is that the integrity of portable wiring systems is highly variable and subject to destructive, wet conditions on construction sites. These requirements apply to portable GFCIs containing one or more receptacles, power supply (extension) cords, and plug units.

The difference between domestic units and those intended for use on construction sites is protection against the open neutral and reverse phasing required in industrial applications. Opening or lifting the line side neutral at a panel creates the condition called the open neutral. Without the neutral connection, the domestic GFCI lacks the power required for operation and no longer provides protection. A faulted tool connected to the unprotected receptacle would create a shock or electrocution hazard. This would be unacceptable in industrial and construction situations. Protection against the open neutral condition occurs when both the hot and neutral circuits on the load side of the GFCI disengage following opening of the neutral or hot lines on the line side (UL 2006).

Under these conditions, breakage of one or more conductors on the supply side of the GFCI could occur (Roberts 1996). A broken hot wire eliminates the power in the circuit needed by the device. A broken grounded conductor (white wire or neutral) would allow the circuit to remain energized although the load would not function. The GFCI would not trip with a ground fault on the load side. The problem would become evident only when someone tested the GFCI. GFCIs protect the open neutral condition through use of a two-pole, normally open relay on the line side of the unit. Power to the relay must exist in order for it to remain closed. Spring pressure opens the relay in the event of an open neutral condition.

In addition, portable industrial GFCIs also provide protection against reverse phasing while GFCI breakers do not. Reverse phasing occurs on reversal of the neutral and phase conductors at the terminals in the receptacle. Protection is achieved by switching both conductors on detection of a ground fault.

GFCIs have earned a position of trust in providing service. This trust, however, is misplaced, as GFCIs are subject to undetected failure (Holt 2001c; Layton 2000). While the device allows the passage of current, the monitoring that is expected to happen may not occur. The cause of failure is damage inflicted by repeated electrical surges. GFCIs contain a Metal Oxide Varistor (MOV) that functions as a surge suppressor. The MOV absorbs the surge and converts the excess voltage to heat. Repeated surges degrade the MOV while still permitting flow of current. Failures are especially high where surges due to lightning strikes are frequent.

### Equipment Leakage Circuit Interrupter

The equivalent of the GFCI at higher voltages is the Equipment Leakage Circuit Interrupter (ELCI). These devices operate from 120 V to 600 V with current ratings to 80 A. Some units have variable trip settings, for example, 6 mA, 10 mA, and 30 mA (TRC 1997). These settings enable the electrician to match the intrinsic current leakage of the equipment and to localize the level of current leakage. Equipment, such as electronic measuring and testing equipment, information technology equipment, and equipment for laboratory use, is permitted to have slightly higher current leakage than normal. This results from use of electromagnetic interference filters.

Even at a trip setting of 30 mA, these units provide considerable protection beyond overcurrent devices and equipment grounding. The trip setting of 30 mA is the norm in Europe. This is intended for equipment protection and will protect about 90% of the population from electrocution (Roberts 1996).

### Other Ground Fault Protective Devices

Ground fault relays or sensors detect ground faults of low magnitude (Cooper Bussmann 2000). When the ground fault current and time exceed the set point, the circuit disconnect opens. Ground fault protection relays involve complex circuitry. Proper function and reliability depend on correct wiring, calibration, and intactness.

The best ground fault protection design is a switch equipped with a ground fault relay, a shunt trip mechanism, and current-limiting fuses. Current-limiting fuses, rather than circuit breakers, are fast acting (within 1/4 cycle), and as a result, protect against damage from high ground fault currents. Circuit breakers can require considerably longer time in which to function. The greatest damage to components by fault currents occurs in the first major loop of the sine wave (first half-cycle).

The (U.S.) National Electrical Code requires ground fault protection for low magnitude, intermediate, and high magnitude ground faults (NFPA 2011a). Ground fault relays open too slowly to offer protection against high magnitude ground faults (Cooper Bussmann 2000). The main or feeder overcurrent devices, such as fuses or circuit breakers, must clear the circuit. Current-limiting fuses substantially limit let-through current from ground faults of high magnitude and thereby offer a higher level of protection. Circuit breakers are not current-limiting protective devices to the same extent and can transmit considerable current before tripping occurs.

## Arc Fault Protective Devices

Arcing can occur in damaged or deteriorated wiring, loose connections, and wiring within wall cavities punctured by nails and staples. Arc faults result from damage to electrical cables, often during installation. This can also result from rubbing together, as occurs during flexing and vibrating. Arcs develop from hot-to-neutral, from hot-to-ground and phase-to-phase (where multiple phases exist). Arcing faults can occur undetected by overcurrent devices because of insufficient current flow and by residential GFCIs. Arc fault devices protect equipment (Roberts 1996).

Electrical cables installed in wall cavities are unseen. Hence, arc faults are undetectable by visual observation. Characteristic of arcing faults are low and erratic current flows (Cutler-Hammer 1997). Arcing faults damage and destroy equipment and can ignite fires in buildings. Sustained arcs require about 375 V in AC systems.

## Insulating Devices and Products

Temporary insulating devices and materials are used to protect linemen and to prevent contact and flashover accidents involving construction and other equipment when work must occur around energized distribution lines. These products include bucket trucks, remote tools, and plastic-rubber products of various shapes, including line hose, blankets, and gloves.

Periodic testing must occur in order to ensure the continuing insulating, and therefore protective capabilities of these devices and products. These devices and products increase the level of safety for individuals who already follow safe work practices. They are likely to be or to become self-defeating when work practices deteriorate.

## Bucket Trucks

Linemen and other aerial workers (telephone and cable television) increasingly work from within insulated boom work platforms (bucket trucks). The insulated boom and bucket normally prevent the transmission of current to metal parts of the truck and ground. This status is effective so long as a conductive path does not develop. A conductive path could develop through longitudinal cracking of the fiberglass in the boom and deposition of conductive windborne material or material carried in rain.

## Remote Devices

Remote devices (Figure F.23) include tools and sensing devices. Tools include cable handling hooks and switch sticks for setting switches and fuses, rescue hooks for retrieving objects and people, and hot sticks.

The hot stick is designed for applying and removing ground clamps and cables from energized conductors while maintaining the user at an extended distance that respects the limits of approach. Voltage detectors and other devices clamp to the end of the hot stick. These devices are fiberglass poles containing a foam core with fittings made from metal or other material at the end.

## Shielding Products

Electrical work on energized conductors places high reliance on the use of temporary insulating devices and personal protective equipment. Safety of the individual relies on concentration on the task and personal discipline, since care in the placement of the temporary insulators (hoses, blankets, and covers) is essential. This means that the insulator must cover all exposed metal surfaces.

Sometimes this is not possible, due to the presence of fittings and attachments, as illustrated by several fatal accidents. The victim was unable to cover all of the exposed metal. The lineman must cover exposed lines in front of the work position, to the sides and behind, and possibly above and below. Again, failure to do so led to contact and several electrocutions.

Shielding products include line hose, fitting shields, covers, and blankets (Figure F.24). These products are manufactured in several configurations in order to fit over various fittings attached to the line. The blankets are configurable to various shapes and as a result can provide protection from contact with bulky components not able to be covered by line hose. Blankets require secure fastening in-place in order not to be displaced by wind.

Figure F.23. Remote devices provide the means to operate equipment and to measure conditions from an extended distance that respects the limits of approach. In this case, the device was used to apply temporary bonding and grounding.

F.24. Shielding products (line hose, fitting shields, covers, and blankets) provide insulation on otherwise bare conductors. Protection depends on insulation value. (Image reproduced with the permission of Salisbury by Honeywell © 2011 Honeywell International Inc. All rights reserved.)

Establishing complete and secure coverage by these products is essential in order to maintain the temporary barrier. Cold weather can severely hamper use of these products as ice and ice can enlarge dimensions beyond what the line hose or protective fittings can accommodate. The cold can stiffen and embrittle the rubber of the blanket.

### Insulating Gloves and Other Personal Protective Devices

Linemen working "hot" wear a combination of an electrically-rated and tested, inner rubber glove and an outer leather glove, as well as covers for the shoulders (Figure F.25). They wear the rubber glove next to the skin for electrical protection and the outer leather glove for protection and abrasion resistance of the inner rubber glove.

During hot weather, perspiration and resultant wrinkling of the skin cause wearing of the impervious rubber gloves to be extremely uncomfortable. Accumulation of sweat causes considerable discomfort due to softening of the skin and sloppy feel inside the glove. Effectively what is occurring is that a highly conductive liquid (sweat) is trapped behind an insulating barrier (the rubber glove), and protected from contact with the energized conductor by the outer leather glove.

Numerous fatal accidents detailed the discomfort experienced by linemen in the use of this system of protection. They failed to use the inner glove, and relied only on the leather outer glove. At least one accident detailed the role of the sweat that accumulated in the rubber glove as the causative agent. One individual raised his arm to drain the sweat from inside the gloves. The stream of liquid provided a conductive path, and dramatically reduced the voltage needed for occurrence of an arc (NIOSH 2000).

### Insulated Hand Tools

Insulated hand tools contain a coating of an insulating material on metal surfaces (where possible to apply). This coating raises the level of protection against unintended contact with energized conductors. The coating provides protection against contact involving potential difference of 1,000 V. The coating covers all nonfunctional metal surfaces.

### **Insulation Testing**

Insulation testing determines the integrity of insulation on new and in-service wiring, as well as other insulated products, such as those mentioned in the previous section. Testing for insulation leakage in new systems should occur prior to energizing and in insulated tools prior to first use. This enables detection and correction of faults and determination of later deterioration due to dirt and moisture, as well as rejection of faulty tools and equipment. Deterioration due to dirt and moisture generally is considered to cause larger leakage currents than initial flaws.

Insulation testing exposes defects not found in any other manner (Palmquist 1988). Insulation testing could have prevented some of the fatal electrical accidents documented in the NIOSH and OSHA reports on fatal accidents (NIOSH 2000; OSHA 1988, OSHA 1990).

Insulation testing determines capacitance charging current, absorptive current, and conduction or leakage current (Palmquist 1988). Capacitance current is the capacitor charging current. This starts out high and decreases to zero as saturation occurs. Absorptive current occurs due to the polarization of the insulating material. This also starts out high and decreases to zero as saturation occurs.

Conduction or leakage current is the biggest concern. This reflects steady-state leakage capabilities over and through the insulation. The cause is dirt, moisture, and normal leakage characteristics of the insulation. The contribution to conduction or leakage current due to moisture and dirt reflect deterioration of the insulation. Moisture and dirt are the largest contributors to leakage current. Normal humidity has no effect on insulating qualities unless condensation has occurred. Hence, the contribution due to moisture actually refers to liquid wetting, as occurs during sweating and external wetting of the surface due to rain.

Insulation tests are performed on wiring, cables and conductors, and windings of motors, transformers, and generators. The testing equipment uses DC voltage. Significant charge accumulation at high voltage can occur, depending on the test. Hence, bleeding off charge accumulated during the test is essential to prevent injury.

Figure F.25. Lineman protection can include insulation-tested inner gloves, abrasion-resistant outer gloves and shoulder coverings. (Image reproduced with the permission of Salisbury by Honeywell © 2011 Honeywell International Inc. All rights reserved.)

Testing is not without cost since this challenge stresses the insulation and equipment with voltages above the nominal maximum service voltage and creates heat. Insulation value halves for every increase of 10° C in temperature. Premature aging of the insulation due to the application of deliberate stress is deemed to be a small price to pay for the acquired knowledge and assurance.

## Electrostatic Discharge

Accumulation of electrostatic charge results mostly from the contact and separation of dissimilar materials and the presence of isolated, ungrounded conductors. To most people, electrostatic discharges pose only an annoyance during winter weather when humidity is low. Uncontrolled electrostatic discharge leading to arcing is a serious hazard in many industries.

Accumulation of electrostatic charge following contact and separation is a common problem in activities such as transport, mixing, pouring, agitation, and filtering of liquids, gases, and flowable solids (Gregg 1996; Lees 1996). The rate of generation of electrostatic charge depends on conductivity of the material, turbulence, interfacial surface area, and presence of impurities. Common problem areas include piping systems, filling operations, filters, and dispersing operations where mixing, thinning, combining, or agitating occur. Electrostatic accumulation and discharge are an especial problem when particulates are present. Where insulating materials are involved, hazardous arcing can occur in mixers.

A common situation occurs where a conductive layer becomes separated by an insulator from another conductive layer. This happens, for example, in plastic piping that is supported by metal brackets and transports conductive material. The structure formed by this geometry creates a capacitor.

Plastic containers, films and linings of containers, and nonconductive fabric bags used to contain bulk quantities of powders (super sacs) can pose electrostatic problems. In the case of fabric bags, electrostatic charge accumulates on the surface of these materials due to friction between particles moving against each other and between the bulk solid and the interior surface (Reardon 2002). Charge accumulation can also occur on exterior surfaces due to contact and rapid separation.

A current issue that illustrates the universal nature of the problem involves flash fires in small containers during refueling at gasoline stations (PEI 1996). These episodes involved both plastic and metal containers transported by vehicles. One hypothesis to explain these occurrences is the generation of electrostatic charge on the container due to windflow and prevention of dissipation by carpets and plastic liners of truck beds. Where metal containers are involved, the insulating material in conjunction with the steel of the vehicle create a capacitor as described previously. An electrostatic arc develops when the metal nozzle of the fuel filler approaches the container. The latter can ignite a flammable mixture of gasoline in air.

The human body also can act as a capacitor when in an electric field and isolated from ground as in the preceding situation. Grounding leads to transmission of current and a small shock. This is the more typical situation that people encounter when charged by accumulation of electrostatic charge brought about by contact and separation of materials and fabrics. Electrostatic charge can accumulate on de-energized conductors through contact with windborne dust, sand and snow, abrasive blasting, and the rotation of helicopter blades (BC Hydro 1998).

Electrostatic discharges are most recognized as a source of ignition in fire and explosion situations. These involve ignitable gases and vapors, and in unusual situations, dusts. Before a fire or explosion caused by electrostatic discharge can occur, several conditions must apply simultaneously (Schmieg 1991). These include a mechanism for generation of electrostatic charge, the ability to store electrostatic charge at a voltage, the minimum amount of stored energy, and creation of an arc in an ignitable environment.

Electrostatic discharge is a form of surge that can affect circuits in microprocessor-controlled equipment (Gorosito 2000). These surges can disrupt transmission of data signals through low- voltage conductors and can destroy equipment.

As with other forms of electricity, bonding and grounding are fundamental to control of electrostatic discharges (Gregg 1996). Bonding maintains all interconnected conductive materials at the same potential. Grounding maintains the interconnected conductive materials at the same potential as ground. Isolated conductors, such as nozzles, screens, rims on nonconductive drums, equipment probes, and high-pressure cleaning equipment, can become sufficiently charged to cause an electrostatic discharge that results in an arc. These require bonding connection to a grounding system.

Cables and attachments must have low electrical resistance. Clamps must penetrate painted surfaces, rust, and surface contaminants to establish metal-to-metal contact (Browne 1996; Gregg 1996). Cables and fittings used in electrostatic protection systems usually are uninsulated. This provides the opportunity for inspection for cleanliness and corrosion as well as mechanical damage. Insulation is an issue where workers can come into contact with bonding conductors potentially carrying current electricity. Fittings include clamps of various designs (Figure F.26). Materials of construction include aluminum, bronze copper, and stainless steel.

As with bonding and grounding systems used for electrical protection, static protection systems must also be interconnected with them in order to maintain a common level of potential with ground. Testing using an ohmmeter of components in the system is essential. This usually means measuring resistance between the component and the grounding bus. Resistance less than $10^6$ $\Omega$ (ohms) is considered acceptable in bonding and grounding systems that dissipate electrostatic charge (NFPA 2007). Hence, the bonding and grounding system used for electrical protection is completely capable of offering protection against electrostatic accumulation and discharge.

Maintaining liquid flow rates below 4.5 m/s (15 ft/s) in piping systems will minimize generation of electrostatic charge (Gregg 1996). This requires attention to piping and component design. Splash filling and other turbulent action involving free fall of liquids are especially capable of generating electrostatic charge. Anti-static additives which are used in part-per-million concentrations can considerably reduce generation of electrostatic charge.

Antistatic bulk bags dissipate electrostatic charges (Reardon 2002). Both groundable and non-groundable products are available. The groundable product is manufactured from an insulating fabric which is interwoven with conductive threads that lead to a bonding connector. Resistivity of the

Figure F.26.   Bonding cables for protection against electrical contact contain clamps capable of penetrating paint and other materials on the surface of metals. (Image reproduced with the permission of Salisbury by Honeywell © 2011 Honeywell International Inc. All rights reserved.)

bag is as low as 1,000 Ω (ohms). The non-groundable product uses a coating on the fabric to dissipate electrostatic charges in a controlled manner.

A variant on the theme of bonding and grounding is the use of antistatic devices during service of electronic circuit boards and other components. These devices include bands worn on the wrist and conductive mats. Both devices connect to ground through the electrical grounding system. Products having very low resistance are contraindicated for use on people as there is the risk of current flow during contact with energized conductors and other surfaces. Grounding connections made to people must have resistance greater than 25,000 Ω  (Schmieg 1991).

## Lightning

Lightning strikes are an important cause of surges in electrical systems. Just prior to a strike, charge neutralization equalizes the voltage between the earth and the bottom of the overhead cloud (Paschal 2000b). This involves current flow through both the earth and the air toward the location of the strike. An electromagnetic pulse in the form of an induced voltage develops in all mutually coupled wiring, pipelines, and other conductors in the vicinity of the strike. This creates an electrostatic pulse in the form of an induced waveform that travels from any conductor above the charged earth down to the earth through a connecting conductor.

Structures 60 m and higher above ground are capable of attracting lightning strikes (Bernstein 1991b). Structures of lesser height attract only lightning that would have struck in their vicinity. This suggests that these structures provide protection to structures of still lesser height. Various geometries are used in the design of zones of coverage in lightning protection systems (Golde 1973).

Current surges produced by lightning have typical rise time of 2 μs (microseconds) and decay to 50% of peak by 40 μs. Median peak current is 30,000 A. Maximum recorded current flow is 270,000 A (Bernstein 1991b). This would suggest that the surges likely exceed the interrupting values of most circuit breakers but are less than the interrupting values of most current-limiting fuses (Cooper Bussmann 2000a). Current-limiting fuses provide a protective barrier to lightning surges. Despite the enormity of the currents, they do not destroy the relatively small-diameter conductors used to convey them. This is related to the very short duration of the surge (100 µs).

The historic method of protection is the lightning diverter system (NFPA 2011a, NFPA 2011b). This system increases the level of protection by a factor of two to three over a non-protected structure according to recent data (Chowdhury 1999). The lightning diverter system uses air terminals (lightning rods), down conductors, and bonding to the electrical grounding system (NFPA 2011a, NFPA 2011b). The conductors and equipment used in the lightning diverter system must be separate from other components. The systems must be bonded together. Side-flash can occur when nearby grounded objects are not bonded to the lightning conductor system. When all systems are bonded together, their individual components tend to have the same potential. This minimizes the tendency for occurrence of side-flash.

Buildings containing a "Faraday cage" of bonded reinforcing bars and structural steel surrounding the structure increase lightning protection by a factor of 100 compared to non-protected structures (Chowdhury 1999).

Recent advances in protection against lightning strikes include lightning avoidance systems. (Paschal 2000b). These are intended for use in applications where lightning must not strike. One example is the NASA space shuttle launch facility. Others include refineries and petrochemical plants.

One type of lightning avoidance system is the dissipation array system. The dissipation array system continuously equalizes earth-to-cloud charge over a long period of time instead of allowing accumulation to the point where a stroke of lightning occurs. The dissipation array in these units utilizes pointed conductors charged at voltages of 10 kV to 15 kV. This voltage ionizes molecules in air creating an ion- and electron-rich cloud. These systems require excellent connection to ground (very low resistance) for collection of electrons from the surrounding area.

Another type of lightning avoidance system is the early streamer emission system (Chowdhury 1999). This type of system stores charge gathered by lightning rods and injects the charge into the atmosphere just before a strike is anticipated. This generates an upward spark from the electrode toward the cloud. This induces a lightning discharge in the opposite direction.

A lightning protection strategy for electrical power systems is the surge protector. Surge protectors contain choke coils. Choke coils contain a number of turns of large-diameter wire (Middleton 1986). Resistance in the coil to normal flow of current is very small. Hence, steady current flows without voltage drop. The choke coil impedes the flow of surges and imposes a time delay on their passage. This delay is sufficient for the surge to follow an alternate path across a spark gap or through a gas tube to ground. The surge has sufficient energy to ionize the air in the spark gap or gas in the tube to form a less resistive conducting path.

A voltage applied suddenly to a coil induces a magnetic flux. As the lines of flux build, they cut the turns of the coil and induce an electromotive force (induced voltage). The induced voltage opposes the applied voltage. This process of inductive opposition to a sudden change in current is known as inductive reactance. Overcoming the inductive reactance to complete passage through the coil requires time. This time is sufficient for the surge to follow the less restrictive path across the air gap or through the gas tube.

Communications equipment also requires surge protection from lightning strikes (Paschal 2000b). Dataline protectors should have a rise time of 5 ns (nanoseconds) and an energy rating of 500 J (Joules). Modern surge protectors can withstand multiple hits without failing, unlike the old generation of Metal Oxide Varistors (MOVs) that failed while converting surge energy to heat. (See also the section on Ground Fault Circuit Interrupters.)

Telephone equipment requires lightning protection (NFPA 2011b). Similar requirements apply to cable systems and television antennas. The protective device is intended to limit voltage to ground that might develop on internal telephone wiring or handsets because of lightning or power surges. These also provide protection from contact between high-voltage lines and telephone lines (Bernstein 1991b).

The protective device is either a carbon block containing a spark gap or a gas tube connected to ground. In either case, the high,voltage surge diverts to the alternate path because of the presence of an inductor (described previously). This causes ionization of air across the spark gap or gas in the tube to create the necessary conductive path. Bonding of telephone protective systems to electrical protective systems is absolutely essential to maintain both at common electrical potential.

Another option for lightning protection is to "harden" electrical equipment (Paschal 2000b). Hardening is used to improve survival of electrical equipment during a lightning surge. This involves increasing insulation by a factor of 2 to 10 to increase flash-over distance of an insulator string and specifying greater amounts of insulation in equipment.

## Electrochemical Corrosion/Cathodic Protection Systems

When two metals are electrically connected in the presence of an electrolyte, such as soil water containing dissolved salts and oxygen, electrons will flow from the more active to the less active metal due to the difference in electrical potential (Shackelford 1988). The latter is referred to as the driving force. The more active metal (anode) gradually disintegrates and dissolves into the electrolyte. The less active metal to which electrons flow becomes negatively charged and protected as a result against corrosion. (See Table F.6.)

Electrolytic corrosion sets up DC current in the metals. Electrolytic corrosion is one reason for the failure of electrical grounding systems, concrete reinforced structures exposed to road salt, and internal as well as external corrosion of ship structures, and ballast tanks exposed to seawater.

Cathodic protection is a system to protect metal structures containing bonded dissimilar metals in contact with electrolytes. Cathodic protection is needed even on coated surfaces because of incomplete coverage and breakdown of the coating. Examples where cathodic protection is used include underground tanks, ships and barges, steel reinforced concrete and steel-framed building structures, among others. Bonding of all metal surfaces is essential.

Cathodic protection is achieved passively by use of sacrificial anodes (Figure F.27) or actively by impressed current systems (Shackelford 1988). In either case, DC current flows through the structure. Sacrificial anodes intended to protect steel are made from zinc, aluminum, and magnesium. Periodic replacement of sacrificial anodes must occur in order to maintain cathodic protection. Impressed current systems couple the metal to be protected with the negative electrode of a DC source. The positive pole couples to an auxiliary anode. In this application, an inactive metal, such as platinum, that has a low rate of dissolution can serve as the anode. Sacrificial anodes can descale rust by charging the underlying steel negatively.

Where electrical systems are involved, electrical devices can block DC currents while allowing passage of AC faults (Paschal 1998b). DC currents can pass through bonded components located outside the ground. DC currents on raceways and other components can cause serious corrosion problems.

## Maintaining and Servicing Electrical Equipment

Previous parts of this Appendix have focused on issues related to operation of electrical and electronic equipment on whose performance in a predictable way depends the safety of workers involved in installation, maintenance, servicing, repair, and renovation.

In order to make appropriate decisions about deactivation, de-energization, and isolation, issues related to real-world performance, as opposed to design performance, must be known and appreciated to the fullest extent possible. This section focuses on activity related to deactivation, de-energization, isolation, lockout, and verification of the state of zero energy in electrical equipment.

At the same time, despite strong advocacy to deactivate, de-energize, isolate and lock out, some work on electrical equipment occurs "hot;" that is, while energized completely or energized on the supply side of an enclosure. This work is justifiable only in extreme circumstances where isolation and lockout could pose extreme hardship and inconvenience to end-users. As pointed out earlier, a considerable proportion of the "hot" work is work that otherwise could proceed, but for the inconvenience, in the de-energized, isolated, and locked-out condition.

Figure F.27. Sacrificial anode used for cathodic protection in a saltwater marine environment. The sacrificial anode undergoes an oxidation process leading to dissolution of the metal

## Assessing Electrical Energy

Little exists in print to enable assessment of sources of electrical energy that exist in electrical equipment during preparation for lockout and tagout. Table 7.6 provides a generalized way of considering this question for different sources of energy. Table F.7 contains a specialized version of Table 7.6 that focuses on sources of electrical energy possibly requiring isolation.

For situations not requiring immobilization, the example presented in Table F.7 is satisfactory for assessing isolation and lockout of equipment, machines, and systems. For situations requiring immobilization, refer to Table G.7.

Complicating the assessment of electrical systems is the existence of supplemental and reliable power systems that maintain base levels of power in circuits following outages in the electrical system.

## Shutting Down Electrical Equipment and Systems

In most installations, the electrical system is not isolated from equipment, machines, and systems. The electrical system supplies the energy that powers all of the other systems. This means that when the electrical system shuts down for service, maintenance, repair, modification, upgrading, and renovation, so also does the entire entity. This constitutes a major disruption to the operation of the entity.

As a result, electrical systems in most entities are designed and constructed for partial isolation to enable work to occur on selected portions. This design evokes hub and spoke imagery and possibly also branches diverging from individual spokes. The effectiveness of the design is reflected in the ability for work to occur on selected parts of the system while at the same time enabling operation of the remainder. As mentioned, interconnection to supplemental and reliable power systems is a requirement for consideration prior to shutdown.

The first level of consideration for performing work on parts of the electrical system is to deactivate, de-energize, isolate, and lock out all loads receiving energy from the feed (Rosenberg 1999). This ensures that the electrical circuit is not under load at the time of opening the disconnect that feeds the

## Table F.7

## Assessing Energy in Electrical Systems

| ABC Company | Hazardous Energy Assessment | Work Activity |
|---|---|---|
| Location: | Equipment, Machine, System: | Assessed by: Tel: Date: |
| Overview: | | |
| Hierarchy of Energy Input | Conversion Energy Output | Equipment, Machine, System Affected |
| | | |

### Circuit Name (Main)

| | | |
|---|---|---|
| Function/Description: | | |
| Electrical Isolation  Strategy: | | |
| Failure/Consequence Analysis: | | |
| Input: | Storage: | Dissipation: |
| Output: | Conversion: | Immobilization/Isolation: |
| Primary Isolation: | Secondary Isolation: | Tertiary Isolation: |
| Location: | Location: | Location: |
| Action: | Action: | Action: |
| Electrical Isolation? | Electrical Isolation? | Electrical Isolation? |
| (Circuit Name) (Main) Electrical Isolation Procedure: | | |

## Table F.7 (Continued)
## Assessing Energy in Electrical Systems

| Verification Test Procedure: | | |
|---|---|---|
| **Circuit Name (Supplemental or Reliable Power)** | | |
| Function/Description: | | |
| Electrical Isolation  Strategy: | | |
| Failure/Consequence Analysis: | | |
| Input: | Storage: | Dissipation: |
| Output: | Conversion: | Immobilization/Isolation: |
| Primary Isolation: | Secondary Isolation: | Tertiary Isolation: |
| Location: | Location: | Location: |
| Action: | Action: | Action: |
| Electrical Isolation? | Electrical Isolation? | Electrical Isolation? |
| Verification Test Procedure: | | |
| Overall Isolation Procedure: | | |
| Overall Verification Test Procedure: | | |
| Notes: | | |

## Table F.7 (Continued)
## Assessing Energy in Electrical Systems

| Hazardous Condition | Occurrence/Consequence | | |
|---|---|---|---|
| | Low | Moderate | High |
| | | | | |
| (circuit name) main circuit | | | |
| (circuit name) supplemental or reliable power | | | |

In this table, **NA** means that the category does not apply in any normally foreseeable situation. **Low** means that exposure is readily identifiable but believed to be much less than applicable limits or that exposure to nonquantifiable hazardous conditions is unlikely to produce injury. **Low-Moderate** means that exceedence of regulatory limits is believed possible or that nonquantifiable exposure could produce minor injury requiring self-administered treatment. Control measures or protective equipment should be considered. **Moderate** means that exposure is believed capable of exceeding regulatory limits or causing traumatic injury requiring first aid treatment or attention by a physician. Protective equipment or other control measures are necessary. **Moderate-High** means that exposure is believed capable of considerable exceedence of regulatory limits or causing serious traumatic injury. Advanced control measures or protective equipment are required. **High** means that short-term exposure is believed capable of causing irreversible injury including death. Advanced control measures or protective equipment are required.

circuit. Locking out equipment and machines prior to opening the disconnect prevents unintended re-activation prior to disconnection from the power supply.

Some machines and processes do not stop immediately after the safety circuit trips. This adds additional complication to incorporation of safety circuits. One possibility is the use of the plugging circuit which reverses current flow to a motor thus stopping it quickly.

Normal regulatory practice is to require electrical lockout along with mechanical and other types of lockout when the work involves a risk of electrocution as well as other types of injury. In the U.S., two possible regulatory requirements (29CFR 1910.147 and 29CFR 1910.333) may apply simultaneously during this work (OSHA 1989, OSHA 2011a). Situations where servicing or maintenance can cause exposure to multiple types of energy (electrical and non-electrical) require adherence to both standards. Only persons who are electrically qualified may perform the electrical work.

The circuits and equipment on which work is to occur require disconnection from all sources of electrical energy. This would include the upstream source so that the supply side of a circuit in an enclosure on which work is to occur is no longer live and capable of posing a hazard. Control circuit devices, such as push buttons, selector switches, and interlocks, are not to be used as the sole means for de-energizing circuits or equipment. Interlocks for electric equipment are not to be used as a substitute for lockout and tagging procedures.

Electrical circuits are complex and often multi-phasic. Phase refers to the time-behavior of the sinusoidal wave of electrical energy flowing in a circuit (Nelson 1983). Single-phase service is used for lighting and small motors. Three-phase service is utilized where heavy loads exist on large motors.

Three-phase service utilizes three separate conductors and sinusoidal waves, each 120° out of phase with each other. Provision of energy in this manner smooths out the irregularities in flow. Three-phase circuits provide three separate surges of power in half a cycle, each separated by one sixth of a cycle. Single phase current provides two surges of power, each separated by half a cycle. Three phase induction motors are reversible. Reversal of direction occurs through interchanging any two of the three supply leads. For this reason, care when reconnecting supply leads to a three-phase motor is essential to ensure that reversal has not occurred.

An additional necessity with complex circuits is to trace the path from the power (energy) source(s) to the load. This activity should ascertain whether shutting down part of the system will eliminate the hazard, as desired, or create an additional one. As discussed previously, some circuits contain uninterruptible power (energy) sources. These include stand-by generators, circuits that receive feed from more than one supply, and storage batteries. Surety of supply in some installations is so critical that they receive electrical feed from two separate generating stations. Following normal deactivation, deactivation, and lockout procedures in these situations provides no assurance that re-energization of the circuit cannot occur. Investigative diligence is required at all times.

Control systems pose additional complications in electrical circuits. Some systems contain multiple start/stop controls. Computer-controlled systems and remotely located, radio-controlled systems may be present and not be readily apparent. The potential for problems from these systems can be especially difficult to recognize because the low voltages involved can activate high-voltage equipment through the action of relays and solenoids. Activation of computer- and radio-controlled circuits can occur without warning.

The action of deactivating, de-energizing, isolating and locking out electrical systems can create hazards. In dusty operations, airborne particulates enter and accumulate in electrical disconnects. Shutting down a circuit under load by opening the disconnect has caused dust explosions because of arcing inside the cabinet or enclosure. For this reason, deactivation of electrical equipment should occur through the local ON-OFF control switch (Table F.8). The next step is to shut off the disconnect/breaker. To minimize injury from a possible explosion, the person shutting off the disconnect should face away from the panel (UAW-GM 1985).

## Lockout Devices for Electrical Equipment

Owing to the dependence of equipment, machines, and systems on electrical energy for their source of power, lockouts involving disconnects form the great majority of lockouts. To put this into perspective, electric motors use about 65% of the electrical power.

Lockout devices for electrical systems require consideration and preplanning. Secondary protective devices, such as START switches, may have components built in to enable lockout. However, as mentioned previously, secondary protective devices do not provide assured protection and should be considered as the starting point in the sequence to ensure that the circuit is deactivated and de-energized at the time that isolation occurs at the disconnect, circuit breaker or removed fuse or removed plug.

Lockout devices for electrical equipment can be subdivided according to design and execution. Intrinsic lockout devices are designed into the equipment as supplied, or are designed into retrofits created by the original equipment manufacturer. This is the most desirable route to follow since the manufacturer is most familiar with the equipment.

Extrinsic lockout devices are temporary add-ons to components of the electrical system. Extrinsic lockout devices for electrical equipment include wall switch lockouts, plug lockouts, and circuit breaker and fuse lockouts.

Wall switch lockout devices (Figure F.28) mount onto the cover plate of the switch using the existing screw positions. The hinged cover to which the lock attaches hides the screws and prevents access to them, and thus movement of the switch. Some designs enable the switch to be locked in the ON or OFF position. Use of these products requires care since switches are installable in either the DOWN = OFF (the normal convention in North America) or the UP = OFF position (the normal convention in Europe). Users must be absolutely certain about the intent of the locked position.

Wall switches are usually part of the control circuit and not the point of isolation of the circuit. Isolation occurs at the circuit breaker or fuse in the panel. The ability to lock out a control switch is useful to prevent unauthorized reactivation of the equipment while isolation is progressing.

## Table F.8
## Generic Procedure for Shutdown of Electrical Equipment

**Step**

Shut down the equipment using the procedure recommended by the manufacturer.

Allow actions controlled/powered by electrical equipment to stop completely.

Open line side breaker or pull fuses using procedure recommended by the manufacturer.

Dissipate residual energy from electrical circuits.

Apply primary and secondary energy isolating device(s) and lockout devices.

Verify using testing equipment that electrical energy is not present in affected circuits.

Source: UAW-GM 1985.

Plug lockout devices (Figure F.29) are useful in situations where the plug is remote from the equipment, and the individual performing the lockout and the work does not have control over it. Plug lockout devices are available in two designs. One type accepts the plug into a surrounding structure. The structure is locked closed thus preventing access to the plug. The second type attaches to the prongs of the plug and locks through the holes in the prongs.

Circuit breaker lockout devices (Figure F.30) attach to the movable switch in the breaker and prevent angular or sliding movement. These devices are available for single pole and multi-pole units as well as voltages up to 600 V. These enable reliable lockout of individual breakers without locking closed the panel cover. The latter practice is unsatisfactory as it prevents access to other breakers that still may be energized and may require rapid de-energizing.

Crafting devices for isolating circuit breakers is an art. The movable part of the circuit breaker is specific to the manufacturer and sometimes not easily retrofitted with a lockout device that is strong enough and securely enough attached to a moveable element that is tapered and pivots to resist the excessive force required by OSHA. Miller (2005) reported on tests of devices for use with circuit breakers. The results were discouraging. The consequences of inability to attach a lockout device to breaker handles of this design is catastrophic to the ability to achieve lockout.

Non-adjustable "universal" devices that utilize the handle holes are ineffective. The holes provide a pivot point which imparts a sliding action to the lockout device as the handle moves. The cavity walls that surround the handles have sufficient clearance to allow movement and in some cases operation of the breaker. In addition, these holes are highly variable between breaker types and models. They do not appear to have the strength necessary to comply with the OSHA requirements. For these reasons, abandonment of the use of side handle holes for lockout purposes should occur.

Multi-pole devices that capture the tie-bar without clamping, or single pole devices with handle cavities large enough to accommodate many models of breaker, contain excessive clearance. This can allow operation of the breaker with the device installed.

Adjustable clamping designs work well in general although specific combinations are not acceptable. The lockout device clamps to the handle and attempts to prevent the rocking motion needed to activate the breaker. Overtightening the clamp can damage the breaker handle.

The report by Miller (2005) highlights the obvious: circuit breaker and lockout device manufacturers must work together to ensure compatibility between their products. To illustrate, tapered handles are incompatible with lockout devices and require redesign. A mushroomed handle allows more effective clamping. Effective grooves or depressions on the the handle are essential to ensure secure clamping and immobilization between the lockout device and the breaker.

Figure F.28.  Wall switch lockout device. Note that a wall switch is a control switch. (Courtesy of Panduit Corporation, Tinley Park, IL.)

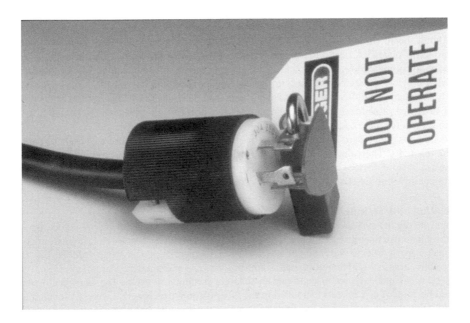

Figure F.29.  Plug lockout device. (Courtesy of Panduit Corporation, Tinley Park, IL.)

Figure F.30. Circuit breaker lockout. (Courtesy of Panduit Corporation, Tinley Park, IL.)

Fuse lockout devices (Figure F.31) prevent reinsertion into position in the holder. Removal of fuses should occur through a dedicated fuse-pulling device. Fuse pullers are manufactured from nonconductive plastic. A fuse puller made from nonconductive plastic provides protection against electrocution compared to metal pliers even those containing insulated handles. Use of the designed fuse puller protects in the event that a cartridge fuse is unexpectedly energized at time of removal. A properly designed fuse puller also protects the fuse against damage during removal and insertion. Fuse pullers, however, provide no protection against electrical arcs caused during removal of the fuse under load conditions from the contacts on the fuse block.

Metal parts of standard lockout devices and padlocks are possible sources of impact sparks and thermite reactivity. These devices can create hazards in hazardous locations. Manufacturers have created plastic lockout devices and locks for these applications (Figure F.32). These devices, however, may not comply with OSHA requirements for excessive force.

### Verification and Testing

Electrical energy in a system is not readily apparent to the senses. The only means for detecting electrical energy is to use diagnostic equipment. Testing the circuit downstream from the lockout is absolutely necessary in order to verify that deactivation and de-energization have occurred. Testing should involve activating ON-OFF controls, and as appropriate, use of diagnostic equipment. Deactivation, de-energization, and isolation must provide an absolute certainty of results. Too many accidents have occurred because of disconnects that failed in the ON position, breakers that were mislabeled, or because of unknown redundancy in the electrical supply system.

Verification devices include noncontact devices that sense through induction of current as well as contact devices. The latter can determine the correctness of wiring as well as ground fault interruption. Equipment used for testing electrical circuits must be suitable for voltages and currents that may be present. Inadequate equipment can become involved in arc faults and cause considerable injury to the

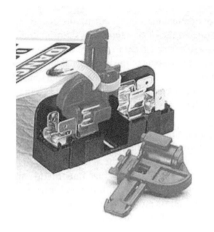

Figure F.31.  Fuse lockout device. (Courtesy of WH Brady Inc., Richmond Hill, ON.)

Figure F.32.  Locks and lockout devices for use in hazardous locations are non-sparking.

user and damage to the surroundings. Explosions involving improperly selected test equipment have caused serious traumatic injury.

## Residual Energy

Residual energy refers to energy retained in a circuit during the deactivation, de-energization, and isolation sequence due to deliberate storage or unintentional entrapment. This situation is akin to fluid pressure trapped in pipes when valves are closed to isolate a section prior to depressurizing.

A typical home electrical circuit consists of the main breaker that connects to the distribution system, a buss that distributes to the circuits and wiring from the panel to form the circuit. Each circuit also contains a circuit breaker or fuse and one or more ON-OFF switches in branching circuits. Circuits involved in lighting especially can have several ON-OFF switches.

In typical fashion, the homeowner locates the appropriate circuit by opening circuit breakers or unscrewing fuses while the circuit is under load. The appropriate circuit breaker or fuse is detectable by extinguishing of lights or shutdown of appliances in the circuit of interest. The homeowner never gives a second thought to possible consequences of the actions involved in locating the correct circuit.

The action of disconnecting a circuit breaker or fuse under load can potentially create an electric arc between the contacts. As mentioned previously, this depends on voltage. The action of the 110 V or 220 V circuit breaker in domestic circuits is rapid and the production of an arc normally is not a concern even though arc production is occurring. In circuits operating at higher voltages and currents, arcing is a serious concern.

An operating electrical circuit is live along its length from supply to return. Simultaneously touching any part of an uninsulated conductor and a metal object that has a path to ground under these conditions creates a circuit. The only regulating factor at this point is the ability of the electrical energy to pass through the body to complete the path.

Electrical circuits that are isolated under load at the circuit breaker, as in the household example, deplete energy in the circuit to the threshold that cannot pass through the equipment. Further, opening of the ON-OFF switch ensures two open switches as a block to further inflow of current, provided that the switch is located on the supply (line) side of the circuit.

By opening the circuit breaker first while under load, energy in the circuit upstream from the ON-OFF switch has dissipated before the second switch opens.

The outcome from this situation is that the only the line side of the circuit in the electrical cabinet or box containing the breaker remains live while the load side is deactivated. This means that work involving the interior of the electrical cabinet still could pose an electrocution hazard.

A different situation arises when the ON-OFF switch is opened first and then the breaker. Electrical energy is trapped in the line between the opened breaker and the opened ON-OFF switch. The extent of possible harm depends on current flow that can result following completion of a path to ground. This situation is not critical at the level of household voltage but can create serious problems in industrial situations where much higher voltages are present.

In industrial situations, the ON-OFF switch is usually part of an indirect control circuit rather than being wired in series with the overcurrent device (breaker or fuse) in the circuit. Control circuits generally operate at low voltage compared to operating circuits. Opening the contactors of the high-voltage circuit by the control circuit under load can cause arcing at the point of disconnection. This situation usually occurs when nobody is present at the electrical disconnect, and while certainly a serious occurrence, constitutes a property damage issue, and not a personal injury one. This distinction is one of the main purposes for the use of control circuits, namely to separate the human operator from the mechanical operation of connection and disconnection of electrical circuits (Hall 1992).

In a typical pushbutton START-STOP system, pushing the START button energizes a relay that engages a contact that maintains the circuit, so that the operator need not further engage the switch (Hall 1992). As well, the relay engages contactors that complete the circuit in the high-voltage or high-current system. Pushing the STOP button breaks the circuit. This type of system also may include limit switches and other sensors that monitor position of components. Failure of any of these switches to engage prevents operation.

In the event that the circuit opens without incident, the disconnect and the wire upstream from the opened contactors remain energized at the same potential. Opening the disconnect under this condition can occur with considerably reduced potential for occurrence of an arc.

Over the years, failures of secondary components in these systems when exposed to the rigors of the operational environment or simply wear and tear through repeated use have indicated that reliance on these components for absolute safety is misplaced (Hall 1992). Despite the complexity of these systems, the fact still remains that the only assured way to prevent electrical energy from entering the system is to disconnect the circuit breaker or remove the fuses at the panel.

Failure modes result from weakened springs, pitted contacts from electrical arcing, chemical corrosion, failure of structural plastics due to attack by ultra-violet energy, erosion from wear and tear, and restrained motion due entry of dirt, grease, and foreign materials into moving parts. There are many potential causes of failure. An additional concern relates to the fact that relays and solenoids are electromagnetically operated.

The actual amount of electrical energy trapped in a circuit under this scenario prior to opening the disconnect depends on components in the circuit. Inductors and capacitors store energy through the magnetic field in the former case and the electrical field in the latter.

An additional problem occurs with isolated circuits. Noncontact induction and capacitance can occur due to nearby active circuits. (See the sections of electromagnetic and electrostatic induction.) This can lead to storage of electrical energy in the isolated circuit.

Where higher voltages are present in circuits involved in distribution, the placing of temporary grounds using a hot stick removes residual energy and prevents unexpected energization. This process is accomplished by securely attaching the ground cable or strap to the ground conductor and then placing or attaching to the conductor of interest. The ground cable or strap is handled using an insulated pole (hot stick) in order to prevent contact between the person applying the ground cable, the ground cable, or the conductor or conductive surface to which it is being connected. Ground cables are applied to several parts of high-voltage circuits. In addition to conducting residual electrical energy to ground to equalize the potential, these cables also carry off electrical energy from lightning strikes and backfeed situations (BC Hydro 1998).

Problems with grounding can result from use of several grounding systems at different potential from each other (Holt 2001d). Current can flow between grounding systems under these circumstances (ground loops). This problem is a potential occurrence when temporary grounding systems are used. Temporary grounding systems are established using rods that are driven into the ground. In some soils, the resistance is too high to enable flow of high current. In situations involving temporary grounding systems and high-voltage transmission systems, explosive expulsion of the ground rod can occur under these conditions.

These problems are controlled using appropriate numbers of ground rods, testing, and use of conductivity enhancing salts mixed into the soil at the time of installation.

### Low-Voltage Equipment

Low-voltage equipment is another approach to safety in the use of electrical equipment (Ericson 1995). Thus far, lighting is the principal industrial application of low-voltage equipment. Low-voltage systems operate at 6 or 12 V AC through transformers that convert from 110 V AC. Some designs shield the low-voltage side from the high-voltage side. The transformer may be located upstream from the light or in the assembly. These units are produced for use in hazardous and nonhazardous locations.

Low- voltage lighting is sanctioned in regulatory statutes as protective devices in conductive and hazardous locations. In the U.S., see for example, 29 CFR 1926.404 on wiring design and protection, components and equipment for general use, and for specific guidelines on protection using low-voltage equipment (OSHA 2011b). OSHA mandates use of hand lamps operating at 12 V or less. Hand lamps designed for use in hazardous locations are specially designed units.

### Battery-Powered Equipment

A number of portable tools (drills, screw drivers, and reciprocating and rotary saws) have become available in battery-powered configurations. These operate at DC voltages ranging from 12 V to 18 V.

While this equipment would seem to match requirements for low energy, DC-powered equipment sold for use in residential applications is prone to arcing, and as a result is not suitable for use in hazardous locations (Figure F.33). This concern may also apply to industrial versions of the same equipment. The only way to establish applicability is rigorous testing.

## Working "Hot"

Fatal accidents reviewed in Chapter 1 indicate that, despite concerted efforts through education and training, electrical and other workers continue to work "hot" on electrical equipment while energized. They do not de-energize electrical circuits (NEMA 1997).

This concern continues to this day. Statistics compiled by the Electrical Safety Authority of Ontario indicate that between 1998 and 2006, 53% of electrical incidents causing injury occurred while work occurred on live electrical equipment (ESA 2008). During the same period, 39% of critical injuries involved burn from an electrical arc. The same statistics indicate that 68% of electrical incidents occur as a result of incorrect procedure and human error.

The first concern about working on live equipment is short-circuit arc faults (NEMA 1997). A short circuit results from contact between conductors of opposite polarity or phase. These contacts have resulted during use of screwdrivers, fish tapes, wrenches, and testing instruments. Only the total impedance (resistance) of the circuit may limit the arc fault. Arcing continues until the protective device opens. Protective devices on the line side include fuses, circuit breakers, and arc-fault or equipment ground-fault protective devices as discussed in previous sections.

Short circuit arc faults of even short duration can result in explosive emission of hot gases, molten metal, shrapnel, and intense ultra-violet, visible, and infra-red energy, described under the terms arc flash and arc blast.

Figure F.33. Arcing from the motor of a low-voltage, battery-powered tool. These products are unsuitable for use in hazardous locations because the arc can act as a source of ignition.

The second concern relates to the occurrence of normal and abnormal switching operations while components are energized (NEMA 1997). Switching operations normally produce electrical arcs, and the resultant products of arc flash and arc blast. These normally are confined to switching cabinets and enclosures. Cabinets and enclosures function partly to contain these emissions although this protection is not total. A switching operation occurring in a cabinet or enclosure opened while under load can cause serious injury up to and including death.

Electrical arcs are widely recognized as major hazards associated with opening circuit breakers and knife switches in older electrical switchgear. Accumulation of dirt or ignitable material from the process following long periods of use is well known as the fuel for explosions inside electrical apparatus. The arc caused by breaking the contacts and the resulting explosion pose serious risk of injury. Electrical maintenance personnel are routinely advised to face away from the switchgear during contact breaking.

Electric shock or electrocution is the third outcome of accidental contact with energized conductors and conductive paths to ground while working hot on equipment.

Working hot on electrical equipment often is a means to save time while performing a job. This approach can result from inadequate or inappropriate training and complacency borne from performing many repetitions of an unsafe practice. Regulators, however, routinely require shutdown of electrical equipment prior to performance of work.

There are circumstances where work on live systems is justifiable. These include creation of hazard or creation of additional hazard as a result of shutdown or infeasibility of shutdown due to limitations of design or operation (NFPA 2004). Given these circumstances, a critically important consideration is that systems that cannot be shut down due to importance of maintaining the service become shut down in the event of an accident.

Circumstances where working hot is a necessity include potential interruption of life-support equipment, deactivation of emergency alarm systems, and shutdown of ventilation equipment used in hazardous locations. In addition, there is the need to perform tests on electrical circuits that can occur only under energized conditions. In all such cases, a qualified person, someone familiar with the construction and operation of the equipment and the hazards involved must perform the work. At the same time, this individual requires protection against the possibility that an arc can occur.

Restart of equipment shut down because of a fault can lead to further hazard especially when loads are not shut down due to the high inrush current and the potential for an arc during closing of the contactors. This is especially the case with the power distribution system following an outage. Much of the load remains because switches do not reset to the OFF position. Homeowners take for granted that equipment and appliances will restart once restoration of power occurs.

Determining that a circuit is deactivated is not always obvious (NEMA 1997). This situation is most likely to occur in circuits containing tie breakers, double throw disconnect switches, automatic transfer switches, and emergency generators. In the latter case, backfeed can re-energize the circuit, as discussed previously. Testing to prove the negative is required.

The risk of causing an electrical arc is not restricted to opening of contacts. Work in electrical cabinets containing switchgear always entails risk especially when the line side is still energized. The load side may be safely de-energized following opening of the disconnect or pulling of fuses. Ensuring that the interior of an electrical cabinet is not energized entails de-energizing the line-side circuit as well as the load-side one. Contact made by tools such as screwdrivers and tape measures, and even inappropriate testing equipment on energized, line-side conductors, can lead to arc faults between phases and between phases and ground.

The only way to eliminate the potential for arc generation is to shut down equipment through normal procedures and to work backwards through breakers and fuses upstream from the work area. That is, equipment must be shut down and completely de-energized prior to starting work.

The electric arc is a graphic example of the power intrinsic to the rapid discharge of electrical energy (Bernstein 1991a). Depending on its size and available energy, the electric arc can act as a source of ignition at currents less than 15 A. The temperature of an electric arc at these current levels ranges from $2,000°$ C to $4,000°$ C ($3,600°$ F to $7,200°$ F) (Cobine 1958). Such arcs can cause damage if sufficient energy is available in them (Baitinger 1995). In electrical switchgear, the temperature at the ends of an arc can reach 20,000° C (35,000° F) or about four times the temperature of the surface of the sun (Cooper Bussmann 2000a).

The voltage that can initiate an arc in air depends on the shape of the electrodes and the waveform. A typical value is 30 kV/cm (Cobine 1958). A fundamental property of arcs is that no matter how small the gap between two conductors, the voltage must exceed 300 V. Breakdown across an air gap cannot develop at smaller voltages. An arc, however, can develop at much lower voltages when electrodes initially in contact are separated as is the case in a circuit breaker.

After establishing an arc, only 20 V/cm is needed to sustain it (Golde 1973). This is the reason that welding equipment operates at low voltage. As well, an arc established in this way can exist over a large air gap. Voltage needed to sustain the arc depends primarily on arc length and is practically independent of current. To illustrate, an arc welder employing open circuit voltage of 80 V cannot initiate an arc across an air gap (Bernstein 1991a). Separating the electrodes after touching initiates the arc. This voltage is sufficient to maintain the arc.

Skin temperature raised above 96° C for 0.1 s results in total destruction of tissue (incurable burn) (Lee 1982). Skin temperature raised to below 80° C for 0.1 s results in just curable burn. This data and the underlying energy needed to produce this level of burn were incorporated into NFPA 70E in an attempt to stop the occurrence of thermal burn due to electrical arcs (NFPA 2004). NFPA 70E requires measures to prevent burn injury including engineering controls and personal protective equipment.

Radiant emissions from the electric arc also include high levels of ultra-violet, visible and infra-red energy. Emissions from electric arc flash have similar characteristics to welding arcs since they originate by the same process. However, the high voltages and currents involved in electric arc flashes are considerably greater than those employed during arc welding.

Electrical arc flashes also expel molten metal and shrapnel from destroyed contacts and supporting structures and other components (Lee 1987). The molten metal can contribute to burn injury on the skin and in the respiratory system, and the shrapnel to ballistic wounds. Inhaled metal vapor and fume (metal oxide) can cause serious lung injury due to toxicological reaction. Metals of particular concern include copper, cadmium (used to plate iron and steel), zinc (galvanizing), and chromium and nickel (plating). Explosive air pressure develops from the rapid expansion of metals from solid to vapor and rapid expansion of air. These pressures have knocked workers off ladders and hurled them across rooms. These pressures also can cause injury to the ear drum, the lung, and nervous system.

The only means to remove the risk implicit in these situations is to shut down the equipment prior to starting. There are many situations where action of equipment that can cause arcing occurs independent of the actions of individuals. This is the case in electrical generating stations with equipment whose function is to protect electrical circuits associated with the generators. Engineering means are the recommended first line of defense in these situations.

One example of these is the breaker used to cut primary output from generators. Breaking the circuit as fast as possible during emergency situations is critical in order to minimize arc generation and thereby to prevent damage to contacts. This is achieved in some equipment through use of sulfur hexafluoride, an insulating gas, in place of air. Other equipment uses a blast of compressed air to flush the space between the contacts. This flushing thereby prevents formation of the arc during opening of the breaker (air blast breakers). This type of equipment employs shielding materials to contain the arc flash. Asbestos was commonly used in this application.

The cabinet in which the breakers are housed provides additional protection against the arc flash and and some protection against the arc blast. Real-world experience indicates that cabinetry is not sufficient alone to provide full protection to workers in the vicinity during one of these episodes. Kolak (2009) offers extensive background commentary on this situation.

Personal protective equipment is available for use against electric arc flash (Figure F.34). The ensemble ranges upward from socks and underclothing to pants, shirt and head covering, to faceshield, gloves and outer jacket and pants, and finally to a jacket containing a hood, and built-in face shield and pants. The difference between this clothing and regular clothing is the rating for fire and thermal resistance. Note that the rating does not consider protection against projectiles and molten metal.

Where fire-resistant clothing does not provide neck and head protection, a faceshield is also required. As well as typical clothing, faceshields intended for impact or splash protection of the face

Figure F.34. High-level arc flash protective clothing incorporates multiple layers and a faceshield. (Image reproduced with the permission of Salisbury by Honeywell © 2011 Honeywell International Inc. All rights reserved.)

are unsuitable for this application. The weakest link in the protective clothing ensemble is the window of the faceshield. Hearing protection is required against noise generated during the blast. Respiratory protection should also be considered because of the potential for high levels of ozone and vaporized metals.

Protective ensembles used in these applications must be fire-resistant (OSHA 2011c). Fire-resistant (FR) clothing self extinguishes after ignition, usually within seconds. As well, OSHA requires clothing worn by workers exposed to electric arcs or flames not to increase the risk of injury. Clothing made from acetate, nylon, polyester or rayon, either alone or in blends, is not to be worn because of risk of melting and intensification of contact with the skin and transmission of heat during a more serious burn. NFPA 70E requires protective apparel worn to protect against electric arcs to prevent against anything more serious than a curable (second degree) burn (NFPA 2004). Note that second degree burn, in and of itself, is a serious injury.

Flame-resistant (FR) clothing is designed to limit or reduce but not to eliminate burn injury (Wulf 2009). Survival, extent of injury, and recovery time all depend on the performance of FR clothing subjected to an electric arc flash. The reason for the concern is that most severe burn injuries and fatalities are caused by ignition and prolonged combustion of clothing with the resultant contribution to the thermal burn. The skin under the burning clothing is often burned more severely than skin exposed to the arc blast.

The entire ensemble requires testing to determine the Arc Rating for use at different levels of thermal energy (ASTM 2010a, ASTM 2010b).

ASTM F1506 details the specifications of textiles to be used by electrical workers for arc protection (ASTM 2010a). This includes the minimum performance requirements for clothing intended to protect against electric arc flash. Fabrics must retain resistance through multiple launderings. Multiple launderings typically mean 25 to 50 wear and cleaning cycles. Manufacturers of some fabrics guarantee fire retardancy for the life of the garment. This reflects intrinsic capabilities of the textile used in these garments rather than a supplemental treatment (Westex 2005).

To qualify as an arc-protective garment, the label must state the tracking identification code, name of the manufacturer, care instructions and fiber content, size and associated standard labeling, the arc rating, and adherence to ASTM F1506 (ASTM 2010a).

ASTM F1959 details the standardized test method to be employed to determine the thermal protective value of a textile in the standardized apparatus used for the test (ASTM 2010b). The method requires at least 20 data points to develop the comparison to the Stoll curve. The Stoll curve predicts human tolerance to heat and onset of a second degree burn. The intersection of the curve derived for a particular fabric with the differential from the Stoll curve leads to the Arc Rating.

The Arc Rating for garments is usually stated as the Arc Thermal Performance Value (ATPV).These ratings are expressed in $cal/cm^2$ (calories per square centimeter of surface area) and are based on exposure to the arc for 1 s. This rating expresses the radiant energy against which the weakest link of the ensemble will provide protection against a 50% chance of occurrence of a second degree burn. The apparatus used in testing measures the amount of energy that passes through the material during exposure to the electric arc used in the test.

To put these values into perspective, an exposure to the hottest part of the flame from a cigarette lighter in 1 s delivers 1 $cal/cm^2$ or 4.2 $J/cm^2$. An exposure of 1 to 2 $cal/cm^2$ will cause a second degree burn on human skin. (Note that the Calorie = 1,000 calories. The Calorie expresses the energy value of food.) Typical lightweight non-FR workwear ignites at 4 to 5 $cal/cm^2$. FR fabrics can at least double this rating. Electric arcs typically emit 5 to 30 $cal/cm^2$, and arcs of 30 to 60 $cal/cm^2$ are not uncommon.

In order to meet the full requirement, faceshields and the window in the hood of protective suits must also meet the requirements for eye and face protection of ANSI Z87.1 (ANSI 2010). The shade of the material increases as the intensity of radiant emissions from the arc increases. As with shades used for protection against welding arcs, the lens becomes opaque and does not transmit enough light to enable the wearer to see properly. This situation applies equally to the windows of hoods providing protection against high levels of emission by electric arc.

Recent studies using super-slow-motion video of electric arc blasts have demonstrated that molten metal droplets and particles ejected during the blast play a significantly greater role in ignition than accommodated by models of incident energy and hazard analysis (Margolin 2010). These models quantify energy due to the arc flash and not energy contained in molten metal that accompanies the blast. The blast projects hemispherically outward thousands of particles of copper in molten and solid form. Copper melts at 1,040° C (1,900° F). Droplets blasted outward are at least at this temperature. Ignition temperature of non-FR (Fire Retardant) fabric is typically about 425° C (800° F). Even with cooling induced by the rapid motion, these droplets still impose significant risk of ignition of fabrics, beyond that considered in models based on emission of thermal energy.

Previous discussion to this point has indicated the complexity of the situation involving arc flash and arc blast. One of the roles undertaken by NFPA 70E was to provide an organized and disciplined approach to the question (NFPA 2004). NFPA 70E continues to undergo an evolution as this question becomes more thoroughly defined and refined.

NFPA 70E took the view and stated clearly as a starting-point the need to deactivate, de-energize, isolate, and to lock out electrical circuits prior to performing work, if at all possible. This reflects the obvious outcome from real-world experience concerning shock, electrocutions, and other electrical injuries. Floyd and Doan (2007) provide background commentary about the importance of this approach.

NFPA 70E introduced the concept of a hazard analysis applied to arc flash and shock hazard as the means to structure an organized approach to assessing and addressing the risks posed by working on energized equipment (NFPA 2004). NFPA 70E restricts this work to individuals who are qualified. A person who is qualified can distinguish live parts from other parts of electrical equipment. This skill also includes the ability to determine the nominal voltage of the live parts and to know minimum safe distances of approach. As well, the qualified person also will know about special techniques, shielding materials, insulated tools, and personal protective equipment required for safe work.

## Table F.9
## Contents of the Electrical Work Permit

Description of the circuit (voltage, current) and equipment on which work is to occur.

Location of the work.

Assigned task.

Justification for performing the work live.

Description of safe working practices to be followed including special precautions and control of energy sources and lockout requirements.

Results of the shock hazard analysis.

Shock hazard protection boundaries.

Results of the arc flash hazard analysis.

Flash protection boundary.

Testing equipment required.

Personal protective equipment required to perform the assigned task.

Means to prohibit access of unqualified persons into the work area.

Summary of job briefing including specific hazards prior to the start of work.

Acknowledgment of Job briefing and understanding of assigned task and required protective measures.

Approval signature.

Signature of qualified person.

Source: NFPA 2004.

A further requirement is the permit for work on energized electrical equipment (Table F.9). The permit captures as much information as possible to enable a person qualified by training and experience to be able to work on energized equipment. The hazard analysis is the critical part of the document and the one most difficult to address.

This situation raises the question: how much radiant energy would be expected in a real-world situation? The hazard assessment must assess the shock hazard and the hazard of the flash to determine the level of energy to which a person performing the specific task might be exposed. The next step is to evaluate engineering options, and the final step, to select protective clothing and other equipment. The level of energy against which the protective apparel will offer protection must meet or exceed the level of energy determined in the hazard assessment.

While this description sounds straightforward, answering the questions posed appropriately is very difficult. Underestimating the amount of energy radiated during an arc flash would lead to under-protecting the worker. Considerable discussion has revolved around this question.

The rationale for the arc flash assessment is to predict the flash protection boundary and the incident arc energy at the position of work for the specific situation.

The flash protection boundary is the distance from the exposed live part at which the incident thermal energy is less than or equal to $1.2$ cal/cm$^2$ in the event of an arc flash. A person outside this boundary would sustain no more than a curable burn, meaning a second degree burn. Note that this boundary presumes that a burn of lesser severity can occur. As well, this boundary offers no protection against the shower of molten metal emitted from the area of the arc, larger projectiles, and effects from the blast (Margolin 2010). As a result, current standards may underestimate the need for performance from fabrics currently in use.

NFPA 70E assigns arc hazard risk categories based on the incident energy at the position of work (NFPA 2004). These determine the extent of protection required for qualified persons potentially affected by the arc and for potentially affected non-qualified persons (Table F.10).

NFPA 70E approaches the question of incident energy at the position of work through two alternatives (NFPA 2004). The first is a generic version. This provides a conservative result in the absence of calculation. NFPA 70E provides tables containing example situations and recommended hazard risk categories. This approach considerably simplifies the selection of protective equipment for situations that lie within the envelope of these recommendations. The decision-making captured in the recommendations is complex. Intrinsic to use of the recommendation in a specific situation is study and acknowledgment of the assumptions listed in the notes that accompany the tables.

Tasks lying outside the envelope that includes those listed in the tables require formal assessment. An additional reason for performing a formal assessment is that the recommendations provided in NFPA 70E may impose considerable and unnecessary burdens of protective and personal protective equipment not justifiable by specific realities. The alternative is to apply a method of calculation that more closely assesses the risk of the situation.

The bases for calculations used to predict the radius of the flash protection boundary are equations for specific equipment and more generalized equations initially developed by Lee (1982). IEEE 1584 and NFPA 70E contain these equations (IEEE 2002; NFPA 2004).

These equations contain as variables: current, voltage, arcing fault clearing time (duration of exposure), distance from the source, distance of the source from the face, arc gap, gap between phase connectors and phase to ground, type of equipment and type of grounding. These equations are complex and require thorough understanding by users. Otherwise, errors of underestimation and overestimation are possible (Marroquin 2008a, Marroquin 2008b; Taylor 2007; Weigel 2006).

Knowledge of clearing time and the device in the circuit that clears the fault are critical concept in the applying the equations and software packages. Clearing time of long duration lasting into seconds or even minutes is possible in low-voltage circuits due to the slow response of upstream protective devices. Small reduction in fault current can cause large increase in clearance time. In these situations, installation of light sensors that detect the arc and trip the breaker in less than two cycles is a possible corrective measure.

Information required to use the raw equations and software packages is complex and detailed. This implies in-depth knowledge about operating conditions, geometry of approach to the arc source, and conditions to be encountered during a fault condition. Making this type of assessment is a difficult undertaking. There is an obvious inclination to obtain the maximum level of protection available rather than to attempt to "play a numbers game." As mentioned, the window in the hood is extremely dark, rendering vision difficult. This could predispose the wearer to other safety-related accidents such as tripping and falling. More appropriate in the circumstances would be to eliminate the cause of arc faults so that this undertaking is not required.

Study of actual industrial installations indicates that many tasks rank at hazard risk category 2, and that 75% are category 2 or less. For those that rank higher, relatively minor change can reduce the level to category 2. This situation supports the growing trend to equipping or requiring industrial electricians to wear category 2 ensemble (cotton underwear (shirt + pants) + FR shirt + FR pants) for all work at all times. This decision provides a starting point for protection against known situations and a base of protection against unforeseen situations involving unexpected arcs from opening of contactors, failure of equipment, dropping tools, use of inappropriate tools, and errors in procedure. With the uncertainty created by the potential role of the spray of molten metal emitted from the arc, current standards may underestimate the need for performance from fabrics currently in use (Margolin 2010). The basis for decision-making may shift from whether FR fabrics are not required during even routine electrical work to what category of protection is required.

## Table F.10

## Hazard Risk Category versus Arc Thermal Protective Value (ATPV)

| Hazard/ Risk Category | Arc Thermal Performance Value cal/cm$^2$ | Ensemble |
|---|---|---|
| 0 | NA | Non-melting ignitable materials (cotton shirt + pants) |
| 1 | ≥4 | FR shirt + FR pants or FR coveralls |
| 2 | ≥8 | Cotton underwear (shirt + pants) + FR shirt + FR pants |
| 3 | ≥25 | Cotton underwear (shirt + pants) + FR shirt + FR pants + FR coveralls or cotton underwear + two FR coveralls |
| 4 | ≥40 | Cotton underwear (shirt + pants) + FR shirt + FR pants + multi-layer flash suit |

Source: adapted from NFPA 2004.

## References

Anonymous: Inrush Current: San Francisco, Wikimedia (Wikipedia), 2011. [website]

ANSI/UL: American National Standard for Leakage Current for Appliances (ANSI C101.1-1992). Northbrook, IL: American National Standards Institute/Underwriters Laboratories, 1992.

ANSI/ISEA: Practice for Occupational and Educational Eye and Face Protection (ANSI/ISEA Z87.1-2010). Arlington, VA: American National Standards Institute/International Safety Equipment Association, 2010.

ASTM: Standard Performance Specification for Flame Resistant Textile Materials for Wearing Apparel for Use by Electrical Workers Exposed to Momentary Electric Arc and Related Thermal Hazards (ASTM F1506-10a). West Conshohocken, PA: American Society for Testing and Materials International, 2010a.

ASTM: Standard Test Method for Determining the Ignitability of Non-Flame-Resistance Materials for Clothing by Electric Arc Exposure Method Using Mannequins (ASTM F1959-99(2010)). West Conshohocken, PA: American Society for Testing and Materials International, 2010b.

Baitinger, W.F.: PC Issues are Hot in the Electrical Utility Business. *Safety Prot. Fab. 4(1)*: January/February (1995). pp. 23-25.

BC Hydro: Temporary Safety Grounding and Bonding (Revision 3). Burnaby, BC: BC Hydro, 1998.

Bernstein, T.: Electrical Systems, Terminology and Components — Relationship to Electrical and Lightning Accidents and Fires. In *Electrical Hazards and Accidents, Their Causes and Prevention*, Greenwald, E.K. (Ed.). New York: Van Nostrand Reinhold, 1991a.

Bernstein, T.: Lightning Protection for Buildings, Equipment, and Personnel. In *Electrical Hazards and Accidents, Their Causes and Prevention*, Greenwald, E.K. (Ed.). New York: Van Nostrand Reinhold, 1991b.

BPA: Electrical and Biological Effects of Transmission Lines: A Review (DOE/BP-524, Revised). Portland, OR: U.S. Department of Energy/Bonneville Power Administration, 1986.

Browne, S.R.: Grounding and Bonding Applications for Control of Static Electricity. Atlanta, GA: Stewart R. Browne Manufacturing Company, 1996.

Cantwell, E.C.: Effective Grounding: The Key to Personnel Protection. *Prof. Safety 26(2):* 16-21 (1981).

Carbone, F.: Double Insulated vs. GFCI Protection. Willoughby, OH: Ericson Manufacturing Company, 7/20/94. 1 pp. [Unaddressed memo].

Carr, M. and L. Sitter: Ensuring Power Quality and Reliability with Gen-Sets. *Elect. Contract. Maint. 100*: March (2001) Suppl. pp. PQ19-PQ20.

CDA: A Primer on Power Quality (A6018-98/98). New York: Copper Development Association, 1998a. [Pamphlet].

CDA: Copper Grounding System Protects Mt. Washington Towers (Copper Applications, A Case Study, A6024-98/00) New York: Copper Development Association, 1998b. [Pamphlet].

CDA: Copper Protects Costly Gear. In *Copper Topics* (No. 92). New York: Copper Development Association, Winter 2001. [Newsletter].

Chowdhury, J.: *Chem. Eng. 106(3)*: 55-57 (1999).

Cobine, J.D.: *Gaseous Conductors*. New York: Dover Publishing, 1958.

Cooper Bussmann: SPD (Selecting Protective Devices), Electrical Protection Handbook. St. Louis, MO: Cooper Bussmann, Inc., 2000a.

Cooper Bussmann: NE99, Overcurrent Protection and The 1999 National Electrical Code. St. Louis, MO: Cooper Bussmann, Inc., 2000b.

Cutler-Hammer: Arc Fault Circuit Interrupters (Pub. No. B31C.03.S.K.). Burlington, ON: Cutler-Hammer Canada, 1997. [Pamphlet].

DeDad, J.A.: Solving Instrumentation Ground Loop Problems. *Elect. Contract. Maint. 98(11)*: 14-20 (1999).

Denton, T.: Radiofrequency Energy Poses Unseen Hazard. *Occup. Hazards. 64(12)*: 45-48 (2002).

Elcon: *Introduction to Intrinsic Safety*. Annapolis, MD: Elcon Instruments, 1989.

Ericson: Safety First, Ground Fault Protection and Confined Space Lighting. Willoughby, OH: Ericson Manufacturing Company, 1995. 7 pp. [Pamphlet].

ESA: Don't Work Live, Respect the Power (Form # 1368 (04/08). Mississauga, ON: Electrical Safety Authority of Ontario, 2008.

Floyd II, H.L. and D.R. Doan: Electric Arc Hazard, Understanding Assessment and Mitigation. *Prof. Safety 52(1)*: 18-23 (2007).

Golde, R.H.: *Lightning Protection*. London: Edward Arnold, 1973.

Gorosito, A.: Simplifying Surge Protection. *Elect. Contract. Maint. 99(11)*: 74-78 (2000).

Gregg, B.: Generation and Control of Static Electricity. *Plant Services 17(6)*: 83-87 (1996).

Hall, F.B.: Safety Interlocks – The Dark Side. Safety Brief 7: No. 3 (June 1992). Niles, IL: Triodyne, Inc., 1992.

Hartwell, F.P.: Don't be Neutral About Grounding. *Elect. Contract. Maint. 98(7)*: 24-35 (1999).

Hartfiel, M. and M. Lamendola: Infrastructure Design Tips Prevent PQ Problems. *Elect. Contract. Maint. 99*: August (2000) Suppl. pp. 10-14.

Hartfiel, M. and M. Lamendola: UPS Cures for Power Quality Problems. *Elect. Contract. Maint. 100*: March (2001) Suppl. pp. PQ10-PQ14.

Hartwell, F.P. and J. Cannatelli: Hazardous Locations: Avoiding the Pitfalls. *Elect. Contract. Maint. 98(7)*: 36-39 (1999).

Holt, M: The Basics of Reducing Touch Potential. *Elect. Contract. Maint. 100(4)*: 120 (2001a).

Holt, M: Equipment Grounding: Know What to Expect! *Elect. Contract. Maint. 100(3)*: 104 (2001b).

Holt, M.: GFCI Basics. *Elect. Contract. Maint. 100(2)*: 112 (2001c).

Holt, M: Ground Rods and Touch Voltage *Elect. Contract. Maint. 100(5)*: 116 (2001d).

Holt, M: Open Service Neutrals. *Elect. Contract. Maint. 101*: November (2002). [Internet Reprint].

Kolak, J.J.: Arc Blast Hazards, The Limitations of Metal-Clad Enclosures to Protect Workers. *Prof. Safety. 54(6)*: 46-51 (2009).

IEC: Part 10-1. Classification of Areas - Explosive Gas Atmospheres (IEC 60079-10.1 ed. 1.0.). Geneva, Switzerland: International Electrotechnical Commission, 2008.

IEC: Part 10-2. Classification of Areas - Combustible Dust Atmospheres (IEC 60079-10.2 ed. 1.0.). Geneva, Switzerland: International Electrotechnical Commission, 2009.

IEEE: Guide for Performing Arc Flash Calculations (IEEE Std. 1584-2002). Piscataway NJ: Institute of Electrical and Electronic Engineers, 2002.

Lamendola, M.: Power Quality Planning: A "Big Picture" Job. *Elect. Contract. Maint. 99:* August (2000) Suppl. pp. 16-18.

Lamendola, M.; A Cure for the Battery Blues. *Elect. Contract. Maint. 100*: March (2001a). Suppl. pp. PQ16-PQ18.

Lamendola, M.: The Earth Is Not a Bonding Jumper. *Elect. Contract. Maint. 100(1)*: 26-32 (2001b).

Layton, J.E.: Why GFCIs Fail. *Occup. Health Safety 69(11)*: 72-74 (2000).

Lee, R.: The Other Electrical Hazard: Electrical Arc Flash Burns. *IEEE Trans. Indust. Appl.IA-18(3)*: 246 (1982).

Lee, R.: Pressures Developed by Arcs. *IEEE Trans. Indust. Appl. IA-23*: 760-764 (1987).

Lees, F.P.: *Loss Prevention in the Process Industries*. Oxford, UK: Butterworth-Heinemann, 1996.

Littelfuse: Basic Fuseology, Overcurrent Protection (2nd Ed.). Des Plaines, IL: Littelfuse, Inc., 1993.

Lyncole: Soil Resistivity Testing, Four Point Wenner Method (LEP-1001). Torrance CA: Lyncole Xit Grounding Solutions, 2005. [www.lyncole.com, Website].

Margolin, S.M.: Arc Flash and Molten Metal – The Hidden Hazard. *Occup. Health Safety 79(5)*: 42-46 (2010).

Marroquin, A.: Assessing Low-Voltage Arc Hazards. *Electrical Source 4(5/6)*: 26-29 (2008).

Marroquin, A.: Assessing the Limitations of Arc Flash Hazard Tables. *Electrical Source 4(7/8)*: 13-16 (2008).

Middleton, R.G.: *Practical Electricity*, 4th Ed. (Rev. by L.D. Meyers). New York: Macmillan Publishing Company, 1986.

Miller, B:D. Circuit Breaker Lockout Devices. Deerfield, IL: B. Miller Engineering, 2005. [www.bmillerengineering.com, Website].

Morgan, M.G.: Electric and Magnetic Fields from 60 Hertz Electric Power: What do we Know About Possible Health Risks? Pittsburgh, PA: Department of Electrical Engineering and Public Policy, Carnegie Mellon University, 1989. [Booklet].

Nelson, C.A.: *Millwrights and Mechanics Guide*. New York: Bobbs-Merrill Company, 1983.

NEMA: Hazards of Working Electrical Equipment Hot. Washington, DC: National Electrical Manufacturers Association, 1997.

NFPA: Standard for Electrical Safety in the Workplace, NFPA 70E (2004 Ed.). Quincy, MA: National Fire Protection Association, 2004.

NFPA: Recommended Practice on Static Electricity, NFPA 77 (2007 Ed.). Quincy, MA: National Fire Protection Association, 2007.

NFPA: National Electrical Code, NFPA 70 (2011 Ed.). Quincy, MA: National Fire Protection Association, 2011a.

NFPA: Standard for the Installation of Lightning Protection Systems, NFPA 780 (2011 Ed.). Quincy, MA: National Fire Protection Association, 2011b.

NIOSH: Worker Deaths by Electrocution (DHHS (NIOSH) Pub. No. 2000-115). Cincinnati, OH: U.S. Department of Health and Human Services/ Public Health Service/ Centers for Disease Control and Prevention/ National Institute for Occupational Safety and Health, 2000. [CD-ROM].

Novotny, D.W. and G.R. Priegel: Electrofishing Boats. Tech Bull. No. 73. Madison, WI: State of Wisconsin Department of Natural Resources, 1974.

OSHA: Selected Occupational Fatalities Related to Welding and Cutting as Found in Reports of OSHA Fatality/Catastrophe Investigations. Washington, DC: U.S. Department of Labor, Occupational Safety and Health Administration (U.S. DOL/OSHA), 1988.

OSHA: "Control of Hazardous Energy Sources (Lockout/Tagout); Final Rule," *Fed. Regist. 54*: 169 (1 September 1989). pp. 36644-36696.

OSHA: Selected Occupational Fatalities Related to Ship Building and Repairing as Found in Reports of OSHA Fatality/Catastrophe Investigations. Washington, DC: U.S. Department of Labor, Occupational Safety and Health Administration (U.S. DOL/OSHA), 1990.

OSHA: Ground-Fault Protection on Construction Sites (OSHA 3007). Washington, DC: U.S. Department of Labor/ Occupational Safety and Health Administration, 1998 (Revised).

OSHA: Selection and Use of Work Practices, Electrical (Subpart S), 29CFR 1910.333. Washington DC: U.S. Department of Labor, Occupational Safety and Health Administration, 2011a. [OSHA website].

OSHA: Wiring Design and Protection, Electrical - Installation Safety Requirements (Subpart K) 29CFR 1926.404. Washington, DC: Occupational Safety and Health Administration, 2011b. [OSHA website].

OSHA: Electrical Power Transmission, Generation, and Distribution; Electrical Protective Equipment; Electrical Safety-Related Work Practices (29 CFR 1910.269). Washington, DC: Occupational Safety and Health Administration, 2011c. [OSHA website].

Palmquist, R.E.: *Electrical Course for Apprentices and Journeymen*. Revised by J.A. Tedesco. New York: Macmillan Publishing Company, 1988.

Paschal, J.: *Understanding NE Code Rules on Hazardous Locations*. (Edited by F. Hartwell). Overland Park, KS: EC&M Books, PRIMEDIA Intertec, 1998a.

Paschal, J.: *Understanding NE Code Rules on Grounding & Bonding*, 2nd Ed. (Edited by F. Hartwell). Overland Park, KS: EC&M Books, PRIMEDIA Intertec, 1998b.

Paschal, J.: Not All Ground Rods Are Created Equal. *Elect. Contract. Maint. 99(9)*: 38-44 (2000a).

Paschal, J.: Don't Let Lightning Catch You Off Guard. Elect. Contract. Maint. 99(6): 36-38 (2000b).

PEI: Portable Container Fueling Incidents at Refueling Sites. Tulsa OK: Petroleum Equipment Institute, 1996. [Website].

Reardon, A.: When to Use a Grounded Antistatic Bulk Bag. *Powder Bulk Eng. 16(10)*: 19-22 (2002).

Roberts, E.W.: *Overcurrents and Undercurrents, All about GFCIs*. Mystic, CT: Reptec, 1996.

Rosenberg, P.: Correspondence Lesson 2: Basic Wired Control Devices. *Elect. Contract. Maint. 98(3)*: 59-64 (1999).

Schiff, N.: Proper Application of Portable GFCIs. *Occup. Health Safety. 68(11)*: 36-40 (1999).

Schmieg, G.: Static Electricity: Causes, Analysis and Prevention. In *Electrical Hazards and Accidents, Their Causes and Prevention*. Greenwald, E.K., Ed. New York: Van Nostrand Reinhold, 1991.

Shackelford, J.F.: *Introduction to Materials Science for Engineers*, 2nd Ed. New York: Macmillan Publishing Company, 1988.

Smoot, A.W. and C.A. Bentel: Electric Shock Hazard of Underwater Swimming Pool Light Fixtures. *IEEE Trans. Power Apparat. Sys. 83*: 945-964 (1964).

Stamper, J.: The Interaction of Power Quality and Your UPS. *Elect. Contract. Maint. 100*: March (2001) Suppl. pp. PQ1-PQ4.

Switzer, K.: Hard Facts about Bonding to Steel. *Elect. Contract. Maint. 99(6)*: 40-44 (2000).

Taylor, M.T.: Regulating Arc Flash Hazards. *Occup. Hazards. 69(5)*: 41-47 (2007).

TRC: HD-PRO Series Application Update. Clearwater, FL: Technology Research Corporation, 1997. [Pamphlet].

UAW-GM: Lockout. Detroit, MI: UAW-GM Human Resources Center, UAW-GM National Joint Committee on Health and Safety, 1985.

UL: Double Insulated Systems for Use in Electrical Equipment (UL 1097). Northbrook, IL: Underwriters Laboratories Inc., 2004.

UL: Ground-Fault Circuit Interrupters (UL 943, 4th Ed.). Northbrook, IL: Underwriters Laboratories Inc., 2006.

Weigel, J.: Their Last Line of Defense. *Occup. Health Safety 75(5)*: 72-76 (2006).

Westex: INDURA® Ultra Soft® and INDURA® Brand Flame Resistant Fabrics vs. Generic/Off-Brand Cotton and Cotton Blend Fabrics, White Paper. Chicago: Westex Inc., 2005 [Booklet.]

Wulf, K.: Flame-Resistant Clothing. *Prof. Safety 54(6)*: 52-57 (2009).

# G  Mechanical Systems

## INTRODUCTION

Mechanical devices change the magnitude, direction or intensity of forces, or the speed resulting from them. Mechanical devices can be as simple as the screwdriver, wrench or hammer, and as complex as a passenger aircraft.

In the industrial context, a machine usually is an assembly of subsystems, each of which can contain simpler subsystems composed of simple mechanical devices. Simple mechanical devices and the subsystems found within industrial equipment can store, transform, and transfer energy. Mechanical action occurs through the motion of levers in straight-line directions and rotating parts (Olivo and Olivo 1984).

Over the years, production machines and equipment have caused many accidents and traumatic injuries, including death, and have long been the subject of efforts in accident prevention. The focus of these efforts was the interface and interaction between the human operator and the point of operation of the machine. The point of operation is the area of a machine where it performs work on material.

At the level of engineering design and development reflected in machinery and equipment of older vintage, the operator was essential for placing materials to be processed into the point of operation in a particular orientation, to command action by the machine through control switches, and to remove the material following action by the machine. The machine could not function without the action of the operator. That is, the human operator literally was part of the machine.

The difference in action between the human operator and the machine is predictability. Whereas the actions of the machine are predictable because of design, manufacture and assembly of components in a precise manner, the actions of the human operator are not totally predictable and beyond control of the designer even after training in operational and machinery hazards. Much of this deviation results from the operation of the human brain and information processing. The action cycle during production mode of placement of material, machine operation, and removal of material could easily occur in less than five seconds. Highly repetitive actions create conditions of monotony and boredom leading to suspension of active information sensing and processing.

These conditions are conducive to inappropriate placement of fingers and limbs for reasons that are inexplicable at the level of normal human consciousness. These movements occur in such a manner that the person does not realize the context in which they have happened. Everyone has experienced similar situations in which the inappropriateness of an action is realized only after its occurrence and often just prior to the inescapable consequence. The person is helpless to change the outcome of the situation after initiation.

This situation is an excellent illustration of the competition for attention between task and conditions. During machine operation, as described, the focus of the operator shifts to task and repetition of particular actions, namely placing material, activating the control and removing material. The conditions under which these actions occur do not distinguish themselves. The operator must consciously shift attention away from the task to conditions. Under pressure, self-induced or otherwise, to maximize production, focus on conditions rapidly disappears from the conscious mind.

No amount of training, knowledge, and supervision can counter the consequence of the loss of active information processing in these circumstances. These idiosyncrasies of human behavior are beyond quantitation and action by designers. The playing of music either by person-portable devices worn by the operator or by "boom boxes" further diverts attention from the tasks to be performed and increases the likelihood of inappropriate action by the operator. Another difference between the human operator and machine parts is fragility and replaceability. Whereas machine parts are substantive and replaceable in the event of destruction brought about by failure of control, human limbs are fragile and not replaceable.

Given the constraints imposed morally and by unreliability of performance of human operators, modern production machinery and equipment utilize complex mechanical mechanisms to position material into the point of operation, to operate the device, and to remove the material following the action. This innovation removes the need for the human operator and the potential for accident and injury during the action phase of production mode. This equipment shifts the potential for accidents to the standby phase of production mode and to maintenance during shutdown.

## Machine Operation and Maintenance

Production machines and equipment illustrate the need and application of the principles and concepts discussed in previous chapters. This equipment experiences repetitive cycles as part of its operating history. These include production mode, standby, and shutdown. This machinery and equipment also experiences production mode and standby phases during production.

Machine operation and maintenance pose considerable challenge to regulators because of overlapping activities. Previous discussion has addressed this situation under production versus maintenance modes. The actions of operation and servicing and maintenance overlap to an extent. Addressing the area of overlap poses the greatest difficulty to regulators and entities affected by regulatory requirements.

What the operator experiences and what the maintenance person experiences are totally different realities. The operator experiences the output function; that is, what the machine does. This is what any observer would experience during operation of the equipment.

The maintenance person, on the other hand experiences what the machine is. This is a completely different environment, namely, how the machine does what it does through the implementation of the design of machine elements and subsystems.

Since the focus of the two different realities is very different, so also must be the approach taken to provide protection. The OSHA documents comment that any machine part, function, or process that may cause injury requires safeguarding. This extends further to elimination or control of hazard when operation of a machine or accidental contact with it can injure the operator or others in the vicinity.

This protection normally is provided partly by the external or boundary surfaces of equipment. The boundary surface of the equipment isolates hazardous components from the exterior environment, and by so doing, protects against contact with them. The maintenance person, however, can experience contact with them as a normal part of performing work. The operator normally would contact them only during unsanctioned and inappropriate activity, the normal focus of the operator being the point of operation.

Mechanical hazards likely to be encountered by the operator at the point of operation occur when the machine performs work on material. This includes cutting, shaping, boring, or forming of stock. Parts of the machine that perform work include flywheels, pulleys, belts, connecting rods, couplings, cams, spindles, chains, cranks, and gears. These parts produce rotation, transverse movement (motion in one direction), and reciprocation (back and forth motion). Hazardous conditions associated with these motions are intrinsic to nearly all types of machinery.

The approach taken by OSHA (the U.S. Occupational Safety and Health Administration) provided guidance to other regulators (OSHA 1989, OSHA 2004, OSHA 2008). OSHA partitioned production activities and maintenance, and allowed certain non-producing activities associated with production to remain in that partition. This overcomes some of the problem that accompanies overly burdensome requirements that would considerably hamper the ability of entities to function in production mode. Management of exposure to hazardous energy during these activities is a critical part of the overall picture.

Safeguarding is the approach taken to ensure worker protection during production activity. Safeguard is the umbrella term for a number of measures that provide workers with effective protection from harmful contact with hazardous moving parts or other harmful conditions. Safeguards include barrier guards, safety devices, shields, awareness barriers, warning signs, and other appropriate means, used singly or in combination (WorkSafeBC 2006). Table G.1 provides OSHA requirements

## Table G.1
## Requirements for Machine Safeguards

Prevent contact of hands, arms, and any other part of the body with dangerous moving parts. A good safeguarding system eliminates possible positioning of parts of the body near hazardous moving parts.

Provide security by preventing easy removal or tampering. A safeguard easily made ineffective is no safeguard at all. Secure guards and safety devices require durable material that will withstand conditions of normal use and secure attachment to the machine.

Prevent contact between moving parts in the machine and external objects. Unexpected and/or unintended and/or inappropriate contact between external objects and moving parts can create high-energy projectiles.

Create no additional hazard. A safeguard that creates an additional hazard defeats its purpose. This highlights the importance of design and attention to details of construction.

Create no interference. A safeguard that impedes performance of the work or creates difficulty is ineffective and at risk of being overridden or disregarded. Well-designed safeguarding can enhance efficiency by relieving apprehensions about injury during work around the machine or equipment.

Allow lubrication without removing safeguards. Locating oil reservoirs outside the guard, with a line leading to the lubrication point, will reduce the need to enter the point of operation.

Sources: OSHA 1992a, OSHA 2007.

for machine safeguards. Any machine part, function, or process that may cause injury requires safeguarding.

OSHA acknowledged additionally that minor servicing activities can occur during normal production operations (Table G.2). OSHA allows the use of alternate methods of control in place of lockout or tagout of a machine where the activity is routine, repetitive, and integral to the production operation. At the same time, the alternative method of control must afford effective protection from the source of hazardous energy. Some of these actions reflect recommendations contained in consensus standards. Careful reading of interpretive letters published by OSHA on their website (www.osha.gov) and the compliance document is essential in order to assist in the decision about whether a particular activity lies within the envelope of production-oriented, minor service (OSHA 2008).

OSHA policy regarding use of consensus standards to establish alternate methods is to possibly deem a violation of an OSHA standard as de minimis. This is the case provided that the employer complies with a consensus standard (not incorporated by reference) rather than the OSHA Standard in effect and that the action clearly provides equal or greater protection to employees. Violations deemed de minimis require no corrective action and result in no penalty (OSHA 2008).

An effective energy control program (lockout or tagout) is necessary to supplement safeguarding during machine servicing and maintenance where safeguards are rendered ineffective or do not protect workers from hazardous energy.

OSHA standard, 29 CFR 1910.147, establishes minimum performance requirements for controlling hazardous energy and complements and augments safeguarding practices (OSHA 1989). The lockout/tagout Standard applies only when employees are exposed to hazardous energy during servicing and/or maintenance activities. An employer may avoid the requirements of the Standard when the

## Table G.2

## Activities Performed on Equipment and Machines

Minor activities performed during production

Lubricating parts

Cleaning assembled machine components and product pathways

Releasing jams and removing jammed material

Making adjustments

Major activities performed during service and maintenance

Setting up for production

Disassembly

Removal and replacement of components and parts

Modification of existing configuration

Source: OSHA 1989.

safeguarding method eliminates exposure to the danger area of the equipment or machine during servicing or maintenance work (using Machinery and Machine Guarding methods in accordance with the requirements contained in 29 CFR 1910, Subpart O).

## Production Machine and Equipment Safeguarding

Accident prevention involving production machinery has received considerable attention over the years and is the subject of many consensus standards and regulatory statutes (ANSI/AMT 2009). The OSHA manual on machine guarding provides a view of hazardous conditions experienced by the operator of machinery and mechanical systems (OSHA 2007).

The classic situation involves close worker interaction with components within the point of operation of production machinery and equipment. These units often contain rotating, translating, and reciprocating components. Many of these machines are used in production of components used in manufacturing.

Every industrial machine has at least one point of operation. The point of operation is the part of the machine where work is performed on the material being processed. Some of the more serious machine accidents occur in the area of the point of operation. Point-of-operation safeguarding prevents access during hazardous machine motion and prevents hazardous machine motion during access.

Safeguarding is a process that pertains to any machine part, function, or process that may cause injury. This extends further to elimination or control of the hazard when operation of a machine or accidental contact with it can injure the operator or others in the vicinity. Machine safeguarding using installed components and equipment is the best way to eliminate injury caused by stationary machinery:

The ANSI/AMT B11 series, some of which have been available since the early 1900s, comprise 24 safety standards for safeguarding metal working machinery (Anonymous 2009). (AMT is the Association for Manufacturing Technology.) The ANSI/MT B11 standards are usually either updated or reaffirmed every five years. This means that the majority of the 24 standards were created, updated, or reaffirmed after the year 2000. The newly created and updated standards are important because some of the technology mentioned in these documents appeared only recently. On the other hand, OSHA

Standards on machine guarding have not changed since 1975, and therefore do not reflect options currently available for machine safety.

The majority of the ANSI/AMT B11 standards are machine-specific. They offer best safety practices for one category of equipment only. ANSI/AMT B11.19-2003 (R2009) serves as an umbrella standard for all machines in the ANSI/AMT B11 series. ANSI/AMT B11.19-2003 (R2009) provides performance requirements for the design, construction, installation, operation, and maintenance of safeguarding when applied to machine tools. This safeguarding includes guards, safeguarding devices, awareness devices, safeguarding methods, and safe work procedures. This standard does not provide the requirements for the selection of safeguarding for a particular application. ANSI/AMT B11.19-2003 (R2009) is not a stand-alone safety standard and makes numerous references to other ANSI/AMT B11 standards.

The process of safeguarding begins with recognition, followed by assessment. Employers should be able to recognize, identify, manage, and control amputation hazards commonly found in the workplace caused by components of machinery, motion that occurs in or near these components, and the activities that workers perform during operation. WorkSafeBC (2006) provides extensive assistance for performing these assessments.

There are many ways to safeguard machines. The appropriate choice in a particular situation reflect consideration of a number of parameters. These include the type of operation, size, shape, and method of handling the stock, physical layout of the work area, type of material, and production requirements or limitations.

Primary safeguarding refers to control methods that protect by preventing contact with hazardous areas of equipment and machines through guarding techniques. Guard and barrier guard refer to a specific type of safeguard. Guards are physical barriers or covers designed, constructed, and installed over moving parts to prevent any contact with them. Guards are the simple solution to protecting workers when access to moving parts, such as belts and drive chains, is not required during operation. They are reliable and cost-effective, and require low maintenance when properly designed and installed.

Safeguarding measures supplement guards to protect from hazards of machines and machine operation. Safeguarding measures are classifiable according to outcome. Primary safeguarding measures prevent access to the point of operation of machines and equipment. This can reflect the operational status of the machine or equipment. Secondary safeguarding measures stop the operation of equipment and machines through circuits that control normal operation. This can occur through activation of detecting devices or through self-operation of stop devices.

Safeguarding devices include a number of alternatives to barrier guards, such as interlocked movable barrier guards, two-hand controls, and electronic presence-sensing devices such as light curtains and pressure-sensitive mats. These solutions are more complex and technical, but may provide the only solution when access to danger areas is required during normal operation such as feeding materials into a machine for processing.

The sources for much of the material on safeguarding that follows are OSHA (2007) and WorkSafeBC (2006).

## Guards

Guards provide physical barriers that prevent access to hazardous areas. Guards usually are preferable to other methods of control because they are physical barriers that enclose dangerous machine parts and prevent employee contact with them. To be effective, guards must be strong and fastened by any secure method that prevents inadvertent dislodgement or removal. Guard attachments typically employ screws, bolts and lock fasteners. Usually a tool is necessary to unfasten and remove them. Generally, guards are designed not to obstruct the operator's view or to prevent employees from doing a job.

Guard openings should be small enough to prevent access to the point of operation using the fingers. ANSI/AMT B11.19-2003 (R2009) updated the OSHA guard-opening scale. The guard-opening scale defines minimum acceptable distance from the hazard as a function of the size of openings in the guard. Minimum distance is 6 mm (0.25 in) regardless of shape of the opening. There is a different version of this measurement scale for square and slotted openings. The guard-opening scale described in

ANSI/AMT B11.19-2003 (R2009) contains smaller dimensions than used in the original scale. This accommodates to the reality of the more slender fingers of smaller women (Also refer to WorkSafeBC (2006) for further assistance in use of measuring scales for determining the acceptability of openings.)

Many manufacturers of single-purpose machines now provide point-of-operation and power transmission safeguards as standard equipment. However, not all machines have built-in safeguards provided by the manufacturer. Guards designed and installed by the manufacturer offer the advantages of conforming to the design and function of the machine. They also can strengthen the machine in some way and possibly to serve additional functional purposes.

User-built and after-market guards are sometimes necessary. There are a variety of reasons for this situation. User-built and after-market guards often are the only practical solution for safeguarding older machines and equipment. As well, they are the only choice for guarding power transmission apparatus in older plants where machinery is powered by a single motor drive. User-built and after-market guards permit options for point-of-operation safeguards designed and built to fit unique and even changing situations. Proper design and construction requires skilled personnel who are familiar with the operation of the machine including die and feeding mechanisms.

Power transmission guards need no opening for feeding stock. As a general rule, power transmission apparatus is best protected by fixed guards that enclose the danger area. Openings are needed in power transmission guards to provide access for lubrication, adjustment, repair, and inspection. These openings should have interlocked covers that require specific tools for removal during service or adjustment.

Point-of-operation guards require an opening for feeding stock. For hazards at the point of operation, where moving parts actually perform work on stock, several kinds of safeguarding may be feasible. The best choice is the most effective and practical in the circumstances.

Metal is the usual best material of construction for guards. The guard framework is usually constructed from structural shapes, pipe, bar, or rod stock. Filler material generally is expanded or perforated or solid sheet metal or wire mesh. Plastic or safety glass is used where visibility is required. Guards fabricated from wood generally are not recommended because of ignitability and lack of durability and strength. However, in areas where corrosive materials are present, wood may be the best choice of material of construction.

## Fixed Guards

A fixed guard is a permanent part of the machine and is not dependent upon moving parts to perform its intended function (Figure G.1). Materials of construction include sheet metal, screen, wire cloth, bars, plastic, or other material substantial enough to withstand impacts and other damage that may occur during prolonged use. This type of guard is usually the most preferable because of simplicity and permanence.

## Interlocked Guards

Opening or removing an interlocked guard deactivates the machine through the tripping mechanism and/or power shut-off. The machine cannot cycle or restart until reinstallation of the guard. An interlocked guard may use electrical, mechanical, hydraulic, or pneumatic power or any combination of these sources of energy. Interlocks should not prevent "inching" or "jogging" by external control, as required. Replacing the guard should not automatically restart the machine. Restart must occur only through the Start control. To be effective, all movable guards should be interlocked.

## Adjustable Guards

Adjustable guards allow flexibility in accommodating various sizes of stock. The operator can adjust the guard to accommodate stock of different sizes. The guard on a bandsaw (Figure G.2) is an example of an adjustable guard. The sheath of the guard is moveable to allow only sufficient height as is needed for cutting by the blade. Adjustment of this type of guard must occur only when the equipment is shut down.

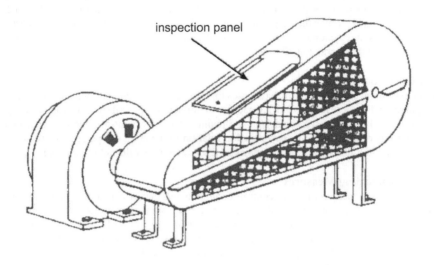

Figure G.1.  Fixed guard. A fixed guard is part of the machine. (Source: OSHA.)

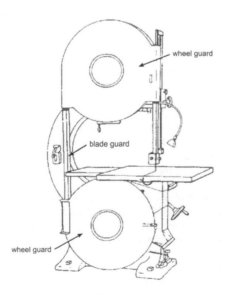

Figure G.2.  Adjustable guard. The dimensions of the stock determine the location of an adjustable guard. When the stock is not in position, an adjustable guard may offer no protection to the operator at the point of operation of the machine. (Source: OSHA.)

## Self-Adjusting Guards

Movement of stock determines the size of openings in these barriers. As the stock moves into the danger area, the guard pushes away, providing an opening that is large enough to admit only the stock. After the stock has passed through the danger area, the guard returns to the rest position. The guard on the blade of a table saw and the jointer planer illustrates the concept. This guard places a barrier between the danger area and the operator at all times.

## Point-of-Operation Guards

The complexity of machines and the different uses for individual machines complicate point-of-operation safeguarding. For these reasons, not all machine manufacturers provide point-of-operation guards on their products. In many cases, the design and manufacture of a point-of-operation guard require thorough analysis of the work to be performed in relation to the point of operation of the machine. Poorly designed, built or installed guards may create a hazard rather than eliminate one.

## **Primary Safeguarding Measures**

Primary safeguarding measures include controls or attachments that, when properly designed, applied and used, prevent inadvertent access to the hazard area of the point of operation of a machine. These types of engineering controls may be used in place of guards or as supplemental control measures when guards alone cannot adequately enclose the hazard. In order for these safeguarding devices to accomplish this requirement, they require proper design and installation at a predetermined safe distance from the danger area of the machine.

## Safe Distance Safeguarding

Safeguarding by location or distance applies the concept that a person cannot occupy two positions at the same point in time. Application of the concept requires a thorough hazard analysis of each machine and particular situation.

To safeguard by location, the dangerous moving part of a machine must not be accessible or present a hazard during normal operation. One means to achieve safeguarding by location is to locate the machine so that the hazardous area is not accessible from operator work stations or other areas where employees walk or work. Positioning the power transmission apparatus of a machine against a wall and performing all routine operations on the opposite side is one way to achieve this requirement. Additionally, enclosures, walls and fences can restrict access to sources of hazardous energy in machines and equipment. Another possible solution is to locate dangerous parts high enough above work surfaces to prevent access by normal reach. WorkSafeBC 2006 provides assistance for making this type of assessment.

Safeguarding by location is applicable to the feeding of stock into machines provided that the activity maintains a safe distance to protect the hands. The dimensions of the stock on which work is occurring may provide the means to ensure adequate safety. To illustrate, for stock that is at least 1 m in length (several feet) long on which work occurs only at one end, the operator may be able to hold the opposite end while the work occurs. Depending upon the machine, protection might still be required for other personnel.

The position of the operator control station provides another potential approach to safeguarding by location. Operator controls may be located at a safe distance from the machine if there is no reason for the operator to tend it. These devices require full cooperation from the operator. Omron STI (2010) provides assistance for assessing location of operator controls.

## Two-Hand Control and Trip Devices

Two-hand control and trip devices require constant, concurrent pressure by the hands on each button to activate the machine to complete the operational cycle. With this type of control, the operator must position the hands on the buttons. This requires the operator to move to the control station which is located at a safe distance from the point of operation of the machine or equipment during the closing cycle. To be effective, the control station must ensure that positioning of the hands into the point of operation during machine action is not possible. As well, the design must ensure that the operator cannot use one hand and another part of the body to activate the machine. These devices work in conjunction with clutches and brakes and monitoring equipment on the machine.

### Gates

A gate is a moveable barrier. Lockable gates can prevent entry into the point of operation. An interlocked, lockable gate can prevent opening during machine operation and can shut down the machine before the cycle starts. In many instances, the gate operates with each machine cycle. To be effective, the gate must be interlocked and closed before the machine can function. The gate may provide protection not only to the operator but to uninvolved bystanders as well.

### Safe Holding Safeguarding (Safe Workpiece Safeguarding)

These devices maintain or move the hands of the operator away from the point of operation during the hazardous portion of the machine cycle.

#### Workpiece Positioning

Workpiece positioning relies on the requirement for both hands to hold or support the workpiece. This also could occur by the requirement for one hand to hold the workpiece while the other hand operates the machine or the control of the machine. To illustrate, work can occur on one end of a long piece of stock while the operator supports or holds the other end while performing the work. This configuration would satisfy this requirement.

The operator's limbs remain out of the point of operation during the hazardous portion of the machine cycle. However, this approach protects only the operator.

#### Pullback Devices

Pullback devices utilize a series of cables attached to the operator's hands, wrists, and/or arms (Figure G.3). This type of device is primarily used on machines with stroking action. When the slide/ram is raised between cycles, the operator is allowed access to the point of operation. When the slide/ram begins to cycle by starting its descent, a mechanical linkage automatically assures withdrawal of the hands from the point of operation.

#### Restraint (Hold-Out) Devices

Restraint (hold-out) devices utilize cables or straps that attach to the operator's hands or wrists and to a fixed point. These devices require adjustment to set the appropriate distance for reach and determination of the zone of safety. Since no extending or retracting action occurs, the operator needs hand-feed tools for moving stock into and out of the danger area.

### Safe Opening Safeguarding

This approach limits access to the hazard area of the machine by the size of the opening or by closing-off the access to the point of operation when the workpiece is in-place. These approaches prevent access to the hazard area during machine operation. Access to the danger area is not adequately guarded when the workpiece is not in place.

#### Automated Feeding and Ejection

Machines containing automated feeding and ejection do not require placement of the hands into the danger area. In some cases, no operator involvement is necessary after the machine is set up and operating. In other situations, the operator manually feeds the stock with the assistance of a feeding mechanism. Some feed and ejection mechanisms may themselves create hazards.

Use of these feed and ejection mechanisms does not eliminate the need for guards and safeguarding devices. Guards and safeguarding devices must be used wherever they are necessary and possible in order to provide protection from exposure to hazards.

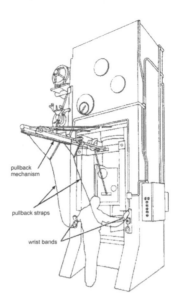

Figure G.3. Pull-back devices maintain the position of the limbs away from the point of operation of he machine. (Source: OSHA.)

## Secondary Safeguarding Measures

Secondary safeguarding measures provide some protection against machinery hazards. Detection devices, awareness devices, safeguarding methods, and safe work procedures are secondary safeguarding measures. These measures provide lesser protection than primary safeguarding measures. Secondary safeguarding measures do not prevent placement of the body into the hazardous area of a machine.

Secondary safeguarding measures are acceptable only when infeasibility prevents installation of guards or primary safeguarding devices. Where primary safeguarding measures are installed, secondary safeguarding measures may supplement them. Secondary safeguarding measures must not replace primary safeguarding measures.

### Probe Detection and Safety Edge Devices

A probe detection device (ring guard) detects the presence or absence of a hand or finger by encircling all or part of the hazard area of the machine. The ring guard provides a warning about intrusion of the hand into the hazard area and usually stops the machine or prevents the occurrence of a cycle or stroke, thereby reducing the likelihood of injury at the point of operation.

This device must be used only on machines in which motion can stop before the worker can reach the danger area. The design and placement of the device depends upon the time required to stop the mechanism and the time needed for the hand to reach across the distance from the device to the danger area at the point of operation. These types of devices are commonly used on spot welders, riveters, staplers, and stackers  because primary safeguarding methods are not possible. Probe detection devices do not prevent inadvertent access to the danger area at the point of operation.

A safety edge device (sometimes called a bump switch) is another type of safeguard that detects the presence of a person when in contact with the sensing edge of the device. A safety edge device initiates a stop command when the sensing surface detects the presence of a person. This can occur when the operator trips, loses balance, or is drawn toward the machine. The position of the bar, trip wire, or mechanical device, therefore, is critical to successful application of the concept. The device must stop the machine before any part of the body reaches the danger area. These devices, when used alone, however, do not usually prevent inadvertent access to danger areas of the machine. These situations require additional guarding or safeguarding devices in order to prevent exposure to a machinery hazard.

Safety trip controls provide a quick means for deactivating the machine in an emergency situation. All of these trip wire, rods, or other safety devices require a manual reset in order to restart the machine. Simply releasing the device to restart the machine will not ensure that the worker has moved out of the danger area when the restart occurs.

## Materials Handling Equipment

Discussion thus far has focused on discrete machines and other equipment used in processing pieces of material typically to create components of products. The operator literally works as part of the machine and has an essential interaction with it at the point of operation.

Other types of machinery create potential hazards for operators and bystanders. These machines act through an an invisible envelope in open space. The movements of machines and other equipment that act in this manner create a large external point of operation. Until and unless enclosed behind a protective enclosure that defines the invisible boundary plus additional distance to provide a safety factor that prevents access, assurance of protection of the operator and bystanders cannot occur.

Examples of fixed equipment include the components in paper mills that handle the large rolls of paper at the end of the papermaking machine and the large conveying equipment used to handle logs and sawn lumber in sawmills. These machines operate repetitively as part of the manufacturing cycle. The strategy for worker protection relies on the use of railings and structural components at the perimeter of building structures to define the limit of the boundary beyond which access is not to occur under routine conditions of operation.

Excavators and cranes that operate in similar manner are examples of semi-mobile equipment that function within an envelope in open space. This type of equipment functions almost totally when stationary on a particular spot. The actions of the equipment, while under the exclusive control of the operator, are unpredictable to individuals who work in close proximity to and often within the point of operation.

The highly coordinated and fluid motion of the heavy steel structural components can easily convince the observer that the equipment is an intelligent being. The equipment is an extension of the operator and subject to the skill of that individual in manipulating the controls and ability to focus on the simultaneous demands posed by task and conditions and the distractions posed by the external environment. The reach of the extendable parts of this equipment plus additional distance needed to provide a safety factor define the boundary of the point of operation. Where within the boundary the point of operation will be present at any moment in time is not predictable.

Forklifts, front-end loaders, and similar equipment are examples of mobile equipment that operate through an almost undefined envelope in open space. While under exclusive control of the operator, the actions of the equipment are unpredictable to a bystander. As a consequence, the envelope that surrounds the point of operation is undefined and undefinable by the bystander. Again, the equipment is an extension of the operator and subject to the skill of that individual in manipulating the controls and the ability of that individual to focus on the simultaneous demands posed by task and conditions and the distractions posed by the external environment.

The reach of the extendable parts of this equipment, plus additional distance to provide a safety factor that defines the upper limit of the boundary of the point of operation. The surroundings define forward and backward and side-to-side boundaries of the point of operation of this equipment. Where within the boundary the active element of the point of operation will be present at any moment in time is not predictable.

## Industrial Robots

The industrial robot is the logical progression from single-purpose human-operated machines and remote manipulation equipment to the next level of advancement. Robotic equipment combines the information processing of the operator with a dextrous manipulator in a self-contained mechanical device. Robots are installed in many industrial applications.

Stated more formally, industrial robots are multi-functional mechanical devices designed to move material, parts, tools, or specialized devices through variable programmed motions to perform a variety of tasks (OSHA 1999). An industrial robotic system includes the robots and the devices and/or sensors required for the robot to perform its tasks as well as sequencing or monitoring communication interfaces. OSHA (1999) forms the basis for much of the following discussion.

Robots generally perform unsafe, hazardous, highly repetitive, and unpleasant tasks. They have many different applications including material handling, assembly, arc welding, resistance welding, machine tool load and unload functions, painting, spraying, and many others.

Robotic units, singly or in groups, are unguarded machines. Robots act through an an invisible envelope (Figure G.4). That is, the actions of the robotic arm and functional peripherals create a large extended envelope within which the point of operation can exist at any time and place, until and unless enclosed behind a protective enclosure that prevents access. The protective enclosure must enclose the invisible boundary plus additional distance to provide a safety factor. Protective enclosure is sometimes needed to protect components of the robotic equipment against the rigors of the environment in which it operates.

Integration of computer technology into control of equipment and machines has made possible industrial robotics. Challenges associated with preventing accidents and injuries in the use of this equipment reflect more than anything the role of the computer as the agent of control. Robotic equipment could not exist without some level of automated control. This equipment illustrates concepts mentioned earlier about operational mode, standby, and shutdown.

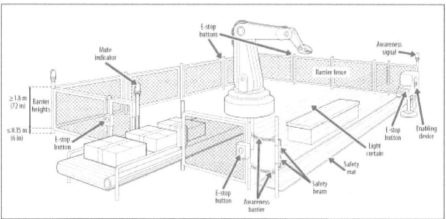

Typical robot work cell safeguarding system.

Figure G.4. The industrial robot operates through a region of space. Employees require protection to prevent entry into the zone of operation of the robot since the point of operation can occur at any point in space and time. (© WorkSafeBC. Used with permission.)

Industrial robots have four major components: the mechanical unit, power source, control system, and tooling.

The mechanical unit includes the manipulative arm as well as the fabricated structural frame, mechanical linkage and joints, guides, actuators (linear or rotary), control valves, and sensors. The physical dimensions, design, and weight-carrying ability depend on application requirements.

The usual primary source of power for these units is electrical. Secondary sources of power for actuators include pneumatic and hydraulic systems, and electrical. Selection for a particular situation reflects application requirements. Pneumatic power (low-pressure air) is used generally for applications involving small forces or weights. Hydraulic systems are used for applications involving medium to high forces or weights, or where smoother motion control than available from pneumatic systems is required. Electrically powered robots are the most prevalent. These use either AC or DC electrical power to supply motor-driven actuating mechanisms and their respective control systems. Electrically powered units offer better motion control and emergency shutdown than either pneumatic or hydraulic units.

Industrial robots contain embedded microprocessors or use an auxiliary computer for control functions. Control occurs through required computational functions as well as interface with and control associated with sensors, grippers, tooling, and other peripheral equipment. The control system performs the necessary sequencing and memory functions for on-line sensing, branching, and integration with other equipment. Programming of the controller can occur at the robot or at remote locations. Self-diagnosis for performance compared to design parameters, troubleshooting, and maintenance greatly improves reliability and predictability, and therefore, safety.

Industrial robots are either servo- or non-servo-controlled. Servo-controlled robots utilize sensors to monitor components continually for position and velocity. The control program compares these readings continually to data recorded during training. Non-servo robots lack the feedback loop, and are controlled through a system containing mechanical stops and limit switches.

Regardless of the configuration of a robot, movement along each axis will result in rotational or reciprocal movement. The number of axes of movement (degrees of freedom) and their arrangement, along with their sequence of operation and structure, will enable movement of the robot to any point within its envelope. Robots have three arm movements (up-down, in-out, side-to-side). In addition, three additional wrist movements can occur on the end of the arm. These include yaw (side to side), pitch (up and down), and rotational (clockwise and counterclockwise).

The operational characteristics of robots can differ significantly from other machines and equipment. Robots are capable of high-energy (fast or powerful) movements through a large volume of space beyond the base dimensions of the unit. The pattern and initiation of movement of the robot is predictable provided that the item experiencing the action and the environment remain constant. Any change to the object, such as a new design, or the environment can affect programmed movements.

Most robots are set up for an operation by the teach-and-repeat technique. In this mode, a trained operator (programmer) typically uses a portable control device (a teach pendant) to teach a robot its task manually. Speed of movement during programming sessions is low.

The program establishes the physical and geometrical relationship between the robot arm and the manipulators and work to be performed. To establish the coordinates precisely within the working envelope requires manual control of the robot. The teaching or programming techniques used include lead-through, walk-through, and off-line.

Lead-through programming or teaching involves use of a proprietary teach pendant. An instructor physically leads the robot through the desired sequence of events by activating the appropriate button or switch on the pendant. The instructor can use the teach pendant solely or in conjunction with other techniques for instruction. This technique may require entry into the working envelope of the robot and deactivation of operational safeguarding devices. The walk-through method of programming puts the instructor in a potentially hazardous position because the operational safeguarding devices are deactivated or inoperative.

Off-line programming involves traditional programming on a computer remote from the robot. This method does not require entry into the operating envelope of the robot. The lead-through or walk-through technique is often employed in conjunction with on-line programming to establish actual positional coordinates and to touch-up the instruction by smoothing out irregularities and jumpiness in movement.

Discussion about training and programming indicates that safeguarding is a key component to ensure safety for the operator and maintenance personnel. Creation of an effective robotic safeguarding system requires an assessment of hazardous conditions created during production, programming, and maintenance. Among the factors to be considered are the tasks a robot will perform, start-up and command or programming procedures, environmental conditions, location and installation requirements, possible human errors, scheduled and unscheduled maintenance, possible robot and system malfunction, normal mode of operation, and all personnel functions and duties.

An effective safeguarding system likely will require a combination of safeguarding methods. Redundancy and back-up systems are especially important when a robot or robot system operates in hazardous conditions or handles hazardous materials. The safeguarding devices employed should not themselves constitute or act as a hazard or curtail necessary vision or viewing by human operators.

Studies indicate that many robot accidents occur during programming, program touch-up or refinement, maintenance, repair, testing, setup, or adjustment, rather than production operations. This differs considerably from the situation discussed earlier involving fixed machinery and operator interaction at the point of operation. The latter situation involved traumatic accidents that occurred during production mode. On the other hand, operator interaction with robots occurs mainly during standby mode or the standby phase during production mode when the unit is energized but production is not occurring. During many of these situations, the operator, programmer, or maintenance worker has entered the working envelope and experienced traumatic injury from unintended or unanticipated movement.

Impact or collision accidents occur through unpredicted movement caused through program changes or malfunction of a component. Crushing accidents have occurred when the arm of the robot pushed the victim against equipment and structures. Entrapping accidents have occurred when a limb or other part of the body becomes trapped between the arm of the robot and equipment and structures.

Some maintenance and programming activities require entry of personnel into the operational envelope while power is available to actuators. The operational envelope of one robot can overlap that of another or that of other industrial machines and equipment. The potential for involvement of the overlap is an important consideration in the hazard analysis to determine possible accident situations.

Accidents involving mechanical parts result from failure of the drive components of the robot, tooling or end-effector, peripheral equipment, or the power source. Mechanical failures include release of parts, failure of the gripper mechanism, or the failure of end-effector power tools such as grinding wheels, buffing wheels, deburring tools, power screwdrivers, and nut runners.

Other accidents have resulted from failure of equipment that supplies power and control to the robot. These situations were expressed through potential electrical and pressurized fluid hazards. Electromagnetic or radiofrequency interference (transient signals) are an additional concern for signal processing.

Software and programming errors and conflicts are other areas of concern in accident causation. A robot operating under the control of two or more resident programs can experience dysfunction and unpredictable behavior in the event of inappropriate cross-loading of modules from the two programs. A software error can lead to calling the incorrect module and loading a program containing different operating parameters including velocity, acceleration, or deceleration, or position. This situation might not be noticeable to maintenance or programming personnel working with the robot. Additional hazards can result from errors in interfacing or programming peripheral equipment.

The Robotic Industries Association is the champion in North America of robotic safety. RIA recently adopted Part 1 of ISO 10218 as an ANSI standard and created and publishes ANSI/RIA R15.06-1999 (R2009).

ANSI/RIA/ISO 10218-1-2007 specifies requirements and guidelines for the inherent safe design, protective measures, and information for use of industrial robots. It also describes basic hazards associated with robots and provides requirements to eliminate, or adequately reduce, the risks associated with these hazards.

ANSI/RIA/ISO 10218-1-2007 provides the basis for adopting emerging robot technologies. One of the important innovations is the wireless teach pendant. Wireless teach pendants eliminate the need for cables and the restrictions to movement that they impose. Cables also pose trip hazards and can limit escape from the enclosure during emergency situations. ANSI/RIA/ISO 10218-1-2007 discusses safety aspects for wireless teach pendants. Another feature of ANSI/RIA/ISO 10218-1-2007 is safety-rated axis limiting.

Many of the new controllers include safety software to contain robot movement internally without any need for external safety sensors or other features. The new safety standard addresses issues that arise when robots and human workers share the same space and work together collaboratively. Lastly, the new standard provides guidelines for situations where simultaneous motion is occurring. This pertains to the movement of multiple manipulators coordinated by one controller.

ANSI/RIA/ISO 10218-1-2007 is a companion to ANSI/RIA R15.06-1999 (R2009). ANSI/RIA/ISO 10218-1-2007 pertains only to robot construction. It is not intended to replace ANSI/RIA R15.06-1999 (R2009). ANSI/RIA R15.06-1999 (R2009) provides requirements for industrial robot manufacture, remanufacture and rebuild; system integration and installation; and methods of safeguarding to enhance the safety of personnel associated with the use of robots and robot systems. This second review further limits the potential requirements for retrofit of existing systems, revises the description of control reliable circuitry, and reorganizes several clauses to enhance understanding.

## Equipment and Machine Shutdown

The manner in which equipment and machines respond to commands to shut down or loss of power is a critical consideration about the approaches available for protecting operators and others (Table G.3). Many machines, such as portable electric drills, can stop at any time after a short run-down and then be restarted without incident or consequence. The operator has complete control over the operation of these machines. This is also true in more advanced equipment such as the microwave oven. An abrupt stop is noncritical to the operation of the equipment and can occur at any time.

In other circumstances, a stop command during an operating sequence leads to a controlled stop. An LCD projector provides an example. Cool-down is an orderly process intended to prolong the life of the bulb. While immediate shut-down is possible, this risks damaging the bulb. In other examples, shutdown can involve slow-down to a gradual stop or braking and use of energy retained within storage components. These provide the operational need for continued supply of energy in order to ensure controlled stoppage. This is the case with hydraulic and pneumatic systems containing accumulators or similar devices. Pressure stored in the accumulator decreases over a finite period following loss of pressure from the power source.

In still other situations, the equipment or machine progresses through a cycle that is unalterable once under way. That is, once the cycle commences, it does not stop until dissipation of energy stored in the equipment or machine has occurred fully. This is usually the case at the normal end-point of the cycle. Punch presses and similar equipment that cycle, operate in this manner.

The manner in which equipment responds to control messages or power failure has obvious implications to the safety of the operator and maintainer.

## Interlocks

An interlock is a device that interacts with another device or mechanism to govern succeeding operations (Krieger and Montgomery 1997). This is a concise description of interlocks and their function, but what it does not say about interlocks is much more important than what it does.

Interlocks generally function through switches or valves to control the action of circuits. Interlocks connected to a safeguarding device can shut down the circuit on removal of the device. The safeguarding device is effective only so long as it occupies the intended position. Without monitoring provided through some means there is no way of knowing whether the device is providing the protection that is presumed to occur.

## Table G.3
## Modes of Machine Shutdown

| Mode | Comment |
|---|---|
| Noncritical | Shutdown can occur abruptly at any time or any time in the cycle |
| Controlled stop | Orderly shutdown required |
| | Energy retained in storage devices expended during gradual shutdown |
| Cyclical | Shutdown occurs only at the end of defined cycle that dissipates energy stored in the system |

Circuits connected to interlocks include electrical, mechanical, hydraulic, and pneumatic circuits, and circuits containing process liquids. Given the generality of the preceding definition, interlocks could include main switches and shut-offs. This, however, is not the case. In fact, defining what an interlock is, is more difficult that defining what it is not.

### Key Interlock Systems

The initial interlocks utilized keys, either trapped or key-actuated. These systems operate on the premise that a single key can occupy only one lock at a time, regardless of the number of available locks. The key interlock system is a physical application of a logic system. The key located in position A or position B, at any time, cannot occupy position A and position B simultaneously. Moving the key from position A to position B can initiate other actions. The ability of occurrence of the other actions is contingent on relocation of the key from position A to position B.

In the simplest example (Figure G.5), the key controls operation of a valve (Kanis 2001). The key locks closed one valve in a pair or larger group at any point in time. Movement of the key in the lock allows the valve to be opened and releases the key at the same time. That is, the lock and the valve are interlocked. Removal of the key from the first lock enables its use in the second lock to close the valve. Closure of the valve traps the key in the second lock. This design prevents closure of more than one valve at any time and ensures that at least one valve remains open. This particular system is totally hand-operated

In a second, more typical example, operation of a piece of equipment requires the presence of the key in position A (Kanis 2001). The availability of the key to be used in position A is contingent on release from position B. Position B is a lock that controls opening of a panel that provides access to the point of operation. Safe opening of the access panel, therefore, is contingent on stopping the equipment, this being a requirement for release of the key from position A. Position A is a control switch in an electrical circuit.

Trapped-key interlock systems are used in high-voltage electrical power systems (Figure G.6). Turning the key from the captive position to the release position deactivates and de-energizes the control circuit in the cabinet. The key is now removable and applicable to the lock in the disconnect. Turning the key in this lock from the free to the captive position releases the disconnect. Moving the disconnect to the OFF position exposes the hole for insertion of the lockout device or the shackle of a padlock to enable lockout.

In more complex situations, key-actuated interlock systems use the insertion of one or more keys in lock mechanisms to cause the occurrence of a defined sequence of actions (Figure G.7). Effectively, the key interlock system requires the following of a shutdown procedure through operation of hardware through the lock and key as the interface. This concept is similar to relay- and microprocessor-actuated shutdown but under hardware control. Operation of key interlock systems requires two critical

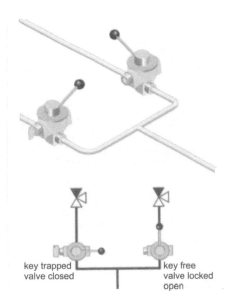

key trapped
valve closed

key free
valve locked
open

Figure G.5. Trapped key interlock. Entrapment of the key in the lock and the actions performed force the logic of the design on the user. (Courtesy of Castell Safety International Ltd., London and Castell Interlocks Inc., Chicago.)

Figure G.6. Trapped key interlock system in an electrical room.

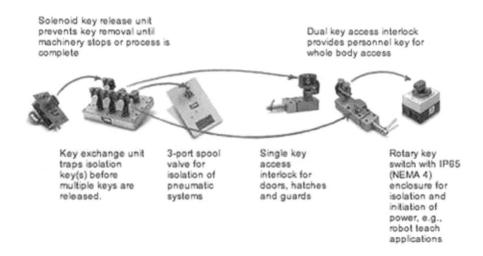

Solenoid key release unit prevents key removal until machinery stops or process is complete

Dual key access interlock provides personnel key for whole body access

Key exchange unit traps isolation key(s) before multiple keys are released.

3-port spool valve for isolation of pneumatic systems

Single key access interlock for doors, hatches and guards

Rotary key switch with IP65 (NEMA 4) enclosure for isolation and initiation of power, e.g., robot teach applications

Figure G.7. Key-actuated interlock systems employ unique keys and rotation to ensure the occurrence of a unique sequence of actions. Rotation of a unique key causes an action and releases one or more keys for additional action. Actions must occur in sequence. (Courtesy of Omron Scientific Technologies, Inc., Fremont, CA.)

actions: insertion of the key into the lock and rotation of the key from the unlocked to the trapped position.

This approach has the advantage that it is independent of the experience and knowledge of the individual who enacts it. This means that the process is person-independent.

The downside of this system is that the person following the key sequence must be able to recognize problems occurring during the procedure. This approach is not suited to situations where access is required frequently or where rapid enactment or undoing of the sequence is required. As with other systems involving keys, one individual holds the key. If this key is unique and no readily sacrificial element, such as the shackle of a padlock is present, obtaining a replacement could be a highly inconvenient activity. This would advocate for putting the unique key under lock and key in a storage box. In this manner, in the event of an unusual problem, the unique key is retrievable following an appropriate procedure, such as described previously.

Additional applications of key interlock systems include pneumatic systems and process flow control through valves, motors, and so on. Valves pose some special challenges in the application of this technology. Emergency override is required in some situations. Some products incorporate a "defeater." Valve status indication is available on some units. Wear on valve seats can lead to over-travel of the interlock device from the position established at commissioning. Some units compensate for this situation.

Insertion requires the use of the correct key. Keys in use in these systems range from standard two-dimensional house-type keys to machined three dimensional keys unique to a particular manufacturer. The problem with house-type keys is the possibility of uncontrolled duplication. This is rendered somewhat difficult, since the keys remain captive in the locks for almost the entire operation. That is, the key is required to be present in a lock at all times. The only exception to this occurs during transfer from one lock to another.

The turning action performed by the operator is the key to functioning of the system. This also differentiates the trapped-key interlock system (Figure G.5 and Figure G.6) from the key-actuated safety interlock (Figure G.7). Both systems utilize key insertion and rotation. Key rotation not only locks the key in position in the lock but also causes or enables the occurrence of some other action. This action can include release of one or more additional keys. These keys are not obtainable in any other way. These keys, in turn, perform additional actions. This hardware-based procedure ensures that the sequence of steps occurs in the manner required. Otherwise, the intended outcome will not occur.

## Electrical Interlocks

Interlocks are widely used in energy control circuits. An example of this is a switch in the main circuit, other than a control switch, disconnect, or breaker. The devices are connected in series. In this configuration, all switches must close to energize the circuit and enable functioning of the equipment. Since this switch is part of the main circuit, it operates at the same energy level.

An example of an interlock in this type of circuit in the home is the switch in the door of a top-loading washing machine. Opening the door during the spin cycle opens the circuit. The tub and agitator, which are rotating rapidly, then coast to a stop. This example highlights the fact that interlocks do not necessarily prevent access to rotating equipment prior to dissipation of all energy.

The home clothes dryer provides a similar, albeit less hazardous example. Opening the door during operation causes the heat source (gas or electric) to shut down and rotation of the drum to stop. In order to restart the machine, the user must reactivate the START switch.

The home blender is another example of equipment containing an interlock in the main circuit. (Cuisinart® is an example of such a unit.) In this case, the user must assemble parts of the plastic feed hopper in order to close switches. The unit will not operate unless these switches are closed, meaning that the intended pieces are positioned in the intended geometry on the unit.

More commonly, interlocks are part of secondary circuits that operate at low voltage or pressure compared to the main circuit. Low voltage or pressure is not the relevant issue with control circuits since these pilot the operation of circuits that operate at higher energy. In this configuration, the interlock indirectly controls energy flow in the main circuit. Opening the interlock cuts power in the secondary circuit, which, in turn, cuts power in the primary circuit. In this manner, the interlock controls the action of the primary circuit. In so doing, the function of the interlock depends on the function of the circuit that interfaces with the primary circuit.

OSHA in the U.S. requires disconnection of circuits and equipment from all sources of electrical energy prior to working on them where there is a chance of electrocution (OSHA 2011). Control circuit devices, such as push buttons, selector switches, and interlocks, are not to be used as the sole means for de-energizing circuits or equipment. Interlocks for electric equipment may not be used as a substitute for lockout and tagging procedures.

Control interlocking is the commonly recognized form of interlocking (Omron STI 2010). An interlock switch attached to the guard detects movement and opens the contacts of the switch when the guard is not fully closed. Some control interlocks cut the power to the device and permit access the moment that the guard is moved to the point of breaking the contact. Control interlocks also can lock the guard in place until rundown has occurred and all motion behind the guard has ceased. This is known as guard locking. Guard locking can incorporate timers or motion sensing.

Safety Interlock Switches
Safety interlock switches link the guard to the control circuit through four types of interaction including key, hinge, cam, and non-contact actuation (Omron STI 2010). While primarily used with electrical circuits, these actuators are adaptable to hydraulic and pneumatic circuits, as well.

The key-actuated design uses a key to operate the actuator. The key, a formed piece of metal that is mounted on the guard, fits into a specifically configured opening located in the body of the switch that is mounted on the frame of the equipment. This configuration is analogous to fitting a key into a lock except that no twisting motion occurs.

Alignment between the actuator and the body of the switch is critical in order to ensure smooth mating. This type of equipment is contraindicated in environments containing oil, grease, product dust, dirt, and liquids that can splash because of potential contamination of the opening that provides access to the interior of the body of the switch. Surreptitious duplication of the key provides a way to defeat this type of device. Since the key is external to the body of the switch, it is subject to mechanical damage. Irregular edges could cause injury to the operator.

The hinge-operated actuator uses motion of the hinge pin to actuate the circuit. The hinge pin and body of the switch form an integrated unit. Rotation of the guard rotates the actuator located inside the body of the switch. This type of unit is sealed, except for the seal on the hinge pin, and as a result, is more resistant to attack by agents in the environment.

This type of switch is insensitive to 3° of rotational movement of the hinge pin. While for many guard doors this is not significant, on long doors, this might provide sufficient gap to enable a hand or arm to penetrate into the restricted area. Alignment of the body of the switch with the door frame is critical so as not to increase the opening angle. Undue strain by the guard on the hinge pin can impair performance and perhaps cause internal damage.

Cam-operated actuators use a linear or rotary cam and a limit switch. Movement of the guard pushes against the cam wheel and depresses the plunger. For fail-safe performance, this type of switch must be installed so that depressing the plunger against the return spring breaks the contact. The cam must operate within a well-defined path.

This system is susceptible to wear. Process dust, dirt, and accumulations of oil and grease can interfere with operation of the cam.

Non-contact actuators are familiar to owners of home intrusion alarm systems. These systems use non-contact proximity devices involving a magnetic field produced by a permanent rod magnet in the contact to establish the closed and open positions. The magnet mounted on the moveable component actuates a reed switch located inside the unit mounted on the stationary component. This type of unit is susceptible to magnetic fields produced by other magnetic and electromagnetic sources, starting with a rod or bar magnet having similar dimensions as the magnet mounted on the guard. Sources of electromagnetic fields include portable tools and installed equipment.

## Light Curtains (Optical Guards)

Light curtains are the modern version of the light beam and photoelectric sensor (Figure G.8). Light curtains use pulsed, sequential emissions of infra-red energy produced by a line of light-emitting diodes to produce a "curtain" of discrete beams (Omron STI 2010). Phototransistors sense the signal, and the electronics determine its validity. This is, in part, aided by the sequencing and pulsing of the components of the beam. These actions form part of the logic for rejecting stray signals. Products use self-checking circuitry to detect internal faults (control reliability). This ensures that failure of a single component within the device, interface, or system will not prevent normal action from occurring. Detection of an internal fault initiates a STOP signal to the equipment.

Light curtains are used to protect operators involved in positioning materials (point of operation safeguarding) and to establish restricted zones around equipment (perimeter safeguarding). Operation of these units offers considerable flexibility. The presence of an obstacle in the path of the beam causes a trip. The system can be set to reset once the obstacle is removed or to require a visual inspection and reset from outside the protected area. When used in perimeter safeguarding, the guarded machine must stop immediately on receipt of a STOP signal from the device. Restart must occur only following visual inspection by the operator from a point external to the protected area.

This type of product is suitable only for machinery that can consistently and immediately stop anywhere in the operating cycle or stroke, and has adequate control devices or mechanisms. This would exclude full-revolution clutched presses and similar machines. Light curtains do not provide protection against flying objects, contrary to implications of the word, curtain. Airborne particulates including smoke, dust, fumes, and mists can degrade performance by the transmitting LEDs and receiving sensors in the device as well as obstructing the path of the beam. Moisture from condensation can coat surfaces in some applications. Mists and sprays can coat surfaces and obstruct the beam.

Reflection by surfaces off-axis to the direction of the beam can cause detection of a secondary reflected beam by the receiving device thus bypassing the obstruction. Reflective surfaces include shiny metal, foil, plastic, and other similar materials. Off-axis reflection can occur to the side as well as

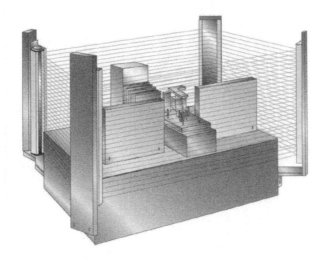

Figure G.8.  Light curtain. Obstructing the path of the beam interrupts the circuit. (Courtesy of Omron Scientific Technologies, Inc., Fremont, CA.)

above and below the axis of the beam. This can lead to false interpretation about the presence of the obstruction.

Multiple sources can interfere with each other. This is an especial concern when pulsing sequences and frequencies are the same. Interference problems can also result when the sending and receiving units are not mounted at the same height and parallel to each other. Mounting should occur so that the receiver of one unit does not "see" the transmitter of another.

Presence-Sensing Mats

Presence-sensing mats (Figure G.9) use the contact or change in capacitance between conductive plates caused by the imposition of force by a prescribed mass to issue a STOP signal to the affected machine (Omron STI 2010). They are not intended to detect removal of force. Presence-sensing mats are used where perimeter access safeguarding is required. When used at the point of access, presence-sensing mats can protect an entire restricted area. A trip caused by the presence of sufficient force on the mat demands a visual inspection and reset from outside the protected area.

The controller unit can detect the presence of broken, missing, or misconnected wires, and in some cases, a corroded or rusted electrode in the mat. Corroded or rusted electrodes change the electrical characteristics of the capacitance or contact relationship used to trigger the STOP signal.

In order to remain effective, pressure-sensing mats require anchoring to the floor so that they cannot conveniently be moved to another position. The mat must cover enough floor area to discourage "jumpers" who would use this technique as a means to circumvent the system of protection.

This type of product is suitable only with machinery that can consistently and immediately stop anywhere in the operating cycle or stroke and has adequate control devices or mechanisms. This would exclude full-revolution clutched presses and similar machines.

Figure G.9. Pressure-sensing mat. Application of pressure interrupts the circuit. (Courtesy of Omron Scientific Technologies, Inc., Fremont, CA.)

Emergency Stop Systems

Emergency stop systems (Figure G.10) provide fast-stop capabilities to prevent serious injury or death by stopping a machine as fast as possible (WorkSafeBC 2006). The emergency stop system occupies the top of the stopping hierarchy. Emergency stopping devices include push buttons, rope-pull switches, and in some cases, presence sensors.

These act through the redirection or removal of energy. Other stopping systems are considered to be variations of machine stops. The emergency stop system halts all movement initiating components of a machine. These include motors, drives, solenoids, and cylinders. The emergency stop system is hardwired in series with all of the emergency stop devices. Following an emergency stop, the equipment must not restart until the operator issues a new START command.

Two-Handed Controls

Two-handed controls (Figure G.11) are an application of interlocks that require a known position of the hands at the start of the machine operation or cycle. That is, the hands must be positioned simultaneously on the switches in order for the machine to actuate. Recent developments utilize capacitive sensing technology (Omron STI 2010). Not only does this technology eliminate the need to apply pressure to depress the switches, which, over time can lead to repetitive trauma injury, it also increases reliability by making defeat more difficult. The defeating mechanism must now match the capacitance of the hand.

False triggering can occur by radiofrequency sources. As result, the device incorporates radiofrequency protection. Control systems must incorporate an anti-repeat feature and a requirement for release of all operator hand controls before resumption of an interrupted cycle can occur.

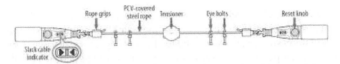

Pull wire system using two emergency stop switches. The switch is activated by a pull from any direction.

Figure G.10.  Emergency stop. (© WorkSafeBC. Used by permission.)

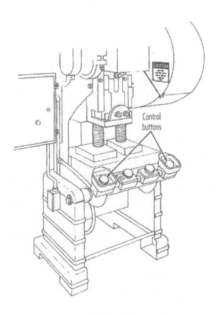

Figure G.11.  Two-handed control. (Source: OSHA.)

**Interlock Operation**

Interlock operations are classifiable by frequency of operation (passive, active, and cycling) (Hall 1992). Frequency of operation has a major bearing on reliability of the interlock and the safety value attached to its use.

Passive (quiescent) interlocks remain in the same position for long periods of time. Long periods can extend into months and years. Equipment containing interlocked guards that operates continuously for long periods between shutdowns is an example where passive interlocks are likely to be present. Testing may require partial or full disassembly of the component in which the interlock is located.

Conditions to which interlocks are subjected make them prone to failure. These include the normal conditions found in the operating environment, including temperature extremes, humidity extremes, vibration, radiation sources, and exposure to chemical agents (gases and vapors, dusts, mists, and fumes). Added to this is possible long-term compression of springs and lack of lubrication of seals through lack of movement.

The ability of the interlock to function is contingent on ability to send the appropriate signal at the appropriate moment. This function depends on the ability to move or to effect some other measure of change in the status quo. Lack of perception of use of the interlock could lead to cannibalization and bypassing (defeating), or even forgetting about its existence in a piece of equipment. As a result, passive interlocks are considered redundant and highly subject to failure.

Active interlocks are actuated frequently. Door interlocks on equipment cabinets and access portals are examples. The door is opened routinely, although not in every cycle. Failure of the interlock is more recognizable.

Cycling interlocks change position at least once per machine cycle. This means that the interlock experiences frequent use. Any deviation from normal performance is readily detectable by the operator. Examples of this type of interlock include the two-handed pushbuttons that the operator must activate simultaneously when using some types of equipment. These indicate the inferred position of the hands. A defect in this type of interlock stops operation of the equipment.

The active interlock and cycling interlock are most likely to be relied on as primary protection. (See later discussion.)

**Dependency**

Barnett and coauthors published a number of articles over the years that are germane to the discussion about interlocks and protection that is inferable from their use.

The first of these articles considered the function of safety devices (Table G.4) irrespective of type (Barnett and Barroso 1981a). Note in Table G.4 the concern of Barnett and Barroso about the use of interlocked guards. This concern reflected their view that interlocks can sometimes increase the danger intrinsic to the system containing the interlock as a protective device. (Danger in this table expresses the measure of likelihood and severity of injury.)

Barnett and Barroso (1981b) continued this exploration in a second article that focused on the functional hierarchy of safety devices (Table G.5). To illustrate the concepts in Table G.5, consider the home washing machine. The manner in which the user interacts with the machine is the zero order safety system. The boundary surfaces that prevent contact with the rotating drum and agitator form the first order safety system. The interlock in the door that shuts down the drive motor that rotates the clothes drum is a second order safety system. The user manual containing information for safe interaction with the machine is the third order safety system. The sign that warns against raising the lid during the wash and spin-dry cycle is the fourth order safety system.

Barnett and Barroso (1981b) note that a first order device is more effective than a second order device, which in turn, is more effective than a third order device, and so on. Further, these authors comment that almost all machines have first order safeguarding, either designed in or retrofitted. This is the result of product liability decisions in the 1960s and implementation of the Occupational Safety and Health Act and rules in the U.S. in the 1970s.

Injury rates have increased despite these actions. Most accidents involving machines occur on machines that have safeguards. Historically, the victim caused about 85% of machine accidents. The

## Table G.4
## Classification of Safeguarding Devices

| Type | Characteristics | Examples |
|------|-----------------|----------|
| I | Always improve safety | Transmission guards |
| II | Improve safety on some occasions and do not affect safety at other times | Awareness barrier |
| III | Do not affect safety | Redundancy in a fail-safe system |
| IV | Sometimes improve safety and sometimes increase danger | Interlocked guard |
| V | Sometimes improve safety, sometimes increase danger, and sometimes do not affect the system | Seatbelts |
| VI | Sometimes increase danger and sometimes do not affect the system | Emergency STOP button in restricted area of equipment geometry that "invites" entry by the operator |
| VII | Always increase danger of the system | "Man cage" used with a mobile crane |

Source: Barnett and Barroso 1981a.

machines caused about 5%. The remainder were not directly assignable. This means that the outlay of money on safeguards (first order safety devices) has not produced the desired results, namely reduction in accident frequency and severity.

Barnett and Barroso (1981b) also commented on the role of the human-machine interaction in this situation (Table G.6). Worker actions reflect uniquely human characteristics (inconvenience) in the assumption of risk. Note the additional category that attempts to optimize productivity while simultaneously not compromising safety.

**Misuse of Interlocks**
Explaining the failure of first order safety devices to reduce accident frequency and severity is the subject of other articles by Barnett and co-authors (Barnett et al. 1983a, Barnett et al. 1983b). These authors contend that incorporation of safety devices into a machine changes the human-machine interaction. These changes can produce unexpected effects that profoundly affect the safety and function of the machine. Barnett et al. (1983a) coined the term "dependency hypothesis" to explain these effects. According to this hypothesis, every safety system produces a statistically significant pattern of user dependence. User dependence leads to misuse as control systems, misuse in kind, and misuse in magnitude.

Misuse as control systems refers to use of the safeguard to control the machine. This results from recognition that operation of the safeguard overrides normal machine operation more conveniently than the main control. Some safeguards freeze motion, shut down rotating equipment, prevent start-up, return a machine to the stand-by position, and temporarily remove the hazard. Shutting down the clothes dryer by opening the access door is one example of the use of an interlock to control a machine. The door interlock on a clothes dryer is not intended to be used as a shutdown device, any more than the

## Table G.5

### Functional Hierarchy of Safeguarding Devices

| Order | Characteristics of Device or Concept |
|-------|--------------------------------------|
| Zero | Independent of safeguarding devices and depend solely on human/machine interaction |
| First | Eliminate or minimize a hazard (enhance zero order systems) |
| Second | Enhance effectiveness of first order devices |
| Third | Enhance effectiveness of second order devices |
| Fourth | Enhance effectiveness of third order devices |
| Higher | Enhance effectiveness of fourth order devices |

Source: Barnett and Barroso 1981b.

door interlock on a clothes washer or a microwave oven or dishwasher. The intended shutdown is the control containing the timer and ON–OFF switch.

As with other actions taken by humans, there is a sound reason why people act in this manner. The control dial or button is less convenient to use, especially on older machines containing mechanical timing mechanisms, or the user has no knowledge about the difference between the intended function of the control devices. In order to operate a control dial, the user must lift or push in and rotate the dial in one direction only, and then to depress or pull it out to make the selection. A simple ON–OFF switch is considerably more user friendly, and therefore more likely to be used. A second consideration with the interlock switch on the door of the clothes dryer is the perception of danger posed by the rotating drum and tumbling clothes. The perception of risk is low, even if the device malfunctioned or failed and rotation of the drum did not stop. The user would be cascaded by hot clothes.

Similar devices are used in industrial equipment, examples being barrier gates, interlocked access doors, and light curtains. Operator practices similar to what occurs with home appliances with this type of protective device could have highly deleterious consequences. One such example is opening an interlocked door or reaching into the restricted area of a machine protected by a light curtain in order to shut it down. Failure of the circuitry to deactivate the machine could lead to serious injury before the operator realized the interlock had failed. Similarly, many machines function like the washing machine during the spin cycle. Opening the barrier gate or interlocked access door cuts energy flow to the machine, but run-down is not instantaneous and contact with rotating equipment can occur for some time thereafter.

Misuse of safety devices includes use of limit switches on industrial equipment as control switches. Using the limit switches on cranes to prevent "two-blocking" (contact between the hoist block and the stationary parts of the hoisting system), and the limit switches on elevator doors illustrate use of safety devices in a manner that completely contravenes their intended function. (Barnett et al. 1983a).

Misuse in kind refers to uses completely beyond the intended or expected function of the protective device (Barnett et al. 1983a). Setting a circular saw containing a rotating blade down on the blade guard while rundown occurs is an action not anticipated by the designer of the guard. The

## Table G.6
## Human–Machine Interaction

| Work Method | Characteristics | Consequence/Outcome |
|---|---|---|
| Maximum output | Piecework or wage + bonus; compensation depends on output | Emphasis on production over safety |
| Least effort | The "easy way" | Ignore safety-related requirements as inconvenient |
| Standard | Specified by method creation specialists | Actions may contravene safety requirements |
| Optimum | More onerous than standard method | Actions maximize safety and minimize impact on production |
| Maximum safety | Minimize danger | Unrealistic perception of risk may hamper production |

Source: After Barnett and Barroso 1981b.

expected function of the guard is to prevent unintended bodily contact with the rotating saw blade. Yet, virtually all users follow this practice.

Operators of presses use the pull-backs as primary safeguards to pull back the hands at the appropriate moment. The expectation in the design is that the operator will pull back the hands without intervention by the device. Reliance on the pull-back device to pull back the hands defeats its purpose as a safety device. Failure of function of the device could lead to injury.

Another example is the safety latch on crane hooks. The intent of the latch is to prevent slipping of the load from the hook. The unintended use of safety latches is to increase the load beyond what the hook can accommodate without the safety device.

Misuse in magnitude refers to overuse of equipment beyond designed-in safety factors (Barnett et al. 1983a, Barnett et al. 1983b). Normal use is the anticipated stress to which the device will be subjected. This includes lifting capacity and rotational speed among other considerations. The rationale for overuse, especially where the safety factor is known, is that the device can tolerate the overload. This certainly is true to a point, but then that rationale defeats the purpose behind the over-design, namely, to minimize risk of use to an acceptable level. Exceeding design capacity of equipment increases the risk that failure will occur in service for reasons other than those anticipated.

### Redundancy

One response to concerns raised about interlocks is to use more of them in the circuit, either redundantly or in response to some other problem (Hall 1992). The concept behind redundancy is that one interlock acts as the back-up to the other. This approach relies on the assumption that the causes of failure are independent from each other, and that the probability of failure is very small. Redundant interlocks installed at the same time could fail at about the same time, given exposure to the same operating environment. Protection offered by redundancy is effective only if failure of the first interlock in the redundant group provides warning. One means to address this concern is the safety-related circuit containing a monitoring function as discussed in Appendix B and Appendix C.

**Actuation Modes**

The manner in which the limit-switch interlock fails is critical to the provision of protection (Hall 1992). Compression-based limit switches contain a spring that pushes against the load and restores the contacts to a resting position. Two designs are available. In the one, the contacts are open in the resting position, and in the other they are closed (Figure G.12).

The switch has two operational positions: compressed and decompressed. The switch also has two transitional phases: compressed to decompressed = decompression, and decompressed to compressed = compression. In order to provide protection, the switch must provide information during as many phases of its operating cycle as possible. With the advent of microprocessor-controlled circuits, information provided by an open circuit is just as valuable as that provided by a closed one.

Protection occurs when the interlock fails toward safety, or fail-safe. To illustrate, consider an interlock containing a pressure switch whose contacts close on compression, and a door that must remain closed to provide safety. The switch can be mounted in two ways. In the first configuration, closing the door compresses the switch. In the second, closing the door decompresses the switch. The question is which mounting strategy provides safety.

The transition from the decompressed to the compressed position is considered more reliable than the transition from the compressed to decompressed position. Hence, failure of a switch to close under compression and to complete the circuit when compressed fails safely. The operator can see that the machine does not operate as expected when the door is closed, as well as open. In the opposite configuration, the return spring fails to break the contact and return the switch to the OFF (decompressed) position when the door opens. In this configuration, the equipment continues to operate, despite the fact that the interlock is supposed to have opened and broken the contact.

Another example is the compression switch whose contacts break when load is applied against the spring and whose contacts close when the spring decompresses. Again, the switch can be mounted

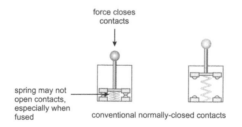

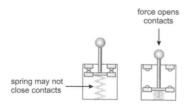

Figure G.12. Fail-to-safe switch contacts. The contacts in compression-based limit switches are either open or closed in the resting position. Failure toward safety occurs when the contacts open against the load. (Modified from Schmersal 2006.)

in two ways. The switch performs more reliably against the load. Hence, failure toward safety occurs when the contacts open against the load; that is, closure of the door.

**Defeat and Failure of Initiators**
Defeat modes refer to activities and conditions that cause interlocks to fail (Hall 1992). These occur due to electrical, mechanical, and environmental conditions to which the actuator is subjected:

## Electrical Factors
Most interlocks contain electrical switches, and as such, are subject to switch-related failures (Hall 1992). For example, jumping a circuit, as is done by electricians when diagnosing a problem, can by-pass a whole series of interlocks, thus completely defeating their purpose.

The second electrical problem is welding of switch contacts by transients. Transients in a circuit result from grounding or a direct short. This can occur in the switch, in wiring to the switch, and in the enclosure of the control cabinet. One possible cause is the formation of a current path following accumulation of conductive dust. Despite overload and blowing of the control fuse, the transient can weld closed switch contacts in a circuit containing many redundant devices. There can be no ready indication of damage. Without indication of damage, the circuit provides no protection since users are not readily aware of the deficiency.

One approach to counter the problem of welding of contacts is the use of 'force-guided contacts' (Omron STI 2010). The two sets of contacts are mechanically connected and move together in this type of relay. The contacts make contact simultaneously. This approach also enables monitoring of the contacts.

The third electrical problem is capacitance caused by long runs of wire and light burden. This can lead to marginal and erratic performance. Inductance caused by the interlocking circuit is also a concern. Magnetic fields produced by other electrical devices, including portable tools, proximity devices and other equipment, can interfere with the action of solenoids.

## Mechanical Factors
Mechanical failure of the interlock can occur due to a broken, stuck, or bent actuator or internal mechanism. Deliberate restraint of the actuator switch or contactor by tying or wedging closed prevents movement. Vibration can affect movement of the actuating arm. Mechanical crushing damage also can occur in the circuits and wiring controlling the interlock.

## Environmental Factors
The initiator is subject to the rigors of the environment to which it is exposed. These begin with accumulations of product dust, dirt, grease, and oil. These can affect electrical function, as mentioned, and can also affect performance of seals. Chemical contaminants include gases and vapors, and particulate dust, mist, and fumes. These substances can be conductive, reactive, or corrosive and excellent solvents to the plastics used in these devices. Chemical attack can embrittle and cause failure of plastics in cases and other components.

Environmental rigors also include extremes of temperature. Low temperatures embrittle plastics leading to susceptibility to cracking. Extremely high temperatures can embrittle plastics through bake-out of plasticizers.

Some environments include exposure to ionizing and nonionizing radiation. Intense sources of penetrating ionizing radiation can, in some cases, activate atoms in the components of the product, making them radioactive. More likely, however is embrittlement of plastic materials brought about by exposure to ultraviolet and intense sources of ionizing radiation.

**Limitations of Interlocks**
The situations described here illustrate consequences that arise from the predilection to design and build the ultimate machine so that nothing can harm the operator regardless of what that person does (Hall 1992). Said another way, these situations highlight the consequences arising from the presump-

tion that strictly technical means can minimize (eliminate) all hazards. The interlock is held to be the device that can achieve this goal.

Hall (1992) considers the perfect interlock to be an unattainable goal for a perhaps unexpected reason, namely impracticality of use. This author highlights the situation where a manufacturer of an interlock for an electrical component had to devise a defeat because the interlock proved to be too restrictive in its use. As well, Hall comments that defeating interlocks is almost a routine part of troubleshooting.

As illustrated in discussion about dependency, interlocks assume in the minds of workers affected by them a role not intended by the designer. The convenience of the protection that they offer is seductive and subtle. Interlock devices easily can assume the role of primary protection. Opening a door or standing on a mat or pushing the stop switch to cut power to a machine is so much easier and faster (more convenient) than shutting it down by opening the disconnect or the breaker in a distant electrical room.

This book has discussed control circuits that operate through interlocks in several places, including this Appendix. The future for interlocks in control of hazardous energy appears to involve two main issues: reliability and redundancy. Reliability relates to the confidence that the device will perform in a predictable manner under all foreseeable conditions. Redundancy and defense in depth rely on the concept of increasing the probability of reliability through the presence of additional devices in series and in parallel. The future cf discussion about isolation must also acknowledge the inconvenience factor and the risk of sudden release of energy during disconnection of circuits still under load that continues to occur as part of isolation.

Probably the circuits containing interlocks of greatest interest are the so-called local disconnects and control circuits. A local disconnect is a remote, low-voltage lockout system. The devices are located conveniently next to control switches beside equipment (Figure G.13) and offer the convenience of local lockout rather than the need to lock out in a distant electrical room. That is, the switches in these devices contain a facility for applying a lock.

In reality, these devices are a type of control circuit. These circuits are dedicated safety systems containing redundant circuits. They use monitoring by safety interface modules to provide control reliable operation. Reactivation cannot occur until release from isolation by ending the lockout.

## Immobilization

Shutdown of mechanical systems creates several possible situations regarding energy. These include rundown of moving components including motors and the components driven by them, cycling to a resting position normally part of the operating cycle, and movement to rest positions not part of the operating cycle. Resting positions are not necessarily located at the low point of the energy profile because this would block accessibility to parts of machines. Any of these actions can retain energy through deliberate or unintended storage.

There are circumstances where removal of energy in a system to the zero state is not possible, is not needed, or will prevent the occurrence of the intended work. These situations arise mainly in mechanical systems where machine parts moved to their position of lowest energy will block access to the area in which work is to occur. Work on these components can occur in safety when the path of movement is blocked and/or the component secured to prevent movement.

The catapult and the guillotine are examples of equipment that store energy for rapid conversion from one form to another. This storage occurs at the top of the energy profile. For the purposes of this discussion, let us say that some maintenance might entail work on the innards of the control mechanism. In order to do this, the worker must insert the head under the path of the moving parts. Of course, the blade of the guillotine, for example, must be positioned in the up position in order to provide access to the section requiring service.

Guillotines and catapults, like most equipment, were designed and built for operation, not for maintenance. Work on the mechanism can occur in safety, only so long as the blade cannot fall and that the hurling arm is not tensioned. However, unforeseen actions, such as burn-through caused by a lighted torch, or the misdirected swing of a sword or axe, could change the circumstances immediately

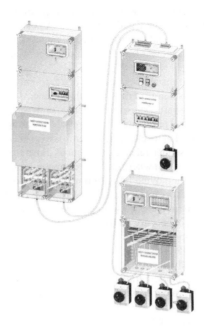

G.13.  Local disconnects offer the convenience of lockout at the equipment or machine rather than in a distant electrical room. The local disconnect is a type of control circuit. (Courtesy of Eaton Corporation - "Safety Lockout System SLS," publication FK8710-1044 GB.)

if these measures were not taken. Again, the potential for these types of accidents is obvious because the system is fully exposed and its actions observable.

The wind-up alarm clock is a more recent example of a device that stores energy by a similar mechanism. The spring mechanism that powers the alarm is essentially similar to the energy storage mechanism in the catapult. The alarm can be armed and triggered at will. Increasing tension in the spring and resistance at the end of rotational travel provides feedback to the operator. A major difference between the wind-up clock and the catapult is that the action produced by rotating the winding key is not visible because the case shields the internal mechanism from view. This situation is consistent with the use of protective enclosures (boundary surfaces) in today's industrial equipment to protect operators and bystanders.

Mechanisms in wind-up clocks fail. The hands fail to turn.  The alarm mechanism fails to function. The cause of these failures is not observable because of the shielding provided by the case. The only way to determine the cause of the problem is to enter the shell of the clock or to dismantle it. Of course, no one can enter the shell of a normal wind-up alarm clock. However, the mechanism in the interior of large clocks, such as "Big Ben" in London, is accessible.

Accessing the interior of a wind-up alarm clock is simply a matter of removing several machine screws from the shell. This exposes a framework containing many intermeshing gears and two flat metal springs. Imagine the shock produced by the rapid unwinding of the mainspring as it flies into a curious young face after loosening one screw too many in the framework.

Stored energy can release rapidly and create a lasting impression. Clocks, as other machines are intended to perform as an operating unit, not to be disassembled by curious young minds. (Instructions accompanying wind-up clocks provided no guidance to the budding young mechanic about how to put them back together, once partly dismantled.)

The sole function of enclosures (actually large guard structures) around many types of equipment is to prevent entry and detrimental contact. That is, by the presence of the enclosure, the manufacturer or installer is stating implicitly that entry for cleaning, servicing, modification, or other purpose cannot occur safely while the equipment is energized and operational. An electrical isolation and lockout can achieve the goal of preventing start-up and operation of equipment, such as a motor, that transfers energy to mechanical subsystems. The same is true for a fluid power system that operates within a machine.

However, these lockouts do not ensure the existence of a zero-energy state within the equipment. Deactivation, de-energization, and isolation of mechanical subsystems within the machine can require the immobilization of freely moving parts following run-down or those that store gravitational energy, or removal of tension in springs. These requirements have evolved through the unfortunate experience of many individuals. Energy retained in mechanical systems is a serious consideration during service and maintenance, especially where sudden release can occur as illustrated by the example of the alarm clock. This release is an especial concern where serious traumatic injury or death can occur.

Momentum is the tendency of a moving body to remain in motion. In mechanical subsystems, momentum is most likely to be present in oscillating and rotating components. Momentum should cease soon after deactivation of external energy sources unless an internal source of stored energy is present. This is the case with components driven by motors.

Components moved during operation of a machine or system are potentially capable of free movement following run-down unless the machinery contains a built-in mechanism for blocking movement. This concern is especially the case with subsystems that operate through the action of levers, and equipment that produces rotation, translation, and reciprocation.

Lever subsystems can amplify minor movement into major motion in large parts of the machine. This is especially the case with weigh scales. Large mass applied to the scale produces small deflection of the rotating dial. On the other hand, small force applied to the rotating dial potentially can produce change in the position of the bed of the scale through movement of the system of levers.

Free rotation potentially can occur in inclined conveyors containing loads at the time of shutdown. Absence of blockage or insufficient strength in the blockage could lead to downward movement of the load on the conveyor. Freely rotating parts can strike or fail to support body weight through misplaced footing or other action that presumes immobility.

Gravity is the most widespread mechanism of energy storage in machines. Parts in machines that move as part of normal operation, such as levers, knives, hammers, and many others, can stop in a position from which downward motion can occur under appropriate conditions. As well, loss of anchorage during dismantlement can lead to pendulum-type motion of many types of functional and structural parts of machines. This can occur suddenly or gradually depending on the circumstance. Sudden conversion of gravitational to kinetic energy leading to unexpected motion and the potential for injury has occurred on numerous occasions during adjustment while workers were within the confines of equipment (NIOSH 1983).

Springs are another type of energy storage device as mentioned previously. Springs are elastic materials. They exert forces or torques and absorb and store energy which is released later. Release of energy stored in a spring can occur rapidly or slowly. Rapid release of energy in conjunction with the compounding action of lever systems can produce unexpected movement in machine parts. Addressing the release of gravitational energy stored in moveable parts and compressional, tensional, or torsional energy stored in springs requires in-depth knowledge of the operation of the equipment. Maintaining the released member in a neutral position could require securement by means of blocks and clamps (UAW-GM 1985).

## Assessing Residual Mechanical Energy
Little exists in print to enable assessment of sources of mechanical energy that exist in equipment, machines, and systems during preparation for lockout and tagout. Typically what are observable are ad hoc configurations created by site personnel in response to the perceived need. Enquiry about engineering input into the design to ensure sufficiency of strength of the structures typically produces negative responses. Table 7.6 provides a generalized way of considering this question for different sources of energy. Table G.7 contains a specialized version of Table 7.6 that focuses on sources of mechanical energy possibly requiring immobilization.

## Table G.7

## Assessing Retained Energy in Mechanical Systems

| ABC Company | Hazardous Energy Assessment | Work Activity |
|---|---|---|
| Location: | Equipment, Machine, System: | Assessed by: <br> Tel: <br> Date: |
| Overview: | | |
| Hierarchy of Energy Input | Conversion Energy Output | Equipment, Machine, System Affected |
| | | |

### Circuit Name

| Function/Description: | | | | | |
|---|---|---|---|---|---|
| Immobilization Strategy: | | | | | |
| Fixed Attachment | Bridging Device | Moveable Attachment | Strength Required | Design Strength | Applicable Standard(s) |
| | | | | | |

| Failure/Consequence Analysis: | | |
|---|---|---|
| Input: | Storage: | Dissipation: |
| Output: | Conversion: | Immobilization: |
| Primary Immobilization: | Secondary Immobilization: | Tertiary Immobilization: |
| Location: | Location: | Location: |
| Action: | Action: | Action: |

## Table G.7 (Continued)
## Assessing Retained Energy in Mechanical Systems

| Immobilization Complete? | Immobilization Complete? | Immobilization Complete? |
|---|---|---|
| (Circuit Name) Immobilization Procedure: | | |
| Verification Test Procedure: | | |
| Notes: | | |
| Overall Isolation Procedure: | | |
| Overall Verification Test Procedure: | | |

**Hazard Assessment (Operational/Undisturbed Space) or (Work Activity)**

| Hazardous Condition | Occurrence/Consequence | | |
|---|---|---|---|
| | Low | Moderate | High |
| | | | | | |

(circuit name)

In this table, **NA** means that the category does not apply in any normally foreseeable situation. **Low** means that exposure is readily identifiable but believed to be much less than applicable limits or that exposure to nonquantifiable hazardous conditions is unlikely to produce injury. **Low-Moderate** means that exceedence of regulatory limits is believed possible or that nonquantifiable exposure could produce minor injury requiring self-administered treatment. Control measures or protective equipment should be considered. **Moderate** means that exposure is believed capable of exceeding regulatory limits or causing traumatic injury requiring first aid treatment or attention by a physician. Protective equipment or other control measures are necessary. **Moderate-High** means that exposure is believed capable of considerable exceedence of regulatory limits or causing serious traumatic injury. Advanced control measures or protective equipment are required. **High** means that short-term exposure is believed capable of causing irreversible injury including death. Advanced control measures or protective equipment are required.

Complicating the assessment of mechanical systems is the absence of pre-engineered and installed immobilization devices that enable compliance with 29CFR 1910.147 (OSHA 1989). The following sections describe the additional terms used in Table G.7. The information collected and considered during this hazard assessment forms the basis for a dialog with the regulator with regard to possible strategies for isolating, immobilizing, and locking out an actual system. Contact with the regulator is strongly advised in order to solicit opinion about whether the proposed approach would be acceptable.

## Fixed Attachment

Fixed attachment refers to the characteristics and particulars of the structure to which the bridging device is to be attached. Often the structural members are substantive enough that strength is not a consideration. This, however, is not always the case. This situation is similar in concept to assessment of existing structures as anchorages for fall protection.

The fixed attachment must have sufficient strength for supporting the bridging device alone and the bridging device attached to the moveable attachment. Assessing these questions requires engineering analysis of masses and forces.

## Bridging Device

The bridging device is the structure that attaches to both the fixed attachment and the moveable attachment. The bridging device and the attaching hardware require sufficient strength to ensure that the attachment will not fail in service.

## Moveable Attachment

The moveable attachment refers to the moveable part of the equipment, machine, or structure attached by the bridging device to the fixed attachment for immobilization.

## Strength Required

The strength required for the structure used to immobilize the moveable part of the equipment, machine, or structure must consider loads that the bridging structure and its attachments may experience.

To illustrate, manuals and warning labels for trucks and mobile equipment routinely warn users, usually in small print against using propping devices used for supporting the dumping body under load, meaning when contents are present. The wisdom intrinsic in such an approach for design of the prop is questionable because operators and mechanics do not always consider the warning at the time of work. The warning is easily forgotten in the urgency of the moment.

Over-design is likely to be a particularly important consideration in such situations. For a truck with a load capacity and lifting capability of the unloading hydraulics of 5 Tonnes (about 5.5 Tons), a prop with a working strength of 2 Tonnes (about 2.2 Tons) that relies on a warning would seem to be an inadequate approach to the realities of work. Inevitably, someone will hoist a load and insert the prop. Ultimately, this places the safety of this situation into the region of strength provided by the safety factor designed into the prop.

## Design Strength

Design strength is the strength calculated for the bridge structure and its attachments to the fixed and moveable structures by the designer. The design strength should equal or exceed the strength required and incorporate a safety factor. Hence, the strength at which failure occurs should be several-fold higher than the actual strength required.

## Applicable Standard(s)

Applicable standard(s) refers to consensus standards and regulatory Standards applicable to the structure created to immobilize the moveable component of the equipment, machine, or structure.

## Blocking and Securement

In all cases where blocking and securement are employed, the following requirements must be met. First, the device must be strong enough to assume the load imposed by the component being immobilized and to prevent expression of energy retained in it. Secondly, the device must be dimensionally stable, so that its position or the position of the load will not shift during or after application of force. That is, the device must provide stability to the load–device combination. The device may require securement to prevent shift in position. Shifting of the device under load can allow the load to move to its zero position, and in so doing, cause needless injury. Securement also must ensure that the position of the load cannot change. The most common direction for blocking and securement is vertical due to the influence of gravity on heavy components.

## Blocking

Blocking is the use of a device to prevent movement of another component. There are many examples of the use of blocking devices. Chocks (Figure G.14) prevent movement along normally traveled horizontal surfaces induced by gravity because of the grade or by collision of a stationary object with a moving one. This problem affects railcars, aircraft, and tractor trailers. Chocks placed on either side of the wheels will prevent most horizontal movement except involving large transfer of impulsive force.

Blocks (Figure G.15) and blocking devices prevent movement of components in machines. Punch presses are prime candidates for blocking of components that move in the vertical direction. Blocking occurs through use of blocks of wood and steel onto which the moveable component is allowed to rest.

Care is required in the selection and use of these devices since they may not withstand the full load against which they are intended to offer protection. This is especially the case with the power stroke of cycling machines where these devices are intended to prevent drift . Wedges ensure a tight fit between the block and the movable surface of the machine or equipment (UAW-GM 1985). Interlock connections ensure that the block remains in place as intended.

Figure G.14.  Chocks placed against curved surfaces of wheels prevent movement along normally traveled surfaces.

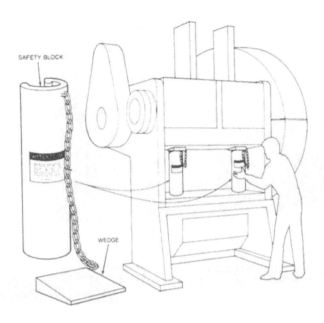

Figure G.15. Blocks prevent expression of energy retained in a machine through deliberate storage. (Source: OSHA.)

Figure G.16. In the absence of support, collapse and crushing injury are an ever-present risk. (Source: OSHA.)

Figure G.17. The prop maintains a structure in a particular position, as in the case of the body of this ice surfacing machine. The prop is intended to support only the empty structure and not the structure plus contents. Often the props are custom-made for a particular piece of equipment.

A prop (Figure G.16 and Figure G.17) supplements the hydraulic system to maintain a structure in position. Props are used with dump trucks and ice-surfacing equipment. The prop for a particular piece of equipment is often custom-designed and manufactured. The prop is intended to support the weight of the structure only, not the structure plus contents.

## Securement
Securement refers to the use of devices to prevent rotation and translational and reciprocational movement in components that are not otherwise immobilized. This occurs through circular tethering against rotational torque and linear tethering maintained under tension or compression. These devices include anchorages and bridging components. Anchorages include clamps and devices attached to frameworks of equipment and machines and the existing infrastructure by screws, bolts, and welded attachments. Bridging components include ropes, chains and cables, and rigid pieces of material and structures. In the case of a securement, the points of attachment must have strength satisfactory to meet the requirements of the load.

Engines of large ships occasionally must undergo service in the harbor. Wave motion from a passing vessel, or in some cases the tide itself can cause rotation of the propeller and movement of components in the interior of the engine. Securement is necessary to ensure immobilization (Figure G.18). This situation poses many unknowns, starting with the force exerted on the components of the engine by the tide and the wake created by passing vessels. The magnitude of the force on components in the engine determines the strength of the attachments and the bridging component needed to ensure immobilization.

Another application of securement occurs with equipment used to dump railcars at loading terminals (Figure G.19). The body of the equipment grasps the car and rotates it while still attached to adjacent cars. (The attachment is a special coupling that enables rotation along the long axis.) The power

Figure G.18.  Securement prevents movement of components in the engine of an ocean-going ship caused by prop wash from a passing vessel or tidal flow.

Figure G.19.  Railcar dumping unit. This device rotates the railcar to dump the contents.

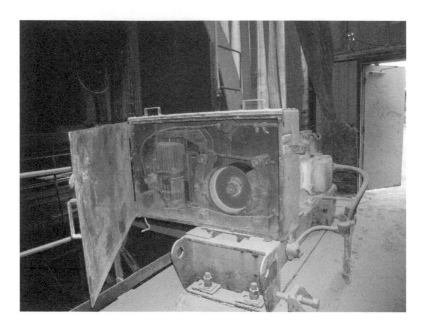

Figure G.20. Braking system for the railcar dumper. The brake acts against the shaft of the drive motor.

unit contains brakes to control position of the dumping unit (Figure G.20). The brake system acts against the shaft of the drive motor. This is an application of a lever system similar in concept to the weigh scale where many rotations of the drive motor are needed to produce a slight change in position of the dumper.

Implicit in the use of internal braking mechanisms applied to geared and direct drive systems is the expectation that the system is not under load, such as a railcar or product, at the time of application. The additional mass may exceed the design parameters of the system and lead to slippage of the brake.

An unusual application of securement occurs during the use of inflatable plugs to isolate flow in sewer systems. The plug holds back the flow of water to enable the performance of work downstream. Pressure exerted by the plug against the walls of the sewer pipeline ensures conformance with the shape to effect a seal. The plugs are shaped like a football. Hence, the contact surface is limited and reflects inflation pressure.

Inflation pressure typically is about 10 to 15 lb/in$^2$ (70 to 100 kPa), depending on the product and its design. One manufacture indicates that the product can resist backpressure of 6 lb/in$^2$ (40 kPa), at which point leakage of water will occur around the sides. Water flow in sewers can exceed capacity and pressurizes the sewer pipeline as a result. The plugs are heavy devices. The product used in one application weighs 280 lb (127 kg) and exerts force downward against the water flowing in the sewer pipeline at time of insertion. Inflation of the plug increases the force exerted by the water against the upstream wall. The upstream pressure will increase to the maximum exerted by the water based on depth and velocity for the particular flow condition.

Fluid flow in a pipe, such as a sewer, exerts force against the walls (static pressure) and force in the direction of flow (velocity pressure). Using the conversion of 0.433 lb/in$^2$/ft enables an estimation of the pressure exerted by the water under flow conditions not including pressurization due to overcapacity. Pressurization is responsible for lifting of manhole covers and spewing forth of water that sometimes occurs under storm conditions. The Owner of sewer lines affected by this condition should know about head pressure created by pressurization.

Use of inflatable plugs without tethering or with inadequate tethering has led to accident situations caused by failure of adhesion to the wall of the sewer pipeline and failure of the tethering (securement) used to maintain position. These accidents have led to near-miss situations and fatal accidents involving rapid flooding and propulsion of the victim along the sewer pipeline. The ability to escape in these situations is limited due to confinement by the pipeline. Manholes in modern sewer design are as much as 100 m (330 ft) apart.

Tethering (securement) and use of redundancy are necessary components to ensure safety during work involving live sewers where the plug is acting as a temporary block to flow.. Considerations about the use of securement include the magnitude of pressurization in the sewer line; strength of the attachment of the inflatable plug before deformational failure of the seal occurs; and strength of available attachments. These considerations will govern the strength needed from the bridging components.

# References

Anonymous: Umbrella Standard for Machine Safeguarding – ANSI/AMT B11.19-2003 (R2009). Rockford, IL: Rockford Systems, Inc., 2009. [www.safetyseminar.org, Website].

ANSI/AMT: Performance Criteria for Safeguarding, ANSI/AMT B11.19-2003 (R2009), McLean, VA: Association for Manufacturing Technology, 2009.

ANSI/RIA: Industrial Robots and Robot Systems – Safety Requirements, ANSI/RIA R15.06-1999 (R2009). Ann Arbor, MI: Robotic Industries Association, 2009.

ANSI/RIA/ISO: Robots for Industrial Environment–Safety Requirements. Part 1 – Robot, ANSI/RIA/ISO 10218-1-2007. Ann Arbor, MI: Robotic Industries Association, 2007.

Barnett, R.L. and P. Barroso, Jr.: On Classification of Safeguard Devices (Part I). Niles, IL: Safety Brief Vol 1, No. 1. Triodyne Inc. April 1981a. [Website reprint].

Barnett, R.L. and P. Barroso Jr.: On Classification of Safeguard Devices (Part II). Niles, IL: Safety Brief Vol 1, No. 2. Triodyne Inc. September 1981b. [Website reprint].

Barnett, R.L., G.D. Litwin, and P. Barroso Jr.: The Dependency Hypothesis (Part I). Niles, IL: Safety Brief Vol 2, No. 3. Triodyne Inc. November 1983a. [Website reprint].

Barnett, R.L., G.D. Litwin, and P. Barroso Jr.: The Dependency Hypothesis (Part II). Niles, IL: Safety Brief Vol 2, No. 3. Triodyne Inc. September 1983b. [Website reprint].

Hall, F.B.: Safety Interlocks – The Dark Side. Niles, IL: Safety Brief Vol 7, No. 3. Triodyne Inc. June 1992. [Website reprint].

Kanis, P.: Controlling the Human Element. *Occup. Health Safety 70(3):* 26, 83-84 (2001).

Krieger, G.R. and J.F. Montgomery (Eds.): *Accident Prevention Manual for Business and Industry, Engineering & Technology*, 11th. ed. Itasca, IL: National Safety Council.

NIOSH: Guidelines for Controlling Hazardous Energy During Maintenance and Servicing (DHHS (NIOSH) Publication No. 83-125). Morgantown, WV: National Institute for Occupational Safety and Health, 1983.

Olivo, C.T. and T.P. Olivo: *Fundamentals of Applied Physics*, 3rd Ed. Albany, NY: Delmar Publishers Inc., 1984.

Omron STI: Engineering Guide. In *Engineering Guide to Machine & Process Safeguarding* (2010-2011 Ed.) Fremont, CA: Omron /Scientific Technologies Inc., 2010.

OSHA: "Control of Hazardous Energy Sources (Lockout/Tagout); Final Rule," *Fed. Regist. 54*: 169 (1 September 1989). pp. 36644-36696.

OSHA: OSHA Technical Manual (OTM), (OSHA Directive TED 01-00-015 [TED 1-0.15A]). Washington, DC: Occupational Safety and Health Administration, 1999.

OSHA: Letter to Edward V. Grund in Reply to His Letter of Enquiry of March 22, 2004 by Richard E. Fairfax. Washington, DC: Occupational Safety and Health Administration, November 10, 2004. [Letter].

OSHA: Safeguarding Equipment and Protecting Employees from Amputations (OSHA 3170-02R). Washington, DC: Occupational Safety and Health Administration, 2007.

OSHA: OSHA Instruction, The Control of Hazardous Energy – Enforcement Policy and Inspection Procedure (CPL 02-00-147). Washington, DC: Occupational Safety and Health Administration, 2008.

OSHA: Selection and Use of Work Practices, Electrical (Subpart S), 29CFR 1910.333. Washington DC: U.S. Department of Labor, Occupational Safety and Health Administration, 2011. [Website.]

UAW-GM: *Lockout*. Detroit, MI: UAW-GM Human Resources Center, UAW-GM National Joint Committee on Health and Safety, 1985.

Schmersal: Man-Machine Safeguarding, Requirements & Techniques. Tarrytown NY: Schmersal USA, 2006.

WorkSafeBC: Safeguarding Machinery and Equipment, General Requirements.(2006 ed.). Richmond, BC: WorkSafe BC, 2006.

# H  Fluid Power, Vacuum, and Water Jetting Systems

## INTRODUCTION

This appendix concerns fluids that are not part of a process. These fluids are contained in circuits enclosed by conduit that links an energy source to one or more energy sinks. The pressures in these systems are similar to and often exceed those used in systems that handle process fluids. The pressure in systems operated under vacuum ranges to –94 kPa (–28 inches of mercury); that is, within one atmosphere. Pressure in systems containing gases ranges to 28 MPa (4,000 lb/in²), while pressure in systems containing liquids extends as high as 690 MPa (100,000 lb/in²). The risks associated with operation of these systems reflect the pressure. These systems are either open, in the case of pneumatic, vacuum, and water jetting systems, or closed, in the case of hydraulic systems. Open systems discharge fluid to atmosphere. Closed systems recycle the fluid in a closed loop.

These systems can contain systems within systems. To illustrate, a low pressure system can increase pressure to a desired higher level for the application in a specific circuit. This occurs in both pneumatic and hydraulic systems through the action of intensifier circuits. Intensification occurs when a larger surface at lower pressure in a primary circuit acts against a smaller surface interconnected to it in an independent, secondary circuit. This interaction boosts the pressure in the secondary circuit containing the smaller surface. Intensification in this manner utilizes energy from a source operating at lower pressure and provides higher pressure in the intensified circuit. This approach minimizes the size of the intensified circuit and need for components and conduit capable of withstanding the high pressure. Intensification occurs in these systems as part of design and sometimes unintentionally and unexpectedly.

Water jetting systems contain externally powered, pressure-boosting pumps. These increase pressure from input level to the pressure at the discharge from the system.

Systems operating at low pressure activate and control systems that operate at high pressure. The low pressure system can be pneumatic or hydraulic. The high pressure system usually is hydraulic.

Systems can retain pressure during operation and when shut down. Intentional retention of pressure occurs in components intended for storage. These include air receiver tanks and accumulators and vacuum tanks. Intentional retention of pressure (or vacuum) provides reserve capacity in the event of unexpected shutdown of the source of power. Discharge of pressure (or vacuum) from these reservoirs enables controlled shutdown of equipment controlled through the circuit. Devices that retain pressure in these systems as part of design are readily identifiable in the circuit in which they are employed.

Intentional retention of pressure (or vacuum) also occurs in isolated circuits. Intentional retention of pressure (or vacuum) in these circuits maintains loads in position during operation. Retention of pressure in these circuits without indication by monitoring devices is not detectable during operation. The observer is aware that the equipment maintains the load in position, but the means that achieve this are not apparent without knowledge of the configuration and operation of the circuit. Unintentional retention of energy can occur in these circuits following shutdown.

## Fluid Power Systems

Fluid power systems include hydraulic, pneumatic, and vacuum systems. Fluid power systems have widespread application across the broad spectrum of human activity. Pneumatic systems sometimes are difficult to distinguish from hydraulic systems. Distinguishing components of pneumatic systems include air filters, regulators, lubricators, and noise silencers/mufflers. Distinguishing components of hydraulic systems include flow control; an electric motor and hydraulic pump; and fluid reservoir. Piping and fittings in pneumatic and hydraulic systems are similar. However, hydraulic systems may contain steel-braided hose and extra heavy-duty pipe and fittings (UAW-GM National Committee on Health and Safety 1985).

Despite the almost ubiquitous application of this technology, fluid power remains a subject about which little appears in print outside the domain of interest of entities involved in use of this equipment. What appears in articles published in trade publications almost always discusses innovations in the industry. These articles, including those written for individuals embarking on a career in fluid power, provide little assistance to the beginner regarding hazards and safety issues. The articles focus on function and operation of hardware.

Hazards and safe operation of this equipment require a totally different focus. Fluid power systems are orphans in the overall spectrum of attention given by safety organizations to equipment. Electrical safety, radiation safety, vehicle safety, machine safeguarding and other areas, by contrast, have received considerable attention. Regulators and other agencies routinely make available information about these concerns. This attention does not apply to fluid power.

While almost nothing appears in print about the safety aspects of fluid power, deactivation de-energization, isolation, and lockout must occur as part of preparation for work on this equipment and these systems. This situation creates difficulty for gaining the knowledge and skills needed to anticipate what could occur, to recognize it when it is occurring, to evaluate the significance of what is occurring and to alert personnel to the need and the means to gain control over specific situations. This knowledge and the ability to make the transference from the generalized information that is available to the specifics of a situation does not appear to be intrinsic in operators of equipment employing these technologies nor to their supervisors and managers. Obtaining information on fluid power requires considerable effort.

In the global sense, risks associated with fluid power reflect the life cycle of the equipment or system (Table H.1). Table H.1 covers the span from design to maintenance and identifies possible causes of failure.

Recognition of issues unable to be addressed during design must translate into action taken throughout the life cycle in order to prevent failure and risk of injury and property damage. This part of the cycle involves the customer who purchases the equipment or system and whose personnel operate it and whose personnel or external contract personnel maintain it. These individuals must be able to make the transition from the generalized information that is available in training materials to the specifics of the equipment with which they are associated.

There are many possible points of failure at this level of the product cycle. There are few influences on quality of work at this level. What influences are available include the quality and comprehensiveness of information provided in the operation and service manual and the ability and determination to follow it. The latter comment reflects on the education, knowledge, skill, and experience of individuals assigned to these tasks.

Table H.2 considers possible accident situations involving fluid power systems during operation and maintenance, in more detail than provided in Table H.1. Many of the accidents resulting from actions and failures involving fluid power components and equipment reflect secondary outcomes. The focus here is the fluid power system and relevant primary outcomes.

In order to appreciate the outlook in Table H.2, some discussion about fluid power is needed. In simplest terms, a fluid power system is an energy source linked by a transmission and distribution system to an energy sink. The energy source and the energy sink convert energy from one form to another. The transmission and distribution system conducts fluid from one location to another through conduits and routing and controlling components. Considering the parts of the life cycle of a piece of equipment or system provides a basis for attempting to predict real-world situations that could lead to injury.

## Design

Design is a critical period in the creation of equipment and systems that incorporate fluid power technology. Design affects selection of components and configuration and geometry of connections. Choices made during design influence outcomes that occur during operation and maintenance.

To illustrate, valve selection can have critical impact on safety. Quarter-turn ball valves operating in a high-pressure system can snap closed or open during the turning motion. The normally expected smooth action experienced in systems designed to operate at low pressure becomes extremely rapid. This situation reflects the action of the pressurized fluid on the surface in the moveable element that be-

## Table H.1
## Factors in Failures Involving Fluid Power Systems

**Design**

Incorrect or inappropriate or incomplete design

Insufficient knowledge about the role of the system

Incompatibility between substances in the environment and materials of construction leading to corrosion

Insufficient knowledge about actual versus planned operation

Unanticipated thermal effects (differential expansion and contraction)

Unanticipated involvement of moisture

**Construction**

Incorrect or inappropriate or inadequate manufacturing practices

Poor (sloppy) workmanship

Substitution of incorrect or inappropriate materials or components of construction

Unauthorized or under-designed modifications

**Operation**

Inappropriate or incorrect operational practices

Failure to operate in the manner anticipated in design

Change in environment from that specified in the design

**Maintenance**

Unauthorized or under-designed modifications

Inappropriate selection of replacement parts and components

Incorrect, inadequate, or inappropriate maintenance techniques

lack of or insufficiency of periodic inspection

Failure to repair/replace failing components

Failure to respond to signs of distress

---

comes exposed as turning occurs. This situation exposes the operator to risk of traumatic injury and the system to risk of structural damage leading possibly to failure due to the destructive impact. Exclusion of this type of valve from consideration reflects knowledge and experience in the design of these systems.

Catastrophic failures of fluid power systems relating to issues addressed in design, while possible, are rare events. This indicates the adequacy of the engineering design and selection of components and assemblies and the equipment and systems created from them. This is not surprising because engineering design involves checks and balances. As well, the manufacturer sustains considerable liability for products defective in design and execution. This shifts the perspective to operation and maintenance.

### Operation

Accidents during operation of fluid power systems occur due to partial or catastrophic failure and consequent loss of pressure in the system. The mode of failure depends on the weakest link in the system at

## Table H.2

## Operation/Maintenance Accidents Attributable to Fluid Power Systems

| Action/Failure | Causative Agent | Outcome |
|---|---|---|
| **Operation** | | |
| Pump operation, fluid flow | Noise | Overexposure |
| Hot surfaces | Heat | Burn |
| | | Ignition of substance/material in the surroundings |
| Failure of component or attachment | Metal or material fatigue | Loss of power to actuator and Loss of control over expected action |
| | Fluid jet | Skin burn |
| | | Eye contact |
| | | Injection |
| | | Fire and possible explosion |
| **Maintenance** | | |
| Inadequate or absent support of component | Gravity | Unexpected movement |
| Deliberate or unexpected release of pressure | Fluid jet | Eye contact |
| | | Skin contact |
| | | Injection |
| | | Fire and possible explosion |
| | | Surface contamination |

time of failure. Most commonly, failure occurs in the conduits that transfer and distribute fluid. The external boundary of the conduit is exposed to the environment and is susceptible to attack by incompatible substances and agents. The interior boundary of the conduit is exposed to the fluid and its contaminants and subject to attack due to incompatibilities. Some of the components of the transmission and distribution system of conduits are rigid and others are flexible.

Partial failure expresses as leakage and escape of fluid. Escape of fluid at high pressure during partial failure creates a jet of liquid and also droplets. Impingement of the jet on a surface can lead to penetration as the jet expends its energy. Injection injury is a major concern during operation of equipment containing fluid power technology. Injection injury has occurred during normal operation as well as diagnostic procedures intended to search for leakage. Leakage occurring at pressure of normal operation can be very difficult to detect, and as a result can cause severe traumatic injury, up to and including amputation.

Amputations have occurred through unknowing contact with a pressurized jet of fluid. Amputations have occurred during operation and diagnostic procedures intended to detect the point of leakage.

Fluid injected into a limb follows an unpredictable path. This results from reflection of the jet on surfaces of bone and tendon. The fluid can inject bacteria into the tissues. Infections resulting from these injections have caused gangrene.

Distribution of the droplets can create ignitable and explosible mixtures even of substances having high flash point (Figure H.1). This results from aerosol formation that occurs during dispersal of the energy contained in the jet. The droplets of liquid can form the ignitable mixture. Droplets of appropriate size behave as vapors.

Catastrophic failure can involve any of the components of the system and can lead to complete and rapid shutdown. Catastrophic failure can occur for several reasons, but principally because of failure of attachment between a coupling and a flexible element. Catastrophic failure can lead to rapid loss of position of the load or speed of movement of a motive element such as a motor. Catastrophic failure also can lead to rapid loss of bulk quantities of fluid. In the absence of emergency contingency devices, such as accumulators, these failures can produce large-scale destruction. These situations are rare. Accumulators provide a cushion to enable orderly shutdown of the system.

### Metal Tubing
Metal tubing provides strength and long service life under tough conditions and accommodates pressures beyond 6,000 lb/in$^2$ (40 MPa) (HP 2003a). Achieving the appropriate shape requires special equipment. Tubing is subject to corrosion unless specially treated.

Immobilized components are intended to remain in a fixed position at all times during operation. Allowing an otherwise intended, immobilized component to move and to move routinely during operation creates a situation of possible metal fatigue. Metal fatigue can lead to abrupt, catastrophic failure. Motion of components and conduit intended to remain immobilized is usually evident to operations and maintenance personnel. This motion should form the basis for immediate service and repair. Immobilization of these components is essential to prevent occurrence of fatigue-related episodes.

Figure H.1. Emission of fluid under high pressure from hydraulic systems has led to formation of clouds of mist and resulting fires and explosions. One accident involved failure of a hose on a bulldozer. (Source: OSHA.)

Another cause of failure of rigid conduit is stress created during attachment of misaligned fittings. Alignment is critical with bent and shaped tubing and the margin for error in alignment is very small.

## Hose

More commonly, failure involves flexible conduit, such as hose. Hose enables movement of parts of the system, otherwise intended to remain immobile. Flexible conduit, such as hose, is essential for the operation and function of the vast majority of equipment that operates through fluid power. Movement by flexible components, despite intentional design, introduces a point of weakness. Points of weakness are susceptible to failure.

Hose combines flexibility with attachment of the flexible component to immobile components. Hose is complex product (Kemper 2011). Regardless of the application, however, all hoses consist of three basic components: inner tube, reinforcement, and cover.

The inner tube, made from synthetic-rubber compounds and composites, must be chemically compatible with the working fluid and to resist corrosion and deterioration. Fire-resistant fluid requires special formulation, as does service at high and low temperature.

In similar fashion to tires on vehicles, reinforcement is essential to enable the hose to handle normal operating pressure and pressure spikes and to prevent premature bursting when properly used. That is, the presence and type of reinforcement determines the working pressure of the hose. Impulse pressure is an important concept in hose function. Impulses result from jolts to the system. These are similar to what occurs in a tire when a vehicle goes over a bump or pothole in the road and sudden and rapid compression occurs. Testing for impulse resistance by manufacturers can range from 600,000 to 1,000,000 impulses. A high rating equates to longer service life.

Hoses with low working pressures normally use textile-fiber reinforcement while those handling higher pressures generally use high-strength steel wire.

Hoses rated for low pressure ($<300$ lb/in$^2$ or $<2$ MPa) usually contain reinforcing of textile material. Some are also rated for suction applications.

Ratings for medium pressure range from 300 to 3,000 lb/in$^2$ (2 to 20 MPa). Hoses rated for medium pressure may have reinforcement from one-wire, steel braid, multiple-wire, and/or textile-braid .

Hose rated for high-pressure applications mainly contains two wire braids with reinforcement by high-tensile steel. Operating pressure ranges as high as 6,000 lb/in$^2$ (40 MPa).

Steel reinforcing in hoses occurs as braid and spiral. Wire-braided hose handles working pressures to 6000 lb/in$^2$ (40 MPa), depending on size, with one or two layers of braid. Spiral hose generally handles high pressures in larger diameters. The wire spirals around the tube on a bias with successive layers laid at opposing angles. These hoses typically contain four or six layers of steel reinforcement. In braid and spiral hose, layers of rubber separate layers of steel wrap to ensure good adhesion throughout the wall.

Spiral reinforcement is particularly suited to high-pressure applications involving impulses because individual wires in each layer are parallel. The thin rubber adhesion layers separate adjacent reinforcement layers and prevent damage to the wires. Spiral construction packs the reinforcement tighter around the tube than does braid reinforcement. Spiral construction provides greater support to the tube by binding more tightly individual ends compared to the over-under gaps present in braiding. However, braided hose generally is more flexible than spiral hose.

The cover protects the tube and reinforcement from heat, abrasion, and corrosion, as well as environmental deterioration from heat, cold, UV light, and ozone. Covers are made from synthetic rubber, fiber braid, or fabric wrap, depending on the application.

Selection of hose for the application is an extremely important decision. Substitution of products containing similar but not identical catalogue numbers can cause catastrophic consequences. This can occur through errors of omission and commission at the interface between the customer and the individual in a stocking entity that provides a product from existing inventory.

Experience indicates that about 80% of hose failure results from external physical damage (HP 2003b). Abrasion is the main cause. Frequent or premature hose failure can be a symptom of equipment malfunction. Prompt corrective action can sometimes avoid serious and costly equipment breakdown. Occasionally, the failure lies with the hose itself. The most likely cause of a faulty rubber hose is old age.

## Couplings and Hose Assemblies

Failure can also occur at points of attachment of the flexible component to immobile components. Certainly, this situation occurs in hydraulic systems. Failure of hoses and their points of attachment is a major cause of accident situations involving these systems. This outcome has caused grave concern to manufacturers of these products. Fittings capable of respecting the physical characteristics of the fluid and impact of pressure are mandatory for successful attachment of components to transmission and distribution conduit. The energy intrinsic to the pressurized fluid causes the destruction and traumatic injury that result from the failure.

Improper assembly and installation are major causes of premature failure of hose assemblies. These involve use of the incorrect fitting and poor routing of the assembly. Manufacturers provide training material to help prevent these problems. Frequent or premature hose failure can be a symptom of equipment malfunction. Considering this factor is essential, because prompt corrective action can sometimes avoid serious and costly equipment breakdown.

A casual observer might view a hose and the end couplings as individual components (Brown 2011). In fact, these parts form a precisely engineered system. That is, a properly constructed and optimized assembly is greater than the sum of its parts. An improperly constructed assembly is less than the sum of its parts.

Fabricating the assembly requires application of the correct amount of pressure during crimping or swaging in order to lock the interface between the hose and the coupling. The interface must hold the end connection, but not damage the hose or collapse the interior of the coupling. The basis for selecting the hose and fittings from the same manufacturer is the proven ability to mate the coupling to the hose. This situation reflects the fact that dimensional tolerances listed in standards are quite broad and that hose and couplings purchased from different sources may not fit together in an optimal manner.

Causes of failure of hose-connector assemblies include twisting out of plane, use below the minimum bend radius, and issues associated with length (HP 2003b).

Bending of reinforced hose should occur only in a single plane. Introducing a twist of 5 ° reduces service life by 70%, and a twist of 7° by 90%. Introducing a twist in an installation is a great temptation when the end fittings do not align properly. Yet, this seemingly insignificant deformation can cause significant future risk and losses in performance of the system.

Minimum bend radius refers to the tightest bend a hose can tolerate per recommendation of the manufacturer. While hose can often tolerate a smaller radius than recommended, use in this manner will likely harm the reinforcement and reduce service life.

Hose length can change between +2% and –4% when pressurized (Brown 2011). Pressure may cause a hose barely reaching the connections to contract and stretch. These actions shorten service life. Hose length should be slightly longer than the actual distance between connections.

## Seals

Containment of the fluid within the system at both fixed and moveable surfaces requires sealing technology. Otherwise, the fluid will escape. Escape of fluid at high pressure from failing seals creates a jet of liquid and also droplets. Impingement of the jet on a surface can lead to penetration as the jet expends its energy. Distribution of the droplets can create ignitable and explosible mixtures even of substances having high flash point.

Leakage in fluid power systems occurs at fittings and seals (HP 2003c). Every connection is a potential point of leakage. Integration of components into a single package is one strategy for reducing the number of connections and points of leakage. Welded fittings instead of threaded fittings also reduce potential for leakage.

The vast majority of O-ring, face seal fitting problems occur due to a problem with the O-ring itself (Haller, and Kleiner 2011). This can include a missing, pinched, or partially extruded O-ring. Often times, the common misconception that O-ring face seal fittings can be finger tightened and will not leak can cause the problem.

Connections with O-ring face seal fittings require the O-ring to be present and properly seated in the groove. Lubricating the O-ring is important to ensure that it remains in place during installation. Proper assembly of O-ring face seal fittings also involves the use of a torque wrench and recommended values.

Seals, primarily rod seals on cylinders, permit controlled leakage of fluid. The fluid coats the surface of the rod. Without the film of lubricant, the seal will experience premature wear. Too thick a film constitutes leakage because accumulation will occur at the base of the cylinder. Leakage of this nature is usually assigned nuisance status.

Leakage beyond nuisance levels can reflect pressure spikes from shock loading and impact and can occur in different applications of the same product in a single piece of equipment.

## Maintenance

Maintenance is the least controlled period in the life cycle of equipment and systems containing fluid power technology. Maintenance poses different issues from operation because the device that pressurizes the fluid is not operating. Maintenance exposes the weakness in the level of knowledge of individuals who dismantle and service equipment containing fluid power technology.

### Fluid Flow Issues

During operation and depending on the availability of storage at operating pressure, the prime mover operates continuously. This operation both pressurizes the system and provides fluid surplus to the needs of the actuators.

In hydraulic systems, the pump moves copious amounts of fluid in order to maintain flow and pressure in the high-pressure lines. Fluid surplus to these needs enters a bypass circuit and returns to the storage reservoir (Vickers 1992).

The quantity of fluid moved by the pump is not always appreciated and has led to accidents during diagnostic procedures when the volumetric flow rate in a disconnected system overwhelmed the ability to collect and retain the fluid in makeshift containers (McLaren 2001a, McLaren 2002a). This situation has led to emission of fluid jets, fire and explosion, and spillage and contamination of surfaces. Hot fluid carries the risk of burns and fire. Hot oil poses a severe burn risk compared to water at the same temperature.

Pneumatic systems cycle on and off in order to maintain the pressure in the receiver tank at the preset level. Cycling is a normal expectation of operation of this type of system.

Vacuum systems can operate continuously or evacuate a reservoir in similar fashion to pneumatic systems depending on mode of operation.

### Pressure Intensification Issues

The hydraulic system in a bulldozer maintains blade height during grading by controlling fluid pressure on both sides of the piston. Changing the pressure on one side relative to the other causes movement of the piston in a double acting cylinder.

The surface area on both sides of the piston is different. This is the result of the presence of the rod in the middle of the piston on one side of the cylinder. This difference in area creates different force for application of the same pressure on both sides of the piston. This situation is a direct result from application of Pascal's law where Pressure = Force/Area.

A larger area for the same pressure creates a larger force. This also means that the rod side of the piston requires larger pressure in order to balance the force applied to the flat side. Without the increase in pressure on the rod side, movement of the piston toward the rod end will increase (intensify) the pressure in that side of the circuit. This situation can create unintentional intensification of the pressure. Said a little differently, if for any reason the flat side of the piston in a double-acting cylinder is pressurized and at the same time fluid some element prevents escape of fluid from the rod side, pressure will increase (intensify) in the rod side of the cylinder until the forces balance, meaning repositioning of the piston or the cylinder fails catastrophically.

Cold weather can gel hydraulic fluid and form a blockage in a line. Blockage of this nature could cause intensification of pressure.

Pressure intensification can create disastrous consequences in bench-level tests to determine leakage involving a piston seal as well as in field locations.

### Residual Energy in Fluid Power Systems

Fluid power systems can retain energy in the form of pressure following deactivation, de-energization, isolation, and lockout. Residual energy includes energy retained intentionally in energy storage devices and energy retained unintentionally in isolated circuits. Of all of the sources of energy considered in these appendices, fluid power systems appear to have the most diverse and problematic issues.

#### *Energy Storage Devices*

Intentional retention of pressure occurs in components installed in circuits for this purpose. Retention of energy occurs in isolating circuits associated with the storage structure (Figure H.2). Absence of knowledge about the retention of energy in fluid power systems has caused many accidents during maintenance.

Energy in accumulators and pressurized lines has caused accidents in situations where this energy is presumed not to be present. As a result, activation of controls powered by energy stored in the accumulator can occur. One incident that highlights this situation involved a tugboat with steerable power units. During service in a shipyard, an electrical foreman showing some visitors how the controls "worked," moved the control unit. Unbeknownst to this individual, this caused the power unit to rotate under energy stored in the hydraulic system. This motion almost struck a surprised shipwright who was working under the stern of the vessel on the power unit. Neither had any idea about energy storage in this system.

This situation highlights the issue of energy retention following shutdown in circuits intended to retain pressure, normally for use during emergency situations to minimize the impact of loss of power.

Figure H.2. Isolated circuit containing an energy storage device and isolating devices.

Air Receiver
The air receiver is the main energy storage device in pneumatic systems. There is no limit to the size of a receiver tank. The air receiver tank is a pressure vessel. A large receiver provides stability to a system, especially one requiring orderly shutdown in the event of power failure. Hence, a receiver tank can continue to supply air at or near system pressure for a period of time after shutdown of the compressor.

The air receiver tank plus related piping and valves comprise the actual energy storage circuit. The circuit comprising these components retains pressure following shutdown of the system. In the absence of a reliable pressure-indicating device in the circuit, the magnitude of the pressure retained will remain unknown.

Pulsation Dampener
The pulsation dampener is a two-chambered device used for removing irregularities in pressure in pneumatic systems. A bladder separates the two chambers. The top section contains compressed air or gas at a pressure lower than the discharge pressure of the pump. A pulse in the system created by the pump (positive displacement) or a valve pushes gas into the low pressure side of the dampener. This pushes against the bladder and the compressed gas on the other side. As the pressure in the system decreases, the force exerted by the compressed gas and bladder push against the gas and force it into the system.

The pulsation dampener plus related piping and valves comprise the actual energy storage circuit. The circuit comprising these components retains pressure following shutdown of the system. In the absence of a reliable pressure-indicating device in the circuit, the magnitude of the pressure retained will remain unknown.

Accumulator
The analogue of the receiver tank and pulsation dampener in hydraulic systems is the accumulator. The accumulator is a fluid storage vessel that maintains pressure in the system through application of force. In parallel with the air receiver in pneumatic systems, the accumulator serves multiple functions including acting as a surge or pulsation dampener.

The accumulator maintains system pressure temporarily following failure or shutdown of the normal supply. This enables orderly completion of operation and shutdown of critical systems. Accumulators can easily range to 750 L (200 gal U.S.) in supply volume and system operating pressures to 140 MPa (20,000 lb/in$^2$).

Accumulators are used in critical circuits such as brake and pilot control circuits. In a brake control circuit, the accumulator provides reserve pressure when the main pump fails or is shut down. An accumulator in a pilot control circuit provides control when the main pump shuts down. This enables a raised bucket or boom to be lowered in a controlled way without operation of the engine and the hydraulic pump. (Nelson 1983; Vickers 1992).

Most accumulators today use a bladder to separate the pressurizing gas from the hydraulic fluid (Bolton 1997). Some use a liquid, such as water, inside the bladder. At equilibrium, the pressure on both sides of the bladder is the same. Sizing the accumulator depends on the application.

The accumulator plus related piping and valves comprise the actual energy storage circuit. The circuit comprising these components retains pressure following shutdown of the system. In the absence of a reliable pressure-indicating device in the circuit, the magnitude of the pressure retained will remain unknown.

Isolated Circuits
Retention of energy both intentionally through storage in devices as described and unintentionally occurs in isolated circuits in fluid power systems. In the absence of a measuring device or pre-installed components that enable testing, these circuits can retain pressure (or vacuum) without indicating about the existence of this condition. Acquisition of this knowledge sometimes occurs through traumatic accidents.

Recognizing these configurations in a particular circuit is essential for gaining the recognition skills needed to anticipate these conditions. This is a difficult task. The word ,"difficult," quite probably understates the magnitude of the difficulty of this challenge. Recognizing how circuits become isolated and where they exist in an overall configuration requires high-level knowledge about hydraulics and expertise in the field of hydraulic components and circuits.

Check (backflow prevention) valves, pilot-operated check valves, relief valves, directional control valves, and counterbalance valves have the ability to form isolated circuits and thereby to trap pressurized fluid (FPSI 2002a). Loss of piloting occurs on loss of pressure in the pilot circuit. Loss of piloting maintains the valve in the position at time of loss of pressure in the pilot circuit unless retraction of the spool to a default position can occur.

The pressure retained in the main circuit is the pressure at which the isolating valve opens to admit additional fluid during operation. This is the pressure in the circuit during operation at the time at which isolation occurred during shutdown.

### Dead-Ended Circuits

Dead-ended circuits contain an isolating valve and an unsupported load (Figure H.3). Open-ended circuits contain two isolating valves (FPSI 2002a).

An unsupported load results when a mass supported by a cylinder remains suspended above the position of zero energy once power in the system is shut down. Shutting down the power to the pressure source eliminates the continuing pressurization and flow of fluid needed to maintain the load in suspension. Suspension continues only so long as the isolated circuit retains the pressure.

Release of pressure either deliberately or through leakage allows the load to move, sometimes rapidly. This situation has led to fatal accidents when service personnel failed to support adequately the load imposed by buckets, booms, and other components of mobile equipment and released the pressure retained in the circuit (FPSI 2002b).

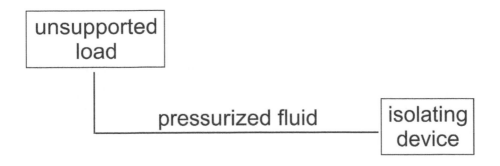

Figure H.3. Dead-ended circuit. A dead-ended circuit contains an isolating device and an unsupported load. Pressure exerted by the mass of the unsupported load pressurizes fluid in the isolated circuit.

*Pressurized Line*
A pressurized line results from entrapment of pressurized fluid between two valves capable of isolating the circuit (Figure H.4). Isolation in this manner is not apparent without careful examination of as-exist circuit drawings by an individual qualified to do so and/or testing of the circuit for retained pressure. Testing of circuits not fitted out with testing points requires careful consideration and extreme caution to prevent against emission by a jet of pressurized fluid.

*Pressurized Hose*
Pressurized hose is a special type of energy-storage component in an isolated circuit. Unlike metal tubing and pipe, hose is flexible. Pressurization causes a hose to swell and to become rigid (Figure H.5). Pressurization can result from isolation devices as well as unsupported loads. The swelling and rigidity are readily apparent to operators of equipment containing hoses.

Depending on type, hoses can elongate up to 2% when pressurized and can contract as much as 4% (Mueller 2002). Radial expansion, which also occurs, depends on the hose and the pressure. Reversible, pressure-induced dimensional changes create the equivalent of a spring that expands and contracts. Expansion and contraction of a spring are expressions of storage and release of energy. Hence, pressure-induced changes in hoses are equivalent to those that occur in accumulators.

Operators learn to check for flexibility in hose as one means to evaluate pressurization of a circuit. This technique is only partially useful because the remaining inflexibility of the hose can mask pressurization at a lower level.

*Piloted Circuits*
Piloted valves can remain active during maintenance when the piloting system remains energized (Vickers 1992). Energy sources for piloted valves can include electrical, hydraulic, and pneumatic cir-

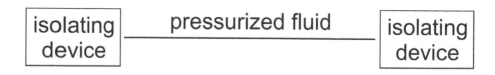

Figure H.4. Pressurized lines occur when valves capable of isolation trap pressurized fluid in a circuit where such occurrence is neither expected or intended.

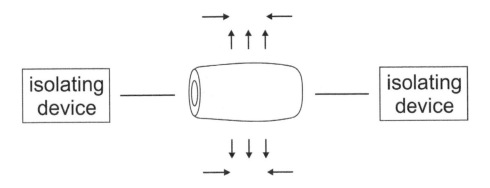

Figure H.5. Pressurized hose. Pressurized hose can occur in circuits containing hose and isolating devices including unsupported loads. During pressurization, the hose expands radially and contracts in length. Rigidity versus flexibility is a rough indicator to the operator about the status of the circuit.

cuits (Figure H.6). Hence, the energy source used for piloting a main circuit hydraulic system need not be the same used in the main circuit.

Energy from energy stored in electrical and electronic systems can power solenoids and relays. Pneumatic actuators receive their energy from the compressed air system. The compressed air system can contain an accumulator or a large receiver tank that remains pressurized long after shutdown of the compressor. Pressurization of the pneumatic circuit by the receiver tank and pressurization of the main hydraulic circuit by one or more accumulators creates an operable circuit even after shutdown of the prime movers in both circuits has occurred.

## Uncontrolled Depressurization

Uncontrolled depressurization is the cause of many accidents involving fluid power systems (FPSI 2002c; McLaren 2001a, McLaren 2002a). These situations resulted from inappropriate actions including deliberate loosening of fittings to relieve pressure in circuits lacking pressure indication. The jet of fluid would emit at circuit pressure compared to atmospheric pressure in the open air. This pressure drop projects the jet into free space. The rate of emission of fluid is fixed when the opening is a "critical" orifice. In some circumstances, large quantities of fluid emit into the surroundings on loss of containment.

Emission in this manner has caused injections, skin burns and contamination and fires. Circuits lacking pre-installed fittings to enable controlled depressurization and drainage of contents pose a serious risk to safety where depressurization must occur.

## Inappropriate Pressurization

Some of the accidents involved inappropriate pressurization of circuits. These typically happened during use of compressed air or portable hydraulic pumps to free seized components of hydraulic cylin-

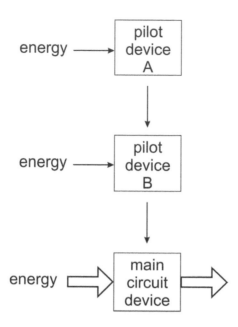

Figure H.6. Piloted circuits. Piloted circuits operating in a hierarchy, such as electric over pneumatic over hydraulic, receive energy from different sources that can remain energized following shutdown of the main circuit. Depending on design, low pressure hydraulic and pneumatic pilot circuits can remain pressurized during shutdown of the main circuit.

ders (FPSI 2002d, FPSI 2002e; McLaren 2001b). The pressure developed overcomes the resistance and the component separates from the cylinder with tremendous force.

Without presence of a pressure gauge in the circuit, the magnitude of the pressure applied by a portable hydraulic pump remains unknown.

## Assessing Hazardous Conditions in Fluid Power Systems

Fluid power systems differ from technologies involved with other forms of hazardous energy. This is rapidly evident to anyone making enquiries to try to learn about and to understand the operation of fluid power systems. Mechanical equipment, for example, shows the action of one element on another. The actions are visible when the equipment is dismantled. The drawings, while providing a static image, still provide enough information to enable an interpretive judgement about movement.

Electrical systems provide no obvious explanation about their actions. An analogy to fluid flow in piping provides some assistance to understanding and comprehension. As well, the electrical industry has made hazard recognition and safety during operation and maintenance an ongoing part of its educational efforts. Education about electricity forms part of the curriculum in the formal education system at the secondary and college levels.

Fluid flow and effects of pressure are not included in the education curriculum. This may reflect familiarity with residential water supply systems, a relatively innocuous system and analogies created from it. Residential plumbing operates at low pressure and relatively low flow. Components are discrete and their functions and action are readily able to be visualized. Conditions arising from domestic systems cause little direct harm, and as a result, receive little attention.

Fluid power systems differ considerably from other systems. Fluid flow does occur in fluid power systems in the same manner as in residential water supply systems. However, the aim of the fluid power system is to provide pressure with suitable flow. The aim of the residential water system, on the other hand, is to provide flow at suitable pressure. Hence, the pressure in a fluid power system is considerably higher than pressure in a domestic water system and the flow is considerably less. This situation is changing with the availability of compressors and equipment for high pressure water jetting aimed at consumers that operate considerably outside the envelope defined by residential water supply systems.

Workers not specifically trained in fluid power are ill-equipped to address the hazardous conditions that can exist. The actions of fluid flow at high pressure are not evident because the equipment is enclosed. The only evidence of activity is the movement of the actuator such as a cylinder. On the superficial level, the equipment seems relatively simple because of relatively few components. Yet, contained within the boundaries are high pressures and unimaginable forces. The components are often highly integrated and multi-function. This produces compactness and small numbers. As well, some components act to compensate for effects of temperature and to tweak performance of core components.

A parallel situation confronts the first-time visitor to a facility containing process piping, valves, and components. A boiler system in a building provides an excellent example. In this situation the critical components are fully exposed for observation and consideration. Given enough time, an outsider can trace the path in all circuits and determine a course of action to effect shutdown and isolation during operation of individual circuits. This is not the case with fluid power systems given the internalization and integration of components.

The logical starting point in the quest for information about any operating system is the operator. The operator normally is the expert regarding the manner of operation of a particular system. The operator "lives" with the equipment and must learn how to gain utmost performance and efficiency from it without endangerment. The operator often knows not only what the equipment is supposed to do and how it does it, but also what the equipment actually does do and also how the equipment fails. This information usually is undocumented. Hence, making contact with and establishing a positive and open relationship with the operator is often critically important for obtaining a successful outcome.

To an extent, the comments in the preceding paragraph also apply to operators of fluid power systems. Complicating this situation is the fact that actuation and control of equipment containing fluid power systems occur indirectly. These units operate through piloted systems one or two levels removed from the actual main power circuit (for example, electric over air over hydraulic). To the operator, actions through control control devices, such as a "joystick," operate the main fluid power circuit, even though expressed through an electrical circuit. That is, the operator knows about the outcome of an action but likely has no knowledge about how the system creates it. This lack of direct communication is another complicating factor in the operation of fluid power systems.

Another difference of critical importance between a steam system and a fluid power system is that the pressure in a steam system depends on temperature. Shut down the source of heat and the pressure throughout the system will revert to an ambient level (which may not be zero). This situation does not apply to fluid power systems. Retention of pressure can occur through entrapment by isolating devices. The mechanism of the entrapment and the involvement of otherwise innocuous components and absence of indication are the key issues.

Compounding the difficulties in the situation are the tools used for communication about system operation. Many years ago, the fluid power industry adopted a unique and counterintuitive set of symbols contained in ANSI Y32.10-1967(R1999) (ANSI 1999). The intent of this action was to foster communication between individuals in the industry and to eliminate jargon. While the symbols may reflect logic and systematic design, they provide little, if any, benefit to individuals lacking formal and in-depth training in fluid power. Diagrams that show flow channels in structures and components are difficult enough to follow and interpret, but diagrams that show no indication of function provide no benefit to anyone lacking formal, in-depth training (Figure H.7 and Figure H.8). That is, the hands-on individuals who work on and service this equipment.

This situation has created barriers to understanding about the nature of operation of this equipment at the hands-on level at which accidents occur. Communication of information needed to assure safety during servicing of this equipment is essential for minimizing accidents.

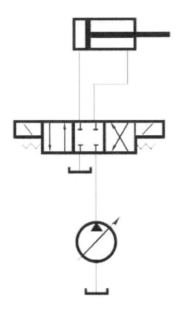

Figure H.7. Fluid power circuit of a two-way cylinder and directional valve described using ANSI symbols.

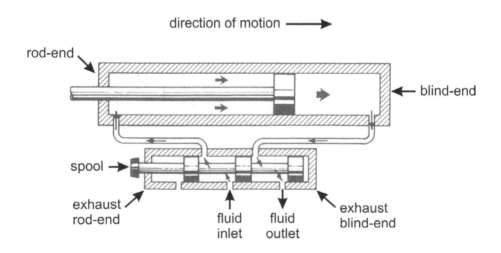

Figure H.8. Fluid power circuit of a two-way cylinder and directional valve depicted in a flow diagram. The position of the spool (moveable element) in the directional valve controls the movement and position of the piston in the cylinder. (Modified from Nelson 1983.)

As indicated previously, these systems may retain pressure intentionally in energy storage devices and unintentionally in isolated circuits. Expression of this energy can occur in unexpected ways at inopportune moments, as also mentioned. Manufacturers typically recommend to customers to toggle the controls and to refer to the pressure gauge after returning loaded structures to a neutral position and shutting down the pressure source.

The idea behind the action is to determine the ability to perform the normal operation through action of the controls. According to the logic of this approach, if toggling the controls produces no action, then de-energization through depressurization has occurred. This is not necessarily the case in all situations. The action produced by toggling the controls depends on availability of the pilot function. As also mentioned, in systems with multiple layers of control such as electric over air over hydraulic, loss of the energy needed for piloting will eliminate this function without necessarily releasing the energy throughout the main circuit.

Toggling the controls functions as a diagnostic tool only when piloting is active after shutdown and all loads are supported and cannot otherwise move. In systems containing articulated components under the influence of gravity, the position of support for all components may not be the position needed in order to provide access to perform the work. Positioning the articulated sections to accommodate the work can create an isolated pressurized circuit. The key questions are what outcome could this produce during dismantling of parts of the system to gain access to other components of the device in order to perform work and what outcome could occur from this situation.

The place to start in formalizing this enquiry is the manufacturer. The components and systems are so highly integrated and as a result, complex, that determining the actual operation can require high-level expertise and in-depth knowledge. The manufacturer has greatest knowledge about parameters involved in the design and operation of the equipment and the ability to perform diagnostic tests to measure temperature, flow, and pressure on circuits under different credible scenarios created during maintenance situations. This information is essential for establishing conditions that truly exist and for answering questions posed in preceding discussion. Personnel servicing this equipment require specific information in order to be able to perform the sequence of deactivation, de-energization, isolation, lockout, and verification as required in regulatory statutes, such as 29CFR 1910.147 (OSHA 1989).

The owner should request a comprehensive circuit diagram from manufacturer for the equipment that indicates pressures and flows. The owner should obtain interpretation of this information from an individual qualified to be able to provide such advice. This situation presents a probable product liability to the manufacturer. As a result, the manufacturer should determine and document the actual status of the equipment that it offers for sale in the marketplace and prepare to offer this information to customers. As well, the manufacturer should provide specific instruction accompanied by documented proof about the efficacy of the procedure to deactivate, de-energize, isolate, and lock out fluid power circuits. This instruction should also include specifics for immobilization and for release of fluid pressure in isolated circuits.

The manufacturer should provide this information in diagrammatic form, along with testing data, so that maintenance personnel can determine exactly what to do and how to do it. This information is a practical expression of requirements for risk assessment contained in consensus standards discussed in Appendix A. Provision of this information in an easily comprehended format is the only way to inform workers who service this equipment and lack high-level, in-depth knowledge of fluid power systems, and thereby to minimize the risk of accidents.

Experience has shown that depressurization points require installation of a needle valve or similar valve that produces highly controlled discharge from a circuit containing fluid at high pressure (FPSI 2002f). Installation of needle valves can occur through permanent fittings or temporarily through quick disconnect fittings appropriate to the circuit and the pressure. Diagnostic devices containing pressure gauges and needle valves for connection to quick-disconnect fittings are available in the marketplace. Depressurization must occur under highly controlled conditions. Otherwise, there is a high risk of accident due to emission of a jet or large quantity of fluid depending on conditions.

Complicating assessment of fluid power systems is the absence of pre-engineered and installed isolation devices that enable compliance with requirements of 29CFR 1910.147 where the decision to perform work on equipment containing isolated pressurized circuits is made (OSHA 1989).

Little exists in print about the means to characterize the presence, movement, and storage of pressurized fluids in equipment, machines and systems to enable assessment in preparation for lockout and tagout. Typically what are observable are ad hoc isolations and lockouts created by site personnel in response to the perceived need. Enquiry about engineering input into the strategy to ensure adequacy of the approach typically produces a negative response.

Table 7.6 provides a generalized way of considering this question for different sources of energy. Table H.3 is a more specialized version of Table 7.6 that focuses on assessment for isolation and lockout of fluid power systems.

The following sections describe the additional terms used in Table H.3. The information collected and considered during this hazard assessment forms the basis for a dialog with the regulator with regard to possible strategies for isolation and lockout of the actual system. Contact with the regulator is strongly advised in order to solicit opinion about whether the proposed approach would be acceptable.

## System Operating Pressure

System operating pressure is the design pressure for systems under construction or actual working pressure for operational systems. This pressure provides a measure of the nature of the design compared to rated working pressure of components selected for use in the application.

## Rated Working Pressure

Rated working pressure is the pressure rating for continuous service recommended by the manufacturer for the component(s) under consideration. The rated working pressure also indicates the closeness of the design pressure or pressure achieved under actual operating conditions compared to the long-term recommended maximum service pressure. Components operating at or near the rated working pressure still have the protection of the safety factor. The safety factor for the component, while not likely to be disclosed by the manufacturer, must exceed the pressure under which hydrostatic stress testing occurs in order to pass that test without failing.

## Residual Pressure

Residual pressure is the pressure retained in piping downstream from pumps or other components capable of pressurizing the system. This pressure reflects retention of pressure due to entrapment of fluid in isolated parts of the system. Residual pressure is capable of creating a jet from a small opening in the flange during intrusive isolation.

## Inflow Isolation

Inflow isolation identifies the component that acts to isolate flow on in the inflow side of the circuit.

## Outflow Isolation

Outflow isolation identifies the component that acts to isolate flow on in the outflow side of the circuit.

## Depressurization Strategy

Depressurization strategy explains the process to effect the depressurization. This process must acknowledge the possibility that depressurization will occur on a pressurized circuit.

## Immobilization Strategy

Immobilization of suspended loads is a possible requirement of this procedure. Articulated loads may require support in unstable positions. This strategy indicates analysis and measures taken to ensure the success of the procedure.

## Table H.3

## Assessing Retained Energy in Fluid Handling Systems

| ABC Company | Hazardous Energy Assessment | Work Activity |
|---|---|---|
| Location: | Equipment, Machine, System: | Assessed by:<br>Tel:<br>Date: |
| **Overview:** | | |
| **Hierarchy of Energy Input** | **Conversion Energy Output** | **Equipment, Machine, System Affected** |
| | | |

### Circuit Name

**Fluid Pathway:**

| System Operating Pressure | Circuit Pressure | Residual Pressure | Inflow Isolation |
|---|---|---|---|
| | | | |
| | | | Outflow Isolation |
| Depressurization Strategy | | Immobilization Strategy | |

**Failure/Consequence Analysis:**

| Depressurization: | Immobilization: | Electrical Isolation: |
|---|---|---|
| Location: | Location: | Location: |
| Action: | Action: | Action: |
| Depressurized? | Immobilized? | Isolated? |

## Table H.3 (Continued)
## Assessing Retained Energy in Fluid Handling Systems

| |
|---|
| **(Circuit Name) Fluid Power Isolation Procedure:** |
| **Verification Test Procedure:** |
| **Notes:** |
| **Overall Procedure:** |
| **Overall Verification Procedure:** |

<table>
<tr><td colspan="4" align="center">Hazard Assessment (Operational/Undisturbed Space) or (Work Activity)</td></tr>
<tr><td>Hazardous Condition</td><td colspan="3" align="center">Occurrence/Consequence</td></tr>
<tr><td></td><td>Low</td><td>Moderate</td><td>High</td></tr>
<tr><td></td><td>|</td><td>|</td><td>|</td></tr>
</table>

(circuit name)

In this table, **NA** means that the category does not apply in any normally foreseeable situation. **Low** means that exposure is readily identifiable but believed to be much less than applicable limits or that exposure to nonquantifiable hazardous conditions is unlikely to produce injury. **Low-Moderate** means that exceedence of regulatory limits is believed possible or that nonquantifiable exposure could produce minor injury requiring self-administered treatment. Control measures or protective equipment should be considered. **Moderate** means that exposure is believed capable of exceeding regulatory limits or causing traumatic injury requiring first aid treatment or attention by a physician. Protective equipment or other control measures are necessary. **Moderate-High** means that exposure is believed capable of considerable exceedence of regulatory limits or causing serious traumatic injury. Advanced control measures or protective equipment are required. **High** means that short-term exposure is believed capable of causing irreversible injury including death. Advanced control measures or protective equipment are required.

## Shutting Down Fluid Power Systems

Among the most complex systems to de-energize and lock out are the fluid power systems. The course on lockout prepared by the UAW-GM (United Auto Workers-General Motors) National Joint Committee on Health and Safety on lockout recommended working on fluid power systems toward the end of the sequence of activities for deactivating, de-energizing, and isolating equipment and machines (UAW-GM 1985). For equipment and machines, the recommended sequence is a hierarchy starting with mechanical momentum sources, followed by gravity, stored mechanical energy, electrical system, pneumatic system, hydraulic system, and process and utility systems.

Shutdown of a fluid power system at the minimum requires return of loads to a supported neutral position and shutdown of the pressure source. As discussed previously, this does not ensure de-energization of all circuits in a system. This can occur only following confirmation through detailed analysis and testing, followed, as necessary by depressurization.

Tracing lines back to their source and examining component parts will assist in this process, along with review of drawings, if available. The investigator should be aware that backup systems could exist in a location remote from the main unit. Also, the investigation should determine whether a hazard could develop as a result of shutting down the pressurizing device. For example, pressure in the system could be essential for supporting some component. Support from a block could be insufficient to maintain safe positioning. Leaving the system energized could be essential.

## Isolation of Fluid Power Systems

The only practical isolation of components in a fluid power system occurs through closure of isolation valves. Closure meets the intent for isolation in regulatory standards, such as 29CFR 1910.147 (OSHA 1989). Note that lockout requires further action, to be discussed in the next section.

Isolation in this manner could apply to energy retained in storage devices such as air receivers, and possibly pulsation dampeners and accumulators. The valves used to effect the isolation must possess reliability and not pose a hazard to the user in their operation. Quarter-turn valves are not suitable in high pressure systems because of the tendency to snap open or closed when pressurized fluid impinges on the moveable element when moved partly from one end position of travel to the other.

Depressurization can occur safely only through appropriate components installed for this purpose as mentioned previously. These devices include permanently installed fittings, such as the valves that bleed pressure to air from components in pneumatic systems. In hydraulic systems that operate at considerably higher pressure, special consideration is necessary. Components for depressurization include quick disconnect fittings on affected circuits and devices containing needle valves and pressure gauges. These devices are readily available in the marketplace.

Depressurization and discharge of fluid should occur from both ends of intensifiers. Movement of moveable element in equipment during depressurization indicates that the achievement of a state of zero energy has not occurred. Blocking, pinning, and physical disconnection also may be required. Once deactivation and de-energization, of the hydraulic system has occurred, the last steps are to isolate and lock out the electrical control system and power supply (UAW-GM 1985).

## Lockout of Fluid Power Equipment

Lockout devices for fluid handling equipment are subdivided according to design and execution. Intrinsic lockout devices are designed into the equipment as supplied or are designed into retrofits created by the original equipment manufacturer. This is the most desirable route to follow since the manufacturer is most familiar with the equipment. Extrinsic devices are add-ons.

Lockout of pneumatic systems can occur at shut-off valves located in lines and incorporated into components (Figure H.9). Testing to determine the effectiveness of the lockout is essential. Tests should include attempts to reactivate the device following application of the lock. Leakage through loosely fitting devices could compromise the safety provided by the isolation. Even slight leakage could produce significant consequences.

Lockout of hydraulic systems, most of which are driven by electric motor, should occur at the hydraulic pump following isolation and depressurization (UAW-GM 1985; Vickers 1992). Lockout in this manner could necessitate shutdown of additional equipment powered by the same pump.

## Verification and Testing

Verification and testing of the effectiveness of isolation and lockout of fluid power circuits should involve determination of pressure in isolated circuits. This can occur through a pressure gauge and toggling controls, given the caveat about the necessity of pressure in piloted circuits.

Figure H.9. A component in a pneumatic system containing a built-in isolation and lockout point.

## Vacuum Loading Equipment

Vacuum loading is a large-scale industrial version of vacuum cleaning as performed in residences and industry using portable and fixed systems. Vacuum loading equipment includes truck- (Figure H.10) and trailer-mounted vacuum suction equipment (Dorman 1993a). Vacuum loading equipment consists of some or all of the following parts: suction tubes and hose, primary separator, debris collection tank, centrifugal (cyclone) separator(s), baghouse, final filter, vacuum blower/sludge pump, discharge air silencer, sludge loading/off-loading pump, and water washdown equipment.

The following material and discussion were obtained from manufacturer's literature and should be considered in their generic context rather than in the context of specific products.

This equipment can accommodate suction hose or metal tubing to 8 in (20 cm) diameter. The suction hose contains the vacuum safety relief. Maintaining suspension of solid particulate material requires a flow velocity of 4,500 ft/min (23 m/s). This is used as a design specification in ventilation systems (McManus 2000).

The primary separator is an optional add-on. This unit consists of a centrifugal (cyclone) separator and associated valves and control equipment. The primary separator intercepts debris in the vacuum line upstream from the vacuum loader. The cyclone separator removes particulates and liquid aerosols from the airstream with 98% efficiency. Intercepted material can be collected into 200 L drums (45 gal UK/55 gal US), bags, dump trucks, or roll-on/roll-off containers. Some units contain a dual discharge. This permits continuous, long-term operation.

For nonhazardous materials, the primary separator can dramatically improve productivity of the vacuum loading equipment since the debris collection tank does not become filled as quickly. In the case of hazardous materials such as asbestos and lead-containing dust or debris, this equipment minimizes contamination of the collection equipment in the vacuum loader. A water injection system is available for dust suppression and cleaning the hopper of the cyclone and eliminating fugitive emissions at the discharge.

Figure H.10. Vacuum loader truck. This unit contains air scrubbing equipment as well as a source of pressurized water and engages in hydroexcavating.

The debris collection tank contains deflectors that guide material flow to effect the initial separation of entrained material from the airflow. These also act to minimize resuspension of material already in the tank. Redirection of flow reduces air velocity in the entrained material. This initial separation reduces particulate loading in air entering the centrifugal (cyclone) separators. A float device prevents overflow of material from the collector body into the downstream air path. Most of the units drain by gravity from a raised position. A cover at the base of the rear panel or the rear panel itself opens to allow escape of collected material. Pneumatic (pressurized), pumped, or augered emptying also are available.

The centrifugal (cyclone) separator(s) in the truck body may occur singly or in multiples. Multiples offer the advantage of smaller diameter and redundancy in the event of damage. These units are rated at 98% efficiency. Liquid and solid debris from the cyclones collects in a storage area located below. This area drains through chutes that empty at the rear of the truck when the debris collection tank is elevated for dumping. A high pressure water injection system is available for cleaning the hopper and chute(s).

Airflow from the cyclones next passes through a baghouse. The baghouse is rated at 99% efficiency for all material larger than 1 μm. Debris that accumulates on the bags is blown off by periodic pulsed blasts of air. Debris from the bags collects in a storage hopper located below. The hopper drains through a chute when the rear of the truck is elevated for dumping. Hatches provide access for washing.

The final filter removes particulates that escape collection in the baghouse because of leakage in a bag or around a seal or through resuspension of debris knocked loose from the bags. The final filter protects the positive displacement vacuum blower. HEPA (High Efficiency Particulate Air) filtration is available as an option.

The air mover in wet/dry vacuum loaders is a high capacity rotary lobe positive displacement blower. These units can generate up to 28 in (94 kPa) of Hg (mercury) vacuum at flow rates up to 6,000 ft$^3$/min (2.8 m$^3$/s). A hydraulic sludge pumping system can be used for vacuum loading and offloading of liquids. Loading and offloading can occur simultaneously. An exhaust air silencer is situated downstream from the blower.

Some units incorporate a water jetting unit for washdown. This is essential where hazardous materials are involved and decontamination of the internals of the unit is required. A recent application of the water jetting unit is hydroexcavation (next section).

Vacuum loading equipment will accept anything that can fit through the opening at the end of the hose and be drawn along its length into the debris collection body. As a result, this capability simultaneously constitutes the greatest strength and the greatest liability of this equipment. Debris can originate from almost any conceivable source. What enters the collection hose reflects decisions made by the operator of the vacuum loader and the organization requesting the service as well as accidents of happenstance. An infinite number of complex mixtures and possibly some reaction products could form inside the collection tank when materials from different sources are combined. Strict attention to requirements of environmental and waste management legislation is required.

Operation of this equipment is not regulated directly in the U.S. at the workplace level (Dorman 1993). As a result, responsibility for safety in operation and application falls to the user. Manufacturers provide some assistance. This has progressed considerably over the last several years to the point where detailed information has become available for some of the equipment. Table H.4 summarizes hazardous conditions associated with operation of vacuum loader trucks.

Contact with overhead electrical utilities is an especial concern during operation of vacuum loaders containing boom extensions. Contact with overhead wires can occur during raising and extending the boom, and similarly during dumping the contents of the debris tank. An additional potential for contact exists when the unit is driven with the boom in the air or the debris tank in the raised position. Even low voltage wires contribute to the risk as insulation often separates from older lines leaving the conductors exposed.

The operator of the boom often cannot see the position of the lines due to obstruction by the body of the vehicle and lack of spatial perspective. (The operator tends to have a two-dimensional view of a three dimensional situation.) An additional contributor is focus by the operator on the position of the tube string at the end of the boom, and not on the surroundings, including the position of the boom itself. Contributing to this lack of visibility is loss of peripheral vision caused by the brim of the hardhat. The brim of the hardhat restricts upward peripheral vision such that overhead wiring is not visible. Positioning of the boom in a smooth, controlled manner depends on the dexterity of the operator and precision of control of position offered by the hydraulic system.

Another complication is the overhead wire itself. Overhead wires are part of the congested visual field. They appear to be completely innocuous, since they provide no warning to the senses about the danger that they pose, other than the passive one of their presence in the visual field and occasionally, buzzing. Warning flags strung near these lines greatly improve visibility but do nothing to reduce the consequences of contact. Shielding devices hung on the line provide some protection against accidental contact.

Insulating devices prevent transmission between the exposed conductor and workers touching the boom tube. Insulated or nonconductive boom tubes and wands must be able to withstand transmission when new and clean, but also when aged and wetted on the inside by conductive substances, such as salty water and mud.

Equipotential bonding and grounding utilize the concept of rendering all components at the same potential (voltage) through interconnection to ground. This concept is used in servicing microcomputers and electronic components. In the latter application, the worker stands on a conductive mat or rests the feet on the conductive mat while sitting and wears conductive wristlets. The electronic component rests on a conductive surface. All of the conductive components are connected together through cabling to a common electrostatic ground. Each component has the same (equi)potential. This concept works satisfactorily only where the work to be performed occurs within a short definable distance, usually a table or desktop, from the connecting point of the ground.

In the case of vacuum loading during hydroexcavating, for example, equipotential bonding is achieved by connecting all components in the system through cables of sufficient gauge to the truck and the truck to ground through a cable and ground rod. The ground rod must establish a predictable

## Table H.4

## Safety Concerns with Operation of Vacuum Loading Equipment

| Concern | Comment |
|---|---|
| Electrical contact | Some units have long extension booms that permit operation over obstacles. Booms of any kind increase potential for contact with overhead electrical power lines. |
| Mechanical contact | Vacuum loaders and their optional add-ons are complex units. They contain subsystems that rotate, articulate, operate under high vacuum, and contain spring-loaded and pressurized mechanisms. These subsystems produce or can produce motion during normal operation, failure mode, or during maintenance. Moving parts create potential for crushing, pinching, cutting, amputation, and decapitation. |
| Exhaust gases | Exhaust gases from diesel engines contain nitric oxide and nitrogen dioxide, carbon monoxide and carbon dioxide, sulphur dioxide, hydrocarbon vapors, and soot. Operation of these units in enclosed areas can confine exhaust gases. This is especially the case during operation in cold weather when exhaust gases stratify at low level. |
| Hot surfaces | Blower casings and associated piping and accessories can become hot enough to cause skin burns and act as a source of ignition. |
| Coupling failure | Failure at a coupling can lead to whipping action. |
| Electrostatic charge | Movement of particulates through vacuum systems at high velocity can cause accumulation of electrostatic charge. Uncontrolled discharge can act as a source of ignition. |
| Spills | Spills have occurred following accidental overfilling due to failure of shut-off mechanism in the collector tank. |
| Noise | Noise is a major operational hazard. Operational noise can mask warning signals and necessitate line-of-sight communication. |
| Decontamination | Interior of debris collection tank and internal surfaces may require decontamination following contact with hazardous material. Debris collection tank, cyclones, baghouse structure, hoppers, and chutes are potential confined spaces. These spaces can trap volatile atmospheric contaminants. Tailgate and access hatches require securement to prevent accidental closing. |

and reliable electrical ground suitable for the anticipated voltage. This can be difficult to achieve in soils encountered in prairie regions and very difficult in rocky terrain. Insufficient grounding renders the level of protection insufficient. When equipotential bonding is employed, the worker stands on a mat containing a woven interior of conductive wires. The mat is connected to the bonding and grounding system.

Realties of this work can considerably complicate use of this technique in routine circumstances. This work is wet and dirty. Mats, clothing, and other equipment become wetted and covered in mud and possibly earth spoil. This can create a conductive path between the mat and surrounding ground. This system is effective only so long as workers remain on the mats. Stepping from the mat to unprotected ground in the event of an electrical contact can expose the individual to a step potential. That is, the voltage experienced by each foot is different. This approach is counter productive to the realities of this work. Workers must have flexibility to move around the work area to move and fetch tools, such as pan shovels for lifting stones and debris from the hole, bars for breaking obstructions, and so on. They must also be free to move around the site to tend to the truck, traffic cones and signage, and interact with bystanders. The equipotential concept requires careful planning and supervision.

API 2219 is a code of practice for the operation of vacuum trucks in petroleum service (API 2005). This document distinguishes between pneumatic conveying equipment and liquid vacuum tankers. The former utilize high airflow created by a positive displacement blower to entrain material. According to API, this approach is not suitable for hydrocarbon service. Indeed, manufacturer's literature for wet/dry vacuum loaders refers to crude oil as the sole hydrocarbon liquid that they handle. Liquid vacuum tankers use vacuum created in the collection tank to draw liquids and sludges and pressure to expel them. The source of the pressure/vacuum is a low flow vane pump. This minimizes vapor formation in the airflow above the liquid during collection.

The major concern of API 2219 with high airflow is formation of ignitable mixtures (API 2005). High airflow across a surface wetted by a volatile liquid that forms ignitable mixtures promotes evaporation. Under suitable conditions of temperature and airflow, ignitable mixtures could form in the airspaces within this equipment. These could include the collector hose, debris collection tank, centrifugal separator, baghouse, collection hoppers, final filter, blower, and exhaust air silencer.

API 2219 documents incidents involving vacuum trucks in which fires have occurred. The accident summaries indicate that diesel engines, despite the limited electrical system compared to gasoline engines, are sources of ignition. Other possible sources include exhaust systems and external wiring. API 2219 points out that shielded ignition systems and muffler-mounted spark arresters do not substitute for a gas-free work area. Locating the truck remote from possible sources of vapor is an important safety precaution.

Dieseling (run-on) can occur when high concentrations of ignitable vapor enter the air intake of a diesel engine. Vapor clouds can develop through discharge of volatile substances by the vacuum pump or blower.

The blower used in vacuum loading units runs hot and can act as a potential source of ignition of ignitable mixtures both internally and externally. Fire and explosion suppression systems are installed in some of these units as an option. Slippage of drive belts can cause pulleys to run hot.

Material moving through the piping and hose can generate high levels of electrostatic charge. Electrostatic discharge can ignite mixtures of gas, vapor and dust in air, causing fires and possibly explosions. This is especially a concern where work is occurring in a hazardous location or where gas or vapor release is occurring from a failed pipeline or even the ground itself.

Bonding and grounding are important safety precautions for ensuring that build-up of static electrical charge does not occur. Continuity testing of the bonding cable is required. API 2219 mentions that petroleum industry experience indicates that electrostatic discharge does not present an ignition problem with either conductive or nonconductive hose. Exposed metal in isolated conductors including non-bonded, conductive hose and flanges and couplings can act as a capacitor and can accumulate considerable electrostatic charge. An arc can occur if this metal touches or is brought close to ground. API 2219, therefore, recommends bonding and grounding all metal parts when any hose is used in applications other than closed systems.

Attention to personal safety is a critical component of operation of vacuum loading equipment. This begins with wearing of standard safety equipment including a hardhat, safety glasses plus sideshields, protective clothing, and gloves. Protective clothing may include a chemical protective suit, and chemical protective gloves, depending on the environment and likelihood of contact. Vibration in the vacuum loading system would necessitate vibration-isolating gloves. Potential for exposure to the substance being vacuumed would necessitate respiratory and possibly other protection.

Of critical importance to safety in the operation of vacuum loading equipment is vacuum relief. Vacuum relief occurs through two main systems: an in-line emergency valve located near the open end of the hose (Figure H.11) and a vacuum release on the truck (Figure H.12).

The in-line vacuum relief valve is configured in the shape of a letter T. The valve is a is a flat disc containing beveled edges to which is attached a perpendicular rod. It is located in a "T"-shaped fitting. The "rip cord" attaches to the rod. During an emergency, the operator pulls on the "rip cord," This dislodges the disk from the seat and admits air into the line, thereby reducing the suction. The ease with which the flap releases indicates the considerably smaller contribution of the negative static pressure to the overall pressure needed to operate the vacuum suction system.

The vacuum release on the truck is a powered valve that releases the vacuum when the engine shuts down while still vacuuming to prevent damage to the blower.

Additional safety issues regarding operation of this equipment concern road safety from shifting cargo, braking distance, and cargo dumping.

## High Pressure Water Jetting

Water jet cleaning, or hydro-blasting has been used for many years. (Summers, 1993) Water jetting is used for loosening soil for vacuum removal during hydroexcavation, cleaning of surfaces, and removal of coatings, and cutting of materials including steel and concrete. Water jetting equipment pressurizes water to levels far above those available from the water supply system. These applications use the jet to loosen and remove residues and intentionally applied coatings. The blast can be applied manually or by remotely controlled delivery systems. Most of the equipment operates at pressures of 7 to 14 MPa (1,000 to 2,000 lb/in$^2$) and flows of 20 to 40 L/min (5 to 10 gal/min).

The Water Jet Technology Association (WJTA, 1994) defines the following ranges for differentiating pressures. Pressure cleaning (powerwashing) or cutting refers to pump pressures less than 35

Figure H.11. Vacuum relief valve. Pulling the cord breaks the suction between the disk and the wall of the valve. This is a practical demonstration of the low level of static pressure present in these systems. (Courtesy of Vactor/Guzzler Manufacturing Inc., Streator, IL.)

Figure H.12. Vacuum relief switch controls a vacuum release valve on the truck. This valve releases the vacuum to protect the blower.

MPa (5,000 lb/in$^2$). High pressure ranges from 35 to 210 MPa (5,000 lb/in$^2$ to 30,000 lb/in$^2$). Ultra high pressure exceeds 210 MPa (30,000 lb/in$^2$). UHP systems currently are achieving pressures close to 100,000 lb/in$^2$ (690 MPa).

Equipment used for low pressure water jetting operates at less than 35 MPa (5,000 lb/in$^2$). This pressure was chosen as an upper limit because considerable energy is required to improve jet performance above this value. The major application of this equipment is industrial cleaning. High pressure water jetting involves pressures exceeding 7 MPa (1,000 lb/in$^2$) at the orifice. High pressure water jetting is a relatively recent development. Table H.5 indicates minimum pressure required for various kinds of cleaning operations.

The basic components of all water jetting system are the same: a pump, pressure-relief system, hose (piping), and nozzle (Figure H.13).

The heart of this equipment is the pump. The most common type for water jetting is a positive displacement plunger pump. A plunger pump has some resemblance to a piston pump (Yared 1994). The plungers are driven through rotation of a crankshaft (Anonymous 1994a). The plunger, having very small surface area compared to a piston, concentrates the motive force in order to produce higher pressure. Achieving these high pressures typically requires use of intensifiers.

Intensifiers are powered hydraulically or by air. Hydraulically-powered intensifiers typically operate in the range of 210 to 410 MPa (30,000 to 60,000 lb/in$^2$) with flows of 4 to 20 L/min (1 to 5 gal/min) (Anonymous 1994a). Air-powered intensifiers can produce pressures as high as 700 MPa (100,000 lb/in$^2$) with a flow of 0.4 L/min (0.1 gal/min). An intensifier uses a large, low pressure hydraulically driven piston to drive small high pressure plungers that contact the water. The plunger displaces the water in the high pressure cavity. An intensifier is equivalent in electrical terms to a step-up transformer.

A critical component in these systems is the pressure-relief system (Williams 1995). These systems utilize two designs: an on/off valve or a bypass. Both are located in the gun and are used to control

## Table H.5
## Pressures Needed for Removal of Embedded Materials

| Pressure | | Application |
|---|---|---|
| MPa | lb/in$^2$ | |
| 10 | 1500 | Silt, loose debris |
| 20 | 3000 | Light marine fouling, light scale, fuel oil residue, aluminum cores and shells |
| 30 | 4500 | Weak concrete, medium marine growth, sandstone and mudstone, light mill scale, limited core removal, loose paint and rust |
| 40 to 70 | 6,000 to 10,000 | Concrete in pipes, severe marine fouling, ferrous casting molds, runway rubber, soft limestone, lime scale, burnt oil deposits, medium mill scale, petrochemicals |
| 70 to 100 | 10,000 to 15,000 | Concrete cutting and removal, most paints, medium limestone, most mill scale, silica cores, burnt carbon deposits, heavy clinkers |
| 100 to 200 | 15,000 to 30,000 | Granite, marble, limestone, marine epoxies, aluminum, lead, rubber, frozen food |

Source: Summers 1993.

flow. The on/off valve is either open or closed. A nitrogen-charged bladder that controls an internal valve modulates pressure in the system. Expansion or contraction of the bladder maintains constant pressure. The bypass design dumps water to atmosphere through an unrestricted port. The bypass is hand- or foot-operated. Pressure and flow can be modulated according to the setting of the valve in the bypass. Each design has its respective advantages and disadvantages.

Flexible, wire reinforced hose is the common way to conduct water from pump to nozzle (Anonymous 1993). With correct sizing, lengths of several hundred feet are possible. A variety of delivery tools are available: hand lances, shotguns, mole heads, flex lances, and stiff lances. Nozzle tips can produce round, needle, or fan sprays (Figure H.14). Nozzles are made from steel, tungsten carbide, ceramic, and sapphire as well as coated and plated materials.

Heating the water after it leaves the pressure pump improves performance when cleaning hydrocarbon and similar coatings of dirt. Chemical substances also are used in some cleaning operations. Long-chain polymers enable pressure jets to deliver greater energy over longer distances by increasing cohesive forces. Bubbles increase the area of coverage without requiring an increase in delivery power. Abrasives added to the water jet assist in removing especially difficult coatings. Hard media such as sand, garnet and slag are used, as well as soluble abrasives, such as sodium bicarbonate (Dorman 1993b, Anonymous 1994b).

High pressure water jetting creates a number of hazardous conditions (Table H.6) (WJTA-IMCA 2005). Water jetting applications and ultimately the hazard posed by pressurized liquids depend on delivery pressure. In practice, there is no pressure at which fluid jets are totally safe. Every fluid jet has the potential to cause human injury. Some of this concern relates to the diameter of the jet. Equipment operating at a wide range of pressures is readily available to consumers from big-box, as well as corner

Figure H.13. Water jetting systems, such as this unit available to consumers, can generate pressures approaching 5,000 lb/in² (35 MPa).

Figure H.14. Nozzle pattern of a high pressure jet of water. Note that the pattern delivers the jet by indirect paths as well as the direct one. This reduces the potential for damage caused by the jet. (Photograph ©2011 by NLB Corp., Wixom, MI. Used with permission.)

## Table H.6
## Hazardous Conditions Created by Water Jetting

Piercing of the skin and injection by the water jet

Flying debris emitted from the point of contact of the jet with surfaces

Release of hazardous materials into the atmosphere from dislodged deposits

Noise which can impede communication and exceed regulatory limits

Difficulty in breathing due to the high concentration of water vapor and mist in the air

Electrostatic charging of nozzles and lances and discharge and potential for discharge

Electrocution caused by damage of electrical cables and wires and creation of a conductive path

Reactive force generated by the discharge of water creating a backward thrust against the nozzle holder

Catastrophic destruction of equipment by prevention of flow by ice in the line and high pressure water injected by the pump

Source: WJTA-IMCA 2005.

hardware stores. Hence, the risk of injury due to use of pressurized fluids is not restricted to workplaces. The same can be said for both air powered and airless spraying equipment which is also available to consumers.

Equation H-1 indicates the parameters involved in water jets that discharge to atmosphere.

$$F_r = Q\rho v = \rho A v^2 \tag{H-1}$$

where:

$F_r$ = reaction force exerted by a flat surface, N.

$Q$ = volumetric flow of water, $m^3/s$.

$\rho$ = density of water, 1,000 $kg/m^3$.

$v$ = velocity, m/s.

$A$ = area of opening, $m^2$.

The maximum recommended counter-reactive force applied by the operator for an extended period of time should be equivalent to less than one third (1/3) of body weight (WJTA-IMCA 2005).

Hydroexcavation is a major application of water jetting. The equipment operates in the range of pressure cleaning. A jet of water loosens soil and a vacuum loading system removes it to a debris tank. Hydroexcavating technology using heated water enables work to occur during the coldest part of the winter.

Hydroexcavation offers major advantages in speed and reduced risk to workers during work involving underground electrical cable, since direct contact with tools such as picks and shovels, does not occur. In work involving unprotected cable, the water jet exposes the cable free from adhering soil. Following repair during winter, these cables reseat and refreeze. The pressure at which the water jetting equipment operates (1,500 to 3,000 $lb/in^2$ or 10 to 21 MPa) is sufficient to cut or damage insulation on wires and cables when used in a narrow jet on prolonged contact. This damage manifests as tearing of the cover. In warmer times, rain penetrates the soil, and in some cases, penetrates the insulation causing a new failure. Some cables are so prone to damage of this nature that hydroexcavating is contraindicated as an approach to expose them.

Unintended contact with underground utilities, especially electrical utilities, is an especial concern during hydroexcavation. Underground utilities include telephone and communications cable, streetlight power cables, electrical distribution, and household power lines and electrical transmission lines. Telephone and communications cables normally do not pose an electrocution risk. The electrical system contains distribution and transmission lines. Voltage in underground electrical distribution lines ranges from household levels to 25 kV. Voltages in underground transmission lines ranges to 250 kV. The latter are contained in concrete-encased plastic duct.

Unexpected change in grade compared to depth of over-cover indicated on as-built drawings is a major cause of unintended contact. Unexpected change in grade can mislead about the actual versus indicated depth profile. Another cause of unintended contact is failure of locates to indicate the presence of underground utilities. This can result from discontinuity in lines due to use of plastics and absent metal tracing in place of metal piping and conduit.

Exposure during high pressure water jetting of conductors carrying large quantities of current at low voltage and contact with an ungrounded system (the wand or boom tubes + the truck) pose an electrocution risk of unknown magnitude to the operators who may simultaneously touch a conductive object to complete the path to ground. Current flow to the truck and its components is limited only by the tolerance of the circuit protective device.

While the hazard of circuits operating at 480 or 600 Volts is intuitively obvious, the hazard posed by circuits operating at 110 or 220 Volts is just as real but not as obvious. Many fatal accidents have occurred due to exposure to circuits operating at 110 Volts. Procedural controls fail to offer the level of protection that is needed during this work. Protection is achievable only through use of insulating devices and/or by the technique of equipotential bonding and grounding as discussed in the previous section.

Movement of fluid from the end of the nozzle and formation of droplets of water can lead to electrostatic charging on equipment. Removal of material from surfaces following contact with the jet of fluid also can lead to electrostatic charging. Accumulation of high levels of electrostatic charge can occur during this process. Arcs created during electrostatic discharge are sources of ignition of ignitable gases and vapors and combustible dusts. Bonding and grounding of this equipment is necessary to prevent this situation. Establishing an electrostatic ground may require driving a ground rod.

An important concern in this industry is freeze-up. Ice in lines can have a serious impact on the safe use of this equipment. Given the small diameter of the tubing used in this equipment, formation of a plug of ice can occur readily during sub-freezing weather. Blockage of flow channels against the high pressure exerted by the fluid could lead to leakage and emission of a jet of water at high pressure or catastrophic failure.

Basic protective clothing and equipment thus are mandatory for every operator and possibly for assistants, depending on proximity to the work area. Adequate precautions are required at all pressures. Katakura and Tsuji (1985), who tested the effectiveness of PVC clothing as protection against injury, concluded that while PVC clothing may be useful, there is no guarantee that it can provide full protection against injury.

Table H.7 outlines protective equipment likely to be needed during this work. Introduction of body armor to this arena is an essential addition to personal protection as service pressure increases (Figure H.15). Protective devices must be able to ensure protection of the operator, the assistant and bystanders against the potential injurious impacts from these jets.

## References

ANSI: Fluid Power Diagrams (ANSI Y32.10—1967 (R1999)). New York: American National Standards Institute, 1999.

Anonymous: Contractor's Program Featured at 7th Water Jet Conference. *Indust. Clean. Contractor 1(5)*: 14-17 (1993).

## Table H.7
## Protective Measures for Water Jetting

### Direct Contact with the Water Jet

| Part Affected | Protective Measure |
| --- | --- |
| Gun | At least one shut-off valve, dump valve or other shut-off device |
| Lance | Lance shield to block back spray |
| Delivery hose | Protection against crushing and other damage |
| Hose connections | Contingency protection to prevent unexpected separation |
| Surrounding area | Barricades or barriers to restrict access |

### Electrical Protection

| Part Affected | Protective Measure |
| --- | --- |
| Whole body | De-energize and lock out or install equipotential bonding and grounding of equipment and work zone |

### Electrostatic Charge Generation and Discharge

| Area Affected | Protective Measure |
| --- | --- |
| Atmosphere | Bond and ground equipment |

### Freeze Protection

| Part Affected | Protective Measure |
| --- | --- |
| Flow channels | Disassemble gun from delivery hose and drain water to dryness |
| | Add anti-freeze to the water tank and circulate through the system |

### Personal Protection

| Area Affected | Protective Measure |
| --- | --- |
| Head | Hard hat |
| Eyes and face | Faceshield + chemical splash goggles or full facepiece respirator |
| Respiratory system | Respirator selected to provide protection against expected levels of hazard predicted to be present |
| Ears | Hearing protection (jetting noise easily can exceed 90 dBA) |
| Torso and limbs | Waterproof and/or liquid- or chemical-resistant suits, body armor |
| Hands | Waterproof and/or liquid- or chemical-resistant gloves Vibration-resistant gloves, body armor |
| Feet | Waterproof boots with steel toe caps + metatarsal guard for jetting gun operators, body armor |

Source: WJTA-IMCA 2005.

Figure H.15. Body armor is now available for protection against high pressure water jets. This clothing allows the operator to maintain close proximity to the action of the jet in order to be able to observe its action on the surface. (Courtesy of Warwick Mills, New Ipswich, NH.)

Anonymous: High Pressure Pumps, Intensifiers and Accumulators. *Indust. Clean. Contractor 2(6)*: 18-19 (1994a).

Anonymous: New Baking Soda System Adds Muscle to Power Washing. *Indust. Clean. Contractor 2(4)*:22-25 (1994).

API: Safe Operation of Vacuum Trucks in Petroleum Service, 3rd Ed. (API Pub. No. 2219). Washington, DC: American Petroleum Institute, 2005.

Bolton, W.: *Pneumatic and Hydraulic Systems*. Oxford, UK: Butterworth-Heinemann, 1997.

Brown, G.:Engineering Hose Assemblies for Safe and Reliable Performance. Cleveland: Penton Publishing Company. Originally published in *Hydraulics & Pneumatics*. www.fpweb.com [website]

Dorman, S.: Technological Developments in Industrial Vacuum Loaders, Part II. *Ind. Clean. Contractor 1(5)*: 24-28 (1993a).

Dorman, S.: High Pressure Water Jetting Systems, Part I. *Indust. Clean. Contractor 1(5)*: 9-13 (1993b).

FPSI: Maintenance Safety Bulletin (MSB-025). Salt Lake City: Fluid Power Safety Institute, 2002a. www. fluidpowersafety.com [Website].

FPSI: Maintenance Safety Bulletin (MSB-012). Salt Lake City: Fluid Power Safety Institute, 2002b. www. fluidpowersafety.com [Website].

FPSI: Maintenance Safety Bulletin (MSB-015). Salt Lake City: Fluid Power Safety Institute, 2002c. www. fluidpowersafety.com [Website].

FPSI: Maintenance Safety Bulletin (MSB-022). Salt Lake City: Fluid Power Safety Institute, 2002d. www. fluidpowersafety.com [Website].

FPSI: Maintenance Safety Bulletin (MSB-026). Salt Lake City: Fluid Power Safety Institute, 2002e. www. fluidpowersafety.com [Website].

FPSI: Engineering Safety Bulletin (ESB-003). Salt Lake City: Fluid Power Safety Institute, 2002f. www. fluidpowersafety.com [Website].

Haller, J. and G. Kleiner: The Four Culprits – Where to Look to Reduce Hydraulic System Leakage. Cleveland: Penton Publishing Company. [www.fpweb.com, Website].

HP: Hose and Tubing Assemblies. In *Fluid Power Directory*. Cleveland: Penton Publishing Company, 2003a.

HP: Hose Installation. In *Fluid Power Directory*. Cleveland: Penton Publishing Company, 2003b.

HP: Leakage Prevention. In *Fluid Power Directory*. Cleveland : Penton Publishing Company, 2003c.

Katakura, H. and S. Tsuji: A Study to Avoid the Dangers of High Speed Liquid Jets. *Bull. Japan Soc. Mech. Eng. 28*: 623-630 (1985).

Kemper, D.: A Hose for Every Job. Cleveland: Penton Publishing Company. Originally published in *Hydraulics & Pneumatics*. [www.fpweb.com,Website].

McLaren, R.: Fluid Power Safety in the Workplace, Part 1. *Hydraul. Pneumat. 54(6)*: 44-46, 62 (2001a).

McLaren, R.: Fluid Power Safety in the Workplace, Part 3. *Hydraul. Pneumat. 54(6)*: 44-46, 62 (2001b).

McLaren, R.: Fluid Power Safety in the Workplace, Part 5. *Hydraul. Pneumat. 55(7)*: 35-37 (2002a).

McManus, N.: *Portable Ventilation Systems Handbook*. London: Taylor & Francis, 2000.

Nelson, C.A.: *Millwrights and Mechanics Guide*. New York, NY: The Bobbs-Merrill Company, Inc., 1983.

OSHA: "Control of Hazardous Energy Sources (Lockout/Tagout); Final Rule," *Fed. Regist. 54*: 169 (1 September 1989). pp. 36644-36696.

Summers, D.: Historical Perspective of Fluid Jet Technology. In *Fluid Jet Technology. Fundamentals and Applications*, 2nd Ed. St. Louis: Water Jet Technology Association, 1993.

UAW-GM: Lockout. Detroit: United Auto Workers-General Motors Human Resource Center, 1985.

Vickers: *Vickers Industrial Hydraulics Manual*. Rochester Hills, MI: Vickers, Inc., 1992.

Williams, D.: Shut Off Gun versus Dump Gun. *Indust. Clean. Contractor 3(1)*: 18-19 (1995).

WJTA: *Recommended Practices for the Use of Manually Operated High Pressure Water Jetting Equipment*. St. Louis: Water Jet Technology Association, 1994.

WJTA-IMCA: *Recommended Practices for the Use of High Pressure Water Jetting Equipment*, 4th Ed. St. Louis, MO: Water Jet Technology Association-Industrial & Municipal Cleaning Association, 2011.

Yared, D.J.: Contractor Waterjetting: Equipment and Applications. *Indust. Clean. Contractor 2(4)*: 10-15 (1994).

# I Fluid-Handling Systems

## INTRODUCTION

This appendix addresses fluids in systems other than fluid power systems (hydraulic, pneumatic, and vacuum). Fluids in this context include liquid and gaseous phase substances used in processes or handled during commercial activities. Systems handling liquids can be unpressurized or pressurized. Unpressurized systems can include partially and fully enclosed conduits including channels, duct, and piping.

## Fluid-Handling Systems

Fluid handling systems pose considerable challenges to individuals attempting to assess risk and to create procedures for deactivation, de-energization, isolation, and lockout. Given all of the piping and valves, pumps, and other components, a person unfamiliar with these systems is likely to experience considerable anxiety and discomfort in addressing these questions.

### Residential Water Supply System

The water supply system in a house provides a straightforward example and a starting point for discussing about fluid-handling systems. The water supply system in a house contains piping, around 15 to 25 valves (depending on complexity), and end-point consumers of water provided by the piping at various temperatures. Viewed on a drawing, a home water supply system is a maze of branches and interconnections and complex structure.

This is the situation that confronts anyone lacking a trained eye who attempts to create procedures for isolation and lockout. Yet, to the weekend plumber, the system, as described, is merely a means to move water at various temperatures to various places and to provide the means to control and isolate the flow. Furthermore, the weekend plumber learns about breakdown and wear and tear on components, and the strategies required to isolate and depressurize lines, and to remove and replace or modify them.

The water supply system in the house contains many isolation valves. Their function is to enable work to occur on various parts of the system without the need for shutting down the remainder. The home plumber shuts the isolation valve and opens the tap to allow the liquid to flow and to depressurize the line. Otherwise, pressure in the line will disappear at the moment cutting occurs when water will spray and wet surfaces of building materials and contents.

The preceding discussion describes the way the residential water supply system is supposed to function. The reality is often considerably different. Isolation valves installed in water supply systems in houses typically are globe valves. These contain a seat of a plastic/rubber material. Normally these valves remain in the fully OPEN position for long periods of time. This condition exposes the seat to water flow. Over the passage of time, the seat erodes due to exposure to the flow of the water and the surface intended to make the seal disappears. Erosion is most evident in taps in parts of the system that pass large quantities of hot water. These include bathtubs and showers, toilets, kitchen sinks and dishwashers. Erosion is also evident in taps in circuits that pass small amounts of water and are subject to quiescence for long periods. These include bathrooms that receive little use.

Erosion of the soft seat of the valve by fluid flow or other mechanism means that the valve no longer functions in one of its intended roles, namely to provide isolation. In this situation, isolation of the line must occur upstream from this valve, possibly at another valve in the line or at the isolation valve for the house. The isolation valve for the home water supply system is is located at the entry of the water supply into the building. If that valve fails to stop flow of water due to age and erosion, another valve is present in the water supply system at the property line. If the latter valve fails to prevent water flow, life becomes considerably more complicated because isolation now must occur at valves located under the surface of the ground in the street in the water distribution system.

Contact with liquid remaining in the line is also possible. Liquid remaining in the line impedes or prevents use of heat for soldering and brazing. Upon strong heating of the pipe, water in the line heats and eventually boils. Boiling water released from discontinuities can cause scald injury. Cutting of isolated, depressurized pipe allows drainage of contents and wetting of surfaces. Wetting can cause considerable damage to building materials and contents.

The hot water side of a residential plumbing system is the more susceptible to leakage and failure. This is especially true for taps that are used regularly for providing and controlling flow of hot water. Washers used in seats are subject to more rapid wear.

Cold weather can freeze cold water in stagnant sections of piping. Expansion of the plugs of ice during freezing has caused splitting of pipe.

While this example may seem elementary in its simplicity, the residential water supply system is a microcosm of systems that handle large quantities of fluid. Sometimes the fluids are water or water-based solutions and slurries, and often are other liquids and gases that are substantially more hazardous than water.

In summary, previous discussion identified that fluid-handling systems contain branches and sometimes multiple and alternate paths that distribute to the same end-point. These systems rely on valves for isolation. Seats in valves exposed to the fluid for long periods in the partially or fully OPEN position are subject to erosion and possibly other mechanisms that induce failure. Failure can occur in multiple valves in a sequence intended for use in isolation. Systems usually contain redundant points for isolation. Sometimes isolation is not possible without extensive and extraordinary measures not anticipated in the original design and execution of the system.

Retention of pressure and fluid can occur in sections of piping following isolation due to deliberate storage or entrapment. The fluid can spray outward following partial or complete severance of the pipe upon creation of a path from containment of the structure. Residual fluid in depressurized piping still can cause exposure and damage upon release from containment.

## Potable Water Treatment and Distribution Systems

Conventional treatment to produce potable (drinkable) water is mostly a gravity flow process (Figure I.1). Water flows by gravity from the source into chambers positioned at increasingly lower elevations. Supplying raw water to the intake may require piping, valves, and pumps, depending on the source. Isolation of the chambers involves use of slide gate valves mounted on the walls. The height of the gate relative to the height of the water passing through the opening in the wall governs the extent of flow. Water treatment plants also contain pressurized systems. The latter are used for transferring process chemicals and operation of the plant.

Some water purification systems employ reverse osmosis involving membranes. The pressure required depends on the concentration of the salt solution on the pressure side of the membrane. Salt removal from seawater requires >800 lb/in$^2$ (>5,500 kPa) gauge pressure. Residential units located under the sink operate at 50 lb/in$^2$ to 70 lb/in$^2$ (345 kPa to 480 kPa).

Potable water distribution systems usually are underground and as a result are not part of normal awareness. Transmission systems utilize piping and sometimes open channels. In the latter case, purification occurs at the receiving end of the transmission system.

Gauge pressure in distribution lines ranges from 120 lb/in$^2$ to 150 lb/in$^2$ (830 kPa to 1,030 kPa) and as high as 250 lb/in$^2$ (1,720 kPa). The lines interconnect with those from alternate sources in order to ensure a continuing supply of water in the event of a major disruption of the infrastructure. Water supply piping and components remain buried and undisturbed for long periods of time. The valves usually remain in the fully OPEN position, and occasionally the fully CLOSED position.

These systems contain single valve isolations. Obtaining complete isolation of flow may require closing more than one valve in a series. Occasionally a failure occurs in piping or a connection. This leads to rapid erosion of surrounding fill and often a geyser of water that shoots into the air. These failures are very rare events.

Figure I.1. Potable water treatment occurs mainly in gravity flow systems. Behavior of fluids in open channels is as important for establishing control as for enabling flow.

Gauge pressure in the residential distribution system ranges from 60 lb/in² to 80 lb/in² (410 kPa to 550 kPa). Supply pressures at the upper level necessitate use of pressure-reducing valves in residences in order to prevent failure due to overpressure in supply piping.

Designers and owners specify valves used in potable water service to meet standards produced by the American Water Works Association (AWWA). Appendix A mentions these standards along with criteria for acceptable leakage. Standards produced by AWWA for individual types of valves specify acceptable rates of leakage, as new and, in some cases, following multiple openings and closings. The latter are intended to mimic conditions experienced by the valves in service. The standards also provide guidance for maintenance and servicing

## Wastewater Collection and Treatment
Wastewater handling operations mostly involve gravity flow between the source of wastewater and the treatment plant in the collection infrastructure. Within the wastewater treatment plant, flow occurs mostly by gravity between units. The most prominent parts of the collection system are the sewers and related components. Sewers usually are underground, except in locations that experience severe winter conditions.

Less noticeable among the components of the wastewater collection system are the lift (pump) stations. These components collect wastewater to a central location and pump it to the sewer at a higher elevation. The discharge pressure of the pumps is typically around 20 lb/in² (140 kPa), and residual pressure in the force main (the discharge line) around 10 lb/in² (70 kPa).

Conventional wastewater treatment is mostly a gravity flow process. Water flows by gravity from the source into chambers positioned at increasingly lower elevations. Supplying wastewater to the intake may require piping, valves and pumps, depending on the source. Isolation of the chambers involves use of slide gate valves mounted on the walls. The height of the gate relative to the height of the water across from the opening in the wall governs the extent of flow. Wastewater treatment plants also contain pressurized systems. The latter are used for transferring process chemicals, tank filling and operation of the plant.

These systems contain single valve isolations, along with some backflow prevention. Obtaining complete isolation of flow may require closing more than one valve in a series. Occasionally a failure occurs in piping or a connection. These failures are very rare events. The valves often remain in the fully OPEN position and occasionally the fully CLOSED position. Valves that meet AWWA standards have known leakage rates when new and possibly in service.

## Combustion Systems

Combustion systems are very common in industrialized societies. These provide the heat needed in boiler systems to produce steam, in cupola foundries and blast furnaces to melt metals, in drying units to remove moisture from materials, and in ovens to roast materials beyond dryness. Combustion systems use various fuels including log and wood waste, coal, oil and waste oil and natural gas burned in a furnace and boiler.

Combustion systems often occur in twos or threes, so that one or two units remain operational during service while the second or third undergoes service, as the case may be. This design often utilizes a single stack and a header (manifold) to channel combustion gases to the stack. These systems also usually contain one or two fans per unit to optimize combustion efficiency. The result is pressurization of the duct that leads to the stack at all times during which one or more units in the system are operational. This situation requires isolation of the discharge duct of the inoperative unit(s) from the discharge duct receiving pressurization from the operating unit.

The usual approach to achieve isolation of flow is installation of a plate across the discharge duct of the inoperative unit(s) at a flange between sections. This is considered superior to closing a damper in the discharge line.

Pressures in ducts created by fans are considerably lower than pressures achieved in pipes by pumps. As a result, the gauge of metal used in the plate is less of a concern in these applications than the gauge of metal used in the plate of a gate valve. Bolts secure together the flanges and the plate. Outflow of hot pressurized combustion gases and possibly particulates is likely during isolation upon opening the flange between sections of duct in the inoperative unit(s). Hence, isolation by both means ensures that spillage of combustion gases into the boiler house due to leakage at the flange is minimal.

## Steam Systems

Steam systems are used for energy transfer in many applications. These include nuclear and fuel-burning thermal generating stations, oil refining and petrochemical processing, manufacturing processes, and heating of buildings.

As the temperature increases and the water approaches its boiling condition, some molecules attain enough kinetic energy to reach velocities that allow them to escape momentarily from the liquid into the space above the surface, before falling back into the liquid (Spirax-Sarco 2011a). Further heating causes greater excitation and the number of molecules with sufficient energy to escape from the liquid increases. When the number of molecules escaping from the liquid surface exceeds the number returning, the water has reached boiling point or its saturation temperature. At atmospheric pressure, the saturation temperature is 100° C. Allowing the pressure to increase will allow the addition of more heat and an increase in temperature.

When steam is able to flow from the boiler at the rate of production, addition of further heat solely increases the rate of production. Excess energy raises the pressure, in turn allowing the saturation temperature to increase, because the temperature of saturated steam correlates to its pressure.

Steam having a temperature equal to the boiling point at a particular pressure is known as dry saturated steam. Production of 100% dry steam in an industrial boiler designed to produce saturated steam is rarely possible, and the steam usually contains droplets of water.

Steam at a condition above saturation is known as superheated steam (Spirax-Sarco 2011b). Temperature above saturation temperature is called the degree of superheat of the steam. Superheat cannot be provided to the steam while water is present. Additional heat evaporates more water. To cause superheating, the saturated steam must pass through an additional heat exchanger. This may involve a second heat exchange in the boiler, or a separate superheater unit. The heating medium can include either hot flue gas from the boiler or a separately fired unit.

Superheated steam powers turbines and process heating in steam plants around the world, especially in oil refining and petrochemical production. This application is more likely to occur in these operations because superheated steam is already available on site for power generation.

Superheated steam must cool to saturation temperature in order to release heat of evaporation through condensation. The amount of heat given up by superheated steam as it cools to saturation temperature is relatively small compared to the heat of evaporation. As a result, superheated steam is not as effective as saturated steam at the same pressure for applications involving heat transfer.

Steam systems contain single valve isolations. Obtaining complete isolation of flow in an operating system may require closing more than one valve in a series. Occasionally a failure occurs in piping or a connection. These failures are very rare events. The valves often remain in the fully OPEN position and occasionally the fully CLOSED position.

Work involving steam systems often occurs overhead from a ladder due to cramped conditions. The surfaces at these locations are often very hot due to absence of insulation. Thermal burn from unintended contact with hot surfaces is an ever-present risk arising from this work. Unintended and unexpected pressurization in a line undergoing isolation is a potential reality acknowledged by operators and maintainers of these systems. Exposure to steam during the opening of flanges has occurred on many occasions. Workers in steam plants take avoidance measures during these activities. Attempts to work on steam lines has led to failure of metals due to embrittlement (OSHA 1985).

## Refrigeration Systems

Refrigeration is a universally applied technology in the industrialized world. The largest installations of refrigeration systems include commercial food preparation and storage, ice arenas, petrochemical installations processing low-boiling products, and gas liquefaction plants.

Ammonia is commonly used in large-scale compression refrigeration systems. A compressor compresses the gas to liquid while a water-based evaporative cooling system removes the heat from both the compressor and the liquefied ammonia. Expansion of the liquefied ammonia to gas chills brine, a solution of calcium chloride, that circulates through piping in the installation. The salt solution has a considerably lower freezing temperature than pure water, a phenomenon known as freezing point depression. Use of brine as the medium for heat transfer minimizes the quantity of ammonia in circulation and localizes its presence to the refrigeration room.

Ammonia is highly toxic and must not escape from containment because of inappropriate measures taken during isolation. Ammonia can dissolve into the oil used for lubricating the compressor and into the brine. Off-gassing can cause considerable overexposure during oil changes (WorkSafeBC 2007).

These systems contain single valve isolations. In emergency situations, ammonia vents to the outdoors. Obtaining complete isolation of flow may require closing more than one valve in a series. Occasionally a failure occurs in piping or a connection. These failures are very rare events. The valves often remain in the fully OPEN position and occasionally the fully CLOSED position.

## Natural Gas Transmission and Distribution Systems

Natural gas has become an important source of energy to industry, institutions, and residential customers. Like the potable water distribution system and the wastewater collection system, the system involved in transmission and distribution of natural gas is a largely unnoticed part of the underground infrastructure.

Transmission of natural gas occurs in high-pressure pipelines. These pipelines can operate at gauge pressures of 750 lb/in$^2$ (5200 kPa). Pressures in the distribution system range progressively downward. Pressure provided in residences is around 0.5 lb/in$^2$ (3 kPa).

These systems contain single valve isolations. Obtaining complete isolation of flow may require closing more than one valve in a series. Occasionally a failure occurs in piping or a connection. This leads to rapid escape and discharge of a jet of pressurized natural gas into the air, followed by ignition. These failures are very rare events.

Repair of the failure requires closing the isolation valve and possibly installing an additional isolation valve in the line through a technique known as a hot tap. This technique involves specialized
equipment for penetrating the pipeline and installing the valve. Operation of this equipment must itself
not create an additional hazard.

## Crude Oil and Products Pipelines

Crude oil is an important source of energy in industrialized economies. Like the system involved in
transmission and distribution of natural gas, the transmission of crude oil is an unnoticed part of the infrastructure. Pipelines handling crude oil tend to be monolithic heading from origin to destination.
There is little, if any, distribution from the main line in the route. These pipelines can operate at gauge
pressures of 750 lb/in$^2$ (5200 kPa).

These systems contain single valve isolations. Obtaining complete isolation of flow may require
closing more than one valve in a series. Occasionally a failure occurs in piping or a connection. This
leads to rapid escape and discharge of a jet of oil into the air, followed by ignition. These failures are
very rare events.

Repairing the failure requires closing the isolation valve and possibly installing an additional isolation valve through a technique known as a hot tap. This technique involves specialized equipment for
penetrating the pipeline and installing the valve. Operation of this equipment must itself not create an
additional hazard.

## The Operator: The Critical Resource

The situation that confronts the first-time visitor to a facility is bewilderment with the complexity of
piping, valves, and components in drawings and in real life (Figure I.2). What is the stranger to think
about a boiler system in a building where there are many valves, and everything is very hot? How does

Figure I.2. The complexity of systems found in buildings and other installations is overwhelming to the
visitor.

one devise a means to shut down such systems? Some jurisdictions require a Qualified Person to create a procedures for isolation and lockout as part of the procedure for entry and work involving confined spaces. Hence, this person could be a stranger to the facility for which the procedures are to be created.

This begs the question about how to approach the problem in a strategic way to gain the needed information in an efficient manner and to create a functional procedure where none exists. Previous discussion has pursued this theme in a general way.

The starting point in the quest for information regarding fluid-handling systems is the operator. The operator is the expert regarding the manner of operation of a particular system. The operator "lives" with the equipment and must learn how to gain utmost performance and efficiency from it without endangerment. The operator often knows not only what the equipment is supposed to do and how it does it, but also what the equipment actually does do. This information usually is undocumented. Hence, making contact with and establishing a positive and open relationship with the operator is critically important for obtaining a successful outcome.

A positive and open relationship is not always easy to establish because of the natural and real concern about the loss of security and control over information that revealing "secrets" could entail. Some operators resent the intrusion into their sphere of influence. This is a natural expression of the territorial imperative. Gaining willing cooperation of these individuals is critical to success and may require considerable patience. This situation also requires demonstration of legitimate interest and previous knowledge of the system to indicate that the visitor has come to learn and to contribute at the same time.

The visitor is not intending or attempting to "reinvent the wheel," but is attempting to capture the knowledge of the operator and to ask intelligent questions at the same time. This is a valuable enquiry that the operator should welcome, because some of the current practices could violate regulatory requirements and possibly safe work practices. Hence, the intent of the exercise is to produce the best possible version of the procedure.

The employer could ask the operator to prepare the procedure. This approach makes a presumption that may not be appropriate. Operators operate. They are hands-on people. Procedure writing that captures complex instructions in simple phrases is an art and a developed skill. Capturing complex thoughts in simple language requires discipline.and practice. The Qualified Person, who has received training in hazard recognition, is often a health and safety practitioner. The outcome from the enquiry is to create a string of instructions that capture in a few thoughts what to do, so that anyone could operate the equipment. This could become necessary, given the sudden death, prolonged recovery from traumatic injury or disease, or retirement of the operator. The "anyone" could include the operator returning to work following prolonged absence, a replacement, or a trainee. A simple expression of how to operate the equipment will guide and not frustrate.

The answer to questions about deactivation, de-energization, and isolation of a complex system may be as simple as moving a switch to the OFF position. In some situations, the "switch" is an area of a computer display activated through the click of a button on a "mouse." A single command executed through action of a switch can initiate a sequence that shuts down the system in orderly fashion. This includes the fuel supply, circulating pumps, feedwater pumps, induced draft fans, forced draft fans, exhaust fans, and so on.

What is required is time for cool-down of fluid that circulates in piping and return to storage reservoir. Forcing cool-down arbitrarily by unsanctioned means can destroy the internal structure of a system. This has happened because of operator error caused by stupidity and arrogance. This has occurred totally independently from the involvement of outside individuals and indicates the requirement for thorough understanding by operators about what must and what must not occur during shutdown of a system.

To illustrate, opening a system for entry by workers prior to cool-down of appropriate duration to ambient temperature can create a life-threatening situation due to induction of heat stress and trauma from thermal burn following contact with hot surfaces. The body swells upon warming. This can prevent movement through the openings in pressure vessels. Thermal conditions that melt the soles of shoes create the conditions for life and death. Operators control the existence of such conditions through their knowledge of operating conditions in these systems.

What the previous example illustrates is that, what at first seems to an outsider to be extremely complex, is actually quite simple in its execution. The action illustrated in this example shuts down the entire system. This approach would be inappropriate where isolation of part of the system must occur while the remainder must remain operational. The example also illustrates the paramount importance of the knowledge and judgment of the operator regarding the safety of the process.

The challenge for the person attempting to capture this information is to determine how to simplify the highly complicated. This challenge exists also for the regulator who is confronted with what seems to be unfathomable complexity. The regulator needs assurance that the system is managed competently, because as mentioned in previous discussions, events are either managed or they are allowed to happen. Events that are allowed to happen reflect loss or absence of control over how an operation operates a system.

A parallel theme in a complex system, or what seems to the casual observer and visitor to be a complex system, is the knowledge gained by operators of this equipment. Operators soon learn that what is supposed to happen during some expression of control may not happen, and that what does happen is not recorded anywhere. The experience gained through operating a system is the reason that operations will continue to need operators, despite what computer-based control can accomplish.

The issue for organizations containing operations employing operators is to capture the experience and knowledge gained by these individuals while they are actively engaged in operating as part of this experience, rather than as an adjunct when they plan to retire or when the organization decides to downsize in the belief that these individuals are not needed or are replaceable by junior individuals.

## Drawings: Another Critical Resource

Visualization is a critical aspect of assessing and forming strategies in a particular situation. First-hand observation and picture-taking, coupled with detailed note-taking and explanation by a knowledgeable individual, are extremely important parts of this strategy. These actions provide only some of the information needed for thorough analysis of the system.

The other important resources that should be available are drawings. Drawings provide key information about the system. Up-to-date as-built drawings are the penultimate resource. Of course, the penultimate is usually not available. Fortunately, the depth of the analysis that occurs during assessment of isolation is shallow compared to what might be needed for other purposes. Drawings may not exist for older systems. This reality complicates this part of the analysis and necessitates alternate approaches to information-gathering.

Facility designers create drawings to capture information about specific concepts. These include civil construction, infrastructure, process, electrical circuits, instrumentation, and flow of liquids due to change in elevation. These drawings are complex and contain considerable information related to the facility and its largest components.

Civil drawings provide information about the overall layout of a site, as well as details about existing facilities, and structures to be constructed, modified.and demolished. In this context, civil drawings provide the "big picture." Sometimes the relationship of various components that create the "big picture" have an importance to the consideration about the details that form the normal focus of this inquiry.

Structural drawings provide details about structures constructed within civil works. The information provided in these drawings is considerably more detailed than what is available from civil drawings. Structural drawings provide the basis for determining size, height, and elevation of structures. Height is often a consideration regarding fall protection and rescue. These considerations are peripheral to the work to be performed.

Process drawings indicate the equipment and structures involved and sequence of flow of fluids within the various components of the process. Process drawings provide information about process structures, piping, valves, pumps, and other components. Process drawings generally are the most useful for gathering the information needed to assess the situation.

Electrical drawings provide details about circuits used for control, as well as wiring information leading to sources of supply. This depth of information sometimes is necessary for understanding the means by which a system is controlled.

Hydraulic drawings (Figure I.3) are especially useful where fluid flow is dependent on gravity and differences in elevation. This is the case in swimming pools, hydroelectric installations, water purification and wastewater treatment plants where gravity and elevation govern pressure exerted against surfaces. Difference in elevation provides the means to calculate pressure exerted against walls of structures, including sluice gates and slide gates.

None of the drawings alone conveys the totality of information needed to be captured in order to create the assessment and procedures for deactivation, de-energization, isolation, and lockout. This effort often requires consultation of several of the drawings mentioned in the discussion.

Deactivation, de-energization, isolation, and lockout focus on individual circuits that produce some action and the circuits needed to regulate or control it. These circuits involve piping; valves, pumps and other critical components; electrical and pneumatic and hydraulic control circuits needed to control this equipment; control switches and disconnects; and switches on consoles and computer displays for electronic circuits acting through Programmable Logic Controllers. Creating a drawing that captures all of this information requires a synthesis of information derived from other drawings and sources of information. Such a drawing would readily indicate the components involved in the process of deactivation, de-energization, isolation, and lockout.

Two-dimensional drawings (Figure I.4) are the norm in design. Two-dimensional drawings, however, provide limited information in a three-dimensional world. The wealth of information provided in system drawings often overwhelms the observer. As a result, such drawings are not the best medium for conveying the information needed to discuss deactivation, de-energization, isolation and lockout of fluid-handling circuits.

Isometric drawings (Figure I.5) provide a way to visualize three-dimensional structures on two-dimensional paper or electronic "paper." The isometric view projects horizontal lines into the

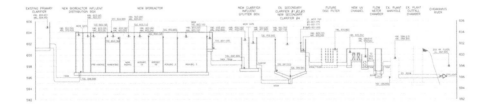

Figure I.3. The hydraulic flow drawing is a key resource in gravity flow systems. Height of weirs governs overflow capability. The latter is an important characteristic in isolation of gravity flow. Differences in elevation govern pressure exerted against surfaces, including slide gate valves.

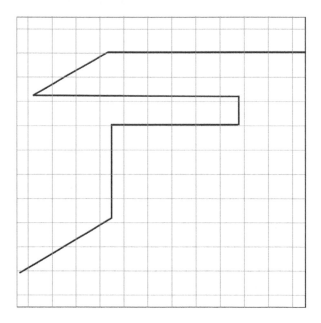

Figure I.4.  A two-dimensional drawing provides a flat version of a three-dimensional world.

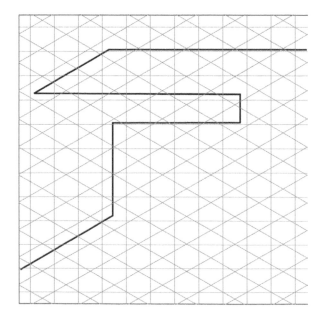

Figure I.5.  The isometric view of the same diagram conveys considerably more information to the brain.

plane of the paper at an angle of 30° to the horizontal. Vertical lines remain vertical. This "trompe l'oeil" fools the brain into perceiving a path into the paper for horizontal lines. This technique provides considerable benefit because of the importance of changes of elevation in the flow of fluids, especially liquids both during operation and following depressurization of lines.

Some drawing programs include isometric views. Free versions of CAD (computer-aided design) programs that provide isometric capability are available from the internet using a search engine. Some of these programs convert "flat" 2D drawings to the isometric format. The result can be quite startling as representation of the structure becomes considerably more realistic to the brain.

## Flow Inducement Strategies in Fluid-Handling Systems

The manner in which fluids flow or are made to flow in channels, duct, and piping systems has critical implications on isolation and lockout. This results from the fact that "textbook" solutions rarely exist in real-world installations and effective isolation must occur using existing equipment and infrastructure. The geometry of the installation is one of the elements potentially critical to understanding the potential for flow after shutdown of equipment, and ultimately, the means to implement an effective isolation.

### Gravity Flow

Flow due to gravity was among the first of the sources of energy to be harnessed during industrial development. Water flow powered ventilation systems in mines, mills for grinding grains, and many other applications.

At this time, flow due to gravity still is a critical part of the operation of many industrial installations, starting with hydroelectric power stations and potable water distribution systems. Flow due to gravity is exploited in potable water and wastewater treatment plants and swimming pools, pulp and paper mills, locks in waterways, and other processes. Gravity flow eliminates the need for pumps and other costly infrastructure and operational costs. Gravity flow is applicable to piping as well as open channels.

Gravity flow depends on the head created by the difference in elevation between two points. In systems having available only a single slide gate for isolation, the safety of the action depends on the security of the installation and its ability to stop water flow. This concept is equally applicable to tanks that empty by gravity flow. The drain line upstream from isolation valves experiences pressure created by the head of liquid remaining in the tank.

### Pumped Flow

Pumped flow refers to the use of pumps as power sources to move fluid from a lower to higher elevation and to create and/or to maintain pressure in a line. That is, the pump transfers energy to the fluid. Without the energy provided by the pump, the fluid does not move in the forward (intended direction), and depending on the design of the pump and the circuit and action taken in preparing for shutdown, may not be able to move in the reverse direction either.

#### Pump Fill

Many systems use pumps to fill structures for storage of fluid. Shipping terminals, for example, unload railcars by pump and fill the storage tanks. This is also the case in chemical processing facilities where pumps fill the storage tanks.

The pump generally is located at the lowest elevation in the system. As a result, residual pressure remains in the line that contains discharge from the pump so long as contents remain in the line and the tank. Emptying the line and the tank removes the backpressure.

#### Pump Empty

Many systems use pumps to empty structures. This process can harness pressure generated by fluid in the tank from bottom unloading or require siphoning during top unloading. The type of pump to be employed depends on placement (DiMatteo and Archambault 2000).

In the former case, the pump is often a submersible unit that is located at the bottom of the structure. This situation is similar in concept to pump filling, described in the previous section. This design is prevalent in sewer lift (pump) stations. The pump pumps liquid to higher level for drainage to the sewer system. Residual gauge pressure in discharge piping is around 10 lb/in² (70 kPa). Since the pump generally is located at the lowest elevation in the system, residual pressure remains in the line upstream from the pump so long as contents remain in the structure. Emptying the structure removes the pressure.

In the case of siphon operation, the pump is located at the high point in the system. As a result, in order to retain residual pressure in the lines on either side of the pump, backflow prevention devices are required to maintain feed in the line on the intake side so as not to allow starvation of the pump.

## Recirculation

Recirculation is the action of maintaining turbulent contact between all components of a liquid through the action of pumping. Recirculation prevents stagnation and settling and maintains uniformity of concentration and contact between all elements of a liquid. Recirculation is used in petrochemical distillation and in wastewater treatment. The system exploits the injection of the jet of pumped liquid into stagnant liquid to induce mixing.

Recirculation occurs in a loop involving the processing unit, the pump, connecting piping, and valves. If the pump is located at a low point in the system, liquid in the processing unit will pressurize piping on both sides of the pump.

## Gas-Pressurized Displacement

In some situations, gas pressurization in a blanket above or below the fluid forces the fluid from storage tanks.

For liquids, the pressurizing blanket of nitrogen or compressed air is located at the top of the tank. Pressure from the gas forces the liquid from the tank. While this approach necessitates use of pressure vessels, it may provide benefits regarding elimination of corrosion concerns for components, such as pumps and valves.

For gases and vapors, the concept is employed where the minimum amount of mixing is desired. A more dense gas introduced at the bottom of the structure will displace a less dense gas or vapor from the top, and vice versa (Figure I.6). Through careful positioning of the injection points and control of injection rate, the displacing gas and the resident gas form stratified layers having different densities. This mode is highly efficient and preserves the composition of the displaced component with minimal mixing (AGA 2001). This is the most efficient means of removal of gases and vapors from a structure both in volume and time. Gases in use include carbon dioxide to displace upward gases of lower density, and hydrogen to displace downward gases of greater density.

## Fans

Fans in furnace systems, cupola foundries, and other situations in which combustion gases and highly hazardous particulates are transported often operate in parallel. This means that these systems converge in a header (manifold) at the stack. Shutdown and entry of structures of one or more of the subsystems requires isolation from the operational system. Fans in the operational system will pressurize the duct connecting to the other systems.

## Blowers

Blowers provide the airflow for propelling solid materials through piping in many industrial facilities. Blowers are used for unloading flowable solid materials, including cement and many other powdered granular and pelleted materials, from highway trailers and railcars and transporting them through piping to storage structures.

Blowers also provide the airflow to convey these products in facilities that produce and consume them. Blowers provide the air for aeration in tanks in wastewater treatment plants. Blowers also create the vacuum cleaning capability of vacuum loading trucks.

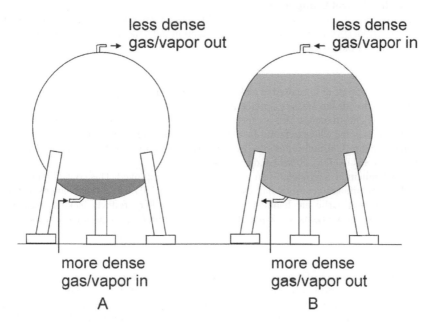

Figure I.6. Displacement by gases less dense or more dense than the displaced gas is a highly effective technique for purging vessels.

### Compressors
Gas pipelines use compressors to create the pressure needed for transport of the product. These systems operate in parallel to provide the redundancy needed to ensure reliability of supply.

## Techniques Used in Isolation
Isolation is the action taken to block passage of fluid within a channel, duct or pipe. Discontinuity or gap, which is satisfactory to isolate electrical circuits, is not functional in fluid-handling circuits. In the latter circumstance, the fluid would escape from the containment of the channel, duct or pipe. This consideration is critically important in actions taken to achieve safe isolation of fluids or to prevent operation of circuits containing fluids. Air acts an effective insulator to prevent escape of electrical current from energized conductors, but this approach is not an option with fluids.

"Textbook" solutions for isolation often do not exist in real-world installations. The effective isolation must occur with existing equipment and infrastructure. Isolation occurs in a number of conditions: full line pressure, contents but no line pressure, and residual contents only. A force exerted in a fluid remotely to a point of isolation can create pressure in the system in residual liquid.

### Lockout The Critical Component
The simplest of the techniques used for isolation is to lock out the critical component, such as a pump, fan, blower, or compressor. These components in many circuits are essential for fluid flow. Electrical lockout (including immobilization) of the pump can prevent movement of fluid within the circuit. This step, coupled with drainage of a storage structure and piping located at higher level in the circuit, prevents movement of fluid.

## Removal of the Critical Component

A simple approach to isolation is removal of the critical component. A pump containing unique fittings in a circulation system is a critical component. Disconnection and removal of the pump from the circuit prevents operation of the system.

## Jumpers

Jumpers are pieces of pipe or hose used to make connections for the purpose of interconnecting components. Jumpers provide the means to interconnect piping and valves without the need for costly installation of permanent structures. This is especially useful where a central hub connects infrequently to different terminuses and where terminuses connect infrequently to each other (Figure I.7). The number of jumpers also limits the number of interconnections at any moment in the operation. This is a simple and elegant means to effect control and isolation.

Cross-contamination is possible when a single jumper is employed. This consideration is a serious issue where contamination of food flavors may occur, especially where flavor-intensive products containing onions and garlic are handled in the midst of other products lacking these flavors. Cross-contamination by allergens is also possible where one product contains allergens and others do not.

## Dedicated Jumpers

Dedicated jumpers are pieces of pipe cut and shaped to fit between two particular connection points. That is, the piping can connect only between these two points. The origin of the concept of dedicated jumpers appears to have been the Manhattan Project in World War II (Thayer 1996). Dedicated jumpers were used in certain of the processing units. The jumpers were prefabricated and test assembled in a mock-up unit prior to shipment and assembly in the final location. This approach ensured the unambiguous fitting and precision sealing that were necessary in these operations.

Dedicated jumpers require precise fabrication in order to ensure exactness of fit and absence of the need for stretching or compression between points of connection. Inexactness of fit can lead to leakage. Hoses obviously provide considerably greater tolerance to inexactness of measurement than steel piping, but can lead to loss of control over the logic intrinsic in the design of the connection.

An additional benefit from the use of dedicated jumpers is that the shape of the pipe can ensure that the connection can occur only under specific conditions of preparation, such as the opening of a hatch or setting the specific position of valves (Figure I.8). The orientation of the piping following connection also can prevent entry into an open hatch.

This design is the hardware analogue of a logic system for isolating structures, such as tanks. This approach is used in food processing where tanks are routinely filled, operated, and emptied. This approach is elegant in its simplicity. The connections that hold the jumpers in place, however, are not lockable, and as a result lie outside the bounds of the normal concept of isolation and lockout.

## Freeze Plug Isolation

Freeze plug technology employs cold to freeze liquid in a pipe to form a solid plug (Figure I.9). The solid plug prevents movement of liquid. This concept is a simple application of the freezing of water pipes that occurs on cold winter nights. This technique appears to be suitable mainly for water or solutions involving water. Note that the water is static and not flowing in the pipe.

This system uses either pressurized liquid carbon dioxide ($CO_2$) or liquid nitrogen as the cooling agent. The liquid $CO_2$ system uses an insulated jacket wrapped around the pipe to contain the product. Upon injection from the cylinder into the space between the pipe and the jacket, the liquid solidifies to the solid form (dry ice). The dry ice sublimes to the gas. Dry ice sublimes at $-78°$ C ($-109°$ F). This temperature rapidly freezes water in the pipe. The ice plug forms only in the section of the pipe covered by the jacket. Hence, the resulting rise in pressure is very small and the pipe suffers no damage.

The nitrogen system operates through tanks containing liquid nitrogen. This system uses a wrap constructed from a coil of flexible copper tubing through which the liquid nitrogen passes.

Both systems vent the gases to the atmosphere. Venting in this manner will enrich the atmosphere in these gases and can lead to overexposure to $CO_2$ and oxygen deficiency. The gases are cold com-

Figure I.7.  Jumpers provide a way to interconnect between storage structures in a highly controlled manner in number and in time. The number of jumpers limits the number of connections.

Figure I.8.  Dedicated jumpers are connectable only in a specific manner. The shape of the jumper controls the logic intrinsic in the connection.

Figure I.9. Freeze plug isolation exploits the ability to solidify liquids and to block flow in piping without causing damage to the pipe. (Courtesy of COB Industries, Melbourne, FL.)

pared to ambient air. Carbon dioxide is also about 1.5 times as dense as air at the same temperature. As a result, gases released from the equipment are likely to subside and to remain in position. Aggressive ventilation to disrupt these accumulations of gas is essential in these applications.

This approach can provide satisfactory results under appropriate conditions. The first condition relates to the concern about splitting or other damage inflicted on the pipe by the forces exerted by the solid plug. This concern appears to be minor, as the technique has had many years of successful application. The second concern reflects possible embrittlement of the pipe caused by the cold. Plastics now are used extensively in piping, whereas previously only metals were present. Copper is potentially more tolerant of extreme cold than cast iron. Plastic pipe is subject to embrittlement, as a result of normal aging as plasticizers migrate from the interior of the product.

The third issue concerns separation of components present in the liquid. These can include components having different freezing points and solids dissolved in the liquid. The freezing agent must be capable of freezing all components of a multi-component liquid. Otherwise, movement of liquid or vapor from the low-freezing component may continue in the pipe.

### Inflatable Plugs
Inflatable plugs include bladders, "football"-shaped devices, and cylindrical devices (Figure I.10). These products often contain multiple layers of vulcanized rubber or other polymeric material reinforced by cordage to increase strength. These products use inflation and sometimes mechanical means, involving a rigid frame and set-screws to increase the strength of the contact between the circumference of the device and the inner wall of the pipe.

Successful application of the concept depends on the security of the adherence of the device to the wall of the pipe. When used to stop water flow during bypass applications involving sewers, these devices require tethering, at the minimum as a back-up measure, to prevent slippage along the pipe due to

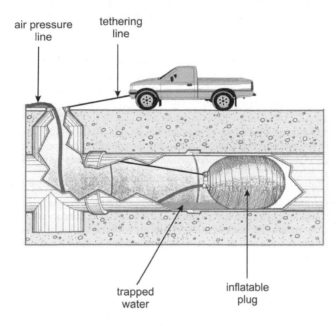

Figure I.10. Inflatable plugs exert pressure against the wall of the pipe in which they are inserted. (Modified from COB Industries, Melbourne, FL.)

the pressure of the fluid. Tethering is critically important because failure of adherence of inflatable devices has occurred, sometimes resulting in fatal injury.

Incompatibility with the environment is an issue in some applications. Manufacturers offer products containing different polymers where petrochemicals are present.

In dry situations, these devices can function as dams to prevent flow of gas at low pressure through the pipe.

These devices require inflation by a compressor. As a result, they are subject to deflation due to loss of power or leakage. They also lack the ability to be protected by lockout by conventional means.

### Flow-Through Plugs

Flow-through plugs combine an inflatable plug with a cylinder located in the middle of the device to enable flow. The cylinder contains a fitting on one end to enable connection to a hose for collection of the flow. This connection enables removal of fluid from the upstream side of the plug through a pump or other means. This concept requires secure adherence of the device to the wall of the pipe in order to prevent leakage and slippage, and also to ensure channeling of the fluid along the intended path through the cylinder located along the center axis.

As with inflatable plugs used for blockage of flow, successful application of the concept depends on the security of the adherence of the device to the wall of the pipe in order to prevent leakage and slippage. These devices require tethering, at the minimum as a back-up measure, to prevent slippage along the pipe due to the pressure of the fluid. Tethering is critically important because failure of adherence of inflatable devices has occurred, sometimes resulting in fatal injury.

Incompatibility with the environment is an issue in some applications. Manufacturers offer products containing different polymers where petrochemicals are present.

In dry situations, these devices can function as dams to prevent flow of gas at low pressure through the pipe.

These devices require inflation by a compressor. As a result, they are subject to deflation due to loss of power or leakage. They also lack the ability to be protected by lockout by conventional means.

## Expansion Plugs

Expansion plugs contain an expandable seal or inflated bladder positioned between two metal plates (Figure I.11). Tightening the nut on the stem in the center of the device compresses the metal plates against the sealing surface. Compression pushes outward the circumference. Radial expansion of the sealing surface makes contact and eventually a tight seal against the inner wall of the pipe.

Small versions of these products can resist gauge pressures up to $100 \text{ lb/in}^2$ (690 kPa). Versions of these products that fit more typical sizes of pipe can resist up to $40 \text{ lb/in}^2$ (280 kPa).

The ability to use these products requires access to the open end of the pipe. This can lead to exposure to the contents of the pipe during depressurization and disassembly or to residual materials remaining in the pipe following drainage.

The seal formed with the surface depends on the smoothness of the interior wall and absence of deposition of corrosion products, such as rust, and roundness of the pipe.

These devices are most effective in the middle of the range of expandability. As compression occurs, elongation of the bladder reduces the area of contact with the wall of the pipe.

As delivered, these devices lack the ability to be protected by lockout by conventional means. Adaptation of the covers used as lockout devices for some types of valves for this purpose may be possible.

## Positive Displacement Devices

Pump designs either allow backflow on cessation of rotation or prevent backflow. Circuits containing pumps not able to prevent backflow incorporate backflow prevention valves either singly or in series to prevent backward migration of fluid during normal operation.

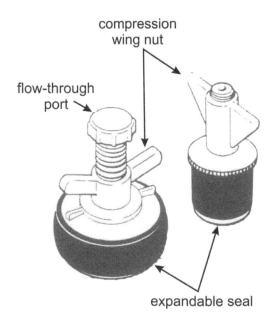

compression
wing nut

flow-through
port

expandable seal

Figure I.11. Expansion plugs utilize compression of an expandable seal to make the seal with the wall of the pipe. Variations on this design utilize an inflated bladder similar in shape to an inner tube. (Courtesy of COB Industries, Melbourne, FL.)

A positive displacement pump causes a fluid to move by trapping a fixed amount on the suction side and forcing (displacing) that volume through the outlet. Positive displacement pumps have an expanding cavity on the suction side and a contracting cavity on the discharge side. Liquid flows into the pump as the cavity on the suction side expands and the liquid flows through the discharge as the cavity contracts.

Positive displacement pumps prevent backward flow through continuous physical contact between movable elements in the pump. Movement of these elements during normal operation pushes or displaces the fluid in the forward direction. If the pump and drive motor can rotate in the opposite direction when not operating, pressure of the pumped fluid acting against the contact surfaces of the pump potentially can create backflow. Backflow prevention valves installed downstream from the discharge and upstream from the intake or a locking mechanism on the pump that prevents reverse rotation when not operational can prevent the occurrence of this situation.

Blowers and compressors are positive displacement devices. Mechanical pumps and blowers used in vacuum systems also are positive displacement devices.

## Blanks

Blanks are flat pieces of steel that fit into the space between the flanges of two adjacent pieces of pipe to block the flow of fluids. API 590 (superceded by ANSI/ASME B16.48-2005) governs the design and selection of materials for use in steel blanks (ANSI/ASME 2005). Selection occurs according to diameter of the opening, pressure class (150, 300, 600, 900, 1500 and 2500), and configuration of the face of the flange. Blanks used without gaskets are subject to leakage.

Use of the blank requires a number of steps. The initial steps include isolating, draining, depressurizing, washing and purging, and cooling to ambient temperature the section of pipe that will contain the blank. The next steps include opening the flange, removing the gasket, rotating or inserting the blank, inserting the original or a new gasket, and reclosing the flange.

## Valves

Valves are the devices designed specifically to control flow while remaining in an intact piping circuit containing fluid. Valves are designed to interrupt, throttle, stop, and allow to restart the flow of fluid. Valves perform this function by the action of a moveable element in its relationship with the body of the device. The moveable element establishes contact with the seat to effect the seal.

Valves are subject to specification and testing of performance, including leakage when new and sometimes following repeated openings and closings intended to mimic conditions and stresses experienced in actual service. Allowable leakage permitted according to standards ranges downward to zero, the latter being the only acceptable performance from valves used in residential applications. Valve manufacturers established standards reflective of performance obtainable from their products.

## Isolation Strategies

The previous section illustrates some of the techniques and devices available for isolating fluids in channels, duct, and piping. Without doubt, there are other examples. Isolation being the action taken to block and to prevent passage of fluid within a channel, duct, or pipe can occur in two main ways: nonintrusive isolation and intrusive isolation. The hazards associated with these strategies and the risks associated with them reflect whether the work occurs at full line pressure, with contents but no line pressure, and with residual contents only. Given the option to choose between the two approaches, the qualified person must clearly understand about the risks and benefits between the choices.

### Nonintrusive Isolation

Nonintrusive isolation occurs external to the channel, duct, or piping structure and creates no possibility for contact with the contents by the individual or component effecting the isolation. That is, the product remains contained at all times within an enclosed system. Contact is not possible in the absence of a catastrophic event.

This approach minimizes risk of exposure to this individual effecting the isolation. This approach forms the basis for isolation of channels, duct, and piping found in consensus standards, such as ANSI Z244.1 and regulatory Standards, such as 29CFR 1910.147 (ANSI/ASSE 2004; OSHA 1989). These approaches allow for closing and locking closed a valve to effect the isolation. Safety in this situation focuses on the individual who performs the isolation.

## Factors Involved with Nonintrusive Isolation

Table 7.4 presents risk factors for use in assessing hazardous conditions in the uncharacterized workspaces in which isolation of fluid handling systems often occurs. These risk factors provide the basis for identifying the risks of performing nonintrusive isolation procedures. The human costs of this activity, as expressed through accident and injury statistics and reports, have yet to be determined.

### Oxygen Deficiency

Purging and inerting of chemical lines often involve the use of nitrogen. Purging is the flushing procedure intended to prepare the line for exposure to air upon removal of hydrocarbon or other chemical contamination resulting from the product handled. Inerting maintains a concentration of an inert gas in the line sufficient to prevent combustion and reaction of the lining with oxygen of the air.

A line remaining pressurized by nitrogen or other gas capable of diluting or displacing the atmosphere that remains closed during the process of isolation potentially can leak at the flange or stem of the isolating valve. This leakage is unrelated to closure. For lines that regularly handle gases, this leakage is expected to be similar to what would occur during in-service operation at the same pressure and temperature. That is, the rate of leakage reflects the environmental condition and not the content of the pipeline. Legally oxygen-deficient conditions (<19.5% oxygen) are impossible to assess without the use of an instrument containing an oxygen sensor.

A valve or flange exhibiting sufficient leakage under conditions of isolation to cause oxygen deficiency in the surroundings during or resulting from isolation by closure behaves in the same way during normal operation. This valve or flange requires service or replacement.

### Oxygen Enrichment

Some pipelines transport oxygen. A line remaining pressurized by oxygen during the process of isolation potentially can leak at the flange or stem of the isolating valve. This leakage is unrelated to closure. Legally oxygen-enriched conditions (>23..5% oxygen) are impossible to assess without the use of an instrument containing an oxygen sensor.

Oxygen-enriched conditions promote combustion. This includes materials that do not burn readily in air. A valve or flange exhibiting sufficient leakage under conditions of isolation to cause oxygen enrichment in the surroundings during or resulting from isolation by closure requires service or replacement.

### Biochemical/Chemical Exposure Issues

A line remaining pressurized by nitrogen or other gas that remains closed during the process of isolation potentially can leak at the flange or stem of the isolating valve. This leakage can contain residues from the original content and is unrelated to closure. During manual closure of the valve, the source of these substances is positioned very close to the breathing zone. When heated to temperatures above the boiling point, these substances will vaporize rapidly when exposed to the atmosphere.

For lines that regularly handle gases, this leakage is expected to be similar to what would occur during in-service operation at the same pressure and temperature, excluding dilution by the nitrogen. That is, the rate of leakage reflects the environmental condition and not the content of the pipeline. A valve or flange exhibiting sufficient leakage under conditions of isolation to cause overexposure to the contents in the surroundings during or resulting from isolation behaves in the same way during normal operation. This valve or flange requires service or replacement.

Estimating exposure to leaking gases and vapors during isolation is essential to ensure worker protection. This can require specialized instrumentation. Work around valves or flanges could necessitate use of chemical protective clothing and high-level respiratory protection, up to and including Self-Contained Breathing Apparatus (SCBA).

## Fire/Explosion Issues

A line remaining pressurized by nitrogen or other gas that remains closed during the process of isolation potentially can leak at the flange or stem of the isolating valve. This leakage can contain residues from the original content and is unrelated to closure. When heated to temperatures above the boiling point, these substances will vaporize rapidly when exposed to the atmosphere.

For lines that regularly handle ignitable gases and liquids, this leakage is expected to be similar to what would occur during in-service operation at the same pressure and temperature, excluding dilution by the nitrogen. That is, the rate of leakage reflects the environmental condition and not the content of the pipeline.

A valve or flange exhibiting sufficient leakage under conditions of isolation to enable formation of an ignitable atmosphere in the surroundings during or resulting from isolation behaves in the same way during normal operation. Many flash fires have resulted under similar conditions. This valve or flange requires service or replacement.

## Ingestion/Skin and Eye Contact

Pipelines handling corrosives require special consideration. Leakage from the flange or stem of a valve of is a possible route for exposure to substances that can irritate the eyes and the skin. This can occur from pressurized contents and also from residues (Grossel 1998a, Grossel 1998b; DiMatteo and Archambault 2000). The source of these substances is positioned very close to the breathing zone.

A jet of fluid emitted at high pressure from the stem of a valve during movement could enter the eye or contact the skin of an unprotected individual. Shields to block these emissions are available for valves and jackets for flanges. A valve or flange exhibiting sufficient leakage under conditions of isolation to enable formation of an irritating atmosphere in the surroundings during or resulting from isolation behaves in the same way during normal operation. This valve or flange requires service or replacement.

Work around valves or flanges confining corrosives necessitates use of full-face protection, at minimum a faceshield + tight-fitting, chemical splash goggles, a respirator appropriate to the circumstances and chemical protective clothing. Establishing the appropriate type of respirator necessitates testing or modeling.

## Fall Hazard

Many of the structures on which isolation points are located are elevated. Access to these locations requires use of a ladder or a personal lifting device or installation of chain wheels. Ladders or chain wheels are the only possible devices for use indoors in boiler rooms and mechanical rooms where isolation points are located. Work at heights and hot surfaces pose risk of accident.

## Hot/Cold Surfaces

Steam systems and similar systems that operate at elevated temperatures contain many hot surfaces where insulation is not present. Thermal burn can occur following exposure of the skin to temperatures exceeding $45°$ C. Contact with hot surfaces poses a risk of burn injury.

## Intrusive Isolation

Intrusive isolation necessitates partial dismantlement of the channel, duct, or piping structure, and as a result creates the potential for contact with the contents and the internal environment. Invasive isolation is the basis for isolation mandated in regulatory Standards, starting with 29CFR 1910.146 (OSHA 1993).

OSHA Standard 29CFR 1910.146 does not allow techniques of lockout or tagout for isolating flowable material (OSHA 2008). The reason in OSHA's view is that compliance with 29CFR 1910.147 (Lockout/Tagout) does not in all cases adequately isolate hazards created by materials such as steam, flammable gases, and flammable and combustible liquids.

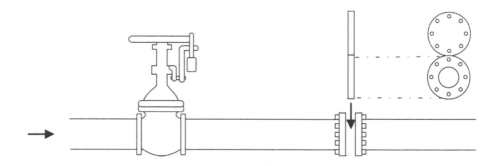

Figure I.12. Blanks. Opening the line to insert the blank exposes the worker to a number of risks.

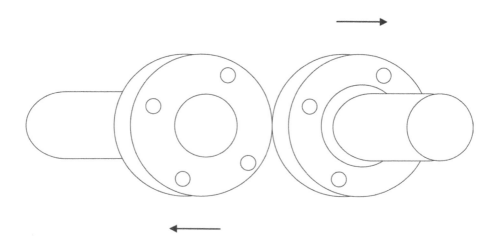

Figure I.13. Misaligning pipe and removal of a section of pipe. Opening the pipe exposes the worker to a number of risks.

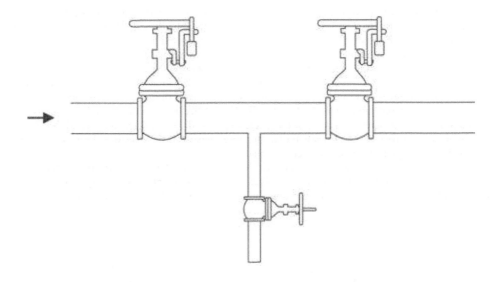

Figure I.14. Double block and bleed isolation. Opening the bleed valve creates a number of risks.

In a permit-required confined space, OSHA considers isolation of hazards associated with flowable materials to have occurred only by the use of blanking (Figure I.12) or blinding; misaligning or removing sections of line (Figure I.13), pipe, or duct; or use of a double block and bleed system (Figure I.14). A double block and bleed isolation system usually utilizes the closure of two valves, the opening of a bleed valve, and the application of lockout or tagout devices. This approach, in OSHA's view, offers complete employee protection, whereas an employer can comply with requirements of 29CFR 1910.147 simply by closing and locking out or tagging out a single valve. Closing and locking out a single valve could create atmospheric hazards in the space.

OSHA has indicated the nature of the process to be followed in assessing the capability of alternate measures in a letter of interpretation to Procter & Gamble (January 5, 1998), which accepted its specific safety disconnect system (inherently fail-safe system) as equivalent to an energy isolating device (OSHA 2008). The equivalency determination was based upon specific facts concerning the process machine and a failure analysis. This example illustrates that possibilities exist for comparing alternative approaches against OSHA requirements through thorough design review and application for obtaining a variance described in 29 CFR 1905 (OSHA 2008).

Intrinsic to all approaches is an element of risk. This applies to the approach adopted by OSHA in 29CFR 1910.146 (OSHA 1993) just as much as to any other approach to regulating behavior. The unknown in the approach mandated by OSHA through regulation is the magnitude of the risk intrinsic in different approaches to the same problem.

### Risk Factors Involved with Intrusive Isolation

Table 7.4 presents risk factors for use in assessing hazardous conditions in the uncharacterized workspaces in which isolation of fluid-handling systems often occurs. These risk factors provide the basis for identifying the risks of performing intrusive isolation procedures. The human costs of this activity, as expressed through accident and injury statistics and reports, have yet to be determined.

## Oxygen Deficiency

Purging and inerting of chemical lines often involve the use of nitrogen. Purging is the flushing procedure intended to prepare the line for exposure to air upon removal of hydrocarbon or other chemical contamination from the product handled. Inerting maintains a concentration of an inert gas in the line sufficient to prevent combustion and reaction of the lining with oxygen of the air.

A line pressurized by nitrogen or other gas capable of diluting or displacing the atmosphere potentially can leak at the flange or stem of the isolating valve prior to isolation. For lines that regularly handle gases, this leakage is expected to be similar to what would occur during in-service operation at the same pressure and temperature. That is, the rate of leakage reflects the environmental conditions and not the contents of the pipeline. A valve or flange exhibiting sufficient leakage under conditions prior to isolation to cause oxygen deficiency in the surroundings behaves in the same way during normal operation. These components require service or replacement.

A line remaining pressurized by nitrogen or other gas capable of diluting or displacing the atmosphere that is opened as part of the activity involved in installing a blank, cap, or plug could create an oxygen-deficient atmosphere in the surroundings around the flange being opened. Legally oxygen-deficient conditions (<19.5% oxygen) are impossible to assess without the use of an instrument containing an oxygen sensor.

## Oxygen Enrichment

Some lines transport oxygen. A line remaining pressurized by oxygen prior to isolation potentially can leak at the flange or stem of the isolating valve. A valve or flange exhibiting sufficient leakage under conditions of isolation to cause oxygen enrichment in the surroundings during or resulting from isolation by closure requires service or replacement.

Opening a pressurized line during disconnection to install a blank, cap, or plug could expose the worker to an oxygen-enriched atmosphere. Legally oxygen-enriched conditions (>23.5% oxygen) are impossible to assess without the use of an instrument containing an oxygen sensor. Oxygen-enriched conditions promote combustion. This includes materials that do not burn readily in air.

## Biochemical/Chemical Exposure Issues

A line remaining pressurized by nitrogen or other gas that remains potentially can leak at the flange or stem of the isolating valve prior to isolation. This leakage can contain residues from the original content. When heated to temperatures above the boiling point, these substances will vaporize rapidly when exposed to the atmosphere.

For lines that regularly handle gases, this leakage is expected to be similar to what would occur during in-service operation at the same pressure and temperature, excluding dilution by the nitrogen. That is, the rate of leakage reflects the environmental condition and not the contents of the pipeline. A valve or flange exhibiting sufficient leakage under conditions of operation to cause overexposure to the contents in the surroundings requires service or replacement.

Opening a pipeline at a flange to install a blank or opening a line to install a cap or plug introduces a number of possible routes for emission and exposure to chemical substances. This can occur from pressurized contents and also from residues. The source of these substances is positioned very close to the breathing zone.

Estimating exposure to leaking gases and vapors during isolation is essential to ensure worker protection. This can require specialized instrumentation. Work around valves or flanges could necessitate use of chemical protective clothing and high-level respiratory protection, up to and including Self-Contained Breathing Apparatus (SCBA).

Asbestos-containing gaskets and sealants pose an exposure risk during disassembly. This exposure may remain unrecognized with the focus on exposure to the product contained in the channel, duct, or pipeline.

## Fire/Explosion Issues

A line remaining pressurized by nitrogen or other gas can leak at the flange or stem of the isolating valve prior to isolation. This leakage can contain residues from the original content and is unrelated to closure. For lines that regularly handle ignitable gases and liquids, this leakage is expected to be similar to what would occur during in-service operation at the same pressure and temperature, excluding dilution by the nitrogen. That is, the rate of leakage reflects the environmental condition and not the contents of the pipeline. When heated to temperatures above the boiling point, these substances will vaporize rapidly when exposed to the atmosphere. Many flash fires have resulted under similar conditions. This valve or flange requires service or replacement.

Opening a pipeline at a flange to install a blank or opening a line to install a cap or plug introduces a further possible route for development of an ignitable atmosphere. This can occur from pressurized contents and also from residues.

### Ingestion/ Skin and Eye Contact
Pipelines handling corrosives require special consideration. Leakage from the flange or stem of a valve of irritating substances is a possible route for exposure to substances that can irritate or destroy the eyes and the skin. This can occur from pressurized contents and also from residues (Grossel 1998a, Grossel 1998b; DiMatteo and Archambault 2000; Pasquariello 2000).

A jet of fluid emitted at high pressure from the stem of a valve during movement could enter the eye or contact the skin of an unprotected individual. Shields to block these emissions are available for valves and jackets for flanges. A valve or flange exhibiting sufficient leakage under conditions of isolation to enable formation of an irritating atmosphere in the surroundings during or resulting from isolation behaves in the same way during normal operation. This valve or flange requires service or replacement.

Opening a pipeline at a flange to install a blank or opening a line to install a cap or plug introduces a further possible route for exposure to chemical substances present on surfaces and in the air. This can occur from pressurized contents and also from residues. These source of these substances is positioned very close to the eyes and the breathing zone.

Work around valves or flanges confining corrosives necessitates use of full-face protection, at minimum a faceshield + tight-fitting, chemical splash goggles, a respirator appropriate to the circumstances, and chemical protective clothing. Establishing the appropriate type of respirator necessitates testing or modeling.

### Noise/Vibration Hazard
Pressurized gases and vapors released from a small orifice as could occur during opening of a pressurized line can pose a considerable noise hazard that can exceed regulatory limits. The source is positioned very close to the ears.

### Biomechanical /Mechanical Hazard
The effort required to open a flange or to remove a valve or other fitting to install a cap or plug can result in musculoskeletal injury. This is especially the case for components that have remained undisturbed for a long period of time.

### Hydraulic, Pneumatic, Vacuum Hazard
Unanticipated pressure in lines poses continuing hazard to individuals attempting to perform isolation involving dismantlement of flanges. Exposure to contents of lines that remain pressurized while thought to be depressurized has occurred during many accidents, as noted previously. Contents include liquids, gases and vapors, and mists.

### Structural Hazard
The effort needed to remove components that have remained undisturbed for a long period of time can subject the components to damage and possible destruction and catastrophic failure (OSHA 1985). This is especially the case with components in steam systems. These components have failed catastrophically during such activity. Metal subjected to high temperature and high pressure in service for long periods embrittles and can fail abruptly and catastrophically.

Pipeline networks in many installations are welded together and surrounded by insulation. Critical points at elbows are usually covered in an insulating cementitious material (mud). Manhandling piping networks with heavy equipment in order to break open a seized flange runs the risk of cracking welds at critical points. Determining the existence of these cracks requires nondestructive testing that can subject the technician to nonionizing or ionizing radiation. Repairing the damage could require extensive dismantlement and reconstruction.

### Fall Hazard

Many of the structures in which isolation points are located are elevated. Access to these locations requires use of a ladder or a personal lifting device. Ladders are the only possible devices for use indoors in boiler and mechanical rooms where isolation points are located. Work at heights, combined with potential exposure to hot, pressurized fluid and hot surfaces poses extreme risk of accident.

### Explosive/Implosive Issues

Exposure to contents of lines that remain pressurized while thought to be depressurized has occurred during many accidents, as noted previously. The lines themselves pose an explosive hazard due to the internal pressurization. Under appropriate circumstances explosive release of the pressure can cause injury and damage.

### Hot/Cold Surfaces

Steam systems and similar systems that operate at elevated temperatures contain many hot surfaces where insulation is not present. Thermal burn can occur following exposure of the skin to temperatures exceeding 45° C. Opening hot lines of operating equipment to install blanks poses considerable risk from the potential for thermal burn resulting from contact with the hot surface.

## Additional Considerations

Previous discussion has examined factors that could apply to any workplace or workspace containing a fluid-handling system. The following discussion considers questions pertaining to fluid handling systems and their safety as expressed ultimately through concepts for isolation expressed in 29CFR 1910.146 (OSHA 1993).

### Leakage

Discussion in the previous section referred to potential leakage from components involved in fluid-handling systems. Table I.1 lists emission factors for components used in fluid-handling systems (EPA 1994). While the data may reflect performance levels since surpassed by improvements in design and materials of construction, they do provide a sense of relativity of performance of different components used in fluid handling systems.

To illustrate, the most problematic emitters are seals in compressors. Flanges are more tightly sealed than valves. Valves that handle gases are less tightly sealed than valves that handle heavy liquids. Some of these observations are intuitive while others are not. These observations provide the data for using models to assess questions raised in discussion in the previous section (Keil et al. 2009; Powell 1982). The data indicate that work around in-service valves and flanges, as encountered during non-intrusive isolation, is considerably less problematic than work around operating compressors.

With regard to valves there are two or three points of leakage: externally through the stem and the flange, and internally through and around the seat. The data provide no information regarding internal leakage. Hence, there is no direct basis for comparison. Often conspicuous by its absence from the list of criteria for specification of valves is leakage.

Leakage is a major concern regarding exposure during disassembly of piping in open environments and during work in confined spaces into which the valve controls flow. In a valve, the stem is the external connection for adjusting the position of the moveable element. In order to perform this function, the stem must pass through the body of the valve. The stem to body path is a route for leakage

## Table I.1
## Average Emission Factors for Fugitive Emission

| Component | Service | Emission Factor kg/hr/unit |
|---|---|---|
| Valves | Gas | 0.0056 |
| | Light liquid | 0.0071 |
| | Heavy liquid | 0.00023 |
| Pump seals | Light liquid | 0.0494 |
| | Heavy liquid | 0.0214 |
| Compressor seals | Gas/vapor | 0.228 |
| Pressure relief seals | Gas/vapor | 0.104 |
| Flanges | All | 0.00083 |
| Open-ended lines | All | 0.0017 |
| Sampling connections | All | 0.0150 |

Source: EPA 1994.

of fluid from the wetted interior surface to the exterior. Manufacturers use various materials to minimize leakage by this route.

Leakage from the connection with the line containing the valve is slow and readily visible. This situation would provide considerable warning to workers affected by this condition. Leakage also occurs due at the interface between the seat and the moveable element. The seat forms the seal between internal moveable parts of the valve and the body of the valve. Leakage also could occur due to catastrophic failure of the geometric relationship between the seat and the movable part or catastrophic failure of either the seat or the movable part. In either case the body of the valve would remain intact as would attachments to piping at the flanges. These occurrences would provide considerable warning to workers about the need to take emergency action.

A number of standards address leakage permissible from valves. The U.S. EPA (Environmental Protection Agency) references performance criteria for leakage mentioned in performance standards published by API (American Petroleum Institute), FCI (Fluid Controls Institute), and MSS (Manufacturers Standardization Society) and discussed in Appendix A. These organizations have produced consensus standards that specify leakage as part of their criteria for performance. Table I.2 summarizes these requirements.

The Fluid Controls Institute created one of the first standards on leakage, ANSI/FCI 70-2-2006. ANSI/FCI 70-2-2006 is a reference to standards created by other groups, such as the American Water Works Association (AWWA) and the American Petroleum Institute (API). ANSI/FCI 70-2-2006 and the similar standard for regulators (ANSI/FCI 70-3-2004) set six classes for leakage (ANSI/FCI 2004, ANSI/FCI 2006)). ANSI/FCI 70-3-2004 applies to pilot-operated and direct acting pressure-reducing, pressure-relieving (backpressure), differential pressure, and temperature regulators. ANSI/FCI 70-2

## Table I.2
## Valve Leakage Standards

| Standard | Summary of Performance Requirements |
|---|---|
| **Primary Standards** | |
| ANSI/FCI 70-2-2006 | Class I: no test required. |
| | Class II: 0.5% of full open capacity at maximum operating differential or pressure drop of 50 lb/in$^2$ (344 kPa), whichever is lower at 50 to 125 °F (10 to 52 °C) for water, or at maximum operating differential or 45 to 60 lb/in$^2$ (310 to 414 kPa) gauge pressure differential, whichever is lower for air at the same temperatures. |
| | Class III: 0.1%, identical test conditions. |
| | Class IV: 0.01%, identical test conditions. |
| | Class V: 0.0005 mL/min of water/inch of port diameter/lb/in$^2$ of pressure differential at the maximum service pressure differential and not to exceed the ANSI rating (ANSI/ASME B16.34) for the valve body and 50 to 125 °F (10 to 52 °C). |
| | Class VI: 0.0005 mL/min of air or nitrogen/inch of port diameter/lb/in$^2$ of pressure differential at 50 lb/in$^2$ or maximum rated differential pressure across the valve plug, whichever is lower, and 50 to 125 °F (10 to 52 °C). |
| ANSI/MSS SP-61-2009 | Gate, globe, and ball valves: 10 mL/hr/inch of pipe diameter. |
| | Check valves: 40 mL/hr/inch of pipe diameter. |
| **Derivative Standards** | |
| API 598 | Fluid Test: |
| | Resilient-seated valves: zero leakage is allowable. |
| | Metal-seated valves: 0 drops/min for valves <2 inches (50 mm) and 12 to 28 drops/min (0.75 to 1.75 mL/min) for valves 2 to 14 inches (50 to 350 mm) in diameter. |
| | Check valves: 0.18 cubic inch (3 mL) per minute per inch of nominal pipe size in the liquid test. |
| | (1 mL = 16 drops. 0 drops = no visible leakage for the minimum specified duration of the test.) |
| | Gas Test: |
| | Metal-seated valves: 0 bubbles/min for valves <2 inches (50 mm) and 24 to 56 bubbles/min for valves ranging from 2 to 14 inches (50 mm to 350 mm) in diameter. |
| | Check valves: 1.5 standard ft$^3$ (0.042 m$^3$) of gas per hour per inch of nominal pipe size. |
| | (0 bubbles means <1 bubble per specified duration of the test.) |
| ANSI/AWWA | Refer to Table A.3. |

Sources: ANSI/FCI 2006; ANSI/MSS 2009; API 2009.

(ANSI/FCI 2006). ANSI/FCI 70-2-2006 is an antecedent to standards created by other groups, such as AWWA and API.

The Manufacturers Standardization Society of the Valve and Fittings Industry (MSS) has long specified testing methods to assess valve performance and acceptable rates of leakage (ANSI/MSS 2009). This standard forms the basis for derivative standards created by organizations including AWWA and API. ANSI/MSS SP-61-2009 establishes requirements and acceptance criteria for shell and seat closure pressure testing of valves other than control valves. This standard forms the basis for derivative standards created by organizations including the AWWA and API.

## Disassembly and Reassembly

Disassembly and reassembly of piping in order to effect isolation runs counter to safety in engineering design. Engineering design focuses on permanency and not disassembly and reassembly. Piping systems that transport hazardous substances must perform consistently over their expected lifetime according to design expectations. Many accidents have occurred during disconnection and reconnection of lines and components (Lees 2001). Unnecessary disconnection and reconnection of components are to be avoided.

These systems often exist in areas containing equipment that contains, handles, and processes hazardous substances. Failure during operation or maintenance would produce catastrophic consequences to the operation and quite possibly the surroundings. Chemical complexes no longer are located in isolated areas. In many cases, they are surrounded by urban development. The tragedy at Bhopal illustrated the consequences intrinsic to the assumption of the availability of distance to offer protection in the event of failure and emission of contents.

Exposure to contents of lines during depressurization of lines not known or believed to be pressurized at time of opening is an occurrence acknowledged and feared by operators. Resealing the line requires considerable effort, including reinstalling of gaskets and leak-testing to establish integrity.

## Double Block and Bleed Isolation

Double block and bleed isolation utilizes two blocking valves and an intermediate bleed valve. The idea behind the concept is that in the event of leakage or outright failure of the upstream valve, leakage will drain from the bleed valve. The second block valve prevents further migration of liquid or gas or vapor along the line. The intent is to prevent entry of the chemical product into confined spaces.

As usually depicted (Figure I. 14), the bleed valve is located in a piece of pipe attached to the main line at a 90° angle. This configuration creates a dead leg. Valve manufacturers are actively attempting to develop products that accommodate the concepts of this paradigm in a single package. Unitary designs that combine the double block valves and the bleed valve into the same body attempt to minimize this concern (Figure I.15). Valves used in isolation of gas pipelines use a variation on this concept to introduce a sealing fluid in the cavity between double block valves (Figure I.16).

Dead legs are stagnant pockets where fluid collects (Schmidt 2001). Dead legs have provided the conditions that have led to serious accidents independent from the intended function as a low point in piping to collect leakage. Freezing in this pipe has led to rupture. Entrapment of condensed liquid and other drainage has led to corrosion and failure of the valve and piping that forms the dead leg.

Liquid can collect and remain in a dead leg even after cleaning to prepare for entry because the cleaning fluid does not flow through this part of the circuit. Vertical dead legs that drain downward are preferable to horizontal dead legs, and horizontal dead legs are preferable to vertical dead legs that collect fluid by downward flow.

Maintaining dead legs as short as possible is critical to minimizing the concerns reflected by this issue. Length/diameter (L/D) ratio is a critical consideration in evaluating the significance of a dead leg. An L/D of 2 is a practical limit to removal of contamination. This reflects the reality of the occurrence of the concentration gradient and eddy and molecular diffusion in a dead leg.

Creation of dead legs in piping circuits involved with confined spaces is an unintended consequence of regulatory requirements stated in 29CFR 1910.146 and adopted by other jurisdictions (OSHA 1993). Double block and bleed isolation is regulation driven, and not engineering driven. This

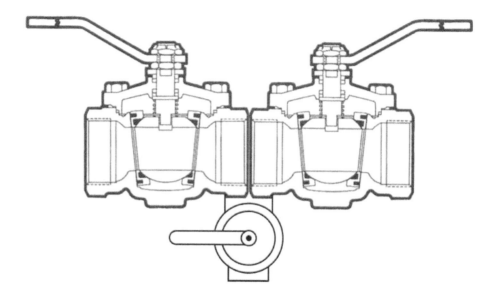

Figure I.15. Double block and bleed valve contained in a single body minimizes trapped fluid.

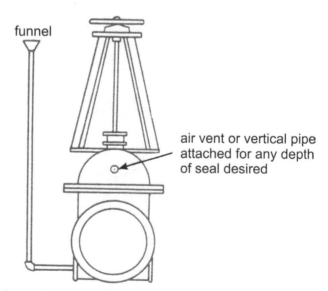

double disc gate valve with water seal between discs

Figure I.16. Fluid sealing in the cavity between two block valves is employed in valves used in gas pipelines. (Modified from AGA 2001.)

is an important distinction. Comments in online engineering forums indicate this to be the case. Some commenters openly express hostility to the concept.

Corollary concerns not mentioned in the discussions are the nature of what would drain from the bleed line in the event of an actual leak, where it would go once free from containment of the piping, and what hazardous conditions would the drainage pose. Discussion in this section has already indicated that this drainage can pose serious risks to personal safety from toxicity, ignitability, and chemical reactivity. A double block and bleed system amounts to single valve isolation with a path for leakage versus a double block system created from the two block valves minus the bleed.

## Air Pollution Concerns

Regulatory requirements for severance of connections or bleeding leakage from the pipe create the likelihood for emissions to the environment. Entities affected by environmental regulations have expended considerable effort to minimize emissions to the environment from known sources, including points of connection at flanges, and ends of pipes that accommodate caps and plugs.

Emission during isolation of piping and emission from bleed valves can constitute major sources of load to the environment. While the overall consequence of knowingly allowing escape of substances from the containment of piping systems designed to confine them may be small, this judgment is less easy to make when the substance(s) is/are hazardous. Hazardous can take its meaning from one or more of the following distinctions: toxicity, ignitability, and chemical reactivity. These properties can affect personnel working in these workspaces, people in the surroundings, and sensitive targets in the environment.

## Process Safety Concerns

Other influences are acting in conflict with the classic paradigm used in isolation and lockout of process systems. There is a trend in process system design to move to welded construction. Welded systems leak less than systems containing bolted construction and are more reliable. These designs offer obvious benefits to process safety and reduction of emissions to the environment. This trend has important implications in preparing confined spaces for entry, since the ability to remove sections of pipe conflicts with other design priorities.

In the U.S., process safety is the subject of 29 CFR 1910.119 (OSHA 1992). The focus of this Standard is maximizing knowledge about processes and the equipment in which they occur for the benefit of control. Part of this focus centers on piping and equipment design for the control of process hazards, such as fire and personal exposure and emissions to the environment.

## Valves versus Blanks

Isolating devices, such as valves and blanks, both reflect performance requirements according to consensus standards for strength. This is the reason that valves and other components rarely fail in service. Valve selection and specification can include performance for internal leakage. Obtaining a tight seal at a flange where a blank is used requires inclusion of a gasket.

Engineering design is very conservative in its approach. The approaches used today reflect considerations that emerged during the reign of Queen Victoria (Petroski 1992). In 1849, Queen Victoria assembled together leading engineering designers to ponder about the use of iron in the design of railway bridges. Safety concerns manifested through fear about collapse of bridges and explosion of boilers were blocking widespread adoption of emerging technologies.

The Royal Commission expressed the view that design should overbuild beyond the perceived greatest load by a multiple (factor) of 4 to 6. As well, the Royal Commission made recommendations for testing of components at a multiple of 1 to 3 beyond perceived greatest load. These deliberations and others led to the modern practice of engineering design.

Valves specified by design engineers are used at a fraction of rated working pressure. Rated working pressure is the pressure assigned by the manufacturer for continuous service. With the additional consideration for failure beyond rated working pressure, system operating pressure, therefore, usually is a small fraction of the failure pressure. This is one of the reasons that failure of valves and other components in engineered fluid handling structures is so infrequent. Environmental assessments are sources of information regarding probability of valve failure determined from engineering studies of operating systems.

This consideration during design accommodates for the many unknowns associated with operation of valves in real-world environments. Often, the valves are located in open air. This means that they are subject to the rigors of the weather. Additional operating environments include burial in backfill and submersion in water or other liquid.

Internal and external surfaces of valves are subject to corrosion, and erosion from within due to fluid flow. Internal and external coatings attempt to minimize the hostility of the internal and external environments. The experience of valve manufacturers and valve users is extremely important in assessing the long-term impact of conditions on the longevity of these products.

Previous discussion raises the valid question about whether a single valve or two or even three valves in series are inferior to, equivalent to, or technically superior to a single plate of metal inserted at potentially great risk and expense in the path of fluid in a line at a flange that joins two separate sections of pipe.

Previous discussion has also indicated reliance on redundancy and defense in depth in risk considerations, especially in Europe. These concepts are based on the presumption that the risk of failure of performance of two or more devices in series is much lower than the risk of failure of a single device acting alone. Failure in this case can include leakage or outright catastrophic failure. To this point, an informed discussion about application of this approach to isolation of fluid-handling systems has yet to occur.

## Maintaining and Servicing Fluid-Handling Equipment

Previous parts of this appendix have focused on issues related to operation of fluid-handling equipment on whose performance in a predictable way depends the safety of workers involved in installation, maintenance, servicing, repair, and renovation.

In order to make appropriate decisions about deactivation, de-energization, and isolation, issues related to real-world performance, as opposed to design performance, must be known and appreciated to the fullest extent possible. This section focuses on activity related to preparing for deactivation, de-energization, isolation, lockout, and verification of the state of zero energy in fluid-handling equipment, machines, and systems.

### Assessing Fluid-Handling Systems

Little exists in print about the presence and movement of fluids in equipment, machines, and systems to enable assessment in preparation for lockout and tagout. Typically what are observable are ad hoc isolations and lockouts created by site personnel in response to the perceived need. Inquiry about engineering input into the design to ensure adequacy of the approach typically produces negative responses.

Table 7.6 provides a generalized way of considering this question for different sources of energy. Table I.3 contains a specialized version of Table 7.6 that focuses on assessment for isolation and lockout of fluid-handling systems.

Complicating assessment of fluid-handling systems is the absence of pre-engineered and installed isolation devices that enable compliance with requirements of 29CFR 1910.146 for work in permit-required confined spaces (OSHA 1993).

Previous discussion indicates that the geometry of the circuit, types of components, and ability to drain fluid from the circuit through normal operating procedures can provide options for consideration for isolation. None of these options may meet the "textbook" description of requirements for isolation. Hence, they require deliberation and consideration in an organized manner in order to determine whether they meet or exceed regulatory requirements for the situation.

The following sections describe the additional terms used in Table I.3. The information collected and considered during this hazard assessment forms the basis for a dialog with the regulator with regard to possible strategies for isolation and lockout of the actual system. Contact with the regulator is strongly advised in order to solicit opinion about whether the proposed approach would be acceptable.

## Table I.3

## Assessing Retained Energy in Fluid Handling Systems

| ABC Company | Hazardous Energy Assessment | Work Activity | | | |
|---|---|---|---|---|---|
| Location: | Equipment, Machine, System: | Assessed by:<br>Tel:<br>Date: | | | |
| Overview: | | | | | |
| Hierarchy of Energy Input | Conversion Energy Output | Equipment, Machine, System Affected | | | |
| | | | | | |

**Circuit Name**

Function/Description:

| System Operating Pressure | Rated Working Pressure | Residual Pressure | Performance Standard(s) | Allowable Leakage | Actual Leakage |
|---|---|---|---|---|---|
| | | | | | |

Fluid/Material Flow Isolation Strategy:

Failure/Consequence Analysis:

| Input: | Storage: | Dissipation/Purge: |
|---|---|---|
| Output: | Conversion: | Immobilization: |
| Primary Isolation: | Secondary Isolation: | Tertiary Isolation: |
| Location: | Location: | Location: |
| Action: | Action: | Action: |

## Table I.3 (Continued)
## Assessing Retained Energy in Fluid Handling Systems

| Flow Isolation? | Flow Isolation? | Flow Isolation? |
|---|---|---|
| (Circuit Name) Flow Isolation Procedure: | | |
| Verification Test Procedure: | | |
| Notes: | | |
| Overall Flow Isolation Procedure: | | |
| Overall Verification Test Procedure: | | |

**Hazard Assessment (Operational/Undisturbed Space) or (Work Activity)**

**Hazardous Condition**                                    **Occurrence/Consequence**

                                                    Low          Moderate          High

                                                     |               |               |

(circuit name)

In this table, **NA** means that the category does not apply in any normally foreseeable situation. **Low** means that exposure is readily identifiable but believed to be much less than applicable limits or that exposure to nonquantifiable hazardous conditions is unlikely to produce injury. **Low-Moderate** means that exceedence of regulatory limits is believed possible or that nonquantifiable exposure could produce minor injury requiring self-administered treatment. Control measures or protective equipment should be considered. **Moderate** means that exposure is believed capable of exceeding regulatory limits or causing traumatic injury requiring first aid treatment or attention by a physician. Protective equipment or other control measures are necessary. **Moderate-High** means that exposure is believed capable of considerable exceedence of regulatory limits or causing serious traumatic injury. Advanced control measures or protective equipment are required. **High** means that short-term exposure is believed capable of causing irreversible injury, including death. Advanced control measures or protective equipment are required.

## System Operating Pressure
System operating pressure is the design pressure for systems under construction or actual working pressure for operational systems. This pressure provides a measure of the nature of the design compared to rated working pressure of components selected for use in the application.

## Rated Working Pressure
Rated working pressure is the pressure rating for continuous service recommended by the manufacturer for the component(s) under consideration. The rated working pressure also indicates the closeness of the design pressure or pressure achieved under actual operating conditions compared to the long-term recommended maximum service pressure. Components operating at or near the rated working pressure still have the protection of the safety factor. The safety factor for the component, while not likely to be disclosed by the manufacturer, must exceed the pressure under which hydrostatic stress testing occurs in order to pass that test without failing.

## Residual Pressure
Residual pressure is the pressure retained in piping downstream from pumps or other components capable of pressurizing the system. This pressure reflects retention of pressure due to entrapment of fluid in isolated parts of the system. Residual pressure is capable of creating a jet from a small opening in the flange during intrusive isolation.

## Performance Standard(s)
Performance standard(s) identifies consensus standards with which components subject to leakage during isolation comply. These standards apply at time of manufacture of the component and possibly in-service obtainable through testing protocols. Performance testing that mimics long-term service provides a basis for estimating potential leakage under actual conditions of operation.

## Allowable Leakage
Allowable leakage refers to leakage stated in the performance standard as being acceptable for a particular component under anticipated conditions of operation.

## Actual Leakage
Actual leakage refers to leakage measured during tests performed on the actual component prior to installation or during operation. Leakage measured during operation could include testing in situ and testing following removal from the installation. Values for actual leakage of a particular component are likely to be rare and difficult to obtain.

### Shutting Down Fluid Handling Systems
Among the most complex systems to isolate and lock out are the process and utility systems and utility feeds into equipment and machines. For the purpose of this discussion utility systems include systems that provide water, steam, natural or fuel gas, and possibly gases, such as hydrogen, helium or argon, carbon dioxide, nitrogen, and oxygen. Process units can have any of these, as well as supply and drain lines and materials storage.

Previous discussion in various places has highlighted operator knowledge and experience, especially for complex systems. Additional sources of information include the operations manual, the vendor, and the manufacturer.

The course prepared by the UAW-GM (United Auto Workers-General Motors) National Joint Committee on Health and Safety on lockout recommended working on process and utility systems last in the sequence of activities for deactivating, de-energizing, and isolating equipment and machines (UAW-GM 1985). For equipment and machines, the recommended sequence is a hierarchy starting with mechanical momentum sources, followed by gravity, stored mechanical energy, electrical system, pneumatic system, hydraulic system, and process and utility systems.

## Isolation of Fluid-Handling Systems

The least complicated isolation occurs through closure of isolation valves. Isolation valves are present in many systems, starting with those that supply potable (drinking) water and natural gas to residences. Closure of a single valve is often sufficient to stop flow to drip tightness in the case of water and bubble tightness in the case of natural gas. Engineering design and codes dictate the requirement for tightness of the seal in these products. Industrial installations often have multiple valves in series. Closure meets the intent for isolation in regulatory standards, such as 29CFR 1910.147 (OSHA 1989). Note that lockout requires further action, to be discussed in the next section.

Germane to process and utility systems are requirements stated in 29CFR 1910.146, the OSHA Standard on permit-required confined spaces in general industry (OSHA 1993). OSHA considers isolation of fluid-handling systems that eliminates the hazard to occur by such means as blanking or blinding; misaligning or removing sections of lines, pipes, or ducts; or a double block and bleed system (Figure I.12 to Figure I.14). The Standard considers isolation to be the process by which a permit space is removed from service and completely protected against the release of energy and material into the space.

The diagrams provided here (Figure I.12 to Figure I.14) illustrate the classic paradigm of isolation involving manually operated valves. Computerization of process control introduces the possibility of using motorized valves and achieving isolation through software. Control through software introduces the potential for tampering.

An issue with the preparation of confined spaces for entry that tends to be overlooked is the nature of the work activity to follow. Many welding and cutting processes performed in confined spaces require the use of gases. These can include helium or argon; acetylene, hydrogen, propane, and other fuel gases; carbon dioxide, and oxygen. These cylinders contain single valve isolations for gauge pressures up to 4000 lb/in$^2$ (27.6 MPa).

Normal practice is to position the cylinder(s) outside the space and to provide the gases to the workspace through flexible hoses. The latter are subject to damage and cutting. Welding units and valves on guns are subject to leakage. In large projects, custody of gas lines is an important issue. Errors in connection can and do happen. The level of care employed in ensuring safe conditions during the work activity should reflect the level in preparing the space for occupancy.

## Lockout Devices for Fluid-Handling Equipment

Lockout devices for fluid-handling equipment are subdivided according to design and execution. Intrinsic lockout devices are designed into the equipment as supplied, or are designed into retrofits created by the original equipment manufacturer. This is the most desirable route to follow, since the manufacturer is most familiar with the equipment.

Extrinsic lockout devices are temporary add-ons to components of the fluid-handling system. The simplest devices for valve lockout are chains or cable devices that prevent movement of the valve wheel (Figure I.17 and Figure I.18). Extrinsic lockout devices for fluid-handling equipment include covers for valve wheels and handles (Figure I.19 and Figure I.20). The wheel cover encloses the valve wheel and allows free rotation of the wheel. A removable plug allows upward and downward movement of the valve stem. Immobilization devices prevent movement of the lever of ball valves. Some types of valve require dedicated wrenches. Placing the wrench in a lockable box prevents its use under uncontrolled conditions. This approach is similar in concept to use of a lockbox to hold the keys used to lock out specific points.

Control valves are more complicated, since they operate through pneumatic, hydraulic, or electrical operators. Isolating and locking out the power source to the operator isolates and locks out the valve.

## Verification and Testing

Verification and testing of the effectiveness of isolation and lockout of fluid-handling circuits can occur in several ways. The first is to determine pressure in isolated circuits. This can occur through a pressure gauge. A bleed valve can release stored pressure if opening and release of contents can occur safely.

Figure I.17. Chain lockout.

Figure I.18. Cable lockout.

Figure I.19. Cover for the wheel of a valve. The cover prevents rotation of the wheel. (Courtesy of Panduit Corporation, Tinley Park, IL.)

Figure I.20. Device for immobilizing the handle of a ball valve. (Courtesy of Panduit Corporation, Tinley Park, IL.)

**Retained Energy**

Retained energy refers to energy retained in a circuit either through deliberate storage or through unintentional entrapment during deactivation, de-energization, and isolation. This situation is akin to fluid pressure trapped in pipes that remain in service when valves are closed to isolate a section prior to depressurizing.

## References

AGA: *Purging Principles and Practice*, 3rd Ed. Arlington, VA: American Gas Association, 2001.

ANSI/ASME: Line Blanks (ANSI/ASME-2005). New York: American National Standards Association/American Society of Mechanical Engineers, 2005.

ANSI/ASSE: Control of Hazardous Energy – Lockout/Tagout And Alternate Methods (ANSI/ASSE Z244.1-2003). Des Plaines, IL: American Society of Safety Engineers, 2004.

ANSI/FCI: Regulator Seat Leakage (ANSI/FCI 70-3-2004). Cleveland, OH: Fluid Controls Institute, 2004.

ANSI/FCI: Control Valve Seat Leakage (ANSI/FCI 70-2-2006). Cleveland OH: Fluid Controls Institute, 2006.

API: Valve Inspection and Testing (API Standard 598, 9th Ed.). Washington, DC: American Petroleum Institute, 2009.

DiMatteo, R. and D.J. Archambault: Bulk Storage Tanks for Acids & Solvents. *Chem. Eng. 107 (11)*: 76-84 (2000).

EPA: Control Techniques for Fugitive VOC Emissions from Chemical Process Facilities (EPA/625/R-93/005, Handbook). Cincinnati, OH: Center for Environmental Research Information, Office of Research and Development, U.S. Environmental Protection Agency, 1994

Grossel, S.S.: Safe, Efficient Handling of Acids (Part I). *Chem. Eng. 105(7)*: 88-98 (1998).

Grossel, S.S.: Safe, Efficient Handling of Acids (Part II). *Chem. Eng. 105(12)*: 104-112 (1998).

Keil, C.B., C.E. Simmons, and T.R. Anthony (Eds.): *Mathematical Models for Estimating Occupational Exposure to Chemicals,* 2nd Ed. Fairfax, VA: American Industrial Hygiene Association, 2009.

Lees, F.P.: *Loss Prevention in the Process Industries*, $2^{nd}$ Ed., Vol. 2. London: Butterworths-Heinemann, 2001.

MSS: Pressure Testing of Valves (ANSI/MSS SP-61-2009) Vienna, VA: Manufacturers Standardization Society (MSS) of the Valve and Fittings Industry, 2009.

OSHA: *Selected Occupational Fatalities Related to Toxic and Asphyxiating Atmospheres in Confined Work Spaces as Found in Reports of OSHA Fatality/Catastrophe Investigations.* Washington, DC: U.S. Department of Labor, Occupational Safety and Health Administration (U.S. DOL/OSHA), 1985.

OSHA: "Control of Hazardous Energy Sources (Lockout/Tagout); Final Rule," *Fed. Regist. 54*: 169 (1 September 1989). pp. 36644-36696.

OSHA: "Process Safety Management of Highly Hazardous Chemicals, Explosives and Blasting Agents," 29CFR Part 1910, Washington: US Department of Labor, Occupational Safety and Health Administration (1992).

OSHA: "Permit-Required Confined Spaces for General Industry; Final Rule," *Fed. Regist. 58*: 9 (14 January 1993). pp. 4462-4563.

OSHA: OSHA Instruction, The Control of Hazardous Energy – Enforcement Policy and Inspection Procedure (CPL 02-00-147). Washington, DC: Occupational Safety and Health Administration, 2008.

Pasquariello, M.: Safe Handling of Caustic. *Chem. Eng. 107(9)*: 76-85 (2000).

Petroski, H.: *To Engineer is Human, The Role of Failure in Successful Design*. New York: Vintage Books, 1992.

Powell. R.W.: Estimating Worker Exposure to Gases and Vapors Leaking From Pumps and Valves. *Am. Indust. Hyg. Assoc. J. 45(11)*: A-7 to A-15 (1982).

Schmidt, M.: Selecting Clean Valves. *Chem. Eng. 108(6)*: 107-111 (2001).

Spirax-Sarco: What is Steam? Cheltenham, UK: Spirax-Sarco Ltd., Steam Engineering Tutorials, 2011a [www.spiraxsarco.com, Website].

Spirax-Sarco: Superheated Steam. Cheltenham, UK: Spirax-Sarco Ltd., Steam Engineering Tutorials, 2011b [www.spiraxsarco.com, Website].

Thayer, H.: *Management of the Hanford Engineer Works in World War II: How the Corps, DuPont, and the Metallurgical Laboratory Fast Tracked the Original Plutonium Works*. New York: American Society of Civil Engineers, 1996.

UAW-GM: Lockout. Detroit: United Auto Workers-General Motors Human Resource Center, 1985.

WorkSafeBC: Ammonia in Refrigeration Systems. Richmond, BC: WorkSafeBC, 2007.

# J Flowable Solid Materials

## INTRODUCTION

The study of solid materials usually considers solids capable of flowing. Flowable solids range from heterogeneous soils to homogeneous powders. Concern about the properties of flowable solids is widespread and affects safety in recreational skiing and snowmobiling in mountainous areas; in construction of buildings set in deep excavations, sewers, and other underground utilities, and dams and other earthworks; and in the handling of processed materials of commercial importance in piles and in storage structures. These issues are the subject of snow, rock, and soil mechanics in the former case, and bulk materials handling in the latter.

Previous information has provided a means to organize discussion about these materials and the manner in which they are stored and handled. They occur in environments in which the energy source is confined, as in the case of material held in handling structures (Figure 1.4); semi-confined, as in a bank of sand undergoing removal by excavation (Figure 1.5); and semi-unconfined, as in a pile of processed material awaiting shipment (Figure 1.6). A hurricane is an unconfined source of energy, and hence, not of interest in this discussion.

Materials confined in handling structures, such as silos, bins, and hoppers, usually possess considerable economic value and are usually processed in some way. Semi-confined environments contain virgin materials in situ, starting with materials of economic significance, such as sand and gravel. These materials also can include soils exposed during excavating. These include undisturbed soils and soils created from backfill during previous construction projects.

Materials in semi-unconfined environments often include materials of economic value, starting with unprocessed materials, such as excavated gravel that has not undergone size classification and materials that have undergone processing. Sulfur is an example of a material stored in open piles that has undergone processing. These materials are sometimes stored outdoors, and sometimes, as in the case of raw sugar, in handling structures that protect it from the weather.

Flowable solid materials in semi-unconfined piles located outdoors and sometimes in buildings can be very large. Materials handling can involve earthmoving equipment, including dump trucks, front-end loaders, bulldozers, draglines, and excavators, to shift the material from one position to another. These movements must always respect gravity because of the possibility of slope failure and cave-in. In the latter case, gravity is the motive force.

Removal occurs at the leading edge of the pile or the bank. As removal occurs, the machine moves toward the center of the pile. Surfaces become steeper than the (presumed) stable initial contour of the pile. At this point the parameters of stability, as discussed previously, become increasingly prominent in explaining what occurs as more and more material is removed. At some point, the pile or bank becomes unstable and slumpage (avalanche) occurs. Occasionally, this has led to burial of the materials' handling equipment or personnel standing in the vicinity.

From a regulatory perspective, flowable solid materials located on the boundary surfaces of trenches, excavations, and earthworks in semi-unconfined structures in the natural environment usually are considered separately from materials contained in semi-unconfined structures, such as storage piles and confined structures, such as bins, silos, and hoppers. The focus of regulatory attention toward trenches, excavations, and earthworks is to prevent collapse as an expression of flow of material. Collapse has led to many fatal injuries and is a major concern in construction and mining.

## Flowable Solid Materials

Flowable solid materials are solids that can flow under the influence of gravity or a disturbance. In this respect flowable solid materials are similar to liquids. Within certain limits, flowable solid materials will adopt the shape of the container into which they are placed. Flowable solid materials include materials of commerce, as well as those of the natural environment. Flowable solid materials of commerce comprise a wide range of inorganic and organic substances and products. Flowable solid materials in the environment include snow, and materials of the Earth's crust, such as soil disturbed during excavating and blasting, and activities involving vibration.

Flowable solid materials of commerce range from minerals to synthetic industrial products to agricultural products. Minerals of interest include sands that occur naturally in geological deposits as discrete unbound crystals, heterogeneous deposits of sand and gravel, and rock and rock-like material of industrial significance. Rock and rock-like material require fracturing with explosives to enable removal from the native material, and crushing to desired size prior to sale for use in industry and industrial processes. Rock includes minerals of direct interest, such as limestone, salts, and other crystallized minerals. Rock also contains minerals that are present in small or trace quantities. This ore must undergo extensive physical and chemical processing to enable extraction of the desired minerals. Rock-like minerals include coal, oil sand, oil shale, diatomaceous earth, clays, mica, and asbestos.

Industrial output occurs in the form of granules, powders, crystals, fibrous materials, flakes, or pellets. Products prepared in these forms include plastics and resins, pigments, building materials, such as cement and gypsum, adhesives, and fertilizers, to name but some of the possibilities.

Agricultural products occur or are prepared in several flowable forms: kernels of dried grain, powders (for example, flour, brewery and bakery waste), crystals (sugars, salts), fibrous materials (peat moss), and pellets.

Flowable solid materials span a wide range of chemical and physical properties. These ultimately are responsible for determining the behavior of these products during handling and storage in large quantities. The properties of the material in the form in which it is present at a particular stage in the process are intrinsic to the manner in which it has been treated, and can be considerably different from those of the starting material. Consider, for example, limestone rock versus crushed limestone, or powdered coal versus chunk coal freshly removed from the native structure in the ground. Some of the physical properties are complicated by the state of subdivision of the material (Wahl 1985).

A classification scheme containing four divisions ranks flowable solid materials according to their ability to flow (Wahl 1985). These classes range from granular, free-flowing materials (Material Class I) to fibrous or flaky materials of low bulk density that tend to interlock and absorb vibration (Material Class IV). Flowability is a measure of the tendency of the material to form a bridge.

## Semi-Confined Materials

Semi-confined materials experience confinement by boundaries on two or more surfaces. These can include the side, top, back, or bottom of the material. These materials are present in natural deposits of material on slopes and in pockets surrounded by boundary surfaces. Similarly, existing boundaries can semi-confine materials that are dumped onto a slope. Sometimes the dumping is intended to increase the usable surface area at the top of the slope and sometimes the slope acts to channel the material for storage and unloading at the base. Within the confines of the boundary surfaces, the material can potentially slide.

Every mass of material located beneath a sloping surface or the sloping sides of an open cut has the tendency to slide downward and outward (Terzaghi et al. 1996). This applies to snow, soil, and piles of mineral and nonmineral materials. This tendency to slump occurs under the influence of gravity and possibly disturbances caused by vibration, both natural tremors (earthquake) and man-made (dynamite).

A shearing resistance counteracts the tendency to slump. Sufficient shearing resistance prevents slumpage. Undercutting the foot of an existing slope or pile of material or excavating with unsupported sides in creating excavations and trenches represents obvious external factors that reduce shearing resistance and cause collapse. Increase in pore water strength and progressive deterioration of soil strength also can decrease shear resistance to the point where slumpage occurs. High pore water pressure is associated with heavy rainfall and snow melt. These conditions can affect any outdoor pile of material.

Failure of a mass of material located at the top of or beneath a slope constitutes a slide. Slides involve the downward and outward movement of the entire mass of material that participates in the failure of the slope. Slides occur suddenly or slowly, and with and without apparent provocation. Slides usually occur due to excavation or undercutting the foot of an existing slope. They can also occur due to gradual disintegration of the structure of the soil, increase in pore water pressure in permeable layers or shock that liquefies the soil beneath the slope. Predicting stability is difficult due to the multiplicity of factors that can contribute to a slide.

Initial formation of tension cracks and steps or benches at the uppermost point or root characterizes almost every slide. During the slide, the root subsides and the lower part or tongue bulges outward. This movement of material forms an "S"-shaped curve in side view. The shape of the tongue depends on the type of sliding material. Constrainment occurs when the flow of material reaches the opposite side of a structure.

## Natural Slopes

Excavating and contouring existing terrain for highway and railway cuts create natural slopes (Terzaghi et al. 1996). Moments of rotation about an imaginary point form the basis for stability computations for slopes. Curvature of the surface of motion by the subsurface material in real-world situations resembles the arc of an ellipse. The arc of a circle substitutes as an approximation for the arc of the ellipse in actual calculations. Calculation of moments occurs for the slice of material that tends to produce failure and the slice that tends to resist it.

The calculation uses the center of mass for each slice and the distance along the perpendicular line drawn from the center of mass to the radius of the circle. In the simplest case, sliding resistance at equilibrium depends on the difference between the moments and the radius of the imaginary circle (the critical circle) and the length of the surface of sliding. A failure that terminates at or above the toe of the slide is a slope failure. A failure that proceeds beyond the toe of the slide is a base failure.

Experience indicates that a slope ratio of 1.5:1 (horizontal:vertical) in the exposed soils of open cuts usually is stable (Terzaghi et al. 1996). Certain types of soils also are less stable and require shallower slope (larger horizontal to vertical ratio). Canals and other flooded cuts require slopes of 2:1 to 3:1 for stability. On the other hand, some conditions permit steeper slopes to as much as 0.25:1.

Government agencies routinely take action against instabilities in natural slopes. These include detonating charges into mountain slopes to trigger avalanches preemptively so as to minimize the magnitude of a naturally occurring situation. Similarly, authorities involved in highway maintenance routinely inspect rock outcrops for instability and "scale" the surfaces to remove loose material that otherwise could roll onto roadways.

## Excavations and Trenches

Excavations and trenches are cuts made into the ground. The terms, "excavation" and "trench", take their specific meaning in regulations and regulatory Standards enacted by various jurisdictions. The so-called OSHA excavation or trenching standard (29CFR 1926.650, 29CFR 1926.651, and 29CFR 1926.652) provides some specific definitions to use as an example (OSHA 1999).

An excavation is a man-made cut, cavity, trench, or depression in the ground that is created by removal of earth. A trench is a narrow excavation (in relation to its length) made below the surface of the ground. Hence, a trench is a subset of an excavation. In general, the depth of a trench is greater than its width, and the width (measured at the bottom) is not greater than 15 ft (4.6 m). If a form or other structure installed or constructed in an excavation reduces the distance between the form and the side of the excavation to 15 ft (4.6 m) or less (measured at the bottom of the excavation), the excavation is also considered to be a trench.

Excavations and trenches, usually being outdoors, share similarities with piles of loose material, and also exhibit some differences. Excavations and trenches are cut into what usually is undisturbed material. Undisturbed is a relative term, since the material could be native material or backfill from previous digging. Refilled, previously excavated material is not undisturbed unless tightly compacted and is subject to lack of cohesion, compaction, and subsidence. It also may have completely different composition compared to undisturbed material.

Excavations and trenches often pass through utility lines containing hazardous and possibly pressurized materials including natural gas, crude oil and petroleum products, potable water and wastewater, steam, and chemical waste. The excavation or trench also could pass through energy conduits, such as underground electrical cable. Breakage exposes workers in the trench and potentially those on the surface to these agents.

As well, the soil into which the cut is made can contain contamination or seepage from previous industrial activity (Figure J.1). This can include heavy metals and volatile hydrocarbons. Vapors from the hydrocarbons can seep into the airspace of the cut. As well, the excavation or trench may pass through nonmineral material, such as shoreline detritus, wood waste, or peat, since filled over. Decaying vegetation or animal material decomposes to form methane and hydrogen sulfide. These gases can diffuse from such soils into the airspace of the excavation.

Soils contain considerable mass. The weight of soil varies with type and moisture content. One cubic meter, 1 m$^3$ (35.3 ft$^3$), of soil can weigh more than 1,350 kg (3,000 lb). A relatively small volume of soil from the failed wall of a cut can easily immobilize or crush an individual.

Excavations and trenches usually have restricted ingess and egress by ladder or sometimes a stairtower. Hence, in addition to potential for cave-in of the soil of the wall of the cut, excavations and trenches can possess many of the same hazards encountered in spaces designated in regulations as confined spaces (McManus 1999). With the presence of contamination and possibly methane from seepage through soil, the distinction between these designations becomes completely blurred.

The first point of consideration in addressing the problem posed by soils that form the walls of trenches and excavations is to establish their relative stability. The system used by OSHA (Table J.1) categorizes soil and rock deposits into four types (OSHA 1999).

Evaluating the type of soil is critical to determining the precautionary measures necessary to provide protection against collapse. Many kinds of devices and methods are used to determine the type of soil present in an area (OSHA 1999). Soil types can range from the highly predictable to the highly un-

Figure J.1. A trench that exposed oil and other contamination resulting from previous industrial activity in the area. Core sampling failed to identify the existence of the contamination. (A geotechnical engineer assessed and approved the slope of the walls.)

## Table J.1
## Classification of Soils (OSHA Criteria)

| Classification | Description/Comments |
|---|---|
| Stable rock | Natural, solid mineral matter |
| | Vertical sides and remain intact when exposed |
| | Usually identified by names, such as granite or sandstone. |
| | Determining stability of a deposit depends on knowledge about cracks and whether they run into or away from the excavation. |
| Type A | Cohesive soils with an unconfined compressive strength of 1.5 tons per square foot (tsf) (144 kPa) or greater |
| | Inclusions: clay, silty clay, sandy clay, clay loam, and in some cases, silty clay loam and sandy clay loam. |
| | Exclusions: soil that is fissured, is subject to vibration of any type, has previously been disturbed, is part of a sloped, layered system where the layers dip into the excavation on a slope of 4 horizontal to 1 vertical (4H:1V) or greater, or has seeping water. |
| Type B | Cohesive soils with an unconfined compressive strength greater than 0.5 tsf (48 kPa) but less than 1.5 tsf (144 kPa) |
| | Inclusions: angular gravel, silt, silt loam, previously disturbed soils unless otherwise classified as Type C, soils that meet the unconfined compressive strength or cementation requirements of Type A soils but are fissured or subject to vibration, dry unstable rock, and layered systems sloping into the trench at a slope less than 4H:1V (only if the material would be classified as a Type B soil). |
| Type C | Cohesive soils with an unconfined compressive strength of 0.5 tsf (48 kPa) or less |
| | Inclusions: granular soils such as gravel, sand and loamy sand, submerged soil, soil from which water is freely seeping, submerged rock that is not stable, and material in a sloped, layered system where the layers dip into the excavation or have a slope of four horizontal to one vertical (4H:1V) or greater. |
| Layered geological strata | Soils configured in layers |
| | Overall classification based on the classification of the weakest layer. |
| | Each classified individually when a more stable layer lies below a less stable layer |

Source: OSHA 1999.

predictable. The latter is true where industrial activity has actively disturbed the soil leading to imprecise knowledge about the situation.

A critical measure of soil stability is the unconfined compressive strength. Unconfined compressive strength is the load per unit area at which soil will fail when compressed. Determination of this parameter can occur by laboratory testing or estimation in the field using various instruments and tests (Table J.2).

The stability of a given cut depends on the balance between forces that produce failure and forces that oppose failure (Terzaghi 1996). An investigation of such forces and conditions leads to a stability determination. In making a stability determination, the position of the potential surface of sliding must be estimated, and resistance against sliding along this surface estimated. Both quantities are subject to considerable uncertainty.

As with piles of loose material, temperature and moisture content of the soil are major considerations in stability computations. Moisture in soils originates from above in the form of rain and snow, and from below in the form of groundwater. Pore water exerts seepage pressure against walls of cuts which, being exposed to atmosphere are unsupported. Being exposed to atmosphere and temperature conditions, soil water at the exposed surface and to a depth in the interior affected by conduction of heat is subject to freezing and thawing.

These influences can cause stresses and deformations in and surrounding an open cut or trench. These stresses and deformations can adversely affect the stability of a trench or excavation and can lead to failure of the wall. Stresses and deformations that affect excavations and trenches include tension cracks, sliding, toppling, subsidence and bulging, heaving or squeezing, and boiling (OSHA 1999). Most of these concepts should be familiar from previous discussion on slopes. Some are unique to trenching and excavation and arise because of the steep walls created during this work.

Tension cracks usually form at a horizontal distance of 0.5 to 0.75 times the depth of the trench, measured from the top of the vertical face of the trench (Figure J.2). Sliding or toppling may occur as a result of tension cracks. Sliding describes collapse of the trench wall at the base. Toppling describes forward collapse of the wall of earth as the vertical face shears along the tension crack line and topples into the cut. Subsidence and bulging occur together. Subsidence refers to sinking of the bench material adjacent to the cut, and bulging to bowing outward of the wall. An unsupported cut can create an unbalanced stress in the soil, which, in turn, causes subsidence at the surface and bulging of the vertical face of the trench. If uncorrected, failure of the wall can occur due to this condition (Figure J.3).

Loads and vibrating equipment placed near or adjacent to the edge of the trench exacerbate conditions that lead to trench failure. Loads include heavy equipment, such as front-end loaders, excavators, dump trucks and ready-mix concrete delivery trucks. Spoil excavated from the cut and piled close to the edge can impose considerable load on the edge of the cut.

Heaving or squeezing refers to bulging of soil in the bottom of the cut caused by downward pressure created by the weight of adjacent soil. Heaving and squeezing can occur even after correct installation of shoring or shielding. Boiling refers to upward flow of water into the bottom of the cut. Boiling produces a "quick" condition in the bottom of the cut, and can occur even when shoring or trench boxes are used. High water table is one of the causes of boiling.

## Slope Failure Protection Concepts

The slopes of temporary cuts, such as trenches and excavations, are made as steep as soil conditions permit without creating undue risk of failure or they are made as close to vertical as possible and supported to prevent collapse (Terzaghi et al. 1996). The slope of the cut may exceed the angle of internal friction of the loose material. In such situations, support is essential to prevent failure.

Failure of a vertical wall in an excavation or trench can occur at the top and bottom, as discussed in the previous section. The data needed to design the support system depend primarily on the depth of the cut. Shallow cuts are considered to be less than 6 m deep (Terzaghi et al. 1996). Bracing used during construction of sewers and water mains at these depths is standardized and applicable to different soil conditions.

### Shoring Systems

Shoring is the provision of a support system to trench faces in order to prevent movement of soil, underground utilities, roadways, and foundations (OSHA 1999). Shoring or shielding is used when the location or depth of the cut renders impractical sloping back to the maximum allowable slope.

## Table J.2
## Tests for Soil Stability

| Test | Comments |
|------|----------|
| Penetrometer test | The pocket penetrometer is a direct-reading, spring-operated instrument used to determine the unconfined compressive strength of saturated cohesive soils. Once pushed into the soil, an indicator sleeve displays the reading. The instrument is calibrated in either tons per square foot (tsf) or kilograms per square centimeter (kPa). Penetrometer error ranges from ±20% to ±40%. |
| Shearvane (Torvane) test | The shearvane (torvane) applies controlled pressure exerted through the blades of the vane to compress the soil. To determine unconfined compressive strength, the blades of the shearvane are pressed into a level section of undisturbed soil, and the torsional knob slowly turned until failure occurs. |
| Thumb penetration test | The tester firmly presses the thumb into the soil. Indentation occurring only with great difficulty indicates that the soil is probably Type A. Penetration no further than the length of the thumb nail indicates that the soil is probably Type B. Full penetration of the thumb indicates that the soil is probably Type C. The thumb test is subjective and is therefore the least accurate of the three methods. |
| Dry strength test: | This test considers properties exhibited by dry soil. Dry soil that crumbles freely or with moderate pressure into individual grains is granular. Dry soil that falls into clumps that subsequently break into smaller clumps that are broken only with difficulty is probably clay in combination with gravel, sand, or silt. If the soil breaks into clumps that do not break into smaller clumps (and the soil can be broken only with difficulty), the soil is considered non-fissured unless there is visual indication of fissuring. |
| Plasticity (wet thread test) | The plasticity (wet thread) test considers properties exhibited by wet soil. This test uses a ball formed from a moist sample containing a thin thread approximately 1/8 inch (3 mm) in diameter (thick) by 2 inches (50 mm) in length. The soil sample is held by one end of the thread. If the sample does not break or tear when suspended, the soil is considered cohesive. |

Source: OSHA 1999.

Trenches at depths typical of utility work require cross bracing to prevent cave-ins. Shoring systems consist of posts, wales (horizontal timbers), struts (vertical timbers), and sheeting (Figure J.4). Cross braces consist of timbers and extensible metal supports (screw jacks or hydraulic or pneumatic cylinders). Wales provide longitudinal support. Sheeting includes vertical support planks or sheet material. Struts (vertical support) are placed at 2.5 m intervals horizontally and 1.5 to 2 m vertically. Installation of all shoring should occur from the top down and removal from the bottom up.

Figure J.2. Tension cracks appear at a horizontal distance of 0.5 to 0.75 times the depth of the trench, measured from the top of the vertical face of the trench  (Source: OSHA.)

Figure J.3. Trench collapse. Notice the S-shaped curve in the pattern of slope failure.

Figure J.4. Engineered shoring system.

Screw jacks contain threaded steel sleeves and pins that extend by unscrewing. The process of setting, adjusting, and removing the struts of a screw jack system is a manual one that requires the worker to be in the trench during the procedure. Uniform "pre-loading" is not achievable with screw jacks because of inability to measure applied pressure in a convenient manner. The weight of these components also creates handling difficulties. Applying pressure by turning the screw requires considerable strength.

The trend today is toward the use of hydraulic shoring, and prefabricated strut and/or wale systems manufactured from aluminum or steel. Hydraulic shoring provides a critical safety advantage over timber shoring because workers do not have to enter the trench to install or remove the hydraulic cylinder.

Most hydraulic systems in this application are light enough to be installed by one worker. These systems contain gauges to enable distribution of pressure along the trench line and are easily adaptable easily to various depths and widths. Trench faces can be "pre-loaded" to use the natural cohesion of the soil to prevent movement. Hydraulic shoring should be checked at least once per shift for leaking hoses and/or cylinders, broken connections, cracked nipples, bent bases, and other damaged or defective parts.

Single-cylinder hydraulic shores are generally used in a water system, as an assist to timber shoring systems, and in shallow trenches where face stability is required.

Pneumatic shoring is similar to hydraulic shoring (Figure J.5). The primary difference is that pneumatic shoring uses gas pressure in place of hydraulic pressure. A disadvantage to the use of pneumatic shoring is possible loss of pressurization. This can necessitate an on-site air compressor.

Deep trenches and excavations require combinations of struts, wales, sheet piles, lagging and tie-backs. Tiebacks are soil anchors. Tiebacks are driven into the soil at an angle of 45° to the horizontal to provide anchorage. Tiebacks are similar in concept to rock bolts used in mining. Rock bolts anchor material in the ceiling. The bolts penetrate into the rock matrix to provide strength and support.

Figure J.5. Pneumatic shoring used in combination with shoring panels in a shallow trench. Note the unstable angle of attachment. One of the pneumatic shores flew through the air like an artillery shell soon after photographing of this picture occurred.

The process of drilling into the soil to place tie-backs can create a pathway for escape of methane, volatile hydrocarbon vapors, and hydrogen sulfide created by decay of buried vegetation, wood debris and contamination of soil by spills of hydrocarbons, as mentioned previously. These emissions can constitute serious safety and exposure hazards. Methane and hydrocarbon vapors can accumulate in sufficient quantity to create ignitable mixtures. These can pose fire and explosion hazards during welding and other hot work. These emissions and also emissions of hydrogen sulfide can constitute serious exposure hazards. The emissions can be high level and very brief in duration, especially in outdoor locations when the wind is blowing. Such situations necessitate use of testing instruments, portable ventilation systems, and possibly high-level respiratory protection.

Underpinning is the process of stabilizing adjacent structures, foundations, and other intrusions that may have an impact on the excavation. Underpinning involves physical reinforcement of the foundation. Underpinning should occur only under the direction and with the approval of a registered professional engineer.

### Shielding Systems

Shielding systems protect against cave-ins and other similar incidents, rather than shoring up or otherwise supporting the trench face (OSHA 1999). Shielding systems involve use of fabricated steel walled and braced structures that are positioned into the trench (Figure J.6). The excavated area between the outside of the trench box and the face of the trench is backfilled as needed to prevent slumpage and collapse at the face of the cut. This minimizes the gap between the metal of the trench box and the face of the cut and prevents lateral movement of the box. Shields are engineered to withstand specific loads and are not to be used in situations that exceed these limits.

Trench boxes are generally used in open areas, but they also may be used in combination with sloping and benching, depending on the actions permitted within a particular jurisdiction. The wall of

Figure J.6. Trench box. The top of the wall of the trench box used where access is possible requires installation of fall protection.

the box should extend at least 18 in (0.45 m) above the surrounding area if there is sloping toward the excavation. This can be accomplished by providing a benched area adjacent to the box.

The top of the wall of the trench box used where access is possible likely also requires installation of fall protection. The depth of allowable unprotected fall distance varies by jurisdiction.

OSHA allows excavation of earth to a depth of 2 ft (0.6 m) below the shield (OSHA 1999). This is allowable only when the shield is designed to resist the forces calculated for the full depth of the trench, and there are no indications while the trench is open of possible loss of soil from behind or below the bottom of the support system. Actions of this type require inspection to determine bulging, heaving, and boiling as well as surcharging, vibration, and effects on adjacent structures. Careful visual inspection of the conditions mentioned here is the primary and most prudent approach to hazard identification and control.

### Sloping
Sloping is a stabilization technique involving excavation to create walls having slopes below which collapse is known with high certainty not to occur (Figure J.7). Maximum allowable depth for excavations using sloping as a failure control strategy is less than 20 ft (6.1 m). Slope angles are based on soil type and angle to the horizontal. These reflect practical experience and built-in safety factor. OSHA criteria for sloped excavations (Table J.3) provide an example for illustration (OSHA 1999).

Slope ratios and angles used in regulatory standards in other jurisdictions could differ from values adopted in the U.S. The variety of factors and processes that lead to slides are extraordinary and complex. As a result, excavations involving these materials require input from experienced civil engineers. Massive flows of material (flow slides) have occurred unexpectedly under natural conditions as well as during construction activities. The former can occur because of poor adhesion between layers of stratified material.

## Table J.3
## Sloping Requirements (OSHA Criteria)

| Classification | Slope Ratio (Horizontal:Vertical) | Slope Angle |
|---|---|---|
| Stable rock | Not applicable | Vertical 90° |
| Type A (short term) maximum depth, 12 ft (3.7 m) | 0.5:1 | 63° |
| Type A (longer term) | 0.75:1 | 53° |
| Type B | 1:1 | 45° |
| Type C | 1.5:1 | 34° |

Source: OSHA 1999.

*Benching*

There are two basic types of benching (Figure J.8), simple and multiple (OSHA 1999). The type of soil determines the horizontal-to-vertical ratio of the benched side. As a general rule, the bottom vertical height of the trench must not exceed 4 ft (1.2 m) for the first bench. Subsequent benches may be up to a maximum of 5 ft (1.5 m) vertical in Type A soil and 4 ft (1.2 m) in Type B soil to a total trench depth of 20 ft (6.1 m). The slope of all subsequent benches must be below the maximum allowable for that type of soil. For Type B soil, trench excavation is permitted in cohesive soil only.

*Spoil Control*

Spoil is the material removed from an excavation or trench. In trenches excavated for placement of utilities, reuse of excavated soil may occur. In these situations, placement of temporary spoil must occur no closer than 2 ft (0.6 m) from the edge of the excavation, measured on the surface from the nearest base of the spoil to the cut (OSHA 1999). Measurement of this distance should not occur from the crown of the spoil deposit. This requirement for distance ensures that loose rock or soil from the temporary spoil will not fall into the trench. Entry of spoil in this manner could injure workers in the trench. Placement of spoil must occur to channel rainwater and other run-off away from the excavation.

Current practice in many situations, especially work involving urban streets, is to remove the spoil from the excavation or trench to a remote location by truck and to bring in sand and gravel for backfill. This can necessitate operation of the truck in close proximity to the side of the cut.

Permanent spoil is the term used to describe excavated material that is not returned to the cut as backfill. Permanent spoil should be placed at some distance from the excavation. Placement of permanent spoil at insufficient distance from the working excavation can negate compliance with the horizontal-to-vertical ratio requirement for a particular slope in the wall of an excavation or trench. Permanent spoil can change undisturbed soil to disturbed soil and dramatically alter slope requirements.

Other Issues

Visual inspection of excavations and trenches is critical to establishing and maintaining site safety. The visual inspection is a qualitative evaluation of conditions around the site and the excavation or trench.

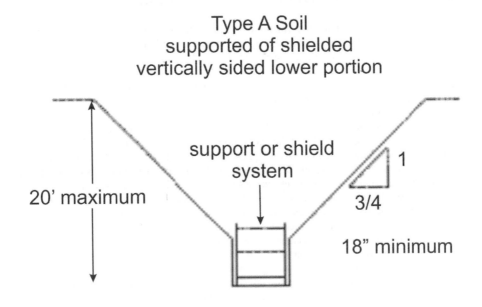

Type A Soil
supported of shielded
vertically sided lower portion

support or shield
system

20' maximum

1

3/4

18" minimum

Figure J.7. Sloping used in conjunction with a trench box. (Source: OSHA.)

Figure J.8. Benching. Open pit mines are primary examples of the application of benching techniques.

Some organizations use an external evaluator to perform this task, as this is a specialized skill and responsibility.

OSHA provides guidance about performance of site inspections (OSHA 1999). The evaluator observes the entire site, including the soil adjacent to the excavation and the soil being excavated. Soil that remains in clumps is cohesive. Soil that appears to be coarse-grained sand or gravel is considered granular. The evaluator checks for signs of vibration and splitting along the failure zone that would indicate tension cracks, and for existing utilities, and indication of any previous disturbance of the soil. The evaluator also observes the open side of the excavation for indications of layered geologic and anthropomorphic structuring.

The evaluator should also look for signs of bulging, boiling, or sloughing, as well as for signs of seepage of surface water from the sides of the excavation or from the water table. If the cut contains standing water, the evaluator should check for "quick" conditions. In addition, the evaluator should check the area adjacent to the excavation for signs of foundations or other intrusions into the failure zone. The inspection should also include a check for surcharging and the spoil distance from the edge of the excavation. Surcharging is instability caused by excessive vertical load on the failure zone. This loading can occur due to spoil piles, overburden, vehicles, equipment, and activities.

Shoring and shielding systems are intended for contiguous use during trench work. Installation of shielding systems in contiguous fashion cannot always occur because of the presence of underground utilities that traverse the trench. This situation creates problems for the competent person responsible for shielding and shoring specification, as well as for workers at the surface. Fall is a serious risk during work under circumstances where the excavation is deep and partially unguarded, or is shallow and partially unguarded and contains protrusions that can cause injury.

This situation requires consideration not just of the potential for cave-in but also for fall. Consideration for the two hazards must occur simultaneously.

## Unusual Situations

One interesting circumstance that is appearing more frequently is the need to trench under structural slabs poured on grade and engineered slabs supported by piles. This situation has arisen with subdivision of low-lying land previously left undisturbed due to lack of desirability for industrial and commercial purposes. These lands pose unique challenges because the water table is close to the surface. There is considerable microbiological activity in these former swamps and bogs with aerobic and anaerobic decay of vegetation, and dumped trash, trees, lumber, and other unwanted items and materials.

Anaerobic decay leads to production of methane and hydrogen sulfide and emission from the soil of these gases. The ground under the slabs of buildings erected on these sites often contains perforated plastic piping used as part of a soil gas collection and venting system. The ground under the buildings is often pre-loaded by the use of a thick layer of soil temporarily applied (and later removed) to compress the soil that will remain in place.

The problem that develops with this construction is that the ground under the slab sinks further following completion of construction, possibly in response to lowering of the water table. Sometimes an air gap develops between the bottom of the entire slab or part of it and the top of the earth. Continued sinking of the ground stresses storm and sanitary sewer lines and lines associated with the gas collection system attached to the bottom of the slab by hangers. The piping fails and leaks into the surrounding soil.

This situation necessitates entry under the slab to access the piping. Concrete used in conjunction with the pilings is especially strengthened and very difficult to penetrate from above. In addition, the length of the piping is often considerable and can involve bends and turns. The buildings are often furnished and contain existing business and commercial operations. Access from the top by removal of concrete can lead to severing of electrical and other services embedded in the slab. Approaching this problem from above is highly disruptive.

A more practical approach is to create a trench under the slab to gain access to the broken sewer piping for removal and replacement. The piles support the slab so that the risk of collapse (unless highly overloaded from above) is not a consideration. An assessment by a registered professional engineer will address this concern.

The trench is shallow, often around 1 m (about 3 ft) deep. Trenching considerations differ from those normally discussed because work occurs on hand and knees. The worker is easily surrounded by the bank created by the wall of the trench, even though the depth is shallow..

Shoring and shielding are impractical in this situation due to inability to move these devices into position and interference posed by existing piping, piles, and concrete beams. Sloping under the direction of a registered professional engineer will ensure safety of the approach.

Techniques employed for soil removal include high vacuum and use of long-handled tools, and sometimes small, portable conveyors to remove the spoil from the area. Rigid tubing at the end of the vacuum line extends the distance from the working end of the trench to the face of the slope. The vacuum system induces considerable airflow along the trench.

These techniques require considerable pre-planning and hazard assessment. The trench and overhead slab form a confined space and require consideration for ventilation, air monitoring, lighting, communication, dust suppression, protection against exposure to microorganisms in wastewater, and emergency extrication as minimum additional considerations for pursuing this work.

## Inappropriate Slope Modification

Circumstances arise where owners whose properties border slopes decide to increase usable land by placing fill onto the top. Similar tactics are employed to level sloping land, also sometimes located at the top of an abrupt slope.

Every mass of material located at the top of a sloping surface has the tendency to slide downward and outward (Terzaghi et al. 1996). This applies to soil placed as fill. This tendency to slump occurs under the influence of gravity and possibly disturbances caused by vibration, both natural tremors (earthquake) and man-made (dynamite). Periodically, failures of such "improvements," along with the consequences of such modifications appears on the news. Such events are a direct consequence of the instability created by upsetting the balance of forces involved in the stability of slopes containing flowable solid material that rests on underlying confining surfaces that are not inclined to fail.

Pushing material over the edge of the slope to restore the original angle of repose followed by dumping at the base can restore stability under such conditions. Slumping indicates that the slope angle near the top is too steep. Bulging at the base indicates that the foundation cannot support the weight of the pile. Bulging is indicative of potential failure of the entire slope (MSHA 1991).

## Semi-Confined Material Stockpiles

Material storage piles provide temporary storage for material awaiting shipping or processing (MSHA 1991). Some of the material storage piles are located on the side of existing embankments. The material slides down the slope formed by the boundary of the underlying stable material. This configuration offers the advantage of gravity flow to a lower level and saves considerable wear and tear on equipment otherwise used to effect this transition.

A related problem concerns the dumping of materials to extend the edge of a pile, as would occur during filling to amend grades on highway construction projects and building of tailings piles. This method of stockpiling involves backing the truck to the edge of the pile and then dumping the load. In order for this procedure to occur safely, berms (piles of material located along the edge of the slope) are necessary to prevent backing too close to the edge. This procedure, combined with removal of material from the toe of the pile at the base, produces an undercut and steepened slope (Figure J.9). This approach is obviously a dangerous method of stockpiling (MSHA 1991). A steepened slope is less stable and cannot support as much weight, and also is subject to failure.

The greatest risk of accident occurs at the edge (or crest) of the pile (Figure J.10). Hence, trucks should deposit loads away from the crest of the pile. This technique uses a bulldozer or front-end loader to push the load to and over the edge. This approach minimizes potential involvement of this equipment in slumpage involving dumped or pushed material and underlying material. By dumping inward from the edge, the truck, which has backed in, is unlikely to be caught in a slide. Similarly, this approach minimizes the potential for the bulldozer or front-end loader to slip over the edge or be caught in a slide. Construction of the stockpile in layers is considered to be a good alternative method. Compaction of previous layers by the movement of heavy equipment strengthens the pile.

Figure J.9.  Undercutting a semi-unconfined stockpile can lead to collapse. (Source: MSHA.)

Figure J.10.  The top of a stockpile poses greatest risk of accident from tip-over situations . (Source: MSHA.)

Warning signs of instability in stockpiles created in this manner include cracks along the crest (Figure J.11), slumping on the slope (Figure J.12), and bulging at the toe (Figure J.13). Cracks along the edge of the slope or slumping along the face indicate inability of the slope to support its own weight. As well, the presence of vibrating mobile equipment in this area further promotes instability until collapse occurs.

Slumpage at the edge (or crest) of a stockpile is a given part of the work routine involving tailings piles and piles of stored materials, such as sand, coal, and minerals. Excavation at the base of a pile of material increases the angle of the slope of the pile. This increase in angle of slope increases the potential for collapse of the pile. Whereas the normal interest in soil mechanics is stabilization of slopes, the interest of equipment operators during movement of bulk material is controlled destabilization. The intent is to collapse the slope in a minor and controlled way, but not to create conditions that resist collapse or to collapse the pile in a way that buries the machine and the operator (Figure J.14). This type of collapse involving burial of the equipment and fatal injury of the operator has occurred in many fatal accidents (MSHA 1988, MSHA 1991).

Working perpendicular to the crest or toe of a pile minimizes risk of injury to operators of materials' handling equipment. As well, the unit should always face the crest or toe of the pile. Perpendicular rather than parallel operation maintains the operator compartment as far as possible from the point of uncontrolled movement of the material. As well, burial of the cab during perpendicular operation requires more material than during parallel operation. Facing the crest during operation at the top of the pile maintains the greater portion of the mass of the equipment as far as possible from the edge. This minimizes potential for causing slope failure and subsequent involvement in a slide. Bulldozers have a lower center of gravity than front-end loaders, and as a result, engender a false sense of security. A unit operated parallel to a slope during failure is highly at risk of roll-over.

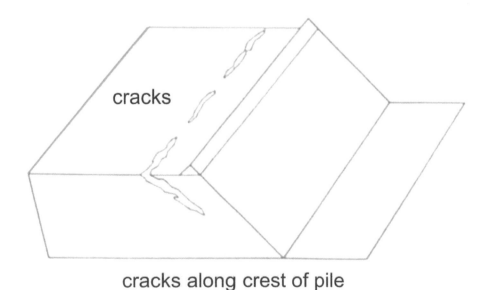

cracks

cracks along crest of pile

Figure J.11. Cracking along the top of a material stockpile is an initial indicator of slope failure. (Source: MSHA.)

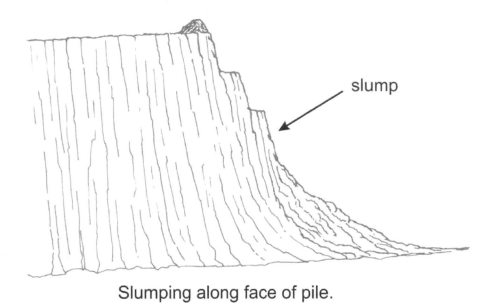

## Slumping along face of pile.

Figure J.12. Slumpage at the edge of a stockpile. (Source: MSHA.)

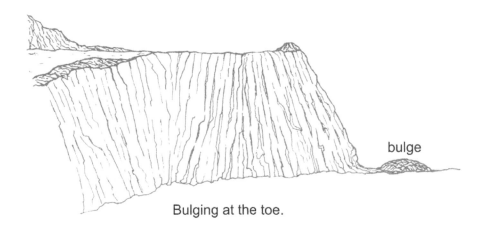

## Bulging at the toe.

Figure J.13. Bulging at the toe of a stockpile. (Source: MSHA.)

Figure J.14. Burial of machines and personnel at the base of a stockpile occurs following collapse induced by instability caused by removal of material. Note the small size of the front-end loader compared to the height of the slope. (Source: MSHA.)

## Semi-Unconfined Materials

Semi-unconfined material experiences confinement on only one surface, namely the base. The base is normally level. Hence, using the base as a constant influence on stability, the pile is constrained in its movement along that boundary only by its intrinsic properties. The latter create the angle of slope of the pile and its outward visual appearance.

Material storage piles provide temporary storage for material awaiting shipping or processing (MSHA 1991). Storage may occur as part of a flow regularization strategy (surge pile) or long term, as part of maintaining inventory available for delivery. A material storage pile constantly changes in shape and size, depending on production and shipping schedules. An outdoor material storage pile is affected by weather, including exposure to sun, rain, and freezing temperatures. Activity by equipment and personnel on foot occurs at the top of the pile where dumping occurs and at the toe of the pile where loading occurs. Conveyors are also used in the deposition of material onto storage piles.

Removal from semi-unconfined storage piles usually involves mobile bulk handling equipment, such as front-end loaders, excavators, and bulldozers. Removal occurs at the leading edge of the pile. As removal occurs, the machine moves toward the center of the pile. Surfaces become steeper than the (presumed) stable initial contour of the pile. At this point the parameters of stability, as discussed previously, become increasingly prominent in explaining what occurs as removal of more and more material occurs. At some point, the pile or bank becomes unstable and slumpage (avalanche) occurs. Occasionally, this has led to burial of the materials' handling equipment or personnel standing in the vicinity (Figure J.15).

During removal of material from the toe of a stockpile, generally the pile collapses slightly, thus maintaining the angle of repose (MSHA 1991). When the material in the pile does not flow easily, over-steepening or undercutting can occur. Pile stability is subject to weather conditions, including wetting by rain, freezing by sub-freezing temperatures, and drying by the sun. Moisture within the pile increases stability and leads to a steeper angle of repose. Freezing also increases pile strength. These

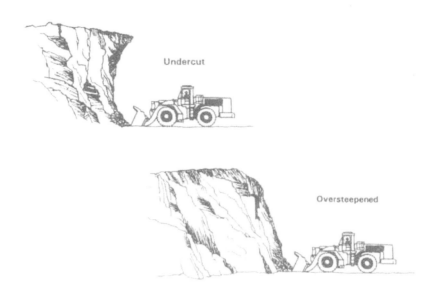

Figure J.15. Slumpage at the base of storage piles has led to burial of personnel and equipment. (Source: MSHA.)

temporary increases in pile strength are subject to change in weather conditions and can rapidly appear and disappear. Pile stability also depends on compaction. Compaction occurs due to passage of heavy equipment and the addition of layers of material to the pile.

Over-steepening results when material slumps from the crest but not the main slope. The crest becomes set back from the main slope. This results in a slope that is steeper than the angle of repose. Another problem with undercut or over-steepened stockpiles is sudden slumpage of compressed or frozen chunks of material. This problem is similar to what occurs in enclosed structures during attempts to remove hang-ups of solidified material from walls and overhead structures, except in the latter case, this results from deliberate action.

Undercutting occurs when the pile does not slump to maintain the angle of repose as removal of material occurs from the base. Undercutting can result in overhangs that can fail and fall onto equipment.

Piles that have become undercut or over-steepened require immediate amendment, as they pose serious safety hazards due to potential for collapse. This can be achieved using bulldozers or more preferably excavators equipped with long-reach booms. These actions constitute emergency measures only, and are not part of the normal routine of operation. The situation should never be allowed to develop to the point where such actions become necessary.

A storage pile usually sits on a firm base. Failure of the slope of the pile following removal of material is known as a slope failure. Base failure also can occur. Failure in a pile of cohesive material usually is preceded by formation of tension cracks. This is followed by sliding along a curved surface. The radius of curvature varies according to the position on the curve. Curvature is greatest in the middle, intermediate at the lower end, and least on the upper end. The curve formed by the slide resembles an ellipse.

A slope, such as a material storage pile, underlain by a dry cohesionless material, such as clean dry sand, is stable regardless of height, provided that the angle between the slope and the horizontal (angle of repose) is equal to or smaller than the angle of internal friction for the material in the loose state (Terzaghi et al. 1996). The angle of internal friction is the maximum angle of stability of a slope of loose material. This is the angle formed by material at the bottom of a pile of sand following excavation by a front-end loader or on the test bench when a spatula is pulled vertically through a pile of material (Ware 1999a). The factor of safety is the ratio of the tangent of the angle of friction to the tangent of the angle of repose. Few natural soils and materials are cohesionless. Stability of a slope is also a function of height.

Some stockpiles hold material prior to loading into an underground feeder. An underground feeder is essentially a large hopper that connects to a conveying system. These configurations are similar in concept to the hoppers located at the base of enclosed storage silos and are subject to the same problems. In unconfined operations, the bulldozer or front-end loader pushes material to the draw hole of the hopper.

Flow of the material into the hopper can also draw in the equipment under some situations. This is similar in concept to entrapment of workers in hoppers and silos containing flowing bulk solid material. In the process, the unit must overcome the increasing slope of the hopper as engulfment occurs, as well as the inflow of material.

In outdoor operation, weather conditions can affect flowability of solid materials, as described previously. Bridges (or arches) result when material coagulates above a draw point. Material subjected to outdoor weather is considerably more susceptible to coagulation. This is due to the presence of moisture and occurrence of freezing conditions. Failure of bridges (arches) has led to complete engulfment of bulldozers and other equipment over the draw point (Figure J.16). Such bridges, and the use of bulldozers and front-end loaders to break them have caused many fatal accidents (MSHA 1988).

Safe breakage of a bridge may require an excavator equipped with a long-reach boom or building demolition equipment containing hydraulic jack hammers and drills. This equipment permits fracturing the bridge without the need to work on it.

## Confined Materials

Storage of materials under confined conditions usually occurs in structures designed and built for such purposes. These include bins, silos, and hoppers. Hoppers can be free standing and form the tapered structures at the base of the cylindrical parts of many bins and silos. The boundary surfaces of storage structures prevent bulk movement of the material into fully extended stable piles. Hence, there is an inherent instability of a pile of material forced to remain within the confines of a storage structure. As a result, the material exerts pressure against the bottom and sides of the structure.

### Storage Structures

Storage structures built to handle and store flowable solid materials appear at first glance to be benign. This view fails to comprehend what is actually under observation. First of all, the storage structure must bear its own weight plus the weight of the contents. These masses are substantial.

About 1000 industrial and farm storage structures fail each year throughout North America (Jenkyn and Goodwill 1987). The term failure in the context used here indicates structural distress (Figure J.17), as well as catastrophic collapse (Figure J.18) (Carson and Holmes 2001). Some indicators of structural distress include crackage in a concrete wall, kinkage in a steel shell, bulges, distortions, and deformations. These signs, while not impeding operation of the structure, indicate that a serious problem has developed. Failure (Carson and Holmes 2001; Carson and Jenkyn 1990; Jenkyn and Goodwill 1987) of storage structures results from many factors (Table J.4). According to this analysis, failure results from inappropriate action taken throughout the life cycle of these structures.

Silos experience loading stresses from flowing and nonflowing bulk solid material (Carson and Jenkyn 1990). Loading conditions reflect flow properties of the material (an intrinsic property), as well as the design of the storage structure. Transient and static loading conditions can exist simultaneously. Loading stresses on storage vessels more closely resemble those involving aircraft more than those on buildings. Loading conditions on aircraft on the ground differ from those in flight, whereas loading

Figure J.16. Entrapment of materials handling equipment in a hopper following collapse of material blocking a grating. (Source: MSHA.)

Figure J.17 Silo failure due to structural distress. (Courtesy of Jenike & Johanson, Inc., Tyngsboro, MA.)

Figure J.18. Silo collapse due to base failure. Note erosion of material from the base. (Courtesy of Jenike & Johanson, Inc., Tyngsboro, MA.)

conditions on buildings do not change. Hence, calculating loading conditions is more difficult than first appears.

Bulk materials exert pressure outward on the walls of the structure (Carson and Jenkyn 1990; Jenkyn and Goodwill 1987). Ideally, the pressure should be uniform around the circumference, or the perimeter of the structure in the case of nonround structures. A single central discharge produces uniform outward pressure. This should be the case with an isolated bin or silo. However, bins and silos are often interconnected and share common walls. Silos often form the walls for bins that occupy the interstitial spaces between them. These interconnections can produce unintended rigidity due to the common walls and the mass of material held behind them. Reduced outward pressure on one part of the wall in a steel silo can cause buckling and the appearance of the characteristic "diamond" shape.

With myriad possible variations of downgrading incidents and the large installed base of storage structures, the number of failures estimated per year is understandable. Perhaps surprising is the small number of structures that fail catastrophically, spilling their contents to the ground and injuring or killing people in the process.

Nonuniform forces in storage structures can result for other reasons, starting with an off-center (eccentric) location of the discharge opening. An off-center discharge opening creates an eccentric flow channel. An eccentric flow channel modifies the load (outward pressure) imposed on walls. An eccentric flow channel that includes a wall adjacent to the opening removes the load from that portion of the circumference/perimeter of the structure. At the same time, the outward pressure on the opposite wall is much higher, if not increased above limits addressed in the design (Carson and Holmes 2001; Carson and Jenkyn 1990; Jenkyn and Goodwill 1987). The designer should recognize the impact of the off-center location of the discharge and incorporate this deviation into the design. If this does not happen, deflection in the wall in the horizontal plane (an outward bulge) can occur in areas of increased pressure and straightening in areas of reduced pressure. In the case of steel walls this is manifested through diamond-shaped buckling, and cracking in the case of concrete walls.

## Table J.4
## Factors in Silo Failure

**Factor**

**Design**

Incorrect or inappropriate or incomplete design

Insufficient characterization of the site and its load-bearing capabilities

Insufficient knowledge about the role of the silo in the process flow at the site

Incompatibility between contents and materials of construction leading to corrosion

Insufficient knowledge about the contents and their flow characteristics

Insufficient knowledge about actual versus planned operation

Failure to anticipate flow channel geometry in conjunction with discharge geometry

Unanticipated thermal effects (differential expansion and contraction of the structure and contents)

Unanticipated transfer of moisture within contents

**Construction**

Incorrect or inappropriate or inadequate construction practices

Poor (sloppy) workmanship

Failure to utilize the specified quantity of material (e.g. reinforcing bars)

Incompatibility between contents and materials of construction

Substitution of incorrect or inappropriate materials or components of construction

Unauthorized or under-designed modifications to the structure and its ancillary components

**Operation**

Inappropriate or incorrect operational practices

Failure to operate discharges in the manner anticipated in design

Change in contents from that specified in the design

**Maintenance**

Unauthorized or under-designed modifications to the structure and its ancillary components

Inappropriate selection of replacement bin liner

Incorrect, inadequate or inappropriate maintenance techniques

Lack of or insufficiency of periodic inspection of the exterior and interior of the structure

Failure to repair/replace the liner (required to maintain frictional characteristics)

Failure to respond to signs of distress (cracking in concrete, creasing in steel)

Sources: Carson and Holmes 2001; Carson and Jenkyn 1990; Jenkyn and Goodwill 1987.

Anchorage of walls by connection to adjacent structures compounded by the mass of material inside can considerably increase stress on flexible walls. This situation can also result from modification to the storage structure through relocation of the discharge opening, as well as through selective use of multiple discharge openings (Carson and Holmes 2001; Carson and Jenkyn 1990). A better design where multiple discharges are required splits the discharge after the flow has left the silo.

Another problem of this type concerns sweep augers located on the floor of the structure. Sweep augers must operate around the complete circumference of the structure rather than in an arc. An arc will cause development of an eccentric flow channel. Development of an eccentric flow channel in a steel silo can lead to toppling due to imbalanced loads on the walls.

Vibratory flow enhancers positioned on one side of the tapered section also can promote development of eccentric flow channels. To prevent this, cycling of these units on and off, rather than continuous operation, should occur. As well, vibratory flow enhancers can also compact the material, thus impeding flow (Ware 1999b).

The pressure exerted by bulk materials against walls of storage structures can change due to local conditions (Carson and Holmes 2001; Carson and Jenkyn 1990). These include migration of moisture from wet or moist particles to dry ones. This migration expands the dry particle and imposes increased radial pressure against the wall of the structure. The material has little freedom to expand upward, and therefore exerts additional pressure outward against the wall. A similar situation develops due to thermal effects. Heating causes slight expansion of the structure. Material flow into this expanded volume cannot be pushed back when contraction occurs at lower temperature.

Prolonged exposure to temperatures <4° C (<25° F) can lead to freezing of wet material or material containing moisture (Ware 1999b). Moisture in stored materials is a serious issue. Moisture can enter the storage structure in the material being stored, in leakage through surfaces exposed to moisture, in humid air admitted through the vent to hygroscopic contents, and in seepage through the base where the water table is high. Resolution of a problem may require consideration of more than one of these possible mechanisms.

Freezing inside the circumference (perimeter) of walls occurs similarly to frost penetration in soil. A ring of frozen material develops. In the transition section and near the outlet, the entire cross-section can freeze. Freezing occurs point-to-point through contact between wet particles or by immersion. Point-to-point freezing is less of a concern because of the air voids between particles of material. Immersion freezing means that the air voids are filled with water, and that the entire mass is frozen. Freezing leads to expansion of the material and increased outward pressure on the wall.

Thermal ratcheting is a phenomenon related to the expansion and contraction of the walls of storage structures. The material of construction of the wall expands during the heat of the day and contracts during the night as temperature decreases (Carson and Holmes 2001). Free-flowing material inside the silo adopts the profile of the structure during expansion. However, once occupying the enlarged cross-section, the material does not return to its original position as contraction occurs. Contraction of the material of construction of the wall against the in-filled material leads to hoop stress that can exceed wall strength. Exceedence of wall strength leads to failure and possible catastrophic loss of containment of the bulk material. Catastrophic failure can lead to rapid expulsion of the contained bulk material. A parallel to settling due to thermal ratcheting is swelling due to absorption of moisture by the bulk material.

Bulk materials exert pressure against walls of structures in the vertical direction. This increases from top to bottom (Carson and Jenkyn 1990; Jenkyn and Goodwill 1987). Vertical forces are caused by accumulation of wall friction effects at any point from the top surface of the material to the bottom, as well as dead loads inherent in the structure. Uniform increases are anticipated and incorporated into the design of the structure.

Vertical pressure profiles unanticipated in design can occur due to flow patterns not occurring according to expectations. These can lead to unanticipated pressure peaks in the vertical flow profile that exceed expectations and design limits. One such cause is an asymmetric flow channel, especially one that includes the wall of the structure as a boundary. This leads to vertical bending moments in the wall.

Lateral pressure exerted against the walls by nonmoving material is static pressure (Ware 1999b). Maximum static pressure occurs at the break line, the point of transition at the base of the cylindrical part of the structure where the tapered section begins. Lateral pressure exerted against the walls by moving material is called dynamic pressure. Dynamic pressure is about three times larger than static pressure. Again, maximum lateral force is exerted at the point at the break line where the slope of the wall of the vessel changes.

Internal components, such as support beams, inverted cone inserts, and gravity bending tubes, can impose large concentrated loads or asymmetric pressures or both onto silo walls (Carson and Holmes 2001). These components impose bending stresses onto the wall of the silo at the point of attachment. The tendency to create a hopper-like convergence at the support point imposes a localized pressure peak on the silo wall. Open cone inserts and gravity-bending tubes can impose large loads. Hence, plugging is a concern, since the vertical load including friction loads, stress field loads, and other loads greatly exceeds the dead weight of material plugging it and the surcharge above it.

Integrity of storage structures is also affected by stresses imposed by natural events such as tornados, high winds, and earthquakes, and anthropomorphic ones, such as nearby explosions (Carson and Jenkyn 1990). Internal and external inspection is the best preventive tool for enhancing longevity of a storage structure to prevent failure (Holmes and Carson 2002). According to these authors, external inspection should occur every two or three months when the structure is operating, and internal inspection, annually. Corrosion and erosion of silo and hopper walls are critical issues, as they are predictive of catastrophic failure. Damage to hopper liners can cause change in the flow pattern.

## Bulk Material Flow

A major focus in the design of storage structures is the means by which contents leave at the base. The base must focus flow from the large cross-section of the interior of the structure to the small cross-section of the opening through which the contents will pass. The success of the design in enabling flow to occur under controlled conditions is essential to the performance and safety of the operation of the structure.

When material first enters a storage vessel, a mound forms. As material continues to enter, large particles flow down the surfaces of the mound to the periphery (Ware 1999b). Small particles remain in the center. Exit from the structure can occur in two primary ways according to whether flow from the discharge involves favored or impartial flow.

Funnel flow involves a first-favored flow channel. Funnel flow produces material movement described as last-in, first-out (Ware 1999b). The characteristic of funnel flow that enables this to occur is development of the first-favored flow stream. The first-favored flow stream is an active flow channel that develops through the material as removal occurs.

The first favored flow channel may develop in the center of the material or eccentrically to one side. The outlet flow channel is circular. Its diameter corresponds to the diameter or diagonal of the outlet. If material around this channel is stable, discharge will occur only at the center. The result is an open channel through the center of the material commonly known as a rathole.

Funnel flow can create a situation where some of the material in the structure moves downward while the remainder remains stationary. An active flow channel develops above the outlet. Non-flowing material usually is present around the periphery. As material level in the structure drops, the layer of non-flowing material may remain in place permanently against the wall (Purutyan et al. 1999).

When the diameter of the first-favored flow stream exceeds the critical core diameter of the stored material, material flows radially inward. Annular shear lines, similar to those observed prior to trench, excavation, slope, or stockpile failure, develop on the top surface of the material. Characteristic of failure of surfaces in the other mentioned environments, material sloughs inward from the location of these cracks. This pattern creates the last-in, first-out characteristic of funnel flow. Entry of material from the top into the first-favored flow channel causes the increase in lateral pressure observed at the break line during material flow. Material in the flow channel pushes the remaining material against the walls of the structure.

The base of a bin in which funnel flow occurs has a shallow taper or flat bottom. A flat-bottomed bin containing one or more outlets always produces funnel flow, since material remains on the sides of each outlet. This is also the case with storage structures having shallow conical or pyramidal outlets. The hopper and not the cylinder of a storage structure exhibiting funnel flow sets the pattern.

Flow of material through the central first-favored flow channel can lead to separation on the basis of size during discharge. The smallest particles will discharge first and the largest, last. In situations where ingredients of blended mixtures are stored in silos exhibiting funnel flow, separation of ingredients based on their different component sizes can occur (Purutyan et al. 1999). This situation can change somewhat where an eccentric flow channel develops.

Funnel flow is adequate for situations where the material is free-flowing and coarse enough to prevent aeration, and in processes not affected by segregation of components in mixtures or by particle size or last in, first out.

Material flow that occurs in a first-in, first-out, non-preferential pattern is known as mass flow (Ware 1999b). Mass flow occurs in vessels having long, steep-walled transition sections compared to the shallower transitions present in vessels exhibiting funnel flow. During mass flow, recombination of separated material occurs at the discharge, thus restoring original composition. Mass flow is essential in some situations to prevent separation of particles and components of mixtures.

Mass flow occurs when all of the material in a bin moves downward when any removal occurs. The cone of a mass flow bin has a steep taper. Flow pattern is first in, first out. As a result, segregation of material does not occur. Walls of a storage structure in which mass flow occurs are smooth and steep.

Pressure distribution in a silo operating in mass flow differs considerably from one operating in funnel flow. Mass flow in a silo designed for funnel flow can lead to collapse of the hopper section. Hence, unexpected operation in the mode opposite from that specified in the design can lead to failure. Additional factors include pressure peaks that develop during flow, off-center flow and unusual flow characteristics unique to the material. Breaking of a bridge can create a flow peak capable of destroying the base of the silo. Storage structures also collapse due to suction on the roof created during collapse of a bridge (Lease 2000).

A serious type of silo failure results from unexpected conversion between mass flow and funnel flow (Carson and Holmes 2001). The expectation of one type of flow over the other is a design consideration. This affects the geometry of the hopper and interior finish of the silo. Mass flow hoppers are less tapered and are longer than those employed in funnel flow. The interior of a silo designed for mass flow has lower frictional resistance.

In the case of silos where the slope of the hopper section is close to the boundary between mass flow and funnel flow (hopper wall angle is close to the limiting angle for mass flow), slight changes in flow properties can change the flow pattern and resulting loads (Carson and Holmes 2001). Polishing occurring in the hopper section during use because of abrasiveness of the material is capable of inducing this type of change in flow pattern. In a silo designed for funnel flow, unintended polishing of the hopper section can indicate mass flow. This is also indicative of a pressure peak at a position not designed to handle it.

In silos where the wall friction angle is lower than that assumed in design, less vertical load is transferred to the walls in the cylinder section (Carson and Holmes 2001). This increases pressure on the walls in the hopper section leading to possible collapse. In the reverse circumstance, increased compressive loads on the walls can cause buckling.

External components, such as discharge gates and feeders can alter flow patterns in the storage vessel (Holmes and Carson 2002; Ware 1999b). Any feeder below the storage vessel (rotary valve, screw feeder, apron, belt or weigh feeder or vibratory feeder) can have its own first-favored flow pattern. This can affect the flow pattern in the vessel above.

## Flow-Related Issues

Flowable solid materials handled in confined structures and facilities possess a wide range of properties. This is evident from the spectrum of sources of these materials. Some examples include agricultural products including grains and seeds, feed and flours; brewery and winery waste; fish and meat meal; plastic resins in flake, powder, and pellet form; inorganic and organic chemical products; wood waste; and mineral products.

The size of the component particles ranges from heterogeneous macro objects to micronized powders. In all cases, the product enters the storage structure from the top and exits from the bottom. That is, the product must flow, and flow freely, in order for this system to function effectively. This is a distinct contrast to other aspects of flowable solid materials where flow is a sign of failure, and failure is to be avoided at all cost.

The type of flow achieved in a storage vessel depends on the flow properties of the material (Craig and Hossfeld 2002). These properties also aid in diagnosis of flow-related problems. Flow-related problems have a huge impact on the safety of these operations.

Cohesive (shear) strength is one of the critically important quantitative properties. A material that gains cohesion because of applied pressure could arch (bridge) and form ratholes (flow channels surrounded by coalesced material). Cohesiveness of a flowable solid is a function of moisture content, particle size and shape, temperature, time of storage at rest, and additives. Moisture increases cohesiveness, as would be expected. Flowable solids containing finer particles are more cohesive, and as a result, more difficult to handle. Angular or fibrous particles are more cohesive than round ones. Increasing the temperature of some bulk solids increases cohesiveness. This is especially true for plastic powders. Materials that sit undisturbed can become more cohesive. This occurs due to settling and compaction, crystallization, chemical reaction, or adhesive bonding.

Materials experience both internal and external friction. Internal friction occurs because of inter-particle contact. External friction refers to the friction experienced by the mass of material sliding against walls of the structure. Factors affecting frictional properties include pressure, moisture content, particle size and shape, temperature, time of storage at rest, and wall surface properties.

Bulk density of a material varies with the consolidating pressure. As a result, the density-to-pressure relationship is required in order to describe the properties of the material correctly.

Permeability refers to the ability of gas to flow through a bed of powder when a pressure differential occurs. During flow of fine powders, expansion and contraction of voids can create an upward gas pressure gradient at the outlet of a mass flow vessel. During discharge, the upward gradient acts counter to gravity to reduce the discharge rate. This phenomenon does not usually occur with coarser materials because air flows freely into and out of voids during expansion and contraction.

Segregation is a property of solids containing particles of varying sizes. Sifting is the most common effect. Smaller particles move through the matrix formed by the larger particles. Sifting causes the separation of fines to the bottom of a cereal box. Segregation on the basis of size occurs during pouring into a pile through the process of sifting. The larger particles flow to the periphery, leaving behind the small ones at the center.

Fluidization refers to the ability of an upward air flow to suspend particles. Fine or light particles that have high permeability are most likely to fluidize. Fluidized particles behave as a liquid in their flow characteristics. Airborne particles are difficult to recover as product and can vent from the vessel or be collected in dust collection systems. In some circumstances, fluidization can develop through banging on the wall of a hopper.

Fluidizing of the material causes flooding (Marinelli 2000). Fluidized solids behave like liquids, and flow accordingly. Feeders are designed to discharge deaerated, nonfluidized material and cannot handle the flood of fluidized material. Flooding is especially likely in situations where segregation on the basis of size has occurred during filling, and the material first out through the first-favored flow stream is the fines.

Formation of hang-ups on side walls, ratholing, and bridging across the cross-section of material are characteristic of storage structures exhibiting funnel flow. Bridging also occurs in storage structures exhibiting mass flow (Carson and Johanson 1985; Purutyan et al. 1999).

Hang-up formation occurs when the diameter of the first-favored flow stream fails to exceeds the critical core diameter of the stored material. The particles of the material coalesce around the sides. Funnel flow can create the situation where some of the material in the structure moves downward while the remainder remains stationary.

Coalescence can result in caking on vertical surfaces of the structure and bridging across the horizontal plane of the material. A bridge is a ring of particles compressed together. This compression provides its strength (Ware 1999a). Bridging can occur at or below the surface of the material. Material that has not coalesced can flow from underneath the bridge, thus creating a hollow. As well, ratholes can develop. Ratholes are vertical channels in the coalesced material. In permitting some flow to occur, the rathole may hide the existence of a bridge. Rathole formation is the more extensive expression of coalescence. In this case, bridge formation occurs but contains one or more flow channels.

Bridging (arching) is the formation of a coalesced mass of material that spans across the cross-section and blocks flow. The bridge may support the weight of an individual, but could collapse at any time. The material under the bridge may flow away, potentially leaving behind a hollow under an arched structure. Unaffected material continues to flow from under the bridge and creates a void space. A bridge can form at or under the surface of material in the storage structure. Bridge formation reflects both storage conditions and properties of the material.

Bridge formation occurs as a result of compression of particles of the material. This can happen as material enters the hopper section and remains undisturbed for a period of time. In this case, the outlet of the vessel is too small to overcome compressive forces and cohesive properties of the material. The minimum outlet size that overcomes the arching problem is directly related to the cohesive strength of the material (Purutyan et al. 1999). Arching can occur in structures designed for mass flow, as well as funnel flow.

Bridge formation also can occur due to the presence of elevated levels of moisture in stored material. Moisture, in combination with sub-freezing temperatures, can lead to the formation of a frozen layer of stored material. Addition of new material to the top and removal of material from the bottom creates the bridge and void space. Moisture in seeds and grains coupled with appropriate temperatures leads to the onset of germination. Swelling of the grain or seed that occurs during the early stages of germination can lead to formation of a solidified mass of material.

Some materials are hygroscopic. That is, they absorb water. Storage in silos vented to atmosphere coupled with long periods of storage, and high and low temperatures coupled with high levels of humidity can create the conditions needed for coalescence. Heating of the structure during the day expands and expels air from the airspace connected to the vent. Cooling during the evening entrains air into the structure in the airspace above the material. Air containing a high humidity provides the moisture needed for reaction with the top layer of the contents of the structure. Hence, failure to maintain humidity in the airspace at low level all the time can lead to swelling and bridge formation.

An appreciation of the nature of problems confronting dry removal is helpful for putting the accomplishments of mechanical cleaning technology into perspective. (See the next section.) Powdered materials can form extremely hard, caked solids. Procedures for removal have included vessel entry and use of picks and shovels. This practice is extremely hazardous, as the intent is to induce flow and collapse material underlying or surrounding the worker. This carries a serious risk of engulfment and striking by falling solidified material. Destruction of coagulated material by blasting with dynamite was tried as a last resort (Anonymous 1989a; Anonymous 1989b).

In ready-mix concrete equipment, for example, concrete sets in the cavity located between the fins and wall of the mixing drum. Periodic removal must occur in order to maintain proper operation. The concrete sets into a curved shape and is removed using a jackhammer. Pieces of concrete can fall from the drum without warning during breakout and are capable of causing serious traumatic injury. Removal requires use of a jackhammer, chipper, or similar equipment. This work exposes the entrant to injury from heavy falling objects and is extremely noisy.

Flow restoration in a storage structure holding flowable solid material can impose considerable loads onto the structure. The sudden collapse or destruction of flow irregularities such as bridges and ratholes and resulting movement of large quantities of material creates significant dynamic loads on silo walls (Carson and Jenkyn 1990). This collapse can occur naturally or through use of flow inducements such as explosives, excessive vibration, or air injection. Collapse of a bridge across the cylinder section of a silo or bin following discharge of material from underneath transfers the full weight of the material above the bridge to the wall below (Carson and Holmes 2001). This can cause buckling of the wall in metal silos. This results because material normally has a restraining effect against the silo wall, increasing pressure where the wall deforms inward and decreasing pressure where the wall deforms outward.

Numerous fatal accidents in flowable material storage structures have occurred following entry to promote flow (Figure J.19). Entrants attempted to promote flow by enlarging flow channels and break bridges or hang-ups of material on walls (MSHA 1988; OSHA 1983).

Most of the accidents occurred due to burial (engulfment) in material that was already flowing. During flow, the level of material decreases irregularly. This can lead to sudden slumping and inflow of

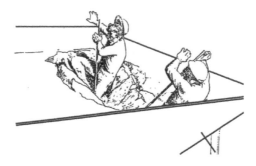

Adapted from MSHA

Figure J.19. Engulfment in flowing material. Attempts to promote flow in first-favored structures have led to engulfment because of the inability to remain on top of the flowing material. (Source: MSHA.)

material around the body. The victim must fight to keep on top of the material by continually pulling out the legs and feet. This action requires considerable strength and rapidly exhausts the victim. Burial can occur despite the wearing of harnesses and retrieval lines. Entry therefore must not be attempted while flow is actively occurring.

Material flows inward around the chest of the victim and maintains the cross-section. The mechanism of breathing requires expansion and contraction of the chest. This, in turn, requires expansion against the flowing material. The outward pressure exerted against the material during inhalation is insufficient to enable expansion of the chest. As a result, the victim is unable to breathe and suffocates. Suffocation can occur when the material is only chest high.

Many of the accidents that occurred in silos, hoppers, and other storage structures resulted from collapse of a bridge during attempts to open a channel or to improve flow. Injury occurred during the sudden release of large coalesced fragments (Figure J.20). Engulfment occurred during release of trapped flowable material and downward suction by the vortex created by the flowing material. The victim was unable to remain on top of the material as it cascaded into the center of the cross-section of the structure.

Some of the accidents resulted from dislodging hung-up material adhering to vertical surfaces. Slumping can occur slowly or suddenly, with or without apparent provocation. The fragments have considerable mass and are jagged and irregular in shape.

## Mechanical Cleaning Systems

Narrowing of a first-favored flow channel can severely restrict, if not completely cut off flow of material from storage structures. As a result, flow restoration is an urgent matter. Flow restoration in the past has occurred through what are most charitably described as novel approaches up to and including the

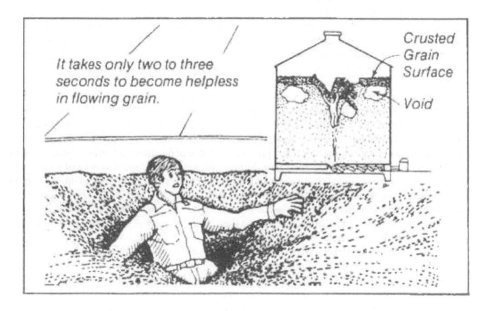

*It takes only two to three seconds to become helpless in flowing grain.*

Crusted Grain Surface

Void

Figure J.20. Collapse of a bridge creates a vortex of flowing material and heavy solidified material that fills the former void space. (Source: NIOSH.)

use of jack hammers and dynamite. Hang-ups on walls and bridges pose similar problems. As a result, a whole industry has developed to correct them.

This industry manufactures equipment to dislodge adherent material as well as to open channels to enable the flow of material. Sending workers into storage structures to perform the same function still is common practice. In these situations, the purpose of the activity is to destabilize the structure of the material in order to induce flow. Techniques are similar to undercutting the foot of the slope or digging an excavation with unsupported sides. These techniques mimic naturally occurring phenomena, including earthquakes. As fatal accidents that have occurred in these structures have shown, the moment of failure of the slope is not always predictable (OSHA 1983, MSHA 1988). The most prudent expectation is that failure will occur and that this type of work poses high risk.

The use of flow enhancers that inject compressed gas into the silo through material or use of dynamite to break apart obstructions can rapidly increase pressure in the headspace. Venting or blow-out panels are required to prevent overpressure under these conditions.

Long-term strategies for preventing flow problems, such as ratholes and bridges, include impelled retrieving, static design, and induced vertical flow (Ware 1999). Impelled retrieving requires equipment that conveys and sometimes forces flow from the vessel. This includes internal and external augers, and vibratory feeders located under the hopper. Impelled retrieving can cause high abrasion and breakdown of particles by degradation or attrition. Static design relies on gravitational flow. Induced vertical flow relies on flow enhancers, such as vibrating cones located in the transition section or an activator that vibrates the entire storage vessel.

Mechanical cleaning systems are used for removing bridged and hung-up materials from storage vessels. These systems are designed for use with products that must not come into contact with water. To illustrate, cement that has caked or bridged in a storage silo is still saleable, so long as wetting has not occurred (Hauck 1993; Laing and Berg 1999).

## Cutting (Attrition) Head Systems

This system contains a mechanism that "chews" at the coalesced flowable solid material. This equipment is proprietary. As a result, no comprehensive description is available. One unit appears to be similar to an industrial version of a garden roto-tiller (Anonymous 1989a). These systems can remove accumulated deposits from a 10 cm (5 inch) pipe to a 15 m (50 foot) diameter silo. Cleaning can occur from top to bottom or bottom to top, and also in corners and chutes, and downward-pointing cones. These systems cannot accommodate sludges.

Application of this technology has occurred in a wide variety of structures utilized in the processing and storage of flowable solid materials. Examples of materials removed successfully include coal, coke, cement, sand, ore, grain and flour, coffee, cottonseed meal, ladle and furnace linings, powdered inks and pigments, lead oxide, sawdust, clays, powdered adhesive resins, and powdered milk. These materials span a broad range of industry. They provide evidence of the many potential applications for this equipment that avoid the need to enter into storage structures to perform this type of work.

Materials used in the cutting (attrition) heads are compatible with the properties of the material to be removed (Anonymous 1989a; Anonymous 1989b; Hauck 1993; Laing and Berg 1999). These materials include hardened steel, fiberglass-reinforced rubber, acrylonitrile-butadiene-styrene (ABS) plastic, and aluminum. Selection of an attrition head for a particular circumstance depends on the hardness and other properties of the material to be removed.

To illustrate, the preferred material of construction in the attrition head for removing coal is fibreglass-reinforced rubber. Fiberglass-reinforced rubber is sufficiently hard to remove the relatively soft coal without creating sparks that could cause fire. (An additional safety measure is to purge the structure with carbon dioxide or nitrogen at the time of this operation.) Cement normally is the hardest of the materials removed by this equipment. This system is also able to remove rust from steel silos.

The operator controls the motion of the cutting head through equipment located outside the space. The opening used to admit the cutting equipment is the only access needed to the space. The cutting heads can be positioned into a bridged area to cut a path or to remove built-up material from walls. The system can accommodate one or more cutting heads. The cutting heads vary in size, depending on the application.

Power sources for this equipment include compressed air, steam, other nonflammable gases, water or hydraulic systems (Anonymous 1989a; Anonymous 1989b).

## Whips

The whip (Figure J.21) is a motorized device that uses rotational motion of chains and flails containing disks and bolts to abrade consolidated bulk materials (Hauck 1993). The composition of the flails, disks, and bolts is compatible with the properties of the material to be abraded. This equipment is positioned into the structure from outside.

Consideration in selection of the abrading material includes hardness of the deposit and underlying structure, and fire and explosibility hazard of the resulting dust cloud. Chain is used against hard noncombustible materials. On the other hand, plastics are used with soft materials, such as grain, coal, or sand to prevent sparking. The flails and attached disks and bolts strike against the deposit, thereby abrading the material. Power sources for the drive motor in the central hub include compressed air or hydraulic systems. An expandable boom is used to position the abrading head.

The whip is effective against deposits on vertical surfaces only, and cannot be used to clean out corners, tops, bottoms or cones, or large pipes or chutes.

## Drills

An adjunct to abrading equipment is the drill (Figure J.22). The drill is used to create a flow channel through the solidified material. Two basic designs are available. In the one case an external hydraulic motor rotates the entire shaft. In the other, the drill motor is located at the drill head. Depending on the needs of the situation, these units can work from above or from underneath the bridged material. The flow channel created by the drill provides a path for drainage, as well as operation of the other equipment mentioned above.

Figure J.21. The whip uses rotating flails and other devices to abrade coalesced material. (Illustration supplied by Cougar Vibration, a division of Martin Engineering (www.cougarindustries.com).)

Figure J.22. The drill creates a new flow channel which the whip can enlarge. (Illustration supplied by Cougar Vibration, a division of Martin Engineering (www.cougarindustries.com).)

## Safety Hazards and Protective Measures

Cleaning and dislodging powdered bulk solids from inside structures that store and handle them can involve a number of health, safety, and fire hazards. Primary among them are fire and explosion. Pulverizing the caked materials and exhausting of compressed air can create a highly concentrated dust cloud and possibly vapors. Without proper design and execution, the equipment and its ancillaries could act as ignition sources.

Measures that minimize this potential include use of steam or other nonignitable gases or hydraulics to power drive motors and inerting the space using inert gas, such as carbon dioxide or nitrogen, to prevent combustion of dust and vapor off-gassed by the material. Use of nonsparking materials for the attrition heads or whip prevent sparking, and bonding and grounding conductive hoses, carbon-based or with interior wire and connections, prevent arcing from accumulation and and discharge of electrostatic charge. Electrical lighting equipment rated for use in hazardous locations can prevent ignition due to hot surfaces and smashed bulbs.

Emission of highly toxic dust from openings in the structure can occur as part of the escape of the compressed air used to power equipment. These emissions could pose exposure hazards to individuals working in the area.

# Maintaining and Servicing Systems Handling Flowable Materials

Previous parts of this appendix have focused on issues related to handling flowable solid materials by equipment on whose performance in a predictable way depends the safety of workers involved in installation, maintenance, servicing, repair, and renovation.

In order to make appropriate decisions about deactivation, de-energization, and isolation, issues related to real-world performance, as opposed to design, actual performance must be known and appreciated to the fullest extent possible. This section focuses on activity related to preparing for deactivation, de-energization, isolation, lockout, and verification of the state of zero energy in equipment, machines, and systems that handle flowable solid materials.

Previous discussion indicates that systems that handle flowable solid materials lack devices commonly used for isolation. Instead, isolation of flow in these systems can successfully occur at components, such as rotary valves, feed conveyors, diverters, and turnheads. Rotary valves (Figure 5.37) and feed conveyors control flow to downstream conveyors and to processing and storage structures. Routing devices, including diverters (Figure 5.38) and turnheads (Figure 5.39), can prevent flow along specific pathways. Isolation and lockout of these devices can occur through process control and electrical lockout and through physical blockage of movement.

Discussion finalized in Appendix I indicated the critical role of the operator in such decision-making and the role of circuit drawings in visualizing the role of elevation change in flow. These resources are important in considerations about isolation and lockout of systems containing flowable solid materials.

Flow promotion and restoration are critically important in systems that handle flowable solid materials. Optimized design and choice of components and materials of construction are critical starting points to minimize these problems. Development of mechanical devices that restore flow has shown that the need for entry by workers into these structures for flow promotion and restoration is unnecessary and creates risks to safety.

That said, there are reasons for entry into these structures. These include inspection, repair, and modification to the existing infrastructure.

## Flowable Solid Material Transporting Systems

Handling of flowable solid materials occurs in a number of ways. The devices used for handling flowable solid materials dictate the means by which isolation and lockout of these systems can occur. Systems that handle flowable solid materials move these materials from place to place. Places include unloading facilities, storage structures, processing units, and loading facilities.

Transport systems empty into storage structures and process units, and similarly, drain structures and process units. The individual pieces are attached together to form integrated systems of interdependent components. These systems often do not contain valves or valve-like structures. Many are constructed to remain assembled at all times and not to be dismantled to achieve isolation. Effective isolation of flow in these systems is achievable. To achieve this end-point necessitates consideration for functions performed by components and the means by which they perform them.

## Gravity Flow

Previous discussion mentioned the role of gravity in movement of materials in a controlled way down the slopes of hills. Gravity flow is exploited in the filling and emptying of storage structures and process units and the handling of flowable solids in conveyors, and the piping and duct in between.

Application of gravity flow to product transport is readily visible in agricultural feed mills where gravity flow through ducts fills bins with product. These operations elevate the product to a high point in the system and use a rotating distributing device (the turnhead) to direct flow to the appropriate duct. The duct contains diverters to direct flow to the appropriate end-point.

The controlling factor in these systems is the device that induces movement of material. This usually is an electric motor. The electric motor operates a mechanical conveyor to move the material from a lower point to the high point from which gravity flow can occur.

Isolation and lockout performed on this component prevents use of the remainder of the distribution system for the duration of the work to be performed. Isolation and lockout performed at the turnhead or the diverters enables use of the distribution system during work on the isolated component.

## Mechanical Conveyors and Components

Material transporting systems utilize mechanical conveyors of varying types to move flowable materials from place to place. Mechanical conveyors connect unloading facilities to storage structures and processing units and loading facilities. These devices move material horizontally and vertically and at all angles in between these extremes. The choice of conveyor reflects the type of service and characteristics of the material. Conveyors empty into storage structures and process units, and collect material from the base of storage structures and process units. Conveyors also empty into other conveyors.

Certain conveyors, known as feeders, can play a critical role in isolation. Feeder conveyors receive and control flow from the base of storage structures. When a feeder conveyor is isolated and locked out, flow from the structure does not occur. When operational, increasing the speed of a feed conveyor increases the throughput of material. Isolating and locking out a feed conveyor prevents transfer of flowable solid material to other devices in a component string.

Disabling the controlling conveyor in a sequence stops the flow of material in the system. The logic intrinsic to flowable material handling systems stops all conveyors and downstream process units as part of the process of isolation and lockout. Stopping the movement of a key conveyor prevents inflow and outflow of material

A type of screw conveyor feeder known as a solids pump is also capable of forming a seal against pressure (Dick 1989). The solids pump contains a chamber without flights, two or more independently moveable augers, and a device to impede or block flow. Rotation of the augers pushes material into the chamber and pulls it from the far end. Stopping rotation of the downstream auger and positioning the impeding device in the flow cause accumulation of material in the chamber. This results in formation of a compacted plug of material, which, in turn, forms the pressure-resistant seal. Impeding devices include horizontal weirs, hinged flaps, and sliding plates.

Screw feeders are enclosed devices. They are suited to situations involving handling of dusty or toxic materials requiring containment, where space is restricted and where particle breakage (attrition) is not a problem (Marinelli 2002). Screw feeders having constant pitch cause the formation of a preferential flow channel at the first flight of the screw. Once the first flight fills with material, there is no capacity for removal of material from further along the flight. Hence, an eccentric flow channel develops in the tapered section of a hopper. This can destroy mass flow in a silo. Screw flights having increasing pitch and tapered shafts overcome this problem.

Belt feeders pose the same problem as screw feeders (Marinelli 2002). Preventing formation of a flow channel requires making flow accessible to the belt from all parts of the opening at the base of the tapered section. The shape of the hopper-to-belt interface should expand in the direction of flow, both in plan view and elevation view so that material can continue to flow evenly.

The rotary valve located at the base of storage structures is another component capable of effecting isolation of flow. Rotation of the rotary valve enables flow of material from the storage structure. Isolation and lockout of this device will prevent outflow from the storage structure. Rotary valves also block outward leakage of gases and vapors that accompany the stored material and help prevent loss of inerting gas used to prevent contact of contents with the atmosphere (Thorn 2008). Inerting gases include nitrogen and carbon dioxide. Leakage of inerting gas from a storage structure could lead to oxygen deficiency or conditions of overexposure in the surrounding area.

Rotary valve feeders are used extensively to interface storage structures to pneumatic conveying systems because they can form a seal against a pressure gradient (Marinelli 2002). The direction of motion of the vanes creates suction against the material in the base of the tapered section. This action differentially draws material downward from one side of the hopper in an uneven manner that can destroy the mass flow pattern. A vertical section of pipe located between the base of the tapered section and the intake to the feeder corrects this problem. A vent line between the bin and the feeder corrects problems due to air pressure in empty wedges of the feeder.

## Pneumatic and Vacuum Conveying
Some operations utilize airflow at high velocity to move flowable materials through the system. These systems can unload incoming material, transfer it to storage structures and processing equipment and transfer it to loading systems. These systems utilize blowers operated in positive- or negative-pressure mode, depending on the design of the system. These systems can contain valves for isolating flow between different segments.

System design has an impact on isolation. In full suspension (dilute phase) systems, flow velocity entrains all material into the suspended phase during transport. Some material will settle onto surfaces when cessation of flow occurs. This can happen during isolation when transport is occurring. The presence of settled material can create difficulty during restart from an abrupt stop.

Partial suspension (dense phase) systems operate at flow velocity insufficient to maintain full suspension of material at all times and in all locations during operation. As a result some material remains on surfaces during normal transport. The quantity depends on the design of the system. Shutdown during operation can create an unexpected burden of material on surfaces and can create difficulties in resuspension on resumption of transport following abrupt stoppage.

Leakage of dust from failed containment is a concern with systems operated under positive pressure (Mills 1999). Aggressive leakage in confined geometry could create a concentrated cloud of dust with sufficient concentration to be explosible. Another concern with systems operated under positive pressure is the presence of vapor and aerosols from lubricating oils emitted during operation of the air mover. The vapor can condense to liquid phase in contact with surfaces of appropriate temperature. The aerosols can coat interior surfaces and impede flow. The material in lubricating oils is known for breakdown into carbonaceous matter and spontaneous combustion.

## Contained Atmosphere Systems
In many operations that handle flowable solid materials, the components of the system are sealed in an enclosed system. These systems are self-contained and utilize inert gases, such as nitrogen or carbon dioxide to form the internal atmosphere and thereby prevent contact between air and the transported material. These enclosed systems can involve all interconnected components or more typically only the storage structures. Enclosed systems also enable collection of toxic or ignitable gases and vapors emitted from surfaces of material experiencing transport.

The movement of solids through interconnected parts of a system has an impact on the pressure of atmospheric or other gas in the storage structure. Venting is essential to maintain internal/external pressure relationships and hence structural integrity. Vessel filling displaces air or blanket gas from the airspace above the material. This air or gas must have a defined route of escape from the vessel. Similarly, emptying the structure creates a partial vacuum. Air or gas compensating for the increase or decrease in pressure must also have a defined route of entry or escape for movement of sufficient volume to maintain pressure in the airspace near atmospheric. Failure to do so can lead to overpressurization or underpressurization and collapse of the structure.

Blow-out panels are one means for ensuring that the pressure in the storage structure does not exceed certain limits. Blow-out panels mechanically compare the difference between pressure inside and outside the structure, and fail when the differential exceeds a preset value. Reversing the panels enables blow-in to occur when the headspace experiences unacceptable under-pressurization.

The air or gas in the airspace above stored materials becomes heavily loaded with dust from fines in the material or the material itself if ground finely enough during loading activity. This action can generate explosible concentrations of dust. One means to control this problem is the use of exhaust systems that withdraw air from the headspace.

An additional concern arises when the dust and bulk material contains solvent or gas that can off-gas into the headspace of the storage structure. These structures may contain inerted atmospheres as a protective measure. Inerting is the process of maintaining the concentration of oxygen below levels capable of supporting combustion in airspaces of the system. Inerting involves use of nitrogen or carbon dioxide. Oxygen concentrations in such systems including safety margin typically range around 6% to 8% (McManus 1999).

## Combustible Dust Issues

Much writing about combustible dust has occurred in recent years. This is an outcome from many industrial explosions involving combustible dust as the causative agent. Dust is present in two main places in industrial facilities: inside systems created to handle flowable solid materials and outside systems created to handle flowable solid materials.

Confinement inside systems created to handle flowable solid material is the place dust is intended to be present. This is true for the vast majority of material present in industrial facilities. About 40% of explosions involving combustible dust in the U.S. and Europe over the previous 25 years involved dust collectors (Johnson 2008). By its design and installation, a dust collection system is likely to transport high concentrations of very small particulates. If these particulates are ignitable, then an explosion is possible. During a dust explosion, the pressure can increase from ambient to 690 kPa (100 lb/in$^2$) in milliseconds (Mills 1999). Whether this will occur depends on design and maintenance and appropriateness of components.

Properly designed, specified, and constructed, structures and systems accommodate the presence of combustible material by eliminating sources of ignition, and incorporating contingencies such as blow-out panels or explosion-suppression systems to protect the equipment from the overpressure that can develop during an explosion. Systems created to handle combustible dust include mechanical and pneumatic conveying systems operated in positive pressure and negative pressure mode, and exhaust systems intended to control emissions of combustible dust.

What is also true is that maintaining containment of materials capable of forming a cloud of combustible dust is extremely difficult, if not impossible (Figure J.23). A dust-free facility that handles flowable solid materials capable of being reduced in size to the range found in airborne dust seems to be an oxymoron. That is, this achievement seems to be a distant goal. This results from the difficulty in controlling day-to-day emissions from leaking fittings to upset conditions in which emission of large quantities of dusty material occurs into the airspace of the building. Dispersion of material into suspension typically occurs at the discharge to a receiving vessel, and during start-up or shutdown of a transient operation (Mills 1999).

Dust that becomes airborne in the airspace of a building inevitably settles onto horizontal surfaces. Unlike gases and vapors, dust can migrate considerable distances from the point of origin to a point of settlement on a surface within a building envelope and not escape. The destination of migration may lie outside the area protected by design against ignition of combustible dust. As well, often surfaces on which settlement occurs are located in areas not normally accessible without considerable effort. This is especially true in roof spaces (Figure J.24).

Settled dust poses only a fire risk. Settled dust made airborne through a mechanism of disruption in the presence of an energetic ignition source poses the risk of a dust explosion. Frank (2004) summarizes information from several dust explosions investigated recently by the U.S. Chemical Safety and

Figure J.23. Accumulation of combustible dust on surfaces reflects loss of containment. Confining dust within containment in operations that handle flowable solid materials seems to be a distant goal. (Courtesy of Jenike & Johanson, Inc., Tyngsboro, MA.)

Figure J.24. Dust on horizontal surfaces in roof spaces usually reflects accumulation over a long period of time. Accumulated dust contains fine particulates that are readily suspended by abrupt movement of the structure. Particles of combustible dust can fuel a fire or explosion.

Hazard Investigation Board. Accumulation of dust on surfaces was a critical element, as was some event that shook the surface and made the dust airborne.

Cleaning and removal of dust from surfaces is a huge undertaking. Preventing the emission is the only practical means to eliminate the issue. This was the strategy employed at the Imperial Sugar Refinery just prior to the explosion that destroyed parts of the plant. Enclosure of a conveying system that had operated for 80 years previously led to the condition that triggered the explosion (Vorderbrueggen 2011). This situation highlighted the critical issue of management of change and anticipation of potential problems that could result from change.

Preventing unexpected discharge of electrical energy and creation of surfaces hot enough to act as ignition sources is paramount to protecting people and facilities. Article 500 of the U.S. National Electrical Code (NFPA 70) provides a means to classify locations based on the substances known to be or possibly present in air at levels capable of providing the fuel for a fire and/or explosion (NFPA 2011a). Article 500 subdivides hazards into classes, divisions, and groups.

Class I pertains to flammable gases or vapors. Class II pertains to dust. Class III pertains to fibers and flyings. Systems handling flowable solid materials may belong to one or more of the classes in the classification of Article 500. Division refers to the frequency of occurrence of the hazardous condition during the life cycle of the operation. Division 1 situations are expected to be present at any time. Division 2 situations are intermittent in their occurrence and normally are not hazardous. They apply to areas that become hazardous in the event of accidental discharge of ignitable materials from confined systems.

Class I contains four groups. Groups contain chemical substances having similar properties. Class II contains three groups.

Explosion-proof equipment generally is required for Class I Division 1 locations. Explosion-proof electrical equipment is not gas-tight. The cast-metal electrical enclosure must prevent escape of flames and operate below the ignition temperature of the ignitable material in the ambient environment. The chief purpose of an explosion-proof enclosure is to prevent initiation of a fire or explosion in the ambient atmosphere by electrical equipment contained within it.

Normally, non-sparking equipment or apparatus containing make-or-break contacts immersed in oil or hermetically sealed is used in Division 2 areas. This equipment can cause ignition only if malfunction occurs at the same time as development of a flammable concentration in the area. This event is unlikely considering the tiny probability of simultaneous occurrence of electrical failure and release of ignitable materials. When non-sparking equipment is not available, explosion-proof equipment or electrical apparatus contained in explosion-proof housings must be used.

Electrostatic discharge is a major consideration in the transport, handling, and storage of combustible dust. Charging occurs during transport in conveying systems and dropping into storage structures. Discharge can involve the bulk material and isolated metal conductors that are insulated from ground (Gregg 1996; Lees 2001). Conductors insulated from ground can include tramp metal located inside vessels, screens, metal rims on nonconductive drums, probes, thermometers, spray nozzles, high pressure cleaning equipment, and wire netting around lagging and insulated metal containers.

One aspect of control of electrostatic events is the type of container in which substances are handled and stored. Classification of containers occurs in order of increasing electrostatic hazard. A container made from conducting material and grounded provides greatest control. A container made from conducting material and insulated from ground provides less control. A container made from insulating material provides least control.

A metal drum containing mixed vacuumed material including metal and nonmetal fragments that is sitting on a wooden pallet is an example of a metal container insulated from ground. In this case a large charge can accumulate in the contents and on the container. Moving the drum to a metal surface could provide a discharge path to ground. A similarly large charge may accumulate in a container made from an insulating material, such as plastic. The conductive material of containers, piping, and other equipment provides a means to control the generation and accumulation of electrostatic charge (Scarbrough 1991). The main technique is bonding and grounding.

Bonding involves electrical connection between conducting objects. Electrostatic charging currents are small. Preventing accumulation of a hazardous amount of charge through unimpeded flow to ground generally can occur by ensuring that resistance in the bonding system is low. The presence of paint, grease and oil, corrosion products, and rust on conductive elements can increase resistance, thus impeding flow and enabling charge accumulation. Bonding across flanged joints ensures better electrical continuity than use of bolts that pass through the flange. This is important to consider during procedures that require opening of flanges to isolate lines from vessels.

Grounding drains away electrostatic charge that develops or accumulates on objects under control or the bonded system. Grounding requires an electrical connection between a conducting object and ground. Ground means the earth. Bonding maintains the connected objects at the same electrical potential. Grounding a bonded system maintains the electrical potential of all interconnected objects in the system at the same level as the ground. Grounding in normal circumstances involves electrical connection to water pipes that pass through the ground or to rods of conductive metal rods that are driven into the ground. Bonding is usually effected using copper strips or stranded wire.

Use of the current-carrying part of the electrical system as a ground should never occur (Gregg 1996). Fires have occurred where the neutral side of the electrical system was used as a ground for controlling electrostatic discharge. Similarly, water pipes are poor candidates for the electrical ground, as are underground systems equipped with cathodic protection. (Cathodic protection systems contain a sacrificial metal that corrodes or impressed voltage to protect the metal of the system against corrosion.) Disconnection for maintenance or modification renders inoperative the grounding capabilities of these systems.

Precautions against accumulation of electrostatic charge on machinery can take several forms (Scarbrough 1991). In some cases, electrical grounding is sufficient to prevent accumulation of electrostatic charge on moving machinery. Electrical grounding is not sufficient for conveyor belts and transmission belts. These can accumulate considerable electrostatic charge. The usual solution is to increase conductivity of the material of the belt by incorporating additives during manufacture or by use of dressing compound during operation. Charge scavenging combs made from conductive metals can provide a discharge path.

Static eliminators based on ionization are used to prevent charge accumulation in many applications. The charge either drains to ground through ionized air or is neutralized by ions from the air. Air can be ionized by heat, ultraviolet light, electrical discharge, or radioactive substances. Equipment used for elimination of electrostatic charge itself must not act as a source of ignition (Lees 2001).

Many storage structures contain little, if any, oxygen as a result of the use of inerting gases, such as carbon dioxide and nitrogen. In these circumstances, the inerting gas must maintain the concentration of oxygen below the Lower Flammable Limit of the ignitable gas or vapor off-gassing from the product or the Lower Explosible Limit of the dust itself (Bodurtha 1980; Wysocki 1991; Zabetakis 1965).

In general, carbon dioxide reduces the height of the ignitability envelope compared to nitrogen per unit of volume. For most ignitable mixtures the inerting concentrations of carbon dioxide and nitrogen are 28% and 42%, respectively. The lesser quantity of carbon dioxide required for fire suppression reduces the oxygen level to a lesser extent. The greater safety and efficiency in fire suppression is the basis for use of carbon dioxide in fire extinguishers. Successful inerting using carbon dioxide requires reduction of oxygen to 11%, and 8% using nitrogen (DI 1964).

The appropriate margin of safety depends on the dust and on the conditions, but normally would necessitate a further reduction of at least 2%. Thus, the system should maintain the oxygen concentration below 9% when the minimum needed to support combustion is 11%. Concentrations reported in the literature normally are measured at ambient temperature.

The appropriate concentration of inert gas for protection against ignition at high temperatures is considerably higher. This also requires confirmatory testing (Lees 2001). Minimum ignition energy and ignition temperature increase as oxygen concentration decreases. Inerting down to the limiting concentration of oxygen of 11% may not be essential when a strong source of ignition is not present. Partial inerting by carbon dioxide and/or water vapor generated in a dust dryer may be entirely satisfactory (Bodurtha 1980).

In addition to suppressing combustion, the inert gas must not react with the product. Certain metallic dusts, for example, can react with carbon dioxide or nitrogen under specific conditions. Helium and argon are more suitable inerting gases in such situations (Schwab 1991).

Careful design and testing of a system relying on inerting as a mode of control are essential. Often this type of plant is totally enclosed. Dust addition and removal occurs through valves. These permit escape of only a small amount of the inert gas. Where recycle occurs, preventing accumulation of fines not removed by gas cleaning equipment is critical. When dead spots and points of in-leakage of air are possible, multiple points of injection of inert gas usually are required. Reliance on inerting as a control requires monitoring of oxygen concentration and mechanisms for shutdown of dusty operations when malfunction occurs.

Inert dust also is used for suppression of fire and explosion involving combustible dust. Limestone dust is utilized on the floor of passageways in coal mines (Lees 2001). The inert powder reduces ignitability by absorbing heat from the source of ignition. However, the amount of airborne inert powder needed to prevent an explosion usually is considerably higher than concentrations normally found or tolerated. To be effective, the inert dust must constitute at least 65% of the total airborne concentration (DI 1964).

Unique strategies, procedures, and equipment are required for handling combustible dust and powders because they are solid materials (Lees 2001). Equipment that produces powders or dusts by size reduction is prone to dust explosions. Grinding elements or foreign bodies, such as tramp metal, can act as sources of ignition. Minimizing quantities of material where dust suspensions may occur is advisable. Several small quantities pose less hazard than one large one.

Over-designing or over-sizing a hopper or a mill creates large void spaces in which a dust cloud can form. This also provides the possibility for long free fall. Removing dust from suspension as soon as possible reduces the possibility and severity of ignition. Removing suspended dust locally is advisable to transferral long distances along ducts to a central cleaning unit. Design that minimizes accumulation of settled dust is highly preferable to reliance on housekeeping. Startup, shutdown, or fault conditions are critical periods during plant operations. Conditions that otherwise provide acceptable control can deteriorate rapidly.

Routine materials handling poses potential challenges with dusts and powders. Scoops used for manual transfer of material should be grounded plastic or metal to prevent electrostatic accumulation and discharge. Other devices for minimizing the potential for generating electrostatic arcs include grounded wristlets and anklets worn by the person performing the transfer. Water sprays provide a simple means of dust suppression (Bodurtha 1980).

In process operations, dust and powder handling occurs by manual methods or by mechanical or pneumatic conveyors. Manual methods should be restricted mostly to bagging or cleanup. Mechanical and other equipment require careful selection in order to minimize potential for electrostatic events (Lees 2001).

Mechanical conveying equipment can include screw and drag-link conveyors, belt conveyors, and bucket elevators. Screw and drag-link conveyors are effective for dusts because of the minimal free volume. The return leg of the drag-link conveyor usually requires explosion protection. A belt conveyor is suitable for dusts only when enclosed and protected against explosion. This equipment is particularly prone to generation of electrostatic charge. The bucket elevator poses severe hazards unless safeguarded. All mechanical conveying methods involve the risk of overheating due to mechanical failure.

As indicated previously, systems that separate dust from air or gas are primary concerns with regard to explosions related to combustible dust. From the perspective of dust explosions, the critical factor in selecting a technology for separation is free volume in the equipment. Settling chambers are unsuitable because of large free volume. Cyclones are widely used for this purpose. The basis for separation in a cyclone is the difference in angular momentum between particles and the airstream. In the cyclone, the concentration of dust increases from the center to the wall. At some distance from the center the concentration of dust could lie within the explosible range.

The baghouse is another common separatory device. An explosible concentration may arise during mechanical shaking to clean the bags. The electrostatic precipitator uses ionization for separating dust from gases. The inlet concentration of dust normally is less than the explosible limit. Mechanical rapping of the electrodes to loosen deposited material can create an explosible concentration within the unit. Electrostatic precipitators are not generally suitable for removal of ignitable dusts.

Drying is another important part of many processes. In some processes, the liquid evaporated from the solid is ignitable. Given appropriate conditions of concentration, temperature, and pressure, the vapor could exist in the ignitable range and thereby could constitute a potential fire or explosion hazard. Ventilation and inerting are the main methods used for maintaining the concentration of vapor below the Lower Flammable Limit and the dust below the Lower Explosible Limit. Low feed rate can lead to overheating of the feedstock. Overheating also can occur when residual material is exposed to hot air on startup. Hot product is more prone to self-heating.

Screening, classifying, and other methods of size classification and separation, as well as mixing and blending, tend to create dust clouds. Operations should occur under enclosure and preferably run under a slight vacuum. A silo or storage hopper also is prone to formation of dust clouds during filling operations.

Vacuum cleaners and mobile sweepers used for spill cleanup and general housekeeping are reservoirs of settled dust removed from horizontal surfaces. Although rare, explosions have occurred in this equipment. All metal parts, including wands and wire in flexible hoses, require grounding to minimize the risk of ignition due to electrostatic discharge. Normal techniques for explosion protection often are not feasible with vacuum cleaners. Minimizing accumulation of powder by prior wash-down and shoveling is preferable to over-reliance on vacuum cleaning in these circumstances (Bodurtha 1980).

Another means of protecting systems that handle ignitable dusts, as well as gases and liquids, is explosion protection (Senecal 1991; Garzia and Guaricci 1995). Explosion protection systems utilize several strategies. They all act to prevent buildup of the pressure wave from the explosion to the level at which the structure fails, and in so doing, to prevent damage and destruction. Dust explosions produce gases. Heat raises the temperature of the air and combustion gases. Rapid expansion of these gases produces the pressure that will act destructively on surrounding enclosures unless vent area is sufficient. Explosion protection systems act actively or passively.

Venting systems are passive. Venting systems release buildup of explosion pressure in a controlled manner by providing an escape path from the equipment. The rate of pressure increase is an important consideration in the design of explosion vents (Schwab 1991). This factor largely determines the area required for the vent and its practicality. Venting involves the use of blow-out panels (Figure J.25). These burst open under slight, predetermined overpressure in the airspace of the equipment or structure and permit the pressure wave and flame to vent into an area where damage cannot occur.

Active systems intervene to suppress the progress of the explosion and thereby its impact through use of sensors that monitor conditions and rapid injection of suppressing agent to prevent buildup of pressure in the system (Figure J.26). One approach utilizes a chemical suppressant and closure of rapidly acting valves to isolate the explosion in one part of the system. Another approach utilizes one or more extinguishers to discharge suppressant into vulnerable parts of the system. The suppressant typically is a dry chemical extinguishing agent, water, or a halogenated methane or ethane gas. The chemical extinguishing agent acts by removing heat from the flame front and by interfering with the chemical chain reaction involved in combustion. Active suppression systems employ one or more detectors and discharge units. These systems act to contain the explosion within milliseconds of the increase in pressure in the system. The detectors and discharge devices are located inside the structure to be protected.

Special tools made from copper-beryllium and other alloys minimize the generation of sparks resulting from energetic metal-to-metal and metal-to-product contact. Such tools cannot, however, wholly eliminate the danger of spark generation in hazardous locations, because a spark may be produced under several conditions.

Reports, such as API 2214 (1989) have considerably reduced concern about the hazard of ignition of vapors or gases by friction sparks. API 2214 indicates that use of non-sparking hand tools in place of steel tools provides no benefit in preventing explosion of hydrocarbons. Aluminum tools also provide little benefit. When struck against iron oxide (rust) with sufficient force, an aluminum tool could initi-

Figure J.25. Blow-out panels provide a path to vent gases produced during explosion and to prevent failure of the structure.

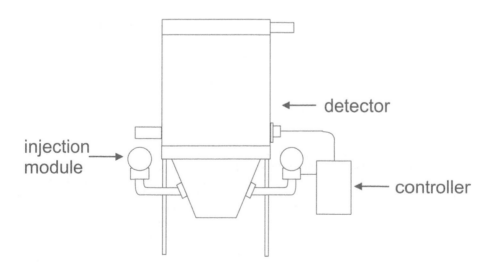

Figure J.26. Active explosion suppression systems rely on sensors to evaluate conditions and rapid discharge of the suppressing agent into the structure to prevent accumulation of pressure.

ate a thermite reaction. Leather, plastic, and wooden tools are free from the issue of friction sparks. However, they may not be practical because of durability and ability to participate in electrostatic charging and discharging events. Nickel, monel, and bronze pose a very slight spark hazard. Stainless steel poses a considerably lower spark hazard than ordinary tool steel (Drysdale 1991).

Protection against lightning strikes is the subject of several standards (API 2008, NFPA 2011b). Maintaining lightning protection during turnaround work on structures containing combustible dust and ignitable gases and vapors is essential to prevention of fires and explosions. Similar to other areas involving potential for electrical contact, lightning protection relies on bonding and grounding (Davis 1991). Turnaround activities raise the possibility of disrupting the continuous path from point of contact on a structure to ground and creating isolated conductors. This could lead to arcing to complete the path to earth. Arcing occurring inside a structure containing an ignitable or explosible atmosphere could create an explosion.

## Assessing Systems That Handle Flowable Solid Materials

Little exists in print about the assessment in preparation for lockout and tagout of flowable solid materials in equipment, machines, and systems. Typically what are observable are ad hoc isolations and lockouts created by site personnel in response to the perceived need. Inquiry about written assessment to ensure adequacy of the approach typically produces negative response.

Table 7.6 provides a generalized way of considering this question for different sources of energy. For situations not requiring immobilization, the example presented in Table F.7 is satisfactory for assessing isolation and lockout of equipment, machines, and systems. For situations requiring immobilization, refer to Figure G.7. Table J.5 provides a version for assessing hazardous energy in circuits that handle flowable solid materials.

Elevation change in the circuit, types of components, and the ability to remove material from the circuit through normal operating procedures prior to shutdown, and immobilization of material conveying devices are critical components to consider about isolation. None of these considerations meets "textbook" description of requirements for isolation. Hence, they require deliberation and consideration in an organized manner in order to determine whether they meet or exceed regulatory requirements for the situation.

OSHA Standard 29CFR 1910.146 does not allow techniques of lockout or tagout for isolating flowable material (OSHA 2008). The reason in OSHA's view is that compliance with 29CFR 1910.147 (Lockout/Tagout) does not in all cases adequately isolate hazards created by materials such as steam, flammable gases, flammable and combustible liquids.

In a permit-required confined space, OSHA considers isolation of hazards associated with flowable materials to have occurred only by the use of blanking or blinding; misaligning or removing sections of line, pipe or duct; or use of a double block and bleed system. This approach, in OSHA's view, offers complete employee protection, whereas an employer can comply with requirements of 29CFR 1910.147 simply by closing and locking out or tagging out a single valve. Closing and locking out a single valve could create atmospheric hazards in the space.

OSHA has indicated the nature of the process to be followed in assessing the capability of alternate measures in a letter of interpretation to Procter & Gamble Company (January 5, 1998) which accepted its specific safety disconnect system (inherently fail-safe system) as equivalent to an energy isolating device (OSHA 2008). The equivalency determination was based upon specific facts concerning the process machine and a failure analysis. This example illustrates that possibilities exist for comparing alternative approaches against OSHA requirements through thorough design review and application for obtaining a variance described in 29 CFR 1905 (OSHA 2008).

Intrinsic to all approaches is an element of risk. This applies to the approach adopted by OSHA in 29CFR 1910.146 (OSHA 1993) just as much as to any other approach to regulating behavior. The unknown in the approach mandated by OSHA through regulation is the magnitude of the risk intrinsic to different approaches to the same problem.

The information collected and considered during this hazard assessment forms the basis for a dialog with the regulator with regard to possible strategies for isolation and lockout of the actual system.

## Table J.5
## Assessing Energy in Systems Handling Flowable Solid Materials

| ABC Company | Hazardous Energy Assessment | Work Activity |
|---|---|---|
| Location: | Equipment, Machine, System: | Assessed by:<br>Tel:<br>Date: |
| Overview: | | |
| Hierarchy of Energy Input | Conversion Energy Output | Equipment, Machine, System Affected |
| | | |

### Circuit Name

| Function/Description: | | |
|---|---|---|
| Material Flow Isolation Strategy: | | |
| Failure/Consequence Analysis: | | |
| Input: | Storage: | Dissipation/Purge: |
| Output: | Conversion: | Immobilization/Isolation: |
| Primary Isolation: | Secondary Isolation: | Tertiary Isolation: |
| Location: | Location: | Location: |
| Action: | Action: | Action: |
| Flow Isolation? | Flow Isolation? | Flow Isolation? |
| (Circuit Name) Solid Material Flow Isolation Procedure: | | |

## Table J.5 (Continued)

## Assessing Energy in Systems Handling Flowable Solid Materials

| Verification Test Procedure: |
| --- |
| Notes: |
| Overall Flow Isolation Procedure: |
| Overall Verification Test Procedure: |

**Hazard Assessment (Operational/Undisturbed Space) or (Work Activity)**

| Hazardous Condition | Occurrence/Consequence | | |
| --- | --- | --- | --- |
| | Low | Moderate | High |
| | &#124; | &#124; | &#124; |

(circuit name)

In this table, **NA** means that the category does not apply in any normally foreseeable situation. **Low** means that exposure is readily identifiable but believed to be much less than applicable limits or that exposure to nonquantifiable hazardous conditions is unlikely to produce injury. **Low-Moderate** means that exceedence of regulatory limits is believed possible or that nonquantifiable exposure could produce minor injury requiring self-administered treatment. Control measures or protective equipment should be considered. **Moderate** means that exposure is believed capable of exceeding regulatory limits or causing traumatic injury requiring first aid treatment or attention by a physician. Protective equipment or other control measures are necessary. **Moderate-High** means that exposure is believed capable of considerable exceedence of regulatory limits or causing serious traumatic injury. Advanced control measures or protective equipment are required. **High** means that short-term exposure is believed capable of causing irreversible injury, including death. Advanced control measures or protective equipment are required.

Contact with the regulator is strongly advised in order to solicit opinion about whether the proposed approach would be acceptable.

Retention of energy in material storage and transfer systems occurs deliberately in storage structure and unintentionally in inclined and vertical conveyors and conveyances, such as elevators, and in ducts of pneumatic and vacuum conveying systems. Deliberate retention of energy in stored materials is part of the design of the structure. Unintentional retention, as may occur in material transport systems, could express at an unexpected moment during dismantlement if not recognized and considered during planning for this work. Isometric drawings of the circuit provide valuable information toward this assessment and deliberation.

## Shutting Down Systems Handling Flowable Solid Materials

Among the less complex systems to isolate and lock out are those that handle and process flowable solid materials. Generally, stopping movement of a material transport system stops the operation unless a critical step is occurring. This occurs because of the interrelationship of operation of interlinked components. All components in a sequence must operate or none can operate. This interlinkage is expressed through hard wired relays and through programming of Programmable Logic Controllers in current systems. The ability of components of an interlinked system to continue to operate following shutdown of any one component should not be possible. The latter would represent catastrophic failure of the logic of the operation.

Previous discussion in various places has highlighted the critical role of the operator in this inquiry. Operator knowledge and experience are essential, especially for situations involving complex systems. Additional sources of information include the design and operations manual, the vendor, and the manufacturer.

The course prepared by the UAW-GM (United Auto Workers-General Motors) National Joint Committee on Health and Safety on lockout recommended working on process and utility systems near the beginning in the sequence of activities for deactivating, de-energizing, and isolating equipment and machines (UAW-GM 1985). For equipment and machines, the recommended sequence is a hierarchy starting with mechanical momentum sources, followed by gravity, stored mechanical energy, electrical system, pneumatic system, hydraulic system, and process and utility systems.

## Isolation of Systems Handling Flowable Solid Materials

The least complicated isolation occurs through shutdown of an electric motor that operates a feeder conveyor or a blower in a pneumatic or vacuum conveying system. Shutdown of the motor prevents further movement of the flowable material.

Shutdown meets the intent for isolation in regulatory standards, such as 29CFR 1910.147 (OSHA 1989). Note that lockout requires further action, to be discussed in the next section.

Germane to systems handling flowable solid materials are requirements stated in 29CFR 1910.146, the OSHA Standard on permit-required confined spaces in general industry (OSHA 1993). OSHA considers isolation of systems handling flowable solid materials that eliminates the hazard to occur by such means as blanking or blinding; misaligning or removing sections of lines, pipes, or ducts; or a double block and bleed system. The Standard considers isolation to be the process by which a permit space is removed from service and completely protected against the release of energy and material into the space.

Figure I.12 to Figure I.14 illustrate the classic paradigm of isolation of fluid-handling systems involving blanking or blinding; misaligning or removing sections of lines, pipes, or ducts; and valves. Some of these concepts may apply to systems handing flowable solid materials. Computerization of process control introduces the possibility of using mechanical operators and achieving isolation through software and electrical lockout.

## Lockout Devices for Systems Handling Flowable Solids

Lockout devices for systems handling flowable solids are subdivided according to design and execution. Intrinsic lockout devices are designed into the equipment as supplied, or are designed into retrofits created by the original equipment manufacturer. This is the most desirable route to follow, since the manufacturer is most familiar with the equipment.

Extrinsic lockout devices are temporary add-ons to components of the fluid-handling system. The simplest devices for lockout of equipment handling flowable solid materials are chains or cable devices that prevent movement.

## Verification and Testing

Verification and testing of the effectiveness of isolation and lockout of circuits handling flowable solid materials can occur in several ways. The first is visual inspection following confirmation of associated electrical lockout to confirm cessation of movement or flow or passage of material into or from a storage structure along a conveying system. Indication of position of components by sensors is a secondary means of confirmation.

## Retained Energy

Retained energy refers to energy retained in a circuit either through deliberate storage or through unintentional entrapment during deactivation, de-energization, and isolation. This situation is a natural consequence from cessation of motion of equipment transporting flowable solid materials prior to emptying as would occur at the end of a production run.

## References

Anonymous: Mechanical Moles Get to the Bottom of Silo Storage Problems. *Powder Bulk Eng. 3(1)*: (1989a). 3 pp. [Reprint].

Anonymous: Silo Cleaning System Moves Cement Terminal's Stock. *Powder Bulk Eng. 3(12)*: 22-25 (1989b).

API: Spark Ignition Properties of Hand Tools, 3rd Ed. (API Publ. 2214). Washington, DC: American Petroleum Institute, 1989.

API: Protection against Ignitions Arising out of Static, Lightning, and Stray Currents, 7th Ed. (API Recommended Practice 2003). Washington, DC: American Petroleum Institute, 2008.

Bodurtha, F.T.: *Industrial Explosion Prevention and Protection*. New York: McGraw-Hill Book Company, 1980.

Carson, J.W. and D.S. Dick: Seals Improve Coal Flow into a Pressurized Environment. *Power Eng.* March 1988. pp. 24-27.

Carson, J.W. and T. Holmes: Why Silos Fail. *Powder Bulk Eng. 15(11)*: 31-43 (2001).

Carson, J.W. and R.T. Jenkyn: How to Prevent Silo Failure with Routine Inspections and Proper Repairs. *Powder Bulk Eng. 4(1)*: 18-24 (1990).

Carson, J.W. and J.R. Johanson: Design of Bins and Hoppers. In *Materials Handling Handbook*, 2nd Ed. Kulwiec, R.A. (Ed.). New York: John Wiley & Sons, 1985.

Craig, D.A. and R.J. Hossfeld: Measuring Powder-Flow Properties. *Chem. Eng. 109(9)*: 41-46 (2002).

Davis, N.H., III: Lightning Protection Systems. In *Fire Protection Handbook*. 17th Ed. Cote, A.E. and J.L. Linville (Eds.). Quincy, MA: National Fire Protection Association, 1991.

DI: Pressure Development in Laboratory Dust Explosions by J. Nagy, A.R. Cooper, and J.M. Stupar (Report of Investigation, 6561). Pittsburgh: U.S. Department of the Interior/ Bureau of Mines, 1964.

Dick, D.S.: Screw Seal (Patent Number 4,881,862). Washington, DC: U.S. Patent Office, November 21, 1989.

Frank, W.L.: Dust Explosion Prevention and the Critical Importance of Housekeeping. *Process Safe. Progress 23(3)*: 175-184 (2004).

Gregg, B.: Generation and Control of Static Electricity. *Plant Serv. 17(6)*: June 1996. pp. 83-87.

Hauck, R.M.: How to Clean Your Storage Vessel Quickly and Safely. *Powder Bulk Eng. 6(11)*: 47-50 (1993).

Holmes, T. and J.W. Carson: Silo Maintenance: It's up to You. *Powder Bulk Eng. 16(11)*: 31-40 (2002).

Jenkyn, R.T. and D.J. Goodwill: Silo Failures: Lessons to Be Learned. *Eng. Digest*. September (1987). pp. 17-22.

Johnson, G.O.: Designing Your Dust Collection System to Meet NFPA Standards – Part I. *Powder Bulk Eng. 22(12)*: 53-59 (2008).

Laing, H.D. and L.A. Berg: Safety First: Hiring a Vessel Cleanout Service to Safely Unblock Your Silo. *Powder Bulk Eng. 13(11)*: 47-54 (1999).

Lease, T.J.: Storage Silos — Steel or Concrete? *Powder Bulk Eng. 14(11)*: 43-49 (2000).

Lees, F.P.: *Loss Prevention in the Process Industries*, 2nd Ed., Vol. 2. London: Butterworths-Heinemann, 2001.

Marinelli, J.: Taking the Guesswork Out of Silo Design. *Powder Bulk Eng. 14(11)*: 35-40 (2000).

Marinelli, J.: A Practical Approach to Bins and Feeders. *Chem. Eng. 109(7)*: 39-42 (2002).

McManus, N.: *Safety and Health in Confined Spaces*. Boca Raton, FL: Lewis Publishers/CRC Press, 1999.

Mills, D.: Safety Aspects of Pneumatic Conveying. *Chem. Eng. 106 (4)*: 84-91 (1999).

MSHA: Think "Quicksand": Accidents around Bins, Hoppers and Stockpiles, Slide and Accident Abstract Program. Arlington, VA: U.S. Department of Labor, Mine Safety and Health Administration, National Mine Health and Safety Academy, 1988.

MSHA: Stockpiling Safety (Safety Manual No. 30). Arlington, VA: U.S. Department of Labor, Mine Safety and Health Administration, National Mine Health and Safety Academy, 1991.

NFPA: National Electrical Code, NFPA 70 (2011 Edition). Quincy, MA: National Fire Protection Association, 2011a.

NFPA: Standard for the Installation of Lightning Protection Systems, NFPA 780 (2011 Edition). Quincy, MA: National Fire Protection Association, 2011b.

OSHA: Selected Occupational Fatalities Related to Grain Handling as Found in Reports of OSHA Fatality/Catastrophe Investigations. Washington, DC: U.S. Department of Labor, Occupational Safety and Health Administration (U.S. DOL/OSHA), 1983.

OSHA: "Control of Hazardous Energy Sources (Lockout/Tagout); Final Rule," *Fed. Regist. 54*: 169 (1 September 1989). pp. 36644-36696.

OSHA: Excavations: Hazard Recognition in Trenching and Shoring (Section V: Chapter 2). In OSHA Technical Manual (Publication No. 9664). Washington, DC: U.S. Department of Labor, Occupational Safety and Health Administration, 1999.

OSHA: "Permit-Required Confined Spaces for General Industry; Final Rule," *Fed. Regist. 58*: 9 (14 January 1993). pp. 4462-4563.

OSHA: OSHA Instruction, The Control of Hazardous Energy – Enforcement Policy and Inspection Procedure (CPL 02-00-147). Washington, DC: Occupational Safety and Health Administration, 2008.

Purutyan, H., B.H. Pittenger, and J.W. Carson: Six Steps to Designing a Storage Vessel That Really Works. *Powder Bulk Eng. 13(11)*: 56-67 (1999).

Scarbrough, D.R.: Control of Electrostatic Ignition Sources. In *Fire Protection Handbook*. 17th Ed. Cote, A.E. and J.L. Linville (Eds.). Quincy, MA: National Fire Protection Association, 1991.

Schwab, R.F.: Dusts. In *Fire Protection Handbook*. 17th Ed. Cote, A.E. and J.L. Linville (Eds.). Quincy, MA: National Fire Protection Association, 1991.

Terzaghi, K., R.B. Peck, and G. Mesri: *Soil Mechanics in Engineering Practice*, 3rd Ed. New York: John Wiley & Sons, Inc., 1996.

Thorn, J.O.: Six Practical Ways to Handle Rotary Airlock Valve Leakage. *Powder Bulk Eng. 22(10)*: 53-61 (2008).

UAW-GM: Lockout. Detroit: United Auto Workers-General Motors Human Resource Center, 1985.

Vorderbrueggen, J.B.: Imperial Sugar Refinery Combustible Dust Explosion Investigation *Process Safe. Progress 30(1)*: 60-81(2011).

Wahl, R.C.: Properties of Bulk Solids. In *Materials Handling Handbook*, 2nd Ed., Kulwiec, R.A. (Ed.). New York: John Wiley & Sons, 1985. pp. 882-900.

Ware, B.: How Vertical Flow Effects and Material Characteristics Affect Discharge — Part I. *Powder Bulk Eng. 13(4)*: 17-24 (1999a).

Ware, B.: How Vertical Flow Effects and Material Characteristics Affect Discharge — Part II. *Powder Bulk Eng. 13(5)*: 27-36 (1999b).

Wysocki, T.J.: Carbon Dioxide and Application Systems. In *Fire Protection Handbook*. 17th Ed. Cote, A.E. and J.L. Linville (Eds.). Quincy, MA: National Fire Protection Association, 1991.

Zabetakis, M.G.: *Flammability Characteristics of Combustible Gases and Vapors* (Bull. 627). Washington, DC: U.S. Department of the Interior/Bureau of Mines, 1965.

# Index